FOREST GENETICS

DEDICATIONS

To the people who have inspired and sustained me throughout my life: Mom, Dad, Mary, Dorothy, Suzie and Antonio. TW

To Cathy, Christine, Michael, Patrick, Neal, Kenda, Halli, Jethro, Ashley, Lily and Clark. TA

To my parents, Ken and Jane, who introduced me to the forests and trees of America. DN

About the Cover

This natural stand of *Pinus tecunumanii* is located in Las Piedrecitas, Chiapas, Mexico. This species occurs in a series of small disjunct populations from southern Mexico to central Nicaragua and was named for the famous Mayan chief "Tecun Uman." *Pinus tecunumanii* is a cornerstone species in many forested ecosystems in Mesoamerica and is also established commercially in plantations outside of its natural range. The person climbing the tree is collecting seed as part of the *ex situ* gene conservation efforts for this species. (Photo courtesy of Dr. Bill Dvorak, Camcore, NCSU, Raleigh, NC, USA. Cover design by farm, Portland, Oregon).

FOREST GENETICS

TIMOTHY L. WHITE, Professor and Director, School of
Forest Resources and Conservation, University of Florida

W. THOMAS ADAMS, Professor and Head, Department
of Forest Science, Oregon State University

DAVID B. NEALE, Professor, Department of Plant Sciences,
University of California, Davis

www.cabi.org

CABI Publishing is a division of CAB International

CABI Publishing
CAB International
Wallingford
Oxfordshire OX10 8DE
UK

Tel: +44 (0)1491 832111
Fax: +44 (0)1491 833508
E-mail: cabi@cabi.org
Website: www.cabi-publishing.org

CABI Publishing
875 Massachusetts Avenue
7th Floor
Cambridge, MA 02139
USA

Tel: +1 617 395 4056
Fax: +1 617 354 6875
E-mail: cabi-nao@cabi.org

A catalogue record for this book is available from the British Library, London, UK.

A catalogue record for this book is available from the Library of Congress, Washington, DC.

ISBN   978 0 85199 083 5

The paper used for the text pages in this book is FSC certified. The FSC (Forest Stewardship Council) is an international network to promote responsible management of the world's forests.
Printed and bound in the UK from copy supplied by the editors by Cromwell Press, Trowbridge.

CONTENTS IN BRIEF

CONTENTS

SECTION II: GENETIC VARIATION IN NATURAL POPULATIONS

Chapter 7: Within-population Variation – Genetic Diversity, Mating Systems and Stand Structure

SECTION III: TREE IMPROVEMENT

Chapter 11: Tree Improvement Programs – Structure, Concepts and Importance 285

Chapter 12: Base Populations – Species, Hybrids, Seed Sources and Breeding Zones 303

Chapter 13: Phenotypic Mass Selection – Genetic Gain, Choice of Traits and Indirect Response 329

SECTION IV: BIOTECHNOLOGY

Chapter 18: Genomics – Discovery and Functional Analysis of Genes 523

Chapter 19: Marker-assisted Selection and Breeding – Indirect Selection, Direct Selection and Breeding Applications 553

About the Authors

Tim White (center), Director of the School of Forest Resources and Conservation, IFAS, University of Florida, is a quantitative geneticist with interests in mixed linear models, breeding theory, tree improvement and international forestry.

Tom Adams (left), Head of the Department of Forest Science, College of Forestry, Oregon State University, is a population geneticist with interests in variation in natural and breeding populations of forest trees, gene conservation and ecological genetics.

David Neale (right), Professor in Plant Sciences, University of California, Davis, is a population and molecular geneticist with interests in genomics, adaptation, complex traits and bioinformatics.

ACKNOWLEDGMENTS

This book would not have been possible without the generous help of many people who gave willingly of their time. Six people deserve special thanks: (1) Claudia Graham who crafted all of the figures; (2) Jeannette Harper who did the layout of the entire manuscript; (3) Rose Kimlinger who obtained all the permissions for using material from other sources and managed references for TLW; (4) Sara Lipow who co-wrote the first draft of Chapter 10; (5) Raj Ahuja who co-wrote the first draft of Chapter 20; and (6) Greg Powell who managed most of the photographs and reviewed multiple chapters.

We are also truly indebted to the many friends, colleagues and students who assisted in so many ways by brainstorming ideas, reviewing chapters, sending photos, providing examples of their work, helping with layout, editing chapter drafts, counseling us and more. We sincerely thank each and every one of them and can only hope that the final product properly reflects the high quality of their input. They are: Ryan Atwood, Brian Baltunis, John Barron, Gretchen Bracher, Karen Bracher, Jeremy Brawner, Garth Brown, Rowland Burdon, John Carlson, Mike Carson, Tom Conkle, John Davis, Neville Denison, Mark Dieters, Rob Doudrick, Gayle Dupper, Bill Dvorak, Sarah Dye, Ken Eldridge, Christine Gallagher, Sonali Gandhi, Salvador Gezan, Rod Griffin, Dave Harry, Gary Hodge, Vicky Hollenbeck, Dudley Huber, Bob Kellison, Eric Kietzka, Claire Kinlaw, Bo-hun Kinloch, Krishna Venkata Kishore, Ron Lanner, Tom Ledig, Christine Lomas, Uilson Lopes, Juan Adolfo Lopez, Pengxin Lu, Barbara McCutchan, Steve McKeand, Gavin Moran, PK Nair, John Nason, John Owens, David Remington, Don Rockwood, Rebeca Sanhueza, Ron Schmidtling, Ron Sederoff, Victor Sierra, Richard Sniezko, Frank Sorensen, Kathy Stewart, Steve Strauss, Gail Wells, Nick Wheeler, Jeff Wright, and Alvin Yanchuk.

PREFACE

Interest in forest genetics began more than two centuries ago when foresters first realized that seed of the same species collected from different geographical locations grew differently when planted together in a common environment. Approximately 50 years ago, pioneers initiated large-scale tree improvement programs to develop genetically improved varieties of some commercially important species. Today, forest genetics is an exciting and challenging field of study that encompasses all subdisciplines of genetics (Mendelian, molecular, population and quantitative genetics) and their applications in gene conservation, tree improvement and biotechnology. Each of these fields has its own terminology and set of concepts; however, all forest geneticists should have a basic understanding of all subdisciplines and be able to integrate across them. Thus, we strive in this book to provide a balanced presentation of the current state of knowledge in each subdiscipline, while also integrating and demonstrating the linkages among them.

The study of forest genetics is important not only because of the unique biological nature of forest trees (large, long-lived perennials covering 30% of the earth's land area), but also because of their social, ecological and economic significance in the world. Trees are the key component of a variety of forested ecosystems, whether they are preserved in their native state or managed for a variety of resources, including forest products. Thus, the most important reason to study forest genetics is to provide insight into the evolution, conservation, management and sustainability of the world's natural and managed forests. For this reason, the intent of this book, *Forest Genetics*, is to describe concepts and applications of genetics in all types of forests ranging from pristine natural forests to monoculture plantations.

The focus of *Forest Genetics* is on genetic principles and their applications. When possible, we have tried to pursue the following, consistent pedagogical style for each principle discussed: (1) Motivate the need for or importance of the principle; (2) Describe the underlying concepts and their applications using a combination of text, equations and figures; (3) Reinforce the principle and its application with examples from forest trees; and (4) Synthesize and summarize the current state of knowledge and main issues regarding the principle. With the focus on principles, there is necessarily less emphasis on species-specific details and on laboratory and field methods associated with implementation of some technologies. To overcome this potential limitation, citations to classical and current literature are provided for interested readers.

In place of a glossary of terms at the end of the text, important words are typed in **bold-face** the first time they are used a meaningful context. Emboldened words are identified in the Index along with the page numbers corresponding to their formal definitions and to other places in the text where these words are used. There are also several other good glossaries for both general genetics (Ridley, 1993; Miglani, 1998) and forest genetics (Snyder, 1972; Wright, 1976; Helms, 1998). Throughout the text, Latin names are used for all species because of inconsistent usage of common names.

Forest Genetics is intended for several audiences as: (1) A first course for advanced undergraduate and graduate students; (2) A reference for professionals working in forest genetics or forest management; (3) An introduction for forest scientists interested in other subdisciplines of forest genetics (e.g. for quantitative geneticists interested in biotechnology or molecular geneticists interested in tree improvement); and (4) A synthesis for geneticists and other scientists working with species other than forest trees. No previous knowledge of genetics is assumed.

Forest Genetics is organized into four major sections. Section I, *Chapters 2-6*, provides a summary of basic genetic principles. Examples from forest trees are used, when possible, to illustrate the principles described, but the concepts are widely applicable to most plant and animal species. Section II, *Chapters 7-10*, focuses on genetic variation in natural populations of forest trees: its description, evolution, maintenance, management and conservation. *Chapters 7, 8* and *9* address these concepts at three distinct levels of organization: among trees within populations (within stand), among populations within species (geographical), and among species, respectively. *Chapter 10* deals specifically with strategies for gene conservation.

Section III, *Chapters 11-17*, relies on the principles developed in previous chapters and discusses the application of these principles in applied genetic improvement programs of tree species. *Chapter 11* is a general overview of the nature of tree improvement, and subsequent chapters in this section are organized around the steps and activities of the breeding cycle common to most breeding programs: defining base populations (12), making selections (13), establishing genetic tests (14), analyzing the data from these tests (15), deploying commercial varieties (16) and developing long-term breeding strategies (17).

Section IV, *Chapters 18-20*, describes genomic sciences and molecular DNA technologies and their applications in forest genetics and tree improvement. *Chapter 18* addresses the technologies used to discover and map genes at the molecular level and to understand their function. *Chapter 19* introduces the concepts and applications of marker-assisted selection and marker-assisted breeding in tree improvement programs. Finally, *Chapter 20*, describes genetic engineering in forest trees.

To use this book as a first course in forest genetics, different chapters may be stressed, highlighted, or omitted completely, depending upon the orientation and objectives of the course. A course in applied tree improvement might feature *Chapters 1, 5-9*, and *11-17* with sections from other chapters highlighted as appropriate. A course on genetics of natural forest populations and gene conservation might rely heavily on *Chapters 1-10*. Lastly, an emphasis on molecular genetics and biotechnology might emphasize *Chapters 1-4, 11* and *18-20*.

Although we have strived for both correctness and completeness in the presentation of topics in *Forest Genetics*, there are necessarily errors of both commission and omission. Further, the examples chosen to illustrate principles reflect our own experiences and biases. Therefore, we hope that readers will alert us to mistakes, make suggestions for improvement, and share their experiences with us.

Tim White
Gainesville, Florida
tlwhite@ufl.edu

Tom Adams
Corvellis, Oregon
w.t.adams@oregonstate.edu

David Neale
Davis, California
dbneale@ucdavis.edu

CHAPTER 1
FOREST GENETICS – *CONCEPTS, SCOPE, HISTORY AND IMPORTANCE*

Genes are the basis for all genetic variation and biodiversity in the world, and genetics is the branch of biology that studies the nature, transmission and expression of genes. **Genetics** deals with heritable variation among related organisms, and studies resemblances and differences among individuals related by descent. **Forest genetics** is the subdiscipline of genetics dealing with forest tree species. In one sense forest trees are not model organisms for studying genetic principles because of their large size and long life spans. However, the study of forest genetics is important precisely because of the unique biological nature of forest trees and also because of the social and economic importance of forests in the world.

Tree improvement is the application of principles of forest genetics and other disciplines, such as tree biology, silviculture and economics, to the development of genetically improved varieties of forest trees. Like breeding programs for crops and farm animals, tree improvement aims to develop varieties that increase the quantity and quality of harvested products. However, unlike agricultural varieties, forest trees are still essentially undomesticated because large-scale tree improvement programs only began in the 1950s and current breeding populations have diverged little in genetic makeup from wild populations of the same species. Therefore, studying the genetics of breeding populations in tree improvement programs can provide many insights into the genetics of natural populations and *vice versa*.

The intent of this book is to describe concepts and applications of forest genetics in all types of forests ranging from pristine natural forests to monoculture plantations. Therefore, this chapter begins with a brief discussion of the different types of forests in the world, and their scope and importance. We then outline the causes of variation in forests, provide a brief history of forest genetics, and conclude with a discussion of the importance of forest genetics in both natural and managed forests.

GLOBAL SCOPE AND IMPORTANCE OF NATURAL AND MANAGED FORESTS

There are 3.4 billion hectares of forests in the world occupying nearly 30% of the earth's total land area (Sharma, 1992; FAO, 1997). Forests are important on every continent and range in coverage from nearly 50% of the total land area in Latin America and the Caribbean, to approximately 30% in North America, Europe and the former USSR, and only 20% in Africa and Asia (FAO, 1995a). The total growing stock of wood is 384 billion cubic meters, with almost half of this accounted for by the combined forests of the former USSR and Latin America (which includes the tropical forests of the Amazon Basin).

The world's forests vary widely in their species composition from temperate and boreal conifer forests composed of relatively few tree species to tropical forests containing literally hundreds of tree species. Forests serve many different functions by providing different products and social values. For example, in developing countries 80% of all wood harvested is used for fuelwood and forests provide a range of indigenous uses (Fig. 1.1a). However, in developed countries, 84% of harvested wood is used for industrial purposes (FAO, 1995a; Fig. 1.1b). In all countries, forests are valued for their conservation and scenic values.

While all forests provide many biological, economic and social benefits, it is some times useful to conceive of a continuum of different types of forests with each link in the continuum providing multiple, yet not identical values (Brown *et al.*, 1997). At one extreme (Fig. 1.2a), undisturbed natural forests are excellent for several biological and social values; however, they often produce low harvest yields and are undesirable sources of commercial wood products for several reasons (Hagler, 1996). At the other extreme, intensively managed plantations grow rapidly, yet sustain lower levels of biodiversity (Fig. 1.2b). Between these two extremes are a number of different types of forests each providing a somewhat different set of values (Fig. 1.3). No single type of forest can provide all possible benefits; therefore, all options are needed (Kanowski *et al.*, 1992).

Fig. 1.1. Forests provide a wide variety of useful products, including: (a) Fuelwood for cooking and heating, which is especially important in developing countries; and (b) Industrial wood harvested for both solid wood and paper products. (Photos courtesy of P.K. Nair and T. White, respectively, University of Florida, Gainesville)

Fig. 1.2. *Eucalyptus grandis* growing in two different conditions that exhibit different levels of phenotypic variability: (a) In a natural stand in Australia, the large variability is caused by both genetic and environmental differences among the trees; and (b) In a plantation of Mondi Forests in South Africa, the trees are more uniform both because the site is uniform, minimizing environmental differences among the trees, and all trees are the same genotype. (Photos courtesy of K. Eldridge, CSIRO Australia, and T. White, respectively)

Knowledge of forest genetics is valuable for understanding the sustainability, conservation and management of all types of forests on the continuum shown in Fig. 1.3. For example, deforestation is reducing the amount of forested land by nearly 1% annually in some countries (World Resources Institute, 1994), which is seriously eroding the genetic base of some tree species. Forest geneticists can help ameliorate this situation in two ways: (1) Gene conservation programs can preserve the genetic diversity of threatened species; and (2) Tree improvement programs can ensure that well adapted, and even genetically improved trees, are used to reforest cut-over lands.

THE ROLE OF PLANTATIONS AS FOREST ECOSYSTEMS

Although the application of tree improvement principles can benefit the quality and yield of forests managed under natural regeneration systems, the great majority of formal tree improvement efforts in the world today depend on artificial reforestation or afforestation to utilize the varieties developed. Therefore, it is appropriate here to briefly discuss the role of plantations in global forestry. **Plantation systems** are defined broadly as any planting regime that contains forest trees, including large-scale commercial (*i.e.* industrial) plantations, agroforestry systems, small woodlots and community forests. The great majority of plantations in the world were established after 1950; today there are approximately 135 million hectares of plantations in the world accounting for 4% of the total forested area (Kanowski *et al.,* 1992). As with natural forests, the fraction of the forested area occupied by plantations varies widely among countries, from 1-3% in Brazil, Canada, Indonesia and the former USSR, to 15-25% in Chile, China, New Zealand, South Africa and the USA, and to 45% in Japan (FAO, 1995b). A low percentage of plantations can indicate either a small area of plantations or a large area of natural forests.

Plantation forests currently provide approximately 10% of the world's consumption of wood, and this may rise to as much as 50% within the next several years, depending on

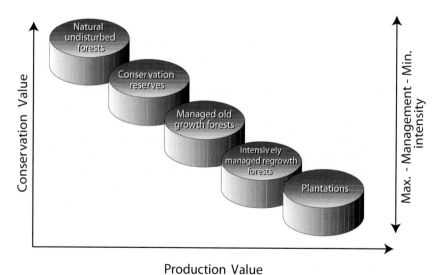

Fig. 1.3. Schematic diagram of several different types of natural and managed forests with each type providing a different set of economic and social values. In general the more intensively managed, faster growing plantations have a higher production value, but lower gene conservation value. (Adapted from Nambiar, 1996). Reprinted with permission from the Soil Science Society of America.

global demand and rates of plantation establishment (Kanowski *et al.,* 1992; Hagler, 1996). Many countries are relying on plantations to supply a significant portion of their wood needs (FAO, 1997). Most early plantations were established for industrial wood production and this is still true in developed countries. However, since the 1970s much tree planting in the tropics has been directed at meeting the indigenous needs described above (Kanowski *et al.,* 1992). For example, China has the largest plantation estate of any single country (36 million hectares) and this is largely due to a government sponsored community planting program.

Plantations offer a number of advantages over natural, undisturbed forests for supplying the world's wood needs (Savill and Evans, 1986; Evans, 1992a):

- Plantations can grow substantially faster than natural forests, especially when fast growing genotypes are intensively managed. For example, plantations in the tropics can average 10 times the growth rate of native tropical forests (Kanowski *et al.,* 1992; Hagler, 1996). This means that wood can be harvested sooner (*i.e.* rotation ages are much shorter) and that less total forested area is required to produce a given amount of wood.
- Plantations produce trees of much higher uniformity (compare Figs. 1.2a and 1.2b) meaning lower costs of harvest, transport and conversion, and higher yield for some products.
- There is great flexibility in the type of land used for plantations and the land can be conveniently located near work forces and infrastructure such as conversion facilities. Abandoned and degraded lands (such as former agricultural lands) are sometimes very suitable for forest plantations.
- Plantations can serve environmental functions such as stabilizing soil to reduce erosion, increasing water quality, providing windbreaks, reclaiming abandoned industrial sites and sequestering carbon to slow global warming.

All of these advantages taken together mean that plantations can play a key role in meeting the rising global demand for wood products (projected to rise nearly 50% in the next 20 years) and therefore, reduce reliance on natural forests for wood production. Under reasonable assumptions, the total amount of plantation area needed to meet the global demand for industrial wood could be as low as 5% of the world's forested area (Sedjo and Botkin, 1997). For example, only 1% of the forested area in Brazil and Zambia is in plantations; yet, these plantations supply over 50% of the total industrial wood produced by those countries. Similarly, the plantations of Chile and New Zealand occupy 16% of the forested area and produce 95% of their industrial wood output (FAO, 1995b).

The higher efficiency of plantations can reduce the total environmental impact of forestry (sometimes called the environmental footprint) since less forested area is needed to meet the global demand for wood. Therefore, reliance on plantations is one way to reduce pressure on and to conserve natural forests. Further, it is important to dispel the notion that plantations are a cause of deforestation. Rather, most deforested land in the tropics is converted to other land uses and less than 1% ends up in forest plantations (FAO, 1995a).

Although plantations also have disadvantages such as less biodiversity, questionable value for gene conservation and questionable long-term sustainability (in some cases), all types of forests on the continuum are needed to provide the range of desired social and economic values. In the last decade, there has been increasing recognition that plantations and other types of forests must be properly managed to ensure sustainable production for many rotations. Intensifying global competition caused by more open world markets has contributed to a trend towards more intensive management of plantations, especially by industrial organizations. Tree improvement programs that produce genetically improved varieties have become part of the operational silviculture in most large plantation programs. When coupled with good plantation establishment and post-establishment silviculture, use of improved varieties can greatly increase plantation productivity and health.

CONCEPTS AND SOURCES OF VARIATION IN FORESTS

Separating Genotypic and Environmental Influences on Phenotypic Variation

Having briefly discussed some aspects of the world's forests, we now address the tree-to-tree variation in those forests and the underlying causes of this variation. The outward appearance of a tree is called its phenotype. The **phenotype** is any characteristic of the tree that can be measured or observed; it is the tree that we see and is influenced both by its genetic potential and by the environment in which it grows. Sometimes the simple equation $P = G + E$ (phenotype = genotype + environment) is used to indicate that the tree's genotype and environment are the underlying causes that together produce the final phenotype (Fig. 1.4). The environmental effects on phenotype include all non-genetic factors such as climate, soil, diseases, pests and competition within and among species.

Genes residing in the genome of every living cell in the tree determine the **genotype**. If the deoxyribonucleic acid (DNA) sequences of two trees are identical for all of the tens of thousands of genes, then their genotypes are the same. Two trees of the same species have more similar DNA sequences than two trees of different species, and two trees with the

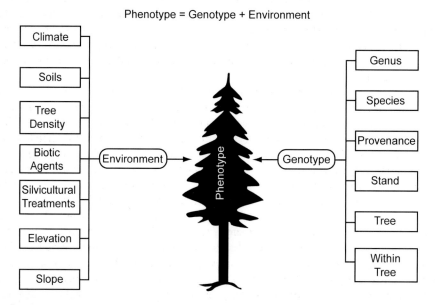

Fig. 1.4. Schematic diagram showing some of the many different environmental and genetic factors that contribute to a tree's phenotype. Differences in any of these factors cause differences among tree phenotypes leading to the abundant phenotypic variation among trees in forest stands.

same parents (*i.e.* in the same family) have more similar DNA sequences than two trees from different families.

No two trees have exactly the same phenotype. Normally, there is tremendous **pheno-typic variation** among trees in a forest (Fig. 1.2a). There is variation from tree to tree in all characteristics including size, morphology, phenology and physiological processes. It is only on a very homogeneous site planted with trees of the exact same genotype (called a **clone**) that the trees may appear to have little phenotypic variability (Fig. 1.2b); even then, phenotypic differences can be found. An important question asked by forest geneticists and that will arise frequently in this book is, "Is the observed phenotypic variation caused mostly by genetic differences among the trees or by differences in environmental effects?" In other words, which is more important, nature or nurture?

Although it is impossible to look at the outward appearance of a tree (its phenotype) and know anything about its underlying genotype, there are two experimental approaches available for separating environmental and genetic effects: common garden tests and molecular genetic approaches.

Common Garden Tests

The first approach for separating environmental and genetic effects on phenotype has been used for more than two centuries and aims to separate environment and genetic influences by holding the environment constant, which isolates the genetic effects on phenotypic variability. To accomplish this, seeds are collected from trees (*i.e.* stands of trees) growing in many different environments, and the progeny formed from these seeds are planted in randomized, replicated experiments in one to many field locations. Under these conditions, the environmental influences are similar for all trees and any differences found between progenies from different trees (or from different stands) are mostly due to genetic causes. These experiments are known as **common garden tests**. The basic premise of common garden tests is that it is not possible to measure or observe trees in nature and use these measurements to determine the relative importance of genetic *versus* environmental influences on the observed phenotypic variability. Rather, it is necessary to establish separate experiments (common garden tests) in order to provide all genotypes with common, replicated environments.

As a simple example, consider the heights of 1000 trees measured in a single even-aged plantation of *Pinus taeda*. Each tree has its own genotype and each has experienced its own microclimate and microsite in the unique part of the stand where it grows. Therefore, the measurements of 1000 phenotypes produce a wide range of tree heights, from short to tall, and this is the phenotypic variation for the particular trait of height growth. A common garden experiment designed to assess whether any of this variation is caused by genotypic differences among the 1000 trees might be established by collecting seed from the tallest five trees and the shortest five trees in the stand. The seed from the five trees in each group could be placed into two bags (treatments) labeled tall and short. The seedlings produced from the two different treatments would be planted in randomized, replicated designs in several field locations; these locations would be the common gardens. If after several years, the trees in the tall treatment (progeny from the five tall trees in the parent stand) are consistently taller than the trees in the short treatment (progeny from the short parents), then at least a portion of the original phenotypic variability among the parents must have been caused by genetic differences. This genetic portion of the phenotypic superiority for height growth of the taller parents was passed on to the offspring, as proven by the superior growth of their progeny in common garden environments.

There are many types of common garden tests such as species trials, provenance tests, and progeny tests that are discussed in detail in later chapters. These tests have different specific objectives and therefore, different names; however, they are all examples of common garden experiments planted to unravel the genetic and environmental contributions that cause natural variation.

Molecular Genetic Approaches

The second approach for separating genetic and environmental effects on tree phenotypes uses the techniques of molecular genetics to measure the genotype directly which eliminates the confounding influence of the environment. For example, it is possible to determine the DNA sequences of many genes of a tree (*Chapter 2*) and to make several types of DNA-based genetic maps (*Chapter 18*). It is also possible to measure some types of gene products (such as terpenes and proteins) that are little influenced by the environment. These techniques have developed rapidly in recent years, and as seen in later chapters are used in a variety of ways to study genetic variation in natural populations and to determine genetic differences among trees.

Except for special cases, it is currently not possible to relate DNA sequences of genes directly to the entire phenotype of the tree or to the phenotypic expression of complex traits. For example, although we can elucidate DNA sequences of two trees that have different phenotypic expressions for several traits (*e.g.* growth rate, crown form, etc), we usually cannot say what it is about the differences in their gene sequences that lead to differences in the phenotypic expression. This is because: (1) We still do not understand gene expression well enough at the biochemical and physiological level; and (2) The interaction of environment and genotype is quite intricate and occurs over the entire life span of the tree. The exciting new field of functional genomics (*Chapter 18*) is progressing rapidly and holds much promise for developing this understanding of gene function and expression. Molecular geneticists, physiologists and forest geneticists are working together on several important gene systems to elucidate the complex interactions of genes and the environment during the ontogenetic development of trees.

In the future, molecular approaches for measuring, understanding, managing and manipulating genetic differences among trees will become increasingly more important. However, common garden tests are currently the main approach for isolating genetic and environmental influences on tree phenotypes.

Environmental Sources of Variation

Many sources of environmental variation (Fig. 1.4) are widely appreciated by foresters and extensively studied in forest ecology and silviculture. Differences in environment cause phenotypic variation at a range of scales. On a small scale, phenotypic variation between neighboring trees in the same stand is caused by differences in microclimate, microsite, competition (between trees of the same species, other tree species and understory plants) and exposure to insects and diseases. Large-scale environmental effects on phenotypic expression include differences in elevation, rainfall, temperature regimes and soils that cause tremendous differences in growth rates, tree form and morphology among forests of the same species growing in different locations. For example, *Pseudotsuga menziesii* grows from British Columbia south through the western USA and into Mexico spanning a latitudinal range from $19°-55°$ L and an elevation range from 0-3300 meters above sea level. Not surprisingly, forests of *P. menziesii* in different locations can vary dramatically.

Some aspects of the environment can be altered by silvicultural treatments in managed forests such as fertilization to improve soil nutrients, bedding prior to planting to improve soil moisture conditions, weed control and thinning to reduce intertree competition. Other environmental factors are difficult, if not impossible, to manipulate such as rainfall patterns, freezing weather and disease epidemics. All sources of environmental variation contribute to the rich pattern of phenotypic variability observed in natural and managed forests.

Genetic Sources of Variation

As with environmental variation, genetic variation occurs on a range of scales. The six sources of genetic variation shown in Fig. 1.4 are nested or hierarchical in the sense that there is a natural progression from bottom to top with each lower source being nested within the source above it (*e.g.* species occur within genera and trees within stands). Generally, the progression from bottom to top is accompanied by larger average differences among the genomes. For example, two genera are more distinct genetically than are two species within the same genus.

The sources of genetic variation described in the next paragraph (species, genus, family) are well known to foresters and are the basis of the Latin names given to different taxa. The four sources of variation below the species level are less well known to foresters, yet contribute to large genetic differences among trees within the same species. Understanding the patterns and importance of the different sources of genetic variation is the central thrust of forest genetics.

Genus and Species

On a large taxonomic scale, trees in different families or genera are usually quite distinct genetically. On a smaller scale, different species within the same genus also differ genetically. Consider two related species of pine from the southeastern USA, *P. elliottii* and *P. taeda*. While resembling each other in several ways, there are also distinct differences in growth habit, morphology and reproductive patterns (*e.g. P. elliottii* flowers approximately one month earlier in the spring). These differences are consistently expressed in a range of environments (*i.e.* common gardens) which demonstrates that they are genetically determined. Proper choice of species in a plantation program is often the most important genetic decision that a forester makes. Use of the wrong species can result in lost productivity or health and can sometimes lead to plantation failure if the species is poorly adapted to the planting environment. For plantation programs where the best species is not clearly known, species trials (common garden tests with many candidate species planted over the range of plantation environments) are established (*Chapter 12*).

Provenance and Stand

The term **provenance** refers to a geographical location within the natural range of a species. Variation associated with different provenances (also called **geographic variation**) is discussed extensively in *Chapters 8* and *12* and only briefly mentioned here. Genetic differences among provenances can often be quite large, especially for wide ranging species occupying many diverse climates. It is very common to find patterns of adaptation that have evolved over many generations making provenances well suited to their local growing environments. For example, *Picea abies* has an extremely wide natural distribution in Europe and Asia that spans many different elevations, climates and soil types. Natural selection during the course of its evolution has led to pronounced genetic differences among provenances. Provenances originating from colder regions (farther north or higher elevations) tend to grow more slowly, begin height growth earlier each spring, end height growth earlier each fall and have narrower crowns with flatter branches than provenances from warmer climates (Morgenstern, 1996). These characteristics of provenances from colder regions are adaptations to the drier snow, shorter growing seasons and higher frequency of frosts in colder climates. That these differences are genetic in origin has been demonstrated by common garden experiments, called **provenance tests**, where several provenances of the same species are compared in randomized, replicated studies.

To underscore the importance of common garden tests in determining the relative importance of environmental and genetic causes of geographic variation in forests, consider the example of wood density in *P. taeda* (Zobel and van Buijtenen, 1989; Zobel and Jett, 1995). One portion of the large natural range of *P. taeda* extends approximately 1300 kilometers in a north-south transect from the much colder temperate climates in the northern parts of the range in Maryland to the warmer subtropical climates of Florida. Along this transect, wood density measured in natural stands increases from north to south (more dense wood in the south). When seeds from many points along the transect are planted in common field locations in randomized, replicated tests, the opposite trend is found (trees from seed collected from southern provenances have lighter wood). Therefore, in this example both environmental and genetic differences contribute to the observed natural patterns of variation in wood density, and the environmental influences are opposite to the genetic trends. Common garden tests are required to properly characterize the genetically caused geographic variation.

Understanding the importance and patterns of geographic variation is significant in both tree improvement and gene conservation. In tree improvement programs, breeders choose provenances that are best adapted to produce the desired yield and product quality (*Chapter 12*). In gene conservation programs, knowledge of geographic variation is important for designing sampling schemes to ensure that genes are conserved from all genetically distinct provenances (*Chapter 10*).

Differences can also exist among neighboring stands within the same provenance. Normally, these differences are much smaller than the provenance differences just discussed, and usually differences among stands are mostly caused by environmental differences associated with different site qualities, slope position, etc. However, there can also be average genetic differences between neighboring stands (see *Chapters 5* and *8*).

Tree and Within Tree

Genetic differences among trees of the same species in the same stand are often large. As with human populations, no two trees growing in a natural forest have the same genotype (unless they are members of a clone). The relative importance of genetic and environmental causes of phenotypic variation among trees in the same stand is different for different traits. For example, wood density has stronger genetic control and is less influenced by environment, while growth rate has weaker genetic control and reflects more environmental influence (*Chapter 6*). Genetic variation among trees forms a portion of the genetic diversity of the species, and knowledge of this variation is critical for gene conservation programs. Tree-to-tree genetic variation also forms the main basis of applied tree improvement programs that use selection and breeding to locate and repackage the existing natural variation into improved genotypes. Study of this level of genetic variation is central to the discipline of forest genetics.

Finally, some traits may even show variation within a tree. For example, wood density in conifers is typically lower nearer the pith of the tree and increases in rings towards the outer portion of the tree (Megraw, 1985; Zobel and Jett, 1995). Even within a given annual ring, the density is usually lower in the wood formed early in the growing season compared to the wood formed later. Are these differences environmental or genetic? The answer is both. The tree genotype interacts with its environment throughout the course of the many years of the tree's development. The tree's genotype (*i.e.* its set of genes) remains essentially constant throughout the course of its life; however, different genes are expressed in different seasons and at different ages. The fact that conifers produce lower density wood near the pith is a result of the expression of a particular set of genes interacting with the environment. Towards the outer part of the tree, different genes are expressed (some new ones turned on, others turned off) resulting in a different effective set of genes influencing wood density. Recognition of the importance of developmental regulation of genetic expression has increased markedly in recent years with the advent of new techniques in molecular genetics (*Chapter 2*), and specific examples are provided in *Chapter 18*.

HISTORICAL PERSPECTIVE ON FOREST GENETICS

General Genetics

The earliest domestication of both plants and animals began about 10,000 years ago in the Late Stone Age in several areas of the world; these domestication efforts accompanied development of other technologies such as cooking, making pottery and weaving fibers into cloth (Allard, 1960; Briggs and Knowles, 1967; Table 1.1). Evidence from several early civilizations indicates that seed from superior phenotypes was saved to use for the following year's crop and that through time this practice was effective in developing improved varieties. By 1000 BC (3000 years ago and still prior to the Historical Period), the majority of important food crops had been domesticated and were phenotypically very similar to their appearance today. It is fascinating that these successful crop improvement programs took place in the absence of any knowledge of genetics.

In the general field of genetics, there was a wealth of important developments before the discovery that DNA was the hereditary material in 1944 (Table 1.1). In 1856, Louis de Vilmorin developed progeny testing as a means to rank parents (Briggs and Knowles, 1967). In 1859, Charles Darwin published his hypothesis on natural selection in *The Origin of Species*. A few years later in 1866, Gregor Mendel studied inbred lines of peas and developed the classical laws of diploid inheritance. In 1908, Godfrey Hardy and Wilhelm Weinberg developed the relationships between allele and genotype frequencies in random mating populations that are the foundation of population genetics. In the first 20 years of the twentieth century, Yule, Nilson-Ehle and East showed that multiple segregating genes with similar effects explain the inheritance of quantitative traits. In the 1920s and 1930s, Sir Ronald Fisher derived the statistical concepts of randomization, experimental design and analysis of variance that became the foundations of all modern experimental methods. Also in the 1930s, Fisher, Haldane and others initiated the field of quantitative genetics and introduced the concept of heritability. Still in the 1930s, Jay Lush published the book, *Animal Breeding Plans*, which contained some of the theory and methods of animal breeding that are still used today in animal, crop and tree improvement programs. Sewall Wright also worked over a period of many years in the field of population genetics and developed concepts of path coefficients, inbreeding and strategies for animal breeding.

In the second half of the twentieth century, the field of molecular genetics was developed and began to have major impacts on all areas of biology (see Lewin, 1997 for more details; Table 1.1). James Watson and Francis Crick's discovery of the double helix structure of DNA in 1953 began the era of studying genetics at the level of DNA. Soon thereafter, the triplet nature of the genetic code was discovered along with how the information encoded in the DNA leads to proteins (*Chapter 2*). The 1970s and early 1980s were very important for the development of molecular methods that serve as the basis for many of the techniques used today. These advances included discovery of restriction enzymes, methods to clone DNA, and development of chemical methods for determining the sequence of nucleotides in DNA molecules (*Chapter 4*). These basic methods make up what is called recombinant DNA technology. These techniques were used to develop new genetic marker technologies that have been used extensively in forestry (*Chapter 4*).

In the 1980s, one very important discovery was the polymerase chain reaction (*Chapter 4*) that is used in nearly all areas of biological research. This technique has made it simple and routine to study DNA without the difficult task of cloning the DNA prior to study. The science of biotechnology was also developed during this period (*Chapter 20*). **Biotechnology** is broadly defined as the array of recombinant DNA, gene transfer and tissue culture techniques used in the study and improvement of plants and animals. Biotechnology research in forest trees began in the late 1980s, with primary emphasis on transform-

Table 1.1. Chronology of some important developments through 1990 in general genetics (G) and forest genetics (F). All developments not referenced at the bottom of the table are in the reference list at the end of the book.

Year	Discovery or development	References
BC	G: Early crop and animal domestication	Many[a,d]
1700s	F: Importance of seed origin	Many[b]
1800s	F: Hybridization, vegetative propagation	Many[c]
1856	G: Progeny testing in plants	Vilmorin[d]
1859	G: Natural selection, evolution of species	Darwin (1859)
1866	G: Classical laws of inheritance	Mendel (1866)
1908	G: Gene frequency equilibrium in populations	Hardy[b], Weinberg[b]
1916	G: Inheritance of quantitative traits	Yule *et al.*[a]
1925	G: Modern statistics: Randomization, ANOVA	Fisher (1925)
1930s	G: Mathematical theory of selection	Fisher (1930), Haldane[e]
1930s	G: Genetics of populations, inbreeding	Wright (1931)
1930s	G: Theory and strategies for animal breeding	Wright (1931), Lush (1935)
1942	G: Reconciliation of Darwin's and Mendel's laws	Huxley[e]
1944	G: Discovery of DNA as hereditary material	Avery *et al.* (1944)
1950s	F: Large-scale tree improvement programs	Many[c]
1953	G: Helical structure of DNA	Watson and Crick (1953)
1960s	G: Isozymes for population genetic studies	Soltis and Soltis (1989)
1961	G: Deciphering of genetic code	Nirenberg and Matthai (1961)
1970s	G: Mixed model analysis in quantitative genetics	Henderson (1975, 1976)
1971	F: Isozymes applied to forest trees	Conkle (1971)
1977	G: Chemical determination of DNA sequence	Sanger *et al.* (1977)
1980	G: RFLP mapping techniques	Botstein *et al.* (1980)
1980	F: CAMCORE gene conservation cooperative	Zobel and Dvorak[f]
1981	G: Transformation by *Agrobacterium*	Matzke and Chilton (1981)
1985	G: Polymerase chain reaction	Sakai *et al.* (1985)
1986	F: Paternal inheritance of chloroplast DNA in conifers	Neale *et al.* (1986)
1987	F: First transgenic forest tree	Fillatti *et al.* (1987)

[a] Allard 1960; [b]Morgenstern 1996; [c]Zobel and Talbert 1984; [d]Briggs and Knowles 1967; [e]Ridley 1993; [f]Dvorak and Donahue 1992.

ing and regenerating pines and poplar. The first transgenic forest tree was reported in 1987 (Fillatti *et al.,* 1987). Another important breakthrough during this period was the first successful demonstration of somatic embryogenesis in a conifer (Hakman and von Arnold, 1985).

Forest Genetics

In forest genetics there were also many accomplishments by early pioneers. Between 1700 and 1850, scientists in Europe recognized the importance of provenance variation, created hybrids between some tree species and developed methods for vegetative propagation of some trees (Zobel and Talbert, 1984). Largely these early pioneers were extremely observant, curious individuals with tremendous foresight and persistence. They succeeded in setting the stage for future forest geneticists.

The first half of the twentieth century saw scattered efforts in tree improvement with provenance testing and selections made in a variety of commercially important tree species. Large-scale tree improvement programs were initiated in the 1950s in more than 14 countries (Zobel and Talbert, 1984). At that time little was known about genetic control of different traits in tree species; the pioneers who began these programs relied on their knowledge of crop breeding and the faith that domestication efforts also would be successful in forest trees. A major thrust of these early programs was development of field methods that became important for successful tree improvement programs, such as selection, grafting, pollen extraction, control pollination and progeny test establishment.

The increasing world population and subsequent pressure to utilize remaining natural forests to sustain this population have heightened the awareness and need to conserve genetic resources of forest trees. Tree improvement programs have normally served this role for their commercially important species, and several countries have foreign aid agencies that have helped in exploratory gene conservation efforts. An excellent example of a formal gene conservation organization is the Central American and Mexican Coniferous Resources Cooperative (CAMCORE) founded in 1980 (Dvorak and Donahue, 1992).

In terms of genetics research, population genetics studies were virtually impossible until recent years because of the lack of single gene traits (*i.e.* genetic markers described in *Chapter 4*). Biochemical markers called allozymes first became available for studying population genetics in the 1960s. Allozymes were quickly adapted by forest geneticists resulting in a large effort directed at describing patterns of genetic diversity in natural and artificial populations of forest trees. Much has been learned from these studies about the distribution of genetic variation between and within tree species, and about the evolutionary forces responsible for causing the observed patterns of genetic diversity (*Chapters 7-9*).

The application of molecular genetic techniques to forest trees was well established by the early 1990s. Restriction fragment length polymorphism (RFLP) genetic markers were applied to studies of the inheritance of organelle genomes, the development of genetic maps and measuring genetic diversity. Neale *et al.* (1986) used RFLP markers to show that the chloroplast genome in conifers is inherited through the male parent; they later showed that the mitochondrial genome in *Sequoia sempervirens* is inherited from the paternal parent as well. These novel modes of organelle inheritance have provided unique opportunities to study genetic diversity and phylogeography in conifers using genetic marker data on maternal and paternal lineages (*Chapters 8* and *9*). Many genetic marker types were also developed for the nuclear genome (*Chapter 4*) and these markers were used to construct genetic maps of forest trees (*Chapter 18*).

In the late 1990s, the genomics era came to forest genetics. The DNA sequencing technology was used to build gene catalogs in several species (*Chapter 18*) and methods to understand the function of genes in trees were first introduced. These technologies will be used to understand regulation and expression of genes in forest trees and will lead to the development of new tree varieties that will help to meet global demands for forest products and also maintain valuable genetic resources. The twenty-first century promises to be an exciting and productive period for all of genetics, including forest genetics.

WHY STUDY FOREST GENETICS?

The primary reason to study forest genetics is to provide insight into the evolution, conservation, management and sustainability of the world's forests. In more detail, specific ecological, scientific and practical reasons for studying forest genetics include:

- Forest trees provide an opportunity to study genetic principles in a unique life form. Compared to other organisms, most forest tree species are very long-lived, highly outcrossing, very heterozygous and highly variable among individuals within a species (Conkle, 1992; Hamrick *et al.,* 1992).
- Forest genetics allows study of natural evolution on a large scale. Some species have natural ranges spanning many millions of square kilometers and exhibit extremely intricate patterns of adaptation to past and current environments.
- Knowledge of general forest genetic principles and of genetic structure of forest tree species is required to develop sound gene conservation strategies.
- Forest genetics can help us understand the implications and guide applications of silvicultural reforestation operations such as seed tree and shelterwood systems in forests managed with natural regeneration.
- Forest genetics principles are central to tree improvement programs that develop genetically improved varieties for plantation systems around the world. Genetically improving the yield, health and product quality of these plantations directly enhances the economic and social value of the plantations.
- Biotechnology, including insertion of novel genes by genetic engineering and use of molecular markers to aid breeding and selection decisions, promises to greatly enhance the development of new tree varieties in the future.
- At the gene and genome levels, forest trees are fundamentally different than other organisms commonly used for genetic research, and thus provide many opportunities in the basic sciences. For example, trees are perennial plants that produce large amounts of secondary xylem and should have unique genes and metabolic pathways, and conifers are gymnosperms that are much older evolutionarily than angiosperms. Therefore, the more ancestral genes of conifers will provide useful information about plant evolution in general and, more specifically, about the evolution of function of plant genes.

The subdisciplines of forest genetics (molecular, transmission, population and quantitative forest genetics) are introduced in *Chapters 2-6*. Work in these areas has provided great insights into genetic principles that are important in forest populations. The problems of today and those of the future are quite complex. Scientists from these subdisciplines must work together as teams more than ever before and also work with social and other biological scientists to understand, conserve and ensure the sustainable utilization of the world's forest resources.

CHAPTER 2
MOLECULAR BASIS OF INHERITANCE – *GENOME ORGANIZATION, GENE STRUCTURE AND REGULATION*

Trees, like most living things, begin from a single cell that contains all the genetic information needed for the entire life of the tree. This information is inherited from the trees' parents. The goal of this chapter is to learn more about the hereditary material. Although Gregor Mendel established the conceptual framework for the existence of hereditary material with his classic studies in peas (*Chapter 3*), he had no idea of the biochemical basis of heredity. It was not until 1944 that Oswald Avery and his coworkers discovered that this material was deoxyribonucleic acid (DNA). Two major topics are discussed in this chapter: (1) The molecular structure of the DNA molecule and its organization in the cell; and (2) The structure and regulation of expression of genes. These topics fall within two major subdisciplines of genetics: cytogenetics and molecular genetics. Further reading on these subdisciplines can be found in excellent texts by Stebbins (1971) and Lewin (1997).

GENOME ORGANIZATION

The DNA Molecule

Even before Avery's demonstration in 1944 that DNA was the hereditary material, it was known that chromosomes within the nucleus were composed of DNA. However, the chemical composition and structure of DNA was not fully understood. By the early 1950s, biochemists knew that DNA was a very large molecule made up of four types of chemically linked **nucleotide bases**: A = adenine, T = thymine, G = guanine and C = cytosine (Fig. 2.1). The nucleotide bases are linked to one another by a sugar-phosphate backbone to form a polynucleotide chain. In *Pinus*, there are at least 1×10^{10} nucleotide bases making up the entire nuclear **genome**, which is the totality of all genes on all chromosomes in the nucleus of a cell.

The final and fundamentally important aspect of the DNA molecule is that it exists as a double helix, which was discovered in 1953 by James Watson and Francis Crick (Watson and Crick, 1953). They showed that two polynucleotide chains (or strands), with the sugar-phosphate backbones in opposite (antiparallel) orientation, are bound to one another through a series of hydrogen bonds. Specifically, adenine always pairs with thymine by two hydrogen bonds and guanine always pairs with cytosine by three hydrogen bonds. This chemical structure of the DNA molecule is known as **complementary base pairing.**

The fidelity of the genetic information is maintained during DNA replication because each of the two DNA strands serves as a template for synthesis of a new complementary strand. This is known as **semiconservative replication** (Fig. 2.2) because each daughter cell receives one of the original strands and one newly synthesized strand. This concept is discussed again in *Chapter 3* as it relates to mitosis and meiosis.

© CAB International 2007. *Forest Genetics* (T.W. White, W.T. Adams and D.B Neale)

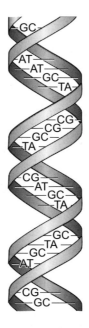

Fig. 2.1. Deoxyribonucleic acid (DNA) is a double-stranded macromolecule composed of four different nucleotide bases (A = adenine, T = thymine, G = guanine, and C = cytosine). The strands are anti-parallel (meaning that the strands' sugar phosphate backbones are in opposite orientation) and the nucleotides are linked together by hydrogen bonds. The structure of the DNA molecule was proposed by James Watson and Francis Crick in 1953.

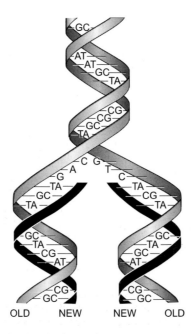

OLD   NEW   NEW   OLD

Fig. 2.2. The DNA molecule is replicated by each of the two strands serving as a template for synthesis of a new strand. This is known as semiconservative replication because each daughter cell at cell division receives one new and one old strand. Complementary base pairing of A with T and C with G ensures fidelity of the genetic code.

Cellular Organization of Genomes

In trees and all higher plants, DNA is found in the **nucleus** and two types of organelles in the cell: the **chloroplasts** and **mitochondria** (Fig. 2.3). Most DNA in a cell is found in the nucleus and this DNA contains the vast majority of genes. The nuclear DNA is divided

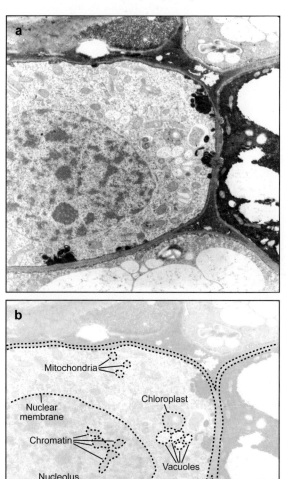

Fig. 2.3. DNA is found in the nucleus, chloroplasts and mitochondria of plant cells. Most of the genetic information is encoded in chromosomes, collectively called chromatin, in the nuclear DNA (nDNA); however, genes related to their respective functions (photosynthesis and respiration) are found in the chloroplast DNA (cpDNA) and the mitochondrial DNA (mtDNA): (a) Original photograph of a cell in the nucellus tissue of a *Pseudotsuga menziesii* ovule; and (b) Dashed lines and labels have been added to highlight important structures. (Photo courtesy of J. Owens, University of Victoria, British Columbia, Canada)

among a number of chromosomes, whose organization and structure is discussed later in this section. The chloroplasts and mitochondria also contain DNA, which encodes a small number of genes related to their respective functions, photosynthesis and respiration. Plant chloroplasts and mitochondria are believed to have descended from cyanobacteria and aerobic bacteria, respectively, and their circular DNA genomes reflect their prokaryotic origin (Gray, 1989). According to the endosymbiont theory, free living bacteria colonized the primitive plant cell and formed a symbiotic relationship with their host; however, over evolutionary time many of the bacterial genes were transferred to the plant nuclear genome. Each chloroplast and mitochondrion contains many copies of the circular DNA molecule and there are several chloroplasts and mitochondria per cell. Therefore, there is a very large number of copies of these genomes in the cell.

Genome Size

The total amount of DNA in the nucleus determines the genome size for a species. Genome size is usually reported as a **C-value**, which is the number of picograms of DNA per haploid nucleus. Genome sizes vary enormously among plant species (Table 2.1) and the evolutionary and practical significance of this variation is a topic of considerable interest (Price *et al.*, 1973). Angiosperm tree species, such as those in the genera *Populus* and *Eucalyptus*, have significantly smaller genomes than gymnosperm species. This raises the concept of the **C-value paradox** observed when comparing the size of eukaryotic genomes: genome size does not increase linearly with apparent evolutionary complexity of the organism (Fig. 2.4). It seems unlikely that amphibians would have more genes than humans and other mammals, and plants would have more genes than most animals, but this is what is observed. This paradox is also apparent between angiosperm and gymnosperm tree species, the latter having more DNA although more evolutionarily primitive. Later in this chapter some possible reasons are discussed as to why plants, and especially gymnosperms, have so much DNA.

Estimation of the DNA contents of conifers began with the work of Miksche (1967), who used a method called Feulgen cytophotometry to show that conifers have among the largest genomes of all higher plants. Deoxyribonucleic acid content can also vary among trees within the same species. Several reports show that DNA content within some tree species increases with increasing latitude (Mergen and Thielges, 1967; Miksche, 1968, 1971; El-Lakany and Sziklai, 1971), although this trend has not been observed in all studies (Dhir and Miksche, 1974; Teoh and Rees, 1976). The positive association between DNA content and latitude led to the hypothesis that increased DNA content in conifers is an adaptation to stressful environments.

Table 2.1. Chromosome numbers, ploidy levels and DNA contents (C-value) for a select list of forest tree species.

Species	Chromosome number (N)	Ploidy level[a]	C-value
Pinus taeda	12	2x	22.0
Pinus radiata	12	2x	23.0
Pinus lambertiana	12	2x	32.0
Pseudotsuga menziesii	13	2x	38.0
Picea abies	12	2x	30.0
Sequoia sempervirens	11	6x	12.0
Eucalyptus grandis	11	2x	1.3

[a]The total number of chromosomes is two times N (*i.e.* diploid) in all cases except *Sequoia sempervirens* where the total is 6N = 66 (hexaploid).

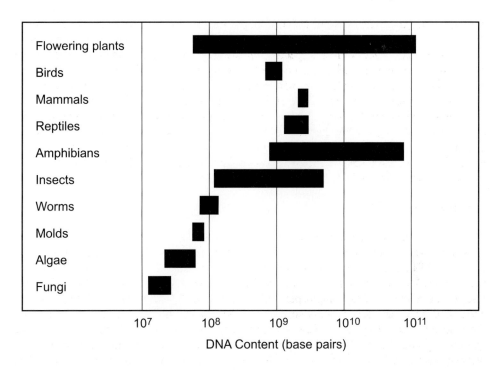

Fig. 2.4. There is not a simple linear relationship between DNA content and evolutionary complexity of an organism, known as the C-value paradox. For example, amphibians often have larger genomes than mammals and many plants have larger genomes than animals. Clearly, many organisms have much more DNA than is needed to encode the structural gene loci necessary for their development and function. Black bars indicate range in size of DNA base pair content among species within the various groupings.

Newton *et al.* (1993) and Wakamiya *et al.* (1993, 1996) have expanded the scope of this hypothesis by showing a positive relationship between the size of genomes of different *Pinus* species and the severity of environmental factors such as temperature and precipitation in the species' native ranges. Their hypothesis is that increased DNA content increases cell volume, a relationship shown earlier by Dhillon (1980), which in turn increases tracheid volume, which then leads to better water conductivity. It seems clear that genome size is an important aspect of plant evolution, and that such differences are, at least in part, adaptive (Stebbins, 1950).

Chromosomes and Polyploidy

The DNA in the cell nucleus is organized into a discrete set of units called **chromosomes** (Fig. 2.5). Chromosomes are usually complexed with DNA binding proteins called histones to form a dense mass called **chromatin**. In many organisms, such as humans and many tree species, chromosomes exist as nearly identical pairs that are called **homologous chromosomes** or **homologous pairs**, and the organisms are said to be diploid. The number of chromosomes can be given as the **diploid (2N)** number, the number in all vegetative cells, or as the **haploid (1N)** number, the number in gametic cells. For example, all species of *Pinus* are diploid and each has a total of 2N = 24 chromosomes. The haploid number is 1N = 12, representing the number of homologous pairs.

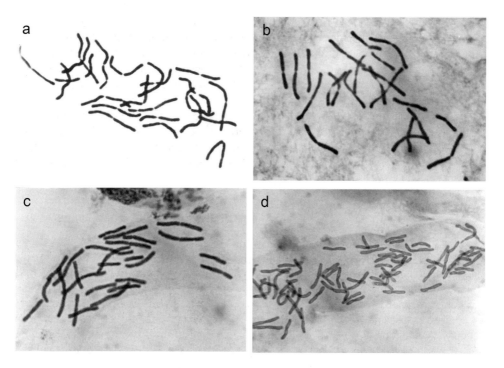

Fig. 2.5. Light microscope photographs of conifer chromosomes prepared from root tips (x1200): (a) *Pinus jefferyi* 2N = 24; (b) *Metasequoia glyptostoboides* 2N = 22; (c) *Pseudotsuga menziesii* 2N = 26; and (d) *Sequoia sempervirens* 2N = 66. (Photos courtesy of R. Ahuja, Institute of Forest Genetics, Grosshansdorf, Germany (retired))

The number of chromosomes in gymnosperms varies little among species (Sax and Sax, 1933; Khoshoo, 1959, 1961; Santamour, 1960) (Table 2.1). Most species have a diploid number of 22 or 24 chromosomes. All members of the family *Pinaceae* have 24 chromosomes with two known exceptions. The first exception is *Pseudotsuga menziesii* with 2N = 26 chromosomes, whereas all other species in the genus have 2N = 24 (Silen, 1978). The type of chromosomal rearrangement leading to the extra chromosome in *P. menziesii* is not clear, but it is hypothesized that one of the chromosomes in a 2N = 24 progenitor broke into two, somehow resulting in the formation of an extra chromosome pair (Silen, 1978).

The second exception is *Pseudolarix amabilis* of the monotypic genus *Pseudolarix*. *P. amabilis* has a total of 44 chromosomes (Sax and Sax, 1933) and is certainly a polyploid species. **Polyploids** are species which have 4, 6, 8 or even higher times the haploid number of chromosomes (Stebbins, 1950). The designation X indicates the base number of chromosomes, *e.g.* in *Pinus* X = 12. For example, species with 4X chromosomes are tetraploids, 6X are hexaploids, and so forth. *P. amabilis* is most likely a tetraploid that was derived from an ancestor of X = 12, although it is not clear how it came to have 44 chromosomes versus the expected number of 48.

Tree species in two closely related families of conifers, *Taxodiaceae* and *Cupressaceae*, all have 2N = 22 chromosomes with two exceptions. The first is *Juniperus chinesis pfitzeriana* that has 44 and is most likely a tetraploid. The second very notable exception is *Sequoia sempervirens* that has 66 chromosomes (Hirayoshi and Nakamura,

1943; Stebbins, 1948), making it the only known hexaploid conifer. Stebbins (1948) argued that *S. sempervirens* is an auto-allohexaploid, meaning that there was a doubling of one common genome to form an autotetraploid species followed by a subsequent hybridization to a different diploid species to form the hexaploid. The progenitor species are not known and are probably extinct, although *Metasequoia glyptostroboides* (2N = 22) may be one of the parent species.

Chromosome number and ploidy levels vary much more among angiosperm forest tree species than among gymnosperms. Nearly all species, however, have X values between 10 and 20. Species in genera commonly used in tree improvement, such as *Populus* and *Eucalyptus*, are nearly all diploids, whereas genera such as *Salix, Betula* and *Acacia* have many polyploid species.

Karyotype Analysis

Cytogenetics includes more than just the study of the number of chromosomes and ploidy levels; it also is concerned with morphological differences among chromosomes. This is known as **karyotype analysis**. The basic approach is to examine metaphase chromosomes (see explanation of mitosis in *Chapter 3*) from actively dividing cells, such as root tips, under a light microscope. The chromosomes must be stained with special chemicals to be seen; the two commonly used are Feulgen and acetocarmine. Several variables are generally measured, but the most common are length of chromosome arms and location of secondary constrictions (Fig. 2.6). The karyotypes are used to distinguish chromosomes from one another within a species and as characteristics in taxonomic analyses (*Chapter 9*).

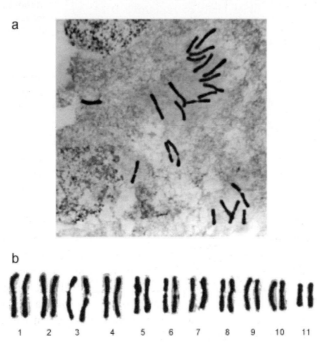

Fig. 2.6. Light microscope photographs of *Metasequoia glyptostoboides* (x1200): (a) Somatic chromosome complement (2N = 22) and (b) Karyotype showing 11 pairs of metacentric chromosomes. (Photos courtesy of R. Ahuja, Institute of Forest Genetics, Grosshansdorf, Germany (retired))

Secondary constrictions are thin regions along chromosomes where nucleoli reside and are also called **nucleolus organizer regions (NORs).** The NORs contain the ribonucleic acid (rRNA) genes that form ribosomes, structures important in the translation of the genetic information coded in DNA sequences into structural proteins. Karyotype analysis has been used extensively in organisms such as *Drosophila* and wheat, but relatively infrequently in forest trees. Most work has been done in *Pinus* (Natarajan *et al.*, 1961; Saylor, 1961, 1964, 1972; Yim, 1963; Pedrick, 1967, 1968, 1970; Borzan and Papes, 1978; MacPherson and Filion, 1981; Kaya *et al.*, 1985); *Picea* (Morgenstern, 1962; Pravdin *et al.*, 1976); *Pseudotsuga menzesii* (Thomas and Ching, 1968; Doerksen and Ching, 1972; De-Vescovi and Sziklai, 1975; Hizume and Akiyama, 1992); and *Sequoiadendron giganteum* and *Calocedrus decurrens* (Schlarbaum and Tsuchiya, 1975a,b). The limited number of karyotype analyses reported in conifers is probably because it is difficult to make distinctions among conifer chromosomes using conventional approaches. Little work has also been done in angiosperm forest species due to the small size of their chromosomes. More powerful cytogenetic techniques, such as fluorescent *in situ* hybridization and confocal microscopy, are leading to a rebirth in cytogenetics of forest trees (Brown *et al.*, 1993; Doudrick *et al.*, 1995; Lubaretz *et al.*, 1996).

Repetitive DNA

The protein-coding and regulatory genes in higher eukaryotic organisms make up only a very small fraction of the total DNA in the genome. This fraction is generally known as the single- or low-copy fraction. The vast majority of the genome is composed of nonprotein-coding DNA or other kinds of **repetitive DNA**, so called because the DNA sequences are repeated in the genome (Britten and Kohne, 1968). **Deoxyribonucleic acid reassociation kinetics** analysis is a technique that has been used extensively to estimate the proportion of single- or low-copy DNA *versus* repetitive DNA (Box 2.1). Miksche and Hotta (1973) were the first to use this technique on conifers and showed that repetitive DNA content in pines is generally much greater than in other plants. They also suggested that repetitive DNA content and genome size might be positively correlated in conifers. Kriebel (1985) determined that 25% of the *Pinus strobus* genome is of single- or low-copy DNA; the remaining 75% is of more or less a continuous distribution of repetitive DNA classes and not just a few discrete size classes as earlier reports had suggested. Kriebel (1985) also estimated that only 0.1% of the pine genome codes for genes that are expressed, so most of the DNA in pines has some function other than coding proteins.

There are many types of DNA sequences that make up the repetitive DNA in plants and all are not known. However, four types have been studied to some extent in forest trees: (1) Ribosomal DNA; (2) Minisatellite and microsatellite DNA; (3) Retrotransposons; and (4) Pseudogenes.

Ribosomal DNA

There are two classes of ribosomal RNA genes, called rDNA, which code for the RNA that forms the two types of subunits of mature ribosomes: (1) The 18S-5.8S-26S rRNA genes (18S-26S rDNA); and (2) The 5S rRNA genes (5S rDNA). These genes are highly repeated along chromosomes and exist in thousands of copies in plant genomes (Flavell *et al.*, 1986; Rogers and Bendich, 1987). Hotta and Miksche (1974) estimated that there were between 10,000 and 30,000 copies of the 18S-5.8S-26S rRNA genes in *Pinus*. Variation in the number of copies of rRNA genes within and between species has been studied in several

Box 2.1. Estimating the proportion of repetitive DNA by the technique of reassociation kinetic analysis.

Deoxyribonucleic acid reassociation kinetic analysis is a technique that can be used to estimate the proportions of DNA in the genome that occur in varying number of copies (Fig. 1). Double-stranded DNA (Step 1) is first fragmented into pieces of approximately several hundred base-pairs (Step 2). Next, these shorter pieces are denatured into single strands by heating (Step 3). The single-stranded DNA is then allowed to reassociate by complementary base pairing (Step 4).

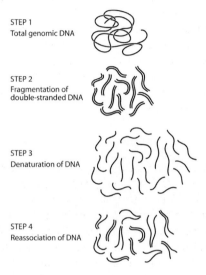

STEP 1
Total genomic DNA

STEP 2
Fragmentation of
double-stranded DNA

STEP 3
Denaturation of DNA

STEP 4
Reassociation of DNA

Fig. 1. Steps in the fragmentation and reassociation of DNA.

The rate of reassociation is plotted in what is known as a **Cot curve**, where Cot is an abbreviation for the product of DNA concentration (C_0) and time of incubation (t) (Fig. 2). Highly repeated DNA sequences have low Cot values and single-copy DNA sequences have high Cot values. The point at which one-half of the DNA in a sample is reassociated ($Cot_{1/2}$) is used as a standard measure of the proportion of repeated DNA.

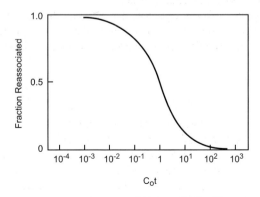

Fig. 2 An example of a Cot curve.

conifer species (Strauss and Tsai, 1988; Strauss and Howe, 1990; Bobola *et al.*, 1992, Govindaraju and Cullis, 1992); however, precise estimates of copy number are technically difficult to obtain. The 18S-5.8S-26S rRNA genes are found at 10 or more locations in the genome and usually correspond to the NORs (Fig. 2.7), whereas the 5S rRNA genes are found at only one or possibly two locations (Cullis *et al.*, 1988a; Hizume *et al.*, 1992; Brown *et al.*, 1993; Doudrick *et al.*, 1995; Lubaretz *et al.*, 1996; Brown and Carlson, 1997). Bobola *et al.* (1992) estimated that rRNA genes might account for as much as 4% of the entire *Picea* genome, which is a considerable proportion but still only a fraction of the total amount of repetitive DNA in the genome.

Minisatellite and Microsatellite DNA

Minisatellite DNA is a type of repetitive DNA where a core sequence of 10-60 base pairs is repeated many times and is dispersed throughout the genome. Minisatellites of variable lengths, called **variable number tandem repeats (VNTRs)**, were first discovered in humans and have been used for **DNA fingerprinting** (Jeffreys *et al.*, 1985a,b). Polymorphic minisatellite sequences have been detected in both angiosperm and gymnosperm tree species (Rogstad *et al.*, 1988, 1991; Kvarnheden and Engstrom, 1992).

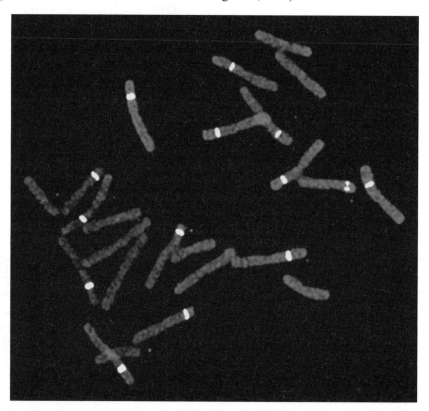

Fig. 2.7. Fluorescent *in situ* hybridization (FISH) in *Picea glauca* (2N = 24). A DNA probe from soybean encoding the 18S-5.8S-26S ribosomal DNA (rDNA) genes was labeled with a fluorescent tag (fluorescein isothiocyanate) and hybridized to *P. glauca* chromosomes. The 14 bright regions correspond to seven homologous pairs of rDNA loci found in *P. glauca*. Each region contains many copies of the rDNA genes. (Photo courtesy of G. Brown and J. Carlson, University of British Columbia, Canada)

Microsatellites, or **simple sequence repeats (SSRs)**, are similar to minisatellite DNA except that the core-repeat length is only 1-5 base pairs. For example, the di-nucleotide sequence AC repeated 10 times would be indicated as $(AC)_{10}$. Microsatellite sequences are distributed throughout genomes and can vary widely in the number of repeats among individuals (Litt and Luty, 1989; Weber and May, 1989). Microsatellites are found in forest tree species and are being developed for use as genetic markers (*Chapter 4*). One study estimated that there were 40,000 copies of the $(AG)_N$ microsatellite sequence in *Picea abies* (Pfeiffer *et al.*, 1997).

Transposons and Retrotransposons

Another class of highly repeated DNA found in forest trees is **transposable DNA**. **Transposons** are mobile DNA elements that can excise from one location in the genome and reinsert at another location. Transposons were discovered first in maize through the pioneering work of Barbara McClintock. **Retrotransposons** are a special type of transposon found in all eukaryotic genomes. These DNA elements are synthesized from messenger RNA molecules (see *Gene Structure and Regulation* section of this chapter) and then integrate into the chromosomal DNA. A retrotransposon in *Pinus radiata* exists in more than 10,000 copies (Kossack and Kinlaw, 1999). Likewise, a retrotransposon in *Pinus elliottii* exists in a very high number of copies and is dispersed throughout the genome (Kamm *et al.*, 1996) (Fig. 2.8). It is clear that retrotransposons are abundant and ubiquitous in *Pinus* and other conifers, although no functional or adaptive significance to these DNA sequences has been postulated.

Pseudogenes

The final known class of repetitive DNA sequences that contributes to the large genomes of conifers is **pseudogenes**, or false genes as the name implies. Pseudogenes are non-functional copies of genes and are created by either: (1) Direct duplication of the DNA sequence of a structural gene during chromosomal replication; or (2) Synthesis from messenger RNA molecules, which are called processed pseudogenes (see *Gene Structure*

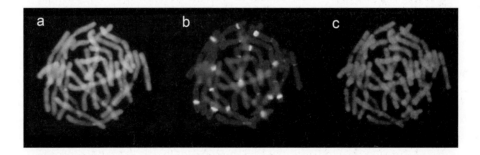

Fig. 2.8. Localization of a Ty1-copia retrotransposon family and the 18S-5.8S-26S ribosomal DNA genes in *Pinus elliottii* by fluorescent *in situ* hybridization (FISH): (a) Metaphase chromosomes stained with DAPI (4',6-diamidino-2-phenylindole); (b) Same metaphase chromosomes stained with a 18S-5.8S-26S ribosomal DNA probe (light spots showing discrete locations of rDNA); and (c). Same metaphase chromosomes hybridized with Ty1-copia DNA probe from *P. elliottii*; staining occurs throughout all chromosomes showing broad distribution of the Ty1 element. (Photo courtesy of R. Doudrick, Southern Institute of Forest Genetics, Saucier, Mississippi)

and Regulation section of this chapter). The first suggestion that pseudogenes might be abundant in conifers came from data on *Pinus* restriction fragment length polymorphisms (RFLP; *Chapter 4*). Restriction fragment length polymorphism analyses revealed that there were significantly more copies of genes in conifers than expected based on the numbers of copies of genes in angiosperms (Devey *et al.*, 1991; Ahuja *et al.*, 1994; Kinlaw and Neale, 1997). It is possible, but unlikely, that all these extra copies represent functional copies of these genes; it is more likely that most are pseudogene copies.

In the alcohol dehydrogenase (ADH) gene family in *Pinus banksiana*, Perry and Furnier (1996) found seven functional ADH genes, whereas no more than two to three genes are found in angiosperms. Further analysis of the ADH gene family revealed additional pseudogene copies of ADH in the *Pinus* genome. Kvarnheden *et al.* (1995) have also found direct evidence for a pseudogene of the p34^{cdc2} protein kinase gene in *Picea*, a gene involved in controlling the cell cycle. Therefore, it is clear that pseudogenes make up a considerable portion of the conifer genome and may in part explain the large size of these genomes.

GENE STRUCTURE AND REGULATION

Before describing how genes are structured and regulated, it is important to first understand exactly what a gene is. The classical (Mendelian) definition of a **gene** is the unit of heredity that resides at a specific location on a chromosome, called a **locus**. There can be varying forms of the gene at a locus, called **alleles**, which contribute to the phenotypic differences among individuals. In diploid organisms, a maximum of two alleles can occur at any given locus in a single matching pair of chromosomes. If the alleles are identical (meaning that they have the same DNA sequence), the individual is said to be **homozygous** at that locus. Likewise, if the alleles are different, the individual is **heterozygous** at that locus. In a population of individuals, there can be many alleles at each locus and individuals can be either homozygous or heterozygous for various combinations of the alleles (*Chapter 5*).

The first biochemical description of a gene was provided in 1941 by George Beadle and Edward Tatum who proposed the **one gene-one enzyme** hypothesis (Beadle and Tatum, 1941). More recently, three classes of genes are recognized: (1) Structural genes that code for proteins; (2) Structural genes that code for rRNAs and transfer RNAs (tRNAs); and (3) Regulatory genes that serve as recognition sites in the genome for various factors that control the expression of genes.

The Central Dogma and the Genetic Code

The mechanism by which the linear sequence of nucleotide bases (A, T, G and C) along a DNA molecule is interpreted and processed to make proteins and other gene products is known as the **central dogma** (Fig. 2.9) and is one of the great marvels in biology. The fundamental aspect of this process is the **genetic code** (Fig. 2.10). Each triplet of nucleotide bases, called a **codon**, codes for a single amino acid. For example, the sequence AAA specifies the amino acid lysine. Because there are 20 amino acids, codons must be made of at least three nucleotides in order to have a unique codon specifying each amino acid. A codon of two nucleotides would have only 16 (4^2) possible combinations of the four bases, whereas 64 (4^3) combinations are possible with a triplet codon sequence. This leads to what is known as **degeneracy** of the genetic code, because nearly all 20 amino acids can be specified by more than one codon. Most often it is the first two nucleotides that specify the amino acid, while the third nucleotide can be any of the four. This lack of specificity in the third nucleotide is known as the **wobble hypothesis**. In the next section, the individual steps leading from the information in the DNA sequence to a polypeptide are outlined.

Transcription and Translation

There are two fundamental steps in producing a polypeptide from the information encoded in the DNA molecule: **transcription** and **translation** (Fig. 2.9). Transcription, which takes place in the nucleus, is the process of synthesizing a single-stranded RNA molecule, called **messenger RNA** (mRNA), from one strand of the DNA template. The enzyme RNA polymerase binds to the DNA template and controls the synthesis of the mRNA. The DNA strand that serves as the template for mRNA synthesis is called the anti-sense strand. The complementary and non-transcribed DNA strand is called the sense strand, because it has the same sequence of nucleotide bases as the mRNA. The small exception to the sense DNA strand and the mRNA being identical is that RNA contains the nucleotide base uracil (U) in the place of thymine used in DNA. As the name implies, mRNA is a copy of the genetic information encoded in the DNA sequence, and the mRNA itself is used as a template to assemble the proper linear sequence of amino acids into a polypeptide chain.

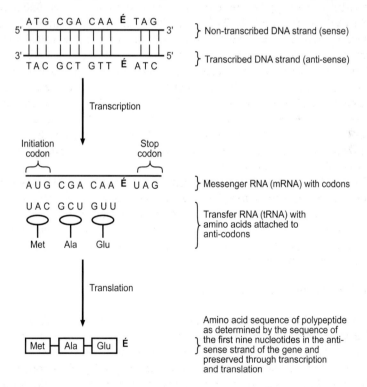

Fig. 2.9. The central dogma refers to the two processes that lead to the synthesis of a polypeptide from the DNA blueprint. In the first process called transcription, RNA polymerase synthesizes a single-stranded RNA molecule, called messenger RNA (mRNA), from the anti-sense DNA template strand. The nucleotide base T found in DNA is replaced by the nucleotide base U in RNA. In the second process called translation, the mRNA joins with a ribosome and amino-acid-carrying transfer RNAs (tRNAs) to build a polypeptide. A polypeptide consists of many amino acids joined together by peptide bonds formed during translation. É represents the remainder of the DNA and corresponding polypeptide sequences that are not shown.

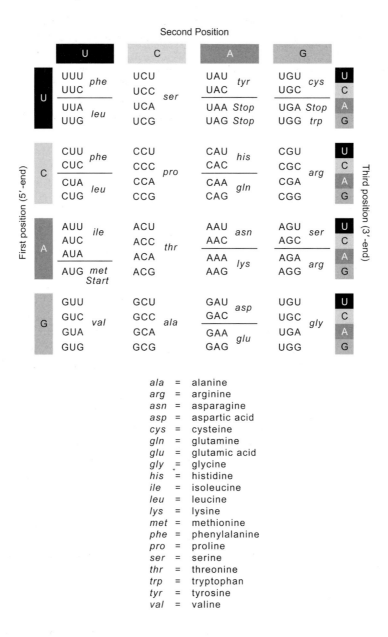

Fig. 2.10. Each of the 20 amino acids (bottom list) is specified by one or more nucleotide triplets (top matrix), which are known as codons. In addition, there is one codon (AUG) that initiates polypeptide synthesis and three codons that terminate synthesis (UAA, UAG and UGA).

Following transcription, mRNAs are exported from the nucleus to the cytoplasm where **translation** occurs (Fig. 2.9). Translation is the process of synthesizing a polypeptide (a polymer made up of amino acids linked by peptide bonds) translated from the information encoded in the mRNA. Translation begins with a ribosome binding to an mRNA. A tRNA carrying the first amino acid then pairs with the mRNA. A tRNA is a small RNA molecule with a three-nucleotide sequence called an anti-codon, which is the complementary sequence to a specific codon of the mRNA. Only a tRNA with the correct anti-codon can pair with each codon on the mRNA and, because there is a specific amino acid associated with each tRNA anti-codon, the order of codons on the mRNA specifies the order of amino acids that are linked to form the polypeptide. Polypeptide synthesis proceeds one amino acid at a time until the complete polypeptide is formed. One or more polypeptides come together to form proteins. There are many types of proteins, but among the most important are enzymes, which catalyze biochemical reactions in the cell.

Structural Organization of a Gene

General Concepts

The simplest model for the structural organization of a gene is one having a linear array of 3X nucleotides coding for a polypeptide of X amino acids. In reality, eukaryotic genes are significantly more complex in their structural organization (Fig. 2.11). Genes rarely are structured in a continuous linear sequence; rather, they are interrupted by one or more non-coding regions called **introns**. The coding regions are called **exons**.

Introns can be extremely large (sometimes many thousands of base pairs) in animal genes; in plant genes, introns are generally no more than a few hundred base pairs in length. During transcription both introns and exons are copied; therefore, the intron sequences must subsequently be deleted. This process, called **RNA splicing**, results in the mature mRNA for translation. This mechanism also enables a process called **alternative splicing** to occur whereby one subset of exons are assembled to code for one polypeptide and a different subset of exons are used to code for an alternative polypeptide.

The starting point for the initiation of transcription along the DNA template is called the **promotor** region (Fig. 2.11). Two DNA sequence motifs that are found in promotors of all eukaryotic genes are the **TATA box** and the **CCAAT box** (Fig. 2.11). The function of the TATA box is to identify the location of the start of transcription by facilitating the denaturation of the double helix. The CCAAT box is involved in the initial recognition by RNA polymerase. Some eukaryotic genes also have **enhancer** elements that act to increase the efficiency of promotors.

Translation of the mRNA template must begin at a precise location if the template is to be read correctly to produce the intended polypeptide. This is specified by an **initiation codon**, AUG, which is the codon for methionine. Therefore, all polypeptides begin with the amino acid methionine. Likewise, completion of translation is specified by one of three **termination codons**: UAG, UAA, or UGA (Fig. 2.10).

Gene Structure in Forest Trees

Gene structure in forest trees has not been thoroughly investigated. Most gene structure studies are based on the DNA sequence of just the coding region of the gene as can be derived from a complementary DNA (cDNA) sequence (*Chapter 4*, Box 4.2). This provides the DNA sequence of the exons, but not the introns or the non-transcribed regions upstream or downstream of the coding regions. There are far fewer examples of gene structure studies derived directly from the entire genomic DNA sequence, although a few reports provide some insight into gene structure in trees. The ADH gene structure in *Pinus taeda* is remarkably similar to that in other plants (Fig. 2.12). The number and position of introns is highly conserved, although lengths of introns vary somewhat and DNA sequences of introns vary significantly among organisms. Promotor sequences such as the TATA and CCAAT

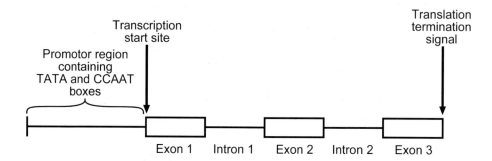

Fig. 2.11. The structural organization of a gene can be very complex. Most genes are made up of both coding (exons) and non-coding regions (introns). Both exons and introns are transcribed, but the intron segments are spliced out before translation. Promotor sequences such as the TATA and CCAAT boxes are in the non-transcribed regions at the 5' end of the gene and are involved in the regulation of gene expression.

boxes are found at their predicted locations. Regions far upstream of coding regions where enhancer sequences are found are rarely sequenced; when they are, there is little similarity to corresponding sequences in other plant genes. This would suggest that a large portion of the phenotypic differences among and within tree species is due to the regulation of genes, rather than differences in the structure or DNA sequence of the coding regions of genes.

Regulation of Gene Expression

General Concepts

Regulation of gene expression can occur at many places in the process that begins with transcription and ends with a mature protein. Steps where regulation commonly occurs include: (1) Transcription; (2) mRNA processing; (3) Transport of mRNA to the cytoplasm; (4) mRNA stability; (5) Selection of mRNAs for translation; and (6) Post-translational modification. Although the expression of genes can be regulated at any one or all of these steps, transcriptional regulation is the most understood and possibly the most important.

Factors that regulate transcription are of two types: (1) **Cis elements**; and (2) **Trans-acting factors**. Cis elements, such as promotors and enhancers, are those found adjacent to genes, whereas trans-acting factors, such as DNA binding proteins, are coded at genetic loci not adjacent to the genes they regulate. The most well known form of trans-acting factors are **transcription factors** which are proteins that bind to promotors and enhancers and effect the rate, developmental timing, and cell specificity of gene expression. Clearly, the regulation of gene expression is as important as gene structure, if not more important in determining the phenotype.

Gene Regulation in Forest Trees

The regulation of gene expression in forest trees has been studied much more extensively than gene structure. Gene expression studies are conducted to better understand the genetic control of: (1) Physiological and metabolic processes such as photosynthesis, nitrogen metabolism, embryogenesis, flowering, and lignin biosynthesis; (2) Responses to abiotic

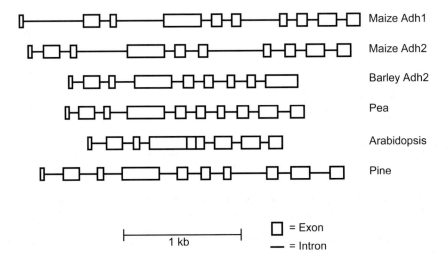

Fig. 2.12. The structural organization of the alcohol dehydrogenase (ADH) gene in *Pinus taeda* is very similar to that in other plants. All ADH genes have 9-10 exons, but are separated by introns of varying lengths. (Figure courtesy of D. Harry, Oregon State University)

stresses such as drought and temperature extremes; and (3) Responses to biotic stresses such as insects and disease. Reviewing the results of these studies is beyond the scope of this book. However, two studies that illustrate unique patterns of gene regulation discovered in trees are discussed.

Light-independent gene expression in conifers. Light energy is captured in plants by the green pigment chlorophyll. In most plants, light is required for chlorophyll biosynthesis; grown in total darkness plants are yellow, or etiolated, because they lack chlorophyll. However, conifer seedlings germinated and grown in complete darkness are green; therefore, chlorophyll biosynthesis must not be dependent upon light.

Molecular geneticists have recently investigated this unique aspect of conifer biology at the level of gene regulation. Studies have shown that a family of genes coding for the light-harvesting complex proteins (LHCP) are transcribed in dark-grown seedlings and that LHCPs are found in the cell (Alosi *et al.*, 1990; Yamamoto *et al.*, 1991; Mukai *et al.*, 1992; Canovas *et al.*, 1993). The LHCP genes have been cloned and sequenced and they are very similar in structure and sequence to their angiosperm counterparts (Jansson and Gustafsson, 1990, 1991, 1994; Kojima *et al.*, 1992). For some reason, these genes are not expressed in dark-grown angiosperms, supporting our earlier hypothesis that differences in phenotype can be due to the differences in the expression of a gene without changes in the amino acid sequence of the polypeptide it codes.

A recent study has shown that the DNA sequences of the promotor region of a LHCP gene from *Pinus thunbergii* cause the gene to be expressed in dark-grown plants (Yamamoto *et al.*, 1994). This was determined using a powerful technique called a **reporter gene assay.** A LHCP gene from *P. thunbergii* was cloned and then the coding region was removed, leaving just the promotor region. The *P. thunbergii* promotor was then fused to a reporter gene called GUS (*B*-glucuronidase) which is an enzyme that produces a blue-staining product if the enzyme is present and substrate is added. Yamamoto *et al.* (1994) genetically engineered (*Chapter 20*) a rice plant that contained the *P. thunbergii* promotor fused to the reporter gene. These genetically engineered rice plants were then grown in complete darkness. Presence of blue staining cells was proof that the

P. thunbergii promotor initiated gene expression in the dark. Although further study is required to identify the exact region of the promotor and DNA sequences that permit gene expression in the dark, discovery of such controlling elements could have practical value for both genetically engineering plants and for understanding the molecular regulation of photosynthesis.

Regulation of genes that control the biosynthesis of lignin. Trees produce large amounts of wood. The three major chemical constituents of wood, or secondary xylem, are cellulose, hemi-cellulose, and lignin. Lignin is the glue that holds the cellulose fibers together. Pulping of wood requires the removal of lignin to obtain pure fibers for making paper. It is not surprising that the biochemistry and genetics of lignin are of major interest in forestry. Reductions in the quantity or changes in quality of lignin in trees grown for pulp and paper could make significant improvements in pulp yield.

The biochemical pathway leading to the biosynthesis of lignin is fairly well known (Fig. 2.13). Recently, many of the genes coding for enzymes in the pathway have been cloned from *Pinus* and *Populus* including genes coding phenylalanine ammonia-lyase (PAL; Whetten and Sederoff, 1992; Subramaniam *et al.*, 1993); O-methyltransferase (OMT; Bugos *et al.*, 1991; Li *et al.*, 1997); 4-coumarate:coenzyme A ligase (4CL; Voo *et al.*, 1995; Zhang and Chiang, 1997; Allina *et al.*, 1998); and cinnamyl alcohol dehydrogenase (CAD; O'Malley *et al.*, 1992; Van Doorsselaere *et al.*, 1995). The cloning of these genes enables the study of their expression, naturally occurring variation, and their relationship to lignin content variation in wood.

Genetic engineering (*Chapter 20*) of *Populus* for the OMT and 4CL genes demonstrates that genetic manipulation of these genes can affect lignin quantity and quality. Hu *et al.* (1998) reported a 40-45% reduction in lignin quantity in *Populus tremuloides* trees that were genetically engineered with an anti-sense copy of 4CL. **Anti-**

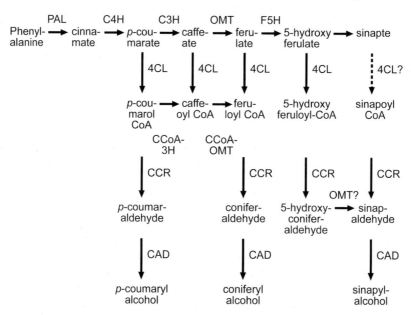

Fig. 2.13. The biosynthetic pathway leading from phenylalanine to the three different lignin monomers involves several enzymatic steps. Most of the gene coding enzymes in the pathway (PAL, C4H, etc) have been cloned from forest tree species and their structure and expression patterns are actively being investigated. (Figure courtesy of R. Sederoff, North Carolina State University, Raleigh)

sense suppression of gene expression occurs because the mRNA produced by the anti-sense copy of the gene, which was put into the plant by genetic engineering, binds to the endogenous sense mRNA through complementary base pairing and effectively reduces the concentration of the mRNA in the cell. In addition to reduced 4CL expression and decreased lignin content, cellulose content was increased and increased growth rates were also observed.

The final enzymatic step in the biosynthesis of lignin monomers is controlled by CAD. Mackay *et al.* (1997) found a naturally occurring mutant in *P. taeda* that had significantly reduced CAD expression and abnormal lignin. Although the genetic engineering of the lignin biosynthetic pathway is still in its infancy, early results such as these demonstrate the potential for modifying lignin content in trees grown for pulp.

SUMMARY AND CONCLUSIONS

The heredity material is encoded in DNA, which is a double-stranded molecule comprised of four nucleotide bases, adenine (A), thymine (T), guanine (G) and cytosine (C). Adenine is paired through hydrogen bonds to T, as is G with C; which is called complementary base pairing and is the foundation for DNA replication and for synthesis of mRNA.

Deoxyribonucleic acid is found in three cellular compartments in the plant cell: (1) Nucleus; (2) Chloroplasts; and (3) Mitochondria. The total DNA complement in the nucleus is called the genome and genome sizes vary enormously, with conifers having among the largest genome sizes in all plants.

The DNA found in the nucleus is organized in the form of chromosomes. In diploid (2N) species, chromosomes are found in pairs called homologous pairs. Some species have more than two sets of chromosomes and are known as polyploids. Chromosomes can be distinguished from one another using various cytogenetic techniques, which collectively are called karyotype analysis.

The genome is made up of various classes of DNA; the smallest proportion is that which encodes structural gene loci such as proteins and ribosomal RNA genes. The largest proportion of the genome codes for various types of repetitive DNA whose function is largely unknown.

The individual hereditary units in the genome are called genes and reside at locations in the genome called loci. Variant forms of genes at a given locus are called alleles. The molecular structure of a gene is complex. Genes are comprised of coding regions called exons, which are interrupted by non-coding regions, called introns. Genes also have non-coding regions that are involved in the regulation of their expression, such as promotor and enhancer sequences.

The process by which the genetic information encoded in the linear sequence of the four nucleotide bases is converted into proteins and other gene products is known as the central dogma. Each triplet of nucleotide bases is called a codon and codes for a single amino acid. The enzyme RNA polymerase synthesizes an RNA molecule (called mRNA) from the DNA template. This process is known as transcription. The mRNA molecule then moves from the nucleus to the cytoplasm of the cell and joins with a ribosome and tRNAs to link the individual amino acids encoded in the mRNA transcript into a polypeptide chain. This step is called translation. There are many places during transcription and translation where the regulation of gene expression can occur. Gene regulation is very important in determining phenotype and causing variation among phenotypes.

Chapter 3
Transmission Genetics – *Chromosomes, Recombination and Linkage*

In *Chapter 2*, we learned that the genetic material is found primarily in the nucleus of the cell and that it is organized in the form of chromosomes. In this chapter, we explore how the genetic material is inherited from generation to generation and the changes in the organization of the genetic material that occur between generations. This subdiscipline of genetics is called **Transmission Genetics** or **Mendelian Genetics** after Gregor Mendel who discovered the laws of inheritance with his experiments on garden peas. The chapter is organized into three major sections: (1) Mendelian Genetics; (2) Transmission and inheritance of chromosomes; and (3) Extensions of Mendel's laws.

Mendelian Genetics

Mendel's Crossing Experiments with Peas

In 1865, the Austrian monk Gregor Mendel began his series of crossing experiments with garden peas. By any standard, Mendel was a very gifted scientist who appreciated the importance of carefully designing and interpreting his experiments. Mendel sought to understand the patterns of inheritance of seven different characters in peas by crossing plants with identifiably different phenotypes of a particular character, and then following the presence or absence of the character in successive generations. The traits Mendel studied were morphological attributes of the seed, pod, flower and stem. The types of crosses Mendel made are still widely used in genetics today. Unfortunately, Mendel's discoveries were largely ignored during his lifetime, and it was not until 1900 that his work was rediscovered and its importance finally appreciated.

Monohybrid Cross

The first crosses Mendel made were between inbred lines of peas that were true breeding for opposite types of a character, for example, tall x dwarf, yellow x green seed, round x wrinkled seed coat, etc. These are called **monohybrid crosses** because the peas were inbred lines that differed for a single character (Fig. 3.1). In the progeny from the hybrid cross, called the F_1 **generation**, he observed that all offspring were always just one of the two alternate types, *e.g.* tall. Never were both types observed or were offspring of some intermediate type.

Mendel then self-pollinated some F_1 offspring to create an F_2 **generation**, whose plants segregated for both characters found in the inbred parents (Fig. 3.1). It was at this point that Mendel was at his scientific best. Not only did he observe the character types in the offspring, but he also counted them and calculated ratios of each type. In the F_2 for all seven of the characters he was studying, he observed approximately three times as many plants of one character type as the other. Furthermore, the type observed in the F_1 was the one in greater number in the F_2.

© CAB International 2007. *Forest Genetics* (T.W. White, W.T. Adams and D.B. Neale)

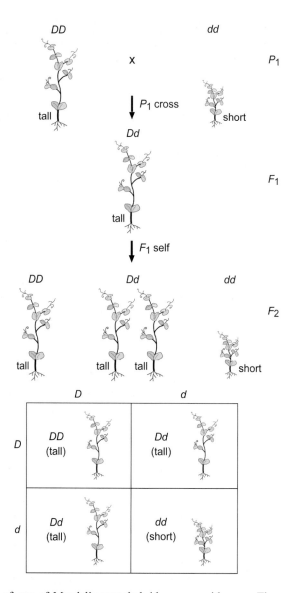

Fig. 3.1. Example of one of Mendel's monohybrid crosses with peas. The experiments involved three generations of plants. First, the true-breeding tall and short parental plants (P₁) were crossed to form the F₁ generation, which were uniformly tall plants. A single F₁ generation plant was then self-pollinated to form the F₂ generation. The Punnett square shows the F₂ offspring resulting from selfing the F₁. There are three possible genotypes (*DD, Dd* and *dd*), but only two phenotypes (tall and short) which segregate in a 3:1 ratio.

Mendel interpreted the results of his monohybrid crossing experiments to develop his first three laws of inheritance. He surmised that there must be one factor for tall and one factor for dwarf, and that these factors are somehow paired (Law 1, Table 3.1). Later this became the basis of the concept of pairs of **homologous chromosomes** with the gene determining a particular character located at the same position (*i.e.* locus) in each homologue. A locus can have different forms of the gene, which are called alleles. For example,

Table 3.1. Mendel's laws of segregation and independent assortment.

Law 1. Each genetic character is controlled by unit factors that come in pairs in individual organisms; these unit factors are now known as alleles.

Law 2. Paired unit factors for a given character are dominant and recessive to one another.

Law 3. Unit factors for the same character segregate independently from one another during the formation of gametes.

Law 4. Pairs of unit factors for two different traits also segregate independently of one another during the formation of gametes.

in Mendel's peas there was one allele coding for tall plants (*D*) and one allele coding for short plants (*d*) and these alleles segregated among the offspring. Plants that have the same allele at a locus on each of the homologous chromosomes are homozygous (*e.g. DD* and *dd*), whereas those with a different allele on each homologous chromosome are heterozygous (*e.g. Dd*).

Mendel further hypothesized that one unit factor (*i.e.* allele) is **dominant** to the other **recessive** factor (*i.e.* the dominant allele masks the effect of the recessive allele), based on the phenotypes he found in the F_1 and F_2 (Law 2, Table 3.1). For example, the allele for tall (*D*) is dominant and the allele for dwarf (*d*) is recessive. This leads to the important distinction between genotype and phenotype. The three possible genotypes are *DD*, *Dd*, and *dd*, although with *D* being dominant to *d*, there are only two phenotypes. The *DD* and *Dd* genotypes are both tall phenotypes and the *dd* genotype is the dwarf phenotype.

A simple method to interpret the results of the monohybrid cross is through the Punnett Square, named after its inventor Reginald Punnett (Fig. 3.1). It can be seen that when a heterozygous F_1 is mated to itself (or crossed to an identical F_1), three genotypes (*DD*, *Dd* and *dd*) are found in the 1:2:1 ratio, respectively. However, only two phenotypes are found, tall and dwarf, in the ratio of 3:1, respectively.

Finally, Mendel proposed that the 3:1 ratio observed in the progeny of selfed F_1 plants is expected if the *D* and *d* alleles are transmitted from each parent to the offspring at random (Law 3, Table 3.1). The frequencies of the offspring genotypes are the products of the frequencies of the alleles transmitted from each parent. The frequency of the *DD* genotype is 1/2 x 1/2 = 1/4, the frequency of the heterozygous *Dd* class is 2(1/2 x 1/2) = 1/2, and the frequency of *dd* is 1/2 x 1/2 = 1/4. Because *D* is dominant to *d* and therefore, *DD* and *Dd* have the same phenotype, the expected proportion of tall offspring is 3/4 and the dwarf offspring is 1/4 (or a 3:1 ratio). Mendel confirmed the expectations of random segregation with another type of cross, called a **test cross**. Here he crossed a *Dd* plant with a *dd* plant and found that approximately half of the offspring were tall and half were dwarf.

Dihybrid Cross

A second type of experiment Mendel performed was the **dihybrid cross**, which was an extension of the monohybrid cross to two characters (Fig. 3.2). Mendel crossed a yellow and round seeded type to a green and wrinkled seeded type. The F_1 offspring were all yellow and round and the alternate types of each character were found in 3:1 ratios in the F_2, as observed in monohybrid crosses involving the same traits.

Mendel also counted the two-character phenotypes and these were in a 9:3:3:1 ratio in the F_2. This is the expected ratio of two-character phenotypes given that both pairs of traits segregate randomly and completely independently of one another (Law 4, Table 3.1). Under independent

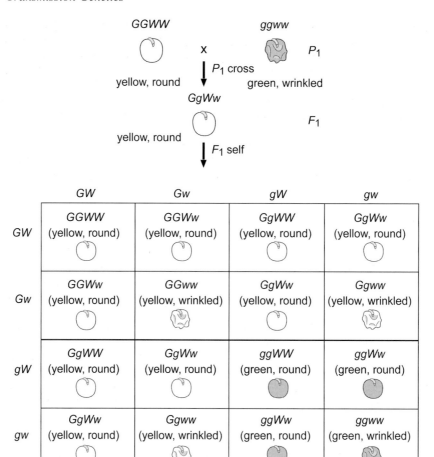

Fig. 3.2. Example of one of Mendel's dihybrid crosses with peas. The Punnett square shows the offspring resulting from selfing the F_1. There are nine possible genotypes but only four phenotypes (yellow and round, yellow and wrinkled, green and round, green and wrinkled) which segregate in a 9:3:3:1 ratio.

assortment, the expected frequency of any two-character phenotype is the product of the frequencies of each component character. Therefore, the frequency of yellow round offspring is f(yellow) x f(round) = 3/4 x 3/4 = 9/16 and the frequency of yellow wrinkled is 3/4 x 1/4 = 3/16. In this case, the genes controlling color and smoothness of seed coat reside on different chromosomes. Later in this chapter, it is shown that when genes controlling different characters reside close to each other on the same chromosome, they do not assort independently (called linkage).

Mendelian Inheritance of Traits in Forest Trees

Simple Mendelian inheritance of morphological traits in forest trees is rarely observed. The reason for this is that forest geneticists and tree breeders have primarily concerned themselves with the genetic improvement of economic traits, such as stem volume and wood quality, which are under the control of many genes (*Chapter 6*). Nevertheless, observant geneticists have identified a small number of traits showing Mendelian inheritance in the progeny of controlled crosses. Some interesting examples of Mendelian inheritance of morphological traits in trees include cone color in *Pinus monticola* (Steinhoff, 1974), seedling foliage color in *Picea abies* (Langner, 1953), chlorophyll deficiencies and other morphological variants in *Pinus taeda* (Franklin, 1969a), diameter growth in *Pinus patula* (Barnes *et al.*, 1987), blister rust resistance in *Pinus lambertiana* (Kinloch *et al.*, 1970), and the narrow-crown phenotype in *P. abies* (Lepisto, 1985). Almost certainly, many other morphological traits under simple genetic control could be identified if geneticists were to invest time in observing such traits in segregating populations.

Another class of traits that show Mendelian inheritance are biochemical traits, such as terpenes, isozymes and molecular markers. There are so many of these types of traits that an entire chapter (*Chapter 4*) is devoted to describing the various types and their applications as genetic markers.

Statistical Tests for Mendelian Inheritance

What if Mendel had observed only a small number of F_2's instead of several hundred or even thousands of offspring? Would he have observed the 3:1 and 9:3:3:1 ratios that led him to formulate the laws of inheritance? Genetic segregation is a random process subject to chance fluctuations. Small numbers of offspring can lead to observed ratios quite different than expected. If Mendel had looked at only ten F_2 plants following the crossing of a tall and a dwarf variety of peas, he might very easily have observed five tall plants and five dwarf plants just due to chance.

Since the early 1900s, geneticists have used statistical methods to test whether genetic segregation ratios conform to expected Mendelian ratios. The test most commonly used is a **goodness-of-fit test** to determine if observed ratios are significantly different from expected ratios. The goodness-of-fit test is performed by calculating a chi-square (χ^2) statistic:

$$\chi^2 = \Sigma \, (o - e)^2/e \hspace{4cm} \text{Equation 3.1}$$

where *o* equals the observed number and *e* equals the expected number in each class and the summation is over all offspring classes. The larger the deviation of the observed number from the expected number, the larger the magnitude of χ^2. A chi-square table *(e.g.* Table A5, Snedecor and Cochran, 1967) must then be consulted to determine the statistical probability of the calculated value. The probability depends on the degrees of freedom (df) in the data, which when testing Mendelian segregation ratios is just the number of offspring classes minus one. The convention is to reject the hypothesis only if the probability of the calculated χ^2 is 0.05 or less. Two examples of Mendelian segregation of traits in *Pinus* are described in Box 3.1, along with the calculation of the χ^2 statistic.

TRANSMISSION AND INHERITANCE OF CHROMOSOMES

In *Chapter 2*, we learned that in the nucleus of every cell of a diploid (2N) organism there are two sets of chromosomes, known as homologous pairs. At the time of fertilization, one set of homologues is contributed by the egg gamete (1N) and the other set by the pollen gamete (1N). The union of these two gametes forms the diploid zygote. This single-celled zygote then undergoes cell division and eventually grows into a fully developed organism. Trees grow from a single cell to very large, multi-cellular organisms by cell division in the meristematic regions of the tree. In order for every cell to have the same composition of chromosomes, there must be a mechanism whereby chromosomes are faithfully replicated during growth. This mechanism occurs through the process of mitosis. A process must also exist whereby during the formation of gametes, each gamete receives a complete haploid set of chromosomes. This process is called meiosis.

Mitosis and Cell Division

Mitosis is the process of nuclear division whereby replication of chromosomes is followed by separation of the replicated products such that each new daughter nucleus receives a copy of the chromosomes found in the mother cell. Before describing the individual steps in mitosis, it is important to review the cell cycle (Fig. 3.3). Cells spend most of the time in a non-dividing phase called **interphase**. Interphase itself has three phases: G1, S and G2. During interphase, the chromosomes consist of extended filaments that are not visible under a microscope. It is during this phase that DNA is transcribed and subsequently translated into gene products that regulate and mediate cell function and growth. The S phase is particularly important because this is when DNA replication occurs.

Box 3.1. Statistical tests for Mendelian segregation of morphological traits in *Pinus*.

Inheritance of cone color in Pinus monticola

Steinhoff (1974) performed a series of crossing experiments to study the inheritance of cone color in *Pinus monticola*. Two color phenotypes were observed, purple and green, and Steinhoff hypothesized that cone color was controlled by a single gene, and that the allele for purple was dominant (*P*) and the allele for green was recessive (*p*). Steinhoff crossed two heterozygous trees (*Pp* x *Pp*) and expected to observe a 3:1 ratio of purple to green cones in the offspring. He observed a total of 109 offspring; 83 progeny with purple cones and 26 offspring with green cones.

Phenotype	Geno-type(s)	Expected ratio	Observed (*o*)	Expected (*e*)	*o-e*	$(o-e)^2/e$
Purple	*PP, Pp*	3/4	83	81.75	1.25	0.02
Green	*pp*	1/4	26	27.25	-1.25	0.06
		Total	109	109.00		0.08

Steinhoff then performed a χ^2 goodness-of-fit test to determine if the observed number of offspring conformed to the expected number based on Mendelian ratios. The expected numbers were calculated by multiplying the total observations by the expected frequency in each class. Therefore, the expected number of offspring with purple cones is 109 x 0.75 = 81.75 and the expected number with green cones is 109 x 0.25 = 27.25. The χ^2 is calculated as $\chi^2 = (83 - 81.75)^2/81.75 + (26 - 27.25)^2/27.25 = 0.08$. The probability of $\chi^2 = 0.08$, with df = 1, is greater than P = 0.75 (Table A5, Snedecor and Cochran, 1967). This means that if the true segregation ratio of purple to green cones is 3:1, then 75% of the samples of 109 offspring would give deviations of the magnitude observed or greater. A goodness-of-fit χ^2 that occurs 75% of the time by chance is certainly not unreasonable, so it was declared that the observed segregation ratio supports single-locus inheritance with purple being dominant to green. (Box 3.1 continued on next page)

Box 3.1. Statistical tests for Mendelian segregation of morphological traits in *Pinus*. (Continued from previous page)

Inheritance of cotyledon number and lethality in Pinus taeda

In this example, the joint segregation of two seedling characters (normal *versus* bright green hypocotyl and normal *versus* cotyledon-stage lethal) was tested (Franklin, 1969a). Franklin hypothesized that the two characters were controlled by two different loci and that there were dominant and recessive alleles at each locus. After crossing two doubly-heterozygous trees, the expected two-locus phenotypic ratios in the offspring were 9:3:3:1.

Phenotype	Genotype(s)	Expected ratio	Observed (o)	Expected (e)	o-e	(o-e)²/e
Normal hypocotyls, non-lethal	*GGLL, GgLL, GGLl, GgLl*	9/16	54	60.19	-6.19	0.64
Normal hypocotyls, lethal	*GGll, Ggll*	3/16	22	20.06	1.94	0.19
Green hypocotyl, non-lethal	*ggLL, ggLl*	3/16	22	20.06	1.94	0.19
Green hypocotyl, lethal	*ggll*	1/16	9	6.69	2.31	0.80
		Total	107	107.00		1.82

Franklin tested this hypothesis by determining the number of offspring expected in each class and calculating the χ^2 goodness-of-fit statistic. The $\chi^2 = 1.82$ (df = 3) is much less than the critical value at the 5% probability level ($\chi^2 = 7.815$) (Table A5, Snedecor and Cochran, 1967), supporting Mendelian inheritance of two independent characters, each controlled by a single locus with dominant and recessive alleles.

The first step in mitosis is **prophase** (Fig. 3.4). The beginning of this stage is marked by the condensation (*i.e.* shortening and thickening) of chromosomes to form visibly distinct, thin threads within the nucleus. Each chromosome is longitudinally split into duplicates called **sister chromatids**, which are the products of replication during the S phase of interphase. The sister chromatids share an attachment point called the centromere. At the end of prophase, the nuclear membrane breaks down and an oval-shaped spindle begins to form, with the poles at opposite ends of the cell and strands attached at the centromere.

The next step is **metaphase**, when the chromosomes move toward the center of the cell until all centromeres line up on a central plane equidistant from the spindle poles. Now the chromatids are positioned to migrate to their respective poles. At **anaphase**, the sister chromatids of each chromosome separate at the centromere and begin migrating towards opposite poles of the spindle. Now each sister chromatid is regarded as a separate chromosome. When this stage is completed, the chromosomes are clustered in two groups at the opposite poles of the spindle, each group containing the same number of chromosomes that was present in the interphase nucleus of the mother cell. Finally, at **telophase** nuclear division is completed. The spindle disappears, a nuclear membrane forms around each cluster of chromosomes, and the chromosomes begin to revert to the extended, diffuse state. Cell division is completed by the process of **cytokinesis**, which divides the contents of the cytoplasm to the daughter cells. The **middle lamella**, which eventually becomes part of the cell wall, is formed to divide the two daughter cells. The process of mitosis is repeated continuously in meristematically active regions of the plant. In fact, all cells in a tree originated from the single-celled zygote.

Meiosis and Sexual Reproduction

Meiosis occurs only in gamete-forming cells of the tree, specifically in the nuclei of the egg and pollen mother cells, and is fundamental to inheritance (Box 3.2). Meiosis differs from mitosis in that two successive divisions (I and II) occur with only one replication of chromosomes, such that four haploid nuclei result. The same stages as in mitosis are used to describe meiosis (Fig. 3.5). Interphase I is essentially the same as interphase in mitosis when DNA replication occurs to double the amount of DNA. Like mitosis, chromosomes shorten and thicken in prophase I, such that sister chromatids are visible. Unlike mitosis, however, pairs of homologous chromosomes pair up (synapse) point for point at this stage. Each pair of chromosomes is called a **bivalent** (or tetrad). During synapsis, the chromatids

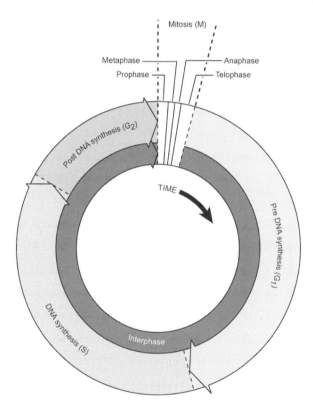

Fig. 3.3. The cell cycle. DNA synthesis occurs during interphase. The mitotic phase is relatively short when replicated chromosomes divide to form two daughter nuclei and daughter cells, each with a complete complement of chromosomes.

are intimately intertwined and exchanges of DNA occur (**crossing over**) between sister and non-sister chromatids (*i.e.* between chromatids of the two chromosomes). Visible evidence of crossovers are characteristic x-shaped configurations called **chiasmata**.

Crossing over between non-sister chromatids is very important because it enables recombination of genes on different chromosomes of a homologous pair. Therefore, crossing over breaks linkages among loci on the same chromosome and is an important mechanism for maintaining genetic variation (Box 3.2).

In metaphase I, the tetrads line up on the central plane; the centromeres of each chromosome in a homologous pair tend to be directed at opposite poles. Homologous pairs separate at anaphase I, with one whole chromosome of each pair moving to each pole. This is called the **reduction division** and ensures that each daughter nucleus has a haploid set of chromosomes. Further, the chromosomes of each pair move randomly to the poles without regard to their parental origin. This is the basis of the random segregation and independent assortment of Mendelian inheritance (Table 3.1).

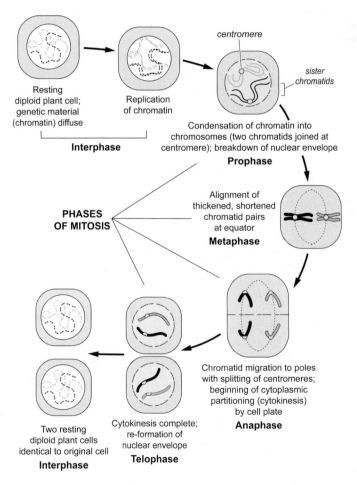

Fig. 3.4. Mitosis is the process whereby replicated chromosomes condense and the nuclei divide to form two new nuclei. The different steps in mitosis are illustrated here for a hypothetical cell having a single pair of homologous chromosomes (2N = 2). Note: Individual DNA strands are not actually visible during interphase, but are shown here to track the chromosome pair of interest.

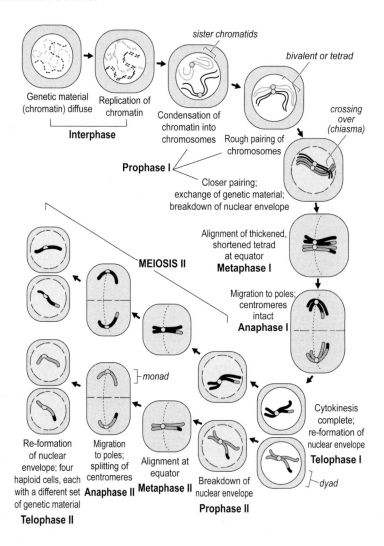

Fig. 3.5. Meiosis is the process of two successive divisions that occur during sexual reproduction to produce four haploid gametes from a single pollen or egg mother (diploid) cell. The series of steps in meiosis are illustrated here for a hypothetical cell with a single pair of homologous chromosomes (2N = 2). There is only a single bivalent or tetrad showing the pairing of homologous chromosomes at metaphase I. Note: Individual DNA strands are not actually visible during interphase, but are shown here to track the chromosome pair of interest.

Interphase II and prophase II together constitute a brief period between the first and second divisions when chromosomes may or may not become diffuse and reform. By the beginning of the second division, the chromosomes are identical with those present at the end of the first division. In metaphase II, the haploid set of chromosomes in each nucleus lines up on the central plane, generally oriented at right angles to the first division. Sister chromatids move to opposite poles in anaphase II. In telophase II, the haploid number of daughter chromosomes arrives at the poles, chromosomes revert to their diffuse nature, nuclear membranes appear and cell walls form that separate the four haploid daughter cells, called gametes. Female gametes are called eggs; male gametes are pollen.

Box 3.2. Significance of meiosis to heredity.

1. Meiosis ensures that each gamete has a complete haploid set of chromosomes.

2. Meiosis results in random assortment of paternal and maternal chromosomes in gametes. Therefore, for each pair of homologous chromosomes, half of the gametes receive the paternal chromosome of the pair and half receive the maternal chromosome.

3. Meiosis results in independent assortment of chromosomes from different homologous pairs. This means that chromosomes originating from the paternal and maternal parents randomly assort to gametes such that one gamete may receive a mix of chromosomes of paternal and maternal origin. Therefore, genes on non-homologous chromosomes segregate independently.

4. Through crossing over, alleles found at different loci on the same chromosome can recombine, further enhancing genetic diversity in offspring of sexually reproducing trees and other organisms.

EXTENSIONS TO MENDEL'S LAWS

Mendel's laws elegantly described the inheritance of a variety of characters in peas and provided the fundamental framework for the laws of inheritance for all characters in eukaryotic organisms. As with most things in biology, however, there are exceptions to the basic rules. In this section, we describe these exceptions and show how all are merely extensions to Mendel's laws.

Partial Dominance

Mendel's experiments established that at every locus there are two alleles, one dominant and one recessive, and that the ratio of dominant to recessive phenotypes in an F_2 is 3:1 when one locus is considered. Subsequent to Mendel's experiments with peas, the inheritance of characters was studied in crosses of many different organisms, and it was found that F_2's do not always segregate in a 3:1 ratio. One explanation for departure from the 3:1 ratio is **incomplete or partial dominance**.

Consider the example of a gene controlling needle color in *P. abies* in a study conducted by Langner (1953). In *P. abies* there is a mutant phenotype, known as the aurea-form, that produces white needles in seedlings homozygous *gg* for the aurea allele. Trees heterozygous (*Gg*) for the aurea allele and the wild-type allele have light-green or golden needles, whereas wild-type trees (*GG*) have normal dark green needles. In this case, it appears that the wild-type allele is not dominant to the aurea allele or else *Gg* trees would have normal green needles. Instead they have a golden needle color that is intermediate between green and white, a so-called blended expression.

Langner established the inheritance of the locus-controlling needle color in *P. abies* by making three different crosses and calculating ratios of needle color phenotypes. The three crosses he made were *GG* x *GG*, *GG* x *Gg*, and *Gg* x *Gg*. As expected, he observed all green needle progeny in the *GG* x *GG* cross. In the *GG* x *Gg* cross and the *Gg* x *Gg* cross, however, he observed 1:1 and 1:2:1 ratios of green to golden and green to golden to white needles, respectively (Fig. 3.6). These phenotypic ratios are consistent only with partial dominance and not complete dominance.

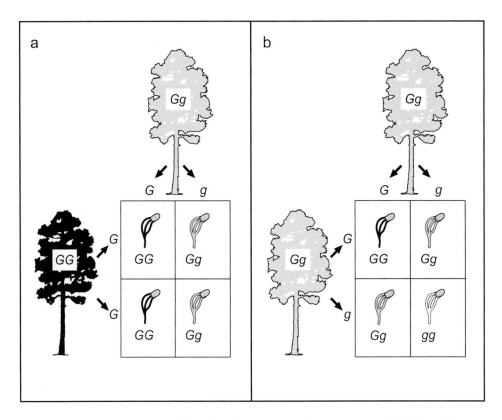

Fig. 3.6. The partial dominance of needle color in *Picea abies* (Langner, 1953). Trees homozygous for the wild-type allele *G* are green and trees homozygous for the mutant aurea allele *g* are white. However, heterozygous trees (*Gg*) are not green as if *G* were dominant to *g*, but rather they are an intermediate color of light green or golden: (a) Crossing *GG* x *Gg* results in an expected 1:1 ratio of green:golden progeny; and (b) Crossing two heterozygotes results in an expected 1:2:1 ratio of normal:golden:white foliage in the progeny.

Codominance

In the previous example, which described the partial dominant inheritance of the needle color locus in *P. abies*, the wild-type (*G*) and aurea (*g*) alleles are said to act additively, because the phenotype of the heterozygote (*Gg*) is a blended result of their individual expression. Alternatively, it is possible to observe a separate phenotype of each of the two alleles at a heterozygous locus, which is known as codominance. A good example of codominance occurs with allozyme genetic markers (*Chapter 4*). At a heterozygous allozyme locus, two distinct bands are observed on gels resulting from the visible and distinct expression of both alleles. The importance of codominance to genetic markers is discussed in more detail in *Chapter 4*.

Epistasis

Deviation from Mendelian ratios can also occur due to the interaction between alleles at different loci. This is known as **epistasis**. Alleles at two loci can interact in antagonistic or complementary ways to modify the expression of phenotypes. When epistasis occurs, two-locus phenotypic ratios do not conform to the standard ratios expected under independent assortment. An example of the epistatic interaction between two loci controlling flower color in pea is shown in Box 3.3; however, there are also many other types of epistatic interactions. The consequences of epistasis are more important when the inheritance of traits are controlled by many genes, as is discussed in *Chapter 6*.

Genetic Linkage

Mendel observed independent assortment among the seven characters he studied in peas, which suggests that all genes assort independently when transmitted from parent to off-spring. However, as we learned in the previous discussion about meiosis, if the chromosome is the unit of heredity, then it would seem that all genes on a chromosome would be inherited as a unit and would not assort independently. This is known as **genetic linkage**, which means that there is a greater association of two or more non-allelic genes than would be expected from independent assortment. Linkage is complete if crossing over does not occur during meiosis and there is no recombination among genes on homologous chromosomes. Linkage is incomplete if crossing over and recombination occur, and the strength of the linkage is a function of the probability that a crossover will occur between two genes. This probability increases with increasing physical (*i.e.* map) distance between genes on a chromosome.

In the early 1900s, British geneticists William Bateson and Reginald Punnett were the first to observe genetic linkage between two flower color genes in peas; however, the understanding of genetic linkage was most developed by Thomas Morgan and Alfred Sturtevant who studied the inheritance of characters in the common fruit fly, *Drosophila melanogaster*. An example of genetic linkage between allozyme loci in *Pinus rigida* is described in Box 3.4.

The standard method for quantifying the amount of linkage between two genes is to estimate the **recombination fraction** (r):

$$r = R/N \qquad\qquad \text{Equation 3.2}$$

where R equals the number of recombinant gametes and N is the total number of gametes in a sample of meioses. Two types of gametes can be distinguished: **parental** and **recombinant gametes.** Before gametes can be classified, however, the **linkage phase** of the two genes on a chromosome must be determined. For example, for two loci with alleles *A* and *a* and *B* and *b*, respectively, **coupling** linkage phase is when *A* and *B* are on one homologous chromosome and *a* and *b* are on the other homologous chromosome. Alternatively, **repulsion** linkage phase is when *A* and *b* and *a* and *B* are on opposite homologous chromosomes. Under coupling linkage phase, *AB* and *ab* are parental gametes and *Ab* and *aB* are recombinant gametes. Under repulsion linkage phase, the opposite is true; *Ab* and *aB* are parental gametes and *AB* and *ab* are recombinant gametes. The concepts of coupling and repulsion and the estimation of the recombination fraction are illustrated in Box 3.4; further details on estimating linkage can be found in Bailey (1961).

Organelle Genome Inheritance

Most of the genetic material is found in chromosomes in the nucleus of the cell (*i.e.* nDNA, *Chapter 2*). However, both chloroplasts and mitochondria have small genomes that encode genes related to their functions. The chloroplast DNA (cpDNA) and mitochondrial

Box 3.3. Epistasis for flower color in pea.

Flower color in pea is a classic example demonstrating a type of epistatic interaction between alleles at two loci controlling a single trait (Fig. 1). There are two loci controlling flower color, the *C* and *P* locus, each with a dominant and a recessive allele. If a doubly-heterozygous individual, *CcPp*, is selfed as in the normal Mendelian dihybrid cross, the expected two-locus genotypes are found as shown in the Punnett Square below. However, the phenotypic ratio of colored flowers to white flowers is 9:7 instead of the 9:3:3:1 predicted by Mendel's law in a dihybrid cross. This is because at least one dominant allele from each locus must be present to have a colored flower. The *C* and *P* alleles from the two loci do not act independently; rather, they have a dependency on one another in determining flower color.

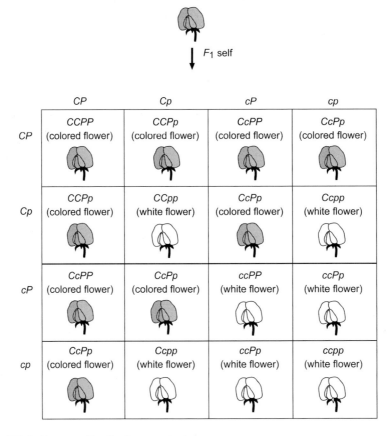

Fig. 1. Dihybrid segregation for flower color in pea.

There is a possible model to explain the epistatic interaction between the *C* and *P* loci. Suppose the pigment producing purple flower color requires two enzymatic steps, one controlled by the *C* locus and the other by the *P* locus. Furthermore, suppose the dominant allele gene product is required to successfully catalyze each reaction. Then, colored flowers result only when at least one dominant allele is found at each locus, because both enzymatic reactions must occur to complete the biosynthesis of the colored pigment.

DNA (mtDNA) in plants are circular DNA molecules just like in prokaryotic genomes, such as bacteria. So how are these genomes inherited? Organelle DNA inheritance studies in plants were traditionally conducted by following the transmission of some mutant phenotype that was assumed to be cpDNA- or mtDNA-encoded, such as chlorophyll-deficient mutants. Studies of this type showed that cpDNA and mtDNA in plants are generally inherited from only a single parent, unlike nuclear chromosomes that are inherited from both parents. In angiosperms, it is nearly always true that both the cpDNA and the mtDNA are inherited from the maternal parent (Fig. 3.7), although there are several cases where cpDNA is inherited from both parents.

More recently, DNA-based genetic markers (*Chapter 4*) have been used to study organelle DNA inheritance in a variety of plants including trees. In angiosperm tree species, such as *Populus*, *Eucalyptus*, *Quercus*, *Liriodendron* and *Magnolia*, inheritance studies show that both cpDNA and mtDNA are inherited from the maternal parent (Radetzky, 1990; Mejnartowicz, 1991; Rajora and Dancik, 1992; Byrne *et al.*, 1993; Sewell *et al.*, 1993; Dumolin *et al.*, 1995), just as is observed in non-woody angiosperms, although inheritance of cpDNA from both the maternal and paternal parent has been observed in a few tree species (Sewell *et al.*, 1993).

In conifers, however, some very surprising differences have been observed (Fig. 3.7). An early indication that organelle DNA inheritance might be different in conifers came from study of the inheritance of a shoot color mutant in a horticultural variety of *Cryptomeria japonica* called *Wogon-sugi* (Ohba *et al.*, 1971). It was found in a series of reciprocal crosses that the mutant phenotype was only transmitted by the paternal parent, suggesting that the mutant was caused by a cpDNA mutation and that the cpDNA was inherited only from the paternal parent. The strict paternal inheritance of cpDNA in conifers has been confirmed by a large number of studies using DNA-based genetic markers (Neale *et al.*, 1986; Szmidt *et al.*, 1987; Wagner *et al.*, 1987; Neale and Sederoff, 1989).

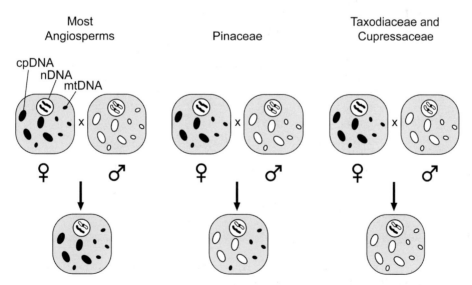

Fig. 3.7. Chloroplast (cpDNA) and mitochondrial (mtDNA) genomes are inherited from one parent, rather than both parents as with the nuclear genome (nDNA). Both cpDNA and mtDNA are inherited through the mother in angiosperms. In all conifers and in Taxodiaceae and Cupressaceae, however, cpDNA is paternally inherited. The mtDNA is also paternally inherited in Taxodiaceae and Cupressaceae, but is maternally inherited in conifers.

Deoxyribonucleic acid-based genetic markers also have been used to study the inheritance of mtDNA in conifers. In the *Pinaceae*, mtDNA appears to be strictly inherited from the maternal parent as in angiosperms (Neale and Sederoff, 1989; Sutton *et al.*, 1991; DeVerno *et al.*, 1993; David and Keathley, 1996; Fig. 3.7), although there is some evidence that mtDNA is occasionally inherited through the paternal parent (Wagner *et al.*, 1991a). A most surprising result, however, is that mtDNA is strictly inherited from the paternal parent in Taxodiaceae and Cupressaceae (Neale *et al.*, 1989, 1991; Kondo *et al.*, 1998; Fig. 3.7). In no other plant or animal species has it been shown that the mtDNA is inherited strictly from the paternal parent.

Box 3.4. Estimation of genetic linkage between two allozyme loci in *Pinus rigida*.

Genetic linkage in forest trees can be illustrated using an example from an allozyme study by Guries *et al.* (1978). Allozymes are codominant genetic markers (*Chapter 4*) and the two alleles at a heterozygous locus are often designated as fast (F) and slow (S), based on their mobility difference on gels following electrophoresis. Guries *et al.* (1978) assayed a number of allozyme loci from seed megagametophyte tissue, which is haploid in conifers and identical to the egg gamete in genetic makeup (*Chapter 4*). Thus, the segregating allelic products of meiosis can be observed directly by assaying megagametophytes in a sample of seeds from a mother tree.

To illustrate the concept of genetic linkage, only two loci are considered in this example, *Got*-1 and *Gpi*-2, each heterozygous for F and S alleles in the mother (Fig. 1). If *Got*-1 and *Gpi*-2 are found on the same chromosome in the mother tree, then two possible configurations (*i.e.* linkage phases) of the F and S alleles at the two loci are possible. In the example below, the F alleles at each locus are on one chromosome of the mother, while the S alleles are on the homologous chromosome: the coupling linkage phase (*i.e.* FF/SS). The opposite configuration (FS/SF), is called the repulsion linkage phase.

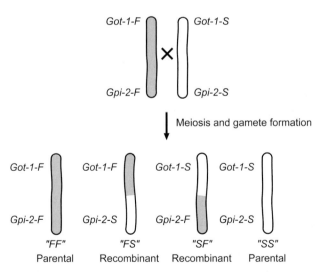

Fig. 1. An example of genetic linkage for two allozyme loci.

(Box 3.4 continued on next page)

Box 3.4. Estimation of genetic linkage between two allozyme loci in *Pinus rigida*.
(Continued from previous page)

Following meiosis in a doubly-heterozygous mother tree, four possible gamete types of the two-locus combinations of fast and slow alleles are expected in megagametophytes (FF, FS, SF, and SS). When alleles in the mother tree are in the coupling linkage phase, FF and SS gametes result if there is no crossover between the *Got*-1 and *Gpi*-2 loci; these are parental or non-crossover gametes. The FS and SF gametes result if there is a single crossover, and are called crossover or recombinant gametes.

   Guries *et al.* (1978) assayed 160 megagametophytes for the *Got*-1 and *Gpi*-2 loci and counted the number of each of the four types. If *Got*-1 and *Gpi*-2 were located on different chromosomes and followed Mendel's law of independent assortment, we would expect each of the four allelic combinations to occur in equal frequency, (*i.e.* 1:1:1:1 ratio or 40 gametes of each type). Clearly, there is a large excess of gametes in the two parental classes (FF and SS) and a corresponding deficiency in the two recombinant classes (FS and SF).

Genotype(s)	Observed (*o*)	Expected (*e*)	$(o-e)^2/e$
FF	90	40	62.5
FS	3	40	34.2
SF	4	40	32.4
SS	63	40	31.2
Total	160	160	142.3

   Guries *et al.* (1978) performed a χ^2 goodness-of-fit test of the 1:1:1:1 ratio and calculated a $\chi^2 = 142.3$ (df = 1), P < 0.001, which supported the hypothesis that these loci do not assort independently and are linked. With seven recombinant gametes out of a total of 160, the recombination fraction is r = 7/160 = 0.043. The recombination fraction x 100 (called centiMorgans; cM) yields a relative measure of distance between genes on a chromosome (*i.e.* map distance). The closer the genes are together (*i.e.* tighter the linkage), the smaller the cM. In this case, the linkage distance between *Got*-1 and *Gpi*-2 is 4.3 cM. Map distances range from 0 (no recombination between two genes that are extremely tightly linked) to 50 (loci that are so far apart on a chromosome that they assort independently).

SUMMARY AND CONCLUSIONS

   The discovery of the laws of inheritance of nuclear genetic material was made in the mid-1800s by the Austrian monk Gregor Mendel, who performed a series of crossing experiments in garden peas. Mendel's monohybrid crosses between varieties of peas differing in phenotype for a single character established his first three laws: (1) Unit factors come in pairs; (2) Unit factors are either dominant or recessive to one another; and (3) Unit factors segregate randomly. Mendel also studied the inheritance of pairs of factors that led to his fourth law of independent assortment. Mendel's discoveries were ignored during his lifetime and were not rediscovered until 1900. Mendel's laws and Charles Darwin's theory of evolution provide the fundamental basis of the science of genetics.

The orderly transmission of chromosomes following cell division and from generation to generation is fundamental to the inheritance of genetic material. Mitosis and meiosis are the cellular mechanisms that control the transmission of chromosomes during cell division and the formation of gametes, respectively. Mitosis occurs in vegetative plant cells during normal growth. Mitosis includes several steps where chromosomes are first replicated and then the nucleus and cell divide to form two new daughter cells each having a complete complement of chromosomes.

Meiosis is similar to mitosis except that it occurs during gamete formation in pollen and egg mother cells. Two successive divisions occur following the replication of chromosomes. The four resulting gametes each have only one homologue of a homologous pair of chromosomes. In addition, recombination can occur between non-sister chromatids, which along with mutation, are the mechanisms for creating genetic variability.

In the years following the rediscovery of Mendel's four basic laws of inheritance, several exceptions were discovered. The alleles controlling characters in peas that Mendel studied acted dominantly or recessively to one another. Incomplete or partial dominance is an exception to Mendel's second rule, where the effects of each of the alleles are blended to produce an intermediate phenotype. Another exception to the second law is codominance, where the phenotype reveals expression of both alleles, such as with allozymes.

Two characters (or loci) located on the same chromosome do not always undergo independent assortment, which is a deviation from Mendel's fourth law. This can be due to genetic linkage. The genetic distance that two loci are from each other on chromosomes is estimated by counting the number of gametes that have undergone crossing over. The proportion of recombinant gametes to the total is called the recombination fraction and is the standard estimator for the degree (*i.e.* tightness) of genetic linkage. Two-locus phenotypic ratios can also be altered by epistasis, where the phenotypic expression of an allele at one locus is modified by an allele at the other locus.

One more exception to Mendel's laws of inheritance is observed with the genetic material encoded in the chloroplast and mitochondrial genomes. These genomes are generally uniparentally inherited from either the maternal or paternal parent. In conifers, chloroplast DNA is inherited from the paternal parent and mitochondrial DNA is inherited from the maternal parent, except in the Taxodiaceae and Cupressaceae where the mitochondrial DNA is paternally inherited as well.

CHAPTER 4
GENETIC MARKERS – *MORPHOLOGICAL, BIO-CHEMICAL AND MOLECULAR MARKERS*

A **genetic marker** is any visible character or otherwise assayable phenotype, for which alleles at individual loci segregate in a Mendelian manner. Genetic markers can be used to study the genetics of organisms, including trees, at the level of single genes. The development of the discipline of genetics would not have been possible without genetic markers such as the visible characters in peas and *Drosophila*. Trees, unfortunately, do not have a large number of visible Mendelian characters (*Chapter 3*) and for many years, this was a limitation in forest genetics research. It was not until the early 1970s that biochemical genetic markers such as terpenes and allozymes were developed for trees. These biochemical markers were applied to an array of research problems, most notably the study of amounts and patterns of genetic variation in natural populations of trees and the characterization of mating systems (*Chapters 7-10*).

A major limitation of biochemical markers, however, is that there are only a small number of different marker loci; therefore, genetic information obtained from such markers may not be very representative of genes throughout the genome. The limitation in the number of markers was overcome beginning in the 1980s with the development of molecular or DNA-based genetic markers. Molecular markers have a wide variety of applications in both basic research and applied tree improvement programs. The goal of this chapter is to briefly describe the basic properties of genetic markers and to introduce some of their uses in forestry. Additional reading on genetic markers can found in Adams *et al.* (1992a), Mandal and Gibson (1998), Glaubitz and Moran (2000), and Jain and Minocha (2000).

USES AND CHARACTERISTICS OF GENETIC MARKERS

Before describing the many types of genetic markers available for use in forest trees, it is first appropriate to consider the various applications of genetic markers and the desired attributes for such applications. Genetic markers are used to study the genetics of natural and domesticated populations of trees and the forces that bring about change in these populations. Some of the more important applications of genetic markers include: (1) Describing mating systems, levels of inbreeding, and temporal and spatial patterns of genetic variation within stands (*Chapter 7*); (2) Describing geographic patterns of genetic variation (*Chapter 8*); (3) Inferring taxonomic and phylogenetic relationships among species (*Chapter 9*); (4) Evaluating the impacts of domestication practices, including forest management and tree improvement, on genetic diversity (*Chapter 10*); (5) Fingerprinting and germplasm identification in breeding and propagation populations (*Chapters 16, 17, and 19*); (6) Constructing genetic linkage maps (*Chapter 18*); and (7) Marker assisted breeding (*Chapter 19*).

We describe several different types of genetic markers in this chapter, each of which has different attributes that make it more or less desirable to use in certain applications.

Some of the desirable attributes of a given type of genetic marker are that it be: (1) Inexpensive to develop and apply; (2) Unaffected by environmental and developmental variation; (3) Highly robust and repeatable across different tissue types and different laboratories; (4) Polymorphic, *i.e.* reveal high levels of allelic variability; and (5) Codominant in its expression.

MORPHOLOGICAL MARKERS

As discussed in *Chapter 3*, very few simple Mendelian morphological characters have been discovered in forest trees that could be used as genetic markers. Many of the identified morphological markers are mutations observed in seedlings such as albino needles, dwarfing and other aberrations (Franklin, 1970; Sorensen, 1973) (Fig. 5.2). Such mutants have been used to estimate self-pollination rates (*Chapter 7*) in conifers. These markers, however, have limited application because morphological mutants occur rarely and often are highly detrimental or even lethal to the tree.

BIOCHEMICAL MARKERS

Monoterpenes

Monoterpenes are a subgroup of the terpenoid substances found in resins and essential oils of plants (Kozlowski and Pallardy, 1979). Although the metabolic functions of monoterpenes are not fully understood, they probably play an important role in resistance to attack by diseases and insects (Hanover, 1992). The concentrations of different monoterpenes, such as alpha-pinene, beta-pinene, myrcene, 3-carene and limonene are determined by gas chromatography and are useful as genetic markers (Hanover, 1966a, b, 1992; Squillace, 1971; Strauss and Critchfield, 1982).

Monoterpene genetic markers have been applied primarily to taxonomic and evolutionary studies (*Chapter 9*). However, they have also been used to a limited extent to estimate genetic patterns of geographic variation within species (*Chapter 8*). Although monoterpenes were the best available genetic markers for forest trees in the 1960s and early 1970s, they require specialized and expensive equipment for assay. In addition, there are relatively few monoterpene marker loci available and most express some form of dominance in their phenotypes. Dominant genetic markers have the disadvantage that dominant homozygous genotypes cannot be distinguished from heterozygotes carrying the dominant allele. Monoterpenes were gradually replaced by allozyme genetic markers because allozymes are less expensive to apply, are codominant in expression, and many more marker loci can be assayed.

Allozymes

Allozymes have been the most important type of genetic marker in forestry and are used in many species for many different applications (Conkle, 1981a; Adams *et al.*, 1992a). Allozymes are allelic forms of enzymes that can be distinguished by a procedure called electrophoresis. The more general term for allozymes is isozymes, and refers to any variant form of an enzyme, whereas allozyme implies a genetic basis for the variant form. Most allozyme genetic markers have been derived from enzymes of intermediary metabolism, such as enzymes in the glycolytic pathway; however, conceivably an allozyme genetic marker could be developed from any enzyme.

Allozyme analysis is fairly easy to apply (Fig. 4.1) and standard protocols for its use in trees are available (Conkle *et al.*, 1982; Cheliak and Pitel, 1984; Soltis and Soltis, 1989; Kephart, 1990). Crude protein extracts are isolated from almost any tissue type and then are separated on starch gels by applying an electrical current (*i.e.* electrophoresis). Isozymes in the protein extract migrate to different positions on the gel depending on the electrical charge and size of the isozyme. Isozymes with different amino acid composition generally have a different charge and/or size, so it is these genetic differences that are revealed as mobility differences on the gel. The location of an isozyme on a gel following electrophoresis is visualized by placing the gel in a solution that contains the enzyme substrate, appropriate cofactors and a dye. The colored bands on the gel are the products of the enzymatic reactions linked to the dye.

Before allozymes can be used for genetic studies, however, the Mendelian inheritance of allozymes must be established. The inheritance of allozymes can be determined by using crosses between trees similar to the types of crosses Mendel used with peas, *e.g.* hybrid crosses and test crosses. In conifers, however, the unique genetic system offered by the seed tissues is most often used. The conifer **megagametophyte** (*i.e.* nutritive tissue surrounding the embryo) is haploid and genetically identical to the egg cell, since they are products of the same meiotic event (Fig. 4.2). Therefore, a sample of seed megagametophytes from a single tree represents a population of maternal meioses. This is a very convenient system because segregation analysis can be performed by simply using open-pollinated seed from mother trees, which eliminates the need for controlled crosses to establish the Mendelian inheritance of allozyme variants (Fig. 4.2).

Assays have been developed for many enzymes and as many as 25-40 different allozyme loci can be detected depending on the tree species and tissue type. Because of their codominant expression and relatively high level of polymorphism, allozymes have been used extensively for estimating genetic variation in tree species (*Chapters 7-8*), to a lesser extent for evolutionary studies (*Chapter 9*), and for monitoring various gene conservation (*Chapter 10*) and tree improvement activities (Adams, 1983; Wheeler and Jech, 1992).

Other Protein Markers

Another type of protein-based genetic marker utilizes **two-dimensional polyacrylamide gel electrophoresis (2-D PAGE)**. Unlike allozymes where single known enzymes are assayed individually, the 2-D PAGE technique simultaneously reveals all enzymes and other proteins present in the sample preparation. The proteins are revealed as spots on gels and marker polymorphisms are detected as presence or absence of spots. This technique has been used most extensively for linkage mapping in *Pinus pinaster* where protein polymorphisms have been assayed from both seed and needle tissues (Bahrman and Damerval, 1989; Gerber *et al.*, 1993; Plomion *et al.*, 1997). Although 2-D PAGE has the potential advantage that many marker loci can be assayed simultaneously on a single gel, assays are more difficult than in allozyme analyses, and the markers are often dominant in their expression.

MOLECULAR MARKERS

A vast array of DNA-based genetic markers has been discovered since 1980 and new marker types are developed every year. There are two general types of DNA markers:

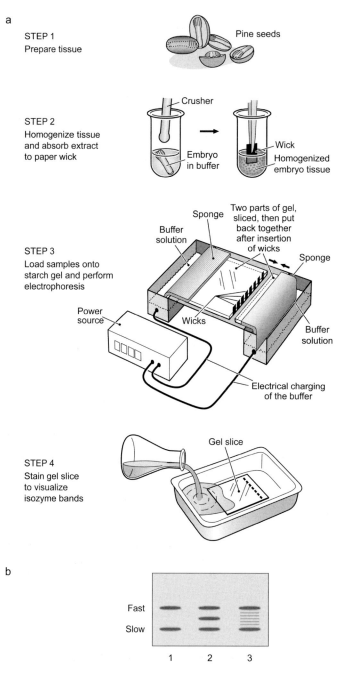

Fig. 4.1. Allozyme analysis in forest trees: (a) Step 1, tissues for assay are prepared such as megagameto-phytes and embryos from conifer seeds; Step 2, tissues are homogenized in an extraction buffer and absorbed onto small filter paper wicks; Step 3, wicks are loaded onto gels and samples are electrophoresed; Step 4, gel slices are stained to reveal allozyme bands; and (b) Three general types of allozyme patterns are observed in heterozygotes carrying slow and fast migrating variants, depending on whether an enzyme is functional as a monomer (lane 1), dimer (lane 2) or multimer (lane 3). Note that in all three cases, both alleles are expressed (*i.e.* are codominant) in heterozygotes.

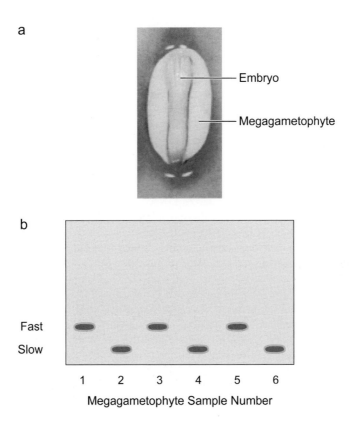

Fig. 4.2. The conifer megagametophyte and embryo genetic system: (a) In conifer seeds, the embryo is diploid (2N), while the megagametophyte is haploid (1N) and genetically identical to the egg gamete; and (b) Allozymes from an individual megagametophyte show the product of just one allele. Megagametophytes from a single seed tree, heterozygous for an allozyme locus, are expected to segregate in a 1:1 ratio of fast and slow allozyme bands (alleles), as shown here for six megagametophytes from seed of a heterozygous mother tree. (Photo courtesy of G. Dupper, Institute of Forest Genetics, Placerville, California)

(1) Those based on DNA-DNA hybridization; and (2) Those based on amplification of DNA sequences using the polymerase chain reaction (PCR). Important technical aspects of these two approaches are discussed in detail in the following sections. More comprehensive reviews of molecular markers in forestry are available (Neale and Williams, 1991; Neale and Harry, 1994; Echt, 1997), so only the marker types most often used in forest trees are discussed here.

DNA-DNA Hybridization: Restriction Fragment Length Polymorphism

Genetic marker systems based on **DNA-DNA hybridization** were developed in the 1970s. Eukaryotic genomes are very large and there was no simple way to observe genetic polymorphisms of individual genes or sequences. The property of complementary base pairing allowed for methods to be developed whereby small pieces of DNA could be used as probes to reveal polymorphisms in just the sequences homologous to the probe. The genetic system derived using this approach is called restriction fragment length polymorphism.

Restriction fragment length polymorphism (RFLP) markers were the first DNA-based genetic markers developed (Botstein *et al.*, 1980). A brief description of the RFLP procedure is shown in Fig. 4.3. To begin, total cellular DNA is digested with a **restriction endonuclease** (Box 4.1), which reduces the genome to a large pool of restriction fragments of different sizes. Hundreds of restriction endonucleases have been discovered that cleave DNA at specific recognition sites of varying length and sequence. However, just a few of these enzymes (*e.g. Hin*dIII, *Eco*RI, *Bam*HI) are routinely used because they generally provide the best size distribution of DNA fragments and are inexpensive. Restriction endonuclease recognition sites are found throughout the genome, both in coding and non-coding regions, and are a powerful way to sample DNA sequence variation in the genome.

The restriction fragments are then separated by their size on an agarose gel by electrophoresis. It is possible to visualize DNA within such a gel by staining it with ethidium bromide; however, because there are typically so many restriction fragments of all possible sizes, discrete fragments cannot be seen. To overcome this problem, the fractionated DNA is transferred and chemically bound to a nylon membrane by a process called **Southern blotting**, named after its inventor E.M. Southern (1975). Specific DNA fragments are visualized by hybridizing the DNA fragments bound to the nylon membrane with a radioactively- or fluorescently-labeled DNA probe. A **DNA probe** is just a small piece of DNA used to reveal its complementary sequence in the DNA bound to the membrane. The DNA probing relies on complementary base pairing; both the DNA fragments on the nylon membrane and the probe are first denatured so that they are single stranded and available to pair with their complementary DNA sequence.

DNA probes have been developed for genetic marker analyses of all three plant genomes: nDNA, cpDNA and mtDNA. The chloroplast and mitochondrial genomes are relatively small, so it has been possible to digest these genomes with a restriction endonuclease and clone individual sections of these genomes using standard plasmid cloning techniques (Box 4.1). The cpDNA and mtDNA probes have been developed for a number of forest tree species by cloning DNA fragments from those genomes (Strauss *et al.*, 1988; Lidholm and Gustafsson, 1991; Wakasugi *et al.*, 1994a,b). Each of these cloned fragments contains several different genes; therefore, when one clone is used for a probe in RFLP analysis, it is possible to reveal genetic variation for a number of organelle-encoded genes at once.

Developing probes for RFLP analysis of nDNA is more problematic because of the large amount of repetitive DNA in the nuclear genome. Two types of probes are commonly used: genomic DNA (gDNA) probes and complementary DNA (cDNA) probes. Probes are isolated from DNA libraries, which are a large collection of cloned fragments resulting from a single cloning experiment. Both cDNA and gDNA probes are equally easy to use and reveal abundant genetic variation in trees (Devey *et al.*, 1991; Liu and Furnier, 1993; Bradshaw *et al.*, 1994; Byrne *et al.*, 1994; Jermstad *et al.*, 1994). The gDNA probe libraries, however, are much easier to construct than cDNA probe libraries because the difficult task of mRNA isolation is not required. The cDNA probes are derived from expressed genes because cDNA is derived from mRNA (Box 4.2), whereas gDNA probes generally are not; therefore, cDNA probes are often preferred for many applications of RFLP analysis in trees.

The genetic interpretation of RFLP banding patterns can be difficult especially in conifers whose large genomes often lead to large numbers of fragments revealed by a single probe. Examples illustrating the molecular basis of several RFLP patterns are shown in Fig. 4.4, as well as the Mendelian interpretations of these band patterns.

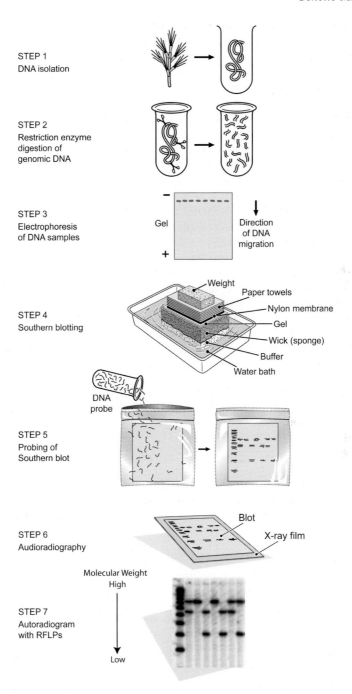

Fig. 4.3. Restriction fragment length polymorphism (RFLP) analysis. Step 1, isolation of DNA from tree tissues; Step 2, restriction enzyme digestion of DNA to cleave DNA into small fragments; Step 3, electrophoresis of DNA samples on agarose gels to separate DNA fragments by size; Step 4, transfer of DNA fragments from gel to nylon membrane by Southern blotting technique; Step 5, hybridization of nylon membrane with specific radioactively labeled DNA probe; Step 6, exposure of nylon membrane to x-ray film (autoradiography); and Step 7, autoradiogram showing RFLP bands.

Box 4.1. Restriction enzymes and DNA cloning.

Experimental methods to study the structure and expression of individual genes were first developed in the 1970s. Together these methods are known as **recombinant DNA technologies.** The essence of recombinant DNA technology is to isolate a specific fragment of DNA (often an entire gene) from the genome and then introduce the fragment into a foreign host genome (usually a bacteria or virus), where large quantities of the recombinant DNA fragment can be amplified and isolated. Recombinant DNA technology was made possible by the discovery of enzymes from bacteria that cleave DNA at specific locations. These enzymes are called **restriction enzymes** or **restriction endonucleases**, because their function in bacteria is to recognize, cleave and destroy foreign DNA invading the bacterial cell. The first restriction enzyme discovered was called *Eco*RI, because it was isolated from *E. coli*. The number of restriction enzymes now available is in the hundreds. Most restriction enzymes recognize a specific DNA sequence at the location where they cleave the DNA. This is called a recognition sequence and for *Eco*RI the sequence is GAATTC (Fig. 1). This sequence is called a palindrome, because it is identical on both complementary DNA strands when read in the same direction (5' to 3' or 3' to 5'). Most restriction enzymes also cleave the DNA at a specific location within the recognition sequence. *Eco*RI cleaves between the G and A.

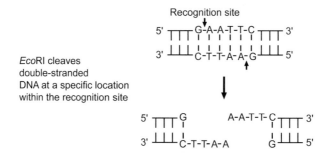

*Eco*RI cleaves double-stranded DNA at a specific location within the recognition site

Fig. 1. Schematic representation of DNA cleaving by *Eco*RI.

Molecular biologists learned that restriction enzymes were not only useful tools for cleaving large DNA molecules into small fragments, but that these enzymes could also be used in the process to clone DNA. The simplest form of DNA cloning involves inserting a fragment of foreign DNA into an *E. coli* **plasmid vector** and then reintroducing the recombinant plasmid back into the bacterial host (called **transformation**). The bacteria can then be grown in culture to produce large quantities of the plasmid containing the foreign DNA fragment to be studied.

The entire cloning process is shown in Fig. 2. In step 1, the plasmid DNA is isolated from the bacteria. In practice, this is rarely done by individual researchers because plasmid DNA can be purchased from suppliers. In step 2, the circular plasmid DNA is linearized by cleaving it with a restriction enzyme. Plasmid vectors have been genetically engineered to include a number of restriction sites that can be used for cloning. In step 3, the foreign DNA fragment to be cloned is cleaved with the same restriction enzyme as was used to cleave the plasmid vector. In step 4, the foreign DNA fragment to be cloned is isolated from the rest of the DNA, usually by electrophoresis and gel purification.
(Box 4.1 continued on next page)

Box 4.1. Restriction enzymes and DNA cloning. (Continued from previous page.)

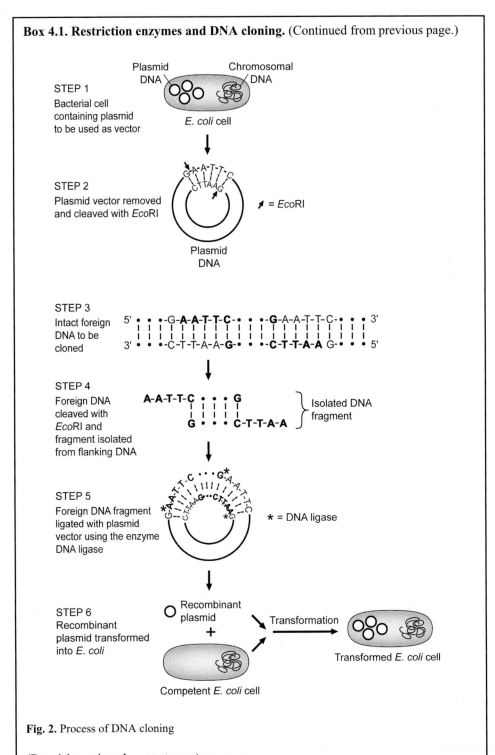

STEP 1
Bacterial cell
containing plasmid
to be used as vector

STEP 2
Plasmid vector removed
and cleaved with *Eco*RI

STEP 3
Intact foreign
DNA to be
cloned

STEP 4
Foreign DNA
cleaved with
*Eco*RI and
fragment isolated
from flanking DNA

STEP 5
Foreign DNA fragment
ligated with plasmid
vector using the enzyme
DNA ligase

STEP 6
Recombinant
plasmid transformed
into *E. coli*

Fig. 2. Process of DNA cloning

(Box 4.1 continued on next page)

Box 4.1. Restriction enzymes and DNA cloning. (Continued from previous page.)

At this point (step 5), it is now possible to join the foreign DNA fragment with the linearized plasmid. The cleaving of both DNAs with *Eco*RI leaves a four-base (AATT) single-stranded end. These are called "sticky ends." Through complementary base pairing, the two DNA fragments join together perfectly. All that remains to reconstitute a circular, double-stranded DNA molecule is to form a bond between the nucleotide bases that were separated by cleaving with the restriction enzyme. This is accomplished by the enzyme DNA ligase and the process is called a ligation reaction. Finally, the recombinant plasmid is re-introduced into the bacterial host (step 6) by transformation. Now the transformed *E. coli* can be grown in culture to produce large quantities of plasmid DNA. Cloning of DNA fragments makes it possible to study aspects of the DNA, such as determining its nucleotide sequence, which would not be possible without cloning because large quantities of DNA are generally needed for such assays.

Restriction fragment length polymorphism analysis has been applied to both chloroplast and mitochondrial genomes to study: (1) Phylogenetic relationships (Wagner *et al.*, 1991b, 1992; Tsumura *et al.*, 1995) (*Chapter 9*); (2) Genetic variation within species (Ali *et al.*, 1991; Strauss *et al.*, 1993; Ponoy *et al.*, 1994); and (3) Modes of organelle DNA inheritance (*Chapter 3*). Restriction fragment length polymorphism analysis of nuclear genomes of forest trees has not been as widely applied, due to the technical challenges of performing Southern blot analysis, especially with the large genomes of conifers. Nuclear DNA RFLP probes are available for a few conifer, *Populus*, and *Eucalyptus* species. These probes have been used to construct genetic linkage maps for a number of species as discussed in *Chapter 18*.

Box 4.2. Complementary DNA (cDNA) cloning.

The ability to obtain millions of copies of individual genes is fundamentally important for molecular genetics research. Molecular biologists developed a very clever method, called complementary DNA (cDNA) cloning, to obtain cloned copies of expressed genes. Complementary DNA cloning usually involves cloning many genes at once, which together form cDNA libraries (Fig. 1). In step 1, mRNA is isolated from one or more tissues. In trees for example, mRNA has been isolated from xylem to obtain a cDNA library of genes expressed in xylem. The mRNA has a tail consisting of many A's (called a polyadenylated tail) which provides a convenient location to attach a primer. In step 2, a poly-T primer is annealed to the poly-A tail. In step 3, the enzyme reverse transcriptase is added, along with free A, T, C, and G nucleotides, to synthesize a DNA strand complementary to the mRNA strand. In step 4, an enzyme called RNAseH is added which digests away the original mRNA template, leaving just the newly synthesized single-stranded DNA copy. In step 5, a DNA strand complementary to the first DNA strand is synthesized by the enzyme DNA polymerase I. Finally in step 6, the double-stranded DNA molecule is inserted into a plasmid or viral vector using the enzyme DNA ligase (Box 4.1).

(Box 4.2 continued on next page)

Box 4.2. Complementary DNA (cDNA) cloning. (Continued from previous page)

Once the cDNA is put into the vector, the vector can be transformed into a bacterial host that can be cultured to produce large quantities of the cDNA.

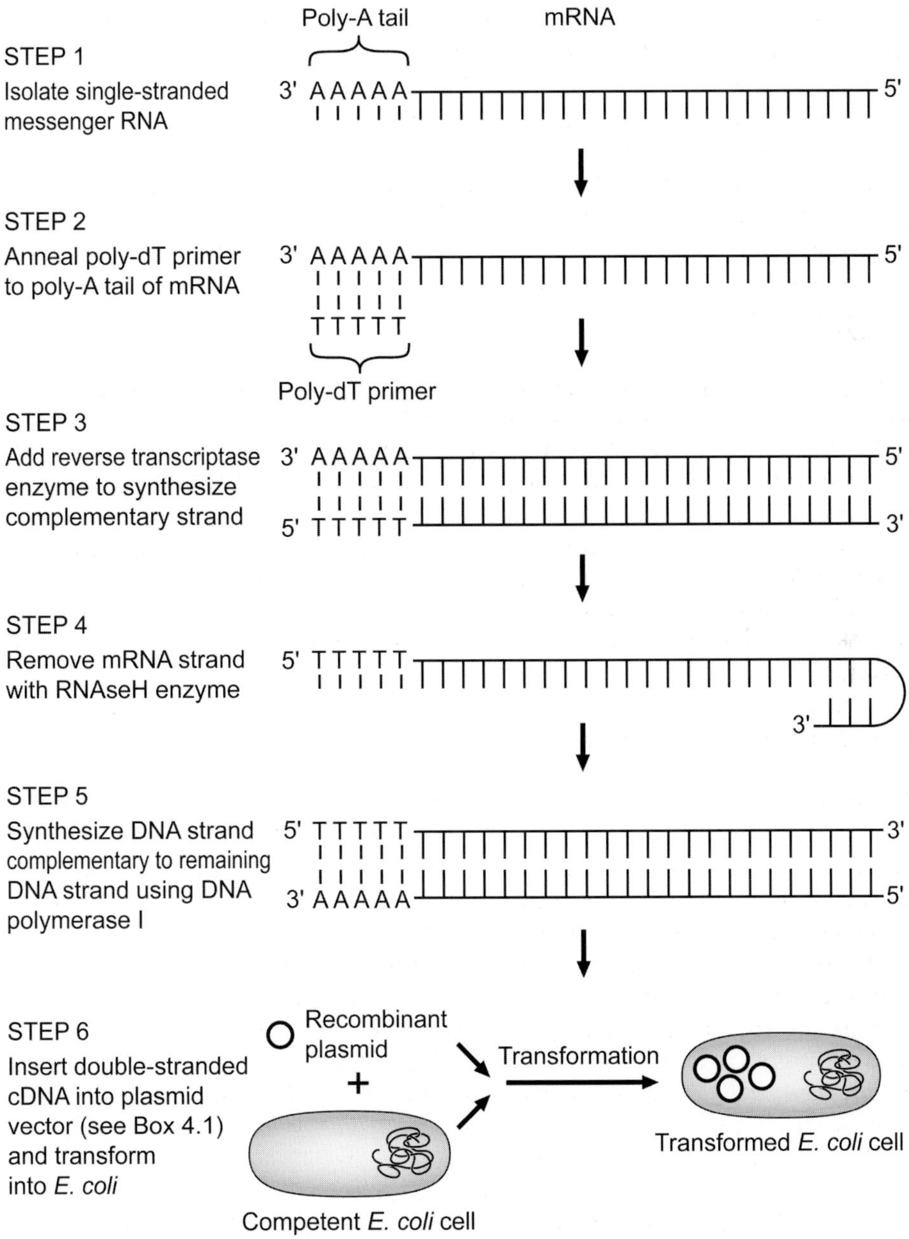

Poly-A tail
mRNA
STEP 1
Isolate single-stranded messenger RNA
3' AAAAA
5'
STEP 2
Anneal poly-dT primer to poly-A tail of mRNA
3' AAAAA
5'
TTTTT
Poly-dT primer
STEP 3
Add reverse transcriptase enzyme to synthesize complementary strand
3' AAAAA
5'
5' TTTTT
3'
STEP 4
Remove mRNA strand with RNAseH enzyme
5' TTTTT
3'
STEP 5
Synthesize DNA strand complementary to remaining DNA strand using DNA polymerase I
5' TTTTT
3'
3' AAAAA
5'
STEP 6
Insert double-stranded cDNA into plasmid vector (see Box 4.1) and transform into E. coli
Recombinant plasmid
+
Competent E. coli cell
Transformation
Transformed E. coli cell

Fig. 1. Steps in cDNA cloning.

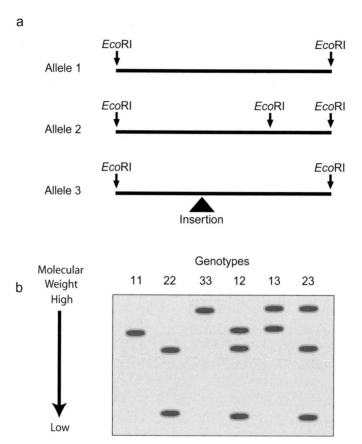

Fig. 4.4. Molecular interpretation of RFLPs resulting from all possible homozygous and heterozygous combinations of three different alleles at a single marker locus: (a) Allele 1 is the wild-type, allele 2 has a mutation that creates a new *Eco*RI restriction site and allele 3 has an insertion of a segment of DNA within the wild-type fragment; and (b) RFLP band patterns of all possible homozygous and heterozygous combinations of alleles 1, 2 and 3. The homozygous 11 genotype has just one band whereas the homozygous 22 genotype has two bands because the third *Eco*RI site creates two fragments from what was once one. The homozygous 33 band is slightly larger than the homozygous 11 band, reflecting the insertion.

Molecular Markers Based on the Polymerase Chain Reaction

The **polymerase chain reaction (PCR)** was one of the fundamentally most important biological discoveries in the 20th century and earned its inventor, Kerry Mullis, a Nobel Prize (Mullis, 1990). Before PCR, the analysis of a specific DNA fragment generally required cloning of the fragment and amplification in a plasmid or comparable vector (Box 4.1). The polymerase chain reaction enables the production of a large amount of a specific DNA sequence without cloning, starting with just a few molecules of the target sequence. One advantage of PCR-based marker methods over DNA-DNA hybridization marker methods is that the latter method requires isolation of large quantities of DNA.

    The polymerase chain reaction has three basic steps: (1) Denaturing of the double-stranded DNA template; (2) Annealing of a pair of primers to the region to be amplified; and (3) Amplification using a heat-resistant DNA polymerase called Taq polymerase (Fig.

4.5). The completed sequence of the three steps is called a cycle and at the end of the first cycle two new double-stranded molecules arise from the original template. These molecules serve as templates during subsequent cycles so that a geometric amplification of molecules occurs. Specialized instruments, called thermocyclers, have been developed to carry out the PCR process. Discovery of the heat-resistant DNA polymerase was critical to

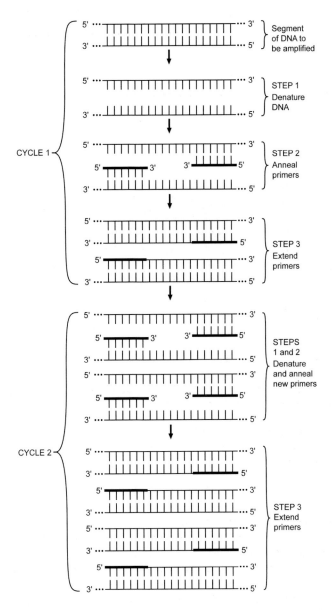

Fig. 4.5. The polymerase chain reaction (PCR) is used to amplify specific segments of DNA from complex genomes. The PCR involves three basic steps: Step 1, denaturing the DNA template; Step 2, annealing primers; and Step 3, synthesizing the complementary DNA strand using a heat resistant DNA polymerase. The three steps are repeated for many cycles to produce large quantities of the specific DNA segment. Only two cycles are shown in this example.

the development of the automated procedure because prior to this, fresh DNA polymerase had to be added each cycle because it was destroyed by high temperature during the denaturing step. The polymerase chain reaction is used not only for DNA marker technology but also for a variety of recombinant DNA assays. Many procedures in molecular biology that previously required cloning of DNA can now be performed by PCR.

Random Amplified Polymorphic DNA

Random amplified polymorphic DNA (RAPD) markers have been the most widely used molecular marker type in forest trees to date. They were the first of the PCR-based markers and were developed independently by Welsh and McClelland (1990) and Williams *et al.* (1990). The RAPD marker system is easy to apply as no prior DNA sequence information is needed for designing PCR primers as is required for other PCR-based genetic marker systems.

In the RAPD marker system (Fig. 4.6), a PCR reaction is conducted using a very small amount of template DNA (usually less than 10 nanograms) and a single RAPD primer. Primers are usually just 10 base pairs long (10-mers) and are of random sequence. There are several thousand primers commercially available, all with a different 10-base sequence, which in theory will all amplify different regions of the target genome. Therefore, the RAPD marker system has the potential to randomly survey a large portion of the genome for the presence of polymorphisms. The small amount of DNA needed is a big advantage of the RAPD technique *versus* RFLPs, because marker analysis can be applied to haploid conifer megagametophytes as was discussed for allozyme markers (Fig. 4.2).

A specific segment of the genomic DNA is amplified when the RAPD primer finds its complementary sequence at a location in the genome and then again at a second nearby location, but in the opposite orientation from the first priming site. If both chromosomes of a homologous pair each have the forward and reverse priming sites (homozygous +/+), PCR amplification products of identical length are synthesized from both homologues and a RAPD band appears on a gel when the amplification product is electrophoresed (Fig. 4.6). Likewise, if both homologues are missing one or both of the priming sites (homozygous −/−), no amplification products are synthesized and no bands are seen on gels. If one homologue has both priming sites, but the other homologue is missing at least one (heterozygous +/−), then an amplification product results from the first homologue. The heterozygous (+/−) band pattern phenotype cannot be distinguished from the +/+ homozygote; therefore, RAPD markers are dominant, di-allelic (*i.e.* only two alleles (+ and −) are expressed at each locus), genetic markers.

The + (band) and − (no band) phenotypes are distinguishable in haploid megagametophytes. Therefore, conifer trees heterozygous for a RAPD marker will segregate for + and − phenotypes in megagametophytes of their seeds, while only + phenotypes will be observed in megagametophytes of +/+ homozygotes. In this manner, +/− heterozygotes can be distinguished from +/+ homozygotes in conifer mother trees.

Since a single RAPD primer can anneal to many locations in the genome, multiple loci are revealed by a single primer. Therefore, it is possible to obtain a large number of RAPD genetic markers in a short amount of time and at relatively low cost.

Carlson *et al.* (1991) first demonstrated the use of RAPD markers in trees by showing the inheritance of RAPD markers in F_1 families of *Pseudotsuga menziesii* and *Picea glauca*. In a subsequent paper, Tulsieram *et al.* (1992) used RAPD markers and megagametophyte segregation analysis to construct a partial genetic linkage map for *Picea glauca*. Random amplified polymorphic DNA markers have since been used for linkage mapping and marker analyses in dozens of tree species (Cervera *et al.*, 2000). However, as the popularity of RAPD markers increased, difficulty in establishing marker repeatability across laboratories slowly manifested itself. Therefore, although RAPD markers are easy and quick to use, they have less overall value than the earlier allozyme and RFLP markers because of the problems with repeatability.

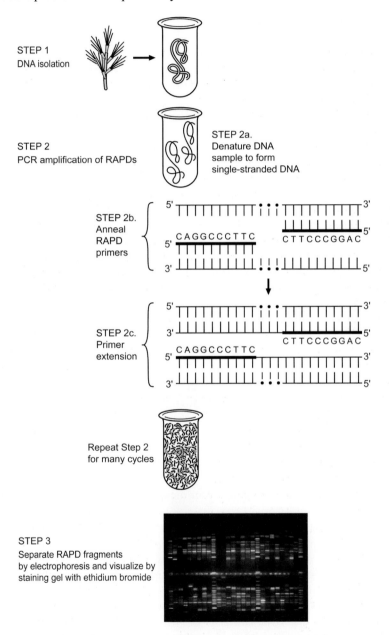

Fig. 4.6. The random amplified polymorphic DNA (RAPD) marker system involves a small number of steps and all are generally easy to apply in forest trees: Step 1, DNA is isolated; Step 2, DNA is amplified by PCR (Fig. 4.5) using single 10-mer primers; and Step 3, RAPD products are electrophoresed and bands are visualized by staining gels with ethidium bromide.

Amplified Fragment Length Polymorphism

Amplified fragment length polymorphism (AFLP) markers are a recent development (Vos *et al.*, 1995). They are like RAPDs in that many markers can be assayed quickly using PCR and they are generally dominant; but, AFLPs appear to be more repeatable than RAPDs. AFLP markers also are similar to RFLPs because they survey the genome for the presence of restriction fragment polymorphisms.

The first report of the use of AFLPs in trees was by Cervera *et al.* (1996) who used this marker system to genetically map a disease resistance gene in *Populus*. Genetic linkage maps based on AFLPs have also been constructed in *Eucalyptus globulus* and *E. tereticornis* (Marques *et al.*, 1998) and in *Pinus taeda* (Fig. 4.7; Remington *et al.*, 1998).

Simple Sequence Repeat

Simple sequence repeat (SSR) markers were first developed for use in genetic mapping in humans (Litt and Luty, 1989; Weber and May, 1989), and are another name for microsatellite DNA (*Chapter 2*). Short, tandemly-repeated sequences of two, three or four nucleotides are found throughout the genome. For example, the dinucleotide repeat AC is commonly found in *Pinus* genomes. Since the number of tandem repeats at a locus can vary greatly, SSR markers tend to be amongst the most polymorphic genetic marker types. For example, one allele might have 10 copies of the AC tandem repeat $(AC)_{10}$, whereas another allele would have 11 copies $(AC)_{11}$, another 12 copies $(AC)_{12}$, and so forth.

Simple sequence repeat genetic markers require a considerable investment to develop. Genomic DNA libraries rich in microsatellite sequences must be created and screened for clones containing SSR sequences (Ostrander *et al.*, 1992). The DNA sequence of these clones must be determined (Box 4.3), because the unique sequence regions flanking the SSR are needed to design PCR primers to amplify SSR sequences from individual samples. Once a pair of primers is developed to amplify the SSR region, it must be determined whether there is polymorphism for the SSR and whether band patterns on gels have simple genetic interpretations (Fig. 4.8).

Some of the first SSR markers developed in trees were from the chloroplast genome (Powell *et al.*, 1995; Cato and Richardson, 1996; Vendramin *et al.*, 1996). Development of these markers was made easier because the complete DNA sequence of the entire chloroplast genome of *Pinus thunbergii* was known (Wakasugi *et al.*, 1994a,b). The SSR sequences were found by a computer search of the entire cpDNA sequence database (see *Chapter 18* for a discussion of database searching and comparison of DNA sequences). Furthermore, since cpDNA sequences are highly conserved among related plant taxa, PCR primers designed from sequences flanking SSR sequences in *P. thunbergii* should easily amplify homologous sequences in other *Pinus* species. The cpDNA SSRs are highly polymorphic relative to other types of cpDNA markers and are useful for many types of studies. For example, because cpDNA is paternally inherited in conifers (*Chapter 3*), cpDNA markers are useful for determining male parentage of offspring (paternity analysis) (*Chapter 7*) and for following the dispersal of pollen in populations where the SSR genotypes of all male trees are known (Stoehr *et al.*, 1998).

Nuclear DNA SSRs have been developed for several forest trees, including species of *Pinus* (Smith and Devey, 1994; Kostia *et al.*, 1995; Echt *et al.*, 1996; Echt and May-Marquardt, 1997; Pfeiffer *et al.*, 1997; Fisher *et al.*, 1998), *Picea* (van de Ven and McNicol, 1996), *Quercus* (Dow *et al.*, 1995) and *Populus* (Dayanandan *et al.*, 1998). Each of these studies describes the isolation and cloning of a small number of SSRs, their inheritance patterns, and their utility for related species.

Molecular weight
standard

Pinus taeda megagametophyte DNA samples

Molecular weight
standard

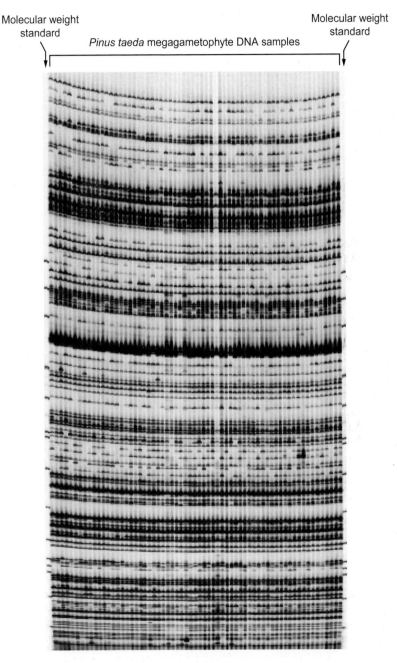

Fig. 4.7. An autoradiogram of amplified fragment length polymorphisms (AFLP) of *Pinus taeda* megagametophyte DNA samples. The first and last lanes are DNA samples of known molecular weight that are used to estimate the molecular weights of the pine DNA samples. All other lanes represent DNA in a sample of 64 megagametophytes from a single *Pinus taeda* seed tree. Each of the horizontal bands represents a different genetic locus. Bands of identical migration on the gel that are segregating for the presence or absence of a band are from polymorphic loci; all other loci are monomorphic in this seed tree. (Photo courtesy of D. Remington, North Carolina State University, Raleigh)

Expressed Sequence Tagged Polymorphisms

Expressed sequence tagged polymorphisms (ESTPs) are PCR-based genetic markers that are derived from **expressed sequenced tags (ESTs)**. Expressed sequenced tags are partial cDNA sequences that have been obtained by automated DNA sequencing methods (*Chapter 18*); therefore, ESTPs are a genetic marker for structural gene loci. The EST databases contain hundreds of thousands of entries from a variety of organisms, most notably *Arabidopsis thaliana*, rice, and maize in plants. In forest trees, there are EST databases for *Pinus*, *Populus*, and *Eucalyptus*. The ESTs are routinely compared to DNA sequence databases to determine their biochemical function. It is also a goal of most genome projects to place the ESTs onto genetic linkage maps. Expressed sequenced tags can be genetically mapped by a variety of methods, all of which rely on detecting polymorphism for the ESTs, hence the name ESTPs for the genetic marker.

Box 4.3. Methods to determine the nucleotide sequence of a DNA fragment.

The ability to determine the nucleotide sequence of a gene or any piece of DNA is fundamentally important to understanding how genes work and the genetic variation found at the DNA sequence level. Chemical methods to determine the nucleotide sequence of a piece of DNA were first developed in the 1970s. In earlier years, it would take months to determine the sequence of just a small piece of DNA. In the 1990s, automated DNA sequencing technologies were developed that allowed individual labs to sequence millions of bases in a single day (*Chapter 18*). These technologies have enabled the determination of the entire DNA sequence of the human genome, drosophila, mice, chicken, *Arabidopsis*, and others. The first tree genome to be completely sequenced is *Populus trichocarpa*, and sequencing of others is likely to be undertaken in the near future. Although DNA sequencing is currently done by automated methods, it is important to understand the basic chemistry of the manual methods that preceeded automated technologies.

There are several different manual methods for determining the nucleotide sequence of a DNA molecule. The most often used approach is the chain-termination method of Sanger *et al.* (1977). The DNA sequence is determined by synthesizing partial complementary strands of DNA. In this example, we wish to determine the unknown sequence of a small fragment of DNA whose actual sequence is ATGCATGC (Fig. 1). In Step 1, a primer is annealed to a single-stranded DNA molecule. The DNA to be sequenced has generally been cloned and the single-stranded copy is derived from the recombinant plasmid-cloning vector (Box 4.1). The primer is complementary to the cloning site within the vector. In Step 2, four different sequencing reactions are set up. To each reaction, a different di-dioxy XTP is added (dd-ATP, dd-TTP, dd-GTP, dd-CTP). When DNA polymerase incorporates a dd-XTP in the synthesis of the complementary strand *versus* a normal dioxy-XTP (d-XTP), the sequencing reaction is terminated because the dd-XTP will not chemically bind to a complementary d-XTP. The dd-XTPs are in low concentration in the reaction mixes, so it is only occasionally and randomly that a dd-XTP is added. In this example, two different chain-termination products are synthesized in each reaction, each corresponding to the base complementary to the dd-XTP in the reaction. In Step 3, the reaction mixes are electrophoresed on polyacrylamide gels by loading one lane for each of the four reaction mixes. The DNA strands produced by chain-termination separate by size. The DNA sequence of the complementary strand is read from the bottom of the gel upwards; in this example, it is TACGTACG. Finally, the complement of this sequence is determined to obtain the sequence of the original single-stranded copy, which is ATGCATGC.

(Box 4.3 continued on next page)

Box 4.3. Methods to determine the nucleotide sequence of a DNA fragment.
(Continued from previous page)

STEP 1

Anneal sequencing primer
to single-stranded DNA template

STEP 2

Set up chain-termination reactions for each base (A, T, G, C) by
adding DNA polymerase, labelled dioxy XTPs and di-dioxy XTPs

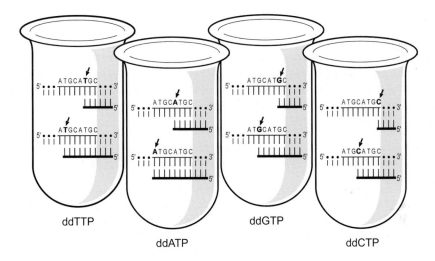

STEP 3

DNA molecules resulting from chain termination reactions are electrophoresed on
polyacrylamide gels. DNA fragments are ordered by their molecular weight.
Read nucleotide order of DNA directly from bottom to top.

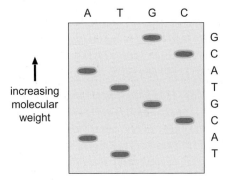

Fig. 1. Illustration of the steps required in determining the nucleotide sequence in a fragment of
DNA.

a

Allele 1 =(ATC)₇

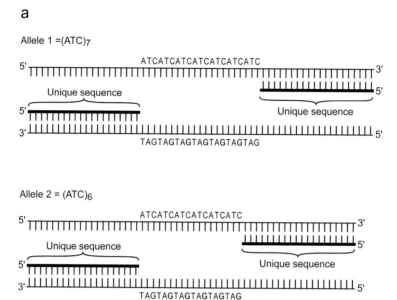

Allele 2 = (ATC)₆

b   Genotypes

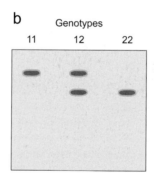

Fig. 4.8. The simple sequence repeat (SSR) or microsatellite marker system requires significant time and cost to develop; however, it is fairly easy to apply once markers are developed: (a) Primers complementary to unique sequence regions flanking the SSRs are used to amplify the SSR sequence by PCR; allele 1 has seven copies of the ATC repeat and allele 2 has six copies of the ATC repeat; and (b) A simplified gel pattern of the two homozygous (11 and 22) and the heterozygous (12) SSR genotypes. Since all three genotypes are distinguishable, this is a codominant marker.

All approaches to ESTP detection require a pair of PCR primers to be designed from the EST sequence. The primers are then used to amplify genomic DNA fragments by PCR. The next step is to reveal polymorphism among amplification products and for this a variety of methods have been used. In *Picea mariana*, Perry and Bousquet (1998) were able to detect length variation among ESTP alleles when amplification products were analyzed on standard agarose gels. This is the fastest and simplest technique for revealing ESTP variation, although it seems unlikely that such length variation can be found for the majority of ESTs. Polymorphisms surely exist as nucleotide substitutions among alleles; however, these polymorphisms are more problematic to detect.

In *Cryptomeria japonica* (Tsumura *et al.*, 1997) and in *Pinus taeda* (Harry *et al.*, 1998), researchers have digested amplification products with restriction enzymes to reveal polymorphism. This approach is similar to RFLP analysis and the markers are sometimes called PCR-RFLPs or cleaved amplified polymorphisms (CAPs). More sensitive techniques for revealing polymorphisms are available such as **simple sequence conformational polymorphism (SSCP)** or **density gradient gel electrophoresis (DGGE)** analysis (Fig. 4.9). Temesgen *et al.* (2000, 2001) have used the DGGE technique in *Pinus taeda* to develop a large number of codominant markers. Cato *et al.* (2000) developed a different method for mapping ESTs that is similar to the AFLP method, except that instead of using two random primers as in AFLP, one primer is complementary to the EST sequence.

An important advantage to ESTP markers over most of the other PCR-based genetic markers, such as RAPDs, AFLPs and SSRs, is that ESTPs most likely reveal variation in

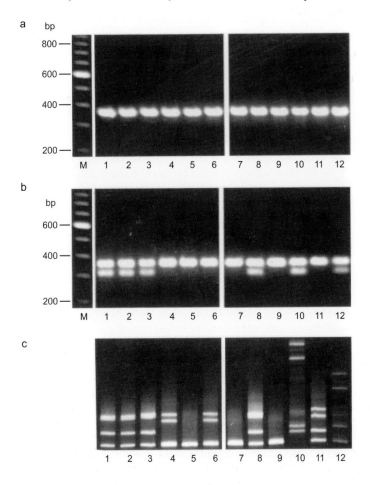

Fig. 4.9. Expressed sequence tagged polymorphism (ESTP) markers can be revealed by several different polymorphism detection methods. An EST from the same 12 individuals (lanes 1-4 and 7-10 are the grandparents of two *Pinus taeda* pedigrees and lanes 5-6 and 11-12 are the parent trees of these pedigrees) is analyzed using three approaches: (a) No polymorphism is revealed in the PCR amplification products separated on agarose gels; (b) Two different alleles were detected in this sample after the DNAs were digested with one restriction enzyme; and, (c) Multiple codominant alleles were revealed when ESTP amplification products were assayed by denaturing gradient gel electrophoresis (DGGE).

structural gene loci whereas the others generally reveal variation in non-coding regions of the genome. Such structural gene markers are called genic markers *versus* non-genic markers. This distinction can be important for some applications of genetic marker analysis, such as candidate gene mapping and the discovery of single nucleotide polymorphisms (SNPs) within candidate genes. SNPs are the most recently developed marker type and are covered in detail in *Chapter 18*.

SUMMARY AND CONCLUSIONS

Genetic markers are essential for a large number of types of genetic investigations in forest trees. Genetic markers are often used with natural populations of forest trees to determine amounts and patterns of genetic variation, to understand mating systems and inbreeding, and to study taxonomic and phylogenetic relationships among species. Genetic markers are routinely used to monitor the efficiency of various tree improvement activities and are necessary for the construction of genetic maps and marker-aided breeding.

Morphological genetic markers are available in forest trees; however, the number of such markers is limited and they are often associated with a deleterious phenotype. Nevertheless, seedling mutant characters have been used to study mating systems in a number of conifer species.

Several types of biochemical markers exist and have been extremely valuable for genetic studies in trees. Monoterpenes were the first biochemical markers in trees and have been mostly used for taxonomic studies in pines. The small number of monoterpene markers and their dominant expression has limited their utility for other applications. Studies based on allozymes, the most widely used of all genetic markers to date, have contributed greatly to our knowledge of the population genetics of forest trees. Allozymes are fairly easy to learn how to assay and apply, and are highly polymorphic and codominant. They have been used to describe patterns of genetic variation both within and among forest tree populations and for estimating mating systems and gene flow. Allozymes are also very useful for genetic fingerprinting and paternity analysis.

Molecular, or DNA-based markers are the most recent to be developed and have many advantages over morphological and biochemical markers. The primary advantages are: (1) There is potentially an unlimited number of molecular markers available; and (2) DNA markers are generally not affected by developmental differences or environmental influences. There are two general classes of molecular markers: those based on DNA-DNA hybridization and those based on the polymerase chain reaction (PCR). Restriction fragment length polymorphism (RFLP) markers rely on DNA-DNA hybridization and have been used for organelle genetic analysis and genetic linkage mapping in forest trees. The three PCR-based molecular marker types used widely in forest trees are: (1) Random amplified polymorphic DNA (RAPD); (2) Amplified fragment length polymorphisms (AFLP); and (3) Simple sequence repeats (SSR). All three of these marker types generally reveal polymorphism in non-coding regions of genomes. The RAPDs and AFLPs are di-allelic, dominant markers, whereas SSRs are codominant and multiallelic. A new PCR-based marker type called expressed sequence tag polymorphism (ESTP) potentially overcomes many of the limitations of the other PCR-based marker types. The ESTPs are codominant, multiallelic and reveal polymorphism within, or in sequences flanking, expressed structural genes. Therefore, ESTPs have significant potential for studies of adaptive genetic variation in forest trees.

CHAPTER 5
POPULATION GENETICS – *GENE FREQUENCIES,*
INBREEDING AND FORCES OF EVOLUTION

Most forest tree species possess considerable genetic diversity. Species vary greatly not only in amounts of genetic variation, but also in how this variation is distributed among and within populations across the landscape. The ultimate goals of **population genetics** are to describe patterns of genetic diversity and to understand their origin, maintenance and evolutionary significance. In simplest terms, population genetics is the application of Mendelian genetic principles to whole populations of organisms and species.

Tree populations may occur naturally or may be formed artificially by actions of foresters or tree breeders. Therefore, population genetics is not only fundamental to understanding evolution and adaptation of forest trees in their native habitats, but also to understanding genetic changes brought about by forest management, including breeding. Although population genetics focuses on traits under single-locus (Mendelian) control, most traits of evolutionary significance (*e.g.* tree size, rate of growth, reproductive capacity) are controlled by many gene loci. These polygenic or quantitative traits are the subject of the next chapter. However, the principles of population genetics apply as well to polygenic traits, since they are the products of cumulative effects of individual gene loci.

In this chapter, we begin by describing how to quantify the genetic composition of populations at single gene loci using genotype and allele frequencies. We show that genotype and allele frequencies remain the same from one generation to the next when large populations conform to a simple model (called the Hardy-Weinberg Principle) which assumes all mating is at random and there is no mutation, no differential selection among genotypes, and no migration of genes from other populations. We then begin to relax the simplified assumptions associated with Hardy-Weinberg populations to examine how failure to meet each of these assumptions results in change to the genetic composition of populations. The second major section in this chapter emphasizes how an important type of non-random mating, inbreeding, impacts the frequencies of genotypes in populations. The third major section describes the impacts of the four forces of evolution (*i.e.* mutation, migration, selection and genetic drift) taken individually. The final section examines the interplay and relative importance of these four forces in natural populations. These concepts of population genetics are explored in greater depth in several excellent textbooks on the subject including Hedrick (1985), Hartl and Clark (1989), Falconer and Mackay (1996), and Hartl (2000).

QUANTIFYING THE GENETIC COMPOSITION OF POPULATIONS

Genotype and Allele Frequencies

A **population** is a group of organisms of the same species living within a prescribed area small enough so that all individuals have the opportunity to mate with all others in the area. Many forest trees, however, are distributed more or less continuously over large regions, such that individuals in one part of a region are unlikely to have the opportunity to mate with distant individuals in other parts of the same region. These widespread species are effectively subdivided into more or less distinct breeding groups called subpopulations, which occupy smaller geographical areas within regions. It is the local breeding group that we generally refer to when we use the word "population" in this chapter.

Genotype and allele frequencies are the key statistics used to quantify genetic composition of populations and we illustrate these concepts by examining allozyme data from a *Pinus ponderosa* population on the east slope of the Rocky Mountains near Boulder, Colorado (Mitton *et. al.*, 1980). Needles were harvested from 64 mature trees and subjected to electrophoretic assay (*Chapter 4*). Three genotypes were observed for genes coding isozymes of glutamate dehydrogenase (*Gdh*; Table 5.1).

Genotype frequencies (*i.e.* the proportion of each genotype in the population) are estimated by dividing the observed number of each genotype by the total number of individuals sampled (*e.g.* the frequency of *Gdh*-11, written f(*Gdh*-11), is 21/64 = 0.328). To calculate **allele frequencies**, remember that each genotype of a diploid tree carries two alleles. Therefore, the total number of alleles is twice the number of individuals sampled, and allele frequency is the proportion of all alleles that are of the specified type. In this example, the frequency of allele *Gdh*-1, written f(*Gdh*-1), is [2(21) + 36]/[2(64)] = 0.609, and f(*Gdh*-2) = [2(7) + 36]/[2(64)] = 0.391. Where there are more than two alleles, the process is the same; sum the alleles of the given type and divide by the total number of observed alleles of all types.

Alternatively, allele frequencies can be derived from genotypic frequencies by summing the frequency of the homozygote for the allele of interest plus one-half the frequencies of all heterozygotes that carry that allele. For allele i:

$$f(i) = X_{ii} + 1/2\sum X_{ij}$$ Equation 5.1

where X_{ii} is the frequency of the homozygote for allele i, and X_{ij} are heterozygote frequencies [*e.g.* f(*Gdh*-1) = 0.328 + 1/2(0.563) = 0.609]. Note that in all cases, genotype and allele frequencies fall between 0 and 1, and the total frequency of all genotypes or alleles must equal 1. When the frequency of an allele is equal to 1, we say that allele is fixed in the population meaning that it is the only allele at that locus; the locus is said to be **monomorphic**.

Allele frequencies calculated from samples of individuals are only estimates of the true allele frequencies in the population. Therefore, sample sizes must be large if the estimates are going to accurately reflect the true values. An estimate of the variance (σ^2) of the frequency for allele i is:

$$\sigma^2_{f(i)} = [f(i)(1 - f(i))]/2N$$ Equation 5.2

where N is the total number of individuals sampled. In our example above, $\sigma^2_{[f(Gdh-1)]}$ = [0.609)(0.391)]/[2(64)] = 0.00186. The square root of $\sigma^2_{f(i)}$ is the standard error of the estimate, so $\sigma_{[f(Gdh-1)]}$ = √0.00186 = 0.043. Generally, sample sizes of at least 50 individuals are required for fairly reliable estimates of allele frequencies, but N ≥ 100 is recommended whenever possible (Hartl, 2000).

Many genes in natural populations of forest trees and other organisms are **polymorphic**, which means they have two or more relatively frequent alleles. Relatively frequent is arbitrarily defined, but the typical convention in population genetics literature is that the most common allele in the population must have a frequency less than 0.95 (sometimes 0.99 is used as the cutoff) for the locus to be called polymorphic.

Table 5.1. Observed genotype frequencies for an allozyme locus, glutamate dehydrogenase (*Gdh*), in 64 trees sampled from a Rocky Mountain population of *Pinus ponderosa* (Mitton *et al.*, 1980).

Genotype	Number	Frequency
Gdh-11	21	0.328
Gdh-12	36	0.563
Gdh-22	7	0.109
Total	64	1.000

Hardy-Weinberg Principle

Various evolutionary forces alter genetic composition of populations from one generation to the next. However, alleles and not genotypes are the link between generations because genotypes are broken up in the process of segregation and recombination during meiosis. Therefore, we need to know the relationship between allele frequencies in one generation and genotypic and allele frequencies in the next. Working independently, Godfrey Hardy and Wilhelm Weinberg worked out these relationships in the early 1900s (*Chapter 1*, Table 1.1) for a simplified set of conditions: (1) The organism is diploid and reproduces sexually; (2) Generations are non-overlapping; (3) Mating is at random; (4) All genotypes are equal in viability and fertility (no selection); (5) Mutation is negligible; (6) There is no movement of individuals (migration) into the population; and (7) Population size is very large.

Single-locus, Two-allele Case

Given a locus with two alleles, A_1 and A_2, with frequencies $f(A_1) = p$ and $f(A_2) = q$, the consequences of the above conditions, as stated in the Hardy-Weinberg Principle, are: (1) Allele frequencies will remain unchanged in future generations; (2) Genotypic frequencies reach constant relative proportions of:

$$f(A_1A_1) = p^2$$

$$f(A_1A_2) = 2pq \hspace{4cm} \text{Equation 5.3}$$

$$f(A_2A_2) = q^2$$

and (3) These constant genotype frequencies are reached after one generation of random mating, regardless of the original allele and genotypic frequencies of the population.

To verify the above results, we first need to define what is meant by random mating. With random mating, individuals mate without regard to genotype; that is, genotypes are paired as often as would be expected by chance. This implies that the zygotes produced in one generation are just a random combination of gametes produced in the previous generation. Therefore, with random mating, genotypic frequencies are calculated by multiplying allele frequencies of the corresponding gametes. Assume that in the parental generation of a population the frequencies of the three genotypes are $f(A_1A_1) = X_{11}$, $f(A_1A_2) = X_{12}$ and $f(A_2A_2) = X_{22}$ (which are completely arbitrary and may or may not conform to Hardy-Weinberg expectations), and allele frequencies are $f(A_1) = p$ and $f(A_2) = q$ (Fig. 5.1).

The frequencies of homozygotes A_1A_1 and A_2A_2 in the offspring generation, X'_{11} and X'_{22}, respectively, are the squares of the frequencies of their corresponding alleles in the

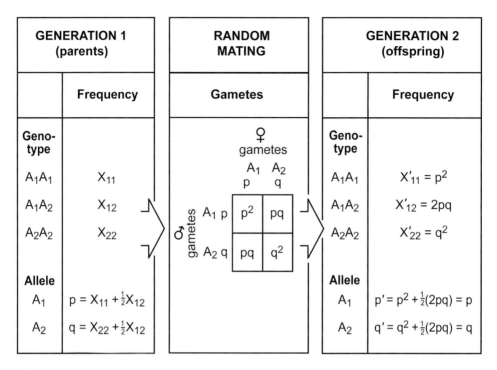

Fig. 5.1. Illustration of the Hardy-Weinberg Principle for a locus with two alleles, A_1 and A_2, having frequencies p and q, respectively. In generation 1, genotypes begin at any frequency (designated X_{11}, X_{12}, X_{22}). However, after one generation of random mating, the genotype frequencies are p^2, $2pq$, q^2. Allele frequencies remain p and q.

gametes: $X'_{11} = p^2$ and $X'_{22} = q^2$. Note, however, that A_1A_2 offspring can be formed from combining gametes in two ways: A_1 (♀) with A_2 (♂) or A_2 (♀) with A_1 (♂), each combination with frequency pq. Considering both possibilities, $f(A_1A_2) = X'_{12} = 2pq$. The frequency of A_1 in the offspring generation (p') (from Equation 5.1) is:

$$p' = X'_{11} + 1/2X'_{12}$$
$$= p^2 + pq$$
$$= p(p + q) = p \quad (\text{since } p + q = 1)$$

The striking conclusion from the Hardy-Weinberg Principle is that in the absence of specific evolutionary forces causing allele frequencies to change, the mechanism of Mendelian inheritance, by itself, preserves genetic variation from one generation to the next. In addition, when mating is random, genotypic proportions can be predicted from allele frequencies using Equation 5.3.

In real populations, genotypic frequencies are never known exactly, but rather are estimated from a sample of individuals representing the entire population. A goodness-of-fit chi-square (χ^2) test is then used to test whether the estimated genotypic frequencies are significantly different from those predicted by the Hardy-Weinberg Principle in Equation 5.3. This is illustrated in Box 5.1.

Genotypic proportions in natural populations of forest trees (and many other organisms) often closely approximate Hardy-Weinberg expectations (*e.g.* Box 5.1). This is because the

Box 5.1. A test of expected genotype frequencies under Hardy-Weinberg equilibrium.

We return to the polymorphism of the gene locus coding *Gdh* isozymes in a Rocky Mountain population of *Pinus ponderosa* (Table 5.1). Estimated frequencies of alleles *Gdh*-1 and *Gdh*-2 were 0.609 and 0.391, respectively. Expected Hardy-Weinberg proportions of genotypes from Equation 5.3 are:

$f(Gdh\text{-}11) = p^2 = 0.609^2 = 0.371$
$f(Gdh\text{-}12) = 2pq = 2(0.609)(0.391) = 0.476$
$f(Gdh\text{-}22) = q^2 = 0.391^2 = 0.153$

Expected numbers, calculated by multiplying each expected frequency by the total number of observations (64), are compared to the observed numbers below.

Genotype	Observed number	Expected number
Gdh-11	21	23.75
Gdh-12	36	30.46
Gdh-22	7	9.79

The observed numbers seem fairly close to those expected and this is tested statistically by using the goodness-of-fit chi-square (χ^2) test introduced in *Chapter 3* as Equation 3.1. Here, $\chi^2 = (21 - 23.75)^2/23.75 + (36 - 30.46)^2/30.46 + (7 - 9.79)^2/9.79 = 2.12$. Recall from *Chapter 3* that the degrees of freedom for χ^2 is generally the number of different classes (three in this case) minus one. In this situation, however, we subtract an additional degree of freedom, because the allele frequencies necessary to calculate the expected genotypic frequencies also were estimated from the data (Snedecor and Cochran, 1967). Therefore, the appropriate degrees of freedom for this χ^2 is $3 - 1 - 1 = 1$. The probability of a particular χ^2 can be looked up in a table of cumulative χ^2 distributions, and 2.12 (with one degree of freedom) lies between the probabilities of 0.25 and 0.15. This probability range means that if the observed genotypic frequencies are in the true population proportions, 15-25% of samples of size 64 would give deviations of the magnitude observed or greater. A calculated χ^2 value that occurs 15-25% of the time by chance is not unreasonable, so we declare that the observed proportions support the hypothesis that genotypic frequencies in this population conform to Hardy-Weinberg expectations. The convention is to reject this hypothesis only if the probability of the calculated χ^2 is 0.05 or less. This means that for a single locus with two alleles and three genotypes, the calculated χ^2 value must exceed the tabular value of 3.84 to be declared significant at the 5% level.

Although there is a reasonable fit of observed genotypic frequencies to Hardy-Weinberg proportions in this case, we cannot conclude that the various evolutionary forces are not acting at this locus. Lack of significant deviation from expectation may simply mean that these forces are weak or that their effects cancel themselves out in combination. Therefore, more information is typically required to evaluate the strength and impact of evolutionary forces. In particular, allele frequencies assessed in different generations, or at different life cycle stages in the same generation (*e.g.* gametes *versus* zygotes *versus* adults), are often required. In addition, spatial patterns of allele frequencies among natural populations are often useful in evaluating the relative strength of the various evolutionary forces (*Chapter 8*).

mating system in forest trees and other higher plants (predominantly self-pollinated species being the major exception) often approaches random mating. In addition, the Hardy-Weinberg Principle is not sensitive to small violations of the assumptions. Therefore, lack of deviation of observed genotype frequencies from expected Hardy-Weinberg proportions cannot, by itself, be taken as evidence that all Hardy-Weinberg conditions are met. On the other hand, the relative lack of sensitivity of Hardy-Weinberg frequencies to violations of the assumptions means that Hardy-Weinberg proportions often roughly approximate the actual genotypic proportions in populations of outcrossing organisms.

It is instructive to examine some mathematical consequences of Equation 5.3 and these are summarized in Box 5.2. While these implications of the Hardy-Weinberg Principle apply strictly only to populations in perfect Hardy-Weinberg equilibrium, the concepts are widely useful, because many natural populations of outcrossing species have genotype frequencies that approximate those predicted by Equation 5.3. Most of the implications mentioned in Box 5.2 arise because when q is small, most of the A_2 allele occurs in heterozygous form. So, for example when p = 0.9 and q = 0.1, then $f(A_1A_2) = 2pq = 0.18$ and $f(A_2A_2) = q^2 = 0.01$ for a ratio of 18:1 (18 times as many heterozygotes as recessive homozygotes). When q is reduced by half (p = 0.95, q = 0.05, 2pq = 0.095 and $q^2 = 0.0025$), the ratio jumps to 38:1 meaning that a larger fraction of the A_2 allele is contained in heterozygotes. When q = 0.001, the ratio is 1998:1.

Box 5.2. Implications of the Hardy-Weinberg Principle in real populations.

1. *Conservation and restoration of genetic variation.* Once Hardy-Weinberg equilibrium genotype frequencies are obtained, they remain constant in future generations as long as Hardy-Weinberg conditions prevail. Therefore, random mating in large populations acts to conserve the genetic variability that exists in the population. Even in highly inbred populations, only one generation of random mating is needed to restore Hardy-Weinberg equilibrium frequencies.

2. *Heterozygote carriers.* Even though homogygotes expressing natural diseases caused by recessive deleterious alleles occur very infrequently in populations, heterozygote carriers of the diseases are fairly common because rare alleles are contained mostly in heterozygotes. An example described in the text is that 1 in 25 Caucasians are carriers of the cystic fibrosis allele, while only 1 in 2500 people get the disease.

3. *Maintaining genetic load.* Rare recessive alleles with detrimental effects can be maintained for many generations in natural populations even though these reduce the population's fitness (called **genetic load**). There is little cost to the population because the large majority of the detrimental alleles are in heterozygotes where the detrimental effects are masked. Therefore, mutations at many gene loci that are currently deleterious can be maintained in large, random mating populations. Ultimately alleles that are neutral or mildly deleterious in current environments form the genetic diversity that allows the species to adapt to changing environmental conditions.

4. *Ridding recessives from breeding populations.* It is difficult to fix a desirable dominant allele by mass selection in a breeding program because as the frequency of the desired dominant allele increases, the unwanted recessive allele becomes rare and exists primarily in heterozygotes where its phenotypic effect is masked.

Also relevant is that even when the A_2 allele is rare, heterozygotes are relatively common. A real example in human populations (Hartl, 2000, p. 33) is cystic fibrosis, a disease expressed only by homozygous recessive individuals. The disease affects 1 in 2500 newborn Caucasians, and assuming Hardy-Weinberg equilibrium, we use Equation 5.3 to calculate $q^2 = 1/2500$, $q = 0.02$, $p = 0.98$ and $2pq = 0.0392$. The latter value means that 1 in 25 Caucasians ($1/25 = 0.04$) are heterozygote carriers of the disease, but only 1 in 2500 are born with the disease.

An important use of the Hardy-Weinberg Principle is to estimate allele frequencies when complete dominance makes it impossible to discriminate phenotypically between homozygotes and heterozygotes expressing a dominant trait. If it can be assumed that genotypes are in Hardy-Weinberg proportions, the frequency of the recessive allele is estimated as the square root of the frequency of homozygous recessive genotypes (X_{22}); that is:

$$q = (X_{22})^{1/2}$$ Equation 5.4
$$\text{and } \sigma^2_q = (1 - X_{22})/4N$$ Equation 5.5

To illustrate, mutant phenotypes (*e.g.* albino needles, dwarfs, fused cotyledons) are often observed among the thousands of germinants in commercial nursery beds (*e.g.* see Franklin, 1969b; Sorensen 1987, 1994). Assume that in a *P. ponderosa* nursery, 20 albino mutants (Fig. 5.2) are found in a seedbed containing 120,000 germinants, and that this mutant is caused by a recessive allele at a single locus. What is the allele frequency (q) of the mutant gene in this seedbed? The frequency of albino mutants is $X_{22} = q^2 = 20/120,000 = 0.0001667$, and $q = (0.0001667)^{1/2} = 0.013$. From this, all genotype frequencies can be estimated from Equation 5.3, but the accuracy of these estimates rests on the a priori assumption that the population is in Hardy-Weinberg equilibrium.

Fig. 5.2. It is common to find albino seedlings in forest tree nurseries, such as this one of *Pinus ponderosa*. The lack of chlorophyll is presumably due to a mutation in a gene important for chlorophyll biosynthesis. If the seedlings die due to lack of photosynthesis, then the mutation is lethal. (Photo courtesy of F. Sorensen, USFS Pacific Northwest Research Station)

Extension to Multiple Alleles and Multiple Loci

The above discussion of the Hardy-Weinberg Principle assumes a single locus with only two alleles. Extension of the Hardy-Weinberg Principle to a single locus with more than two alleles is straightforward. With any number of alleles, expected genotypic proportions after a single generation of random mating are p^2_i for A_iA_i homozygotes and $2p_iq_j$ for A_iA_j heterozygotes, and allele frequencies are the same as those in the previous generation, as long as all other Hardy-Weinberg conditions hold. Therefore, all of the implications and uses of the Hardy-Weinberg Principle mentioned above hold for the case of multiple alleles.

The case of multiple loci is more complex. Although equilibrium in genotype frequencies is attained after a single generation of random mating for a single locus, this is not true of genotypic combinations at two or more loci considered together. Multilocus equilibrium is best examined at the level of haploid gametes because it is simpler; and because under Hardy-Weinberg conditions, if allelic combinations are in equilibrium in gametes they are also in equilibrium in the diploid genotypes formed by the gametes. Assume there are two loci with two alleles at each locus, A_1, A_2 and B_1, B_2, having frequencies of p, q and r, s, respectively. The frequency of each two-locus gametic type (*i.e.* A_1B_1, A_1B_2, A_2B_1, A_2B_2) at equilibrium is just the product of the frequencies of the alleles it contains [*e.g.* $f(A_1B_1)$ = pr and $f(A_2B_1)$ = qr]. If, however, the frequency of any one gametic type is not equal to its equilibrium value, these loci are said to be in **gametic disequilibrium** (also called **linkage disequilibrium**).

Gametic disequilibrium is caused by a variety of mechanisms including by chance if only a small number of parents contribute offspring to the next generation, by mixing of populations with different allele frequencies, or by selection favoring some combinations of alleles over others (sometimes called co-adapted gene complexes). Nevertheless, unless loci are tightly linked, gametic disequilibrium is expected to dissipate in relatively few generations of random mating (Falconer and Mackay, 1996). Therefore, like in other predominantly outcrossing plants, gametic disequilibrium has been observed infrequently in natural populations of forest trees (Muona and Schmidt, 1985; Epperson and Allard, 1987; Muona, 1990).

Gametic disequilibrium studies in forest trees have been based on allozymes, for which few loci (generally 20-30 or less) are generally available and linkage information is incomplete. Interest in gametic disequilibrium in natural or artificial populations is likely to increase with the large numbers of genetic marker loci available through molecular DNA methods. Of particular practical significance is gametic disequilibrium between marker gene loci and genes at closely linked loci that control polygenic traits of economic importance (*i.e.* **quantitative trait loci** or **QTLs**) (*Chapter 18*). An example might be a DNA marker locus closely linked with a QTL involved in growth, disease resistance or wood quality. If gametic disequilibrium is very strong, such markers could greatly facilitate selection and breeding to improve these traits through marker-assisted breeding (*Chapter 19*).

MATING SYSTEMS AND INBREEDING

A key assumption of the Hardy-Weinberg Principle is random mating. In this section, we address deviations from random mating. By itself, the mating system (*i.e.* pattern of mating among individuals) does not alter allele frequencies, but does affect the relative proportions of different genotypes in populations, which under some circumstances profoundly influences the viability and vigor of offspring. Even though genotypic frequencies in natural populations of forest trees often approximate those expected under random mating,

mating systems that depart from random mating do occur and have important implications. Individuals of most temperate forest trees are bisexual and have the capacity for self-fertilization. In addition, nearby trees may be related (*e.g.* siblings originating from seeds of the same mother tree), providing opportunity for mating between relatives. Therefore, forest trees typically have mixed-mating systems (*Chapter 7*), whereby many and perhaps most mates are paired essentially at random (among all possible reproductive adults in the population). There is also some mating between genetically-related individuals that occurs more often than expected from random pairings (called **inbreeding**).

Other departures from random mating include preferential mating between like pheno-types, called **phenotypic assortative mating** (PAM), such as might occur among indi-viduals of similar flowering phenologies (*e.g.* among early flowering or among later flow-ering individuals in a stand); and **phenotypic disassortative mating**, such as might occur in a monoecious species when trees with mostly female flowers mate with trees with a preponderance of male flowers. Because phenotypes are at least partially under genetic control, assortative and disassortative mating have genetic implications. For example, PAM leads to weak inbreeding when the similar phenotypes that are mating are also rela-tives; so, the consequences of inbreeding discussed below apply to PAM.

We emphasize inbreeding in the remainder of this section because of its great signifi-cance to the genetic makeup of both natural populations and breeding populations of forest trees. Inbreeding has two major consequences: (1) In comparison to random mating, in-breeding increases the frequency of homozygous offspring at the expense of heterozy-gotes; and (2) Mating between close relatives is usually detrimental to the survival and growth of offspring (called **inbreeding depression**). Therefore, the magnitude of inbreed-ing among parent trees used to produce seed for reforestation, such as in seed production areas or seed orchards, is of great practical concern (*Chapter 16*).

Influence of Inbreeding on Genotypic Frequencies

The effect of inbreeding on genotypic frequencies is most dramatic when all mating is by self-fertilization, which is nearly the case in many annual crop plants (*e.g.* oats, barley and wheat) and in natural populations of some annual grasses (Schemske and Lande, 1985). Most forest trees produce few selfed offspring, but some species with mixed mating sys-tems produce an appreciable frequency of selfed offspring (*Chapter 7*). An extreme, but rare example is *Rhizophora mangle*, for which more than 95% selfing has been reported in populations in Florida and San Salvador Island (Lowenfeld and Klekowski, 1992).

To illustrate the effects of selfing on genotype frequencies, assume that a population has two alleles, A_1 and A_2, with frequencies $p = 0.40$ and $q = 0.60$, respectively, and that in gen-eration 0 the three genotypes occur in Hardy-Weinberg proportions: homozygotes, $p^2 = 0.16$ and $q^2 = 0.36$, and heterozygotes, $2pq = 0.48$ (Fig. 5.3). Under complete self-fertilization, A_1A_1 and A_2A_2 homozygotes produce only A_1A_1 and A_2A_2 offspring, respectively, as shown by the arrows in Fig. 5.3. Therefore, if the population consisted of 1000 trees, there would be 160 A_1A_1 genotypes (0.16) that self fertilize and produce only A_1A_1 offspring. The A_1A_2 het-erozygotes also self-fertilize and produce offspring of all three genotypes in frequencies: $1/4(A_1A_1)$, $1/2(A_1A_2)$, $1/4(A_2A_2)$. Because only heterozygous adults produce heterozygous offspring and only 1/2 of the offspring are heterozygotes, the frequency of heterozygotes is reduced by one-half in one generation of selfing [$f(A_1A_2) = 0.24$ in generation 1 in Fig. 5.3]. The frequency of heterozygotes is again halved with a second round of selfing such that in generation 2, $f(A_1A_2) = 0.12$. With continued self-fertilization, the frequency of heterozy-gotes in the population continues to be halved each generation.

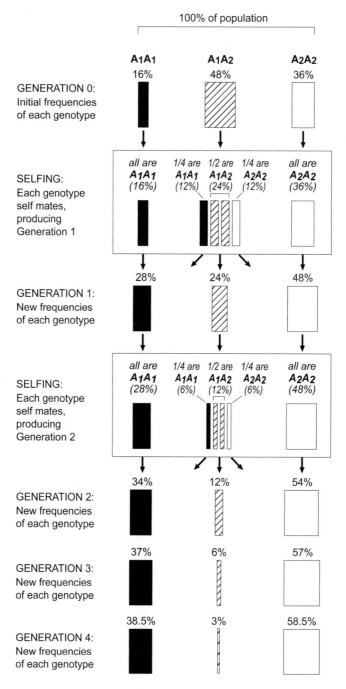

Fig. 5.3. Example of the impact of four generations of complete selfing on genotypic frequencies at a single locus. The population begins with Hardy-Weinberg equilibrium genotype frequencies in Generation 0. Then, all genotypes mate only with like genotypes (selfing), and the relative contributions of each type of mating to the genotypes of the subsequent generation are shown by the arrows. Note that allele frequencies are not altered throughout the three generations remaining constant at p = 0.6 and q = 0.4 as can be calculated from Equation 5.1.

Despite the steady decrease in frequency of heterozygotes, the frequencies of alleles A_1 and A_2 remain the same over generations at $p = 0.4$ and $q = 0.6$. These allele frequencies can be calculated each generation by applying Equation 5.1 to the genotype frequencies. The lack of change in allele frequencies in this example illustrates a key concept that is completely general to all forms and intensities of inbreeding; by itself, inbreeding affects only relative proportions of genotypes, not allele frequencies. Allele frequencies change only if the population is being affected by one of the four forces of evolution described in the next major section (mutation, migration, selection or genetic drift).

While the above example is for selfing, less intense forms of inbreeding (such as mating between siblings and cousins) also lead to reduction in heterozygosity, and in a natural population, there may be several forms of inbreeding all occurring. Also, the effects of inbreeding are cumulative in the sense that the effects build up over generations leading to further and further reduction in the frequency of heterozygotes. A convenient way to quantify the cumulative effects of inbreeding on population structure is in terms of the reduction of heterozygosity that occurs. The magnitude of inbreeding can be measured by comparing the actual proportion of heterozygous genotypes in a population (H) with the proportion that would be produced by random mating in an idealized Hardy-Weinberg population (expected heterozygosity, H_e). This measure, called the **inbreeding coefficient**, is symbolized by F, and defined as:

$$F = (H_e - H)/H_e \qquad\qquad \text{Equation 5.6}$$

Therefore, F measures the fractional reduction in heterozygosity relative to a random mating population with the same allele frequencies. Note that H in Equation 5.6 is the observed frequency of heterozygotes which has been earlier symbolized as both X_{12} and $f(A_1A_2)$ when there are only two alleles per locus. However, we use the symbol H here to be consistent with the literature and to indicate that there could be multiple alleles per locus (in which case H is the sum of genotype frequencies of all heterozygotes). In fact, Equation 5.6 is also utilized to combine information across multiple loci to obtain a single estimate of F, as illustrated in Box 5.3.

When there are two alleles at a single locus, $H_e = 2pq$, $F = (2pq - H)/2pq$, and by rearranging terms, the frequency of heterozygotes is defined as:

$$H = 2pq(1 - F) \qquad\qquad \text{Equation 5.7}$$

The frequency of A_1A_1 homozygotes (X_{11}) in an inbred population can also be expressed in terms of F. Given $p = X_{11} + 1/2H$ (Equation 5.1), and substituting for H in Equation 5.7, $p = X_{11} + pq(1 - F)$, and by rearranging terms:

$$
\begin{aligned}
X_{11} &= p - pq(1 - F) \\
&= p - p(1 - p)(1 - F) \\
&= p^2(1 - F) + pF
\end{aligned}
$$

To summarize for inbreeding, Hardy-Weinberg genotypic expectations are modified as follows:

$$
\begin{aligned}
f(A_1A_1) &= p^2(1 - F) + pF = p^2 + Fpq \\
f(A_1A_2) &= 2pq(1 - F) \quad = 2pq - 2Fpq \qquad\qquad \text{Equation 5.8} \\
f(A_2A_2) &= q^2(1 - F) + qF = q^2 + Fpq
\end{aligned}
$$

The above formulas are general and apply to natural and artificial breeding populations after one or many generations of inbreeding of various forms. The only caveat is that these formulas apply strictly only to large populations with no forces of evolution affecting gene and genotype frequencies. Note that when F = 0, there is no inbreeding and genotypic frequencies are the same as given by the Hardy-Weinberg Principle. When F = 1 (such as after many generations of complete selfing), there is complete inbreeding and the inbred population consists entirely of A_1A_1 and A_2A_2, in frequencies of p and q. This again highlights the dramatic effect of inbreeding on genotype frequencies, while allele frequencies are unaffected.

A common practice in analyzing the genetic makeup of actual populations is to calculate F from estimated genotypic frequencies (Box 5.3). It needs to be recognized, however, that factors other than inbreeding, such as selection or random genetic drift, also cause heterozygote frequencies to deviate from Hardy-Weinberg expectations. For this reason, F in this situation is usually referred to as the **fixation index**.

Box 5.3. Effects of inbreeding in a *Pinus sylvestris* seed stand.

Allozymes were used to characterize the genetic makeup of three age classes occurring simultaneously within a *Pinus sylvestris* seed stand in Sweden: seeds, young trees from 10-20 years old and adults approximately 100 years old (Yazdani *et al.*, 1985). In total, 122 seeds (one from each adult), 785 young trees and 122 adult trees were sampled and assayed at 10 loci. Based on the four loci polymorphic in all age classes, mean observed and expected frequencies for heterozygotes, as well as corresponding inbreeding or fixation indices (F) were estimated.

Age Class	Mean heterozygosity		F
	Observed (H)	Expected (H_e)	
Seed	0.295	0.349	0.155
Young trees	0.441	0.450	0.020
Adults	0.469	0.451	−0.040

The fixation index is calculated from Equation 5.6 as $F = (H_e - H)/H_e$, which for seed is $(0.349 - 0.295)/0.349 = 0.155$. The modest deficiency of heterozygotes in seeds indicated by this estimate of F (relative to Hardy-Weinberg expectation) is consistent with a predominantly outcrossing mating system that includes at least some selfing, and perhaps other types of inbreeding. Indeed, the proportion of seeds resulting from self-fertilization in this stand was estimated in a separate study to be approximately 0.12. The estimated fixation indexes for young trees and adults, however, are close to zero, indicating the frequencies of genotypes in these age classes are nearly as expected for Hardy-Weinberg equilibrium. The lower F in young trees and adults has two likely explanations: (1) The proportion of selfs in the seed crops that produced the young trees and adults was less than in the current crop of seeds; and/or (2) The well-known inferiority of inbreds (due to inbreeding depression) resulted in their higher mortality in seedlings, eliminating the excess homozygosity due to selfing by the time the trees reached the older age classes. The latter explanation is consistent with observations in a number of allozyme studies of forest trees (Muona, 1990). That is, the majority of offspring produced by self-fertilization in natural stands die before reaching reproductive age.

Inbreeding Coefficient and Regular Systems of Inbreeding

An alternative development of the inbreeding coefficient in terms of probability is very useful for calculating F from pedigrees for crosses completed or those being planned. Then, the inbreeding consequences of alternative mating schemes can be evaluated before any breeding actually occurs. To express the inbreeding coefficient in terms of probability, consider the two alleles at a particular homozygous locus in an inbred individual. These two alleles could be derived by replication from a single allele in a common ancestor; that is they are identical by descent, or **autozygous**. Alternatively, the two alleles, although chemically identical, are not identical by descent, but rather come from unrelated ancestors (**allozygous**). Because production of autozygotes is responsible for the increased frequency of homozygotes after inbreeding, F is defined as the probability that two alleles at a locus are autozygous. When formulated in this way, it is clear that F is a relative measure of inbreeding; relative to some specified or implied base population where F is presumed to be zero. All alleles at a locus are identical by descent, if one is willing to go back far enough in time, but they have differentiated due to mutation. As a practical matter, therefore, the base population typically refers to a population existing only a few generations prior to the one of interest.

To show that the definition of F based on probability of autozygosity is equivalent to F based on genotypic frequencies, consider a population where the average probability of autozygosity is F, and there are two alleles, A_1 and A_2, with frequencies p and q, respectively. The probability of drawing one A_1 allele from this population is p. The probability of drawing another A_1 allele that is identical by descent is F, by definition. Therefore, the probability that genotype A_1A_1 is autozygous is pF. It is also possible that two A_1 alleles may be drawn that are not identical by descent. The probability of this is $p^2(1 - F)$, where p^2 is the probability of drawing two A_1 alleles, and $1 - F$ is the probability that they are allozygous. Therefore, the total frequency of A_1A_1 genotypes in this population is $X_{11} = p^2(1 - F) + pF$, which is just as we derived earlier (Equation 5.8). This formulation identifies clearly the two sources of homozygotes: those that are allozygous and occur even when all mating is at random $[p^2(1 - F)]$, and the additional homozygotes that are due to inbreeding (autozygotes, pF).

To determine the inbreeding coefficient of an individual (I), we need to calculate the probability that the alleles at a particular locus are autozygous. This calculation is simplified by first drawing the pedigree in the form shown in Fig. 5.4, where only parents linking the individual with common ancestors are shown. In this case, there is only one common ancestor (A). Let γ_1 and γ_2, symbolize the alleles carried by ancestor A at any single locus. The probability that I has genotype $\gamma_1\gamma_1$ is 1/64, because in each of the six parent-offspring pairs linking A with I (*i.e.* AB, BD, DI, EI, CE, AC), there is a 1/2 chance that a parent carrying γ_1 transmits this allele to its offspring; and the probability that γ_1 is transmitted through all six parent-offspring pairs is the product of the individual probabilities $(1/2)^6 = 1/64$. Likewise, the probability I is $\gamma_2\gamma_2$ is also $(1/2)^6$ so the probability I is either $\gamma_1\gamma_1$ or $\gamma_2\gamma_2$ is $2(1/2)^6$ or $(1/2)^5$.

Common ancestor A may itself be autozygous through previous inbreeding, in which case I is also autozygous if it has genotype $\gamma_1\gamma_2$ or $\gamma_2\gamma_1$. The probability that A is autozygous is its inbreeding coefficient, F_A. Therefore, the additional probability of autozygosity is $(1/2)^5 F_A$, and the total inbreeding coefficient of I is $F_I = (1/2)^5 + (1/2)^5 F_A = (1/2)^5(1 + F_A)$. Note that the exponent 5 is the number of parents in the path connecting I with its common ancestor. If the original parent A is not inbred, $F_A = 0$ and $F_I = (1/2)^5 = 1/32$.

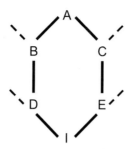

Fig. 5.4. Simplified pedigree for a half first-cousin mating. The particular sexes of the individuals are not shown because they are irrelevant to calculating the inbreeding coefficient of I. Unrelated ancestors are not shown, but are indicated by the dashed lines.

In more complicated pedigrees, parents may be related to each other through more than one ancestor, or by the same ancestor through different paths. Each common ancestor or path contributes an additional probability of autozygosity, such that the inbreeding coefficient is obtained by adding together the separate probabilities for each path through which parents are related (Hedrick, 1985; Falconer and Mackay, 1996). This is illustrated for the case where two full-sibs (*i.e.* individuals with both parents in common) are mated (Fig. 5.5). Here, there are two common ancestors, with three parents linking I with the common ancestor in each path, such that $F_I = (1/2)^3(1 + F_A) + (1/2)^3(1 + F_B)$. If F_A and F_B are both zero, $F_I = 2(1/2)^3 = 1/4$. A general formula for calculating F_I from pedigrees is:

$$F_I = \sum_{i=1}^{m} (1/2)^{n_i} (1 + F_i)$$

Equation 5.9

where m is the total number of paths, F_i is the inbreeding coefficient of the common ancestor in the ith path, and n_i is the number of parents in the ith path.

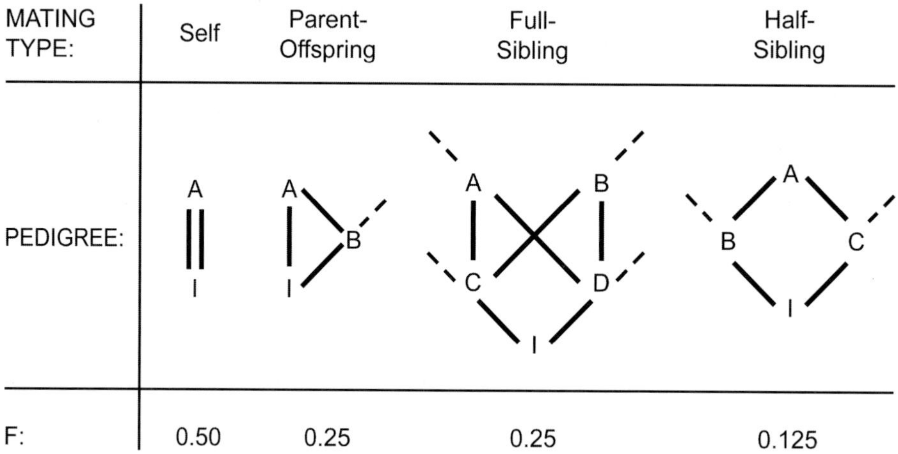

MATING TYPE:	Self	Parent-Offspring	Full-Sibling	Half-Sibling
PEDIGREE:	$\begin{array}{c} A \\ \| \\ I \end{array}$	$\begin{array}{c} A \\ I \end{array} B$	A B C D I	B C I A
F:	0.50	0.25	0.25	0.125

Fig. 5.5. Pedigrees for frequently encountered types of related matings in native stands and artificial breeding populations of forest trees, and inbreeding coefficients (F) of offspring (I) assuming no inbreeding in common ancestors.

It is often of interest to know in breeding programs how rapidly the inbreeding coefficient would increase if a regular system of inbreeding, such as self-fertilization, sib mating or backcrossing to a particular parent, was used repeatedly over generations. Such increases in F could be calculated from pedigrees, but this becomes cumbersome with increasing numbers of generations. It is simpler to calculate F from recurrence equations that relate the inbreeding coefficient in one generation to those of previous generations (Falconer and Mackay, 1996). For example, recall in the case of selfing that heterozygosity is reduced by 1/2 each generation, such that heterozygosity in generation k, $H_k = 1/2H_{k-1}$ (Fig. 5.3). Substituting for H using Equation 5.7, $1 - F_k = 1/2(1 - F_{k-1})$, and $F_k = 1/2(1 + F_{k-1})$. Values of F_k for four mating systems of potential interest in forest tree breeding are plotted over generations in Fig. 5.6

The inbreeding coefficient in each generation can be interpreted in two ways: (1) As the decrease in average heterozygosity over many loci within a single inbred line; or (2) As the decrease in heterozygosity at a single locus across many inbred lines in the same population, each with the same mating regime. Average genotypic frequencies across all lines are those in Equation 5.8. If inbreeding is repeated over many generations, F approaches 1 and all loci become fixed (homozygous) for one allele or another within any one line. At a single locus across many lines, p lines become fixed for the A_1 allele and q lines become fixed for the A_2 allele, where p and q are the allele frequencies (remember allele frequencies are unchanged by inbreeding alone). Only one generation of random mating among unrelated lines, however, will return F to zero. Then at any given locus, genotype frequencies return to those predicted by Equation 5.3 (Hardy-Weinberg equilibrium).

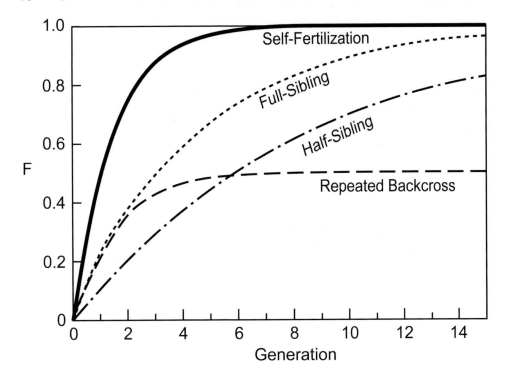

Fig. 5.6. Values of the inbreeding coefficient (F) over generations for four regular systems of inbreeding (Table 5.1 in Falconer and Mackay, 1996).

Inbreeding Depression

Causes of Inbreeding Depression

Increased homozygosity due to inbreeding has detrimental consequences for forest trees, as it does for most outcrossing organisms. The harmful effects of inbreeding are most evident in offspring of close relatives; they can occur in any stage of development and reduce embryo viability, seedling survival, tree vigor, and seed production (Charlesworth and Charlesworth, 1987; Williams and Savolainen, 1996). Inbreeding depression due to selfing is usually so severe that the effects are visually obvious as shown in Fig. 5.7.

Fig. 5.7. Inbreeding depression due to selfing often results in visually apparent reduction in growth and vigor. This photo illustrates the effects of selfing on survival and tree size of *Pseudotsuga menziesii* at age 34. The row of smaller trees in the center are selfs, while trees in the rows immediately to the left and right of the selfs are outcrossed offspring (from deliberate crosses to unrelated males) (OC) and open-pollinated offspring (seed resulting from natural pollination) of the same mother tree, respectively. At age 33 average survival of selfs in this test was only 39% of OC offspring (total of 19 self-OC family pairs); average DBH of surviving selfs was 59% of surviving OC siblings. (Photo courtesy of F. Sorensen, USFS Pacific Northwest Research Station)

Two main hypotheses have been proposed to account for inbreeding depression and the reciprocal phenomenon of hybrid vigor: the dominance and the overdominance hypotheses. According to the **dominance hypothesis**, inbreeding depression results from the increased frequency of genotypes that are homozygous for rare, deleterious, recessive alleles. Recall that rare recessive alleles occur primarily in heterozygotes in random mating populations and are infrequently expressed. For example, an allele with a frequency of $q = 0.01$ is homozygous in $q^2 = 0.0001$ offspring of randomly mated parents, but occurs 50 times more frequently as a homozygote if all parents are self-fertilized [*i.e.* $q^2(1 - F) + qF = 0.0001(0.50) + 0.01(0.50) = 0.005$; Equation 5.8]. If a large number of loci carry rare deleterious recessives, many individuals express one or more of these harmful alleles upon close inbreeding, which under random mating remain largely hidden in heterozygotes. Alternatively, if two unrelated, highly inbred (*i.e.* highly homozygous) individuals are crossed, the deleterious alleles homozygous in one parent may be masked by normal alleles in the other parent, such that the offspring will be superior to either parent. This is the converse of inbreeding depression called **hybrid vigor**.

The **overdominance hypothesis** states that heterozygotes at individual loci are superior to either alternative homozygote because the presence of complementary alleles results in optimum function. In other words, A_1A_2 is superior to either A_1A_1 or A_2A_2 because having both alleles is better than having two copies of either one (see the later section, *Balancing Selection*). If the overdominance hypothesis were true, it would be impossible to develop superior pure-breeding (*i.e.* completely homozygous) lines, because superior genotypes always segregate for alternative alleles. Pure-breeding superior lines, however, are theoretically possible under the dominance hypothesis, because deleterious alleles could be made "visible" by close inbreeding and removed from the breeding population.

Although considerable research has been directed at testing these alternative hypotheses, discriminating between them has proven difficult. While overdominance cannot be ruled out in some cases, the general consensus is that inbreeding depression is primarily due to the expression of recessive deleterious alleles (Charlesworth and Charlesworth, 1987; Husband and Shemske, 1996; Williams and Savolainen, 1996). Strong support for the dominance hypothesis is the negative association in plant species between selfing rate and inbreeding depression. Inbreeding depression in predominantly selfing species is less than 50% of that in predominant outcrossers; yet, selfing species are well adapted to the habitats they occupy. That is, high homozygosity has not resulted in reduced adaptation as expected under the overdominance hypothesis (Husband and Shemske, 1996). Deleterious recessive alleles are readily exposed and purged from populations of predominantly self-fertilized species.

Inbreeding Depression in Forest Trees

In forest trees, as in many other plants, inbreeding depression is most often evaluated by comparing the performance of progeny derived through self-fertilization with the performance of progeny derived by deliberate outcrossing among unrelated parents. Inbreeding due to selfing is of much practical significance in forestry because selfing is a component of the mating system of many tree species. In addition, selfing is a convenient means of studying inbreeding because self-fertilization is readily carried out and inbreeding depression is often substantial at this high level of inbreeding ($F = 0.50$). Although there are exceptions (Andersson *et al.*, 1974; Griffin and Lindgren, 1985), inbreeding depression at other levels of inbreeding is roughly proportional to F, for example, inbreeding depression at F = 0.25 is about half that at F = 0.50 (Sniezko and Zobel, 1988; Woods and Heaman, 1989; Matheson *et al.*, 1995; Durel *et al.*, 1996; Sorensen, 1997). Here we emphasize the results from selfing studies.

Inbreeding depression in seed set has been thoroughly investigated, with self-fertility (R) expressed as the ratio of the proportion of full seeds upon selfing to full seeds upon outcrossing. For many species, such as *Larix laricina* and *Pseudotsuga menziesii*, few viable seeds are produced from self-fertilization (Table 5.2). Presumably, empty seeds result from the expression of lethal or semi-lethal alleles that cause embryo abortion. Species that have more lethal alleles at more loci suffer more embryo abortions and have higher fractions of empty seed, and this is quantified through a concept known as lethal equivalents explained in Box 5.4. When a species has a high number of lethal equivalents, this is indicative of a high genetic load.

Inbreeding depression following selfing has also been investigated for a variety of traits in seedlings and older trees. In these cases, inbreeding depression is often quantified as:

$$\delta = [1 - (W_s/W_c)]$$
Equation 5.11

where W_s and W_c are means of traits following selfing and outcrossing, respectively (Williams and Savolainen, 1996). As an example of Equation 5.11, $\delta = 0.25$ reflects an inbreeding depression of 25%, so the mean of inbred individuals is 75% of the trees resulting from outcrossing. Note that Equation 5.11 can be used for forms of inbreeding other than selfing by substituting the mean for the progeny resulting from inbreeding for W_s.

In recent reviews for conifers, δ for progeny from selfing ranged roughly from 0.10 to 0.60 for survival and tree height, depending on species and age (Sorensen and Miles, 1982; Williams and Savolainen, 1996; Sorensen, 1999) indicating a 10-60% reduction in survival and height growth in progeny from selfing (Fig. 5.7). Inbreeding depression in height for two *Eucalyptus* species 3.5 years after planting was near the lower end of the range observed in conifers (0.11 and 0.26); (Griffin and Cotterill, 1988; Hardner and Potts, 1995). Plants respond to competition by favoring height increment at the expense of diameter and root growth; therefore, slow-growing inbreds are expected to show more depression in diameter and even more in stem volume, than in height (Matheson *et al.*, 1995; Sorensen, 1999).

Trends in inbreeding depression of tree size with age are confounded by mortality, because smaller, weaker (*i.e.* inbred) trees have lower survival, so that measurements made at earlier and later ages are not based on the same set of individuals (Williams and

Table 5.2. Estimates of filled seeds after self- and cross-pollination, self-fertility ratio (R), and number of lethal equivalents (α) for a variety of coniferous and angiosperm tree species.

Species	Filled Seeds (%)		R[a]	α[b]	Source
	Self	Cross			
Abies procera	36.0	51.0	0.706	1.4	Sorensen *et al.*, 1976
Larix laricina	1.6	21.6	0.074	10.4	Park and Fowler, 1982
Picea abies	13.0	45.9	0.283	5.0	Skrøppa and Tho, 1990
Pinus resinosa	71.0	72.0	0.986	0.1	Fowler, 1965a
Pseudotsuga menziesii	7.9	69.2	0.114	8.7	Sorensen, 1971
Eucalyptus globulus	60.0	80.0	0.750	1.2	Hardner and Potts, 1995
Acer saccharum	17.4	35.1	0.496	2.8	Gabriel, 1967

[a] Percent filled seeds after self-pollination divided by percent filled seeds after cross pollination.
[b] $\alpha = -4\ln R$ from Equation 5.10 in Box 5.4 (Sorensen, 1969).

Box 5.4. Lethal equivalents as estimated from self-fertility ratios.

A lethal equivalent is defined as a group of genes, which on average cause one death when homozygous (Morton *et al.*, 1956). This may be one lethal mutant allele or two mutant alleles at different loci, each with 50% probability of causing death, etc. A simple relationship for estimating the number of lethal equivalents (α) from self-fertility values (R) (Sorensen, 1969) is:

$$\alpha = -4\ln R \qquad\qquad\qquad \text{Equation 5.10}$$

This relationship assumes that lethal alleles are completely recessive, act independently, and that each empty seed corresponds to an individual embryo death (Savolainen *et al.*, 1992). Therefore, for parent trees of *Picea abies* from a breeding program in Norway, mean R was estimated to be 0.283 (Table 5.2). The number of lethal equivalents corresponding to this value of R is 5.0. In other words, if a parent tree is heterozygous for lethal recessive alleles at five independent loci, we expect the self-fertility to be approximately 0.28, meaning 28% as many full seeds are produced after selfing versus after outcrossing.

Perennial plants including forest trees typically have more lethal equivalents for embryo viability than are found in annual plants or animals (Ledig, 1986; Williams and Savolainen, 1996). While α often exceeds 5.0 in forest trees (Williams and Savolainen, 1996; Table 5.2), ranges reported for agronomic crops and animals (humans and insects) are 1.2 to 5.2 and 1.0 to 4.0, respectively (Levin, 1984). The high number of lethals observed in forest trees is probably due, in part, to higher cumulative mutation rates in long-lived plants (*Chapter 7*).

The number of lethal equivalents varies greatly among species and individuals within species. The lowest number reported for a forest tree appears to be *Pinus resinosa*, with α estimated at nearly zero. This species is also highly unusual in that it is nearly lacking in genetic variation (Fowler and Lester, 1970; Fowler and Morris, 1977; Mosseler *et al.*, 1992). Tree-to-tree differences in number of lethal equivalents can also be striking. For example, in *Pseudotsuga menziesii* (Sorensen, 1971) and *Larix laricina* (Park and Fowler, 1982), α ranged from 3.0 to 19.0 or more among mother trees within individual stands. Parent trees that have few lethals or other deleterious recessives, yet perform exceptionally well for traits of economic importance, may be of great value for advanced generation breeding (Williams and Savolainen, 1996).

Savolainen, 1996). In a study comparing three conifers from outplanting to ages 25-26, Sorensen (1999) found that inbreeding depression for height steadily decreased after about age 6 in *P. menziesii*, which had the greatest inbreeding depression in survival in the first few years following planting. In *Abies procera*, the species with the least depression in survival after planting, inbreeding depression for height steadily increased with age. The third species, *P. ponderosa*, had somewhat intermediate age trends in inbreeding depression compared to the other two species, in both survival and height. Regardless of the age trends in inbreeding depression, the ultimate impact of inbreeding on productivity was similarly catastrophic in all three species, with δ for total stem volume per unit area averaging 0.80 (range 0.74-0.83) meaning 80% yield reduction for progeny from selfing. These results illustrate why foresters must take care not to use seed resulting from close inbreeding for reforestation (*Chapter 16*).

Reports on inbreeding effects in traits beyond the seed stage, other than for survival and stem growth, are relatively rare. In conifers, sexual maturity seems to be delayed and seed production diminished in individuals resulting from self-fertilization (Durel *et al.*, 1996; Williams and Savolainen, 1996). Some traits appear to be little influenced by inbreeding including stem form in *Pinus elliottii* (Matheson *et al.*, 1995) and *Pinus pinaster* (Durel *et al.*, 1996), rust resistance in *P. elliottii* (Matheson *et al.*, 1995), and spring cold hardiness in *P. menziesii* (Shortt *et al.*, 1996). Much still needs to be learned about the impacts of inbreeding on fitness and economically important traits in forest trees. This is especially true for angiosperm species because inbreeding studies to date have emphasized conifers.

FORCES THAT CHANGE ALLELE FREQUENCIES

In this section, we relax more of the assumptions of the Hardy-Weinberg Principle in order to evaluate the impacts of four major forces of evolution: mutation, migration, selection and genetic drift. **Evolution** can be defined as "cumulative change in the genetic composition of a population" (Hartl, 2000). Such cumulative evolution occurs primarily through changes in allele frequencies over time and space. Among the forces that alter allele frequencies, both mutation and migration introduce new alleles into populations although mutation acts only weakly in bringing about allele-frequency change. Selection and genetic drift also change allele frequencies, the former in a systematic direction and the latter randomly. For simplicity, we discuss each of the evolutionary forces individually in this section. However, since it is the combined effect of all the forces acting simultaneously that ultimately shapes the genetic composition of populations, we end this chapter with a brief discussion of their combined effects.

Mutation

In the broadest sense, mutation is any heritable change in the genetic constitution of an individual. Types of mutation range from substitutions of single bases in DNA to rearrangements of chromosome structure, and to additions or deletions of parts of chromosomes, whole chromosomes, or chromosome sets. While polyploidy has been particularly important in the origin of new species of higher plants (Stebbins, 1971; *Chapter 9*), the large majority of genetic variation within populations of a given species has its origin in mutations that occur in single genes (Hartl and Clark, 1989). For this reason, we focus on the effects of single-gene mutations.

Although mutation is the ultimate source of genetic variation in populations, and is an essential process in evolution, it only weakly influences allele frequencies from one generation to the next. This lack of significant short-term influence is because spontaneous mutation rates are very low. Studies in a range of plant and animal species for a variety of single-gene traits, show that rates of detectable mutations are typically only 10^{-5} to 10^{-6} mutations per gene per generation (Hartl and Clark, 1989; Charlesworth *et al.*, 1990; Lande, 1995). For any single locus, spontaneous change from one allelic state to another that is detectably different (*e.g.* $A_1 \rightarrow A_2$ or $A_2 \rightarrow A_3$) occurs in only about 1 in 100,000 to 1 in 1,000,000 of the gametes produced in a generation. We add the qualifier that these rates apply to detectable mutations, because some changes to DNA sequences may not be detectable (*e.g.* those occurring in non-coding regions and those that do not change protein function).

Most newly arising mutations have harmful effects on the phenotype. This is true because presumably the wild type allele is well adapted to perform its function. Therefore, only a small proportion of mutations that occur at individual loci are likely to be favorable to the organism and to contribute to evolution. An example of a deleterious (in fact, lethal) mutation is shown in Fig. 5.2 in which the albino seedlings of *P. ponderosa* eventually die due to a mutation that blocks chlorophyll biosynthesis.

Polygenic traits are more susceptible to change due to mutation than single gene traits because mutation at any one of the many loci affecting the trait has an effect. It has been estimated that the cumulative mutation rate over all loci producing mutations influencing a particular quantitative trait is typically around 0.01 per gamete per generation (Lande, 1995). Since about 50% of these mutations have highly detrimental effects, about one out of every 200 gametes carries a neutral or less detrimental mutation influencing an individual quantitative trait.

To examine the effect of mutation on allele frequencies, we assume there are two types of alleles: wild type alleles (that code normal phenotypes) and detrimental alleles, and that mutation is reversible. Mutation can occur from wild type alleles to detrimental alleles (forward mutation) and from detrimental alleles to wild type alleles (backward mutation). Forward mutations are expected to occur more frequently than backward mutations because there are many more ways to cause a malfunction in a gene than to repair it (Hedrick, 1985; Hartl and Clark, 1989).

Given that the rate of forward mutation from wild type alleles (A_1) to detrimental alleles (A_2) is u, and the rate of backward mutation is v, then the change in the frequency of A_2 (q) per generation due to mutation alone is:

$$\Delta q = u(1 - q) - vq$$
$$= u - q(u + v)$$

Equation 5.12

The maximum positive value of Δq is u, when q = 0, and the maximum negative value is v, when q = 1. Therefore, Δq is very small, supporting our earlier statement that mutation has only a minor influence on changing allele frequency at any particular locus in any single generation.

Although its impact is small in any single generation, mutation is the ultimate source of genetic variation in natural populations. The large amount of genetic variation common to species of forest trees has accumulated over many, many generations spanning evolutionary time frames (*Chapter 7*). This highlights the importance of conservation efforts to preserve this rich genetic diversity because once lost, it would be impossible to re-create this rich diversity in the short term.

Migration

Migration, also called **gene flow**, is the movement of alleles among populations. There are two important effects of migration on the genetic structure of populations: (1) New alleles are introduced, thereby increasing genetic variability (much like mutation); and (2) Continued migration over generations leads to reduced genetic divergence among populations.

To quantify the effects of migration, assume a population that is large enough that genetic drift can be disregarded. Then, let the proportion of migrants moving into the population each generation be m, so the proportion of non-immigrants (natives) is (1 − m). If the frequency of allele A_2 in migrant individuals is q_m and the frequency of A_2 among natives before migration is q_o, then the allele frequency after migration is:

$$q = (1 - m)q_o + mq_m$$
$$= q_o - m(q_o - q_m)$$

Equation 5.13

The change in allele frequency after one generation of migration is $\Delta q = q - q_o = - m(q_o - q_m)$. Therefore, the rate of change in allele frequency depends on the migration rate and the initial difference in allele frequency between the immigrants and the natives. An illustration of the use of Equation 5.13 to estimate m in a real population is given in Box 5.5.

Migration always narrows the allele frequency difference between the recipient population and migrants, because when $q_o < q_m$, Δq is positive and when $q_o > q_m$, Δq is negative. If migration is one-way (*i.e.* always from the donor population(s) to the recipient population) and q_m is constant over generations, then Δq eventually reaches zero when $q = q_m$; that is, when the allele frequency in the recipient population equals the allele frequency in the donor population. When migration is reciprocal between two or more populations, gene flow eventually leads to all populations having the same allele frequency, intermediate to the original frequencies of the populations (Hartl and Clark, 1989). In this manner, migration can be a potent force in preventing both genetic differentiation among populations and loss of diversity within populations (*Chapters 7* and *8*), such as might occur under genetic drift or selection.

Propagule dispersal, which occurs primarily through pollen and seeds, can be extensive in forest trees, whether it occurs by wind or is mediated by animals (*Chapter 7*). In addition, when geographic patterns of genetic variation are examined using allozymes, typically only a small proportion (often < 10%) of the genetic diversity in a region can be attributed to differences among populations (Hamrick *et al.*, 1992; *Chapter 7*). Taken together, these observations indicate that gene flow is substantial in forest tree species (Ellstrand, 1992; Ledig, 1998).

Box 5.5. Estimating levels of pollen gene flow with genetic markers.

Pollen gene flow in a group of five isolated individuals of *Tachigali versicolor*, a tropical bee-pollinated forest canopy species, was studied with the aid of alleles at an allozyme locus (Hamrick and Murawski, 1990). These five trees, separated by 500m from other trees of *T. versicolor* on Barro Colorado Island in the Republic of Panama, were monomorphic for a common allele at this locus. Among the offspring of these trees, 0.04 were heterozygous for an alternative allele that occurs at a frequency of 0.16 in the surrounding population. Because the alternative allele is absent in the isolated trees, offspring heterozygous for this allele must be the result of fertilization by males outside the group. This method of identifying pollen migrants is called **paternity exclusion**. Not all migrants are detected, however, because pollen gametes from the surrounding population also carry the common allele, and pollen gametes with this allele are not distinguishable from those produced by trees within the group. To estimate pollen gene flow (m), we can use Equation 5.13. Designating the alternative allele as A_2 and only considering pollen gamete frequencies, $q_o = 0$, $q = 0.04$, and $q_m = 0.16$. With $q_o = 0$, the equation simplifies to $q = mq_m$ and $m = q/q_m = 0.04/0.16 = 0.25$. Therefore, it is estimated that 25% of the seed produced by trees in the isolated group is the result of fertilization by pollen gametes from males outside the group.

Selection

Natural selection is the central force in evolution because it is the mechanism by which organisms adapt to their environments. The concept of **natural selection**, as proposed by Darwin (1859), is based on three premises: (1) More offspring are generally produced than can survive and reproduce; (2) Individuals differ in the ability to survive and reproduce and some of these differences are genetically controlled; and (3) Genotypes that enhance survival or reproduction contribute disproportionately more offspring to the next generation, such that their alleles increase in frequency at the expense of alleles produced by less adapted genotypes. In this manner the adaptation of a population is progressively improved over generations. In addition to natural selection, **artificial selection** is the primary tool used by breeders to make genetic improvement in traits of interest in tree improvement programs (*Chapter 13*). Natural and artificial selection act on the same principles and are described here together.

Although selection acts on phenotypes that are the result of many gene loci, the consequences of selection are best understood by focusing on how it changes allele frequencies at a single locus. In the absence of other evolutionary forces to counter the steady loss of alternative alleles, **directional selection** eventually leads to fixation of the favored allele in a population. Directional selection is responsible for progressive adaptation to new environments by natural selection and is the type of artificial selection applied by breeders in the vast majority of improvement programs. **Balancing selection**, on the other hand, promotes polymorphism by selecting for diverse alleles at the same locus. Therefore, balancing selection could be an important force for maintaining genetic variation in populations, genetic variation that is vital to continued adaptive evolution when environments change.

Directional Selection

As an example of directional selection, assume that after a wildfire destroys a forest stand, genotypic frequencies among naturally regenerated seedlings are as shown in Table 5.3. The **viability** of each genotype is its probability to survive until reproductive age. The absolute viabilities of 0.80, 0.80 and 0.20 indicate that none of the three genotypes has a 100% chance of survival to reproductive age, but that the A_2A_2 genotype is least viable.

Selection, in this case, depends only on the relative magnitudes of the viabilities, and the viability of each genotype is expressed in relative terms by dividing all the absolute viabilities by that of the most viable genotype. Therefore, genotype A_2A_2 has a relative viability of $0.20/0.80 = 0.25$, or 25% of the other two genotypes in the population. Differential ability of genotypes to mate and produce offspring once they reach reproductive age (*i.e.* **fertility**), also influences their contribution to the next generation (*i.e.* their **fitness** or **adaptive value**); for simplicity, we assume all genotypes are equally capable of reproduction. Relative viability, in this case, therefore, determines relative fitness.

The proportional contribution of each genotype to the next generation is calculated as its initial frequency (before selection) times its relative fitness. Note, however, that the sum of the proportions $(0.36 + 0.48 + 0.04 = 0.88)$ is less than one because selection removes some individuals. To calculate relative frequencies of each genotype after selection, the individual proportions must be divided by their sum. Therefore, the frequency of A_2A_2 in the next generation is $0.04/0.88 = 0.045$. The influence of selection on the frequency of allele A_2 (q) can be quantified as the change in q from one generation to the next. Given that the initial frequency of q was 0.40 and is 0.318 after selection (Table 5.3), then $\Delta q = 0.318 - 0.400 = -0.082$. There is less of the A_2 allele after selection because it reduces relative viability of the A_2A_2 genotype.

Table 5.3. Hypothetical example of allele frequency change in one generation of directional selection. Assume that the initial genotype frequencies are those in the natural seedling regeneration after a wildfire (with p = 0.6 and q = 0.4). Due to differential ability to survive to reproductive age, A_2A_2 genotypes have a lower viability. The result is that genotype and allele frequencies at reproductive age (*i.e.* following selection; p = 0.682 and q = 0.318) are different from those in the seedlings.

	Genotypes			
	A_1A_1	A_1A_2	A_2A_2	Total
Initial frequency	0.360	0.480	0.160	1.000
Absolute viability (fitness)	0.80	0.80	0.20	
Relative viability (fitness)[a]	1	1	0.25	
Proportion after selection[b]	0.360	0.480	0.040	0.880
Frequency after selection[c]	0.409	0.546	0.045	1.000

Initial frequency of A_2 (q) = 0.160 + 1/2(0.480) = 0.400
Frequency of A_2 after selection (q_1) = 0.045 + 1/2(0.546) = 0.318
$\Delta q = q_1 - q = 0.318 - 0.400 = -0.082$

[a] Relative viability is obtained as absolute viability divided by 0.80 which is absolute viability of the most viable genotype.
[b] Proportion after selection is initial frequency times the relative viability.
[c] Frequency after selection is proportion after selection divided by the sum of those proportions (*i.e.* divided by 0.88).

To formulate a general expression for genotype and allele frequency change under directional selection, the Hardy-Weinberg model is modified to allow the fitness of the genotypes to differ (Table 5.4). Genotype fitness is defined using two parameters: **the selection coefficient** (s) and the **degree of dominance** (h). The selection coefficient is the selective disadvantage of the disfavored allele and is calculated as one minus the relative fitness of the homozygote for the disfavored allele. In our numerical example, the selection coefficient against allele A_2 is s = 1 − 0.25 = 0.75. Because selection acts on phenotypes, we also need to account for the level of dominance in the expression of fitness. When h = 0, relative fitness of A_1A_1, A_1A_2, and A_2A_2 is 1, 1, and 1 − s, respectively, so that A_2 is recessive to A_1 in fitness (Table 5.4). When h = 1, the relative fitness is 1, 1 − s, and 1 − s, and A_2 is dominant to A_1. When h = 1/2, there is no dominance in the expression of fitness. Note that dominance in terms of fitness is not necessarily equivalent to dominance with respect to visible effects of the same gene.

Table 5.4. General model for allele frequency change under directional selection.

	Genotypes			
	A_1A_1	A_1A_2	A_2A_2	Total
Initial frequency	p^2	2pq	q^2	1.0
Relative fitness	1	1 − hs	1 − s	
Proportion after selection	p^2	2pq(1 − hs)	$q^2(1 − s)$	$T = 1 - 2pqhs - sq^2$
Frequency after selection	p^2/T	2pq(1 − hs)/T	$q^2(1 − s)/T$	1.0

s = selection coefficient against A_2
h = degree of dominance of A_2 (0 = no dominance, 1 = complete dominance)
Frequency of A_2 after selection (q_1) = $(q - hspq - sq^2)/(1 - 2hspq - sq^2)$
$\Delta q = q_1 - q = -spq[q + h(p - q)]/(1 - 2hspq - sq^2)$

Following the same procedures outlined in the numerical example, a rather complicated-looking formula for Δq after one generation of selection is derived (Table 5.4). This expression simplifies to:

$$\Delta q = - sq^2(1 - q)/(1 - sq^2) \text{ (when A}_2 \text{ is recessive)} \qquad \text{Equation 5.14}$$

$$\Delta q = - sq(1 - q)^2/[1 - sq(2 - q)] \text{ (when A}_2 \text{ is dominant)} \qquad \text{Equation 5.15}$$

$$\Delta q = - 1/2sq(1 - q)/(1 - sq) \text{ (when there is no dominance)} \qquad \text{Equation 5.16}$$

Regardless of the degree of dominance, the effect of directional selection against A_2 is to decrease its frequency in the population. The magnitude of change in allele frequency, however, depends not only on the selection intensity (s), but also on the initial allele frequency (q).

The impact of allele frequency and dominance on the effectiveness of directional selection is illustrated in Fig. 5.8 for the three cases of dominance above, when s = 0.20. For example, assume that the deleterious allele is recessive but nearly fixed in the population (*i.e.* q is near 1). This figure shows that it would take about 22 generations of directional selection to reduce the frequency of this allele to 0.50 and another 15 generations to reduce it to 0.25. Selection for or against a recessive allele is, however, very ineffective (*i.e.* little change in allele frequency) when the allele is rare (*i.e.* to go from q = 0.05 to 0.025 would take approximately 100 generations). As shown earlier in this chapter, when a recessive allele is rare it occurs mostly in heterozygotes, and therefore, is not exposed to selection. On the other hand, selection against a dominant allele is least effective when it is common (*i.e.* when the favored recessive allele is rare).

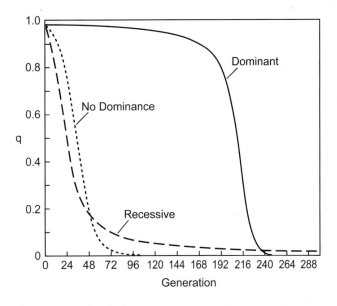

Fig. 5.8. Frequency of a deleterious allele (q) undergoing directional selection in a large, random mating, population when its effect on fitness is dominant, recessive, and when there is no dominance. All plots assume that the deleterious allele is nearly fixed in the population at generation 0 (*i.e.* q near 1) and that the selection coefficient against this allele is 0.20. Values were calculated using the *Populus* computer program (Alstad, 2000).

How effective is selection against an unfavorable allele if s = 1, which is the case of a lethal allele in a natural population, or the selective removal of all individuals expressing the allele in a breeding program? Only a single generation is required to remove the allele when it is dominant, but when recessive, it is increasingly difficult to remove as it becomes rare. Substituting s = 1 in Equation 5.14, $\Delta q = -q^2/(1 + q)$, so that when q = 0.01, the change in q expected from removing all homozygous recessive individuals would be only 0.0001. Therefore, increasing the selection intensity, even to the maximum, has little additional impact on removing rare, recessive alleles when mating is at random.

Removal of deleterious alleles in breeding populations can be accelerated greatly through progeny testing (Falconer and Mackay, 1996). By crossing parents with the recessive phenotype to all trees in the breeding population, those trees that are heterozygous for the deleterious allele can be identified by the presence of recessive homozygotes in their progeny. Using the progeny test data to eliminate heterozygous trees along with parental phenotypes to eliminate the recessive homozygotes, the recessive allele could be removed in a single generation (barring misclassification of any genotypes).

Balance between Mutation and Selection

As stated earlier, inbreeding depression is largely due to the expression of rare deleterious alleles in homozygous progeny of related individuals. Under random mating, selection removes these alleles very slowly, but can never eliminate them entirely because recurring mutation of normal alleles creates a continuous supply of new mutants. Eventually, an equilibrium is reached whereby the removal of rare recessives by selection is counterbalanced by mutation. The frequency of the recessive allele at equilibrium (q_e) can be derived by equating the change in allele frequency due to mutation with that due to selection, and solving for q (Falconer and Mackay, 1996). An approximate solution is:

$$q_e = (u/s)^{1/2} \hspace{4cm} \text{Equation 5.17}$$

where u is the rate of mutation from the normal to the deleterious allele, and s is the selection coefficient against the deleterious allele. With mutation rates normally observed, only very mild selection is necessary to keep q_e very small. For example, if $u = 10^{-5}$, a selection coefficient of only 0.10 results in $q_e = 0.01$. Even lower selection intensities are required to keep q_e small when the deleterious allele is dominant or intermediate in dominance, because heterozygotes are also vulnerable to selection. It is clear that the joint effects of selection and mutation can readily explain the low frequency of deleterious mutants in natural populations.

Balancing Selection

The joint effect of mutation and selection explains the presence of deleterious alleles at low frequency, but it does not account for the majority of genetic variation that is commonly observed in populations of outcrossing organisms, including forest trees. In typical population surveys of forest trees based on allozymes, for example, at least 40% of the loci have two or more widespread alleles at frequencies too high to be accounted for by mutation-selection equilibrium (Ledig, 1986; Hamrick *et al.*, 1992). Although factors other than balancing selection are probably responsible for much of the variation in allozymes and other molecular genetic markers (Box 5.6), balancing selection certainly plays a significant role in maintenance of genetic diversity influencing polygenic traits of adaptive significance (*Chapter 7*).

Box 5.6. Allozyme polymorphism: selection versus neutrality.

Considerable debate has ensued in the last 20 years over the adaptive significance of allozymes and other molecular variants that often express high levels of polymorphism in native populations, especially in forest trees (Hamrick *et al.*, 1992). Little is known about the phenotypic effects associated with particular alleles of molecular variants in trees, but two observations have been cited as evidence supporting an adaptive role for this variation. The first observation is the finding in a number of studies that rate of stem growth increases with increasing proportion of heterozygous allozyme loci, leading some authors to conclude that at least some of the loci are expressing overdominance (Bush and Smouse, 1992; Mitton 1998a,b). Only a small proportion of the variation in stem growth, however, is accounted for by allozyme heterozygosity in these studies, and there are even a greater number of reports of no association between allozyme heterozygosity and stem growth (Savolainen and Hedrick, 1995). Others have concluded that positive association between allozyme heterozygosity and stem growth is more likely due to partial inbreeding in populations (Ledig, 1986; Bush and Smouse, 1992; Savolainen, 1994; Savolainen and Hedrick, 1995). That is, variation in allozyme heterozygosity is merely a reflection of variation in levels of inbreeding and slow growth of individuals with low heterozygosity results from the expression of normally deleterious recessives in inbreds.

A second observation supporting an adaptive role for allozymes in forest trees is when allele frequencies are repeatedly correlated with the same environmental variables in geographical surveys. As such, this is evidence that these alleles are responding to directional selection; however, such correlations are only rarely reported in the literature (*e.g.* Grant and Mitton, 1977; Bergmann, 1978; Furnier and Adams, 1986a). Rather, most allozymes appear to be weakly associated, at best, with environment even when environments change relatively dramatically over geographical or topographical gradients (Muona, 1990).

An alternative explanation for the high degree of polymorphism observed in molecular markers is that these variants have little or no effect on fitness, and polymorphism results primarily from a balance between the creation of new variants by mutation and their loss through genetic drift called the **Neutrality Hypothesis** (Falconer and Mackay, 1996; Kreitman, 1996; Ohta, 1996; Hartl, 2000). It is not proposed that these loci lack important adaptive function, only that most allelic differences code for molecules that are equivalent, or nearly equivalent, in function. Observed levels of polymorphism are generally consistent with expectations based on the neutrality hypothesis. Nevertheless, expected levels of polymorphism can vary greatly depending on the assumed rates of mutation and effective population size. Therefore, the degree to which the neutrality hypothesis accounts for molecular polymorphism is still unclear. As a practical matter, it may make little difference because even if allozymes or other molecular variants are subject to selection, it is rarely strong enough to be detected experimentally at individual loci. Molecular polymorphisms, therefore, usually provide little information on adaptation of populations. These markers, however, are excellent for investigating mating systems and the effects of gene flow and genetic drift.

One example of balancing selection is heterozygote superiority. In directional selection, the heterozygote is intermediate in fitness between the two homozygotes or equal in fitness to one of them. With **heterozygote superiority** (or **overdominance**), the fitness of the heterozygote is greater than either homozygote and neither allele is eliminated by selection. Given that the relative fitness of A_1A_1, A_1A_2, and A_2A_2 is $1 - s_1$, 1 and $1 - s_2$, respectively:

$$\Delta q = pq(s_1p - s_2q)/(1 - s_1p^2 - s_2q^2)$$ Equation 5.18

An equilibrium (*i.e.* $\Delta q = 0$) that results in both alleles remaining in the population is achieved when $s_1p = s_2q$. The frequency of A_2 at this equilibrium is:

$$q_e = s_1/(s_1 + s_2)$$ Equation 5.19

As long as the population is not fixed for one allele or the other, selection will move allele frequencies toward the equilibrium value (Fig. 5.9). Even a slight superiority of the heterozygote can result in a balanced polymorphism (meaning that both alleles remain at appreciable frequencies) as long as population size remains large.

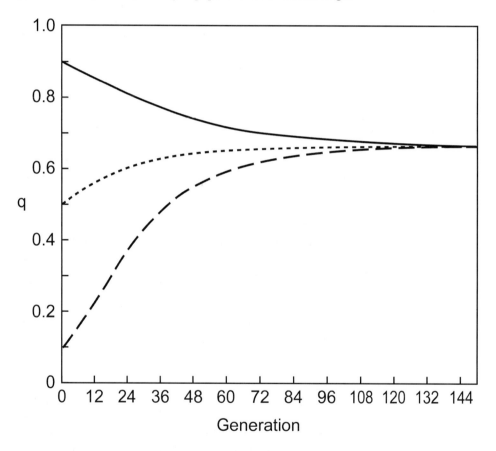

Fig. 5.9. Frequency of allele A_2(q) over generations in a larger, randomly mating, population when selection favors the heterozygote (overdominance). The plots assume three starting values of q (0.10, 0.50, 0.90) at generation zero and that selection coefficients against alleles A_1 and A_2 are $s_1 = 0.10$ and $s_2 = 0.05$, respectively. As long as q is between 0 and 1, overdominance eventually leads q to converging to the equilibrium value, which in this case is 0.66 (as calculated from Equation 5.19). Values of q were generated using the *Populus* computer program (Alstad, 2000).

A variety of mechanisms may be responsible for overdominance (Ledig, 1986; Savolainen and Hedrick, 1995; Falconer and Mackay, 1996). At the molecular level, for example, heterozygotes for a structural gene produce two or more forms of the enzyme, compared to only one in homozygotes. If these forms have different functional properties, including different optima for environmental factors such as pH and temperature, heterozygotes may be better adapted to a range of habitat conditions. Overdominance can also arise if alternative alleles at the same locus are favored at different stages of development, different seasons of the year or in different microhabitats in the same site (also called **marginal overdominance**). For example, if growth is favored by one allele in the seedling and at early competitive stages, but by an alternative allele after stand closure, the heterozygote may be the most fit overall. Despite the potential role of heterozygote superiority in the maintenance of genetic diversity, there are few examples where true overdominance has been demonstrated (Box 5.6).

Two other forms of balancing selection are **frequency-dependent selection** and **diversifying selection** (Dobzhansky, 1970; Hedrick, 1985). If having a rare phenotype is an advantage, selection may depend on gene frequency such that when an allele is in low frequency it is favored; when at high frequency, it is selected against. One of the best-known examples of balanced polymorphism maintained by frequency-dependent selection is self-sterility alleles in flowering plants (Briggs and Walters, 1997). Because pollen grains carrying a given self-sterility allele cannot germinate and grow on stigmas of plants with the same allele, pollen grains with a low-frequency allele have an advantage in mating. Such systems result in large numbers of self-sterility alleles being maintained in populations. Interplant competition for growing space among genotypes of the same species may also lead to frequency dependent selection. Some studies in crop plants and forest trees have found that the yield of genotypes (*i.e.* seed production in crop plants and stem growth in trees) can be influenced by the genotypes of neighboring plants (called **intergenotypic competition**; Allard and Adams, 1969; Adams *et al.*, 1973; Tauer, 1975). If two or more genotypes each have greater yield in mixture than when grown in pure stands, stable equilibria of multiple genotypes (and multiple alleles) will likely result.

Environments in which trees grow vary greatly in space due to patchiness of soil type, water availability, competition, etc, and over time, due both to changing climate and to stand development. If one allele is favored in one microhabitat and another allele in another microhabitat, stable polymorphisms can result, even if there is random mating among genotypes in the different microhabitats (Hedrick, 1985; Falconer and Mackay, 1996). The net effect of such diversifying selection may be that heterozygotes are the most fit overall, but stable polymorphism can also result from diversifying selection when marginal overdominance is not involved (Hedrick, 1985).

Genetic Drift

The offspring in any generation are the result of sampling the gametes produced by parents in the previous generation. Given that parents are equally fertile (no selection) and the number of offspring is large, allele frequencies in the offspring closely resemble those in the parents (and this is the result of the Hardy-Weinberg Principle described earlier for populations of infinite size). If the number of offspring is small, however, allele frequencies in the offspring are likely to differ from the previous generation. This is because randomly drawing a small number of gametes is less likely to be representative of the genes in the parents than drawing a large number. Therefore in small populations, allele frequencies can change from generation to generation in a random, unpredictable fashion just due to the vagaries of sampling; this is called **genetic drift**.

The major features of genetic drift are illustrated with a simulated example (Table 5.5). There are 20 replicate populations meaning that we have created 20 identical starting populations in the computer so that we can follow the course of genetic drift. Each population contains 20 diploid individuals, with an equal number (20) of A_1 and A_2 alleles in generation 0. All replicate populations begin with Hardy-Weinberg equilibrium frequencies ($p^2 = q^2 = 0.25$, and $2pq = 0.50$) so there are five individuals of each homozygote and 10 heterozygotes. A computer program is used to simulate random sampling of gametes from each replicate over 20 consecutive generations, with constant population size (N = 20), random mating and Mendelian segregation (Alstad, 2000).

Within only a few generations, the number of A_1 alleles begins to differ among replicate populations. Changes in the number of A_1 alleles in individual populations are haphazard; some populations increasing in A_1 and others decreasing. With more generations, genetic differentiation (*i.e.* drift) among populations increases; by generation 20, the A_1 allele is completely lost from one population (Population 1) and fixed in another (Population 4). Despite changes in allele frequency in individual populations, the mean frequency of A_1 over all replicates ($\bar{p}$) is approximately 0.5 in all generations. With a large number of replicate populations, $\bar{p}$ is expected to be constant over generations, because random increases in allele frequency in some replicates are balanced by random reductions in others. Mean heterozygosity, on the other hand, is expected to steadily decrease with time as allele frequencies within populations move closer to 0 or 1. Over the 20 generations (Table 5.5), H decreased from 0.50 to 0.32. Because mating between relatives occurs more frequently in small populations than in large populations, even when mating is at random, another implication of small sample size is inbreeding. Thus, the reduction in heterozygosity

Table 5.5. Monte-Carlo simulation of genetic drift in number of A_1 alleles in each of 20 populations over 20 generations. Each population has a constant population size of 20 individuals, with frequencies of both A_1 and A_2 starting at 0.5 in generation 0. Mean frequency of allele A_1 (p) and heterozygosity (H) over all populations are also given for each generation. Simulated by the *Populus* computer program (Alstad, 2000).

Gen.	1	2	3	4	5	6	7	8	9	10	11	12	13	14	15	16	17	18	19	20	p	H
0	20	20	20	20	20	20	20	20	20	20	20	20	20	20	20	20	20	20	20	20	0.50	0.50
1	13	26	16	26	18	19	9	16	20	24	16	22	21	19	20	24	23	13	22	21	0.49	0.48
2	8	26	17	24	18	18	7	18	21	22	17	16	23	19	19	29	31	13	18	23	0.48	0.46
3	8	33	15	29	21	18	7	18	26	20	17	19	26	19	20	26	35	18	19	21	0.52	0.44
4	5	33	11	26	18	20	8	15	24	21	20	15	27	14	18	25	33	19	22	19	0.49	0.44
5	4	32	15	23	20	22	12	18	26	20	18	22	18	13	17	26	31	15	28	17	0.50	0.45
6	5	35	18	28	22	27	10	21	17	23	16	21	16	17	17	25	33	16	27	13	0.51	0.43
7	4	37	20	31	26	28	11	19	18	24	21	17	15	16	18	25	30	13	28	9	0.51	0.42
8	7	39	21	28	26	33	9	18	19	22	20	16	12	14	19	25	28	8	30	10	0.51	0.41
9	9	35	28	32	26	30	7	19	17	21	27	18	7	15	15	24	31	5	35	9	0.51	0.38
10	5	35	27	33	19	30	9	23	22	22	33	19	5	18	15	28	32	8	36	7	0.53	0.37
11	4	38	24	35	17	28	12	26	22	24	30	21	4	19	20	29	29	7	37	8	0.54	0.37
12	3	39	24	38	13	22	11	27	23	26	30	21	5	15	20	28	30	6	35	5	0.53	0.35
13	2	38	23	38	13	22	10	24	24	26	29	27	6	15	17	32	25	8	30	9	0.52	0.37
14	4	39	20	37	15	24	11	18	26	22	24	26	5	17	17	29	31	10	29	12	0.52	0.39
15	3	37	22	37	15	26	16	15	28	18	24	25	5	16	17	27	34	12	32	12	0.53	0.38
16	2	36	16	37	14	25	13	16	28	16	19	29	6	27	15	30	32	11	37	10	0.52	0.37
17	1	37	15	40	17	25	11	18	29	12	24	30	8	29	8	34	34	8	36	12	0.54	0.33
18	1	35	16	40	20	18	13	23	28	15	26	30	8	28	7	35	36	9	37	11	0.55	0.34
19	0	36	16	40	20	16	11	16	26	10	28	33	13	24	7	35	35	9	38	9	0.53	0.32
20	0	36	16	40	20	16	11	16	26	10	28	33	13	24	7	35	35	9	38	9	0.53	0.32

accompanying genetic drift can also be expressed in terms of the inbreeding coefficient, F (see Hedrick, 1985; Hartl and Clark, 1989). Eventually, all populations will become fixed for either A_1 or A_2, and because $\bar{p}$ remains the same, the proportion of populations fixed for A_1 is p and for A_2 is $1 - p$. Although this example is for only a single locus, all loci in the genome are similarly affected by random genetic drift in small populations.

In summary, genetic drift has three major impacts on the genetic makeup of populations: (1) Reduced genetic diversity (due to loss of alleles); (2) Reduced average heterozygosity (*i.e.* increased inbreeding) within populations; and (3) Increased genetic differentiation among populations. The magnitude of these impacts depends on population size, with impacts being particularly great when populations are very small. This is illustrated in Fig. 5.10, where the proportion of simulated populations fixed for either allele A_1 or A_2 is plotted over generations for three population sizes, N = 12, 24, 48 (p = 0.50 in generation 0). When N = 12, more than 80% of the populations are fixed for one or the other allele by 50 generations, but when N is four times larger, less than 20% are fixed over the same period of time. Time to fixation for a locus in any one population also depends on initial allele frequency. When p = 0.1, the average time to fixation is 1.3 N generations, and when p = 0.5, the average time is 2.8 N generations (Hartl and Clark, 1989). Therefore, when an allele starts at an intermediate frequency, it takes longer, on average, to lose the allele from the population.

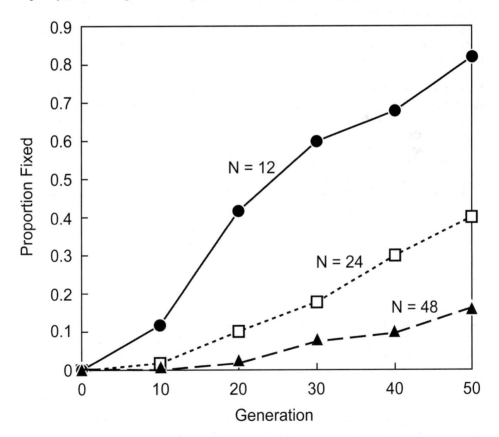

Fig. 5.10. Proportion of populations fixed for either allele A_1 or A_2 at 10-generation intervals (up to generation 50) from Monte-Carlo simulation of genetic drift (Alstad, 2000) in populations of three sizes (N = 12, 24 or 48 individuals). The initial frequency of A_1 (and A_2) in generation 0 is 0.50.

Although forest trees typically exist in rather large, continuous populations, there are a number of situations where genetic drift may be a major factor influencing genetic makeup. Often, populations on the ecological or geographical margins of species' distributions are small and isolated from other populations of the same species. These populations are particularly vulnerable to random changes in allele frequency due to genetic drift (*Chapter 8*). In addition, when sites are colonized by a species expanding its range, only a few individuals may be responsible for establishing new populations. These individuals constitute the founder population, and genetic changes (relative to the progenitor population) resulting from the limited number of founders, is called the **founder effect**. Finally, normally large populations can be drastically reduced in size due to natural or human-caused catastrophes such as glaciation, fire, hurricanes, or shopping malls. Such drastic bottlenecks in population size have profound, long-term effects, even when populations regain their original size within one or many generations.

Extreme population bottlenecks in the recent past are the likely explanation for the very low levels of genetic diversity reported in some endemic tree species that are restricted to few populations and low numbers of individuals. Two examples of species that fall into this category are *Pinus torreyana* in southern California (Ledig and Conkle, 1983) and *Eucalyptus pulverulenta* in southwest Australia (Peters *et al.*, 1990). Although uncommon, some species with wide geographic distribution are also limited in genetic diversity. We have already mentioned that *P. resinosa*, which ranges 2400 km along the Canada-United States border in eastern North America, is extremely limited in genetic variation. It has been hypothesized that this species experienced one or more severe bottlenecks in population size when glaciers covered most of its present range during the Pleiostocene Epoch about 10,000 years ago (Fowler and Morris, 1977). Low genetic diversity in *Thuja plicata*, a species that ranges broadly in the Pacific Northwest of North America, may also be due to historical bottlenecks, although the source of the bottlenecks is unknown (Copes, 1981; Yeh, 1988).

Because changes in allele frequency in any one population are not predictable in direction, the effects of random genetic drift are usually quantified in terms of mean heterozygosity or variance of allele frequency over many population replicates of the same size (or over many replicate loci within a single population). A number of simplifying assumptions are made in deriving relationships between mean heterozygosity, or allele variance, and population size. These assumptions are that within populations, the Hardy-Weinberg conditions of random mating, no mutation, no selection, and no migration are met, and that there are equal numbers of males and females.

No real population conforms to all these assumptions. First, to evaluate the evolutionary consequences of small population size, it is the number of breeding individuals and not the census or total number that is important; the number of breeding individuals is usually fewer than the total number in forest tree populations. Secondly, even the number of breeding individuals may poorly reflect population size in the sense of an idealized population, because factors such as unequal number of males and females, variation among parents in the number of gametes or offspring, and fluctuating numbers over generations are not accounted for. To more accurately evaluate the potential magnitude of genetic drift in a real population, the effective number of breeding individuals or effective size of the population is estimated and used (Hedrick, 1985; Falconer and Mackay, 1996). **Effective population size**, N_e, is the size of an idealized population that would give rise to the same amount of genetic drift as the actual population under consideration. The effective size is almost always less than the number of breeding individuals, and sometimes considerably so (Frankam, 1995). Frankel *et al.* (1995) suggested that N_e is typically only 10-20% the number of adults in forest tree populations (Box 5.7).

Box 5.7. Example of effective population size, N_e, in *Gleditsia triacanthos*.

A population of this dioecious angiosperm tree species in Kansas contained 914 individuals when it was sampled in 1983 for an allozyme study (Schnabel and Hamrick, 1995); only 249 were adults including 174 male and 75 female trees. When the number of males and females is unequal, effective population size, N_e, is reduced because each sex must contribute one-half the genes to the next generation. Assuming all trees in each sex contribute the same to offspring, N_e is equal to twice the harmonic mean number of the two sexes (Crow and Kimura, 1970). In this case, $N_e = 210$ and the harmonic mean is 105, which is calculated as $\{[(1/174) + (1/75)]/2\}^{-1}$. Therefore, N_e is 84% of the number of adults and 23% of the total individuals in the population. Effective size is further reduced if the adults do not contribute equal numbers of gametes to the next generation, or if mating is restricted spatially such that near neighbors mate more often than trees further away in the same population (Crow and Kimura, 1970; Frankel *et al.*, 1995).

JOINT EFFECTS OF EVOLUTIONARY FORCES

With the exception of balance between mutation and selection, we have discussed the impacts of each evolutionary force independent of the others. Although this simplification was convenient for describing the forces, the genetic makeup of real populations is the product of all evolutionary forces acting simultaneously. It is the magnitude of each force, relative to the others, that ultimately determines its influence on genetic composition. Interactions between the various evolutionary forces are quite complex, and their relative magnitudes vary over space and time. Nevertheless, it is instructive to examine some simple cases. Here we address the joint effects of selection, migration and genetic drift on allele frequency. Mutation is ignored, because relative to the other forces, it has little impact on allele frequency change.

Relative to selection and migration, genetic drift has a predominant impact on allele frequencies only when effective population size is quite small. If, for example, the selection coefficient against a mutant allele is s (and there is no dominance), the relative roles of selection and random genetic drift in influencing the frequency of the mutant is a function of $4N_es$, where N_e is the effective size of the population (Hartl, 2000). When $4N_es > 10$, selection predominates; that is, selection tends to reduce the frequency of the allele, with only recurring mutation maintaining the allele in the population. When $4N_es < 0.1$, the frequency of the mutant allele is mostly due to the whims of random sampling. When $0.1 < 4N_es < 10$, both selection and random genetic drift have roles in influencing allele frequency. Weak selection counters genetic drift in all but the smallest populations. For example, if N_e is 25, genetic drift is the predominant force influencing the frequency of a mutant allele only if $s < 0.001$.

Surprisingly little migration between populations curtails the dispersive effects of genetic drift. For example, assume that allele frequency differentiation among a large number of island populations of size N_e is countered by gene flow (m) into the islands from a single large continental population. Eventually these opposing forces lead to an equilibrium with the amount of dispersion among populations being a function of $1/N_em$, where N_em is the number of immigrants into each island population each generation (Hedrick, 1985). The rate of migration (m) required to hold population dispersion at a particular level, therefore, is less in a larger population than in smaller one, which makes sense because larger populations are less susceptible to genetic drift. When N_em is less than 0.5 (*i.e.* one migrant every-other generation), allele frequency differentiation among populations will be large at equilibrium, with many loci fixed, or nearly fixed, for one allele or another within each of the different island populations (Wright, 1969). The amount of dispersion, however, drops off rapidly with increasing $N_e m$, such that it is greatly curtailed when $N_e m > 1$.

Migration can also be a powerful force in limiting differentiation due to selection (Hedrick, 1985; Hartl, 2000). Assume that one of the island populations in the above example is large enough that genetic drift can be ignored. Also assume that an allele, A_2, is selected against in this population, but that its selective removal is countered by migration from the continent where A_2 is fixed in frequency. The expected change in frequency of A_2 (q) over generations, for two levels of migration (m = 0.02, 0.002) and two selection coefficients (s = 0.2, 0.02), is shown in Fig. 5.11. With little immigration of A_2 alleles from the continent (and no dominance), relatively strong selection is expected to rapidly reduce the frequency of A_2 to a very low frequency in the island population (*e.g.* curve for s = 0.2, m = 0.002). Nevertheless, even low rates of migration may be sufficient to maintain A_2 at substantial frequencies (*e.g.* curve for s = 0.2, m = 0.02). When the magnitude of m is greater than or equal to s, migration completely dominates. Therefore, in the face of gene flow, selection is expected to lead to little differentiation of allele frequencies among populations, unless s is much larger than m.

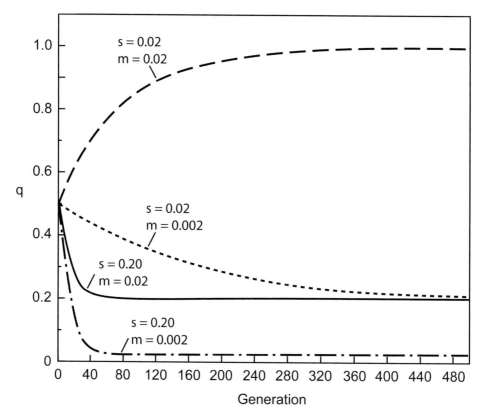

Fig. 5.11. Frequency of allele A_2 (q) over generations in an island population when selection against this allele (no dominance) [s = 0.20 or s = 0.02] is countered by migration (m = 0.02 or 0.002) from a continental population that is fixed for A_2. All plots assume that q = 0.50 at generation 0. Values were calculated using the *Populus* computer program (Alstad, 2000).

SUMMARY AND CONCLUSIONS

As a means of summarizing this chapter, we consider the influence of the five evolutionary forces on genetic differentiation between two populations (Fig. 5.12). We suppose for the purpose of illustration that these are natural populations growing on different sites, although we could just as well be referring to genetic differentiation over generations in a single natural population or in an artificial breeding population. Assuming initially that the populations are identical in genetic composition, we know from the Hardy-Weinberg Principle that they will remain identical as long as the populations are large, mating within each is random, and there is no mutation, selection, genetic drift, or migration from other populations in the region. Furthermore, under these conditions, genotypic frequencies at any one locus correspond to Hardy-Weinberg expectations in Equation 5.3.

Mating patterns within populations (*i.e.* mating systems) do not affect allele frequencies, but do alter the distribution of alleles among genotypes. Although most forest trees are predominantly outcrossing, some mating between relatives, at rates above those expected under random mating (inbreeding), does occur (called mixed-mating). Inbreeding increases the frequency of homozygotes and decreases the frequency of heterozygotes, relative to Hardy-Weinberg expectations. Inbreeding between close relatives (*e.g.* self-fertilization, mating between siblings) often has detrimental consequences (called inbreeding depression). This is because rare deleterious recessives that usually occur only in heterozygotes are expressed as homozygotes in inbreds, resulting either in embryo abortion or poor vigor of surviving progeny. Therefore, large differences between our two populations in levels of inbreeding might not only be reflected in different frequencies of heterozygotes (and homozygotes), but also in differences in survival and growth rates. Most highly inbred individuals, however, are likely to die young, contributing little to the genetic composition of adults in these populations.

Because the process of DNA replication is not error free, mutations occur in offspring, but generally at extremely low frequencies at any one locus. Therefore, mutation is typically disregarded as a force causing differentiation between populations, especially in the short term (*i.e.* over a few generations). Nevertheless, mutation is the ultimate source of genetic diversity, without which evolution could not proceed, and very long time frames are required to accumulate the rich levels of genetic diversity common in most species of forest trees.

Both directional selection and genetic drift lead to allele-frequency differentiation between populations and reduced genetic variation within them, but with different consequences. If the environments inhabited by our two populations are sufficiently different, directional selection favors alternative alleles in each. If, on the other hand, either or both populations go through a severe bottleneck in size, the sampling of alleles in surviving individuals results in random differentiation of allele frequencies between the two populations (genetic drift). There are three major differences in the consequences of directional selection versus genetic drift: (1) Directional selection results in increased adaptation of populations, whereas genetic drift probably does not and may lead to lower adaptation; (2) The effects of directional selection on allele frequencies are predictable in the sense that the same alleles are favored in similar environments replicated in time or space (*i.e.* allele frequencies are associated with environments), while no association of allele frequencies and environments are expected with genetic drift; and (3) Selection affects allele frequencies only at loci influencing fitness (or at tightly linked loci), whereas genetic drift influences all loci in the genome simultaneously.

If both directional selection and genetic drift are operating, genetic drift has the predominant influence on patterns of allele-frequency differentiation only if selection coefficients and/or population size are very small. Populations are generally large in forest trees,

POPULATIONS TEND TOWARD DISSIMILARITY

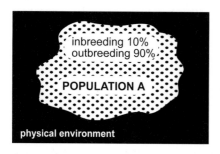
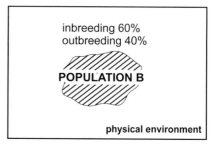

When:

- Levels of inbreeding (the **mating system)** differ between them. This leads to differences in genotypic frequencies (*i.e.* proportions of heterozygotes versus homozygotes) and, perhaps, to different rates of survival and growth.

- Environments differ to the extent that the genotypes most favored in one population are not the most favored in the other (*i.e.* differential **selection** occurs), resulting in allele-frequency differentiation and adaptation.

- Populations are small. Severe reduction in the size of either population results in **genetic drift**, the random (stochastic) differentiation of allele frequencies.

- There is no **migration**. The complete absence of migration facilitates allele frequency differentiation caused by directional selection and genetic drift.

- **Mutation** rates are elevated in one or both populations. Although mutation is the ultimate source of all genetic diversity, allele frequency differentiation due to difference in mutation rates is expected to be negligible in the short term because mutation rates are typically very small.

POPULATIONS TEND TOWARD SIMILARITY

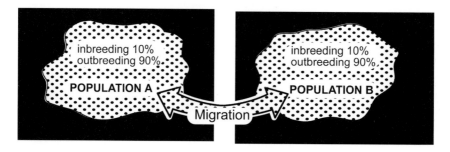

When:

- The mating system is similar in the two populations.

- Environments are similar.

- Populations remain large.

- Migration occurs between them. Even a small amount of gene flow between populations will mitigate the effects of directional selection or genetic drift.

Fig. 5.12. Schematic diagram summarizing the influence of the four major evolutionary forces (mutation, selection, migration, genetic drift) and the mating system on genetic composition, including similarity and dissimilarity, of two populations.

but it is effective population size (the equivalent number of breeding individuals under Hardy-Weinberg conditions) that determines the magnitude of genetic drift and not actual population numbers. In young stands (*i.e.* with few flowering adults) or in poor seed years, for example, effective population size can be substantially less than the number of trees present.

Even modest migration between populations does much to reduce or prevent allele frequency differentiation due to directional selection or genetic drift. Pollen and seed dispersal are the primary modes of migration in forest trees. Pollen dispersal, in particular, can be extensive, occurring in mass over long distances. Therefore, if our two populations are in relatively close proximity to each other, we would expect little differentiation unless one or both has gone through a substantial recent bottleneck or directional selection is particularly strong. If our two populations occupy similar environments, directional selection can be a powerful force preventing differentiation due to genetic drift because selection favors the same alleles in both.

In addition to curtailing allele-frequency divergence between populations, migration helps to maintain genetic variation within populations. This is because migration introduces new alleles into a population and/or replenishes alleles that otherwise would be lost due to selection or genetic drift. Several forms of selection, called balancing selection (*e.g.* heterozygote advantage, diversifying selection, frequency-dependent selection) also promote polymorphism within populations. Unlike directional selection that, in the absence of other evolutionary forces ultimately results in the fixation of one allele at a locus, balancing selection favors two or more alleles simultaneously. Allelic variants having little or no effect on fitness can also be maintained in populations for long periods because these alleles, created anew each generation by mutation, are lost only through genetic drift and this will take a long time if populations are large. This mechanism, called the neutrality hypothesis, probably accounts for much of the extensive polymorphism observed in allozymes and other molecular genetic markers that is frequently observed in populations of forest trees and other outbreeding organisms.

CHAPTER 6
QUANTITATIVE GENETICS – *POLYGENIC TRAITS, HERITABILITIES AND GENETIC CORRELATIONS*

The focus of previous chapters has been the study of simply inherited traits in which different genotype classes are readily distinguishable from each other and can be easily separated into discrete categories. These traits are controlled by one or at most a few gene loci. However, there are many important traits in trees, such as stem height, in which phenotypes are not easily categorized into genotypic classes because the traits vary continuously from one extreme to the other (*e.g.* from short to tall). Often, the phenotypic expression of these traits is influenced by many gene loci, as well as by environmental effects. These traits are called polygenic traits or quantitative traits.

The study of polygenic traits, called quantitative genetics, involves partitioning (*i.e.* subdividing) the measured phenotypic variation in natural or breeding populations of forest trees into the portions caused by genetic and environmental influences. There are many benefits from understanding the genetic control of polygenic traits, and this chapter introduces this field of quantitative genetics. The chapter begins with an overview of the characteristics and methods of studying polygenic traits. Next, we discuss how to model phenotypes and to partition measured phenotypic variance into genetic and environmental components. These sections include the important concepts of breeding values, clonal values and heritability. Since geneticists are often interested in associations among multiple polygenic traits, the next major section introduces genetic correlations, including the correlation between the same trait measured at two different ages, called age-age or juvenile-mature correlations. The next major section on genotype x environment interaction addresses the differential performance of genotypes when grown in different environments, and the implications of these interactions in tree improvement programs. The last major section presents a brief introduction to the use of genetic experiments, specifically clonal tests and progeny tests, for estimating some of the genetic parameters defined in previous sections of the chapter.

The concepts presented in this chapter are used throughout the remaining chapters of this book, especially in designing, implementing and quantifying the gain from tree improvement programs (*Chapters 13-17*). For more detailed accounts of quantitative genetics, the reader is referred to the following books listed in increasing order of mathematical complexity: Bourdon (1997), Fins *et al.* (1992), Falconer and Mackay (1996), Bulmer (1985), and Lynch and Walsh (1998).

THE NATURE AND STUDY OF POLYGENIC TRAITS

Characteristics of Polygenic Traits

A **polygenic trait** is one controlled by many gene loci and therefore, each gene locus has a small effect on the phenotypic expression of the trait. Most complex morphological traits

© CAB International 2007. *Forest Genetics* (T.W. White, W.T. Adams and D.B. Neale)

are thought to be polygenic, and some examples in forest trees include growth rate (measured as bole height, diameter or volume), phenological traits (*e.g.* dates of bud burst, bud set, and flowering), wood specific gravity, stem form (*e.g.* straightness, branch thickness, forking), rooting ability of cuttings, flowering precocity (age to first flowering) and seed fecundity (amount of seed produced by a tree). In many cases, a polygenic trait is affected by several underlying traits. For example, growth rate is affected by many physiological parameters (*e.g.* rates of photosynthesis and respiration), phenological patterns (timing of bud burst, bud set and cambial growth) and organ growth rates (root growth patterns, timing of leaf senescence, etc.).

Characteristics of polygenic traits that distinguish them from traits controlled by one or two major genes (such as discussed in *Chapters 3* and *5*) are: (1) Each single gene locus has a small, unobservable effect on phenotypic expression of the trait; (2) At a given locus, each alternative allele has a small effect on phenotypic expression (Fig. 6.1a); (3) The environment in which a tree grows also affects the phenotypic expression, confounding even more the ability to observe genetic differences directly; (4) The combination of points 1, 2 and 3 means that phenotypic differences usually occur on a continuous gradient (polygenic traits are sometimes called metric or quantitative traits for this reason, but see exceptions below); (5) Individuals with the same phenotype almost always have different genotypes, differing perhaps in specific alleles at many loci; and (6) The number of loci controlling the trait and the allele frequencies at those loci are almost never known.

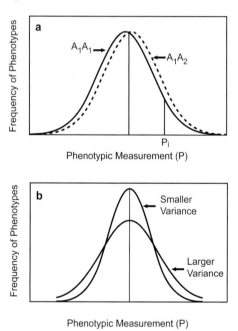

Fig. 6.1. (a) Frequency distribution of phenotypic measurements for a hypothetical polygenic trait such as bole height with a population mean $\mu = 20$ m and phenotypic variance $\sigma^2_P = 25$ m^2 showing: (1) The distribution is continuous with a large fraction of the phenotypes having intermediate values near the population mean, μ; (2) Any single allele has a very small effect on the population mean (curves A_1A_1 *versus* A_1A_2 show two populations that differ only in the substitution of the A_2 allele in all A_1A_1 individuals of the first population); and (3) Individuals with the same measured phenotype, such as P_i, almost always have different genotypes and environmental influences. (b) Frequency distributions of two different populations that have the same mean, but different variances.

For some polygenic traits, the distribution may be more discrete than continuous, and this can happen for natural or artificial reasons. Sometimes traits such as stem form, branch characteristics or crown form are conveniently assessed (*e.g.* visually scored) in discrete categories (such as 1, 2, 3, and 4), even though the trait is actually expressed on a continuous scale. Other traits naturally form a discrete distribution, for example, traits that are measured as 0 = absence or 1 = presence, such as disease resistance (healthy or diseased), survival (alive or dead), or frost damage (yes or no). These are binomial or Bernoulli traits because they only have two categories; but, when their phenotypic expression is controlled by many genes, they are referred to as **threshold traits** (Falconer and Mackay, 1996).

Even though there are many gene loci controlling expression of a polygenic trait, and we do not know the allele frequencies at each locus, these loci follow the same laws of inheritance as for major gene traits discussed in *Chapter 3*. Further, the concepts and effects of selection, mutation, migration, drift and inbreeding discussed in *Chapter 5* for major gene traits also apply to polygenic traits. For example, artificial and natural selection cause changes in allele frequencies at many or all of the gene loci controlling a polygenic trait (but the change at any single locus is very small in any single generation).

Studying Polygenic Traits

Because we cannot identify specific genotypes nor allocate them into specific genotypic classes for polygenic traits, statistical methods are used to: (1) Quantify the amount of phenotypic variation for each measured trait; (2) Estimate the relative contributions of different types of genetic and environmental influences to this observed phenotypic variation; and (3) Predict the overall underlying genetic value, accumulated over all loci, of each individual tree, family or clone that is measured. We introduce these concepts in this section, and then develop each step in more detail later in this chapter and in *Chapters 13, 14* and *15*.

Quantifying the Amount of Phenotypic Variation

Suppose we measure total height of 1000 trees growing in a natural even-aged forest stand of a pure species. The distribution of measured phenotypes might look like one of the curves in Fig. 6.1a. There are N = 1000 phenotypic measurements (P_i) in this population of trees, and we would use the following standard formulae to calculate the **population mean**, μ, and **phenotypic variance**, σ^2:

$$\mu = \sum(P_i)/N \qquad\qquad\qquad \text{Equation 6.1}$$
$$\sigma^2 = [\sum(P_i - \mu)^2]/N \qquad\qquad \text{Equation 6.2}$$

The mean describes the relative placement or location of the distribution of phenotypes on the x-axis and its units are the units of measure; for example, let's say the mean height of the example population is 20 m. The population variance measures the dispersion or variability of phenotypes around the mean, and its units are the square of the units of measure; let's say $\sigma^2 = 25$ m^2 for the current example. The square root of the variance is called the **standard deviation**, σ, and is also a measure of variation but in the units of measure. Populations that are more variable have larger variances (and standard deviations). In Fig. 6.1a, the two populations have slightly different means, μ and μ', and the same variance. In Fig. 6.1b, the means of the two populations are equal, but one population has a larger variance or amount of spread in the phenotypic measurements than the other.

Estimating the Importance of Genetic and Environmental Influences

The relative importance of genetic and environmental contributions to the observed phenotypes and to the phenotypic variance for any trait cannot be deduced from phenotypic measurements of that trait in natural stands. To demonstrate this concept, consider a simple **linear model** to describe the 1000 phenotypic values for height in Fig. 6.1a:

$$P_i = \mu + G_i + E_i$$ Equation 6.3

where P_i is the phenotypic value (height measurement in our example) of a given tree labeled i and i ranges from 1 to 1000; $\mu = 20$ m is the population mean calculated with Equation 6.1 using all 1000 phenotypic values in the population; G_i is the underlying genotypic value of tree i; and E_i is the cumulative effect of all environmental influences on the phenotypic value of tree i. G_i and E_i are defined as deviations from the population mean; they can either be positive or negative and sum to zero over all trees in the population. That is, the genotype (or environment) of a particular tree either increases or decreases its phenotypic value relative to the mean of that trait. In our example, G_i and E_i can be thought of as deviations above or below a mean of 20 m. Therefore, a given tree's height is modeled as the sum of three components: the population mean, and the tree's genotypic and environmental deviations that add to or subtract from this mean.

Since the mean, μ, is a constant, the model in Equation 6.3 specifies that a tree's phenotypic measurement is controlled by two underlying variables (G_i and E_i), and these two underlying variables are confounded. That is, for any number of phenotypes measured, there is twice the number of unknowns as measurements; therefore, G_i and E_i cannot be separated (further illustrated in Box 6.1). For this reason, forest geneticists must plant progeny (either clonal or sexual offspring) in randomized, replicated trials to separately measure genetic and environmental effects on phenotypes (as illustrated in the last section of this chapter). These tests are called common garden tests, progeny tests or genetic tests. There are many mating designs and field designs for genetic tests considered in *Chapter 14*. The mating designs specify the pedigree control and family structure, while field designs ensure proper randomization, replication and representation of edaphoclimatic conditions.

In quantitative genetics, the relative importance of genetic and environmental influences is quantified in terms of their contribution to the observed phenotypic variance. With reasonable assumptions (*e.g.* the variance of the population mean is zero since it is a constant), the phenotypic variance is derived from Equation 6.3 as:

$$\text{Var}(P_i) = \text{Var}(\mu + G_i + E_i) = \text{Var}(\mu) + \text{Var}(G_i) + \text{Var}(E_i)$$ Equation 6.4
$$\sigma^2_P = \sigma^2_G + \sigma^2_E$$ Equation 6.5

meaning that the measured variance among phenotypes can be expressed as a sum of the variance among genotypes (σ^2_G) and variance among environments (σ^2_E). To interpret the meaning of Equation 6.5, consider two extreme, hypothetical examples. First, if all 1000 trees are from the same clone (*i.e.* have the same genotype), then $\sigma^2_G = 0$ (since there can be no genetic variability if all genotypes are identical) and $\sigma^2_P = \sigma^2_E$. Conversely, if the environment is completely uniform (which never happens in reality), such that all trees experience the same microenvironment; but individuals vary genetically, $\sigma^2_E = 0$ and $\sigma^2_P = \sigma^2_G$. These extreme examples are meant to illustrate that it is possible to quantify the relative importance of genetic and environmental variability in terms of their respective contributions to the phenotypic variance. Later in this chapter, this approach leads to the concept of heritability.

Box 6.1. Confounding of genetic and environmental influences in phenotypic measurements.

Consider three tall trees in the stand of trees depicted by the data in Figure 6.1a. Trees i, j and k all have the same height of 30 m (10m above the population mean). Shown below are three of the many possible combinations of genotypic and environmental effects that would result in the same phenotypic value:

$$30 = P_i = \mu + G_i + E_i = 20 + 6 + 4$$
$$30 = P_j = \mu + G_j + E_j = 20 + 12 - 2$$
$$30 = P_k = \mu + G_k + E_k = 20 + 0 + 10$$

(from Equation 6.3)

In this hypothetical example, tree i is tall due to both its underlying genotypic value, G_i (which is 6 m above the average genetic value of 0), and because it is growing in a favorable microenvironmental location that contributes 4 m more growth than the average microenvironment in the stand. Tree j is an outstanding genotype growing in a slightly below average microenvironment, while tree k is an average genotype growing in an outstanding microenvironment.

This example demonstrates that measurement of phenotype alone does not allow separation of the underlying influences of genetic and environmental effects. We say that the genetic and environmental influences are confounded in the phenotypic measurements. Therefore, progeny tests must be used to separate the effects of genotype and environment (*Chapter 14*).

Even after creating offspring from the original parent population and planting these progeny in field trials replicated over several environments, it is still not possible to specify the exact genetic constitution of a parent for a polygenic trait. For example, tree j is an outstanding genotype in terms of height growth, but this superiority could arise from any number of allelic combinations at the many gene loci controlling height growth. Other trees with an identical genetic value of 12 would almost assuredly be different genotypes (*i.e.* with superior alleles at distinct loci from j).

MODELING PHENOTYPES OF PARENTS AND OFFSPRING

Clonal Value and Breeding Value

Before further discussing the partitioning of phenotypic variance into the relative contributions of genotypic and environmental influences, it is necessary to more completely develop the concept of genotypic value and the relationship between parents and their offspring for quantitative traits. If rooted cuttings or tissue culture plantlets taken from a selected tree are used for reforestation, all plantlets are the exact same genotype and therefore, have the same genotypic value, G_i, which is also the genotypic value of the selected tree. The selected tree is called the **ortet**, while plantlets produced through asexual propagation of the same tree are called **ramets**. Taken together, the ortet and ramets form a **clone** with all members sharing the same G_i. For this reason, the genetic value, G_i, is also called the **clonal value** of a tree, because it is the genetic value of all plants of the clone formed by vegetatively propagating that tree. The clonal value of the tree for a particular trait predicts the performance for that trait when its asexually propagated plantlets are used for reforestation. To illustrate this, if two different clones are planted, the one with the higher clonal value for growth rate results in a plantation that grows faster. Clonal reforestation programs, such as for *Eucalyptus grandis* in many countries, use predicted clonal values (labeled G_i and synonymously called **genotypic values**) to rank clones and to choose the best ones to use for operational planting (*Chapters 16* and *17*).

It also is important to be able to predict the performance of a parent's sexually produced offspring. For example, when seedlings (offspring) from different selected trees (parents) are available for reforestation, it would be useful to know which parents' offspring will perform better. Also, when a selected tree is used in a breeding program, interest again centers on the performance of its offspring formed from the matings with other parents. The simple model of phenotype presented in Equation 6.3 does not allow this prediction and must be expanded to include a term called **breeding value**:

$$P_i = \mu + A_i + I_i + E_i$$

Equation 6.6

where A_i is the breeding value of the tree (also called the **additive value**) and I_i is called the **interaction** or **non-additive portion** of the genotypic value. Note that $G_i = A_i + I_i$ and that A_i and I_i are defined as deviations from the population mean so that they both average to zero across the population of parent trees. An individual's breeding value is defined as the portion of the genetic value that is transmitted to offspring created by mating the individual randomly to all other individuals in the population. An individual tree in the parent population has a positive breeding value for height if its offspring grow faster than the mean of progeny produced from all possible matings among all parents in the population.

To fully understand the concept of breeding value, it is necessary to understand the concept of **average allele effect** (as illustrated in Box 6.2 and derived in Cockerham, 1954; Falconer and Mackay, 1996). In sexual reproduction, parents contribute gametes, not their whole genotype, to their offspring. Therefore, parents transmit a random sample of one-half of their alleles to each and every offspring; they do not transmit both alleles from the same locus nor do they transmit intact combinations of alleles from many loci. The Mendelian processes of independent segregation and random assortment coupled with crossing over between linked loci ensure that a vast array of new allelic combinations (at each locus and across all loci) is created among offspring in each generation (*Chapter 3*). Therefore, each allele has an average allelic effect that depends only on that allele as it increases or decreases performance in the offspring, and is independent of the other alleles in the genome. The breeding value of a tree for a specific trait, A_i, is the summation of the 2n average allelic effects influencing that trait (2 alleles from n loci).

Because the breeding value of an individual is determined solely by the average effect of each allele it possesses, part of a tree's clonal value for any trait, G_i, due to interactions with other alleles in the genome (*i.e.* caused by dominance of one allele over another at a specific locus or by the epistatic interactions of specific allelic patterns at several loci) does not influence the mean of offspring produced by random mating. This is called the non-additive portion or interactive portion of the genotype, I_i, to highlight the fact that it arises from specific interactions of alleles that are not conveyed to offspring. Therefore, although each parent has a clonal value, G_i, and a breeding value, A_i, for each trait, these are not identical. The clonal value is the total genotypic value passed on to asexual plantlets (ramets of the clone), while the breeding value of a parent is the portion of its superiority (or inferiority) that is transmitted to sexual offspring under random mating (Box 6.2).

Just as genotypic values cannot be measured directly, neither can breeding values. Breeding values are estimated from progeny tests (*Chapters 14* and *15*) in which the off-

---

Box 6.2. Illustration of breeding value *versus* genotypic value.

Consider a hypothetical one-locus model in which a particular locus, L, is one of many affecting the polygenic trait of tree height growth. Suppose that the average allele effect of the L allele is to add 0.1m to tree height, while that of the l allele is to decrease height by 0.1 m. These average effects are the values of the two alleles if we could measure them independently from the effects of dominance (interactions of alleles at the same locus) and epistasis (interactions among alleles at different loci).

A tree's breeding value for this locus (A_L) is defined as the sum of the average effects of both its alleles, and there are three possibilities, one for each potential genotype (LL, Ll and ll): $A_{LL} = 0.1 + 0.1 = 0.2$ m; $A_{Ll} = 0.1 - 0.1 = 0.0$ m; and $A_{ll} = - 0.1 - 0.1 = -0.2$ m. The tree's total breeding value for height is the sum of its breeding values across all n loci affecting height (*i.e.* the summation of the 2n average effects of 2n alleles, 2 alleles at each of n loci). Note that the contribution of any particular locus to the overall breeding value may be positive, negative or zero depending on the summed average effects of the two alleles at that locus.

Ignoring the other loci influencing height growth, a tree whose genotype is LL has the highest breeding value (0.2 m) since it always contributes an allele with a positive effect to its offspring, while a tree that is Ll contributes a favorable allele half of the time and is therefore, intermediate in its value as a parent (breeding value = 0.0 m). If the L allele is completely dominant over the recessive l allele, then the genotypic values of LL and Ll must be the same. This exemplifies the case where the clonal values and breeding values are different. If asexual propagules are planted, then the same performance is expected from clones that are LL or Ll since the L allele is dominant to l. This is why the clonal values are identical for these two genotypes. However, when LL is used as a parent, its sexually created offspring (generated by random mating) grow faster on average than the offspring of parent Ll because some offspring of Ll are slower-growing ll genotypes. Therefore, while clonal values are identical ($G_{LL} = G_{Ll}$), the breeding value is greater for the homozygous dominant parent ($A_{LL} > A_{Ll}$).

---

spring of many parents are planted together in randomized, replicated field experiments. Parents whose offspring grow faster but have a high disease incidence, have a favorable estimated breeding value for growth, but an unfavorable estimated breeding value for disease incidence. Even with progeny test data, breeding values are only estimates because we measure the phenotypes of a limited sample of offspring from each parent.

Estimating the Average Performance of Offspring

The above section describes conceptually how the breeding value of an individual is that portion of the individual's genotypic value transmitted to offspring under random mating. In this section, we develop this further in a more quantitative manner to estimate the mean value of the progeny from half-sib and full-sib families grown in progeny tests. These concepts are used to estimate gains from different options of selection and deployment of genetically improved material in *Chapter 13*.

By definition, if two parents are mated to form a full-sib family and both have large, positive breeding values for height growth, then the offspring produced grow faster than offspring from matings among parents with smaller breeding values. However, each of the offspring is a different genotype, just as brothers and sisters are different genotypes in human families. Each parent contributes a random sample of half of its alleles to each gamete formed, and genetic recombination (sometimes called **Mendelian sampling**) ensures that there are literally an infinite number of gametic (haploid) genotypes produced by each parent. Some samples are better than others, so some offspring receive a better than average (and some are poorer than average) sample of the alleles from one or both parents. Therefore, it is impossible to know the genotypic value or breeding value of each and every offspring in the family, and we must settle for predicting the mean of the family.

Average Performance for Half-sib Families

Mating one parent tree at random with all other parents in the population produces a **half-sib family**. Normally in forestry, the common parent is the female parent and the male parents supply pollen at random from the population. Each gamete produced by each parent contains a random sample of n alleles from that parent that influence a particular trait, and each of these alleles has a positive or negative average allele effect on the trait in the progeny. Just like in the parents, the average allele effects are additive in the progeny, and the breeding value of each offspring produced is the sum of the 2n average allele effects affecting a given trait. Since each offspring receives half of the alleles (and their corresponding average effects) from each parent, half of the offspring's breeding value is received from each parent. Assuming the male gametes (*i.e.* pollen grains) come at random from the population, the mean value of the male parents' allele effects is zero when averaged across a large number of male parents contributing to a half-sib family. This is because breeding values are defined as deviations from the population mean, and across the entire population of male parents, the mean breeding value is zero for every trait. In equation form for a given trait, the mean breeding value of the offspring in a half-sib family, $\overline{A}_{O,HS}$, is derived as follows:

$$\overline{A}_{O,HS} = [\Sigma(A_F + A_M)/2]/m \quad \text{(the average of all male (A_M) and female (A_F) parental}$$
breeding values of m offspring)
$$= \Sigma(A_F)/2m + \Sigma(A_M)/2m \quad \text{(distributing the summation)}$$
$$= 1/2\ \overline{A}_F + 1/2\ \overline{A}_M \quad \text{(rewriting the mean breeding values)}$$
$$= 1/2\ \overline{A}_F \quad \text{(since the male breeding values average to zero)}$$
$$= 1/2\ A_F \quad \text{(the female breeding value is constant for one family)}$$
$$= GCA_F \quad \text{Equation 6.7}$$

where GCA_F is the **general combining ability** of the female parent (the parent common to all offspring). In other words, the mean or expected breeding value of an offspring in a half-sib family is half of the female parent's breeding value; this is synonymously called the general combining ability of the female parent ($1/2\ A_F = GCA_F$). This relationship means that, on average, a parent passes on one-half of its breeding value to its offspring. This should be intuitively understandable since the parent's breeding value is the sum of the average allele effects for the 2n alleles influencing that trait, and the parent passes on one-half of those 2n alleles to each offspring.

We can also develop a linear model to explain the mean **phenotypic value**, $\overline{P}_{O,HS}$ for the offspring of a half-sib family:

$$\overline{P}_{O,HS} = \mu_O + \overline{G}_{O,HS} + \overline{E}_{O,HS}$$
$$= \mu_O + \overline{A}_{O,HS} + \overline{I}_{O,HS} + \overline{E}_{O,HS} \quad \text{(because } G = A + I)$$
$$= \mu_O + \overline{A}_{O,HS} \quad \text{(because } \overline{I}_{O,HS} = \overline{E}_{O,HS} = 0)$$
$$= \mu_O + 1/2\ A_F \quad \text{(from Equation 6.7)}$$
$$= \mu_O + GCA_F \quad \text{Equation 6.8}$$

where μ_O is the population mean of offspring from all possible matings, $\overline{G}_{O,HS}$ is the mean genotypic value of all offspring in this half-sib family, $\overline{E}_{O,HS}$ and $\overline{I}_{O,HS}$ are the means of all environmental and non-additive genetic effects, respectively, averaged over all trees in this half-sib family, and all other terms are defined similarly to those defined in Equations 6.3, 6.6 and 6.7. The O and HS subscripts are used to highlight the fact that the phenotypes measured are offspring from a half-sib family with female F as the common parent.

To understand Equation 6.8, consider that each offspring in a single half-sib family has a unique phenotypic value (assuming a continuous scale), because each offspring has its own genotype and each one experiences a different microenvironment. As a result, there is a large amount of variation among trees within the same family. However, if a large number of offspring are measured from a given half-sib family, the phenotypic mean of the family reduces to approximately the population mean plus half of the parental breeding value of the common parent. This is because: (1) The mean value of environmental effects, $\overline{E}_{O,HS}$, approaches zero when a large number of trees of the family are planted in randomized, replicated field designs so that the trees experience the range of positive and negative microenvironmental effects; (2) The interactive effects, $\overline{I}_{O,HS}$, due to dominance and epistasis are recreated each generation and these also average to zero; and (3) The mean breeding value of the trees in the half-sib family ($\overline{A}_{O,HS}$) can be expressed as half the breeding value of the female, common parent ($1/2\ A_F$) as derived in Equation 6.7.

Equations 6.7 and 6.8 together form one of the most important concepts in this chapter. Not only is the mean breeding value of a half-sib family equal to half the breeding value of the common parent ($\overline{A}_{O,HS} = 1/2\ A_F$ from Equation 6.7), even more importantly, this mean breeding value of the offspring is a prediction of the mean performance of the family above or below the population mean (μ_O) of offspring produced by random mating of all parents in the population (Box 6.3). Therefore, by having estimates of the breeding values of all of the parents in a population (such as a breeding population in a tree improvement program), it is possible to predict the performance of the offspring from the half-sib families formed from these parents. If many top-ranking parents are randomly intermated to form seed for operational reforestation (such as in a seed orchard, *Chapter 16*), then the seedlings are predicted to perform better than seed from lower ranking parents. Further, the predicted genetic gain of the seedling offspring is obtained by averaging the breeding values of the selected parents (*Chapters 13* and *15*). The use of half-sib families to estimate genetic parameters, such as heritabilities, and to predict breeding values is illustrated in the last major section of this chapter and in *Chapter 15*.

Average Performance for Full-sib Families

To predict the average performance of a **full-sib family**, $\overline{P}_{O,FS}$, created by crossing a specific female parent (F) with a specific male parent (M), recall that each parent passes on one-half of its breeding value to the offspring. Then, proceeding similarly as in Equation 6.8:

$$\begin{aligned}
\overline{P}_{O,FS} &= \mu_O + \overline{G}_{O,FS} + \overline{E}_{O,FS} \\
&= \mu_O + \overline{A}_{O,FS} + \overline{I}_{O,FS} + \overline{E}_{O,FS} \\
&= \mu_O + 1/2\ A_F + 1/2\ A_M + SCA_{FM} \\
&= \mu_O + GCA_F + GCA_M + SCA_{FM}
\end{aligned}$$

Equation 6.9

where terms are defined similarly as in Equation 6.8; FS indicates that this is a full-sib family and SCA_{FM} is the **specific combining ability** of the particular full-sib family created between parent F and M. The similarities with Equation 6.8 (for half-sib families) are: (1) The mean value of environmental effects, $\overline{E}_{O,FS}$, approaches zero since the large number of trees of the family planted randomly experience the range of positive and negative microenvironmental effects; and (2) The mean breeding value of the trees in the full-sib family ($\overline{A}_{O,FS}$) is expressed as the sum of half of each of the parental breeding values since each parent contributes half of its alleles to each offspring. Therefore, the mean breeding value of all trees in a full-sib family is equivalent to the average of the two parental breeding values and to the sum of the two general combining abilities, *i.e.* $\overline{A}_{O,FS} = 1/2(A_F + A_M) = GCA_F + GCA_M$.

Specific combining ability, SCA, is a little more complicated and corresponds to a small portion of the interactive effects ($\overline{I}_{O,FS}$) that does not average to zero across all the trees forming a full-sib family. Even though all of these non-additive, interactive effects are recreated anew each generation, the array of unique genotypes generated in the offspring is limited to the specific alleles present in the two parents, unlike the offspring of a half-sib family where the sampling of alleles from male parents is large and reflective of

---

Box 6.3. Example of mean phenotypic values of half-sib and full-sib families.

1. *Half-sib families.* As an example of Equation 6.8, consider the predicted average performance of the half-sib offspring from parent tree F (where F is a tree chosen from the stand in Fig. 6.1a). Suppose that tree F has a breeding value of 6 m ($A_F = 6$ m and $GCA_F = 3$ m) above the mean height of all trees in the stand of 20 m. The anticipated mean of the half-sib family of progeny having F as the common parent is calculated from Equation 6.8 as: $\overline{P}_O = \mu_O + 1/2A_F = 20 + 3 = 23$ m.

   Note that for a half-sib family, the expected mean is the population mean plus half of the parent's breeding value. The population mean is conceptually both: (1) The mean of all possible half-sib families created from random matings of all parents in the original population; and (2) The mean of a sample of offspring produced from randomly collected seed in the parent population of trees (assuming random mating and equal fecundity of all parent trees). Therefore, the half-sib offspring from parent F are expected to average 3 m taller than seedlings from seed randomly collected from the parent population when planted back in the same environment as the parent stand of 1000 trees.

2. *Full-sib families.* As an example of Equation 6.9, suppose that two trees F and M with breeding values of 6 m and 12 m (and GCA values of 3 m and 6 m, respectively) are chosen from the stand of 1000 trees in Fig. 6.1a and mated together. A large number of their offspring is planted in the same environment as the parent stand randomized with full-sib families from all other possible pair-wise matings among the 1000 parents in the population. Then, if the mean of the full-sib family for height is 25 m, the SCA must be − 4 m (rearranging Equation 6.9: $SCA_{FM} = \overline{P}_{O,FS} − \mu_O − GCA_F − GCA_M = 25 − 20 − 3 − 6 = − 4$). For some reason, the offspring of this family grew 4 m less than predicted by the parental breeding values (or synonymously by parental GCA values), and this is due to the specific dominance effects of the alleles in the offspring that are not accounted for by the average effects of those alleles.

the whole population. In full-sib families, one-quarter of all offspring will share the same (*i.e.* identical by descent; *Chapter 5*) genotype at any one locus, and if one allele is dominate over another at this locus, all trees with the same genotype will share a genotypic value that depends on the dominant gene action and not only on the average allele effects. This leads to the concept of SCA, which refers to the fact that the mean performance of a full-sib family can be different than predicted from the average of the two parental breeding values. If all gene loci have only additive effects (no alleles are dominant to any others), then the SCA contribution to the family mean is zero.

While GCA values are properties of individual parents, SCA is specific to the offspring of two parents. The SCA represents the difference between the anticipated performance of a family based on the parental breeding values and the family's true mean performance for a given trait (Box 6.3). Therefore, SCA values are expressed as deviations and sum to zero over all possible parental pairs in the population. Further, GCA and SCA values are uncorrelated. In other words, if a given female parent is mated with 100 male parents to produce 100 full-sib families, we expect half of the full-sib families to have positive SCA values and half to have negative SCA values; these 100 SCA values will not be correlated with the 100 GCA values of the male parents.

GENETIC VARIANCES AND HERITABILITIES

Definitions and Concepts

The phenotypic variance among individual trees in a reference population for a given trait, σ^2_P, is derived from Equation 6.6 as:

$$Var(P_i) = Var(\mu) + Var(A_i) + Var(I_i) + Var(E_i)$$
$$\sigma^2_P = \sigma^2_A + \sigma^2_I + \sigma^2_E$$

Equation 6.10

where $Var(\mu) = 0$ since the population mean is a single constant value, σ^2_A is the **additive variance** (the variance among true breeding values of the trees in the reference population), σ^2_I is the interaction or **non-additive genetic variance** (the variance caused by differences in allele interactions among individuals), σ^2_E is the variance among trees caused by microenvironmental effects, and all effects in the model are assumed uncorrelated. The importance of Equation 6.10 is that the underlying causal components (σ^2_A, σ^2_I and σ^2_E) add up to the phenotypic variance (Fig. 6.2). We never know the true values of any of the three variances in Equation 6.10, but rather estimate these variances of the causal components from genetic experiments containing offspring from the reference population of parents (as illustrated in the last section of this chapter and in *Chapter 15*). Sometimes it is important to subdivide the non-additive genetic variance, σ^2_I, into two components such that $\sigma^2_I = \sigma^2_D + \sigma^2_\varepsilon$ where σ^2_D is the **dominance variance** arising from dominance effects at individual loci affecting the trait and σ^2_ε is the **epistatic variance** arising from allelic interactions among different loci.

Because breeding values represent the portion of the genotypic value that is passed on to offspring after random mating, it is logical that the variance among breeding values can only be estimated from experiments with sexually produced offspring. If after experimentation, only a small portion of the observed phenotypic variance is due to the variance among breeding values (Fig. 6.2), then the phenotypic variance must be composed primarily of either variance due to environmental effects or interaction effects. In other words,

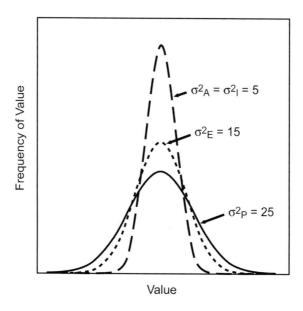

Fig. 6.2. Total phenotypic variance, σ^2_P, and its underlying causal components: additive variance, σ^2_A, non-additive variance, σ^2_I and variance due to microenvironmental effects, σ^2_E. For this hypothetical trait, $\sigma^2_P = 25$, $\sigma^2_A = 5$, $\sigma^2_I = 5$, $\sigma^2_E = 15$, $h^2 = 0.20$ and $H^2 = 0.40$.

when the additive variance is small, each phenotypic measurement is affected more by environmental and interaction effects than by the tree's breeding value. This is exemplified in Fig. 6.2 where the variance among true breeding values of trees in the population ($\sigma^2_A = 5$ m^2) is much smaller than the variance among phenotypic values ($\sigma^2_P = 25$ m^2).

Two common measures of the relative amount of genetic control for a given trait in a population are the **broad-sense heritability** and **narrow-sense heritability**. Broad-sense heritability, H^2, is the ratio of the entire genetic variance to the total phenotypic variance:

$$H^2 = \sigma^2_G/\sigma^2_P$$
$$= (\sigma^2_A + \sigma^2_I)/(\sigma^2_A + \sigma^2_I + \sigma^2_E) \qquad \text{Equation 6.11}$$

The narrow-sense heritability, h^2, is the ratio of additive genetic variance to total phenotypic variance:

$$h^2 = \sigma^2_A/(\sigma^2_A + \sigma^2_I + \sigma^2_E) \qquad \text{Equation 6.12}$$

There are three equivalent interpretations of narrow-sense heritability (Falconer and Mackay, 1996; Bourdon, 1997). First, h^2 is a ratio of the additive variance to the total phenotypic variance; as such, it expresses the fraction of the phenotypic variance that is accounted for by the variance among breeding values of trees in the reference population. Second, h^2 is the regression coefficient, b, for the regression of breeding value on phenotype (Fig. 6.3); thus, it measures the amount of change in breeding value for a one-unit change in phenotype (slope = b = h^2 = σ^2_A/σ^2_P). For example, if $h^2 = 0.20$, then a tree that is 1 m taller than its neighbor is predicted to have a breeding value that is only 0.2 m greater than its neighbor. Finally and perhaps most informative, h^2 measures the degree to which sexual offspring will resemble their parents. If h^2 is near 1.0 for a given trait, then each phenotypic measurement

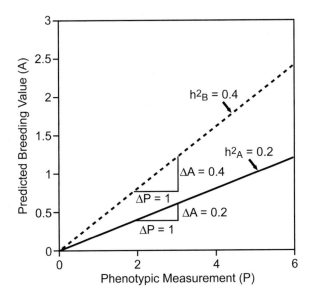

Fig. 6.3. Graphs of the regression lines of breeding values on phenotypic values for two hypothetical traits, A and B, with $h^2_A = 0.2$ and $h^2_B = 0.4$. Note that phenotypic values are closer in magnitude to the underlying breeding value for trait B with its higher heritability. Two trees that are different by one unit in their phenotypic value are expected to have breeding values that differ by 0.2 and 0.4 units for traits A and B, respectively. Selection of outstanding phenotypes for trait B is more likely to result in trees that have higher breeding values and therefore, produce better-performing offspring.

completely reflects the underlying breeding value of that tree. If that tree is then used as a parent, the mean of its half-sib offspring will perform similarly to it (above or below the population mean), since breeding value is the portion of the genotype passed on to offspring following random mating. Broad-sense heritability also has these three interpretations when additive variance, breeding value, and sexual offspring are replaced by total genetic variance, clonal value, and clonal offspring, respectively.

Both narrow-sense and broad-sense heritabilities are trait-specific, population-specific and also greatly influenced by the homogeneity of the environment containing the genetic test (Box 6.4). Further, true heritabilities are never known, but must be estimated from field experiments employing offspring (clonal "offspring" for estimation of H^2 and sexual offspring for estimation of h^2). Two experiments are briefly described in the last section of this chapter and more detailed methods of estimation are discussed in *Chapter 15*. In general, heritabilities are difficult to estimate and require large numbers of clones and/or families for precise estimates. Estimates from small experiments are often very imprecise and have large standard errors and confidence intervals.

An interesting paradox associated with heritabilities is that they cannot be directly estimated in wild stands (such as the stand of 1000 trees used for the examples in Figs. 6.1 and 6.2). This is because a replicated, randomized common garden experiment is required to separate genetic and environmental effects on phenotypes. The experiment contains offspring (sexual or clonal) from only a sample of trees in the original wild stand, and cannot experience the exact same set of environments and annual climates experienced by the parental stand. For any trait, the degree to which the heritability estimated from a genetic test mimics the heritability in the parental stand depends on the sample of genotypes and the similarity of the environments in the test to those in the parental stand.

Box 6.4. Properties of heritabilities.

1. Heritabilities are trait specific meaning that two polygenic traits can have very different heritabilities if they are influenced by different underlying gene loci.

2. For any trait, both broad-sense (H^2) and narrow-sense (h^2) heritabilities range between 0 and 1, and H^2 is greater than h^2 if there is any non-additive genetic variance (compare Equations 6.11 and 6.12).

3. Heritabilities are functions of the genetic composition of the population. When allele frequencies at all loci influencing a polygenic trait are at intermediate levels (say 0.25 to 0.75), then heritabilities (H^2 and h^2) are higher. If natural or artificial selection has acted at many loci to nearly fix certain alleles (frequencies approaching 1.0) and almost eliminate others (frequencies near 0), then genetic variability and hence, heritabilities are reduced.

4. Heritabilities also are functions of the environment in which the population is grown. If field experiments are planted on very homogeneous sites and carefully managed to reduce spurious environmental effects and measurement errors, then observed heritabilities are larger because the environmental variance is smaller (*i.e.* smaller σ^2_E in the denominators of Equations 6.11 and 6.12). This means that observed heritabilities apply to a specific set of environmental and experimental conditions.

5. The heritability of any stress-related trait (such as resistance to disease, drought or cold) must be assessed in an experimental environment in which the stress agent is present. Genetic variation to a stress can be present in the population, but is expressed only in the presence of the stress.

6. For untransformed binomial traits, heritability is a quadratic function of the mean incidence of the trait in the population. When the population mean is very high or very low, there is less phenotypic variation measured and the observed heritabilities are lower (Sohn and Goddard, 1979).

Estimates of heritabilities obtained from a field experiment planted on a single field location are always biased upward (*i.e.* overestimated). This is discussed in the section of this chapter on *Genotype x Environment Interaction*. To obtain unbiased estimates, the field experiment must be replicated over multiple sites in the geographical or edaphoclimatic zone of interest.

Even when field experiments employ large numbers of genotypes planted at several locations, there are several assumptions that must be satisfied for some uses and interpretations of the heritability estimates (Cockerham, 1954). Some problems in forest trees that can seriously affect the estimation or interpretation of narrow-sense heritability estimates are: (1) The tree species is polyploid; (2) Selfs or other inbred offspring occur in some or all families planted in the field experiments (Griffin and Cotterill, 1988; Sorensen and White, 1988; Hardner and Potts, 1995; Hodge *et al.*, 1996); and (3) The reference population is in serious linkage disequilibrium (*Chapter 5*), such as when two geographically distant populations have recently merged or when two species have been hybridized. Another important assumption for valid interpretation of broad-sense heritability estimates is that the propagation system (*e.g.* tissue culture or rooted cuttings) used to vegetatively propagate ramets of a clone has no differential effects on some clones than others. These non-genetic effects caused by the propagation system are called **C-effects** and can upwardly bias heritabilities estimated from vegetatively propagated offspring if some clones perform better than others due to their favorable response to the propagation system (Libby and Jund, 1962; Burdon and Shelbourne, 1974; Foster *et al.*, 1984) (see *Chapter 16*).

There are other types of heritabilities beyond the two types introduced in this chapter (Hanson, 1963; Hodge and White, 1992a; Falconer and Mackay, 1996). The broad-sense and narrow-sense heritabilities discussed here are sometimes called individual tree heritabilities because the individual tree is the unit of measurement, analysis and selection. These individual tree heritabilities are the most useful when attempting to quantify the underlying genetic control of a polygenic trait, and should not be confused with family heritabilities, plot-mean heritabilities or clonal repeatabilities that have other uses and interpretations. These latter types of heritability are most useful in specific applications of selection in tree breeding programs, and are not discussed further in this book.

Estimates of Heritabilities for Forest Trees

In forest tree populations, heritability estimates have been made for many traits such as those reflecting physiological processes, phenology of annual growth and flowering, flowering precocity, and stem growth and wood quality. In an extremely thorough review of estimates of narrow-sense heritability (h^2), Cornelius (1994) reviewed 67 published papers containing more than 500 estimates of h^2 for different species (conifers and hardwoods), traits and ages. Several conclusions can be drawn: (1) For any trait, the estimates range widely depending primarily on the size of the experiment and homogeneity of the experimental environments; (2) The median estimate of h^2 is between 0.19 and 0.26 for most traits of commercial interest except wood specific gravity (these traits include bole height, diameter, and volume, stem straightness, and some branching characteristics); (3) The median h^2 estimate for wood specific gravity is 0.48 indicating that this trait is under significantly stronger genetic control than traits in the former group; and (4) No trends in h^2 estimates due to species groups or age of measurement were noted by the author. There have also been summaries for several species of *Eucalyptus* (Eldridge *et al.*, 1994, p. 175) and for *Pinus radiata* (Cotterill and Dean, 1990, p. 19); these are in general agreement with the above conclusions.

It is important to note that many of the published estimates are based on a single test location, which means they are upwardly biased (see later section on *Genotype x Environment Interaction*). However, it is safe to conclude that narrow-sense heritabilities for many stem growth and form traits range from 0.10 to 0.30, while for wood specific gravity they range from 0.3 to 0.6 (for details on the genetics of wood properties, see Zobel and Jett, 1995).

Many authors have compared narrow-sense heritability estimates for growth rates at different ages in the same species, and some have developed models to explain changes with age in terms of the different stages of stand development (Namkoong *et al.*, 1972; Namkoong and Conkle, 1976; Franklin, 1979). It now seems that heritabilities can change or can stay relatively constant with tree age (Cotterill and Dean, 1988; King and Burdon, 1991; Balocchi *et al.*, 1993; St. Clair, 1994; Johnson *et al.*, 1997; Osorio, 1999), and there is not a single model of stand development that explains age trends across a wide range of species (*e.g.* see comparisons in Hodge and White, 1992b; Dieters *et al.*, 1995).

There are far fewer estimates of broad-sense heritabilities owing to the fact that a relatively small number of tree species are planted commercially as vegetative propagules (Shelbourne and Thulin, 1974; Foster *et al.*, 1984; Foster, 1990; Borralho *et al.*, 1992a; Farmer *et al.*, 1993; Lambeth and Lopez, 1994; Osorio, 1999). There are even fewer studies in which both narrow-sense and broad-sense heritabilities have been estimated for the same population and tree species (Table 6.1). These studies are particularly informative, because the ratio h^2/H^2 measures σ^2_A/σ^2_G when both heritability estimates are based on the same set of genotypes grown in the same set of test environments. Values of h^2/H^2 near one imply that the amount of non-additive variance is very small and that clonal values are similar to breeding values. Values near one also mean that the advantages of clonal forestry are less compared to use of seedlings (this is developed further in *Chapter 16*).

Several observations come from comparing h^2 and H^2 for growth rates in Table 6.1: (1) For both types of heritabilities, there is a wide range in estimates among experiments as is commonly encountered; (2) Estimates of broad-sense heritabilities are larger than those for narrow-sense values in all experiments; and (3) The ratio h^2/H^2 ranges from 0.18 to 0.84 with a mean of 0.49. C-effects can sometimes be large and inflate H^2 values when clonal differences in response to the propagation system are confounded with genetic differences among clones. With this in mind, tentative conclusions for growth traits indicate that non-additive genetic variance is appreciable and may be approximately as large as additive variance, meaning that a true ratio of h^2/H^2 of 0.50 would not be an unreasonable assumption in the absence of firm experimental data. The amounts of non-additive variance, impacts of C-effects and the ratio of h^2/H^2 are areas in need of more research in quantitative genetics.

Uses and Importance of Heritability Estimates in Forest Tree Populations

Heritability estimates are important for understanding the genetic structure of both natural populations of forest trees and breeding populations in tree improvement programs. In natural populations microevolution is a process that shapes genetic structure, and causes populations to differentiate genetically one from another in different environments. Only if a trait displays variation, is heritable (*i.e.* $h^2 > 0$) and is under differential natural selection will large populations diverge genetically for that trait (Lande, 1988a; Barton and Turelli, 1989). Therefore, phenotypic and additive genetic variances along with their ratio, the heritability, are useful measures for evaluating whether a trait has potential to differentiate in populations exposed to differential natural selection pressures in distinct environments (*Chapters 5, 8* and *9*).

Tree species can also be greatly affected by bottlenecks, which are drastic reductions in population sizes caused by ice ages or other events. These bottlenecks cause genetic drift in allele frequencies (*Chapter 5*). If a trait has a large amount of non-additive variance, there is some evidence that the trait will be more responsive to selection following a bottleneck (Goodnight, 1988, 1995; Bryant and Meffert, 1996). While beyond the scope of this book, this field of evolutionary quantitative genetics has received increasing emphasis in recent years (Roff, 1997).

Another implication of heritabilities in natural populations relates to Fisher's (1930) Fundamental Theorem of Natural Selection, which hypothesizes that any trait having a direct and important influence on the organism's overall fitness should display lower genetic variance and therefore, lower heritability (Falconer and Mackay, 1996, p. 339). The theorem rests on the notion that traits closely related to fitness should be under extreme selection,

Table 6.1. Narrow-sense (h^2) and broad-sense (H^2) heritabilities estimated for growth traits from the same experiments of six conifer species. The traits are DBH = diameter at breast height, HT = height and VOL = bole volume. The number of families (Fam) and clones in each experiment are shown.

Species	Fam, clone (#, #)	Trait, age (yrs)	Narrow-sense h^2	Broad-sense H^2	Ratio of h^2/H^2	Reference
Picea mariana	40,240	HT, 5	0.08	0.11	0.73	Mullin and
		HT, 10	0.05	0.13	0.38	Park, 1994
Pinus radiata	16,160	DBH, 7	0.08	0.32	0.25	Carson *et al.*,
		DBH, 10	0.10	0.35	0.29	1991
Pinus radiata	60,120	DBH, 8	0.16	0.38	0.42	Burdon and
		HT, 8	0.20	0.30	0.67	Bannister, 1992
Pinus taeda	30,514	DBH, 5	0.10	0.13	0.76	Paul *et al.*,
		HT, 5	0.21	0.25	0.84	1997
Populus del-toides	32,252	HT, 4	0.16	0.32	0.50	Foster, 1985
		VOL, 4	0.07	0.39	0.18	
Pseudotsuga menziesii	60,240	DBH, 6	0.15	0.38	0.39	Stonecypher
		HT, 6	0.19	0.34	0.56	and McCul-lough, 1986

and that eventually (over many generations of natural selection) selection pressure should lead towards allele fixation at many of the loci affecting the trait (Barton and Turelli, 1989). Conversely, traits only distantly related to fitness are hypothesized to have higher heritabilities because natural selection is not acting strongly to reduce existing genetic variability. There have been no empirical studies of this hypothesis in forest tree species; however, a thorough survey of 1120 estimates from 75 animal species (Mousseau and Roff, 1987) and an intensive analysis of the many heritability estimates in the fruit fly (*Drosophila* spp.; Roff and Mousseau, 1987) support Fisher's hypothesis. There are theoretical explanations other than Fisher's theorem for the relationship between fitness and heritability of a trait (Roff, 1997, p. 64); however, the observed trend suggests that traits linked directly to fitness have lower heritabilities in a range of species. However, this is only a general association and there are certainly exceptions.

There are many uses of genetic variances and heritabilities in tree improvement programs, and these are discussed in some depth in *Chapters 13-16*. In this chapter, we only highlight these uses to provide perspective on the nature of heritability as a concept: (1) All other things being equal, traits with higher heritabilities should receive more emphasis in tree improvement programs because the potential for genetic progress is higher (traits with low heritabilities do not respond as well to selection because phenotypic measurements provide little information about underlying genetic values); (2) The magnitudes of heritabilities of the most important traits in tree improvement programs help determine the amount and type of field experiments and selection that are appropriate (*i.e.* traits with low heritabilities require more offspring spread over more field locations or replications to adequately estimate the breeding values of the parents in the breeding population); (3) Estimates of genetic variances and heritability are used to predict the amount of genetic gain expected for each trait in a tree improvement program (*Chapter 13*); (4) Further, these estimates are used in mixed model methods of data analysis to both predict underlying breeding and clonal values from experimental data and to create selection indices that aim to maximize genetic gain from selection (*Chapter 15*); and (5) A small ratio of h^2/H^2 indicates that there is substantial non-additive genetic variance and that extra genetic gain can be expected from a clonal forestry program in which commercial plantations are established with vegetative propagules of tested clones (*Chapter 16*).

GENETIC CORRELATIONS

Definitions and Concepts

When two different traits are measured in a population of forest trees, there may be an association or correlation between the measurements. For example, suppose that both bole height and diameter are measured on the 1000 trees in the hypothetical stand shown in Fig. 6.1a. We might expect a tendency for trees above average for height also to be above average for diameter. This would imply a phenotypic association between the two traits, which is quantified by the **phenotypic correlation**, r_p, which measures the strength of the association.

We do not present the formula for calculating r_p (Falconer and Mackay, 1996; Bourdon, 1997), but rather highlight the nature of phenotypic correlations: (1) Like all correlations, phenotypic correlations are standardized, unitless measures of association between two traits; (2) Phenotypic correlations range within the theoretical limit from negative one to positive one; (3) A positive correlation indicates the tendency for a tree that is above average for one trait to also be above average for the second trait, and similarly for trees below average; (4) A negative correlation implies that a tree that is above (or below) average for the first trait is likely on the opposite side of the mean for the second trait; (5) A correlation near zero indicates little correspondence between the measurements of the two traits; and (6) When $r_p = 1$ or -1, there is a perfect linear association between the two traits meaning that knowing one trait's value allows complete prediction of the other trait.

An observed phenotypic correlation between two polygenic traits may be due to genetic or environmental causes, and just as with phenotypic variances, there is a need to understand these underlying components that give rise to a phenotypic correlation. The main cause of a **genetic correlation** between two traits is **pleiotropy**, which means that a given gene locus influences the expression of more than one trait (Mode and Robinson, 1959). Consider two polygenic traits that are both influenced by many gene loci (Fig. 6.4). If many of these gene loci are pleiotropic and influence the expression of both traits, then there is a measurable phenotypic correlation between the traits caused by the underlying genetic correlation.

Just as with heritabilities, we can define a broad-sense genetic correlation (r_G) and a narrow-sense genetic correlation (r_A). Intuitively, these are best understood by assuming that we know the true clonal values and breeding values for all individuals in the population of trees (of course, we never know these true underlying genetic values, but rather have to estimate them with field experiments). The **broad-sense genetic correlation** is the correlation of true clonal values for the two traits X and Y: $r_G = \mathrm{Corr}(G_{xi}, G_{yi})$ where G_{xi} and G_{yi} are the true clonal values for traits X and Y for all trees, $i = 1, 2, \dots N$, in the population. The **narrow-sense genetic correlation** is the correlation of true breeding values for the two traits X and Y: $r_A = \mathrm{Corr}(A_{xi}, A_{yi})$ where A_{xi} and A_{yi} are the true breeding values for traits X and Y. These correlations measure, respectively, the association of clonal values (r_G) and breeding values

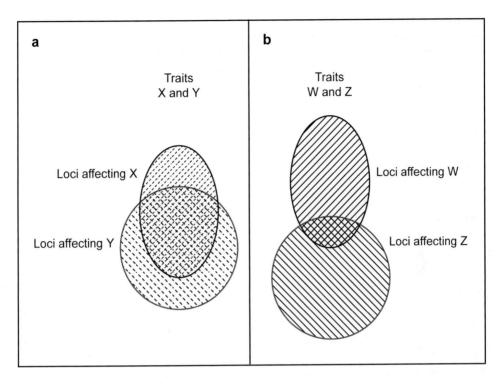

Fig. 6.4. Schematic diagram of pleiotropy for two different pairs of traits: (a) Traits X and Y share many pleiotropic gene loci that influence both traits which results in a strong genetic correlation (which could be negative or positive); (b) Traits W and Z share few pleiotropic gene loci which results in a weak genetic correlation.

(r_A) for the two traits in the reference population. Therefore, a value of r_A near negative one indicates a very strong tendency for a tree with an above-average breeding value for one trait to have a below-average breeding value for the second trait.

An observed phenotypic correlation between two traits may also be caused by an underlying **environmental correlation**, r_E, if environmental effects that influence one trait also influence the second trait. For example, consider a hypothetical conifer stand in which favorable microsites in the stand (*e.g.* locations with moister, deeper soils) lead to faster spring stem diameter growth. That is, the amount of spring wood is increased on favorable sites, but growth later in the season is unaffected (and therefore, the amount of late wood is similar on all microsites). Trees on these favorable microsites tend to have faster diameter growth and lower wood specific gravity, since they have a larger fraction of spring wood, which has lower specific gravity than late wood. In this case, there is a negative environmental correlation between diameter growth and wood specific gravity (faster diameter growth associated with lower specific gravity). This does not imply any relationship between the gene loci affecting the two traits, and there may or may not be a genetic correlation.

In all facets of quantitative genetics, the challenge is to unravel directly observable phenotypic relationships into their underlying causes. The phenotypic correlation between two traits is a complex function of the genetic and environmental correlations and the heritabilities of the two traits (Hohenboken, 1985; Falconer and Mackay, 1996). Therefore, it is not possible to deduce the size or the sign of the underlying genetic and environmental correlations from a phenotypic correlation. This has two implications: (1) The phenotypic correlation may have little utility in making genetic inferences (see below and *Chapter 13*), so we must estimate the genetic correlation; and (2) Randomized, replicated field experiments employing offspring from many parents are required to estimate the underlying causal correlations (*i.e.* the genetic and environmental correlations). With regard to the latter point, genetic correlations are even more difficult to estimate precisely than heritabilities, and require extremely large experiments.

Two other concepts about correlations warrant mention. First, the sign (positive or negative) is somewhat arbitrary and reflects the scale of measurement. For either trait, if the measurements are multiplied by negative one, the sign of the correlation is reversed. Therefore, it is important to know the exact scale of both traits, and it is sometimes preferable to state the correlation as favorable or unfavorable, rather than positive or negative. Second, all types of correlations express tendencies at a population level and do not apply to a given individual tree. Therefore, even with a negative correlation between two traits, it is possible to find individuals that are above average for both traits (these trees are called **correlation breakers**).

There are three different types of genetic correlations that are of interest in forest trees: (1) The genetic correlation between two distinct traits, such as between growth and wood specific gravity; (2) The genetic correlation between the same trait expressed at two different ages, called **age-age correlations** or **juvenile-mature correlations**; and (3) The genetic correlation of the same trait expressed in two distinct macroenvironments. Types 1 and 2 are discussed below, while type 3 is considered subsequently in the section on *Genotype x Environment Interaction*.

Trait-trait Correlations

Genetic correlations have been estimated for many pairs of traits in a wide variety of animals (Falconer and Mackay, 1996; Bourdon, 1997), plants (Hallauer and Miranda, 1981; Bos and Caligari, 1995) and trees (Cotterill and Zed, 1980; Dean *et al.*, 1983; Van Wyk, 1985a; Dean *et al.*, 1986; King and Wilcox, 1988; Volker *et al.*, 1990; Woolaston *et al.*, 1990). Almost all of these estimates are for narrow-sense genetic correlations (r_A), which are nearly always referred to as simply genetic correlations. As clonal programs become more common, more estimates of broad-sense genetic correlations will appear in the literature (*e.g.* Osorio, 1999), and it will be important to distinguish more clearly between the two types.

One consistent conclusion is that different measures of body size in animals and plant size in crops and trees have strong positive genetic correlations. For example in trees, bole diameter, height and volume are strongly inter-correlated with r_A ranging from 0.70 to nearly one in different studies. This is logical since these different traits are functionally related. The genetic correlation between wood specific gravity and stem growth rate has been the subject of many studies summarized by Zobel and Jett (1995). They conclude that for many species (especially most conifers and species of *Eucalyptus*), there is little association between the two traits; however, there are many reports of both positive and negative correlations. In general, it is difficult to draw conclusions about any pair of traits simply by looking at the literature, and therefore, trait-trait correlations must normally be estimated for the traits and species of interest.

Knowledge of genetic correlations among pairs of traits can be important both for the study of natural populations of forest trees as well as in tree improvement programs. If two traits have a strong positive or negative correlation, then either natural or artificial selection on the first trait causes genetic change in the second. This is called a **correlated response to selection** or **indirect selection** (*Chapter 13*). Consider the narrow-sense genetic correlation, and remember from earlier in this chapter that only additive allele effects are passed on to offspring following random mating. With a strong positive genetic correlation (r_A) between two traits, a parent with a high breeding value for the first trait tends to have a high breeding value for the second trait, and this parent produces offspring superior for both traits.

In tree improvement programs, genetic correlations between two traits can be important for several reasons: (1) If two traits have a strong favorable correlation, then selection and breeding on the first trait results in genetic gain for both traits; (2) If two traits have a strong unfavorable correlation, then it is more difficult to make progress in both traits simultaneously; and (3) If genetic correlations are unknown, it is possible for a selection program to produce unexpected results. The last example is termed **inadvertent selection response** in which a trait that is not measured or included in the tree improvement program may still change (favorably or unfavorably) due to its genetic correlation with a selected trait.

Age-age Correlations

It is rare that tree breeders wait until rotation age to make selections of the best trees to include in a tree improvement program. Consider bole volume growth in which selection is done at a young age, while the goal is to improve volume production at rotation age. It is apparent that these may be two different polygenic traits (growth at younger ages *versus* growth at older ages). If nearly the same set of gene loci influence volume growth at all ages, then the genetic correlation between the two traits is high. If a substantial fraction of loci important at younger ages is less influential or new ones become important at older ages (say when the trees dominate the site and begin to flower), then the genetic correlation between the two ages is smaller. Clearly, the success of selection at younger ages in producing gain in volume at rotation age depends at least in part on the age-age correlation of the two ages in question (see further discussion in *Chapter 13*).

Because of the importance of age-age correlations in tree improvement programs, they have been widely estimated for forest tree species. Lambeth (1980) used published estimates of phenotypic correlations of tree heights at several ages from several conifer species to develop the following empirical relationship:

$$r_P = 1.02 + 0.308(LAR) \hspace{4cm} \text{Equation 6.13}$$

where r_P is an estimate of the phenotypic correlation and LAR is the natural logarithm of the ratio of the younger age to the older age. For example, to estimate the phenotypic correlation between ages 5 and 20, $r_P = 1.02 + 0.308*\ln(5/20) = 0.593$. Note that this equation does not depend on the younger or older age, but only their ratio. Therefore, the predicted correlation is 0.593 anytime the younger age is 25% of the older age. The estimates for age ratios of 0.50 and 0.75 (50% and 75% of final age) are 0.81 and 0.93, respectively.

If, as Lambeth (1980) suggested, age-age correlations for growth traits can be roughly approximated by age-age phenotypic correlations, then the empirical relationship for height in Equation 6.13 could be useful for estimating genetic responses at later ages from selection at younger ages. For this reason, many investigators have compared their estimated genetic correlations for height, diameter and volume growth in several species with the predictions from Equation 6.13. Several observations have been made: (1) The age-age genetic correlation for any growth variable (bole height, diameter or volume) is most often slightly larger (by 0.05 to 0.20) than the phenotypic correlation predicted by Equation 6.13, and the difference is larger when the ages are more different (Lambeth *et al.*, 1984; Cotterill and Dean, 1988; Riemenschneider, 1988; Hodge and White, 1992b; Dieters *et al.*, 1995); (2) The genetic correlation is not always linear with LAR or may change across edaphoclimatic regions, thus a more complex model form may be required for some species (King and Burdon, 1991; Matheson *et al.*, 1994; Johnson *et al.*, 1997); (3) Diameter at breast height (DBH) and bole volume may need different equations than height, because DBH and volume are more sensitive to spacing and competition differences that become more pronounced as stands age (Lambeth *et al.*, 1983a; King and Burdon, 1991; Johnson *et al.*, 1997); (4) At one or two years of age, smaller correlations with older ages are more often observed than those predicted by Equation 6.13 (Lambeth, 1980; Riemenschneider, 1988; Matheson *et al.*, 1994); and (5) Whether genetic or phenotypic correlations are estimated with Equation 6.13, the approach has been questioned on theoretical grounds, because stem size at the older age is a function of stem size at the younger age which leads to an autocorrelation between the ages (Riemenschneider, 1988; Kang, 1991).

Based on these observations, it is always theoretically desirable to estimate age-age genetic correlations separately for each trait, species, age combination and set of environments. However, this requires very large experiments (hundreds of families in several environments measured over many years). In the absence of these experiments, Equation 6.13 provides conservative estimates of age-age genetic correlations for stem growth traits.

Investigations of wood specific gravity have shown higher age-age genetic correlations than stem growth variables for the same ages (Loo *et al.*, 1984; Burdon and Bannister, 1992; Vargas-Hernandez and Adams, 1992). For example in *Pinus taeda* (Loo *et al.*, 1984), the genetic correlation for wood specific gravity between ages 2 and 22 was estimated as 0.73, whereas Equation 6.13 predicts 0.28 for growth traits at these very different ages.

GENOTYPE X ENVIRONMENT INTERACTION

Definitions and Concepts

The essence of genotype x environment interaction (hereafter called G x E interaction) is a lack of consistency in the relative performance of genotypes when they are grown in different environments. This may mean that the relative rankings of genotypes change in the different environments (called **rank change interaction**) or that, even in the absence of rank changes, differences in performance are not constant in all environments (called **scale effect interaction**). An example of scale effect interaction is when bole volume growth is greater in one species compared to another, but the difference is large when the species are planted in one soil type, but small in another soil type. A real example of interaction involving change of rank is shown in Fig. 6.5 in which two southern pine species change rank for early height growth when grown in two different silvicultural management intensities (Colbert *et al.*, 1990).

When discussing G x E interaction, the terms genotype and environment are used very broadly. The genotypes compared could be different species or different seed sources, provenances, families or clones of the same species. The planting environments can be different soil types, elevations, climates, fertilizer treatments, planting densities or any combination of these or other environmental or silvicultural factors.

Statistically, G x E interaction is a two-way interaction in an analysis of variance that occurs when true differences between genotypes for any trait are not constant in all environments in which they are planted. This means that the linear model defined in Equation 6.3 is no longer adequate to explain the observed variation and must be amended to include another term (GE) to help explain the interaction:

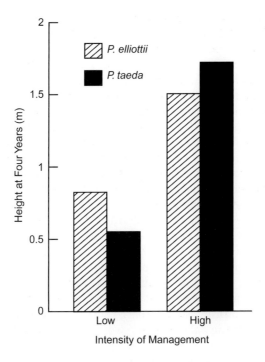

Fig. 6.5. Height after 4 years for *Pinus elliottii* and *Pinus taeda* planted in well-replicated experiments on the same site with two different management intensities (low intensity with no fertilizer or weed control, and high intensity of management with both fertilizer and weed control). Statistical tests indicated a significant species x treatment interaction manifested by the change in ranks of the two species across treatments. (Copyright 1990, Society of American Foresters. Reprinted with permission from Colbert *et al.*, 1990)

$$P_{ij} = \mu + E_i + G_j + GE_{ij} \hspace{4cm} \text{Equation 6.14}$$

where P_{ij} is the true phenotype of the jth genotype when grown in the ith macroenvironment, μ is the population mean of all genotypes averaged over all environments, E_i is the mean effect of a given macroenvironment, G_j is the true mean of genotype j averaged over all macroenvironments and GE_{ij} is the interaction term that measures how much better or poorer genotype j performs in environment i than is expected based on its average performance across all environments. Equation 6.14 is a conceptual model, and for a real experiment the linear model includes terms for experimental design factors (such as blocks) and an error term (*Chapter 15*). Equation 6.14 implies that the difference between any pair of genotypes j and j' in a given environment i depends not only on the average values of the two genotypes across all environments (G_j and $G_{j'}$), but also on the interactions of the two genotypes with that specific environment:

$$P_{ij} - P_{ij'} = (G_j + GE_{ij}) - (G_{j'} + GE_{ij'}) \hspace{3cm} \text{Equation 6.15}$$

where $P_{ij} - P_{ij'}$ is the difference in true phenotypic performance for a given trait for genotypes j and j', $G_j + GE_{ij}$ is the performance of genotype j in environment i and $G_{j'} + GE_{ij'}$ is the performance of genotype j' in environment i.

If the GE terms in Equation 6.15 are zero for all genotypes in all environments, then the difference between any two genotypes in any single environment depends only on the genotypic values averaged across all environments (*i.e.* $P_{ij} - P_{ij'} = G_j - G_{j'}$); hence, the difference between any pair of genotypes is constant across all environments, and this means that true phenotypic performances would form parallel lines when plotted as in Fig. 6.6a. Of course in a real experiment, the genotype lines would never be exactly parallel due to experimental error ignored in Equations 6.14 and 6.15. Therefore, statistical tests are performed to determine whether G x E interaction is significant, and declaring G x E as non-significant is equivalent to both of the following: (1) All of the GE terms in Equation 6.14 are zero (or more properly cannot be declared significantly different than zero); and (2) The genotype lines in a plot such as Fig. 6.6a are parallel (cannot be declared nonparallel).

When G x E is declared statistically significant, this implies that at least some of the GE terms in Equation 6.14 are not zero. Each genotype may have a specific positive or negative reaction to a given environment that makes it perform better (a positive GE term for that genotype in that environment) or worse than predicted by its overall genotypic value averaged across all environments (G_j). In the case of significant G x E interaction: (1) The difference in performance between any two genotypes in any single environment depends on the size and sign of the two interaction terms (*e.g.* the difference in heights between *Pinus elliottii* and *P. taeda* in Fig. 6.5 is 0.27 m in the low intensity environment, but −0.15 m in the high intensity environment); (2) Differences between any pair of genotypes can vary widely among environments; and (3) Plots like Fig. 6.6 can take on any form that depends on the exact set of GE terms encountered. There are several related concepts of G x E interaction discussed in Box 6.5, and three of the many methods of quantifying and studying G x E interactions are outlined in Box 6.6.

Importance of G x E Interaction in Forest Trees

Genotype x environment interaction occurs at many genetic levels (species, provenance, family and clone) and we address these briefly in turn. Many species are planted over a wide range of climatic and edaphic conditions (*e.g. Pseudotsuga menziesii* in northwest portions of USA and Canada; *Pinus taeda* in southeast portions of USA; *P. sylvestris* and *P. contorta* in Europe; *P. radiata* in temperate regions of the southern hemisphere; and *E. grandis* in many tropical and subtropical regions). The fact that a single species is favored over such a large area implies that **species x environment interaction** is minimal within this area (*i.e.* that the chosen species is best on a wide range of sites within this area). With species such as these, strong species x environment interaction is found only when there is a large change in climatic or edaphic conditions. For example, when *E. grandis* is moved to colder areas in the subtropics, it suffers and other species perform better. Likewise, when *P. radiata* is planted in subtropical regions with summer rainfall, it can suffer disease problems, and other species are more appropriate.

While the species cited above have wide amplitudes in terms of the areas where they grow and perform well, other species have narrower limits. For example, *Platanus occidentalis* and several species of *Eucalyptus* are known to be very site-specific with respect to soil quality. In these situations, species can change rank across sites or management intensities (*e.g.* Khasa *et al.*, 1995; Butterfield, 1996). Therefore, when the limits and performance of potentially useful species are not well understood, they must be tested in all edaphoclimatic zones in the area to be planted (Saville and Evans, 1986; Evans, 1992b) (*Chapter 12*).

In a very thorough review of **provenance x environment interaction**, Matheson and Raymond (1986) conclude that strong rank changes in provenance performance are

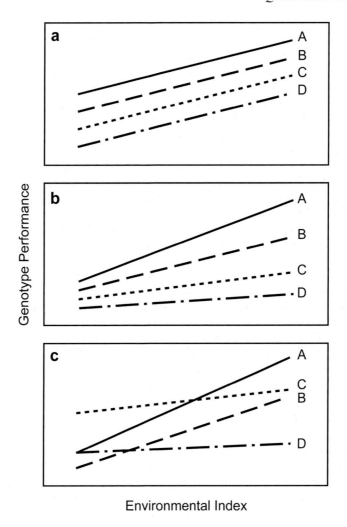

Fig. 6.6. Schematic diagrams of the performance of four genotypes (A, B, C, and D) plotted against an environmental index calculated as the average of the four genotype performances in each environment: (a) No genotype x environment interaction; (b) Interaction due to scale effects; and (c) Interaction due to rank changes.

normally minimal unless associated with large changes in climate or soils (similar to many species x environment interactions). For example, *Pinus contorta* is native to the USA and Canada, and widely planted in Europe. The provenances originating from coastal portions of Oregon, Washington and British Columbia are preferred in the more maritime climates of Ireland, Britain and western Norway, and appear to be widely adapted to planting sites throughout this broad region. On the other hand, provenances from the colder, interior portions of British Columbia perform better in the colder continental climates of northern Scandinavia. In some species, provenances may be strongly adapted to local edaphic or climatic conditions (Campbell, 1979; Adams and Campbell, 1981). In these cases, there may or may not be strong rank changes among provenances planted at different locations within the region, and testing of provenances on many sites is required.

Box 6.5. Concepts of G x E interactions.

1. In data sets from real experiments, differences in genotype performance are never exactly constant across all environments. Analysis of variance techniques (ANOVA as described in many textbooks on statistics) are used to determine if the GE terms are statistically significant or more likely due to experimental noise. An ANOVA must be conducted for each trait that is measured, and G x E interaction may be important for one trait and not important for another.

2. Even when G x E interaction is statistically significant for a trait, it may or may not be biologically important. That is, the statistical significance may be due to relatively small interaction terms (GE terms) as compared to the average genetic values (G terms in Equations 6.14 and 6.15). In this case, genotype differences are still nearly constant across environments meaning that a genotype that is good in one environment is still good in others.

3. If G x E interaction is both statistically and biologically important, then the relative superiority of a genotype clearly depends on the environment in which it is grown. Then, we must test genotypes across all environments of interest in order to estimate their relative performances (rankings) in each environment.

4. When experiments are planted in a single test environment, G and GE are completely confounded and inseparable. Therefore, the variance among genotypes tested in a single site ($\sigma^2_G + \sigma^2_{GE}$) is larger than if estimated from genotypes tested in many environments (σ^2_G). This means that narrow-sense and broad-sense heritabilities estimated from experiments on a single site are biased upward (too large), due to the confounded presence of both genetic variance and G x E variance in the numerators of Equations 6.11 and 6.12.

5. The presence of G x E interaction for a specific trait implies that different sets of gene loci are operating in different environments. In this sense, we can think of the single trait measured in two different environments as being two different, but genetically correlated traits (Burdon, 1977).

Large changes in environmental conditions can often elicit distinct G x E interactions; however, interactions of families and clones with their environment can also be associated with more subtle changes in the environment. When many families are planted across several sites within a region and G x E is statistically significant for growth rate: (1) Most often the G x E interactions are due mainly to scale effects (Fig. 6.6b) as opposed to dramatic rank changes (Li and McKeand, 1989; Adams *et al.*, 1994; Dieters *et al.*, 1995; Johnson, 1997); (2) If there are rank changes among the families in different locations, these **family x environment interactions** may be caused by a small fraction of the families being tested (Matheson and Raymond, 1984a; Li and McKeand, 1989; Dvorak, 1996; Stonecypher *et al.*, 1996); and (3) Sometimes one or two site locations can be very different and cause much of the observed interaction (Woolaston *et al.*, 1991).

Box 6.6. Three methods of quantifying and understanding G x E interactions.

1. *Contributions of individual genotypes and environments.* It is often useful to see whether certain genotypes or environments are causing most of the interaction by calculating the contribution of each genotype and each environment to the interaction sum of squares in an analysis of variance (Shukla, 1972; Fernandez, 1991). If one or a few genotypes or environments cause most of the interaction, then a biological explanation may be sought. For example, perhaps the coldest test site is causing 60% of the interaction in an experiment with 20 test locations (each location contributes 5% of the interaction if all contribute equally). This leads to the biological inference that differential cold tolerance is a plausible explanation for the interaction. Perhaps, more cold-resistant genotypes perform better in the colder environment causing a substantial change in ranking as compared to warmer sites.

2. *Stability analysis.* In stability analysis (Finlay and Wilkinson, 1963; Mandel, 1971; White *et al.*, 1981; Li and McKeand, 1989), the mean of all genotypes in each environment is calculated as an index of the environmental quality for that environment; a separate regression line then is developed for each genotype using the genotype's performance as the Y variable and the environmental index as the X variable (Fig. 6.6). If there is no G x E interaction, regression lines for different genotypes have slopes that are not significantly different from each other (*i.e.* lines are nearly parallel reflecting nearly constant differences among all pairs of genotypes in all environments, as shown in Fig. 6.6a).

   Stability analysis can clearly highlight the two different types of interactions (scale effects and rank changes shown in Fig. 6.6b and c, respectively). When scale effects are the predominant form of G x E interaction, genotypes are more divergent in some environments, but rank changes are not important. This could mean that some genotypes take better advantage of the increasing environmental quality and thus, increase their relative superiority (*e.g.* Li and McKeand, 1989). On the other hand, rank changes are a more serious type of interaction in which decisions about which genotype is best change in different environments.

3. *Type B genetic correlations.* This approach assumes that a single trait, such as bole volume growth, is expressed genetically as a different trait in each different environment. Then, the genetic correlation is estimated for this trait between each pair of environments (Dickerson, 1962; Yamada, 1962; Burdon, 1977, 1991). When the correlation is near one for a given pair of environments, the relative expression of the trait is nearly the same among genotypes in both environments, and the interaction is small and biologically not important. When the genetic correlation for a pair of environments is small, this indicates that genotypes do not rank the same in both environments, and that trait expression is influenced by different (non-overlapping) sets of gene loci in the two different environments. In this case, a genotype that excels in one environment because of specific favorable alleles at certain loci does not excel in another environment if these loci are less influential in trait expression.

Type B genetic correlations (defined in Box 6.6) are commonly used to quantify family and clone by site interaction in tree breeding programs (Johnson and Burdon, 1990; Hodge and White, 1992b; Dieters *et al.*, 1995; Haapanen, 1996; Johnson, 1997; Osorio, 1999; Atwood, 2000; Sierra-Lucero *et al.*, 2002). A particularly good example that illustrates the use of the type B genetic correlation (Box 6.6) as a measure of G x E interaction comes from CAMCORE (Central America and Mexico Coniferous Resources Cooperative), which collected open-pollinated seed from many trees of each of three pines native to Mexico and Central America: *P. caribaea*, *P. patula* and *P. tecunumannii*. The open-pollinated families were then planted in exotic environments in many countries in South America, southern Africa and Indonesia, with volume growth measured at age 5 years (CAMCORE, 1996, 1997). Consistently for all three species, the type B genetic correlations for families within provenances were much higher for pairs of planting sites within the same country (averaging 0.67, 0.60 and 0.66 for the three species, respectively) and much smaller for pairs of planting environments located in different countries (0.50, 0.39 and 0.33, respectively). These values indicate larger differences in family rankings for all three species when planted in two different countries compared to among sites within the same country.

There is some evidence that traits with higher heritabilities show less family x environment interaction. For example, family x planting location interaction for wood specific gravity seems small to nonexistent for several species (see the two examples in the next section of this chapter and Byram and Lowe, 1988; Zobel and van Buijtenen, 1989; Barnes *et al.*, 1992).

The practical importance and implications of family and clone interactions with environments are discussed in *Chapters 12, 14* and *16*; however, there are three important concepts to retain: (1) It is nearly always best to test families and clones over the range of sites in which the winners will be planted; (2) Sometimes there are certain sites where genetic expression and therefore, discrimination among genotypes are better for certain traits (Carson, 1991; Magnussen and Yanchuk, 1994); (3) If genotypes are tested in only a single environment, heritabilities and estimates of genetic gains are upwardly biased as described earlier; and (4) When large rank change interactions are found, the tree improvement strategy must account for this (*e.g.* by dividing the planting zone into multiple breeding units or deployment zones, see *Chapter 12*).

ESTIMATING GENETIC PARAMETERS

Genetic parameters for any natural or breeding population must be estimated from randomized, replicated genetic tests containing vegetative propagules or sexually produced offspring from the population of interest. The steps involved in planning, implementing, analyzing and interpreting genetic tests are outlined in Box 6.7. Our intention here is only to illustrate some key concepts and hence, we do not provide enough information to allow readers to design, execute or analyze genetic experiments. Each of these steps is described in more detail in *Chapters 14* and *15*.

For this section, we chose two real experiments from the literature to demonstrate the principles involved in estimating heritabilities and type B genetic correlations (the latter measures genotype x environment interaction as described in Box 6.6). In these experiments, there were other parameters estimated, but we focus on these two for clarity and simplicity. The first experiment contains 65 clones of *E. grandis* planted as rooted cuttings on five sites representing the typical range of soils and climates on which this species is planted in Colombia (Osorio, 1999). The second experiment consists of 113 open-pollinated families of *P. taeda* measured on 13 sites in Florida, Alabama and Georgia in the lower coastal plain of the southeastern USA (Atwood, 2000). The subsections below contain a brief summary of each of the first six steps in Box 6.7 as applied to these two genetic experiments. Step seven (placing confidence intervals on parameter estimates) is addressed in *Chapter 15*. The results of the analyses are presented in Table 6.2.

Mating Design

For the clonal experiment, the mating design consists of 65 clones obtained from 65 unrelated, first-generation selections of *E. grandis*. The selections were made in plantations in Colombia as part of the first-generation tree improvement program of Smurfit Carton de Colombia (Lambeth and Lopez, 1994). These 65 clones are considered to be a random sample of the total of 400 selections. Therefore, all parameter estimates from this experiment apply to the first-generation breeding population of 400 selections.

For the second experiment, 113 open-pollinated (OP) families were obtained by collecting seed from 113 selections of *P. taeda*; these selections were part of the first-generation program to improve Florida provenances of this pine. Branch tips (*i.e.* scion material) from the selections were grafted into seed orchards (*Chapter 16*) and open-pollinated seed was collected from these seed orchards. Therefore, the maternal parents of the 113 OP families are genetically identical to the 113 first-generation selections, while the pollen came from both other clones in the seed orchards and from external sources outside the orchards. The selections were made only in Florida (a very small part of the natural range of *P. taeda*, see *Chapter 8*), so the parameter estimates apply strictly only to this provenance and to this breeding population.

Field Design

The field design for the clonal experiment of *E. grandis* at each test location was a randomized complete block design (RCB) with a variable number of blocks ranging from two to eight depending on the site. Within each block, each clone was represented by a single 6-tree row plot (meaning six ramets of the same clone planted in a row) and all clonal row plots were randomly assigned to plot locations within each block. There were a total of five test locations established within the main plantation zone of *E. grandis* in Colombia. At each site, the number of clones ranged from 29 to 65 with 27 clones common to all five sites. Nearly all 65 clones were established in at least three locations. The unbalanced nature of the experimental design (variable numbers of blocks and clones per site) was due to the differences among clones in the numbers of plantable ramets.

The field design for the experiment with 113 OP families of *P. taeda* consisted of a RCB with six blocks at each of 11 test locations and three blocks at two additional sites. Within each block, each OP family was represented by a single row plot ranging in size from six to 10 trees depending upon the site. The trees within one row plot had the same maternal parent and were thus related as OP siblings. The families were randomly assigned to plot locations within a block. The number of families at each location ranged from 32 to 72. Not all families were in all tests, but most families were in eight of the 13 locations. Again, the reason for the unbalanced nature of the experiment was that families produced different numbers of plantable seedlings in the nursery.

Study Implementation, Data Cleaning and Standardization

For both studies, field site locations were chosen to be representative of the plantations where the species are planted. The sites were carefully prepared to reduce environmental noise due to woody and herbaceous competition. Rooted cuttings for the clonal experiment and seedlings for the OP family experiment were grown in nurseries and

Box 6.7. Experimental procedures for estimating genetic parameters.

1. *Mating design.* All experiments to estimate genetic parameters (such as heritabilities and genetic correlations) consist of groups of genotypes, which can be different clones, half-sib families, full-sib families or other types of relatives. The mating design specifies how to form these genotypic groups (*Chapter 14*).

2. *Field design.* To obtain unbiased, precise estimates of genetic parameters, plants from the genetic groups must be planted in the field (or other location such as greenhouse) in a randomized, replicated field design that is statistically sound (see Box 14.3).

3. *Study execution.* Several procedures must be followed during the implementation of the study to ensure the accuracy and validity of the data obtained including careful planting, monumentation, maintenance and measurement of the plants.

4. *Data cleaning and standardization.* After measurements are obtained on the traits of interest, the data are edited and cleaned to ensure that all invalid measurements are removed (*e.g.* Fig. 15.1). Sometimes the data for a particular trait are standardized by dividing each tree's measurement by an appropriate standard deviation; this aims to equalize the variance across all environments to remove scale effects that can cause genotype x environment interaction.

5. *Data analysis.* A linear statistical model (*Chapter 15*) is specified declaring some effects as random and other as fixed; genetic groups must always be random to have a "variance." Then, the data for each measured trait are analyzed to partition the total phenotypic variability (*i.e.* the phenotypic variance) into variance components associated with all of the random effects in the model (*e.g.* variance among the genetic groups such as among half-sib families). Sometimes, covariance components are also estimated to partition the phenotypic covariance for two traits.

6. *Parameter estimation.* Genetic parameters are estimated by forming functions of the estimated variance components. This step requires that a genetic model be superimposed on the statistical model. For example, the narrow-sense heritability for each trait is estimated as a ratio of the additive genetic variance in the numerator and the phenotypic variance in the denominator. If the genetic groups are unrelated half-sib families, then the genetic model specifies that the variance among the families is one quarter of the additive variance. Thus, the numerator is estimated as four times the variance component for families.

7. *Confidence intervals.* Finally, various methods are used to place confidence intervals around the estimates of heritabilities and other parameter estimates (Becker, 1975; Bulmer, 1985; Searle *et al.*, 1992; Hallauer and Miranda, 1981; Huber, 1993). Unless the experiments are very large with hundreds of genetic groups (clones or families), the genetic parameters are poorly estimated (meaning wide confidence intervals).

well-labeled as to their genetic identity. The trees were planted on the field sites in the randomized, replicated experimental designs and all plots were properly labeled with block number and genetic identity. In addition, the corners of all blocks were monumented, and border rows of trees were planted around exposed (outside) edges of blocks to ensure that all experimental trees had equal competition. After planting, the sites were maintained with various applications of weed control.

Trees in the clonal experiment of *E. grandis* were measured at 3 years of age, which is half way to the harvest age (rotation) of 6 years. At this point, the trees averaged more than 15 m in height although this varied across the five sites. Bole volume for each tree was calculated from height and diameter measurements of that tree. Increment cores were extracted from each tree and the volumetric displacement method (American Society for Testing and Materials, 2000) was used to determine wood specific gravity. In total, more than 8000 trees were measured across all blocks planted on all five sites.

Trees in the OP family experiment of *P. taeda* were measured at 17 years of age, which is more than half way to the harvest age (rotation). At each site, only two trees in each 6- or 10-tree family row plot were measured and trees were measured in only five of six blocks. Bole volume for each tree was calculated from height and diameter measurements of that tree. Increment cores were extracted from each tree and the volumetric displacement method was used to determine wood specific gravity. In total, more than 3000 trees were measured across all blocks planted on all 13 sites.

For both experiments, the data were checked very carefully to remove recording errors. Then, data were analyzed individually by site to determine if any sites were extremely different in data quality or variance structure. Finally, data for both traits (bole volume and wood specific gravity) were standardized by dividing each tree's measurement by the phenotypic standard deviation from that block. These standardized data were used in all analyses that combined data across sites to ensure nearly equal phenotypic variance across the sites.

Data Analyses

The linear model for the clonal experiment of *E. grandis* had fixed effects for sites and blocks within sites, and these effects have no associated variance components since they are considered as fixed constants (*Chapter 15*). The model included random effects for clones, clones x site interaction, clone x block interaction and within-plot error; the variance components associated with these effects are σ^2_c, σ^2_{cs}, σ^2_{cb} and σ^2_e, respectively. The linear model for the OP experiment of *P. taeda* had fixed effects for sites and blocks within sites. Random effects were family, family x site interaction, family x block interaction and within-plot error; the variance components associated with these effects are σ^2_f, σ^2_{fs}, σ^2_{fb} and σ^2_e, respectively. The SAS software system (SAS Institute, 1988) was used to analyze the standardized data for bole volume and specific gravity in both experiments; variance components were estimated using a Restricted Maximum Likelihood Procedure (REML) in Proc Mixed of SAS.

To appreciate the meaning of each estimated variance component (Table 6.2), consider first the total phenotypic variance for each trait (σ^2_P), which is estimated as the sum of the four estimated variance components for that trait (*e.g.* $\sigma^2_P = \sigma^2_c + \sigma^2_{cs} + \sigma^2_{cb} + \sigma^2_e$ for the clonal experiment). Heuristically, the phenotypic variance for each trait is the variance among trees measured within each block of the experiment (using a formula similar to Equation 6.2) and then averaged over all blocks and all sites. This "within-block" phenotypic variance excludes the variability due to blocks and test sites having different mean values, because these effects were included in the model as fixed effects. The fact that the data were standardized prior to analysis means that the total phenotypic variance is nearly one (Table 6.2); if the variables had not been standardized, then the phenotypic variance would be expressed in units of measure squared (*e.g.* cubic meters squared for volume). With standardized variables, each variance component estimate can be viewed as a proportion of the total phenotypic variability, as demonstrated below.

For the clonal experiment, the following are interpretations of the four variance components: (1) The variance component due to clones, σ^2_c, measures the variance among true means of the 65 clones averaged over all sites as a proportion of the total phenotypic variance (*e.g.* for bole volume, the estimated variance among clonal means is approximately 15% of the total phenotypic variance, $100*0.1503/1.0107$ from Table 6.2); (2) The variance component due to clone x site interaction, σ^2_{cs}, measures the variability associated with the relative performance of clones or their ranking changing across the five sites (*e.g.* for bole volume, the estimate of this component is 0.1007 which is on the same order of magnitude as the previous clonal component indicating appreciable clone x site interaction); (3) The variance component due to clone x block interaction, σ^2_{cb}, measures the variability associated with clonal means changing their rankings across the various blocks within a site; and (4) The variance component due to within-plot error, σ^2_e, measures the variance among trees within a row plot, and since trees in the same 6-tree row plot are ramets of the same clone, σ^2_e measures the amount of microenvironmental variance among neighboring trees (*i.e.* there is no genetic variance among trees within a row plot since ramets of the same clone are identical genotypes).

For the OP family experiment, the interpretation of the four variance components is analogous to that just explained for the clonal experiment (substituting the word "OP family" for "clone" in the explanations above). For example, the variance component due to families, σ^2_f, measures the variance among the true means of the 113 OP families averaged across all 13 sites. Note that for both traits, the family component is only approximately one-fifth as large as the corresponding value in the clonal experiment; for bole volume, the variance among clones accounts for 15% of the total phenotypic variance, while the variance among OP families accounts for only 3% of the total phenotypic variance ($100*0.0307/1.0153$). For both traits, the residual component, σ^2_e, is considerably larger as a fraction of the phenotypic variance in the family compared to the clonal experiment. For bole volume and wood specific gravity, the variance among trees within the same row plot accounts for more than 80% of the phenotypic variance in the OP family experiment, whereas in the clonal experiment, these values are 62% and 15%, respectively. This is because neighboring trees in the same OP family are different genotypes such that this residual variance is a combination of both genetic and microenvironmental variability.

Parameter Estimation and Interpretation

Equations for Estimating Heritability and Type B Genetic Correlation

After the variance components are estimated using a computer program (such as Proc Mixed of SAS in these examples), the analyst must interpret the genetic meaning of these components and then use them to calculate genetic parameter estimates. This requires certain genetic assumptions about the reference population and the type of genotypes being tested. Detailed description of the assumptions is beyond the scope of this book (Falconer and Mackay, 1996); rather, we briefly outline the rationale used to estimate the two parameters in each experiment (Table 6.2): heritability (H^2 or h^2) and type B genetic correlation (r_{Bc} or r_{Bf}).

Table 6.2. Estimates of variance components, heritabilities and type B genetic correlations for bole volume and wood specific gravity from two experiments: (1) Data from 65 unrelated clones of *Eucalyptus grandis* at age 3, tested in 5 experimental sites representing the range of soils and climates in which this species is planted in Colombia (Osorio, 1999); and (2) Data from 113 unrelated open-pollinated (OP) families of *Pinus taeda* at age 17, tested in 13 sites representing the range of soils and climates in which this species is planted in the lower coastal plain of the southeastern USA (Atwood, 2000).

1: Clones of *E. grandis*			2: OP Families of *P. taeda*		
Parameter	Bole volume	Specific gravity	Parameter	Bole volume	Specific gravity
Variance Component Estimates[a]			Variance Component Estimates[b]		
σ^2_c	0.1503	0.5385	σ^2_f	0.0307	0.1110
σ^2_{cs}	0.1007	0.0558	σ^2_{fs}	0.0393	0.0117
σ^2_{cb}	0.1428	0.2777	σ^2_{fb}	0.1333	0.0197
σ^2_e	0.6169	0.1489	σ^2_e	0.8120	0.8586
Calculated parameters[c]			Calculated parameters[d]		
σ^2_P	1.0107	1.0209	σ^2_P	1.0153	1.0010
σ^2_G	0.1503	0.5385	σ^2_A	0.0921	0.2937
H^2	0.150	0.527	h^2	0.091	0.333
r_{Bc}	0.599	0.906	r_{Bf}	0.439	0.905

[a] For the clonal experiment, the four random terms in the statistical linear model are clones, clone x site interaction, clone x block within-site interaction and within-plot error; the corresponding variance components for these effects are: σ^2_c, σ^2_{cs}, σ^2_{cb} and σ^2_e, respectively. Both bole volume and wood specific gravity were standardized so that the total phenotypic variance, $\sigma^2_P \approx 1$.

[b] For the OP family experiment, the four random terms in the statistical linear model are families, family x site interaction, family x block within-site interaction and within-plot error; the corresponding variance components for these effects are: σ^2_f, σ^2_{fs}, σ^2_{fb} and σ^2_e, respectively. Bole volume and wood specific gravity were standardized so that the total phenotypic variance, $\sigma^2_P \approx 1$.

[c] For the clonal experiment, the total phenotypic variance (σ^2_P) is estimated as the sum of the four variance components; the total genetic variance is estimated as $\sigma^2_G = \sigma^2_c$; the broad-sense heritability is estimated as $H^2 = \sigma^2_G/\sigma^2_P$; and the type B genetic correlation is estimated as $r_{Bc} = \sigma^2_c/(\sigma^2_c + \sigma^2_{cs})$.

[d] For the OP family experiment, the total phenotypic variance (σ^2_P) is estimated as the sum of the four variance components; the additive genetic variance is estimated as $\sigma^2_A = 3*\sigma^2_f$; the narrow-sense heritability is estimated as $h^2 = \sigma^2_A/\sigma^2_P$; and the type B genetic correlation is estimated as $r_{Bf} = \sigma^2_f/(\sigma^2_f + \sigma^2_{fs})$.

In the clonal experiment of *E. grandis*, the fact that a random sample of 65 unrelated clones from the eucalypt breeding population is tested in the experiment means that the variance among clones (σ^2_c) can be interpreted as the total genetic variance in the first-generation breeding population (which is the reference population to which the parameter estimates apply). Each clone is a distinct genotype with its own clonal or genotypic value (*i.e.* its own G_i value from Equation 6.3), and σ^2_c is the variance among these clonal values. Therefore, the estimated variance component from the experiment estimates the total genetic variance, which is σ^2_G in Equation 6.5 and also the numerator of H^2 in Equation

6.11. If the sample of clones were related, then this would need to be taken into account.

As mentioned in the previous section, the total phenotypic variance is estimated as the sum of all four variance components, and the total phenotypic variance is the denominator of H^2 in Equation 6.11. Substituting the estimated variance components for the conceptual components, Equation 6.11 becomes:

$$H^2 = \sigma^2_G / \sigma^2_P \qquad \text{(Restatement of Equation 6.11)}$$
$$= \sigma^2_c / (\sigma^2_c + \sigma^2_{cs} + \sigma^2_{cb} + \sigma^2_e) \qquad \text{Equation 6.16}$$

where all terms of the first line are defined following Equation 6.11 and terms in the second line are defined in Table 6.2. The important point is that the conceptual statement of the broad-sense heritability in Equation 6.11 has been re-expressed in terms of the variance components that are estimated in the clonal experiment.

In the OP family experiment of *P. taeda*, the variance among OP families, σ^2_f, measures some, but not all, of the genetic variance in the pine breeding population. As one way to explain this concept, note that individuals within an OP family are genetically distinct and may even have different fathers. Therefore, the total genetic variance in the pine breeding population is partitioned into two parts: among families and among individuals within families. Theoretical derivations (Falconer and Mackay, 1996) show that for half-sib families, the variance among families equals one-quarter of the additive variance (*i.e.* $\sigma^2_f = 0.25 * \sigma^2_A$ where σ^2_A is the additive genetic variance defined in Equation 6.10). If OP families are truly half-sib families, an estimate of the additive genetic variance is obtained as $4 * \sigma^2_f$. For several reasons, OP families are likely to differ more from each other (and hence, have a larger among-family variance) than half-sib families (Squillace, 1974; Sorensen and White, 1988) such that $4 * \sigma^2_f$ will overestimate the true additive genetic variance. Therefore, many investigators estimate the additive genetic variance using a multiplier less than four (Griffin and Cotterill, 1988; Hardner and Potts, 1995; Hodge *et al.*, 1996); the value of three was used in the pine example in Table 6.2 such that:

$$h^2 = \sigma^2_A / \sigma^2_P \qquad \text{(Restatement of Equation 6.12)}$$
$$= 3 * \sigma^2_f / (\sigma^2_f + \sigma^2_{fs} + \sigma^2_{fb} + \sigma^2_e) \qquad \text{Equation 6.17}$$

where all terms in the second line are defined in Table 6.2. Equation 6.17 is the computational formula actually used in the pine experiment to estimate the conceptual parameters defined in Equation 6.12.

The type B genetic correlations (Burdon, 1977) (Box 6.6) were estimated using the following formulae:

$$r_{Bc} = \sigma^2_c / (\sigma^2_c + \sigma^2_{cs}) \qquad \text{Equation 6.18}$$
and
$$r_{Bf} = \sigma^2_f / (\sigma^2_f + \sigma^2_{fs}) \qquad \text{Equation 6.19}$$

where all terms are defined in Table 6.2. These parameters measure the importance of clone x site interaction (Equation 6.18) and OP family x site interaction (Equation 6.19). For example, if the 65 clones have the same performance relative to each other at all 5 test locations, then there is no interaction and no variance due to interaction (*i.e.* $\sigma^2_{cs} = 0$). In this case, $r_{Bc} = 1$ and this can be interpreted to mean that the correlation of clonal rankings is one for any and all pairs of sites. Conversely, if r_{Bc} is near zero, then the interaction variance is large relative to the clonal variance; this is equivalent to saying that the clonal rankings do not correlate well across sites (*i.e.* clonal rankings change across sites). The meaning of r_{Bf} is analogous to that for r_{Bc}, except that the interaction applies to OP families and not clones.

Interpreting Genetic Parameter Estimates

For two reasons, it is not uncommon to see genetic parameter estimates vary widely in the literature even from experiments on the same species. First, genetic parameters are specific to the trait, population and environmental conditions of the experiment. Even when measuring the same trait in the same species, two investigators can obtain dramatically different estimates if: (1) The samples of families or clones come from different populations (this affects the estimates of genetic variance); or (2) The experiments are planted in different environmental conditions (in particular, more "noise" due to environmental variability decreases heritability by increasing the residual environmental variance).

The second reason is that each of the variance components and the parameter estimates is subject to experimental error. Therefore, parameter estimates from large experiments (many clones or families planted on many sites) are more precise and have smaller confidence intervals than parameters estimated from small experiments. Even estimates from large experiments are not perfect, so identical estimates are never obtained by different investigators.

With these two caveats in mind, it is still informative to compare estimates from any set of experiments to those obtained by other investigators. In both of our case studies, the heritability estimates are larger for wood specific gravity than for bole volume growth (Table 6.2), and this is certainly consistent with other workers (see previous discussion in this chapter and review by Cornelius, 1994).

The type B genetic correlations are much higher for specific gravity (approximately 0.9 in Table 6.2) compared to those for bole volume. The value from the family experiment (r_{Bf} estimated as 0.439) is particularly low and indicates a significant amount of family x site interaction for volume. That is, the 113 *P. taeda* families do not rank consistently for volume growth across the 13 test locations; this trend also was found in another study of *P. taeda* from Florida (Sierra-Lucero *et al.*, 2002). This high level of G x E in families originating from Florida and planted in the lower coastal plain, is much larger than estimated for other families from other provenances of this species planted in other regions (Li and McKeand, 1989; McKeand and Bridgwater, 1998). This highlights the need to test families from all provenances of interest on the specific types of sites intended for future planting.

SUMMARY AND CONCLUSIONS

The phenotypic expression of polygenic traits is influenced by many gene loci and by environmental effects. Each gene locus has a very small effect on phenotypic expression, and it is impossible to identify the effects of specific alleles. Phenotypic measurements from many genotypes (such as different families or different clones) planted in randomized, replicated experiments in different test environments are analyzed (*Chapter 15*) to estimate: (1) Heritabilities for each trait that express the fraction of the total phenotypic variance due to genetic causes; (2) Breeding and/or clonal values for each trait to quantify the portion of phenotypic superiority that will be passed on to offspring of a parent or ramets of a clone, respectively; (3) Genetic correlations among traits to understand whether certain pairs of traits are influenced by many of the same gene loci; (4) Genetic correlations among different ages to understand whether performances at early and later ages are similar traits; and (5) Extent of G x E interaction. Some important conclusions follow:

- The breeding value of a parent for a specific trait quantifies the portion of the phenotypic superiority (or inferiority) of the parent that is passed on to its offspring formed by random mating with other parents in the population. Only average allele effects are passed on to offspring from a parent, and a parent contributes, on average, half of its breeding value to its offspring.

- Most polygenic traits in forest trees have narrow-sense heritabilities (h^2) in the range of 0.10 to 0.30. Estimates of h^2 for wood specific gravity are higher for most species (0.30 to 0.60). Little is known about the importance of non-additive genetic variances for most polygenic traits in forest trees. However, as a first approximation, non-additive variance appears to be similar or smaller than additive variance for growth traits (so the ratio of narrow-sense to broad-sense heritabilities approximates 0.5 or greater).

- All heritability estimates apply to a single trait measured for a certain population in a given set of environments. The scientist can increase the heritability for any trait (and therefore, increase gain from selection) by installing well-designed experiments that are implemented with care to reduce experimental error. Broad- and narrow-sense heritabilities estimated from a single test environment are always inflated due to inclusion of G x E variance in the numerator; thus, experiments should be installed on a range of site locations.

- Genetic correlations are very difficult to estimate and require extremely large experiments to estimate precisely. Therefore, estimates from a single set of modest-sized experiments should be viewed cautiously. Although genetic correlations among growth traits (such as stem diameter, height and volume) are usually strongly positive, it is more difficult to draw conclusions for other sets of traits.

- Age-age correlations are very valuable for helping to assess the best age at which to make selections. For growth traits, a first approximation of correlations between measurements at an early age and rotation age are 0.60, 0.81 and 0.93, when the early age is 25%, 50% and 75% of the rotation age, respectively. Age-age correlations for wood specific gravity are higher than these estimates for growth traits.

- Genotype x environment interaction is always more important when the climatic and edaphic conditions among test environments are more diverse. Interactions of species and provenances with environments are often minor within a region, but more pronounced when there are large changes in climates or soils. Family and clone interactions across planting locations within a region can be important; thus, to accurately quantify relative performance of genotypes over a planting zone, thorough testing is required across the range of sites representative of the zone.

CHAPTER 7

WITHIN-POPULATION VARIATION – *GENETIC DIVERSITY, MATING SYSTEMS AND STAND STRUCTURE*

In this and the next two chapters, we address genetic diversity and patterns of genetic variation in natural populations of forest trees. Genetic variation is conveniently considered at three hierarchical levels: (1) Among trees of a single species in individual populations or stands; (2) Among geographical areas or sites within a single species; and (3) Among species. We address genetic variation within populations in this chapter, address patterns of geographical variation within species in *Chapter 8*, and address variation among species, speciation and natural hybridization in *Chapter 9*.

Genetic diversity is essential to the long-term survival of species; without it, species cannot adapt to environmental changes and are more susceptible to extinction. The amount of genetic variation available within species also determines the potential for improving species through breeding programs. Patterns of genetic variation on the landscape reflect the responses of species to evolutionary forces operating within current and past environments, and can tell us much about how species have evolved and may continue to evolve in the future. In addition, many decisions made by foresters in harvesting, reforestation, and other silvicultural practices influence the historical patterns of genetic variation by altering the relative magnitudes of the evolutionary forces. Therefore, knowledge of natural patterns of genetic variation and their evolutionary bases also are of great practical significance. For example, the magnitude of genetic variation, spatial distribution of genotypes and patterns of mating within populations influence the genetic composition and quality of seed collected for reforestation, tree improvement and gene conservation purposes.

Most forest tree species possess considerable genetic variation, much of which can be found within individual populations. This chapter: (1) Describes how genetic variation is quantified; (2) Summarizes observations on the magnitude of genetic diversity in forest trees and the distribution of genetic variation within and among populations; (3) Addresses the characteristics of forest trees that typically promote high levels of genetic diversity within populations, and also the conditions under which population diversity may be limited; (4) Describes how mating systems of forest trees vary dynamically among and within species; (5) Considers patterns of cross-fertilization among trees within populations; (6) Examines spatial and temporal structure of genetic variation within populations; and finally, (7) Discusses practical implications of genetic variation within populations for silviculture, in particular for genetic improvement under natural regeneration regimes and for seed collection in wild stands. Other reviews of these concepts include Ledig (1986), Adams *et al*. (1992b), Mandal and Gibson (1998), and Young *et al*. (2000).

QUANTIFYING GENETIC VARIATION

Measuring levels of genetic variation within and among populations is an important first step in evaluating the evolutionary biology and tree improvement potential of a species.

There are basically two classes of traits used to measure genetic variation: (1) Quantitative traits controlled by many gene loci (*i.e.* polygenic traits, *Chapter 6*); and (2) Traits controlled at the single-gene level (*i.e.* genetic markers, *Chapter 4*). Quantitative traits are generally of greatest interest to evolutionary biologists and forest geneticists because of their adaptive and practical significance. Examples include germination rate, tree height, stem volume, root to shoot ratio, dates of bud burst and bud set, and cold hardiness. However, detailed genetic analysis of quantitative traits is difficult because the expression of individual genes cannot readily be identified or quantified (see analysis of quantitative trait loci, QTL, in *Chapter 18*), and because common garden experiments are required to evaluate the genetic influence on phenotypes (*Chapters 1* and *8*).

The availability of biochemical markers (*e.g.* isozymes, monoterpenes) and molecular markers (*e.g.* RFLPs, RAPDs, SSRs) for identifying genetic variation at specific gene loci allows assessment of genetic variation at the level of individual genes (*Chapter 4*). At these marker loci, more exact estimation of genetic diversity is possible than for quantitative traits. The meaning of variation in genetic markers, however, is obscured because their influence (if any) on the observable phenotype is generally unknown.

Because each individual tree has many thousands of genes its collective genotype can only be partially characterized by any particular set of marker genes or quantitative traits. Therefore, it should not be surprising that estimated levels and patterns of genetic variation depend on the particular markers or quantitative traits chosen to sample the collective genotype. Variation among populations in quantitative traits is easier to detect statistically, and is often of greater magnitude than gene frequency differences at marker loci (Lewontin, 1984; Muona, 1990). Genetic variation in quantitative traits also is expected to recover more rapidly than at individual marker loci after a population has gone through a population bottleneck (*i.e.* extreme reduction in population size), because mutations are manifested faster in quantitative traits (*Chapter 5*). Finally, geographic patterns of variation in quantitative traits often are associated with spatial patterns of environmental variation, suggesting that these patterns are primarily the result of selection (*Chapter 8*). In contrast, geographic patterns of variation in genetic markers often are independent of, or at best, weakly associated with changes in environment, suggesting that these markers are mostly neutral to selection pressure (*Chapter 5*, Box 5.4). For this reason, genetic markers are ideal for studying the effects of genetic drift and gene migration on spatial and temporal patterns of variation, and for evaluating mating systems. However, common garden studies of quantitative traits are usually more informative for describing patterns of adaptive variation. Therefore, to fully characterize levels and patterns of genetic variation in a species, and their evolutionary causes, studies involving both classes of traits are necessary.

Measures of Genetic Variation Based on Genetic Markers

Three statistics obtained from studies employing genetic markers are commonly used to quantify genetic variation within populations: (1) Proportion of polymorphic loci (P); (2) Mean number of alleles per locus (A); and (3) Mean expected heterozygosity ($\overline{H}_e$) (Hedrick, 1985; Nei, 1987). The **proportion of polymorphic loci**, P, is estimated as N_p/r, where N_p is the number of polymorphic loci and r is the total number of loci sampled. Usually the most frequent allele must have a frequency less than 0.95 or 0.99 (95% and 99% polymorphism criterion, respectively) for the locus to be considered polymorphic (*Chapter 5*). However, often all loci with two or more alleles, regardless of frequency, are counted as polymorphic. The **mean number of alleles per locus**, A, is estimated as $\Sigma m_j/r$, where m_j is the number of alleles at the jth locus. Both P and A are sensitive to the number of individuals sampled in the population, because the number of low frequency alleles that are detected increases with increasing sample size. This makes valid comparisons among populations difficult unless sample sizes are equivalent.

The **expected heterozygosity** at any given locus (H_e, also called **gene diversity**) is the probability that any two alleles randomly drawn from individuals in a population will differ, and is equivalent to the heterozygosity (*i.e.* sum of the frequency of all heterozygotes at a locus) expected given Hardy-Weinberg equilibrium (*Chapter 5*). The expected heterozygosity is estimated as $1 - \Sigma x_i^2$, where x_i is the observed frequency of the ith allele at the locus and is summed over all alleles at the locus. For example, when there are only two alleles, H_e can be alternatively calculated as $1 - p^2 - q^2$ or as $2pq$ where p and q are the allele frequencies.

The **mean expected heterozygosity**, $\overline{H}_e$, is the expected heterozygosity averaged over all loci sampled, and is the most widely used measure of genetic variation employing genetic markers. Because low frequency alleles contribute very little to $\overline{H}_e$, it is relatively insensitive to sample size (as compared to A and P). The mean proportion of observed heterozygotes ($\overline{H}_o$) also could be used to measure genetic variation, but the magnitude of $\overline{H}_o$ depends on the level of inbreeding (*Chapter 5*). Therefore, $\overline{H}_e$ is more useful for comparing genetic variation among different species, or populations of the same species, where mating systems may not be the same.

Populations of forest trees often differ in allele frequencies (especially when they are separated geographically, *Chapter 8*), and it is often of interest to determine the degree to which genetic variation in a region is distributed within and among populations (called **population genetic structure**). This information is useful for understanding the degree to which migration counters population subdivision due to selection or genetic drift (*Chapter 5*). It also has practical value when planning seed collections for breeding or gene conservation purposes (*Chapter 10*).

One approach often used to quantify the partition of allele frequency variation within and among populations is **gene diversity analysis** (Nei, 1987). For a single locus, recall from our illustration of genetic drift in *Chapter 5*, that one consequence of drift-caused divergence among replicate populations is that average heterozygosity within replicates decreases and is less than the heterozygosity calculated from the mean allele frequencies over all populations (Table 5.6). In fact, whenever populations differ in allele frequencies, the average heterozygosity within populations (called H_S) is less than heterozygosity based on mean allele frequencies over all populations (H_T, total gene diversity), regardless of the cause of population differentiation (called the **Wahlund effect**) (Hedrick, 1985). This concept is the basis of gene diversity analysis. For any one locus, total gene diversity can be partitioned into two components, $H_T = H_S + D_{ST}$, where H_S is the gene diversity due to variation within populations and $D_{ST} = H_T - H_S$ is the gene diversity due to differences among populations. The ratio of D_{ST} to H_T, called G_{ST}, measures the proportion of total gene diversity due to allele frequency differences among populations, while the proportion of H_T found within populations, on average, is $1 - G_{ST}$. Gene diversity analysis is generally applicable to the entire genome; therefore, a more general definition of G_{ST} is:

$$G_{ST} = \overline{D}_{ST} / \overline{H}_T \qquad \text{Equation 7.1}$$

where $\overline{D}_{ST}$ and $\overline{H}_T$ are obtained by taking averages of D_{ST} and H_T over all loci sampled (usually 20 or more). Note that the average H_S over all loci in a gene diversity analysis (*i.e.* $\overline{H}_S$) is equivalent to the average $\overline{H}_e$ over all populations sampled. However, gene diversity analysis is often conducted with only polymorphic loci because monomorphic loci do not influence the magnitude of G_{ST}. In this case, $\overline{H}_S$ overestimates actual mean heterozygosity within populations, and $\overline{H}_S > \overline{H}_e$. The principles of gene diversity analysis are illustrated with a numeric example in Box 7.1.

Box 7.1. Gene diversity analysis of population structure in *Pinus radiata*.

Although *P. radiata* is an extremely important plantation species of world economic importance (Balocchi, 1997), its native range is restricted to five small, isolated populations: three on the mainland coast of California south of San Francisco, and two on islands (Cedrus and Guadalupe) in the Pacific Ocean off Baja California. Thirty-one allozyme loci were used to examine the distribution of genetic variation within and among these five populations (Moran *et al.*, 1988). Six progeny (seed or seedlings) resulting from wind pollination were assayed electrophoretically from each of 50 or more parent trees in each population. Estimated allele frequencies and expected heterozygosities (H_e) for a single locus, the *Pgi-2* locus, are presented below. Note that allele 2 was the most prevalent in four of the five populations, while allele 3 was most common in the Cedros population ($x_3 = 0.74$ where x_i is the allele frequency, i = 1, 2 or 3).

Allele	Año Nuevo	Monterey	Cambria	Cedros	Guadalupe	Mean
1	0.01	-	-	-	-	0.002
2	0.81	0.76	0.80	0.26	0.88	0.702
3	0.18	0.24	0.20	0.74	0.12	0.296
H_e	0.311	0.365	0.320	0.385	0.211	0.318

H_S is equal to the average H_e within populations (0.318) and $H_T = 1 - \Sigma \overline{x}_i^2$, where $\overline{x}_i$ is the frequency of the ith allele averaged over all subpopulations: $H_T = 1 - (0.002)^2 - (0.702)^2 - (0.296)^2 = 0.420$. Gene diversity due to allele-frequency differentiation among populations at this locus is $D_{ST} = H_T - H_S = 0.420 - 0.318 = 0.102$. The proportion of total gene diversity at the *Pgi-2* locus due to population differences, therefore, is estimated as $G_{ST} = D_{ST}/H_T = 0.102/0.420 = 0.243$.

Over all 31 loci sampled in this study, the means were $\overline{H}_S = 0.098$, $\overline{H}_T = 0.117$, and $\overline{D}_{ST} = 0.019$. Therefore, the mean estimate of G_{ST} over all loci is $G_{ST} = \overline{D}_{ST}/\overline{H}_T = 0.019/0.117 = 0.162$, which is interpreted to mean that 16.2% of total genetic diversity in radiata pine (as measured by allozyme markers) is due to differences among populations. Similarly, 83.8% of the measured diversity is among trees within populations.

Note that when the number of individuals sampled from each population is less than 50, gene diversity estimates should be corrected for bias due to small sample size (Nei, 1987). Investigators often calculate gene diversity statistics using only the polymorphic loci in a sample (27 of the 31 loci in the *P. radiata* study were polymorphic). This increases the values of $\overline{H}_S$ and $\overline{H}_T$, but should have no effect (other than rounding error) on the magnitude of G_{ST}.

The estimated G_{ST} for *P. radiata* is relatively large compared to other conifers (Table 7.1). Genetic drift, either at the time the populations were established (founder effect) or during population bottlenecks, is the most likely explanation for the large differences in allele frequencies among populations. The isolation of the populations from each other also limits migration as a factor that could reduce population differentiation. Because much of the genetic variation is distributed among populations, gene conservation strategies for this species need to consider ways to conserve genetic diversity in all five populations (Moran *et al.*, 1988).

Gene diversity analysis can readily be extended to include further levels of population hierarchy (Nei, 1987). For example, if a species is subdivided into regions and subpopulations within regions, total gene diversity can be partitioned as follows: $H_T = H_S + D_{RT} + D_{SR}$, where H_S is the gene diversity within populations, and D_{RT} and D_{SR} are diversities between regions and between subpopulations within regions, respectively.

In *Chapter 5*, the inbreeding coefficient, F, was defined as the fractional reduction in heterozygosity due to inbreeding relative to heterozygosity expected in a random mating population. Another way to quantify allele frequency differentiation among populations is to view the reduced expected heterozygosity within populations, relative to the expected heterozygosity combined over all populations, as an inbreeding-like effect (Wright, 1969, 1978). Wright defined the "inbreeding coefficient" due to population divergence at a single locus with two alleles as $F_{ST} = (H_T - H_S)/H_T$. Nei (1987) subsequently expanded the formulation of F_{ST} for loci with more than two alleles and showed that F_{ST} is essentially identical to G_{ST} when calculated for a single locus.

Measures of Genetic Variation Based on Quantitative Traits

Most studies of quantitative genetic variation within and among natural populations of forest trees are based on family analysis. Seeds resulting from natural pollination mediated by wind or animals are collected from several to many wild trees in each of a number of locations covering the region of interest, producing as many open-pollinated families as there are mother trees. The seeds are then sown into one or more nursery or greenhouse environments (*i.e.* seedling common gardens). The resulting seedlings may also be subsequently outplanted into one or more field sites for longer-term evaluation. Whether such studies end at the seedling stage or the trees are grown for many years in the field, forest geneticists refer to such experiments as provenance or seed source studies (*Chapter 8*). Similar common garden experiments are widely used to evaluate progenies in tree improvement programs (*Chapters 12* and *14*).

For each trait measured, the variation among all families in the experiment is partitioned by analysis of variance into variation due to differences among populations in the region and due to families within populations (Box 7.2). Patterns of variation among population means are used to examine whether populations are genetically associated with geography or environment (suggesting the patterns may be the result of adaptation). Variation among families within locations estimates only a portion of the total genetic variation within populations. This is because genetic variation among individuals within families and within individuals (*i.e.* heterozygosity) is not accounted for.

GENETIC DIVERSITY IN FOREST TREES

Estimates of Genetic Diversity from Genetic Markers

A great deal has been learned in the past few decades about levels and distribution of genetic variation in forest trees from studies based on allozyme markers. Among recent surveys of allozyme results are those of Hamrick and Godt (1990; all plants), Hamrick *et al.* (1992; woody plants), Loveless (1992; tropical trees), Moran (1992; Australian trees), Muller-Starck *et al.* (1992; European trees), Ledig *et al.* (1997; spruces), and Ledig (1998; pines). In general, individual populations of trees possess much higher levels of genetic diversity than found in animals (Hamrick and Godt, 1990; Hamrick *et al.*, 1992) or other types of plants

(Table 7.1). Within an average forest tree population, about 50% of allozyme loci are polymorphic (*i.e.* at least 2 alleles detected), with 1.75 alleles per locus and mean expected heterozygosity of 15%. Expected heterozygosity is nearly 50% greater in a typical forest tree population than in average populations of annual plants or short-lived perennials.

Box 7.2. Genetic variation for tree height within and among populations of *Pinus palustris*.

Seeds resulting from open pollination were collected from each of three trees in 24 populations (72 total families) of *P. palustris* from a four-state region in the southeastern USA (but mostly in Georgia; Wells and Snyder, 1976). Seedlings were planted at Gulfport, Mississippi, in a randomized complete block design with three replications of five-tree row plots of each family. Survival and mean tree height at age 12 years were 55% and 9.2 m, respectively. The following analysis of variance of row plot means revealed that age 12 tree heights differed significantly (p < 0.05), both among populations and among families within populations.

Source of variation	df	Mean square Observed	Mean square Expected[a]
Blocks	2	0.42	
Populations	23	3.72[b]	$\sigma^2_e + 3\sigma^2_{f/p} + 8.64\,\sigma^2_p$
Families/populations	46	2.11[b]	$\sigma^2_e + 3\sigma^2_{f/p}$
Error	124	1.21	σ^2_e

[a] σ^2_p = variance due to differences among populations, $\sigma^2_{f/p}$ = variance due to differences among families within populations, and σ^2_e = error variance.
[b] Significant at 0.05 probability level.

To partition the height variation among the 72 families into portions due to differences among populations and among families within populations, Wells and Snyder (1976) equated observed mean squares with their theoretical expectations (*i.e.* expected mean squares). For example, the observed mean square for families within populations has two components, that due to differences among family means within populations ($\sigma^2_{f/p}$) and that due to experimental error in plot means (σ^2_e). Because σ^2_e is estimated from the bottom line of the table (1.21), $\sigma^2_{f/p}$ can be estimated by equating the observed and expected mean squares for families/populations by substituting 1.21 for σ^2_e, and solving for $\sigma^2_{f/p} = (2.11 - 1.21)/3 = 0.30$. Following the same procedure, σ^2_p is estimated as 0.19. Therefore, the total variance among families is 0.30 + 0.19 = 0.49, and the proportion of the total due to differences among families within populations is 0.30/0.49 = 0.61. Using similar calculations for three other traits (DBH, survival and plot stem volume), Wells and Snyder (1976) estimated that the average proportion of total family variation due to differences among families within populations (over all four traits) is 66%. This indicated that more than half of the variation among families sampled over a four-state region was found, on average, within a single population. In addition, ranges among the three family means within populations frequently exceeded the range across all 24 population means. Therefore, the estimated genetic variation among families within individual populations was considerable in this study.

Table 7.1. Allozyme diversity within and among populations for five categories of plant species.[a]

Category	P[b]	A[b]	$\overline{H}_S$ [b]	G_{ST}[c]
Annuals	0.29 d	1.45 d	0.101 d	0.355 d
Short-lived perennials	0.29 d	1.40 de	0.098 d	0.245 e
Long-lived perennials				
Herbaceous	0.22 e	1.32 e	0.082 d	0.278 e
Woody[h]				
Gymnosperms	0.53 f	1.83 f	0.151 e	0.073 f
Angiosperms	0.45 g	1.68 g	0.143 e	0.102 f

[a] Derived from Tables 1, 3, and 4 in Hamrick *et al.,* 1992. Values for different categories in each column are statistically different (p < 0.05) if they are accompanied by different letters.

[b] Genetic diversity within populations: P = mean proportion of polymorphic loci (*i.e.* presence of two or more alleles in a sample); A = mean number of alleles per locus; $\overline{H}_S$ = mean expected heterozygosity (includes monomorphic loci). Based on data averaged over 96-226 studies in each category (except long-lived herbaceous plants is based on 24 studies).

[c] Mean proportion of total gene diversity due to differences between populations. Based on polymorphic loci only and data averaged over 73-186 studies in each category (except long-lived herbaceous plants is based on 25 studies).

[h] Trees and a few long-lived shrubs and treelets.

When all loci, including monomorphic loci, are used in gene diversity analyses, the magnitude of genetic diversity within populations ($\overline{H}_S$) is highly correlated with the total genetic diversity ($\overline{H}_T$) in species (Hamrick *et al.,* 1992). Therefore, total diversity in forest trees is also generally higher than found in other plants. However, only a small proportion of the total gene diversity in trees is due to differences among populations (Table 7.1); G_{ST} in trees is frequently 10% or lower, which is only one-half to one-quarter of the G_{ST} estimates typically found in annuals or other herbaceous species. The lower G_{ST} in trees is most likely because most tree species are outcrossing, while a large proportion of annuals and herbaceous plants are either self-pollinated or self-pollination features prominently in their mating system. High levels of self-pollination not only promote inbreeding (*Chapter 5*), but also limit pollen migration between populations. Both pollen and seed migration between populations of forest trees can be extensive (see the section on *Strong Migration Between Populations*).

Although the magnitude of allozyme diversity within forest tree populations is typically high, the range among species is striking. In Table 7.2, allozyme variation estimates are listed for 20 species that fall into four classes of mean within population gene diversity ($\overline{H}_S$), ranging from very high ($\overline{H}_S > 0.20$) to very low ($\overline{H}_S < 0.05$). Here, as in Table 7.1, both monomorphic and polymorphic loci were used in calculating $\overline{H}_S$. All four $\overline{H}_S$ classes include gymnosperms and angiosperms, with a wide variety of genera represented throughout the table. The five pine species listed range from among the highest levels of $\overline{H}_S$ reported (*Pinus sylvestris*, $\overline{H}_S = 0.303$) to the lowest (*P. torreyana* and *P. resinosa*, $\overline{H}_S = 0$). Given that *P. torreyana* has the smallest distribution of any pine (two small stands in southern California, one on the mainland near San Diego, and one 280 km north on an island in the Pacific Ocean off Santa Barbara), the lack of genetic diversity within populations is, perhaps, not surprising. Both *P. sylvestris* and *P. resinosa*, on the other hand, have extensive ranges, one across northern Europe and Asia (*P. sylvestris*), the other across eastern North America

Table 7.2. Allozyme diversity within and among natural populations of 20 forest tree species that span a wide range of levels of average expected heterozygosity within populations ($\overline{H}_s$).

Species	GR[a]	PD[b]	Population sample				Genetic diversity[d]					Reference
			Coverage[c]	No.	Loci	P	A	$\overline{H}_s$	$\overline{H}_T$	G_{ST}	N_em[e]	
High $\overline{H}_s$ (> 0.20)												
Pinus sylvestris	W	C	Scotland	36	16	0.84	2.6	0.303	0.311	0.026	8.8	Kinloch et al., 1986
Robinia pseudo-acacia	W	C	Rangewide	23	40	0.71+	2.6	0.291	0.333	0.125	1.6	Surles et al., 1989
Quercus petrae	W	C	France	32	15	—	2.4	0.277	0.282	0.017	13.6	Kremer et al., 1991
Castanea sativa	W	C	Italy	18	15	0.60+	1.7	0.231	0.255	0.097	2.1	Pigliucci et al., 1990
Larix laricina	W	C	Rangewide	36	19	0.50+	1.8	0.220	0.233	0.055	4.1	Cheliak et al., 1988
Moderate $\overline{H}_s$ (> 0.10 to 0.20)												
Gleditsia triacanthos	W	C	Rangewide	8	27	0.62	2.2	0.198	0.210	0.059	3.0	Schnabel and Hamrick, 1990
Melaleuca alternifolia	L	C	Rangewide	10	17	0.64+	2.0	0.164	0.186	0.119	1.5	Butcher et al., 1992
Pinus palustris	W	C	Alabama west to Florida	24	19	0.54	2.9	0.150	0.160	0.062	3.5	Hamrick et al., 1993b
Taxus brevifolia	W	D	Oregon, Washington, N. California	17	17	0.15	1.5	0.124	0.138	0.104	1.9	Wheeler et al., 1995
Picea abies	W	C	Rangewide	70	22	0.73	1.6	0.115	0.121	0.050	4.6	Lagercrantz and Ryman, 1990

Low $\bar{H}_S$ (> 0.05 to 0.10)

Species	GR	PD	Sampling region			P	A	$\bar{H}_S$	$\bar{H}_T$	G_{ST}	$N_e m$	Reference
Pinus radiata	L	D	Rangewide	5	31	0.46[+]	1.8	0.098	0.117	0.162	0.8	Moran *et al.*, 1988
Picea chihuahuana	R	D	Rangewide	10	24	0.27	1.4	0.093	0.124	0.248	0.6	Ledig *et al.*, 1997
Populus trichocarpa	W	C	W. Washington	10	18	0.32	1.5	0.090	0.096	0.062	3.1	Weber and Stettler, 1981
Pentaclethra macroloba	W	C	Costa Rica	12	14	0.31	1.4	0.074	0.095	0.219	0.8	Hall *et al.*, 1994
Eucalyptus caesia	L	D	Rangewide	13	18	0.29[+]	1.3	0.068	0.176	0.614	0.1	Moran and Hopper, 1983

Very low $\bar{H}_S$ (< 0.05)

Species	GR	PD	Sampling region			P	A	$\bar{H}_S$	$\bar{H}_T$	G_{ST}	$N_e m$	Reference
Thuja plicata	W	C	S. British Columbia	8	19	0.16	1.2	0.039	0.040	0.033	5.6	Yeh, 1988
Magnolia tripetala	W	D	Rangewide	15	17	0.11[+]	1.1	0.033	0.052	0.371	0.4	Qiu and Parks, 1944
Acacia mangium	W	D	Rangewide	11	30	0.13	1.1	0.017	0.025	0.311	0.5	Moran *et al.*, 1989
Pinus torreyana	L	D	Rangewide	2	59	0	1.0	0	0.017	1.0	0	Ledig and Conkle, 1983
Pinus resinosa	W	C	N.E. Canada	5–11	9–20	0	1.0	0	0	0	—	Fowler and Morris, 1977; Simon *et al.*, 1986; Mosseler *et al.*, 1991

[a] GR (Geographic Range) – Distance between most widely separated populations within the bulk of the species range (*i.e.* not including outlier populations): W (Widespread) > 1000 km, R (Regional) 200-1000 km, L (Localized) < 300 km.

[b] PD (Population Distribution): Continuous (populations relatively large and continuous in distribution over much of the species range), Disjunct (populations relatively small with disjunct distribution).

[c] Rangewide means that the sampled populations are distributed over much of the species range; otherwise, the geographic location of the sampling region is given.

[d] P = Mean proportion of polymorphic loci within populations ([+] most common allele must have a frequency < 0.99 for a locus to be counted as polymorphic. In all other cases, any locus with two or more alleles in the sample is considered polymorphic).
A = Mean number of alleles per locus within populations.
$\bar{H}_S$ = Average expected heterozygosity within populations.
$\bar{H}_T$ = Mean expected heterozygosity (gene diversity) over all populations.
G_{ST} = Proportion of total gene diversity due to differences among populations.

[e] $N_e m$ = Estimated number of immigrants migrating into populations each generation (Slatkin, 1987).

(*P. resinosa*). As seen later in this chapter, while levels of genetic variation are often positively associated with current population size and distribution range of the species, extreme bottleneck effects in the past also may have been important in determining current levels of diversity.

Genetic differentiation among populations (G_{ST}) also varies widely among tree species (Table 7.2), ranging from low values in species that have more or less continuous distributions, to high values in species with disjunct population distributions. One hundred percent of the gene diversity in *Pinus torreyana* is due to population differentiation (*i.e.* $G_{ST} = 1.0$); of the 59 loci sampled, the only two that were polymorphic were fixed for alternate alleles in the two widely separated populations of this species.

Levels and distribution of genetic diversity revealed by allozymes are largely consistent with more recent observations based on DNA genetic markers in the nuclear genome, when direct comparisons between these marker types are made (*e.g.* Mosseler *et al.*, 1992; Isabel *et al.*, 1995; Szmidt *et al.*, 1996; LeCorre *et al.*, 1997; Aagaard *et al.*, 1998). This implies that results from allozymes are roughly representative of most genes in the genome, at least genes that are fairly neutral to selection.

Estimates of Genetic Diversity from Quantitative Traits

Even casual observation of trees in even-aged natural stands or plantations reveal large differences among individuals in size, stem form, growth phenology, foliage color, pest resistance, etc. If only a small proportion of these differences is under genetic control, the extent of genetic variation for quantitative traits within individual stands of most forest tree species must be considerable (Zobel and Talbert, 1984). Results of partitioning quantitative-trait variation among open-pollinated families are illustrated for three forest tree species in Fig. 7.1. Gene diversity statistics for these species across comparable geographic ranges are summarized in Table 7.2. Given the broad geographical distribution of populations in these studies, substantial variation among population means is expected. However, nearly one-half to two-thirds of the total variation among families (averaged over a variety of traits) in these studies is due to differences among families within populations. Nevertheless, although within-population diversity is substantial at the family level, a larger proportion of the total family variation is due to population differences than was found for gene diversity using allozyme loci (Table 7.2).

Genetic variation among open-pollinated families represents only part of the within-population genetic variance. Much genetic variation also exists among individuals within the same family. Unfortunately, genetic variation among individuals within families usually cannot be estimated in family studies, because these genetic differences are confounded by environmental effects (*Chapter 6*). However, in the *Populus trichocarpa* example, individuals within families were clonally replicated by rooted cuttings, which makes it possible to estimate genotypic values for individual trees, and to partition total genetic variation among individual trees into components (Rogers *et al.*, 1989). Doing this partition, it was estimated that only 9% of the total genetic variation among individual trees sampled in this study was due to population differences, while 91% was found within populations; 12% was due to families/populations and 79% due to differences among individual trees within families. Therefore, in this case, the distribution of quantitative genetic variation between and within populations at the individual tree level is similar to that observed for allele frequency variation at allozyme loci (Table 7.2).

FACTORS PROMOTING GENETIC DIVERSITY WITHIN POPULATIONS

A number of factors contribute to the high levels of genetic diversity typically found in forest tree populations: (1) Large population size; (2) Longevity; (3) High levels of out-crossing; (4) Strong migration between populations; and (5) Balancing selection (Ledig, 1986, 1998; Hamrick *et al.*, 1992). We briefly address each of these in the next sections.

Large Population Size

Many forest trees occur in large stands and have more or less continuous distributions over broad areas. Large population size reduces the susceptibility of populations to random genetic drift, which leads to loss of genetic diversity within populations and increased diversity among them (*Chapter 5*). In their survey of allozyme variation in forest trees, Hamrick *et al.* (1992) found that size of the geographic range of a species was the best predictor of levels of genetic variation within populations. On average, species with widespread distributions have four times the gene diversity within populations ($\overline{H}_S = 0.228$), and one-quarter of the diversity among populations ($G_{ST} = 0.033$), than geographically limited endemic species (*e.g. Pinus torreyana* and *Eucalyptus caesia*, Table 7.2), which occur in small isolated stands ($\overline{H}_S = 0.056$, $G_{ST} = 0.141$).

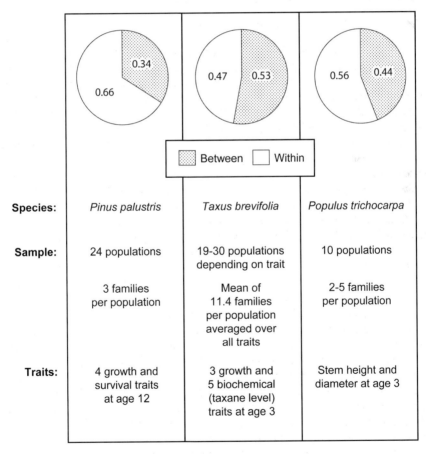

Fig. 7.1. Mean proportion of the total variation among open-pollinated family means for quantitative traits due to population differences and to differences among families within populations, for three tree species: *Pinus palustris* (Wells and Snyder, 1976), *Taxus brevifolia* (Wheeler *et al.*, 1995), and *Populus tricho-carpa* (Rogers *et al.*, 1989). Although populations ranged widely across geographical regions, a substantial proportion (~50%) of the total variation among families is found within individual populations.

Some narrow endemics, however, appear to possess levels of gene diversity within populations that are comparable to species with broad ranges. For example, $\overline{H}_S = 0.121$ in populations of *Abies equitrojani* (Gulbaba *et al.*, 1998), which is restricted to four small disjunct stands (totaling ~3600 ha) in northwest Anatolia, Turkey; and, $\overline{H}_S = 0.130$ in the largest stand (11 ha) of *Picea omorika*, a species found on a total of only 60 ha in the border region between Serbia and Bosnia (Kuittinen *et al.*, 1991). Although the current distribution of these species is very small, individual population sizes may never have been reduced to low enough sizes for extensive genetic drift to occur. In the absence of mutation or selection, and assuming constant effective population size (N_e, *Chapter 5*), gene diversity is expected to decline at the following rate over t generations:

$$H_{et} = [1 - 1/2N_e]^t H_{e0}$$

Equation 7.2

where H_{e0} is the original gene diversity (Hedrick, 1985). With $N_e = 50$, it would take nearly 70 generations for H_e to decrease from 0.150 to 0.075. For typical forest tree species with generation lengths between 20 and 100 years, it would require from 1400 to 7000 years for gene diversity to halve when starting with an effective population size of 50 individuals. Therefore, population size must be reduced drastically for drift to have a large impact on levels of genetic variation within populations.

On the other hand, some species with widespread geographic ranges have extremely low levels of genetic diversity within populations and within the species as a whole (*e.g. Pinus resinosa, Acacia mangium, Magnolia tripetala*, Table 7.2). Low levels of genetic variation in these species are attributed to extreme population bottlenecks that are hypothesized to have occurred as their distributions were displaced and narrowly confined by one or more Pleistocene glaciations or by raised water levels during intervening warm periods (*Chapter 5*). Extreme bottlenecks in the evolutionary history of a species impact genetic diversity for many subsequent generations, because only mutation (which occurs only very rarely, *Chapter 5*), or possibly hybridization with closely related species, can replenish the lost variation.

Within-population gene diversity is expected to decrease, on average, with decreasing population size, as long as populations are relatively small to begin with. For example, Ledig *et al.* (1997) studied *Picea chihuahuana*, a Mexican conifer with an extremely disjunct distribution of small populations (Fig. 7.2a); the number of trees varied from 15 to 2441 trees in the populations sampled. They found that $\overline{H}_e$ decreased linearly with decreasing population size (Fig. 7.3). Similar trends have also been observed in some narrow endemic species of Australian *Eucalyptus* (Fig. 7.2b) (Moran, 1992). However, in other *Eucalyptus* species with narrow, disjunct distributions of small populations, like *E. pendens*, there is no apparent association between population size and $\overline{H}_e$ (Moran, 1992; Fig. 7.3). In these latter cases, migration between populations may counter the loss of genetic diversity due to drift, or the current small population sizes may be relatively recent in origin (*e.g.* due to clearing for agriculture), so there has not been enough time for genetic drift to occur. Therefore, the expected relationship between population size and genetic diversity within populations is not likely to be found unless populations have remained very small and isolated for long evolutionary periods.

In summary, mean gene diversity, $\overline{H}_e$, is high in the great majority of tree species, regardless of their geographical range. Only in species with very small, disjunct populations is $\overline{H}_e$ likely to be reduced greatly, and only if population size has been small for many generations. In rare cases, $\overline{H}_e$ is small in species with wide geographical range and large populations because of one or more extreme range-wide population bottleneck events that occurred in their recent geological history.

Fig. 7.2. Genetic drift in small isolated populations of forest trees can result in greatly reduced levels of genetic diversity within populations and increased diversity among populations: (a) *Picea chihuahuana* in Mexico; and (b) *Eucalyptus caesia* in Western Australia. (Photos courtesy of F.T. Ledig, USDA Forest Service Institute of Forest Genetics, Placerville, CA and G. Moran, CSIRO Forestry, Canberra, Australia, respectively)

Longevity

Long generation intervals mean that trees are less frequently subjected to potential population bottlenecks during reproduction (either due to limited parentage or poor survival of offspring) than species with short life cycles. In addition, because individuals can live a long time, stands may contain cohorts of different age classes. Older cohorts may differ genetically from younger cohorts because environments, and thus selection regimes, differed during their respective establishment periods, or because of chance (Roberds and Conkle, 1984). Therefore, when genetic diversity is assessed across all cohorts, the aggregate diversity is increased.

Another source of variation related to longevity is mutation. We saw in *Chapter 5* that compared to plants with short life cycles, trees possess high numbers of lethal equivalents (*i.e.* deleterious mutants). Because of their long life and great size, trees are expected to accumulate more mutations per generation than short-lived plants (Ledig, 1986; Klekowski and Godfrey, 1989). In plants, spore mother cells are derived from vegetative tissues. Therefore, the older the individual, the more cell divisions that occur between the zygote and gamete production, and the greater the opportunity for DNA copy error and accumulation of mutations during mitosis. Since meristematic tissue gives rise to gametes, these somatic mutations are transmitted to offspring. On the other hand, animals have a germ line in which there are relatively few divisions between the zygote and gametes. For example, there may be as few as 50 cell generations between the zygote and gametes in humans, while gametes produced in cones at the top of a 12 m pine tree may be the result of as many as 1500 cell generations (Ledig, 1986).

Support for higher mutation in trees comes from estimates of the total frequency of genome-wide (diploid) chlorophyll mutants (*i.e.* albino or yellow foliage) accumulated per generation in seedlings (Table 7.3). These mutants are almost always lethal and collectively, they are controlled by a large number of loci. Because the number of loci influencing chlorophyll production is probably similar among higher plants (~300), estimates of total mutation per generation are comparable across species (Klekowski, 1992). Forest trees appear to have about a 10-fold higher level of accumulated mutations per generation for chlorophyll deficiencies than annual plants. Recent studies based on measures of diversity at the nucleotide level, however, suggest that rates of mutation in conifers may be no greater than in plants in general (*Chapter 9*). Certainly, additional research is needed on this issue.

High Levels of Outcrossing

A high level of **outcrossing** (*i.e.* mating among unrelated individuals) is particularly important to maintaining genetic diversity within populations. Outcrossing promotes heterozygosity and maintenance of recessive alleles that might otherwise be exposed to selection (*Chapter 5*). Furthermore, mating between heterozygous individuals leads to genetic recombination and creation of vast arrays of different genotypes in offspring. On the other hand, inbreeding results in less recombination and increased homozygosity, leading not only to loss of variation, but to exposure of deleterious, recessive alleles, reducing average viability of offspring (inbreeding depression).

The proportion of inbreeding versus outcrossing (called the **mating system**), and the factors influencing this proportion, are very important elements of the biology of trees, because of the detrimental effects of inbreeding. The mating system is affected by mechanisms promoting outcrossing, as well as by population structure, tree density and the spatial distribution of relatives. As we will see, most species of forest trees are predominantly outcrossing and this is true for species having different pollination mechanisms and for tropical, temperate and boreal species. However, at least some self-fertilization, the most extreme form of inbreeding, occurs in many forest tree species. Therefore, the degree of selfing is often of particular

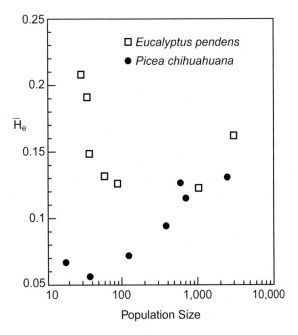

Fig. 7.3. Relationship between mean gene diversity ($\overline{H}_e$) for allozyme loci and population size in two narrow endemic tree species with disjunct distributions of populations. The pattern in *Picea chihuahuana* (Ledig *et al.*, 1997) is consistent with genetic drift having the most impact on smaller populations, which have lower $\overline{H}_e$. The lack of a similar pattern in *Eucalyptus pendens* (Moran, 1992) may be because migration between populations counters the loss of genetic variation due to drift, or because the populations have only recently become small so that there has not been enough time for genetic drift to have had an effect.

Table 7.3 Estimates of total frequency of genome-wide (diploid) mutations per generation for chlorophyll deficiency lethals in forest trees and annual plants.[a, b]

Trees	Estimate	Source
Picea abies	$2 - 5 \times 10^{-3}$	Koski and Malmivaara, 1974[c]
Pinus sylvestris	$1 - 2 \times 10^{-3}$	Koski and Malmivaara, 1974[c]
	$1 - 3 \times 10^{-2}$	Kärkkäinen *et al.*, 1996
Pinus contorta	2.2×10^{-3}	Sorensen, unpublished [c, d]
Pinus ponderosa	3.5×10^{-3}	Sorensen, unpublished [c, d]
Rhizophora mangle	1.1×10^{-2}	Lowenfeld and Klekowski, 1992
	1.2×10^{-2}	Lowenfeld and Klekowski, 1992
	4.1×10^{-3}	Klekowski *et al.*, 1994
Annuals		
Inbreeders (9 species)	< 1.5–5.0×10^{-4}	Klekowski, 1992
Outbreeders		
Zea mays	4.0×10^{-3}	Crumpacker, 1967
Fagopyrum esculentum	2.3×10^{-4}	Klekowski, 1992
Mimulus guttatus	9.7×10^{-4}	Willis, 1992

[a] The sum of all mutations per individual per generation causing chlorophyll deficiency.
[b] Most estimates are based on the assumption that allele frequencies in populations are in mutation-selection equilibrium (see Ohta and Cockerham, 1974; Charlesworth *et al.*, 1990)
[c] These estimates were calculated from data provided by the sources.
[d] Sorensen, F.C., USDA Forest Service, Pacific Northwest Research Station, Corvallis, Oregon, USA.

interest in mating system studies. Despite the generally deleterious results of selfing, retaining the capacity for self-fertilization means that reproduction is always possible, and this is important to population regeneration in some instances: (1) After a catastrophe such as wildfire, where one or only a few widely scattered individuals survive; and (2) During invasion of new habitats by a few members of a colonizing species (Ledig, 1998).

Mating system estimates are usually based on the **mixed mating model** that assumes all mating is the result of either self-fertilization (with probability s) or random outcrossing to unrelated individuals in the population (with probability $t = 1 - s$) (Fyfe and Bailey, 1951). Two general estimation procedures have been utilized: (1) Comparison of proportions of filled seed after open pollination with filled proportions under controlled self and cross pollination (Franklin, 1971; Sorensen and Adams, 1993); and (2) Applying expectations from the mixed mating model to the segregation of genetic markers in offspring resulting from open pollination (Box 7.3). It is important to recognize that in both procedures, selfing and outcrossing are assessed at the filled-seed stage or in young seedlings,

---

Box 7.3. Estimating proportions of self and outcross progeny using genetic markers.

Mating system estimates are often obtained by applying the Mixed Mating Model (Fyfe and Bailey, 1951) to describe the frequencies of marker genotypes observed in the offspring of mother trees (Ritland and Jain, 1981; Shaw *et al.*, 1981). The Mixed Mating Model states that all mating is due either to self-fertilization (with probability s) or to random outcrossing to other individuals in the population ($t = 1 - s$), so that the probability of observing a particular marker genotype, $P(g_i)$, in the offspring of mother tree M is:

$$P(g_i) = tP(g_i|O) + sP(g_i|M)$$

where $P(g_i|O)$ is the probability of g_i given outcrossing and $P(g_i|M)$ is the probability of g_i given selfing. In conifers, where pollen gamete genotypes can be inferred by comparing the genotype of the diploid embryo with the haploid genotype of the megagametophyte (*Chapter 4*), $P(g_i)$ is the probability of pollen gamete g_i in the offspring of mother tree M, $P(g_i|M)$ is the probability mother tree M produces gametes with g_i and $P(g_i|O)$ is the probability of g_i in the random outcross pollen pool.

The simplest application of the mixed mating model to estimating s (or t) is when a mother tree carries a unique marker gene not found in any other tree in the population. In this case, $P(g_i|O) = 0$ and s can be estimated as:

$$s = P(g_i)/P(g_i|M)$$

An example comes from one of the earliest applications of allozyme markers to mating system estimation in forest trees (Muller, 1976). In this study, open-pollinated cones were collected from 86 *Picea abies* mother trees in a 91 year-old stand in Germany. One mother tree was heterozygous at the *LapB* locus, with genotype *LapB-2/LapB-3*; allele *LapB-3* being found in no other tree in the population. In a sample of 987 seeds from this mother tree, 57 pollen gametes had allele *LapB-3*. Thus, an estimate of $P(g_i)$ is $57/987 = 0.058$. Assuming Mendelian segregation, one-half of the mother tree's pollen gametes are expected to carry *LapB-3*, so that $p(g_i|M)$ is 0.50. Therefore, the proportion of self-fertilized offspring from this tree is estimated to be $s = 0.058/0.50 = 0.116$, and $t = 1 - s = 0.884$.
(Box 7.3 continued on next page)

Box 7.3. Estimating proportions of self and outcross progeny using genetic markers. (Continued from previous page)

When marker genes are not unique to individual mother trees, population estimates of t (or s) can be obtained with individual marker loci by applying the Mixed Mating Model to the pooled offspring data of many mother trees using Maximum Likelihood procedures (Brown and Allard, 1970; Shaw and Allard, 1982; Neale and Adams, 1985a). Additional assumptions required in these procedures are: (1) The probability of outcross offspring is independent of maternal genotype; and (2) Allele frequencies in the outcross pollen pool are distributed uniformly over the population of mother trees.

It is better to use many polymorphic marker loci simultaneously in mating system estimation (Shaw *et al.*, 1981; Ritland and Jain, 1981; Cheliak *et al.*, 1983; Neale and Adams, 1985a; Ritland and El-Kassaby, 1985). Multilocus estimates are more precise than those based on a single locus and are less sensitive to violations of the assumptions of the Mixed Mating Model (Ritland and Jain, 1981; Shaw *et al.*, 1981). In particular, with a single marker locus, offspring resulting from mating with close relatives other than selfing (*i.e.* bi-parental inbreeding) may be indistinguishable genetically from true self-fertilization, resulting in downwardly biased estimates of t (and inflated estimates of s). Therefore, estimates based on many marker loci more readily distinguish true selfs from other forms of close inbreeding and are less biased. In addition, when many marker loci are utilized, multilocus estimates of t (t_m) can be compared to mean estimates of t based on each locus assessed individually (t_s). If close bi-parental inbreeding has occurred, t_m is expected to be greater than t_s.

---

which reflects the proportion of viable offspring resulting from self-fertilization (*i.e.* after losses due to embryo abortion), and not the frequency of self-fertilization *per se*. Because s + t = 1, mating systems are described in terms of either the proportion of selfed offspring (s) or the proportion of offspring resulting from random outcrossing (t). For consistency, we will use t for the remainder of this chapter.

Generally, estimates of t are high (> 75%) in most forest trees (Table 7.4). As illustrated with species of *Pinus* and *Picea*, outcrossing approaches or exceeds 0.90 in conifers, where pollination is mediated by wind. There does not seem to be a trend, however, for animal-pollinated trees to be less outcrossed than those that are wind pollinated, or for angiosperm species to be less outcrossed than conifers. Tropical angiosperm species, for example, are animal pollinated, but t estimates also often exceed 0.90 (Nason and Hamrick, 1997) (Table 7.4).

Although most trees are predominantly outcrossed, t ranges widely among species. For example, t averages 0.75 in *Eucalyptus*, but ranges from 0.59 to 0.84 among species in the genus (Table 7.4). The tree species with the highest level of selfing reported in the literature (but not included in Table 7.4) is *Rhizophora mangle*, red mangrove (Lowenfeld and Klekowski, 1992; Klekowski *et al.*, 1994). This small tree, which forms dense forests in the tropical intertidal ecosystem of the New World, has bisexual (*i.e.* perfect) flowers whose anthers begin shedding pollen before the flower buds open. The result is nearly complete (95%) self-fertilization in some populations, although outcrossing as high as 29% also has been observed.

While the emphasis in most mating system studies is on estimating proportions of offspring due to selfing, inbreeding due to bi-parental mating between close relatives (*e.g.* brother-sister or father-daughter mating) is also evident in forest stands. Bi-parental inbreeding is often detected when multilocus estimates of t (t_m) are compared to the mean of estimates based on individual marker loci (t_s). Because t_s estimates are downwardly biased by inbreeding other than selfing, and t_m estimates are much less subject to this bias, the difference between t_m and t_s is a measure of bi-parental inbreeding (Box 7.3). For example, among the pine species included in Table 7.4, comparison of single with multiple-locus estimates of t was possible in 15 cases. The mean t_m in these studies was 0.86 and the mean t_s was 0.84, suggesting that 14% of offspring are due to self-fertilization, while bi-parental mating between relatives results in inbreeding equivalent to an additional 2% of selfing. The difference between t_m and t_s estimates is usually small, which suggests that in most cases, inbreeding other than self-fertilization plays a relatively limited role in the mating system of forest trees.

A variety of mechanisms promote the high outcrossing typically observed in forest trees. These mechanisms, as well as factors influencing variation in t among and within species, are discussed later in this chapter in *Mating System Dynamics*.

Strong Migration between Populations

Migration (or gene flow), the movement of alleles between populations, contributes to the maintenance of genetic diversity within populations in two ways: (1) By contributing new genetic variants; and (2) By counteracting the loss of variation due to genetic drift or directional selection (*Chapter 5*). Gene flow in forest trees is largely accomplished by movement of pollen or seeds from one population into another, although dispersal of broken branches by rivers or streams, or during flooding, also contributes to gene flow in species that readily propagate by stem rooting (*e.g.* species of *Populus* and *Salix*).

Migration via Pollen Dispersal

The total output of pollen from stands of wind-pollinated trees can be considerable. Large masses of pollen may rise high above stands and be carried by wind over long distances. For example, high densities of *Pinus banksiana* and *Picea mariana* pollen were detected at 300 m above ground level at the edge of a seed orchard in northern Ontario (Di-Giovanni et al., 1996). Based on estimated settling rates of these pollens, the authors concluded that pollen could drift as far as 50-60 km under mild (5 ms^{-1}), steady winds. Other authors have reported observations of mass pollen dispersal by wind up to several hundred kilometers from source stands (Lanner, 1966; Koski, 1970).

Table 7.4 Estimates of outcrossing (t) in four groups of forest trees.[a]

Type	No. of species	t Median	t Range	References
Pinus	17	0.90	0.47-1.12[b]	Changtragoon and Finkeldey, 1995; Mitton et al., 1997; Ledig, 1998
Picea	7	0.84	0.08-0.91	Ledig et al., 1997
Eucalyptus	12	0.77	0.59-0.84	Eldridge et al., 1994
Tropical angiosperms	13	0.94	0.46-1.08[b]	Hamrick and Murawski, 1990; Murawski and Hamrick, 1991; Murawski et al., 1994; Boshier et al., 1995a; Hall et al., 1996; Kertadikara and Prat, 1995

[a] Proportion of random outcrossed versus self-pollinated offspring.
[b] Due to estimation error, t estimates may exceed 1.0, especially when the true value of t is near one.

Given the extensive pollen dispersal observed in wind-pollinated species, a relatively large proportion of the pollination within any stand may be attributable to immigrant (background) pollen sources that are external to the stand. Based on pollen densities in target stands relative to pollen densities of the same species in nearby stands that did not contain the species, Koski (1970) estimated levels of background pollination for several species in Finland. These estimates ranged from 15% for a species with relatively low abundance in the region (*Alnus incana*) to more than 60% for *Pinus sylvestris* and *Picea abies*, both species of high abundance.

While pollen can be dispersed in mass over large distances, it is effective in migration only if it is viable upon arrival in the target stand and local females are receptive to fertilization. Therefore, the above observations on pollen dispersal and relative pollen densities only indicate the potential for pollen gene flow between populations. Estimates of actual gene flow by pollen (called **pollen flow**) are obtained from studies using genetic markers (Box 5.5). The number of such studies in natural stands of wind-pollinated species is not large, but estimates indicate that pollen flow can be extensive, accounting for at least 25%-50% of viable offspring produced within stands (Table 7.5). Genetic markers also have been used to investigate levels of pollen flow into conifer seed orchards (called pollen contamination, *Chapter 16*), estimates of which often exceed 40% (Adams and Burczyk, 2000). What is not clear is the extent to which pollen flow observed in both natural stands and seed orchards is due to pollen dispersed from adjacent or nearby stands versus distant populations. In six conifer seed orchards separated by 0.5-2 km from the nearest background sources, estimates of pollen contamination ranged widely, from 1% to 48% (Adams and Burczyk, 2000).

Long distance movement of pollen by animal vectors also can be considerable. Dispersal of pollen by animals is primarily determined by their foraging behavior (Levin and Kerster, 1974). Although pollen may not be distributed in mass over long distances as observed in wind-pollinated species, individual animals have the capacity to forage widely between distant trees (Janzen, 1971; Nason and Hamrick, 1997; Nason *et al.*, 1997). Particularly impressive effective pollen dispersal distances of 300-500 m and more have been reported for tropical rainforest species where adult trees typically occur at densities of less than one tree per hectare, and where pollination is mediated primarily by insects (Table 7.5). Therefore, the extremely low population densities in these species appear to be associated with extensive pollinator foraging ranges and wide pollen dispersal (Nason and Hamrick, 1997; Nason *et al.*, 1997). The long-distance record for effective pollen dispersal in forest trees is held by tropical strangler figs (*Ficus* spp.). Flowers in these trees are visited by tiny (1-2 mm), short-lived (2-3 day), wasp pollinators specific to each fig species (Fig. 7.4). In a genetic marker study of offspring in three fig species in Panama, it was estimated that effective pollination occurs routinely over distances of 5.8-14.2 km between widely spaced trees (Nason *et al.*, 1996, 1998).

Estimates of pollen flow into stands of insect-pollinated species are similar in magnitude to those observed for wind-pollinated species (Table 7.5). Therefore, based on the few estimates of actual gene flow available, we cannot assume that pollen flow mediated by animals is any less effective than pollen flow mediated by wind.

Migration via Seed Dispersal

Most information on seed dispersal in forest trees comes from point-source studies where seeds are collected in traps at varying distances from either isolated trees or from trees whose seeds have been marked in some fashion prior to dispersal (Issac, 1930; Boyer, 1958; Boshier *et al.*, 1995b). These studies typically show that seed dispersal is highly leptokurtic; that is, the great majority of seed is dispersed relatively close to the mother (typically within 50 m), while only a small percentage is distributed greater distances away. Therefore, the potential for gene flow by seeds (seed flow) appears to be much less than by pollen flow.

The distribution of seeds from point sources, however, may underestimate the importance of seed flow for two reasons. First, since each seed carries both paternal and maternal genes from the source population, each immigrant seed has twice the effect on gene flow as each immigrant pollen grain, which carries only paternal genes (Nason *et al.*, 1997). Second, maximum seed dispersal distances are truncated in point source studies because of limited sampling distances. Therefore, seed dispersal may be much greater than

Table 7.5. Estimated effective dispersal of pollen within study plots of forest trees and from unknown sources outside the plots (Immigration %); based on analysis of male-parentage in offspring of mother trees using genetic markers.

Species	Location	Study plot			Pollen dispersal (m)		
		Size (ha)	Adults [a] (#)	Average [b] spacing (m)	Within-plot mean	Long-est [c]	Im-migr. [d] (%)
Wind pollinated							
Pseudotsuga menziesii [e]	S.Oregon	2.4	84	17	55	---	27
		2.4	36	26	81	---	20
Pinus attenuata [f]	N.California	0.04	44	3	5.3	11	**56**
Quercus petraea [g]	N.W. France	5.8	124	22	42	165	69
Quercus robur [g]	N.W. France	5.8	167	19	45	140	65
Animal pollinated							
Gleditsia triacanthos [h]	E. Kansas	3.2	61[d]	18	---	240	22
		4.2	124[d]	15	---	120	24
Pithecellobium elegans [i]	Costa Rica	19	28	82	142	350	29
Calophyllum longifolium [j]	Panama	14	4	183	---	382	62
Platypodium elegans [k]	Panama	84	59	119	388	1000+	31
Spondias mombin [l]	Panama	3	10	55	---	200+	44
		4	22	43	---	100+	**60**

[a] For *G. triacanthos*, a dioecious species, this is the number of male trees. All other species are bisexual or monoecious, and this is the total number of adult trees in the study plot.

[b] Average spacing between trees calculated from the plot size and number of adults within the plot.

[c] Longest detected dispersal of pollen gametes within the plot or longest distance of a sampled mother tree from the plot border.

[d] Estimates in bold (*P. attenuata* and *S. mombin*) are for actual pollen flow; *i.e.* total proportion of offspring fertilized by immigrant pollen. In all other cases, this is the observed frequency of offspring sired by males outside the study plot (apparent pollen flow) and is a minimum estimate of pollen flow. When calculable, estimates of actual pollen flow are typically two to three times the apparent pollen flow (Nason and Hamrick, 1997; Adams and Burczyk, 2000).

[e] Adams, 1992. [f] Burczyk *et al.*, 1996. [g] Streiff *et al.*, 1999. [h] Schnabel and Hamrick, 1995. [i] Chase *et al.*, 1996. [j] Stacy *et al.*, 1996. [k] Hamrick and Murawski, 1990.
[l] Nason and Hamrick, 1997.

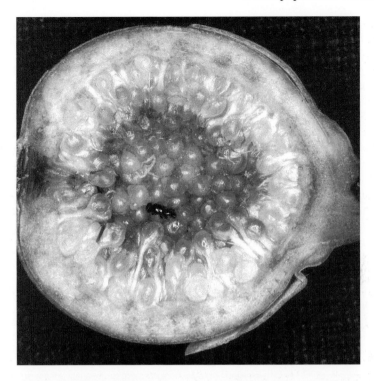

Fig. 7.4. Interior of the fruit of a strangler fig (*Ficus columbiana*) from Panama containing a dead female wasp (genus *Pegoscapus*). These tiny wasps (approximately 2 mm long) typically carry pollen five or more kilometers between trees, and so are extremely effective in promoting long-distance gene flow. (Photo courtesy of J. Nason, Iowa State University)

point-source studies indicate. The potential for long-distance seed dispersal by wind, for example, is illustrated by the ability of exotic pine species to colonize new habitats in New Zealand. Pioneers of *Pinus radiata* and *Pinus contorta*, respectively, have apparently established open areas by seeds dispersed from plantations at least 3 km and 18 km away (Bannister, 1965).

Animals are even more effective in moving seeds. A particularly striking example is Clark's nutcracker (*Nucifraga columbiana* Wilson, Fig. 7.5), a bird species in the western United States that can carry up to 95 *Pinus edulis* seeds at one time in its sublingual pouch (Vander Wall and Balda, 1977). This species has been observed to carry seeds up to 22 km from collection areas, where they bury the seeds in clumps. However, the most effective seed dispersers are humans. In ancient times, nomadic peoples were probably instrumental in spreading large-seeded pine species collected and traded for food (Ledig, 1998). More recently, forest tree seed has been moved extensively for reforestation and for establishing fast-growing exotics on continents far from their origin. Clearly, long-distance seed movement by wind and animals plays an important role in colonization of tree species in new habitats. However, the impact of sporadic long-distance seed dispersal on gene flow between populations is not well understood.

Tracking pollen and seed dispersal, and/or actual gene movement with genetic markers, is necessary to estimate current levels and patterns of gene flow in tree species. Another means of investigating migration, however, is to evaluate the impact of past gene flow on the genetic structure of populations (Box 7.4). Only when gene flow is limited, is it possible for populations to diverge in allele frequencies (*Chapter 5*). The small values of G_{ST} observed in most tree species (Table 7.2) indicate that past gene flow must have been extensive. Comparison of allele frequency divergence among populations for maternally inherited markers *versus* those inherited bi-parentally or through pollen (*Chapter 4*) suggests that historically, pollen flow has had a greater influence on limiting population divergence than seed flow (Box 7.4).

Balancing Selection

Several forms of selection lead to maintenance or "balance" of genetic polymorphisms within populations including heterozygote advantage or overdominance, and diversifying selection (*Chapter 5*). Few examples of heterozygote superiority have been demonstrated, and while it may account for occasional polymorphisms at individual loci, it is an unlikely explanation for the bulk of genetic variation found within populations (Powell and Taylor, 1979) (*Chapter 5*).

Forest trees occupy environments that are extremely heterogeneous in both space and time. For example, changes in soil moisture and fertility, surface roughness and shading, and the presence of diseases, predators and competitors occur over very short distances within a site and have profound influence on survival of seedlings and subsequent tree growth (Campbell, 1979). If different genotypes are favored in different microhabitats within populations, theory suggests that the resulting diversifying selection can lead to stable within-population genetic diversity over time, even when there is extensive intermating among genotypes from different microhabitats (Dickerson and Antonovics, 1973; Hedrick *et al.*, 1976; Powell and Taylor, 1979).

Fig. 7.5. Clark's nutcracker (*Nucifraga columbiana*) harvesting seed from limber pine (*Pinus flexilis*) cones. This bird harvests seeds from a number of high-elevation pines in the western USA burying the seeds in caches, often many kilometers from their source. (Photo courtesy of R. Lanner, USDA Forest Service Institute of Forest Genetics, Placerville, CA)

Box 7.4. Historical levels of gene flow in forest trees.

Recall from *Chapter 5* that for genes relatively neutral to selection (*i.e.* allozymes and molecular markers), the distribution of allele frequency variation between and within populations is primarily a function of the balance between genetic drift within populations and gene flow between them. Wright (1931) showed for a simple model of migration among many populations of equal size, that $F_{ST} = 1/(1 + 4 N_e m)$, where N_e is effective population size and m is the average rate of immigration into each population, each generation. Given that G_{ST} estimates F_{ST} for a set of sampled populations, the number of migrants each generation can be estimated by substituting G_{ST} in the above relationship and solving for $N_e m$, such that $N_e m = 1/4(1/G_{ST} - 1)$.

Slatkin (1987) and Slatkin and Barton (1989) show that this equation provides reasonably accurate estimates of the average level of gene flow even when model assumptions are not strictly met. $N_e m$ is an indirect measure of the amount of gene flow that has occurred historically, resulting in the current population structure. Genetic drift will result in substantial allele frequency differentiation among populations only if $N_e m < 1$ (Slatkin, 1987; Slatkin and Barton, 1989; *Chapter 5*).

Estimates of $N_e m$ in forest trees are typically greater than one, indicating that gene flow is sufficient to prevent substantial differentiation by genetic drift (Table 7.2). Further, these estimates are consistent with the high levels of gene flow observed in gene dispersal studies. Estimates of $N_e m$ in some cases, however, are less than one (Table 7.2), suggesting that gene flow is absent or extremely limited among populations (*e.g. Picea chihuahuana*, *Eucalyptus caesia*, *Magnolia tripetala*). These exceptional cases involve species consisting of small populations that are more or less isolated from each other.

The impacts of pollen and seed dispersal on gene frequency differentiation among populations can be compared by examining G_{ST} values for genetic markers in different plant genomes (*Chapter 4*). For nuclear and for paternally inherited markers (*i.e.* chloroplasts in conifers), gene flow occurs through both pollen and seeds. For maternally inherited markers (*i.e.* chloroplasts in angiosperms and mitochondria in conifers), gene flow occurs only through seed. The ratio of pollen to seed migration rates can be estimated by comparing gene frequency differentiation in the nuclear and maternally inherited genomes (Ennos, 1994). The few ratios reported for trees consistently have been greater than one, indicating, as expected, that rates of gene flow by pollen exceed those by seed (Ennos, 1994; El Mousadik and Petit, 1996). Relative rates of pollen/seed gene flow were particularly high (84-500) in three angiosperms with large nuts (*Quercus petraea*, *Quercus robur*, *Fagus sylvatica*), and somewhat less so (7-44) in lighter-seeded conifers. Two species with particularly low ratios (< 3) are *Eucalyptus nitens* and *Argania spinosa*. Pollen flow in *E. nitens* may be limited because of selfing and because of limited pollen dispersal by insects. *A. spinosa*, an endangered angiosperm tree endemic to Morocco, is also insect pollinated. However, seed dispersal in this species may be quite high because its fleshy fruits are consumed by goats and camels and are subsequently regurgitated later in the day or evening when the animals ruminate, perhaps after traveling many kilometers from source trees.

Year-to-year fluctuations in weather, and habitat changes with stand development may favor different genotypes at different times, leading to genetic differentiation among age-class cohorts within the same stand. Therefore, temporal environmental heterogeneity also can lead to diversifying selection, although stable polymorphisms are more likely to result from environmental heterogeneity in space than over time (Hedrick *et al.*, 1976).

The extent to which balancing selection contributes to the total genetic diversity within forest tree populations is difficult to quantify because of the inability to separate the effects of balancing selection from other factors promoting diversity, such as migration between populations and outcross mating systems. Nevertheless, three observations support the conclusion that balancing selection has a significant role in maintaining genetic variation within populations. First, the large reproductive capacity of forest trees allows for intense selection to act in the replacement of trees from one generation to the next. For example, based on demographic data, Campbell (1979) estimated that it takes more than 350,000 seedlings to replace the 74 mature trees per hectare at age 450 in an old growth *Pseudotsuga menziesii* stand; this is 5200 seedlings for each survivor at age 450 (Fig. 7.6). Similarly, Barber (1965) estimated that after a wildfire, 20,000 seedlings are required to replace each *Eucalyptus regnans* breeding tree that will occupy the regenerated stand at age 400. Even assuming that a large proportion of seedling mortality (say 50%) is due to chance alone, there is still considerable opportunity for selection among the remaining individuals.

Second, when stands within a local area occupy environments that are substantially different, adaptive genetic variation between stands may be extensive, even if gene flow between stands is strong. This has been observed in a number of studies where populations of forest trees span steep topographic gradients or marked discontinuity in soils within a few kilometers (*Chapter 8*). If adaptive variation can exist among local stands in the face of extensive gene flow, it is not unreasonable to assume that diversifying selection acting at a smaller scale (*i.e.* among microhabitats within a stand) also contributes to genetic diversity within populations.

Fig. 7.6. Intense competition occurs among the many thousands of seedlings necessary to replace each mature Douglas-fir (*Pseudotsuga menziesii*) through natural reproduction and subsequent stand development. This competition provides ample opportunity for natural selection to influence the genetic composition of mature stands. (Photo by W.T. Adams)

Final support for the adaptive significance of genetic variation within populations is the increased susceptibility of crop plants and forest trees to pests and climatic extremes when genetically uniform varieties have been planted (Kleinschmit, 1979; Ledig, 1988a). Both the Irish potato famine in the 1840s and the 1970s corn blight in the United States were the consequence of epidemics that occurred in extremely uniform crop varieties planted over large areas. Plantations of individual clones of forest trees are also vulnerable to damaging pest attacks (Libby, 1982; Kleinschmit *et al.*, 1993). The presence of genetic diversity within populations increases the likelihood that at least some individuals are resistant to pest infestation or to damage from climatic extremes that otherwise might completely destroy a population with only one or a few genotypes. Genetic diversity, however, may not be essential to the survival of populations or even species. A number of species listed in Table 7.2 are extremely limited in genetic variation (*e.g. Acacia mangium* and *Pinus resinosa*), yet are apparently very successful, with large geographical ranges. However, the extremely low diversity in these species is exceptional among forest trees, and although these species may be thriving at present, their ability to adapt to future changes in environment may be greatly hampered by the limited genetic variation they possess.

MATING SYSTEM DYNAMICS IN FOREST TREES

Earlier we saw that high outcrossing rates are common in most forest tree species, and this is an important factor contributing to the high levels of genetic diversity in most trees as compared to annual plants (Table 7.1). In this section, we extend our earlier discussion of mating systems to explore: (1) The biological mechanisms responsible for the high rates of outcrossing in forest trees; (2) The factors that lead to unusually low levels of t (*i.e.* high self-fertilization) in some circumstances; and (3) The patterns of cross-fertilization that occur among trees within forest stands.

Mechanisms Promoting High Levels of Outcrossing

A variety of factors in the reproductive biology of forest trees prevent or limit self-fertilization or the viability of selfed offspring and thus, favor outcrossing. Because tropical lowland rain forests often consist of assemblages of many tree species, resulting in wide separation between any two individuals of the same species, early investigators assumed that selfing must predominate in tropical trees (Loveless, 1992). However, later studies revealed that about 25% of tropical angiosperm tree species are **dioecious** (male and female flowers occur on separate individuals), and therefore, are incapable of self-fertilization (Bawa, 1974; Bawa and Opler, 1975; Bawa *et al.*, 1985). Further, although the majority (65%) of tropical angiosperm species have bisexual flowers, most (80%) of these species are self-incompatible. **Self-incompatibility**, which is common in angiosperms, is due to chemical interactions between self pollen grains and female reproductive tissues that: (1) Prevent pollen penetration of the stigma or growth of pollen tubes down the style (early self-incompatibility); or (2) Inhibit the development of the ovule or embryo (late self-incompatibility) (Owens and Blake, 1985; Seavey and Bawa, 1986; Radhamani *et al.*, 1998).

The remaining tropical angiosperm forest tree species (10%), and most conifers and temperate angiosperm trees are **monoecious**; that is, male and female reproductive structures occur on the same individual tree but in separate floral structures (Krugman *et al.*, 1974; Owens and Blake, 1985; Ledig, 1986). Monoecism promotes outcrossing. In some conifers, physical separation of the sexes is enhanced (*e.g.* in *Abies*, *Pseudotsuga*, *Pinus*) because of the tendency for female strobili to occur at the ends of branches and in the upper crown, while male strobili occur on inner shoots in the mid to lower crown. In addition, male floral structures may begin to shed pollen before or after female structures start to become receptive. However, neither spatial nor temporal separation of the sexes in conifers completely prevents self-fertilization, and often both forms of separation are weak to non-existent. Temporal separation of the sexes may also occur in species with bisexual flowers. For example, **protandry** (*i.e.* the stigma is not receptive until some days after pollen begins shedding from anthers in the same flower) helps promote cross-fertilization in *Eucalyptus* species where pollination is mediated by insects and birds (Eldridge *et al.*, 1994). **Protogyny** is when the stigma is receptive before pollen is shed from the anthers of the same flower.

In species lacking self-incompatibility (*i.e.* conifers and many angiosperm trees), the most important mechanism promoting outcrossing is abortion of selfed embryos resulting from the expression of homozygous lethal genes (Orr-Ewing, 1957; Hagman, 1967; Sorensen, 1969; Franklin, 1972; Owens and Blake, 1985; Ledig, 1986; Muona, 1990). Because of embryo abortion, the proportion of filled seed after self-fertilization relative to cross-pollination (*i.e.* self-fertility) is usually small (Table 5.3). Therefore, even when a large percentage of pollen around a mother tree is self-pollen, the frequency of selfs in viable offspring may be very low (Franklin, 1971; Sorensen, 1982). For example, Sorensen and Adams (1993) estimated that in *Pinus contorta* stands in Oregon, the proportion of viable self-fertilized offspring was only one-third to one-tenth the frequency of embryos resulting from self-fertilization.

An important phenomenon reducing the frequency of nonviable (empty) seeds when there is mixed selfing and outcrossing is **polyembryony**. In conifers, several archegonia usually develop within an ovule (Fig. 7.7), each with an egg cell (Krugman *et al.*, 1974; Owens and Blake, 1985). The number of archegonia per ovule varies; in *Pinaceae*, the number ranges from 1-10, but most ovules have 3-5 (Owens and Blake, 1985). Although more than one egg may be fertilized, rarely does more than one develop into an embryo, indicating that competition occurs among embryos within an ovule. Generally, outcrossed embryos are favored in competition with selfed embryos, such that fewer selfed seeds are produced than expected if there was only a single egg fertilized (Barnes *et al.*, 1962; Franklin, 1974; Smith *et al.*, 1988; Yazdani and Lindgren, 1991). Analogous mechanisms to polyembryony also occur in angiosperm tree species (Owens and Blake, 1985). In *Quercus*, for example, the ovary originally has six ovules, but acorns usually contain only one mature embryo (Mogensen, 1975).

Factors Leading to Unusually Low Levels of Outcrossing

While high rates of outcrossing are commonly found in forest trees, unusually low rates of outcrossing are found in situations where the female flowers (or strobili) of individual trees are physically or temporally isolated from outcross pollen sources. As the distances between stands and between trees within stands increase in species with bisexual or monecious flowering, a higher proportion of the available pollen around individual trees will be produced by the individuals themselves, resulting in higher rates of self-fertilization. *Picea chihuahuana*, for example, an endangered spruce species (Fig. 7.2a), has a highly fragmented distribution of mostly small populations in the high mountains of central Mexico. In two small populations (each with less than 40 spruce trees in mixed stands), t was estimated to be 0 in one and 0.15 in the other indicating 100% and 85% selfed progeny, respectively (Ledig *et al.*, 1997). *Pinus merkusii* is the only pine species with a native range in the southern hemisphere. Among 11 populations in Thailand, where

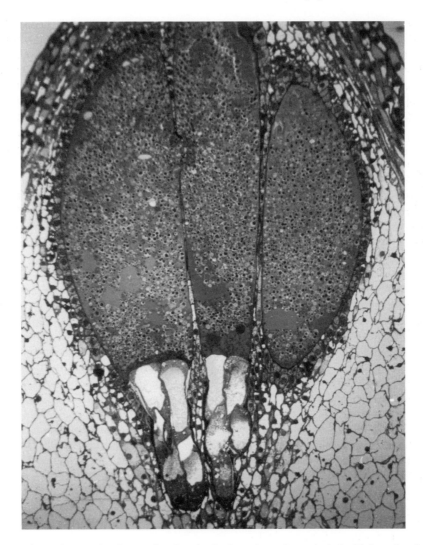

Fig. 7.7. Photomicrograph of a median longitudinal section of a resin-imbedded ovule of *Pseudotsuga menziesii* showing three archegonia, two having young embryos and the third (on the right) being unfertilized. Although ovules of this species typically contain 3-4 archegonia of which two or more are fertilized (resulting in polyembryony), mature seeds seldom contain more than a single viable embryo. (Reproduced with permission from Owens *et al.* 1991)

the species is found in scattered relics of forest in otherwise deforested areas, or in forests dominated by hardwoods, estimates of t ranged from 0.02 to 0.84, and averaged 0.47 (Changtragoon and Finkeldey, 1995). In both of these species, seed collections from populations with unusually low rates of outcrossing also had high frequencies of empty seed, indicating that the high levels of inbreeding may be threatening the capacity of these populations to reproduce.

Tree density also is a factor influencing variation in levels of outcrossing among populations within a species (Farris and Mitton, 1984; Knowles *et al.*, 1987; Murawski and Hamrick, 1992a; Sorensen and Adams, 1993; Hardner *et al.*, 1996). For example, Murawski and Hamrick (1992a) studied the mating system in *Cavanillesia plantanifolia*, a tropical canopy tree with bi-sexual flowers, whose pollination is mediated by insects, hummingbirds and perhaps, monkeys (Murawski *et al.*, 1990). In two populations on Barro Colorado Island in the Republic of Panama, density of flowering trees ranged five-fold over sites and years. Estimates of t decreased linearly with decreasing flowering-tree density (Fig. 7.8). Given that this species is self-fertile, it is amazing that any outcrossed seed were produced in the lowest stand density where there was only about one flowering tree every 10 hectares.

Although outcrossing is expected to decrease with flowering-tree density, this is not always observed (Furnier and Adams, 1986b; Morgante *et al.*, 1991; El-Kassaby and Jaquish, 1996). The occurrence and density of other tree species in the stand and the degree of pollination by pollen sources outside the stand are confounding factors. In addition, when self-fertility is low, even relatively large changes in frequency of self-fertilization may have only limited impact on the proportion of selfs observed in seedlings (Sorensen, 1982). This is certainly evident in tropical angiosperm trees that sometimes have only one or fewer individuals per hectare, yet have very high outcrossing rates due to well-developed self-incompatibility systems (Stacy *et al.*, 1996).

Outcrossing rates can vary widely among individual trees in populations and over years in the same trees. For example, estimates of t ranged from 0.25 to 1.02 among 9 trees in a stand of *Thuja occidentalis* (a wind-pollinated conifer) in northeast Ontario (Perry and Knowles, 1990) (Note that due to estimation error, t estimates can exceed 1.0, especially when the true value of t is near one.). In *Ceiba pentandra*, a tropical angiosperm with bi-sexual flowers, t ranged from 0 to 1.34 among 11 trees in a population in Panama (Murawski and Hamrick, 1992b). In one individual mother tree of *C. plantanifolia*, estimated t ranged from 0.79 in one year when it flowered simultaneously with 6 other trees to zero in the next year when it was the only tree in the cluster to flower (Murawski *et al.*, 1990).

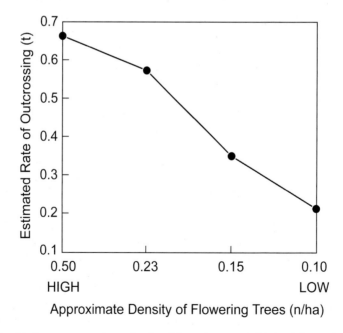

Fig. 7.8. Relationship between approximate density of flowering trees and estimated rate of outcrossing (t) in *Cavanillesia plantanifolia* on Barro Colorado Island, Republic of Panama (Murawski and Hamrick, 1992a). Clearly, a higher density of flowering trees means higher levels of outcrossing in this self-fertile species.

Factors influencing tree-to-tree variation in t are basically the same as those influencing differences in t between stands within species. Trees that are isolated spatially, isolated phenologically because of particularly early or late female receptivity, have high levels of pollen production relative to surrounding trees, or are highly self-fertile will likely have lower levels of outcrossing (Sorensen, 1982; Shea, 1987; El-Kassaby *et al.*, 1988; Erickson and Adams, 1990; Murawski and Hamrick, 1992a; Sorensen and Adams, 1993; Hardner *et al.*, 1996).

Finally, because of the tendency in some conifers for female cones to be borne on branches in the upper portion of crown, and males lower down, one might expect female cones in the top of the tree to be more outcrossed than mid- or lower-crown cones that are closer to the source of self-pollen. This expectation has been confirmed in a number of cases. For example, estimates of t in the upper versus lower crown were 0.90 *versus* 0.76 in *Picea sitchensis* (Chaisurisri *et al.*, 1994), 0.93 *versus* 0.86 in *Pseudotsuga menziesii* (Shaw and Allard, 1982), and 0.87 *versus* 0.74 in *Pinus banksiana* (Fowler, 1965b). In other cases (some involving the above species), crown level was found to have little or no effect on t (Cheliak *et al.*, 1985; El-Kassaby *et al.*, 1986; Perry and Dancik, 1986; Sorensen, 1994). Tree age, size and density influence the distribution of male and female cones, the amount pollen produced, and pollen dispersal characteristics, and therefore, are all factors influencing rates of outcrossing in different parts of the crown. In particular, conditions that favor the development of long crowns and heavy pollen production should favor more outcrossing in the upper than in the lower crown (Sorensen, 1994).

Patterns of Cross-fertilization within Populations

We learned earlier in this chapter that self-fertilization typically accounts for about 10%-20% of the viable offspring produced in forest stands, while pollen flow from nearby or distant stands accounts for another 25%-50%. In this section, we briefly address the remaining component in the breeding structure of forest trees: patterns of cross-fertilization among individuals within populations, and the factors influencing these patterns.

In both wind- and animal-pollinated species, dispersal of pollen from individual trees follows a leptokurtic distribution, similar in shape to the frequency distribution of seed dispersal from point sources, but with a longer tail (*i.e.* greater total dispersal distance; Wright, 1952; Wang *et al.*, 1960; Levin and Kerster, 1974). This pattern of pollen distribution is reflected in the gene dispersal observed in two tree species, where rare allozyme marker alleles were used to track pollen from male source trees in the offspring of females located increasing distances away (Fig. 7.9). In both cases, the source trees were located in mature, relatively pure stands, but while *Pinus sylvestris* is wind pollinated, insects mediate pollination in *Cordia alliodora*.

The rapid decrease in effective pollination (*i.e.* effective pollen dispersal) with increasing distance from the male marker tree shows that nearby males have a greater probability of mating with a specific mother tree, than individual males further away (see also Hamrick and Murawski, 1990; Burczyk *et al.*, 1996; Streiff *et al.*, 1999). This is the expected relationship between mating success and distance between mates as long as tree-to-tree differences in floral timing or pollen productivity are not great (see below). Nevertheless, in both examples, the marker male never accounts for more than 15% of offspring, even in the nearest females (Fig. 7.9). Therefore, the proportional contribution of the nearest male is diluted by the collective contribution of pollen produced by more distant males within the population and by pollen flow immigrating from other populations (Bannister, 1965; Adams, 1992).

Genetic markers also can be used to study cross-fertilization patterns by inferring the male parentage of open-pollinated offspring of mother trees located within designated study plots (paternity analysis; see references in Table 7.5). These studies often show that pollen gametes effective in fertilizing offspring come from numerous and sometimes distant males. Mean effective pollen dispersal distance is frequently estimated as 2-3 times the average spacing between trees, with some pollen gametes coming from trees up to 8 times this distance away (Table 7.5). Therefore, the number of males mating with each female may be considerable. For example, Burczyk *et al.* (1996) estimated that the extensive cross-fertilization in a *Pinus attenuata* stand (due both to mating within the stand and external pollen flow) was equivalent to each female mating at random to 59 males (*i.e.* the effective number of mates for each female was 59). Likewise, Nason *et al.* (1996, 1998) estimated that the number of effective males contributing to the single fruit crops of individual females, ranged from 11-54 (mean 24) among seven *Ficus* species. In addition, because only a small proportion of the total *Ficus* trees in an area flowered during a particular flowering period, the effective male population contributing to each female over several years is at least 10 times these numbers.

A range of factors including pollen productivity and floral phenology differences among males, and the spatial arrangement of trees within stands, influence the relationship between mating success and distance between mates. For example, in conifers, male reproductive success increases with increasing pollen productivity and with increasing overlap of the timing of pollen dispersal and conelet receptivity of mother trees (Shen *et al.*, 1981; Schoen and Stewart, 1986; Erickson and Adams, 1989; Burczyk *et al.*, 1996). Therefore, a neighboring male

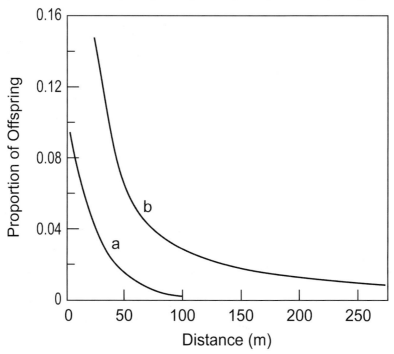

Fig. 7.9. Relationship between distance of mother trees from a male marker-tree, and the proportion of offspring sired by the marker-tree, after pollination mediated by: (a) Wind in a 120-year-old *Pinus sylvestris* stand in Germany (Muller, 1977); and (b) Insects in a *Cordia alliodora* stand in Costa Rica (Boshier *et al.*, 1995b). For both species, mating success of the marker-tree decreased rapidly with increasing distance from females.

may be responsible for siring the majority of offspring when it is in complete floral synchrony with the mother tree, especially if it is a heavy pollen producer. On the other hand, a nearby male may contribute little to a female's offspring if it is out of phase with the receptivity of the mother tree, regardless of the amount of pollen it produces.

Spatial arrangement of trees has a profound influence on cross-fertilization patterns in low-density, insect-pollinated, tropical angiosperms. In the tropical species presented in Table 7.5, individual trees were widely distributed over the study areas and wide effective dispersal of pollen was observed. In two other species, with highly clustered spatial distributions, *Spondias mombin* and *Turpinia occidentalis*, neighboring males in the 2- to 3-tree clusters sired the great majority (72%-99%) of offspring, although small amounts of effective pollen came from 200-300 m away (Stacy *et al.*, 1996).

It is clear that gene dispersal by pollen can be extensive in both wind- and animal-pollinated tree species. It is also clear that insect vectors are highly efficient in promoting cross-fertilization, even in tropical angiosperms where the nearest neighbors may be separated by hundreds of meters. What is not clear, however, are the conditions under which cross-fertilization might be strongly curtailed, as observed in *S. mombin* and *T. occidentalis*, and how frequently this occurs. There is still much to learn about the breeding structure of forest trees.

SPATIAL AND TEMPORAL GENETIC STRUCTURE WITHIN POPULATIONS

The distribution of genotypes over space (spatial genetic structure) and over age classes (temporal genetic structure) within stands influences the mating system and the propensity of populations to become subdivided due to genetic drift or selection (Levin and Kerster, 1974; Epperson, 1992). Spatial clustering of genotypes can arise from limited gene dispersal, vegetative reproduction, historical disturbance, and microenvironmental heterogeneity that favors different genotypes in different parts of the stand (Epperson and Allard, 1989; Schnabel and Hamrick, 1990; Epperson, 1992; Streiff *et al.*, 1998). Genetic variation among age classes within stands can result from yearly variation in mating patterns or in the subset of parents involved in mating, or from selection that differs among age cohorts (Schnabel and Hamrick, 1990).

Spatial Genetic Structure

Three types of evidence have been used to infer the extent and patterns of spatial genetic structure within forest tree populations: (1) Detection of bi-parental inbreeding; (2) Detection of inbreeding depression in offspring after controlled mating of neighboring trees; and (3) Observed clustering of genotypes within stands (Epperson, 1992). It was pointed out earlier in this chapter (*High Levels of Outcrossing* and Box 7.3) that the mean of single locus estimates of outcrossing is often marginally less than the multilocus estimate based on the same marker loci, suggesting that at least some bi-parental inbreeding occurs in forest populations. Such bi-parental inbreeding is likely if relatives are spatially clustered (Ledig, 1998), but can also occur if related individuals are clustered in terms of flowering times (Epperson, 1992).

In most tree species, mating between close relatives results in inbreeding depression (*Chapter 5*). Therefore, if stands consist of clusters of close relatives, mating between neighboring trees is expected to result in reduced seed set and seedling growth, at least relative to offspring of crosses between trees from distant locations. This expectation was observed in studies of two boreal conifers in Canada, *Picea glauca* and *Larix laricina*

(Table 7.6). Based on the degree of inbreeding depression in offspring of near-neighbors, relative to offspring produced by self-fertilization and crosses between distant (and presumably unrelated) trees, the average coefficient of relatedness (r) between neighboring trees was estimated. The **coefficient of relatedness** is the probability that alleles sampled at random from two individuals are identical by descent (Falconer and Mackay, 1996). This value is equal to twice the inbreeding coefficient in the offspring (*Chapter 5*), and is 0 when the two parents are unrelated, 0.125 when they are first cousins, 0.25 when half-sibs, 0.50 when full-sibs and 1.0 when they are clones of the same individual. In *P. glauca*, the average estimate of r (r = 0.26) was consistent with neighboring trees being open-pollinated offspring of the same mother (*i.e.* half-sibs), while the mean relatedness was somewhat less in *L. laricina* (r = 0.17). In contrast to these two boreal conifers, little or no inbreeding depression was observed in the offspring of neighboring trees in a *Pseudotsuga menziesii* stand in Oregon, suggesting that, in this case, there is no spatial clustering of close relatives (Sorenson and Campbell, 1997).

A variety of approaches have been used to directly evaluate spatial clustering of genotypes within stands. The simplest is to plot the location of genotypes on a map and to visually assess the degree of clustering (Knowles, 1991; Furnier and Adams, 1986b). A second approach is to subdivide stands into subplots of various sizes and examine allele-frequency differentiation among subplots using gene diversity analysis (*i.e.* G_{ST} or F_{ST}) (Hamrick *et al.*, 1993a,b; Streiff *et al.*, 1998). However, the most common procedure is to quantify similarity between pairs of genotypes or subplots (based on allele frequencies) within a specified distance and evaluate whether the pairs are more similar than expected by chance under random spatial arrangement (*e.g.* spatial autocorrelation analysis) (Epperson, 1992). This is repeated for pairs at varying distances apart. If a patchy distribution exists, it is expected that pairs will be more similar than by chance at shorter distances, but less similar at larger distances.

Although spatial clustering analyses sometimes reveal little or no spatial structure within stands (Roberds and Conkle, 1984; Epperson and Allard, 1989; Knowles, 1991; Leonardi *et al.*, 1996), more commonly, at least some clustering of genotypes is evident. In nearly all cases, however, cluster sizes are quite small (5-50 m across), suggesting that they are primarily the result of limited seed dispersal, and that clusters are made up of close relatives. This latter result is consistent across a wide variety of tree species and seed sizes, for example, in conifers (Linhart *et al.*, 1981; Knowles *et al.*, 1992; Hamrick *et al.*, 1993b); temperate angiosperms (Schnabel and Hamrick, 1990; Berg and Hamrick, 1995; Shapcott, 1995; Dow and Ashley, 1996; Streiff *et al.*, 1998); and tropical angiosperms (Hamrick *et al.*, 1993a). Differential selection due to microenvironmental heterogeneity is usually not

Table 7.6. Mean percent filled seed and mean seedling growth of viable offspring, after crossing mother trees in stands of two boreal conifers with self pollen, pollen from neighboring trees, and pollen from distant (unrelated) trees.[a]

Pollen type	Picea glauca[b]		Larix laricina[c]	
	% filled seed	Epicotyl length (mm)	% filled seed	Two-year height (cm)
Self	6.1	51.1	1.6	72.0
Near-neighbor	31.2	69.0	18.4	85.2
Distant trees	43.3	73.8	21.5	87.6

[a] Distant trees is a mixture of pollen from stands at least 1km away from mother trees.
[b] Near neighbors were within 100 m of the mother tree (Coles and Fowler, 1976).
[c] Mean distance to near neighbors was 22 m (Park and Fowler, 1982).

offered as a likely explanation for the observed spatial clustering of genetic markers because: (1) Associations between marker genotypes and microenvironments is not evident; and (2) Genetic markers are generally considered to be neutral to selection pressure, or nearly so (Box 5.6).

Spatial clustering of genotypes also can occur when seeds are dispersed in clumps (Furnier *et al.*, 1987; Schuster and Mitton, 1991; Schnabel *et al.*, 1991; Vander Wall, 1992, 1994). For example, two subalpine pine species in North America, *P. albicaulis* and *P. flexilis*, often are found in tight clusters of several individual trees that result from groups of seeds being harvested and later cached by the bird species Clark's nutcracker (Fig 7.5), perhaps many kilometers from source trees. Trees within clumps are more similar genetically than trees from different clumps. In *P. flexilis*, for example, it was estimated that trees within clusters were related, on average, slightly less than half-sibs (r = 0.19; Schuster and Mitton, 1991).

If restricted seed dispersal is the primary mechanism promoting spatial clustering of genotypes, one might expect cluster size to increase over successive generations, especially when clustering is reinforced by mating between relatives within clusters (Levin and Kerster, 1974; Streiff *et al.*, 1998). Three factors contribute to limited cluster size: (1) Extensive pollen dispersal counters the effect of restricted seed movement; (2) Clusters become less distinct as stands age and trees within clusters are lost due to competition-induced mortality; and (3) Disturbance during regeneration.

The differential effect of seed and pollen dispersal on spatial genetic structure is illustrated in a study conducted in a single, 1.5 ha, pure stand of *Pinus ponderosa* in Colorado (Latta *et al.*, 1998). Spatial structure was investigated using genetic markers inherited through two haploid organelle genomes: mitochondrial DNA (mtDNA), which is inherited maternally in *Pinus*, and chloroplast DNA (cpDNA), which is inherited through pollen. Therefore, spatial structure of mtDNA markers is due entirely to seed dispersal, whereas spatial structure of cpDNA markers is the combined result of dispersal, first by pollen, and then by seeds, but should be primarily determined by pollen dispersal. Mapping a rare marker haplotype of each genome shows obvious clusters of the mtDNA haplotype (circled), while the cpDNA haplotype appears to be randomly distributed across the site (Fig. 7.10). Based on an analysis of allozyme genotypes, it was estimated that trees within the mtDNA-haplotype clusters had an average coefficient of relatedness of r = 0.27, suggesting that the clusters represent groups of half-sibs related through their maternal parent. It is clear from Fig. 7.10 why spatial genetic clustering is difficult to detect in forest trees when based on spatial distributions of diploid genotypes, which are the combined result of both seed and pollen dispersal patterns.

When spatial genetic structure is compared among age classes within the same stand, clustering that is clearly evident in seedlings or young trees diminishes, often substantially, in adults (Schnabel and Hamrick, 1990; Hamrick *et al.*, 1993b; Dow and Ashley, 1996; Streiff *et al.*, 1998). This is probably due to differential selection against inbreds within clusters or because most trees within patches are eliminated by intense competition during stand development.

The amount and pattern of disturbance prior to reproduction (*i.e.* stand history) can also greatly influence subsequent spatial genetic structure. Knowles *et al.* (1992), for example, compared spatial genetic clustering in two stands of *Larix laricina* in Ontario, Canada. The stand naturally regenerated after a clearcut, presumably by a few remnant individuals scattered within the stand, shows significant spatial clustering of genotypes. No clustering, however, was observed in a nearby old-field stand colonized by trees along the edge of the field. Similarly, Boyle *et al.* (1990) found significant genetic clustering in a lowland site of *Picea mariana*, where there was little evidence of a major disturbance, while no genetic clustering was found at a second location where high tree density and uniformity of tree size suggested that the site had been regenerated after a fire.

In conclusion, at least some spatial genetic structure appears to be relatively common in forest tree populations, although it is likely to be limited primarily to tight clustering of siblings around mother trees and to younger age classes. The presence of clustering is highly dependent on stand history, especially patterns of disturbance during reproduction.

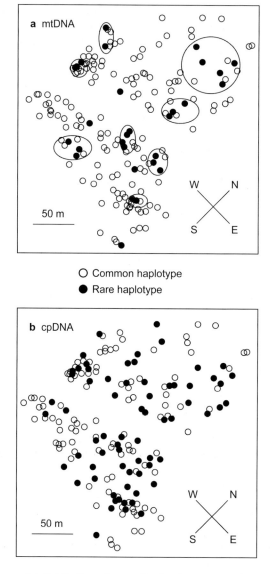

Fig. 7.10. Maps of the spatial distribution of organellar haplotypes among trees in a *Pinus ponderosa* stand in Colorado: (a) Spatial arrangement of the rare mtDNA haplotype in clusters (circled) reflects the maternal inheritance of mitochondria in this species and limited seed dispersal; and (b) In contrast, the lack of clustering in the rare cpDNA haplotype reflects the paternal inheritance of chloroplasts and wide dispersal of pollen. (Reproduced with permission from Latta *et al.*, 1998)

Temporal Genetic Structure

Allele frequency differentiation among age-class cohorts within stands is usually small (Lin-hart *et al.*, 1981; Knowles and Grant, 1985; Plessas and Strauss, 1986; Schnabel and Hamrick, 1990; Leonardi *et al.*, 1996), which is expected given strong gene flow and large population sizes. An exception to this trend is an old-field stand of *Pinus taeda* in North Carolina, where significant allele-frequency differentiation was observed among three fairly distinct age classes at four of the eight allozyme loci investigated (Roberds and Conkle, 1984). Because this stand was probably regenerated by a few nearby scattered trees or small groups of trees, it is not unreasonable to conclude that different small subsamples of parents contributed to each cohort. Therefore, genetic drift (*i.e.* founder principle, *Chapter 5*) may be largely responsible for the allele-frequency differentiation among ages.

Although allele frequencies may differ little between age classes, a common observation is that homozygosity decreases as stands age (Shaw and Allard, 1982; Muona, 1990; Morgante *et al.*, 1993; Ledig, 1998). Typically, the fixation index (F), which measures the frequency of homozygotes relative to that expected given Hardy-Weinberg equilibrium (*Chapter 5*), is positive in seedlings (homozygosity excess), but is near zero in adults. Excess homozygosity at the seedling stage is most likely due to: (1) Inbreeding among parent trees (primarily selfing); and/or (2) Spatial clustering of offspring genotypes due to limited seed dispersal (Wahlund effect). As trees age, however, inbred progeny are selected against, especially in young trees (Box 5.2), and spatial genetic clustering is reduced by competition-induced mortality (see previous section).

PRACTICAL IMPLICATIONS OF WITHIN-POPULATION GENETIC DIVERSITY

Genetic Improvements under Natural Regeneration Systems

The large amount of within-population genetic diversity typically observed in forest trees includes variation for such important economic traits as stem size and form, wood quality, drought hardiness, and disease resistance. Therefore, large improvements in the economic value of trees can be made by selecting and breeding the best genotypes (*i.e.* genotypes that produce offspring with desirable characteristics; *Chapter 13*). This is particularly true in intensive tree improvement programs that involve progeny testing of selected parents and operational deployment of improved offspring (*Chapters 13-17*).

Stands of trees also can be genetically improved in natural regeneration systems by leaving individuals with the most desired characteristics as parent trees. As described in *Chapter 13*, the amount of improvement in offspring that can be attained by selection of parent trees depends primarily on two factors: (1) The heritability of the traits being selected (*Chapter 6*); and (2) The selection intensity. The higher the heritability and proportion of poor phenotypes culled prior to regeneration (*i.e.* greater the selection intensity), the greater the improvement expected in offspring. Therefore, the potential for improvement is greatest in shelterwood and seed tree systems, where relatively small numbers of parents are left for regeneration.

In most cases, genetic improvement through natural regeneration is expected to be modest at best, especially when compared to what can be achieved in more intensive tree improvement programs. The reasons for this are as follows: (1) Traits of economic importance (*e.g.* stem growth rate) typically have low heritability, so it is difficult to judge superior genotypes based on phenotypes alone; (2) Selection intensity is constrained because in order to achieve uniform seed dispersal across the site, parent trees must be uniformly spaced, which somewhat restricts the choice of selected individuals; and (3) Genetic improvements achieved by selecting better genotypes are diluted in the offspring because much of the seed is likely to be sired by unimproved males in nearby stands (*i.e.* by pollen gene flow).

Despite these caveats, growth superiority of naturally regenerated offspring from faster-growing parent trees has been demonstrated (*e.g.* Wilusz and Giertych, 1974; Ledig and Smith, 1981), and even greater improvements documented in offspring of parents selected for disease resistance (Zobel and Talbert, 1984), which has higher heritability than growth traits. Therefore, the potential for genetically improving the value of stands through positive selection of parent trees and natural regeneration should not be overlooked by foresters. It is an inexpensive means of achieving rapid, although modest improvements, while ensuring that offspring are well adapted to the site.

The large amount of genetic variation within forest tree populations also provides the "opportunity" for negative selection (*i.e.* dysgenic selection) or high grading, whereby the best trees are harvested and the poorest are left to regenerate the stand. This practice increases profit from a harvesting operation, but reduces the genetic quality (and value) of the stand for the next and future generations. The problem is exacerbated if only a very few culls are left on site, because in addition to negative selection, levels of inbreeding (*i.e.* selfing) may be substantially increased. High grading is a short-sighted and wasteful process that should be strictly avoided.

Another potential negative consequence of natural regeneration regimes is that the reduced stand density resulting from selective thinning or harvesting may lead to increased proportions of selfed offspring, although this is not likely to be a problem unless stand density is reduced drastically. For example, in two pairs of adjacent old-growth stands of *Pseudotsuga menziesii* in southwest Oregon, one stand of each pair was uncut (~100 trees/ha), while the other was thinned to a shelterwood density (15-35 trees/ha). Multilocus estimates of outcrossing (t_m) showed thinning had little or no effect on the proportion of selfed offspring (*i.e.* mean t_m in the uncut stands was 0.98 and was 0.95 in the shelterwoods) (Neale and Adams, 1985a). Likewise, levels of outcrossing were nearly identical between a thinned seed production stand ($t_m = 1.01$) of *Pinus caribaea* var *caribaea* in Cuba and a nearby wild population ($t_m = 0.98$) (Zheng and Ennos, 1997). However, the proportion of outcrossed progeny in a stand of *Shorea megistophylla*, a bee-pollinated, canopy tree in Sri Lanka that had been thinned to 2 trees/ha was appreciably lower ($t_m = 0.71$) than in an adjacent undisturbed forest (10 trees/ha, $t_m = 0.87$) (Murawski *et al.*, 1994).

Seed Collections in Natural Populations

Whether collecting seed from wild stands for operational reforestation, *ex situ* gene conservation (*Chapter 10*), or for progeny testing of selected parents (*Chapter 14*), it is desirable to avoid inbreeding, particularly self-fertilization (Sorensen and White, 1988; Sorensen, 1994; Hardner *et al.*, 1996). For most applications, it is also desirable that open-pollinated offspring from each mother tree be the result of a large number of male parents. Based on our current knowledge of mating systems and pollen dispersal in forest trees, the following recommendations for wild-stand seed collections can be made:

- Collect seed only in larger stands, where the density of individuals is normal for the species, and only from seed crops resulting from good flowering years. Avoid collecting seed in small, isolated stands, and in stands with particularly low densities of flowering trees, unless these stands are the only stands available in the region of interest.

- Avoid collecting seed from isolated trees that are likely to have elevated proportions of selfed offspring.
- Collect seed only from cones or fruits in the upper one-third of the crown to maximize the potential for cross-fertilization.
- Collect seed only from trees that are at least 50-100 m apart if the goal is to minimize relatedness among seed lots from different mothers (remembering that neighboring trees may be closely related).

SUMMARY AND CONCLUSIONS

Forest trees typically possess levels of genetic variation that are among the highest observed in all living organisms. At the single gene level, most diversity is found within individual populations (often 90% or more of the total diversity in a species), with only limited allele frequency differences between populations (*i.e.* $G_{ST} < 10\%$). Although more genetic differentiation among populations is observed for quantitative traits, genetic variation within populations is still considerable.

The characteristics of forest trees that foster high levels of genetic diversity within populations, but little differentiation between them, include:

- *Large population size.* Large populations are less susceptible to loss of genetic diversity due to random genetic drift.
- *Longevity.* Long generation intervals mean that: (1) There is less opportunity for loss of diversity during bottlenecks that may occur during reproduction, at least relative to short-lived species; and (2) Forest trees accumulate more variation per generation by mutation than short-lived plants.
- *High levels of outcrossing.* The proportion of viable offspring due to outcrossing (*versus* self-fertilization) is usually greater than 75% and often exceeds 90%. This is true for both conifers and angiosperm species, and whether wind or animals mediate pollination.
- *Strong migration between populations.* Gene flow by pollen is extensive and may account for 25%-50% of viable offspring produced within a stand. Most seed is dispersed relatively close to the mother tree (often within 50 m) and, in most cases, appears to have much less of an effect on gene flow that pollen dispersal.
- *Balancing selection.* High environmental heterogeneity within forest sites no doubt leads to balancing selection for genes controlling adaptation to different microenvironments within a stand. Therefore, some of the genetic variation within populations is adaptively significant.

Species with limited ranges and small populations typically have lower levels of genetic diversity than species with broad ranges and larger populations, but there are notable exceptions. For example, if reduction in range size has been relatively recent in a species, little loss of variation due to genetic drift may have occurred (*e.g. Abies equitrojani* and *Picea omorika*). On the other hand, species that currently have broad ranges, but in recent geological history went through one or more extreme bottlenecks, may be severely limited in diversity (*e.g. Pinus resinosa* and *Acacia mangium*).

Within species, small isolated populations are expected to have lower genetic diversity than larger populations that are not isolated, especially if the populations have been small and isolated for several generations.

The mechanisms preventing or limiting the production of self-fertilized offspring in different tree species, in increasing order of effectiveness, include: physical or temporal separation of floral organs of different sexes on the same tree, abortion of self-embryos, self-incompatibility, and dioecy. Unusually high levels of selfed offspring can occur in self-compatible species when mother trees are physically or temporally isolated from outcross pollen sources. This is most likely when individuals of the same species are relatively isolated from each other (*i.e.* occur at very low densities), and especially in poor flowering years.

Because of high levels of pollen gene flow and extensive cross-fertilization among trees within populations, the effective number of outcross males mating with each female is expected to be relatively large. Only in unusual circumstances is it likely that only a few males dominate mating, such as when a female is neighboring one or two trees producing very heavy pollen with which it is in complete floral synchrony.

Weak genetic structure resulting from limited seed dispersal and subsequent spatial clustering of siblings in the offspring generation, appears to be relatively common within forest tree populations, especially in young age classes. However, the presence of spatial genetic structure in any particular stand is difficult to predict because it is influenced by stand history (*e.g.* the amount of disturbance prior to reproduction). Spatial clustering of close relatives is the most likely cause of bi-parental inbreeding (*e.g.* mating between siblings) that is frequently detected in forest stands.

In silvicultural systems relying on natural regeneration, the practical ramifications of the high levels of genetic diversity found within most forest stands are: (1) Modest genetic gains can be made by selecting the best trees to leave as parents in seed tree or shelterwood harvests; and (2) Conversely, leaving phenotypically inferior trees to serve as parents (called high grading or dysgenic selection) can decrease the genetic value of the regenerated stand and future stands.

CHAPTER 8
GEOGRAPHIC VARIATION – *RACES, CLINES AND ECOTYPES*

Some of the best information we have about genetics of forest trees relates to geographic patterns of genetic variation within species. There is a rich history of experimentation beginning in the mid-1700s when seed of *Pinus sylvestris,* obtained from various natural stands in Scandinavia and Europe, was used to establish comparative plots in France to determine the best seed sources for timber production (Turnbull and Griffin, 1986). Systematic provenance testing of several species in Europe and North America was initiated in the early 1900s, and today the literature is replete with reports describing patterns of geographic genetic variation for hundreds of angiosperm and gymnosperm tree species.

The study of geographic variation within the native range is a logical first step in genetics research and/or domestication of any tree species, because understanding the amount and patterns of this variation is important for: (1) Learning about the interplay and significance of evolutionary forces (*i.e.* natural selection, migration and genetic drift) that caused the observed patterns (*Chapters 5* and *7*); (2) Making reforestation decisions about how far seed can safely be transferred from the site of seed collection to a distant planting site and still ensure adequate adaptation; (3) Delineating breeding and deployment zones for applied tree improvement programs that are both genetically sound and logistically efficient (*Chapter 12*); (4) Deciding which geographical sources of seed will give the highest yields in a particular planting region (*Chapter 12*); (5) Designing selection and genetic testing programs that span the appropriate edaphoclimatic regions (*Chapter 14*); and (6) Formulating gene conservation strategies that capture the natural genetic diversity that exists within species (*Chapter 10*).

This chapter describes the concepts, study and importance of geographic variation within species of forest trees. It also focuses on native species, while studies involving the testing of seed sources for planting a species in non-native (*i.e.* exotic) environments are addressed in *Chapter 12*. The first section introduces terms and concepts central to the study of geographic variation. The next section outlines experimental methods used to study, quantify and understand natural geographic patterns of variation in trees. The third section summarizes observed patterns of natural geographic variation in tree species and discusses the underlying evolutionary causes of these patterns. The final section discusses the implications of geographic variation on seed transfer among locations within the species' native range. Because geographic variation impacts nearly all phases of tree improvement programs, *Chapters 12-17* contain repeated mention of some of the concepts introduced in this chapter. Other general reviews of the concepts and applications of geographic variation include Wright (1976), Zobel and Talbert (1984), Turnbull and Griffin (1986), Morgenstern (1996), and Ladrach (1998).

DEFINITIONS AND CONCEPTS RELATED TO GEOGRAPHIC VARIATION

Provenances, Seed Sources and Races

All experiments aimed at quantifying and describing patterns of geographic variation compare plant material originating from different locations. Two important terms are used as labels to describe the location where the material is obtained: (1) **Provenance** is the geographic location of the native population where the plant material originated; and (2) **Seed source** is the geographic location from which the seed was obtained, regardless of whether or not the parent trees are located in their native population For example, *Tsuga heterophylla* seed obtained from parent trees growing in a native stand in Coos Bay, Oregon, USA would be labeled the "Coos Bay" provenance or the "Coos Bay" seed source (Box 8.1). If seed is collected in England from a plantation of *T. heterophylla* established with seed collected from Coos Bay, the seed source is "England" to reflect the parent trees from which the seed was actually obtained, and the provenance is "Coos Bay" to indicate the initial origin of the material within the native range.

We prefer to use the terms provenance and seed source only as labels to identify seedlots or other plant material for purposes of experimentation, seed movement or seed purchase. With this usage, there are no genetic differences implied between provenances or sources. When experiments have demonstrated that genetic differences do exist, then provenances or sources are classified as different **races**. For example, suppose seed of *Tsuga heterophylla* is collected from two nearby coastal locations in Oregon: Coos Bay and Waldport. If experiments comparing seedlings of these two provenances demonstrate genetic differences for any traits measured, then the two provenances are declared to be different races (Box 8.1). If no genetic differences are found, they are still two different provenances that label the two seed collection locations, but they are not different races.

A **geographic race** is defined as a subdivision of a species consisting of genetically similar individuals occupying a particular territory to which they have become adapted through natural selection (Zobel and Talbert, 1984, p 81). The definition is not meant to imply that there is no genetic variation within a geographic race, but rather that the trees within a race are more genetically similar to each other than they are to trees in a different race. These genetic differences between races are found through experimentation as described in *Experimental Methods Used to Study Geographic Variation* later in this chapter. When experiments demonstrate that geographic races have evolved in response to differences in elevations, climates or soils, then they are sometimes called elevational races, climatic races or edaphic races, respectively. The important point is that geographical races differ genetically, and that these adaptive differences evolved in response to differential natural selection in distinct environments.

This chapter focuses on natural geographic variation within the species' native range; however, for completeness, we introduce a term associated with adaptation of species in exotic planting environments (see *Chapter 12* for more explanation). A **local land race** forms when a species is introduced into an exotic planting environment and adapts through natural (and sometimes artificial) selection to the edaphoclimatic conditions of the planting zone. Naturalists and merchants were very active in moving seed of tree species and introducing these species into new countries. Seeds from species of trees from western North America, such as *Pseudotsuga menziesii*, *Picea sitchensis*, and *Pinus contorta*, were imported into Europe over 250 years ago, and these species are now grown widely as exotics there (Morgenstern, 1996, p 154). *Pinus radiata* was introduced into several countries in the 19th century from its extremely narrow range in coastal California, and these original introductions formed the basis of extensive plantation and tree improvement programs in Australia,

Box 8.1. Provenances, races, clines and ecotypes in *Tsuga heterophylla*.

A genecology study by Kuser and Ching (1980) aimed to characterize geographic patterns of genetic variation within *Tsuga heterophylla,* a wide-ranging tree species in the northwestern USA. They collected seed from a total of 20 provenances: (1) 14 coastal provenances located along a north-south transect spanning the nearly continuous part of the native range along the Pacific Coast from northern Alaska (58°N latitude) to northern California (38°N); (2) Three provenances from the Cascade Mountains in Oregon and Washington; and (3) Three provenances from the Rocky Mountains in Idaho and British Columbia. The Cascade and Rocky Mountain provenances are disjunct (*i.e.* isolated) from the populations along the coast and from each other. At any given latitude, the mountain climates are much harsher and colder than the milder climates on the coast.

They maintained the 20 provenances separated throughout the study, growing seedlings from each provenance as a separate treatment in a randomized, replicated experiment in outdoor shade frames in Corvallis, Oregon (44°N latitude). They measured several traits during the course of the experiment including: (1) The Julian Date (number of days after January 1) in autumn of the first growing season when 50% of the seedlings in a provenance had ceased height growth as evidenced by presence of a terminal bud (*i.e.* date of growth cessation); and (2) Survival of seedlings to cold outdoor winter temperatures in the subsequent winter.

Results indicated that considerable racial variation exists within the native range of this species. Several traits changed continuously for the 14 coastal provenances along the north-south transect (called a clinal pattern of variation or a cline). Two of these clines are shown in Fig. 8.1 (a,b) for date of growth cessation and survival, respectively (examine the regression lines for the 14 coastal provenances, there is one cline for each variable). Moving south to north, there was a gradual tendency of seedlings from more northerly sources to cease height growth earlier in the first growing season, and to survive better to cold temperatures in the subsequent winter. Kuser and Ching (1980) concluded, "it is reasonable to assume that evolutionary response to environmental factors varying gradually with latitude has been clinal. The factors causing divergent selection pressure between north and south are probably dates of first fall and last spring freezes and severity of midwinter temperatures."

The study also identified three distinct ecotypes: coastal, Cascade Mountain and Rocky Mountain, for *T. heterophylla* (Fig. 8.1). These three ecotypes differ in many measured traits associated with adaptation to the different environments in the milder coastal habitat and harsher montane regions. There is virtually no gene migration among the three ecotypes due to the large geographical separation, which helps to maintain the adaptive genetic differences caused by natural selection.

To summarize the terminology used in the context of this study: (1) 20 provenances of *T. heterophylla* were tested, and the word provenance specified the 20 areas of seed collection without implying genetic differences among them; (2) Common garden studies showed distinct genetic differences among the provenances, called racial differences; and (3) There were two distinct patterns of racial variation: coastal provenances displayed clinal patterns for two adaptive traits; and the coastal, Cascade Mountain and Rocky Mountain populations were classed as three ecotypes.

Chile and New Zealand (Balocchi, 1997). There are many other examples of local land races. As soon as a species is introduced, natural selection begins to operate, and sometimes humans also effect genetic change by selecting which seed trees are left for reforestation after harvesting. Therefore, the local land race begins to evolve in its new edaphoclimatic and silvicultural conditions.

With all three terms introduced above, there can sometimes be imprecision about the expanse and location of the provenance, seed source or race. For example, does the Coos Bay provenance imply that seed was collected from trees growing within the city limits? For this reason, it is important to accurately specify the location of the parent trees from which seed or other plant material is collected (see *Experimental Methods Used to Study Geographic Variation*). Even still, there can sometimes be confusion about how large an area should be considered the same provenance, seed source or race. Normally, this slight confusion has little practical or biological significance.

Clines and Ecotypes

Many terms are used to describe observed genetic patterns of geographic variation, but two are especially useful. A **cline** is a continuous genetic gradient in a single measurable trait that is associated with an environmental gradient (Langlet, 1959; Zobel and Talbert, 1984, p 84; Morgenstern, 1996, p 70). Since a cline is defined for a single trait, there are potentially as many clines as there are traits measured. Clines describe a certain type of racial (*i.e.* genetic) variation that is continuous. When this continuous genetic variation (called clinal variation) is associated with a corresponding gradual change in an environmental characteristic (such as elevation, latitude, rainfall, day length or annual minimum winter temperature), this is evidence that the species has adapted through natural selection to the environmental gradient (*Chapter 5*). Table 3.3 in Morgenstern (1996) presents examples of clines for several traits and many different tree species, and two examples for *Tsuga heterophylla* are shown in Fig. 8.1 (regressions of date of growth cessation and survival to cold temperatures on latitude of origin).

The most likely underlying genetic explanation for clinal variation is that differential natural selection has acted along the continuous environmental gradient to change allele frequencies in a continuous manner at loci controlling traits associated with adaptation and fitness. Therefore, in the *Tsuga heterophylla* example (Box 8.1, Fig. 8.1), allele frequencies at loci influencing height growth cessation and cold tolerance are presumed to vary in a continuous manner from north to south along the Pacific Coast at the loci influencing these two measured traits. When genetic variation is clinal, the distinction of geographic races is somewhat arbitrary and a matter of viewpoint. The most northerly and southerly provenances of *T. heterophylla* are distinctly different races, but such a declaration is less clear for two provenances separated by only a degree of latitude. Still, clinal variation is an example of racial variation that is often quite useful for helping to explain underlying evolutionary causes of variation.

The term **ecotype** is used to describe a race made up of genotypes adapted to a particular habitat or ecological niche. Unlike the term cline that refers to a single trait, ecotype refers to many traits or characteristics of the race that distinguish that ecotype from others. As with clines, natural selection is implied as the cause of genetic differences between ecotypes. Ecotypes are most likely to develop when environments change abruptly and populations are more isolated from each other (*i.e.* gene migration among populations is limited). An example of three ecotypes (Coastal, Cascade Mountain and Rocky Mountain) in *Tsuga heterophylla* is presented in Box 8.1, and further examples are described later in this chapter (see *Patterns of Geographic Variation in Forest Trees*).

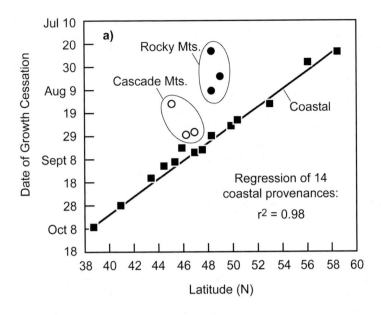

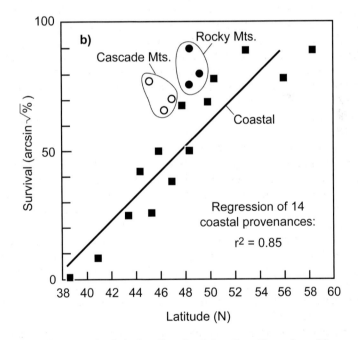

Fig. 8.1. Relationships between the latitude of seed origin of seedlings from 20 provenances of *Tsuga heterophylla* grown in outdoor shade frames in Corvallis, Oregon (44°N latitude) and: (a) Date of growth cessation in the first growing season measured as the Julian date (days after January 1) that 50% of the seedlings had ceased growth; and (b) Percent survival in the subsequent winter. The 20 provenances came from three ecotypes: (1) Fourteen provenances on a latitudinal transect along the Pacific Coast from northern California (38°N latitude) to northern Alaska (58°N); (2) Three provenances from the Cascade Mountains of Oregon and Washington; and (3) Three provenances from the Rocky Mountains of Idaho and British Columbia. (Copyright 1980, Society of American Foresters. Reprinted with permission from Kuser and Ching, 1980)

Varieties and Subspecies

The term **variety** is used in two very different ways by forest geneticists and both may imply genetic differences within a species: (1) In plant breeding and tree improvement programs, variety refers to the genetically-improved cultivar or breed that is planted operationally, and the variety may be composed of a single clone, single family or many families; and (2) Taxonomists use variety as a subdivision within a species such that two different varieties within the same species have different morphological characteristics, occupy distinct ranges and are given different Latin names. It is this latter use of variety that is important in this section and is similar to the word **subspecies** that is also used to indicate taxonomically distinct subdivisions within a species.

One example is *Pinus contorta*, an extremely wide-ranging species in North America, whose geographic variation has been well studied in field experiments planted both in its native range and in Europe (Morgenstern, 1996, p 122). There are four named subspecies (Fig. 8.2): (1) *P. contorta* subsp. *contorta* near the Pacific Coast; (2) *P. contorta* subsp. *bolanderi* in the Mendocino White Plains of California; (3) *P. contorta* subsp. *murrayana* in the mountains of Oregon, California, and Mexico; and (4) *P. contorta* subsp. *latifolia* in the Rocky Mountain and Intermountain areas from Utah to Canada. These subspecies are genetically different in growth patterns, cold hardiness and morphology; further, there is also substantial genetic variation within subspecies (Savill and Evans, 1986, p 69; Rehfeldt, 1988; Morgenstern, 1996, p 124).

Another example of large and consistent varietal differences is *Pinus caribaea*, which has three recognized varieties (*Pinus caribaea* var. *caribaea,* var. *hondurensis,* and var. *bahamensis*). Within subtropical environments, var. *hondurensis* grows faster than the other varieties, but is less straight and more susceptible to wind damage (Gibson *et al.,* 1988; Evans, 1992a; Nikles, 1992; Dieters and Nikles, 1997).

It is important to remember that taxonomists usually assign subspecies and varietal designations within a species based on observations in natural stands (*i.e.* in the absence of common garden experiments to prove genetic differentiation). Most often, large genetic differences are found among varieties or subspecies as is true for the examples of *P. contorta* and *P. caribaea* mentioned above. However, taxonomic classification in the absence of genetic data can sometimes result in: (1) Varietal designations in which the genetic differences between varieties are relatively unimportant; and (2) Lack of varietal designations in a species exhibiting tremendous racial differences (*e.g. P. taeda*). Therefore, well-designed experiments are required to quantify the magnitude and patterns of genetic differentiation between and within different varieties or subspecies.

In most instances of species with varietal (or subspecies) subdivisions, there also is large genetic variation within varieties (or subspecies). In both the examples mentioned above, there are genetic differences among provenances within a variety and also genetic differences among trees within provenances. This is another reason why data from genetic tests are required to properly quantify the relative importance of differences among varieties compared to other levels of hierarchy (*e.g.* among provenances within variety).

Provenance x Environment Interaction

Provenance x environment interaction is a type of genotype x environment interaction (*Chapter 6*) that occurs when provenances do not perform consistently for a measured trait across a range of experimental environments. The environments may be different field locations or artificially imposed treatments in field, greenhouse or growth room studies (*e.g.* different fertilization regimes, different day lengths, distinct moisture regimes). As with any genotype x environment interaction, two types of provenance x environment interactions are observed: (1) Interaction due to rank changes in which provenances change their relative ranking for a particular trait in different environments; and (2) Interaction due to scale effects in which provenances retain their rank order across environments, but the magnitude of the differences among provenances changes.

When important rank changes occur for a particular trait, the differential response of provenances is often interpreted to mean that different gene loci are controlling performance of the trait in the different environments. Morgenstern (1996, p 86) provides an example using data from a study on *Pinus echinata* reported by Wells (1969; Fig. 8.3). Seeds collected from seven different provenances in the southeastern USA were planted in randomized, replicated experiments established at 10 locations. Both the provenances and the plantation locations varied widely from milder, southern locations in Louisiana and southern Mississippi to colder, more northerly regions of Missouri, northern Tennessee and New Jersey. Ten-year-old heights of *P. echinata* demonstrated substantial provenance x plantation interaction, and this was most clearly visualized as distinct clinal patterns that depended on the environment of the planting location (Fig. 8.3).

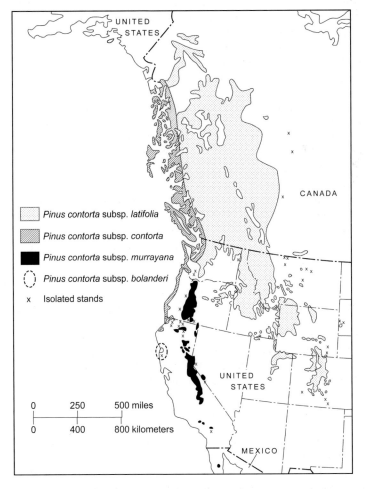

Fig. 8.2. Natural ranges of the four subspecies of *Pinus contorta*: *P. contorta* subsp *contorta, murrayana, latifolia* and *bolanderi* (Wheeler and Guries, 1982).

In the southern and middle-latitude test plantations (lines a and b of Fig. 8.3), seedlings originating from warmer climates (*i.e.* southern provenances) grew faster as evidenced by clines with positive slopes. Conversely, in the northern test plantations (line c of Fig. 8.3), the cline had a negative slope indicating that provenances from more northerly origins were taller. Perhaps gene loci associated with cold hardiness are influencing growth in the northern plantations, but not in southern plantations. If northern provenances have evolved more cold hardiness, this would explain their superior performance in the northern plantations. Conversely, cold hardiness is likely of little adaptive value in the southern plantations. Perhaps southern provenances grow more rapidly in southern environments due to increased frequency of alleles associated with longer growing seasons.

Although other examples of provenance x environment interaction can be found (Morgenstern and Teich, 1969; Conkle, 1973; Campbell and Sorensen, 1978), important provenance x environment interactions involving rank changes are not normally observed unless there is substantial genetic variation among the provenances, and also important differences in the climatic or edaphic conditions among the environments. When test environments or provenances are similar, large interactions are not expected (Matheson and Raymond, 1984a).

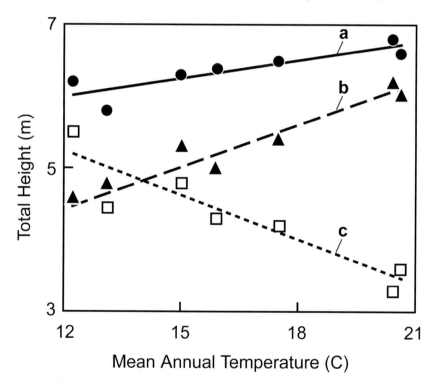

Fig. 8.3. Provenance x location interaction for 10-year-old heights in *Pinus echinata.* Mean total heights (m) of provenances are plotted against mean annual temperature at the seed collection sites for three different series of planting environments: (a) Five middle-latitude experiments in Mississippi, Tennessee and the Carolinas; (b) Three southern experiments planted in the southern parts of Louisiana, Mississippi and Georgia; and (c) Two northern experiments established in Missouri and New Jersey (Adapted from Morgenstern, 1996, based on data from Wells, 1969). (Reprinted with permission of the Publisher from Geographic Variation in Forest Trees: Genetic Basis and Application of Knowledge in Silviculture by E.K. Morgenstern. Copyright University of British Columbia Press 1996. All rights reserved by the Publisher)

EXPERIMENTAL METHODS USED TO STUDY GEOGRAPHIC VARIATION

Early studies of geographic variation were based on measurements of trees growing in natural stands within the native range. However, as was discussed in *Chapter 1*, studies that measure phenotypes in natural populations reveal little, if anything, about the underlying genetic patterns of geographic variation, because phenotypes in different stands are influenced by both environmental and genetic differences between stands. In other words, genetic and environmental effects are confounded when phenotypes are measured in natural populations. Therefore, these types of investigations should not be utilized when the goal is to determine genetic differences among provenances.

To isolate genetic differences among provenances and to characterize genetic patterns of geographic variation, three types of studies are conducted: (1) Studies in which plant material is obtained from natural stands and genetic markers (*Chapter 4*) are measured directly with this material; (2) Studies in which seeds obtained from different provenances are sown in growth room, greenhouse, nursery or other artificial environments and grown for a short time (few months to a few years) followed by measurement of seedling traits; and (3) Studies in which seeds from different provenances are used to establish one or more field experiments and planted trees are measured for an extended length of time (up to rotation age or longer). Genetic markers can be assayed directly from tissues collected in the field because their expression is generally unaffected by the environment of the collection site. On the other hand, the last two types of experimental approaches are common garden experiments that involve growing different provenances together in randomized, replicated trials (*Chapter 1*).

All three types of experimental approaches have been used to: (1) Determine patterns of genetic variation within a species' natural range; (2) Relate these patterns to physical, edaphic and climatic patterns of variation within the range; (3) Understand the relative importance of the evolutionary forces that have shaped the observed patterns; and (4) Construct seed transfer guidelines within the natural range based on inferred patterns of adaptive genetic variation. Each approach is described below, and a related discussion in *Chapter 12* addresses seed source testing for long-distance seed movement and purposeful planting of non-local sources aiming to increase yield.

The three approaches employ different sampling and experimental methods and vary in which of the above four objectives they address most effectively. However, for all three experimental approaches, it is important to maintain good records of provenance locations and mother tree positions within each location. Modern global positioning systems (GPS) facilitate recording of latitude, longitude and elevation of each tree. During the course of the experiment, soil characteristics are sometimes measured at each provenance location (depth and texture of each horizon, moisture holding capacity and cation exchange capacity). Further, climate data can be obtained from the nearest weather station to each location or from predictive maps of the area to determine rainfall patterns (amounts and seasonality) and temperature regimes (*e.g.* maximum summer, minimum winter and annual mean temperatures). These edaphoclimatic variables are subsequently used as predictive variables in regression models. If observed patterns of genetic differences among provenances are associated with any of the edaphoclimatic variables (*e.g.* slower growth rates of provenances from colder climates), then this is taken as evidence of differential natural selection having resulted in adaptive differences among provenances.

Genetic Markers for Studying Geographic Variation

In concept, any type of genetic marker (*Chapter 4*) can be used to study geographic variation within species. Plant material collected from natural stands of many provenances is assayed in the laboratory to test for differences among provenances and patterns in these genetic differences. Historically, most studies of population differences in forest trees employing genetic markers have utilized allozymes, and we primarily focus on the allozyme approach (*e.g.* Moran and Hopper, 1983; Li and Adams, 1989; Surles *et al.,* 1989; Pigliucci *et al.,* 1990; Westfall and Conkle, 1992; Rajora *et al.*, 1998).

Methods Used in Allozyme Studies of Genetic Patterns of Geographic Variation

The number of provenances studied in allozyme studies varies from less than 10 to more than 100 different locations. The locations may sample the entire natural range of the species or portions of the range of specific interest. Usually 20 to 30 trees from each provenance location are sampled. Seed or vegetative plant material (usually leaves or dormant buds) is collected from each tree at each provenance location. The plant material is subjected to electrophoresis to determine allozyme genotypes at, typically, 20 or so loci (*Chapter 4*). If vegetative material is assayed, then the genotype of each tree can be determined directly at all sampled loci. Another approach is to use the genetic composition of a sample of 5-12 offspring (germinants or young seedlings, or megagametophytes in conifers; *Chapter 4*) to infer the trees' genotypes. In some cases, seeds collected from all trees from each provenance are bulked and genotypes of this single sample are used to determine the genetic makeup of the provenance.

In terms of data analysis and interpretation, the basic data used to compare the genetic makeup of different provenances are genotypic and allele frequencies at the sampled loci (*Chapter 5*). The first step is often to partition average allele-frequency variation using gene diversity analysis or F-statistics (*Chapter 7*). The goal with this analysis is to estimate and compare the proportions of total genetic diversity within provenances to that among provenances at various hierarchical levels (*e.g.* among populations within regions and among regions).

The next step is to examine patterns of allele-frequency variation among provenances at individual loci, or across multiple loci using multivariate statistics (*e.g.* principal components or discriminant analysis; Westfall and Conkle, 1992). A more common approach is to evaluate the magnitude and patterns of allele-frequency differentiation between pairs of provenances using one or more measures of genetic distance (*e.g.* Wright, 1978; Gregorius and Roberds, 1986; Nei, 1987). The most used statistic for **genetic distance** is Nei's (1987) standard genetic distance, D, which is calculated for each locus separately and then averaged over all loci. If any pair of provenances has identical allele frequencies, then D = 0. Theoretically, if two provenances have no alleles in common, D has no upper limit. D can be interpreted as the average number of allelic differences per locus. Therefore, if D = 0.05, there would be one allele difference per locus, on average, for every pair of 20 gametes sampled at random from the two provenances.

Once genetic distances have been calculated between all pairs of provenances [*i.e.* if there are 25 provenances, there are (25 x 24)/2 = 300 pairs], the total matrix of estimated distances is used in a software program to cluster provenances and create a cluster diagram (sometimes called a phenogram; Hedrick, 1985; Nei, 1987). Provenance pairs with the smallest genetic distances cluster first and those with the greatest pairwise D values cluster last. Inferences can then be made about patterns of genetic similarity among provenances on a landscape by examining how populations from different locations or environments cluster (*e.g.* Box 8.2).

Box 8.2. Geographic patterns of allozyme variation in *Pinus jeffreyi*.

Allele frequencies were estimated at 20 allozyme loci for 14 provenances of *Pinus jeffreyi* (Furnier and Adams, 1986a). Seven provenances came from the Klamath Mountains in southern Oregon and northern California from 43°-41°N latitude (labeled K1-K7) and seven came from the Sierra Nevada Mountains and further south to northern Mexico (40°-31°N latitude; labeled S1-S7). The Klamath populations are disjunct from the remainder of the range, and here the species grows exclusively on infertile serpentine soils high in chromium, magnesium and nickel. In the rest of the range, the species typically grows on more fertile sites.

Data analysis took place in three steps. First, allele frequencies at each locus were used to estimate a genetic distance (D) for each pair of provenances, according to the methods of Nei (1987). There were 3640 D values calculated: 20 allozyme loci x 182 pairwise combinations of 14 provenances, where 182 = (14x13)/2. Second, the D values were averaged over all 20 loci giving a single average distance estimate for each of the 182 provenance pairs. Finally, the 182 average D values were subjected to a clustering procedure (UPGMA method) (Sneath and Sokal, 1973) to group the 14 provenances by genetic similarity.

Results showed that the seven populations from the Klamath Mountains differ genetically from the seven provenances obtained from the rest of the species' range (Fig. 1). The Klamath provenances (K1-K7) clustered into a group based on smaller average D values; likewise, the seven provenances from the rest of the range clustered into a second group. In addition to these results, the average expected heterozygosity ($\overline{H}e$ from *Chapter 7*) was significantly smaller for the seven Klamath provenances (0.185) compared to the seven other provenances (0.255). The genetic differentiation of the Klamath region from populations in the rest of the range and its low average $\overline{H}e$ may reflect adaptive selection to the harsh soil conditions, genetic drift due to a drastic reduction in the size of an ancestral population (*i.e.* population bottleneck), or both.

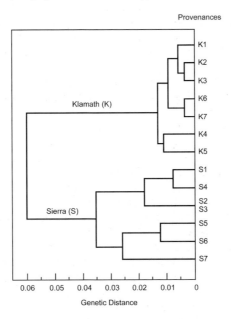

Fig. 1. Clustering of 14 provenances of *Pinus jeffreyi* from the Klamath Mountains (K1-K7) and the rest of the range in the Sierra Nevada Mountains and south (S1-S7). (Adapted from Furnier and Adams, 1986a)

Advantages and Disadvantages of Allozyme Studies of Geographic Variation

A major advantage of using genetic markers (*e.g.* allozymes) compared to common garden experiments is that genotypes can be determined directly from tissues collected in wild populations, without the need to grow progeny and measure traits over a period of time. This means that genetic marker studies can often be completed in 1-2 years and often at lower cost than common garden studies. Another advantage of genetic marker studies of geographic variation is that inferences can be made about genetic differences at the genome level, *e.g.* differences in allele frequencies or in levels of heterozygosity among provenances. If these differences can be related to evolutionary forces causing them, we potentially understand more about the species' genecology (*e.g. Pinus jeffreyi*; Box 8.2).

Because allozymes (and other genetic markers) appear to be largely neutral to selection pressure (see Box 5.6), geographic patterns of genetic variation detected by allozymes are those primarily determined by evolutionary forces other than selection, *i.e.* the interplay between genetic drift causing populations to differentiate and gene flow restricting differentiation. Therefore, a disadvantage of geographic patterns of variation based on allozymes is that they usually explain little about adaptation of populations on the landscape, although there are some exceptions (see *Patterns of Geographic Variation in Forest Trees* later in this chapter).

Short-term Seedling Tests in Artificial Environments

A very common approach to both characterizing natural geographic variation and developing seed transfer guidelines is the use of short-term, common garden experiments planted in artificial (non-field) environments (Fig. 8.4). The interest often centers on the extent to

Fig. 8.4. Short-term common garden studies of seedlings from two provenances of *Pseudotsuga menziesii* in their second growing season in a bareroot nursery in the Willamette Valley, Oregon, USA. Seedlings on the right originated from a provenance in Gardner, Oregon (coastal fog belt with annual rainfall of 2100 mm); seedlings on the left are from a provenance near Elkton, Oregon approximately 50 km inland (dry valley with annual rainfall of 1300 mm). (Photo by T. White)

which differences in provenance means for individual traits, or clusters of traits, are associated with environmental differences across the landscape in order to describe adaptive patterns of variation. Detailed methods for provenance sampling, test implementation and data analysis vary among short-term seedling studies, and a few examples spanning a range of species, test environments and traits measured include Fryer and Ledig, 1972; Campbell and Sorensen, 1973, 1978; Kleinschmit, 1978 (Fig. 8.5); Kuser and Ching, 1980 (Fig. 8.1); Ledig and Korbobo, 1983; Rehfeldt, 1983a,b; Rehfeldt *et al.*, 1984; Campbell, 1986; White, 1987a; Ager *et al.*, 1993; and Kundo and Tigerstedt, 1997.

Methods Used in Short-term Seedling Studies of Genetic Patterns of Geographical Variation

The number of provenance locations sampled in short-term seedling studies varies from fewer than 10 to more than 100. The locations sampled typically span the range of edaphoclimatic conditions in the region of interest, and most appropriately are chosen from an environmental sampling grid of the area. Alternatively, sampling may be restricted to provenances located along a specific environmental transect (*e.g.* the latitudinal gradient of 14 provenances of *T. heterophylla* in Fig. 8.1). Open-pollinated seed is collected from randomly chosen mother trees in each provenance; 10 or more trees are recommended to get a reasonable estimate of the provenance mean. This seed is usually kept separate by mother tree. Bulking seed by provenance reduces the number of treatments in the experiment, but at the expense of not being able to evaluate the relative importance of provenance variation to variation among families within provenances. When the goal is to characterize clinal patterns of geographic variation, it is better to sample more locations (especially if several environmental gradients are involved), rather than more mother trees per location (*e.g.* as few as 1 or 2 trees from 100 or more provenances; Campbell, 1986). Although specific provenance means are not well estimated, the many locations facilitate development

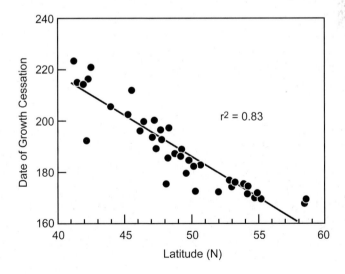

Fig. 8.5. Regression of mean Julian date of growth cessation (bud set in days after January 1) on latitude of origin of 43 *Picea sitchensis* provenances tested at age 3 in a nursery in Germany (from Morgenstern 1996, p 77, based on data from Kleinschmit, 1978). (Reprinted with permission of the publisher from Geographic Variation in Forest Trees: Genetic Basis and Application of Knowledge in Silviculture by E.K. Morgenstern. Copyright University of British Columbia Press 1996. All rights reserved by the publisher)

of robust clinal regression models used to relate provenance differences to environment, and to develop seed transfer equations.

Throughout the study, the identity of seed collected from each mother tree is kept separate by provenance and family within provenance. These identified seedlots are genetic treatments in a randomized, replicated experiment (see *Chapter 14*), and there may be 200 or more such treatments (OP families from different provenances). Seed is germinated and seedlings are established in containers or soil in an artificial environment such as a growth room, greenhouse, or nursery. In addition to the genetic treatments, cultural treatments may be imposed such as different nutrient regimes, photoperiods or stress environments (cold, drought or disease). Seedlings are measured for a few months to a few years, and traits assessed may include growth rates, phenology (dates of growth initiation and cessation), morphological attributes (leaf size, stomatal density, root/shoot ratio), physiological processes (photosynthesis, transpiration, nutrient uptake) and survival or other responses to imposed stress (drought survival, cold hardiness or disease resistance).

After data collection and editing (*Chapter 15*), analysis of variance is usually conducted for each measured trait to determine statistical significance and relative importance of design factors (environments, blocks, cultural treatments), genetic treatments (provenances and OP families within provenances) and all possible interactions. Variance components are often estimated to quantify the relative amounts of genetic variation among provenances compared to among families within provenances (see Box 7.2).

To test for clinal patterns of adaptive genetic variation, regression analyses are conducted for each measured trait. In each multiple regression, the dependent variable is one of the measured traits, while the regressors include values for environmental variables measured at each provenance location (*e.g.* latitude, elevation, minimum and maximum annual temperatures, annual and growing season precipitation, soil characteristics) along with their squared values and first-order interactions. The goal is to obtain parsimonious models in which all regressors are statistically significant, an important fraction of the differences among provenances is explained, and the models have biologically plausible interpretation (*e.g.* Figs. 8.1 and 8.5). Sometimes multivariate statistical methods are used in which principal components obtained from all the measured traits are the dependent variables in the regressions (Campbell, 1979). These multivariate methods are especially useful in developing seed transfer guidelines, but are more difficult to interpret in terms of underlying biological causes of observed patterns.

While statistical analyses are important, it is also valuable to examine the data using less formal methods, such as: (1) Examining provenance means, minimums and maximums for each trait; (2) Tabulating provenance means by region or other smaller unit (*e.g.* to compare means for disjunct populations from different locations); (3) Constructing graphs of the data (with a measured trait on the y-axis and any of the environmental variables on the x-axis); and (4) Conducting correlation or graphical analyses to understand relationships among measured traits. The latter is particularly important and may detect, for example, that provenances from colder climates exhibit higher levels of cold hardiness, slower growth rates and earlier dates of fall cessation of growth.

Advantages and Disadvantages of Short-term Seedling Studies of Geographic Variation

The specific advantages of short-term tests in artificial environments are: (1) Many provenance locations can be studied since the studies are conducted on young seedlings; (2) A large amount of data can be collected in a very short time on morphological, phenological and physiological traits that often effectively discriminates among provenances; (3) The experiments are very powerful (*i.e.* have low experimental error due to homogeneous artificial environments) and are therefore very effective for demonstrating and modeling adaptive variation among provenances; and (4) The experimental environments can be manipulated to assess adaptive variation among provenances in resistance to stresses such as frost, drought or disease.

These advantages make short-term tests especially appropriate for: (1) Characterizing genetic patterns of natural geographic variation (clinal, ecotypic or both, *e.g.* Figs. 8.1 and 8.5); (2) Understanding the differential selective forces that have caused observed patterns of adaptive genetic variation (*e.g.* selection for earlier growth cessation in high elevations); (3) Developing preliminary seed transfer guidelines within a region for later verification by long-term field provenance tests (Rehfeldt, 1983a,b; 1986; Campbell, 1986, 1991; Westfall, 1992); and (4) Narrowing down the number of promising provenances for a reforestation program to be subsequently tested in long-term field trials.

Disadvantages of short-term tests are: (1) The environments are artificial, and therefore observed patterns of variation may not mimic those in field plantations; (2) Only a very short portion of the tree's life cycle is assessed; and (3) It is not possible to determine which provenances actually yield more or have better product quality at rotation age across the range of edaphoclimatic zones and silvicultural regimes employed in an operational plantation program. For these reasons, long-term provenance trials in field experiments are required to develop definitive seed transfer guidelines and to make final decisions about the best provenances for reforestation programs.

Long-term Provenance Trials in Field Experiments

As mentioned above, long-term field experiments (often called provenance trials) are needed for making final decisions about seed transfer guidelines and for choosing the best provenances or seed sources for reforestation. There is simply no replacement for these types of trials if the highest priority objective is to determine which provenances perform best in the edaphoclimatic and silvicultural conditions of a plantation program. These types of field trials have been conducted for literally hundreds of different forest tree species. Many excellent examples are described in Morgenstern (1996) and Ladrach (1998); other studies include Squillace and Silen (1962), Squillace (1966), Callaham (1964), Wells and Wakeley (1966, 1970), Wells (1969), Conkle (1973), Teich and Holst (1974), Morgenstern (1976, 1978), Kleinschmit (1978), Morgenstern *et al.* (1981), Park and Fowler (1981), White and Ching (1985), and Stonecypher *et al.* (1996).

Two major goals of long-term provenance testing are: (1) To characterize natural patterns of geographic variation, which is the same goal as most short-term seedling tests; and (2) To choose the best specific provenance or provenances for an operational planting program. The latter goal may subsequently lead to making selections from the field trials for an applied tree improvement program. Sampling of provenances and field test designs can be quite different depending on which goal has precedence, and these differences are noted below in the discussion of experimental methods.

Sometimes, provenance trials are conducted in exotic plantation environments outside the species' native range. Examples include: (1) Several North American conifers tested in Europe (Morgenstern, 1996); (2) Several species of *Eucalyptus* tested in countries where the species are used commercially (Eldridge *et al.*, 1994); and (3) Mexican and Central American pines tested in South America and South Africa by the Central America and

Mexico Coniferous Resource Cooperative (CAMCORE; Dvorak and Donahue, 1992). Even though these trials are outside the species' native range, they can be very effective at characterizing natural genetic patterns of geographic variation. In other words, the patterns observed in exotic field locations often mimic the patterns observed on sites established within the native range.

Methods Used in Long-term Provenance Studies

In concept, few (less than 10) to many (more than 100) provenance locations may be sampled, but there is a tendency for fewer provenances to be represented than in short-term tests due to the cost and size of long-term field trials. When the primary objective is characterizing patterns of natural geographic variation (goal 1 above), sampling guidelines described for short-term trials are appropriate (more locations with few mother trees per location). When the goal is to extensively test and rank candidate provenances for an operational reforestation program, fewer provenances are usually included and often these are considered the most promising for operational use (*i.e.* not chosen at random or according to an environmental gradient). Each provenance location should be represented by seed collected from at least 20 mother trees to precisely estimate provenance means if those means will be used to choose which provenances to plant operationally. If selections for a tree improvement program are planned from the provenance tests, then mother trees should not be chosen at random, but rather selected based on superior phenotype (mass selection, *Chapter 13*). Whenever possible, mother tree identity should be retained, and the mother trees separated by at least 50 m to minimize relatedness among parent trees (*Chapter 7*). These factors are especially important if subsequent selections within provenances are planned for a tree improvement program.

Once seed is collected from natural stands in the various provenance locations, it is generally germinated and grown in a greenhouse or bareroot nursery for one to two years prior to outplanting in the field tests. In the nursery and at each field location, a statistically sound experimental layout is established (*Chapter 14*). Randomized complete block (RCB) designs are common, but incomplete block designs should be considered when a single block of an RCB design in the field would be larger than 0.1ha or within-block heterogeneity is expected to be large. In the field tests, provenances are generally established in rectangular plots within replications (see Fig. 14.9c) such that each provenance plot within a block contains 25-100 trees from that provenance. When mother tree identities are retained, a split-plot field design is employed in which families from a given provenance are nested within each rectangular provenance plot and identified by mother tree. We recommend using noncontiguous or single-tree plots (*Chapter 14)* instead of row plots for each family.

Use of large, rectangular provenance plots is especially important when big differences are expected among provenances, because they minimize bias that might arise if provenances are mixed together in the same plot (*e.g.* slow-starting provenances could be disadvantaged early and never catch up), and permit competition to occur among trees of the same provenance as would occur operationally in a plantation program. For example, suppose that each provenance whole plot contains 49 trees (7 rows x 7 columns). If competition between provenances becomes important, the outer row of each whole plot can serve as a buffer between plots, with the inner plot of 25 trees becoming the measure plot.

Long-term provenance trials should be replicated on several field sites (preferably, at least six). If characterizing patterns of natural geographical variation is the main objective, the field sites should be chosen to span the range of edaphoclimatic conditions in the zone of interest. If the goal is to identify the best provenances for operational use, then the field sites should be representative of the range of environmental conditions expected in the plantation program. Different cultural treatments (such as fertilizer treatments or different silvicultural treatments) can also be imposed at each location to test for provenance x treatment interactions. Choosing uniform sites and doing a good job of site preparation to eliminate heterogeneity due to uneven competing vegetation are imperative for precisely estimating provenance means. Two complete border rows are commonly planted around the entire test so that all experimental trees are exposed to uniform lighting and competition (*Chapter 14*).

Traits associated with adaptation (*e.g.* early growth, shoot phenology, damage by frosts and drought, and survival) are often assessed at young ages, while stem yield and quality traits (*e.g.* bole straightness, wood quality) are measured in older trees. Provenance field tests are typically measured for at least one-half of the commercial rotation age (Zobel and Talbert, 1984; Morgenstern, 1996). This helps to ensure that provenances that rank the best for growth, yield, survival and other traits are truly adapted to the environmental conditions of the planting region. Results from young tests can sometimes be misleading; and even at one-half rotation age, there is no assurance that occasional climatic extremes that occur in the test region have been adequately experienced during the life of the test. Therefore, experimenters must also use their judgment about the appropriate final age of assessment.

After data collection and editing (*Chapter 15*), approaches to data analysis and interpretation depend on the goal. If the goal is to characterize natural patterns of geographic variation, then the analytical methods outlined previously for short-term experiments are appropriate. If the goal is to identify the best provenance(s) for use in an operational reforestation program, then analysis of variance is used to test for statistical significance of provenance differences, and mean comparison techniques or contrasts are used to separate and rank provenances. If inferences are meant to apply only to the specific set of provenances included in the experiment, then provenances are treated as fixed effects in these analyses (*Chapter 14*); however, they are treated as random effects if the goal is that inferences should apply to a larger population of provenances of which only a sample is included in the experiment. In any case, incomplete blocks, family effects within provenances, and all of their interactions should be treated as random effects to fully capture the benefits of incomplete blocks and to ensure a broad inference space (*Chapter 15*).

Advantages and Disadvantages of Long-term Provenance Studies

Long-term provenance trials in field environments are the only definitive method to rank provenances for operational reforestation programs and are also the most effective way to characterize natural patterns of geographic variation. The main disadvantages of long-term field provenance trials are the high costs, long time interval needed to obtain the data, and limitations in the number of provenances that can be tested. In particular, very large field trials are necessary to accommodate rectangular plots, even with relatively modest test designs. For example, to accommodate a field trial with only 10 provenances, 6 replicates, 49-tree rectangular plots and 3 m spacing between trees, more than 2.5 ha are required, not including border rows around the plantation. Clearly, the total number of provenances and families that can be reasonably tested in rectangular plots is severely constrained. When a large number of provenances are being tested, it may be more appropriate to use single-tree plots (*Chapter 14*), such that there are no discrete rectangular plots for each provenance; this is especially true when the goal is to characterize natural variation. For example, single-tree plots were employed in a provenance trial of 1100 provenances of *Picea abies* in order to fit all sources into a field block of reasonable size (Krutzsch, 1992).

Genetic marker (*e.g.* allozyme) and short-term seedling experiments can provide useful information to aid the design of efficient long-term field trials and to develop interim seed transfer guidelines. However, these types of studies cannot identify which provenances grow best across the range of climates, soils and management regimes. For this, there is no substitute for long-term trials. As one example, several short-term seedling tests of *Pseudotsuga menziesii* using provenances from the west side of the Cascade Mountains in Oregon and Washington have clearly indicated steep clines and strong adaptation to local environments (Campbell and Sorensen, 1978; Campbell, 1979; Campbell and Sugano, 1979; Sorensen, 1983). However, long-term field trials have shown different patterns in which the local provenance is not always superior (White and Ching, 1985; Stonecypher *et al.*, 1996). This reinforces the need for data from long-term provenance trials to guide seed transfer decisions and is why so many plantation programs establish long-term provenance trials in the field.

PATTERNS OF GEOGRAPHIC VARIATION IN FOREST TREES

Although much is known about patterns of geographic variation for temperate species of forest trees, there have been fewer studies of tropical species, and this is an area that needs further research (Ladrach, 1998). By far, the large majority of species studied show important amounts of geographic genetic variation; in other words, there are significant genetic differences among provenances, meaning that races have developed (Zobel and Talbert, 1984, p 82). The patterns of geographic genetic variation observed in any species reflect both the effects of past evolutionary forces and ongoing evolution in current environments. That is, the genetic structure of tree species is not static, but ever changing.

Three main evolutionary forces interact to shape the patterns of geographic variation that are observed today (see *Chapter 5*): (1) Natural selection whereby species become adapted to edaphoclimatic conditions that are ever-changing through geological time scales; (2) Genetic drift influencing genetic variation both within and between provenances if species go through drastic reductions in population size (*i.e.* bottlenecks) and then expand through recolonization; and (3) Gene migration among populations (or the lack of it between disjunct populations). The impacts of these forces on geographic variation are summarized in Box 8.3 and discussed in Gould and Johnson (1972), Turnbull and Griffin (1986), Morgenstern (1996), and Ledig (1998).

As a result of differences in both geologic histories and current adaptive processes, different species show markedly distinct patterns of geographic variation. This makes the study of geographic variation an exciting and challenging field. Scientists use empirical data from genetic markers, short-term common garden experiments and long-term provenance trials, along with deductive reasoning based on their knowledge of biogeographical history (*e.g.* obtained from fossil records) to explain observed patterns for a given species. Often, however, a number of alternative explanations are plausible such that the underlying causes of patterns are not completely understood. For wide ranging species, in particular, patterns can be very complex and variable in different parts of the range; this is illustrated for *Pseudotsuga menziesii* (Box 8.4), *Pinus taeda* (Box 8.5) and *Eucalyptus camaldulensis* (Box 8.6).

In this section, we describe some observed patterns of geographic variation in forest trees, with specific illustrations in Boxes 8.4-8.7. We do not catalogue a large number of species, because a number of recent summaries are found in the literature (Turnbull and Griffin, 1986; Zobel *et al.*, 1987; Eldridge *et al.*, 1994; Morgenstern, 1996; Ladrach, 1998; Ledig, 1998). Rather, we strive to span the range of types of patterns that are observed and speculate on the interplay of evolutionary forces that cause these observed patterns.

Box 8.3. Evolutionary forces that shape genetic patterns of geographic variation.

Natural selection. The most important force shaping geographic variation in forest trees is natural selection. There is a tendency for the fittest individuals in any population to survive and leave more offspring to the next generation. So over generations, a population becomes genetically adapted to the local edaphoclimatic conditions; that is, alleles having favorable influence on adaptation increase in frequency, while disfavored alleles decrease (*Chapter 5*). If environmental gradients exist, then differential selection pressures cause populations in distinct localities (*e.g.* higher and lower elevations) to become differentiated from each other as they adapt to local environments. If environmental gradients vary continuously, clinal patterns of genetic variation are most commonly observed. Steep environmental gradients (*e.g.* large elevational differences) tend to produce steep clines (meaning large genetic differences between provenances). If environmental differences are discrete (such as dramatically different soil types), then ecotypes are more likely to form. Clines are more common than ecotypes, because most environmental characteristics vary continuously over a region.

Migration. Gene flow in the form of seed and pollen dispersal tends to reduce genetic differences among populations (*Chapters 5 and 7*). If nearby populations freely exchange alleles on a continuous basis, potential genetic differentiation due to natural selection can be drastically slowed. At the other extreme, effectively disjunct populations are free to differentiate more rapidly, and as genetic differences develop they are sustained. The balance between selection and migration depends on rates of migration and intensity of differential selection pressure.

Gene flow among species in the form of hybridization and introgression also affects patterns of geographic variation within species (*Chapter 9*). When provenances in one part of a species' range hybridize with another species (or when introgression has occurred in the distant past from another species), then those provenances are genetically different from provenances in parts of the species' range not experiencing hybridization.

Genetic drift. When a population experiences severe reduction in size, genetic drift occurs resulting in loss of genetic variation (reduced heterozygosity and loss of alleles, *Chapter 5*). For example, periods of glaciation during the Pleistocene caused severe reduction of population sizes in many tree species in the Northern Hemisphere, which affected future patterns of geographic variation in two ways. First, during glaciation, the ranges of many species were pushed south, and some species were reduced to several small, disjunct populations, called refugia. Because the effects of genetic drift are haphazard (*Chapter 5*), provenances existing today that are derived from distinct refugia may differ genetically in a manner not associated with differences in the environments where the provenances reside. Second, during interglacial periods of warming, species again expanded their ranges northward (called recolonization). When recolonization occurs in sporadic jumps of a few seed to previously unoccupied territory (called saltation), the founders of new populations carry only a sample of the genetic variation of the parental population. Two potential consequences of saltation are: (1) Current provenances derived from different saltation events will differ genetically due to the founder principle (*Chapter 5*); and (2) A clinal pattern of gradual reduction in heterozygosity from south to north may result, reflecting many sequential saltation events during recolonization (each step northward creating the opportunity for loss of more genetic variation).

Racial Variation Associated with Environmental Differences

Often when provenance differences are examined in short-term seedling tests or long-term field trials, patterns of racial variation are associated with differences in environments. Clinal variation is the most common pattern and tends to occur when environmental gradients are continuous; less commonly, ecotypes can occur when environmental differences in the natural range are sharply discontinuous. There are four concepts relevant to both of these patterns: (1) Both patterns are assumed to be adaptive in the sense that provenances have become adapted through differential natural selection to the varying environments (Box 8.3); (2) Provenances from milder environments (lower latitudes, lower elevations, more mesic areas) tend to have evolved the capacity for more rapid growth, but are generally less hardy to harsh environments than provenances originating from harsher environments; (3) Both ecotypic and clinal patterns are observed on broad geographic scales and on finer, local scales; and (4) Wide ranging species often exhibit quite complex patterns of racial variation that include clines and ecotypes on a range of scales. These concepts are illustrated for four different species: *Tsuga heterophylla* (Fig. 8.1) and *Pseudotsuga menziesii* (Box 8.4) from the western USA, *Pinus taeda* in the eastern USA (Box 8.5) and *Eucalyptus camaldulensis* in Australia (Box 8.6). In all four cases, both clines and ecotypes occur

---

Box 8.4. Patterns of geographic variation in *Pseudotsuga menziesii*.

Pseudotsuga menziesii, a species of great economic importance worldwide, has a very broad range in the western USA. It is found in both the coastal forests of the region (Coast Ranges and west slopes of the Cascade and Sierra Nevada Mountains), as well in the Rocky Mountains to the east (Fig. 8.6). Environmental heterogeneity within this range is tremendous, from the relatively moist, mild climates in the Pacific coastal forests to harsh (colder, drier) climates in the Rocky Mountains. In addition, the species' range spans elevations from sea level near the coast to 3500 m in the southern Rockies. Geographic variation has been studied extensively in this species, including investigations based on measurements of quantitative traits in common garden experiments and allozyme studies.

 Quantitative traits. Ecotypic differentiation between Pacific Coast and Rocky Mountain populations is so great that taxonomists give populations from the two regions varietal status, with the coastal ecotype designated var. *menziesii* and the Rocky Mountain (or interior) ecotype designated var. *glauca* (Fig. 8.6). Variety *glauca* grows slower than var. *menziesii*, but is better adapted to the harsher climates in the Rocky Mountains, being hardier to both cold and drought (Silen, 1978; Kleinschmit and Bastien, 1992). The interior variety can be further divided latitudinally into two subdivisions (at around 44°N latitude) (Fig. 8.6), with provenances of the northern race having slower growth rates, less drought resistance, and greater cold hardiness than provenances of the southern race. Patterns of geographical variation within the varieties and subdivisions are strongly clinal. For example, in seedling common garden tests involving seed collected from a large number of locations in northern Idaho and adjacent areas (*i.e.* northern race of var. *glauca*), geographical variation was found to largely follow elevational clines, with rate of seedling height growth decreasing with

(Box 8.4 continued on next page)

Box 8.4. Patterns of geographic variation in *Pseudotsuga menziesii*. (Continued from previous page)

increasing elevation of provenance (Rehfeldt, 1989). In a seedling common garden study involving 40 populations in Oregon and Washington between the Pacific Ocean and the west slopes of the Cascade Mountains (var. *menziesii*), both length of the growing period and seedling height at age 2 decreased clinally with increasing latitude, elevation, and distance from the ocean (Campbell and Sorensen, 1978). Changes in elevation and distance from the ocean had much larger effects on these traits than changes in latitude.

Allozymes. Li and Adams (1989) investigated range-wide patterns of variation based on seed samples from 104 locations using 20 allozyme loci. Excluding one Mexican population that may be another species of *Pseudotsuga*, genetic clustering of populations based on genetic distances (D from Nei, 1987) resulted in three groupings corresponding to the three major subdivisions of this species (*i.e.* coastal variety and northern and southern races of the interior variety). Allele-frequency differentiation among populations (G_{ST} from *Chapter 7*) accounted for 23.1% of the total gene diversity, with about 50% of this due to differences between varieties, 25% due to differentiation between the northern and southern races of the interior variety, and the remainder due to population differences within the three groupings. Average gene diversity within populations of the southern race of the interior variety ($\overline{H}_S = 0.08$) was only one-half that found in the northern race and in the coastal variety, but genetic diversity among populations within the southern race ($G_{ST} = 0.12$) was two to three times greater than in the other species' subdivisions.

Integrating these empirical observations with biogeographical information, the two varieties probably diverged at least a half million years ago. During the Pleistocene, cycles of subsequent cold periods, when glaciers pushed down from the north, periodically separated the varieties and reinforced their separate evolution. Since the most recent contact between the varieties occurred at least 7000 years ago in southern British Columbia, there has been limited opportunity for gene exchange between them. The three major subdivisions appear to have been repopulated after the retreat of the glaciers from different refugia populations. The large total genic diversity and limited differentiation among populations in the coastal variety and the northern race of the interior variety suggest they were each repopulated from one or more large refugia. The southern interior race, with its limited gene diversity, however, appears to have been repopulated from smaller refugia. The large genetic differentiation among populations in this race is probably a reflection of genetic drift occurring among its many small, isolated populations. It is clear from the common garden and allozyme studies in this species that the complex patterns of geographical variation in Douglas-fir are a product of both its past evolutionary history and more recent adaptations to current environmental conditions.

within the same species, and most of the genetic differences among provenances are thought to be adaptive and genetic differences occur across both broad and fine geographic scales. Also in these species, provenances originating from milder climates grow faster unless they are planted in extreme environments that challenge their adaptability (cold, dry or other stress conditions). In addition, provenance x environment interaction is sometimes observed. This is illustrated for *P. taeda* in Box 8.5 and for *P. echinata* in Fig. 8.3; southern provenances surpass those originating from the north when planted in most locations, but the reverse ranking is observed in the most northern test location.

In a number of species, large-scale geographic patterns of differentiation based on allozyme frequencies mirror those observed for quantitative traits in common garden studies, especially differences between subspecies and varieties. Examples include *Pinus contorta* (Wheeler and Guries, 1982), *P. monticola* (Box 8.7), *Pseudotsuga menziesii* (Box 8.4), and *Picea abies* (Lagercrantz and Ryman, 1990). Although these allele-frequency differences among provenances are associated with large differences in environment, they are probably of little adaptive significance; rather, they most likely reflect the geologic history of the species, especially genetic drift associated with population bottlenecks and subsequent recolonization (Lagercrantz and Ryman, 1990; Boxes 8.2, 8.4, 8.7).

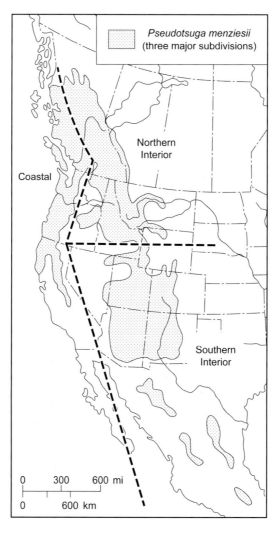

Fig. 8.6. Native range of *Pseudotsuga menziesii* and approximate locations of the three major subdivisions of the species: the coastal variety (var. *menziesii*) and the northern and southern races of the interior variety (var. *glauca*). (Adapted from Little, 1971; Li and Adams, 1989)

On a finer geographical scale, common garden studies show that the strong effects of differential natural selection on patterns of geographic variation are sometimes manifested by steep, local clines (*i.e.* rapidly changing performance of different populations with environmental gradients across small distances) or ecotypes that have arisen across short distances. Two examples of steep, local clines are given in Box 8.8: (1) An elevational cline in *Abies balsamea* showing strong genetic differentiation for CO_2 assimilation rate among seed collection sites separated by a total distance of 3.2 km; and (2) An elevational cline in *Pseudotsuga menziesii* showing strong differentiation for several traits among provenances located in a 10 x 24 km area. A striking example of local ecotypes was also observed in the *Pseudotsuga menziesii* study, where populations from north and south slopes separated by as little as 1.6 km were found to differ genetically (Box 8.8). In both species, little variation was observed among the local provenances in allozyme frequencies, reflecting the considerable gene migration that must be occurring among seed collection sites through both pollen and seed transfer. Therefore, differential selection can overcome the counter-balancing effects of migration when environmental differences are particularly strong.

Although little or no association is usually found between allele frequencies at individual allozyme loci and environmental gradients within geographical regions, there are exceptions (Bergmann, 1978; Furnier and Adams, 1986a; Mitton 1998b). For example, allele frequencies at an acid phosphatase locus in *Picea abies* showed consistent clinal patterns across both latitudinal and elevational transects (Bergmann, 1978). These exceptions appear to be cases where allozyme loci are not selectively neutral (see Box 5.6), but rather where allelic substitutions at particular allozyme loci impact fitness.

Sometimes multivariate statistical methods, such as principle components or discriminant analysis, are used to examine geographic patterns of variation across many allozyme loci simultaneously (Guries, 1984; Yeh *et al.*, 1985; Merkle and Adams, 1988; Westfall and Conkle, 1992). These methods can sometimes improve associations between allozyme patterns and geographic variables, even within regions. Westfall and Conkle (1992) argue that multilocus variables can mark co-adapted complexes of genes and reveal adaptive patterns of variation across many loci simultaneously. It also is possible that these multilocus associations reflect geologic history of the species, and therefore, have little or nothing to do with adaptation.

Racial Variation Not Associated with Environmental Differences

Although not commonly observed, sometimes there are strong genetic differences among provenances that are apparently unrelated to changes in the environment. Two examples are: (1) An Australian angiosperm, *Eucalyptus nitens* (Box 8.6); and (2) A conifer from the western USA, *Taxus brevifolia* (Box 8.7). In both species, differences among provenances are large, yet these differences do not correspond to any known climatic gradients or soil patterns. Since an ecotype is defined as a race adapted to a specific ecological niche, the different races in these species cannot be labeled as ecotypes. There are at least two possible explanations for this pattern of racial variation: (1) Natural selection during the species' evolutionary past resulted in adaptive differences, but environments have since changed so that observed patterns do not relate to today's climates and soils; and (2) The racial variation arose from evolutionary causes other than natural selection (Box 8.3), so that the differences among provenances are not now, and never were, adaptive in nature. In the latter case, a plausible cause is genetic drift due to severe reductions in population size during the evolutionary past. One possibility is that different provenances originated from distinct refugia each affected differently by genetic drift.

Box 8.5. Patterns of geographic variation in *Pinus taeda* and *Pinus elliottii*.

Pinus taeda and *P. elliottii* are the two most important commercial timber species in the southeastern USA and are sympatric (*i.e.* their ranges overlap) in the lower coastal plain regions of Alabama, Georgia, and Florida (Fig. 1). *P. taeda* has an extremely large range spanning many different environments. Large racial differences have been observed in *P. taeda*, and there are complex patterns involving both clines and ecotypes observed range-wide and on more local scales. Growth potential, which can vary by 30% or more among provenances in field tests, is usually highest for provenances originating from southern, coastal areas with high summer rainfall (Wells, 1969; Dorman, 1976; Sluder, 1980; Schmidtling, 1999). For example, in eight field test locations of the Southwide Seed Source Study, a clinal pattern of genetic variation was observed for height growth at ages 5, 10 and 15 years; provenances originating from locations with higher rainfall grew faster as shown by a linear regression of growth on rainfall (Wells and Wakeley, 1966; Wells, 1969; Dorman, 1976, Fig. 70). Significant provenance x location interaction was also observed, because in the most northerly test location (in Maryland), there was no clinal pattern observed for growth and summer rainfall; rather, northern sources survived and grew best.

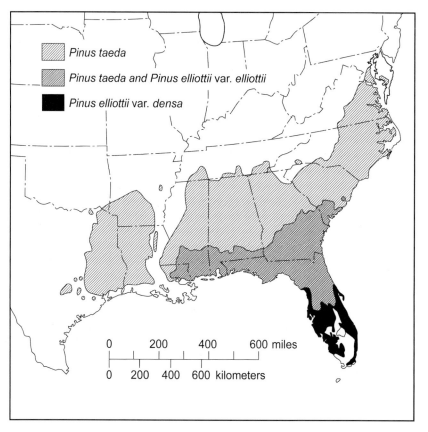

Fig. 1. Native ranges of *Pinus taeda* and *Pinus elliottii* in the southeastern USA (Little, 1971).
(Box 8.5 continued on next page)

Box 8.5. Patterns of geographic variation in *Pinus taeda* and *Pinus elliottii*. (Continued from previous page)

Discontinuous, ecotypic differences also exist in *P. taeda* on rangewide and local scales. Rangewide, western provenances (seed collected from west of the Mississippi River, Fig. 1) are slower growing, more drought resistant and more resistant to the fungal disease caused by the rust *Cronartium fusiforme* than are ecotypes collected from eastern provenances at similar southerly latitudes (Dorman, 1976; Schmidtling, 1999). Possible evolutionary forces contributing to this observed pattern are: (1) Natural selection for drought hardiness and, hence, slower growth in the drier western regions; (2) Evolution of the current western and eastern sources from two distinct Pleistocene refugia (perhaps one in southern Texas and the other in southern Florida, respectively) that were isolated for up to 100,000 years before recolonization began (Schmidtling *et al.*, 1999); (3) Introgression of genes into western provenances through interspecific hybridization (*Chapter 9*) in the distant past with *P. echinata* which is more rust resistant and slower growing (Hare and Switzer, 1969; Dorman, 1976); and (4) Little gene migration between distantly-separated eastern and western locations.

On a local scale, small disjunct populations of *P. taeda* (called the Lost Pines) are located approximately 75 km west of the western edge of the natural range (Fig. 1). When compared in common garden tests to nearby sources, trees from the Lost Pines grow more slowly, are more drought hardy and have different needle morphology associated with their better drought resistance (Dorman, 1976; van Buijtenen, 1978; Wells and Lambeth, 1983). The evolutionary forces effecting differences between the Lost Pines and provenances 75 km to the east could be similar to those mentioned in the previous paragraph, but acting on a much finer geographic scale.

Compared to *P. taeda*, *P. elliottii* has a much smaller (though still appreciable) natural range (Fig. 1) that is relatively homogeneous throughout in terms of similar rainfall and temperature patterns (note that we consider only the main variety of *Pinus elliottii* var. *elliottii* and not the more southern variety, *densa*). Genetic differences among provenances are much less important in the main variety of *P. elliottii*, and in fact racial differences are almost negligible for most traits when compared to the amount of genetic variation existing among individual trees within provenances (Squillace, 1966). In terms of evolutionary forces, we hypothesize that: (1) The relatively homogeneous natural range means that little differential natural selection is acting to effect differentiation among provenances; and (2) Differences that might develop are minimized by high migration rates due to the nearly continuous distribution of *P. elliottii* throughout its range and long-distance pollen dispersal of pines.

While not related to the original cause, restricted gene migration among provenances helps maintain genetic differences among provenances once differences develop. For example, both *E. nitens* and *T. brevifolia* occur in small populations that are effectively disjunct from one another. So, once genetic differences among provenances developed, the lack of gene migration between them helped to perpetuate these differences.

Species with Little or No Racial Variation

There are two types of species that show little, if any, genetic differentiation among provenances. First, there are some tree species that are nearly or completely devoid of any kind of genetic variability: *Acacia mangium* (Moran *et al.*, 1988), *Pinus resinosa* (Fowler and

Box 8.6. Patterns of geographic variation in *Eucalyptus camaldulensis*, *E. nitens* and *E. globulus*.

Eucalyptus camaldulensis, *E. nitens* and *E. globulus* are native to Australia (Fig. 1), are in the same subgenus (*Symphomyrtus*), and are all used as exotics in reforestation programs outside Australia. Information here about these species from field provenance studies is summarized from Turnbull and Griffin (1986) and Eldridge *et al.* (1994). *E. camaldulensis* is the most widely distributed of all species of eucalypts ranging over more than 5 million km^2 from wet, frost-free, subtropical environments in the Northern Territory (12°S latitude) to mild, temperate climates of southern Victoria which experience seasonal frosts (38°S). Amazingly, annual rainfall varies from 200 mm to 1200 mm within this range, with predominately summer rains in the north and winter rains in the south. In hot, dry inland areas, the species is confined to riverbeds and other watercourses.

It is not surprising that *E. camaldulensis* is genetically highly variable with large racial differences expressed in complex patterns. Genetic differences have been found in many field studies and for many traits (*e.g.* growth rate, wood properties, leaf oil content, coppice ability and tolerances to drought, salinity and lime); these results are only highlighted here: (1) There are extremely large ecotypic differences between northern and southern provenances, with northern provenances adapted to tropical growing environments and southern provenances to temperate environments (*e.g.* in one field test in Nigeria, bole volume yield near rotation was 300% more for

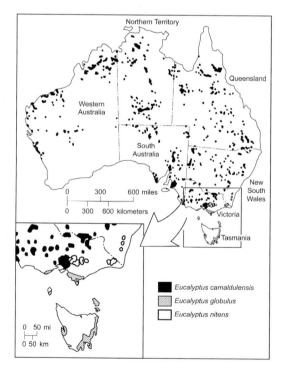

Fig. 1. Native ranges of *Eucalyptus camaldulensis, E. globulus* and *E. nitens* in Australia (Eldridge *et al.*, 1994).

(Box 8.6 continued on next page)

Box 8.6. Patterns of geographic variation in *Eucalyptus camaldulensis*, *E. nitens* and *E. globulus*. (Continued from previous page)

the fastest-growing northern provenance compared to the slowest-growing southern provenance); (2) There are large provenance x location interactions among northern and southern provenances characterized by large rank changes (northerly provenances are better in more tropical environments in Africa, Asia, and South America, and southerly provenances are better in warm temperate climates of the Mediterranean and California); (3) Clinal patterns, associated with decreasing rainfall, are found among provenances along transects from coastal regions to inland areas; (4) Finer-scale genetic differences among provenances are associated with different river drainage systems and even among stands within the same region; and (5) There is evidence of hybridization and introgression with other eucalypt species in some portions of the range (*e.g.* the widely-planted Petford provenance from northern Queensland naturally hybridizes with *E. tereticornis*). Clearly, different evolutionary forces (Box 8.3) have been important in molding the patterns of geographic variation observed in *E. camaldulensis*, but the interplay and relative importance of these forces in different portions of the natural range are poorly understood.

E. nitens occurs in several disjunct populations in a relatively small natural range in the mountains of southeastern Australia from 30°-38°S latitude (Fig. 1). The six provenance areas recognized in this species are located in four regions beginning in the south: (1) Three provenances located in the Central Highland mountains of Victoria (called Toorongo, Rubicon and Macalister); (2) Errinundra located on the border of Victoria and New South Wales (NSW); (3) Southern NSW; and (4) Northern NSW which is approximately 400 km north of stands in southern NSW. Even in the main part of the distribution in Victoria, the species occurs in stands that are effectively disjunct from each other.

For a species with such a small natural range, genetic differences among provenances are extremely large, and some observations include: (1) Important genetic differences among the four regions for several traits (including growth, frost tolerance and others) with provenance x planting location interaction important for some traits; (2) Consistent superiority in growth rate of the Central Victorian provenances (region 1 above) across many test locations in various countries; and (3) Strikingly large differences in juvenile leaf shape and age of transition from juvenile to adult leaves of the Errinundra provenance compared to all others (sometimes this provenance is given varietal status as *E. nitens* var. *errinundra*). Clearly, the small disjunct populations sustain genetic differences, but the original causes of such large genetic differences in a species with such a small natural range are unknown. Hybridization with other species of *Eucalyptus* in different parts of *E. nitens'* range may have played a role.

For *E. globulus* we use the taxonomic nomenclature of Eldridge *et al.* (1994) that recognizes *E. globulus* as a single species with no subspecies (but with four closely related species which we do not consider). *E. globulus* has a natural range that is similar in size to that of *E. nitens* occurring slightly to the south of *E. nitens* in southern Victoria, the islands in Bass Strait and Tasmania (Fig. 1). Climates in its range are generally more similar than within the range of *E. nitens,* yet *E. globulus* is distributed widely across elevations (sea level to more than 500 m) and annual rainfall classes (550 mm to more than 1000 mm).

(Box 8.6 continued on next page)

Box 8.6. Patterns of geographic variation in *Eucalyptus camaldulensis*, *E. nitens* and *E. globulus*. (Continued from previous page)

Many stands throughout its distribution are genetically disjunct from neighboring stands. Racial differences appear to be much less pronounced in *E. globulus* compared to the other two eucalypts discussed here. In many field tests, no large, consistent genetic differences among provenances have been found for growth rate, survival and cold hardiness, but differences in susceptibility to insect attack have been observed (Rapley *et al.*, 2004a, b). In all cases, however, genetic differences among individuals within provenances are extensive. This low level of geographic variation for *E. globulus* contrasts sharply with *E. nitens,* perhaps reflecting either the smaller range of climates occupied by *E. globulus* or their differing (but unknown) geologic histories.

Morris, 1977; Mosseller *et al.*, 1991, 1992), *P. torreyana* (Ledig and Conkle, 1983) and *Washingtonia filifera* (McClenaghan and Beauchamp, 1986). Today's populations may have arisen from a single, very small population that was lacking in genetic variability, such as from a single glacial refugium. These species may not have had sufficient time to recuperate genetic diversity, which might take 10,000 generations (more than 100,000 years; Nei *et al.*, 1975; *Chapter 7*). All of the species mentioned, except *P. torreyana,* have relatively wide natural ranges spanning a range of climates and soils in which we might expect differential natural selection to result in adaptive genetic differences among provenances. However, selection can only act to mold existing genetic variability, not to create genetic variation. Therefore, these genetically depauperate species have possessed insufficient genetic diversity to allow natural selection to operate.

A second group of species exhibits ample genetic variability among trees within populations, but little or no racial variation. Examples are *Pinus elliottii* (Box 8.5) and *Eucalyptus globulus* (Box 8.6). For both of these species, although there are active tree improvement programs employing artificial selection to exploit the considerable tree-to-tree genetic variation that exists (Lowe and van Buijtenen, 1981; White *et al.*, 1993; Tibbits *et al.*, 1997), the genetic differences among provenances are small to non-existent. Characteristics shared by these species that could combine to explain the low levels of racial variation are: (1) The natural range is small in geographic expanse with relatively homogeneous edaphoclimatic conditions throughout, which means that differential natural selection among provenances may not be strong; and (2) Populations are nearly continuous throughout major portions of the range, which would enhance gene migration and limit genetic differentiation among provenances.

A final example in this category is *Pinus monticola* (Steinhoff *et al.*, 1983; Rehfeldt *et al.*, 1984; Box 8.7). This species is subdivided into two major ecotypes that are quite different. Further, although there is substantial genetic variability within ecotypes, most of this variation is among trees in populations, not among populations. The low level of racial differentiation is particularly striking in the northern ecotype, which spans a broad range of more than 10° in both latitude and longitude (Fig. 8.7). Over a similar broad range, *Pseudotsuga menziesii* exhibits considerable adaptive differentiation across the landscape (Box 8.4). Because the range of the northern ecotype of *P. monticola* was covered by glaciers in the late Pleistocene and was colonized rapidly, there may not have been time for racial differentiation to develop (Steinhoff *et al.*, 1983). Alternatively, adaptive differentiation among provenances may be absent because individuals in this species are broadly adapted

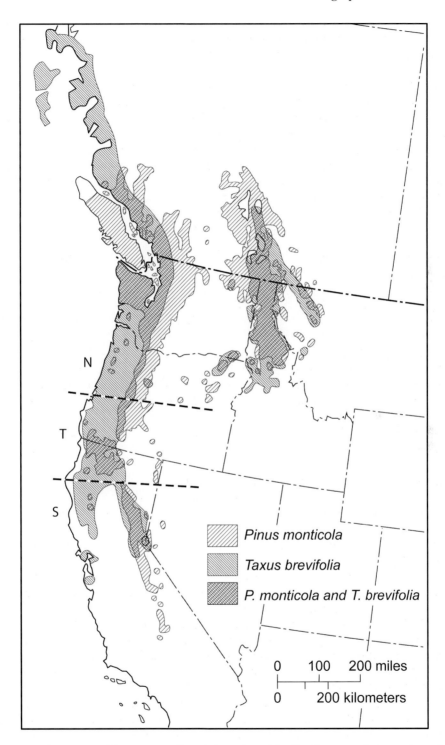

Fig. 8.7. Native ranges of *Pinus monticola* and *Taxus brevifolia*. For *P. monticola*, approximate locations of the northern (N) and southern (S) ecotypes of the species and the transition zone (T) between the ecotypes are shown (Steinhoff *et al.*, 1983).

to environmental heterogeneity. Such **phenotypic plasticity** of individuals could have evolved through selection of heterozygotes (*Chapter 5*) or through selection of alleles that confer wide environmental tolerances (Rehfeldt *et al.*, 1984).

Geographic Patterns of Genetic Diversity

Patterns of racial variation mentioned in the above sections all involved genetic differences in provenance means, *e.g.* mean differences in growth, phenology or allozyme allele frequencies. Provenances can also differ in levels of genetic diversity, a measure of variation instead of averages. Historically, allozyme studies have been widely used to assess genetic diversity in forest trees, and the most widely used measure of genetic diversity is the average expected heterozygosity ($\overline{H}_e$), also called gene diversity. Expected heterozygosity is calculated for each locus as one minus the frequency of homozygotes and then averaged over all loci (*Chapter 7*); it increases with increasing number of alleles per locus and is maximum when all alleles are in equal frequency (*e.g.* with 2 alleles per locus, the maximum value is 0.5 when p = q = 0.5 at all loci). Forest geneticists can sometimes interpret the patterns of $\overline{H}_e$ values, and usually genetic drift is implicated as the evolutionary cause of differences among provenances in their genetic diversity.

Box 8.7. Patterns of geographic variation in *Pinus monticola* and *Taxus brevifolia*.

In tree species with broad geographical ranges, genetic variation for quantitative traits is typically associated with environmental heterogeneity on the landscape in the form of clines or ecotypes (*e.g. Pseudotsuga menziesii*, Box 8.4). Here we describe geographical patterns for two conifer species in the western USA, with distributions that overlap much of the broad range and elevation amplitude of *P. menziesii*, yet have patterns of geographic variation that are quite different.

   Pinus monticola, an important timber tree in the western USA, has a similar distribution to *Pseudotsuga menziesii*, but does not extend as far south in the Rocky Mountains (Fig. 8.7). In most of its range, *P. monticola* is found at elevations below 1800 m, but in the Sierra Nevada Mountains it is typically found from 1800 to 2300 m (Graham, 1990). Rehfeldt *et al.* (1984) grew seedlings from seed collected in 59 populations from throughout the range in greenhouse and nursery environments. Regression analysis was used to relate provenance means to environmental variables. They found that this species is subdivided latitudinally into two ecotypes, with a transition zone in between (Fig. 8.7). The northern ecotype includes both the Rocky Mountains and coastal forests south to about 44°N latitude. The southern ecotype is restricted to the Sierra Nevada Mountains from about 40°N latitude southwards. The northern ecotype is characterized by faster seedling growth, a longer growing season, and lower cold hardiness than the southern (higher elevation) ecotype. There is a steep, latitudinal cline for these traits in the transition zone between ecotypes. Within each ecotype, there is little genetic variation among populations and no relationship between population means for seedling traits and environmental variables, including elevation. That is, within ecotypes *P. monticola* does not exhibit adaptive variation such as clines as evident in *P. menziesii* (Box 8.4), *Tsuga heterophylla* (Box 8.1) and many other conifers in the region (Rehfeldt *et al.*, 1984).

(Box 8.7 continued on next page)

Box 8.7. Patterns of geographic variation in *Pinus monticola* and *Taxus brevifolia*.
(Continued from previous page)

Geographic patterns of allele-frequency differentiation in *P. monticola* based on al-lozymes are nearly identical to those found with seedling quantitative traits, and show additionally that total genic diversity in the northern ecotype ($\overline{H}_T = 0.14$) is only one-half that in the southern ecotype, with the transition zone being intermediate in $\overline{H}_T$ (Steinhoff *et al.*, 1983). These observations suggest that an ancestral species split into two non-contiguous populations, and during the time of separation, perhaps in the late Pleistocene when glaciers covered much of the range of *P. monticola*, the ancestor of the northern ecotype was reduced drastically in population size. After the retreat of the glaciers, the two populations apparently came back into contact in the region of the transition zone, where mixing or hybridization between the ecotypes has occurred.

Taxus brevifolia, a small, shade tolerant conifer, with wind-dispersed pollen and dio-ecious flowering, has a range similar to *Pinus monticola* (Fig. 8.7), but extends further north along the Pacific coast to southeast Alaska. This species typically occurs in small, isolated populations, at low densities (Bolsinger and Jaramillo, 1990). Recently it has re-ceived much attention because its bark contains taxol, used in the treatment of ovarian cancer. A study of variation over most of its range, for both seedling quantitative traits and allozymes, revealed little genetic differentiation between geographic regions (even be-tween the Cascade and Rocky Mountains); yet, extensive differences among nearby popu-lations within regions were observed (Wheeler *et al.*, 1995). While allele-frequency dif-ferentiation among populations (G_{ST}) accounted for 10.4% of the total gene diversity in the species, less than 10% of the G_{ST} was due to differences among regions and 90% was due to differences among populations within regions. Furthermore, variation among popu-lations in both quantitative traits and allozyme frequencies appeared to be totally unrelated to geographic or environmental variables. Therefore, no adaptive patterns of geographic variation are evident in this species; patterns of variation appear to be primarily the result of stochastic processes. Populations of *T. brevifolia* may be particularly susceptible to genetic drift because they are small and isolated, and because their occurrence in the un-derstory hampers gene flow by pollen dispersal (Wheeler *et al.*, 1995).

An example of provenance differences in genetic diversity is *Pinus jeffreyi* (Box 8.2) in which the seven populations from the Klamath Mountains had lower levels of $\overline{H}_e$ than populations from the Sierra Nevada Mountains. Another example occurs across the broad range of *Pseudotsuga menziesii* (Box 8.4). The 24 provenances sampled from the southern race of the interior variety (Fig. 8.6), had a mean $\overline{H}_e = 0.077$, as compared to a mean $\overline{H}_e = 0.15$ for the 36 provenances from the northern race of this variety, and a mean $\overline{H}_e = 0.16$ for the 43 provenances from the coastal variety (*P. menziesii* var. *menziesii*). Interest-ingly, there also were larger allele-frequency differences among the southern provenances of the interior variety than among provenances of the northern interior race or the coastal variety. These patterns are consistent with the effects of genetic drift (less genetic diversity within provenances and more between, *Chapter 5*), which has likely had more of an im-pact on the southern portion of the interior variety which is composed of many small iso-lated populations (Li and Adams, 1989) (Box 8.4).

Another pattern common to several conifers in the western USA is a gradually de-creasing amount of genetic diversity from the southern to the northern parts of their ranges (review by Ledig, 1998). A particularly interesting example is *Pinus coulteri* in which southern provenances from Baja California in Mexico have $\overline{H}_e$ values that are approximately double those of northern provenances near San Francisco, California (Fig. 8.8). In addition to this pattern of $\overline{H}_e$, some alleles present in Baja California disappear at various points along the south-to-north transect, and never reappear. A likely explanation for these observations is that all existing populations are derived from ancestral populations in Baja California through a series of progressive northward long-distance dispersal events following the retreating ice after the last glacial period (Box 8.3). If recolonization occurred in sporadic jumps of a few seed to previously unoccupied territory (called saltation), the founders of new populations carried only a sample of the genetic variation of the parental population. Therefore, each subsequent spread of the range northward resulted in less genetic diversity. The pattern may be especially strong in *P. coulteri* because its fragmented range and very large seed reduce the potential of gene migration among populations.

An analogous pattern of decreasing gene diversity along a single geographic gradient also was observed in *Pinus palustris* in the southeastern USA (Schmidtling and Hipkins, 1998). Western provenances in Texas had $\overline{H}_e$ = 0.14 and $\overline{H}_e$ values gradually decreased from west to east in a linear fashion with provenances from the far eastern portions of the range in Florida having $\overline{H}_e$ = 0.07. The investigators interpreted these results as evidence for: (1) A single Pleistocene refugium of this species located in south Texas or northern Mexico; and (2) Gradual recolonization from this single refugium in an eastward direction to form the current natural range (losing genetic diversity in the process). Interestingly, one stand in southern Louisiana had much lower gene diversity values than predicted by the regression line of $\overline{H}_e$ on longitude. This stand arose from two seed trees remaining after the original forest was clearcut in the 1920s. The reduced diversity values are consistent with the founder principle impact of genetic drift (*Chapter 5*).

Box 8.8. Examples of adaptive differentiation among local stands in forest tree populations.

When forest tree populations span sharp environmental gradients, genetic differentiation among nearby stands (subpopulations) may reflect adaptation to local environments, despite strong gene flow between them. This is illustrated with examples from three North American conifer species.

Fryer and Ledig (1972) studied changes in CO_2 assimilation rates in *Abies balsamea* seedlings grown under uniform conditions. The seedlings were from seed collected from parent trees growing in a continuous population on a single mountain slope in New Hampshire. Sampled stands ranged along an elevational transect from 731-1463 m in altitude. Despite the short distances between sampled stands (the whole transect occurred within 3.2 km), the temperature of optimum photosynthetic CO_2 fixation decreased linearly with increasing altitude (at a rate of -2.7^0 per 305 m), suggesting that the photosynthetic system in this species is precisely adapted by natural selection to the local temperature regime.

Hermann and Lavender (1968) investigated adaptation of *Pseudotsuga menziesii* to changing altitude and slope aspect within a 10 x 24 km area on the west slope of the Cascade Mountains in southern Oregon. Seeds were collected from 2-4 trees on north- and south-facing slopes at 152 m intervals of altitude, ranging from 457-1524 m.

(Box 8.8 continued on next page)

Box 8.8. Examples of adaptive differentiation among local stands in forest tree populations. (Continued from previous page)

The distance between stands sampled in different aspects at each elevation was as little as 1.6 km. Results of seedling tests in both bareroot nurseries and in growth rooms showed, as expected, that growing season length and tree size decrease linearly with increasing elevation of parent trees. Even more striking differences were observed between seedlings from the two aspects. Seedlings originating from south-facing slopes set buds earlier, grew less, and had greater root to shoot dry-weight ratios than seedlings from north-facing slopes, reflecting their adaptation to the hotter, drier growing conditions on south slopes.

Millar (1983, 1989) found consistent association between foliage color of *Pinus muricata* and distinct soil types in northern coastal California. Differences in the shape and waxiness of the chambers above the stomata determine foliage color in this species, which at one extreme results in a bluish cast to foliage and in the other, green (Millar, 1983). These foliage color differences are expressed in common garden experiments, and thus are under genetic control, perhaps by only a few gene loci. Changes in the monoterpene fraction of xylem resin are associated with the foliage color differences. On patches of forest soil characterized by extreme infertility, low soil pH, and mineral toxicities (called pygmy-forest soil because of stunted tree growth), most trees have blue foliage. However, the majority of stands in the area, some only 0.5 km from the blue foliage stands, grow on deep, fertile soils and have green foliage. Although the mechanisms by which blue foliage and the traits associated with it enhance survival on pygmy-forest soil are not understood, the repeated occurrence of the blue foliage type on this extreme soil, and green foliage in stands elsewhere, is strong circumstantial evidence for the adaptive significance of the two foliage types.

In each of the above examples, genetic variation among stands was also investigated using allozymes (Neale and Adams, 1985b; Millar, 1989; Moran and Adams, 1989). In each case, allozyme diversity among stands was extremely weak, indicating that gene migration between these stands has been strong historically. Therefore, selection pressures have been strong enough to counter the effects of gene migration and result in adaptive differences among nearby populations.

IMPLICATIONS OF GEOGRAPHIC VARIATION FOR SEED TRANSFER

In many regions of the world, there is a shift away from uncontrolled exploitation of natural forests to sustainable forestry through silvicultural systems emphasizing natural regeneration or artificial regeneration in plantations. Often, an artificial reforestation program involves planting the same species in several edaphoclimatic zones of varying sizes that differ to a greater or lesser extent in elevation, climate, soils and topography. We have seen in previous sections that genetic differences among provenances can be large, and that frequently these genetic differences are adaptively significant. This means that choosing the proper provenance to use in a particular location is critical to optimum survival, productivity and health of the planted forest. Therefore, one of the main applications of the knowledge gained about geographic variation within a species is the use of this information to choose appropriate provenances for reforestation.

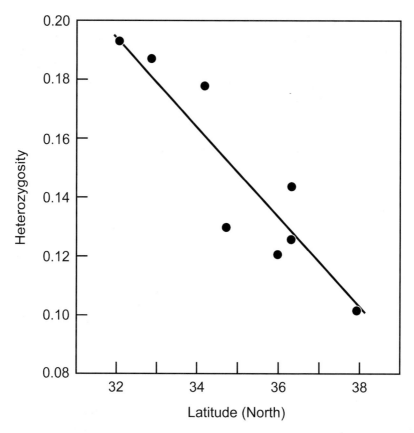

Fig. 8.8. Expected heterozygosity ($\overline{H}e$) as measured in allozyme loci of eight provenances of *Pinus coulteri* sampled along a south to north transect from Baja California (32° N latitude) to San Francisco (38°N latitude). $\overline{H}e$ in the southern populations is twice that of the northern provenances, likely reflecting the effects of genetic drift as the species expanded its range northward by long-distance seed transport from southern refugia at the end of the Pleistocene (see Box 8.3). (Adapted from Ledig, 1998, reproduced with permission of Cambridge University Press)

Before proceeding, it is necessary to differentiate between the **seed collection zone** (also called the provenance, seed procurement zone or donor zone depending upon the author and situation; van Buijtenen, 1992; Morgenstern, 1996) where the seed is collected and the **planting zone** (also called plantation zone, receptor zone, and deployment zone) where the seed is planted. When the seed collection zone and planting zone are similar ecologically and are located in close geographic proximity, then we say that a local provenance is used for reforestation. When some geographical or ecological distance separates the two zones, then a non-local provenance is planted. Sometimes, the seed collection zone and planting zone are combined into a single geographical or ecological unit called a **seed zone** or **seed transfer zone**, which implies that seed collected and planted within the same zone would be considered "local" seed.

The focus of this section is on choosing provenances for operational reforestation programs within the species' native range in the absence of a tree improvement program (or before a tree improvement program has produced genetically improved seed for reforestation). Ultimately, this means defining seed transfer guidelines that specify how far and on what basis seed can be transferred from its seed collection zone to the planting zone. This section covers several topics: (1) Defining objectives that guide selection of provenances for reforestation; (2) Lessons learned from previous provenance studies; (3) A decision tree for choosing provenances and making seed transfer decisions; and (4) Types of seed transfer guidelines and the logistics of their implementation. Discussions of choosing appropriate species and seed sources in exotic planting locations and defining base populations for tree improvement programs are covered in *Chapter 12*.

Setting Explicit Objectives of Provenance Selection

The first step in choosing appropriate provenances for reforestation is to delineate the criteria upon which the decision should be based; that is, how do we judge what is an appropriate provenance? This step is often overlooked or the criteria are only implied. Three possible objectives are: (1) Maintain natural patterns of genetic variation in the planting zone and provide for gene conservation; (2) Ensure long-term adaptation in terms of vegetative vigor and reproductive fitness; and (3) Maximize potential growth, yield and product quality of the planted forest.

If objective one is the highest priority, then the best approach is to first characterize natural patterns of racial variation within the species using short-term seedling tests or long-term field tests, and then design seed transfer guidelines that perpetuate these existing natural patterns. If gene conservation *in situ* is an important part of the operational reforestation program, then this approach works to ensure that extant patterns of racial variation in natural stands are maintained in planted forests. In other words, if objective one is most important, then only the local provenance should be used to plant any particular area, and seed zones are relatively small.

Most reforestation programs that place highest priority on objective two begin with the presumption that the existing patterns of racial variation are adaptive, having resulted from natural selection, and therefore, "local is best". The most common approach to developing seed transfer guidelines employs data from short-term seedling studies, and these guidelines generally: (1) Define seed transfer zones that are as large as possible while still ensuring long-term adaptation; or (2) Specify distances (in km or along ecological gradients such as elevation) that seed can be moved from its point of collection (*i.e.* the local provenance area) to a planting zone to minimize any risk of maladaptation. With the expressed priority of maximizing long-term vegetative and reproductive adaptation to the planting site, seed transfers are usually conservative, aiming to maintain nearly local seed zones by moving seeds short geographic or ecological distances.

If maximum product yield and quality are the highest priorities (objective three) and gene conservation efforts are being conducted separately from the reforestation program, then the goal is to develop guidelines that allocate the best performing provenance to each planting zone subject to an acceptable level of risk (a blend of objectives two and three). Almost always this involves balancing risk and reward: that is, finding the provenance with optimum production potential (the reward) that is well enough adapted to have a high probability of excellent survival and growth for the entire rotation (minimizing the risk of maladaptation). To adequately assess both production potential (reward) and adaptation (risk), there is no substitute for good data from long-term field trials established in the planting zone. When these data are available, guidelines are developed that aim to maximize product yield and quality for all edaphoclimatic regions in the planting zone. Sometimes this can involve moving provenances long geographic and/or ecological distances from point of seed collection to point of planting.

Lessons Learned from Previous Provenance Studies

Good judgment by the reforestation forester plays an important role in choosing appropriate provenances for plantation programs because rarely, if ever, will it be possible to base decisions on perfect information (such as large amounts of data from many well-replicated, long-term provenance trials established in the planting zone). Judgment can be enhanced by developing an appreciation for lessons learned from previous studies conducted with a wide range of tree species.

Local versus Non-local Provenances

In the absence of data from long-term provenance trials established on field sites in the planting zone, it is common to assume that use of local provenances is the best alternative. While planting seed from a local provenance is almost always the safest decision to ensure long-term adaptation to the edaphoclimatic conditions at the planting site, the local provenance may or may not produce the maximum growth, yield or product quality. We consider two lines of reasoning: empirical evidence from field trials and theoretical arguments.

In some long-term provenance studies, the local provenance has performed better than non-local provenances. Examples from three pine species (see review by Ledig, 1998) are: (1) At 12 years, stem volume of *Pinus rigida* was negatively correlated with distance between the seed collection zone and the planting site (Kuser and Ledig, 1987); (2) At 29 years, provenances of *P. ponderosa* collected along an elevational transect in the Sierra Nevada mountains of California performed best when planted at an elevation similar to that of seed collection (Conkle,1973); and (3) At 10 years, the local northern sources of *P. echinata* grew best in the most northern planting locations (line c of Fig. 8.3).

There are as many, if not more, instances where the local source is not optimum for growth, but rather is outperformed by another provenance originating from a slightly milder environment. Ledig (1998) and Schmidtling (1999) each cite several experiments in which provenances from more southerly latitudes or lower elevations performed better than the local provenance. As one specific example, provenances of *P. echinata* from farther south in the range did best in all test plantations scattered across 1000 miles (Fig. 8.3); it was only in the most northerly, harshest plantation that the local, most northerly provenance was superior.

Theoretical arguments also indicate that local provenances may not be optimum for plantation programs (Namkoong, 1969; Mangold and Libby, 1978; Eriksson and Lundkvist, 1986). First, natural selection results in adaptation to meet the requirements of the entire life cycle: both vegetative fitness (ability to survive and grow to reproductive age) and reproductive fitness (ability to produce seed and leave surviving offspring). In a plantation program, critical portions of the life cycle are sometimes obviated and therefore, elements of fitness involved with natural reproduction are less important in plantations: (1) The ability to flower and reproduce is not essential in plantations because seed can be produced or collected elsewhere; (2) Critical phases of germination and seedling establishment in natural conditions are bypassed in plantations when seedlings are raised in nurseries; and (3) Environmental and competitive conditions may be very different in plantations due to silvicultural treatments, application of fertilizers, etc. Eriksson and Lundkvist (1986) differentiated between domestic fitness and natural Darwinian fitness to indicate that even if local provenances are best adapted to natural conditions, they may not be superior in plantations. Therefore, it is often possible to find non-local provenances that express superior vegetative vigor (growth) in the plantation environment, even though they may not be as completely adapted in terms of reproductive fitness or other portions of the life cycle not critical to success in plantations.

Another reason that local provenances may not perform best in plantations is that evolution is conservative such that natural selection may have adapted the local provenance to extreme climatic events not likely to be experienced during most plantation rotations (Ledig, 1998). In addition, there are instances when natural selection has not produced a perfectly adapted local population. We have seen many cases in which current patterns of geographic variation mainly reflect the effects of severe population bottlenecks, rather than adaptation. For instance, Morgenstern (1996, p 143) suggests that several northern species are not utilizing the entire growing season for growth either because of a lag in adaptation following glaciation or due to the conservative nature of evolution.

Finally, other evolutionary factors, such as genetic drift, sometimes strongly influence patterns of genetic differences among provenances. In these cases where selection and adaptation to current environments have not been the principle causes of observed differences among provenances, there is no reason to expect that local provenances should be better adapted to local planting conditions.

The discussion above does not mean that non-local provenances are always best, and in fact the local provenance is the safest to use in the absence of any other information. However, if growth and product yield are the main criteria for choosing appropriate provenances, use of a non-local provenance may result in rapid and inexpensive genetic gains, as discussed below.

General Guidelines for Choosing Provenances

When no field data exist from long-term provenance trials, it is useful to have a set of general guidelines developed from previous studies of other species (Zobel *et al.*, 1987, p 79). These can also be useful for deciding which provenances to include or sample most heavily when planning provenance trails for a plantation program. The most important rule of thumb is: within a species, provenances that have evolved in harsher (colder, drier, etc.) portions of the natural range usually exhibit slower growth potential, but better stress resistance, than provenances from milder portions of the range. Therefore, transferring seed from a provenance evolved in a harsher environment (*e.g.* higher elevation) to a milder planting zone (*e.g.* lower elevation) is expected to result in less growth and yield than use of a local provenance.

Conversely, there may be some gain in growth expected from moving seed from milder to harsher climates. An example would be planting a provenance from a lower elevation at a slightly higher elevation. This potential gain in growth is accompanied by the risk that if the movement is too extreme, the transferred provenance may suffer losses in growth or survival from stress in the harsher environment. The risk/reward decision is never easy and there are two types of errors: (1) Make the conservative decision (use a local provenance) and potentially lose inexpensive and rapid genetic gains that might have been obtained if a non-local provenance from a slightly milder environment had been planted; and (2) Make the aggressive decision and potentially suffer symptoms of maladaptation if seed is moved too far.

Balancing these factors, the following is a general set of guidelines to employ in the absence of good field data (Zobel *et al.*, 1987, p 79): (1) Do <u>not</u> transfer seed from a harsher to a milder environment as a growth loss may result; (2) Do <u>not</u> transfer seed from acidic to basic soils or *vice versa*; (3) Do <u>not</u> transfer seed from summer rainfall climates to winter rainfall climates or *vice versa*; (4) Consider transferring seed from milder to slightly harsher environments to increase growth performance; and (5) Be more aggressive with seed transfer when rotation length is shorter and ecological distance of seed movement is small (even if the physical distance of the movement is large).

A Decision Tree to Guide Seed Transfer Decisions

If objective one (maintain natural patterns of genetic variation in the planting zone) or objective two (ensure long-term adaptation in terms of both vegetative vigor and reproductive fitness) have the highest priority, then seed zones and seed transfer guidelines often emphasize use of local or near-local seed for reforestation. In this section, we assume that the main goal of seed transfer guidelines is to maximize expected product yield from the plantations (objective three), while still ensuring adequate adaptation to the plantation environment. In this situation, the decision tree in Fig. 8.9 can be useful for guiding choice of appropriate provenances for reforestation. As mentioned earlier, the focus here is on reforestation within the native range of a species; discussion of exotics is found in *Chapter 12*.

The most desirable situation is that the reforestation forester has a large amount of information from long-term provenance trials established in the planting zone. However, this is often not the case, and yet decisions still must be made about which provenances to use. The case of least information is that no seed transfer guidelines or provenance data of any kind exist. In this situation, it is usually prudent to plant local provenances or try to choose provenances such that the edaphoclimatic characteristics of the seed collection zone closely match those of the planting zone (Fig. 8.9a). This conservative approach to choice of provenance has the highest probability of selecting a provenance that has long-term adaptation to the planting site and will provide adequate growth and yield. A more aggressive option would be to move seed from milder seed collection zones to slightly harsher planting zones as discussed previously.

Even when no data exist from long-term provenance trials, there may be existing seed transfer guidelines (Fig. 8.9b) developed from climatic information, short-term seedling tests or genetic marker studies. A common first step is to use climatic, topographic and soils information to define ecologically small seed transfer zones. These seed zones are created such that variation of climate, soils and elevation within each zone is arbitrarily small. Original seed zones in the native ranges of several species were developed this way. For example, in British Columbia there were 15 coastal seed zones originally delineated based on this type of edaphoclimatic information and these were the same for all species (Morgenstern, 1996, p 38). Later, genetic data from field provenance trials were used to enlarge each zone and reduce the number of zones from 15 to 4. Similarly, seed zones in Oregon and Washington were originally crafted using climate and elevation information, then were revised mostly based on data from short-term seedling tests (Randall, 1996).

Data from short-term seedling tests in artificial environments are commonly used to develop seed transfer guidelines (Campbell and Sorensen, 1978; Rehfeldt, 1983a,b, 1988; Campbell, 1986; Westfall, 1992) and guidelines based on multilocus patterns of allozyme variation have also been proposed (Westfall and Conkle, 1992) (Fig. 8.9b). Clines of seedling traits revealed in greenhouse, growth room or nursery bed studies are evidence of adaptive differentiation across environmental gradients. Presumably, the steeper the cline, the greater the risk that seed transferred to a non-local planting site along the gradient may be maladapted. Hence, seed transfer guidelines are developed to minimize risk of maladaptation and this usually implies planting local or nearly local provenances. Therefore, although these types of seed transfer guidelines provide conservative rules that minimize risk, they may largely ignore the opportunity for making any gain from seed transfer. For the reasons previously noted in the section on *Local versus Non-local Provenances*, using the local provenance does not always maximize product yield. A more aggressive decision would be to use the guidelines to move seed from its seed collection zone to a slightly harsher planting zone.

When there is ample field data from several mid-rotation or older provenance trials established in the planting zone, then the provenance allocation decision can be based on the most important inferences from the data (Fig. 8.9c, d, e). If there are no important provenance differences in the field tests, then there could be a single, large seed transfer zone with seed obtained from any or all provenances planted anywhere within the planting zone (since all provenances perform approximately the same). An example is *Pinus elliottii* (Box 8.5) which exhibits little provenance variation.

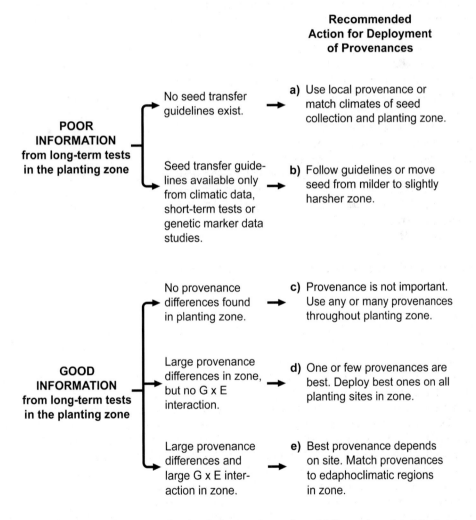

Fig. 8.9. Schematic decision tree for developing seed transfer guidelines when the objective is to maximize yield and product quality. Decisions are based on whether or not good information is available from long-term provenance trials and the nature of the provenance variation observed in those long-term trials.

If there are large provenance differences in important traits, but no provenance x location interaction, then the same one or few provenances are best over all edaphoclimatic zones in the plantation program. This situation is common when soils and climates do not vary dramatically over a region (Matheson and Raymond, 1984a; Zobel *et al.*, 1987, p 37). In this case, it is logical to include all lands in a single seed transfer zone and plant the best provenance or provenances across the entire zone. An example from *Pinus taeda* is the consistent performance of the Livingston Parish provenance from Louisiana that has been widely planted across a wide range of sites in the lower coastal plain of the southeastern USA (Wells, 1985; Lantz and Kraus, 1987).

Finally, when provenances dramatically change rank for growth or adaptability traits across edaphoclimatic conditions within the planting zone (large provenance x location interactions), then it is logical to define multiple seed transfer zones that match provenances to appropriate planting zones based on long-term yield, product quality and adaptability. An example is Weyerhaeuser Company that tested *Pinus taeda* seed from a North Carolina provenance (a mild, coastal climate) in many different field tests located 1500 km away in Arkansas (an inland, more temperate climate; Lambeth *et al.*, 1984). The North Carolina provenance consistently outperformed the Arkansas provenance in tests planted on sites with deeper soils and higher water holding capacity, but suffered some mortality on more shallow soils (presumably due to drought stress). Weyerhaeuser divided their Arkansas lands into two seed zones. The North Carolina provenance is moved 1500 km and planted on sites with high soil moisture holding capacity (Zone 1) and the Arkansas provenance is allocated to Zone 2 (shallow soils).

As mentioned previously, there will rarely, if ever, be perfect genetic information upon which to base seed transfer decisions. Therefore, judgment of the reforestation forester is always important. The decision tree in Fig. 8.9 is meant to help guide the decision by focusing on the type and quality of genetic data available and on the main inferences from those data.

Types of Seed Transfer Guidelines and Logistics of Implementation

Whether developed from edaphoclimatic information, short-term seedling studies, genetic markers or data from long-term provenance trials, guidelines for seed movement to match provenances to the appropriate planting zones can take three forms (Morgenstern, 1996, p 143): (1) Map units drawn on a map of the planting region that show zones within which seed can be safely moved; (2) Written guidelines that specify how far a provenance may be moved or in which types of edaphoclimatic zones it performs best; and (3) Equations developed from multiple regressions or other methods that can be solved for certain latitudes, elevations or precipitation parameters to guide seed transfer. All these methods have useful applications and can be combined together to provide a detailed series of maps, guidelines and equations for a given planting zone. Westfall (1992) reviews the statistical methods used to construct seed transfer guidelines.

There are many examples of seed transfer guidelines blending written guidelines with seed zone maps (Lantz and Kraus, 1987; Morgenstern, 1996; Randall, 1996), and we provide a recent example that is general for all southern pines (Schmidtling, 1999) (Fig. 8.10). Using data from several series of long-term provenance trials of different species of pines from the southern USA (*P. echinata, P. elliottii, P. palustris* and *P. taeda*), multiple regressions were developed to relate provenance performance within a planting zone to climate of the collection zone. The most important explanatory variable was the average yearly minimum temperature of the provenance's location. Analyses showed that moving seed northward from seed collection zones having minimum temperatures 2.8° to 3.9°C (5° to 7°F) warmer than the planting zone resulted in maximum yield gain over the local source. Movements northward of more than 5.5°C (10°F) resulted in growth that was slower than the local source. Since the U.S. Department of Agriculture's Plant Hardiness Zones are based on isotherms of 2.8°C (5°F) average minimum temperatures (Fig. 8.10), Schmidtling (1999) recommends that seed can be safely transferred northward one hardiness zone, but not two. Using seed within a zone is considered safe, but is expected to result in slower growth than movement one zone northward.

Another example is given by Rehfeldt (1983c) for *Pinus contorta* in northern Idaho (Fig. 8.11). Three test sites at elevations of 670, 1200 and 1500 m were planted with seedlings from 28 provenances originating from seed collection elevations of 650 to 1500 m in a small region of Idaho. Using data for tree height at age four in the field and freezing injury to needles tested for damage in a laboratory screening process, Rehfeldt developed

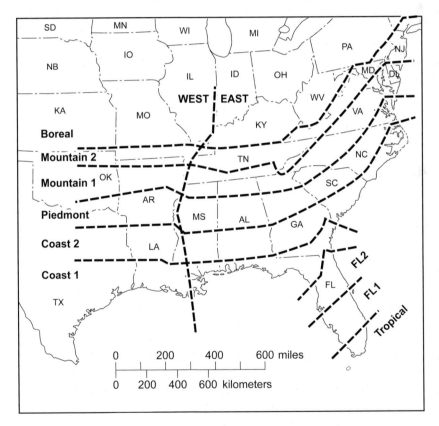

Fig. 8.10. General seed zones for pines in the southern USA based on a series of long-term provenance trials of *Pinus echinata, P. elliottii, P. palustris* and *P. taeda*. The zones are identical to the US Department of Agriculture's Plant Hardiness Zones that are isotherms of 2.8°C (5°F) based on the average minimum temperature. Seeds may be safely transferred within a zone and transfers one zone northward results in improved growth over the local source. Seed transfers of more than one zone northward are discouraged. East-west transfers across the Mississippi River also are discouraged for *P. taeda.* (Adapted from Schmidtling, 1999)

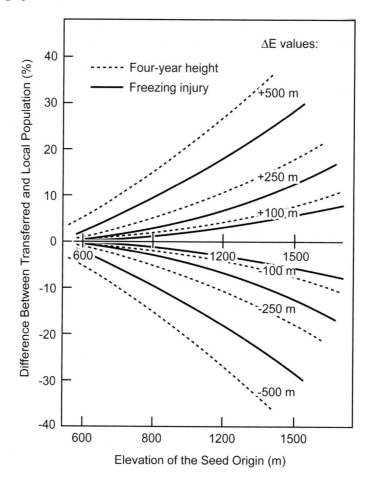

Fig. 8.11. Expected differences in 4-year-old height growth and freezing injury, as determined by multiple regression models, between local provenances of *Pinus contorta* and provenances transferred various intervals of elevation (ΔE). For example, the two curves labeled $\Delta E = + 500$ m indicate the height and freezing injury difference between the local population and provenances transferred upward 500 m. (Copyright 1983, NRC Research Press, Canada. Reprinted with permission from Rehfeldt, 1983c)

multiple regression equations relating provenance performance of these two response variables at different elevations to elevation of seed collection (regressor or predictor variable). The regressions indicated that both height and freezing injury were strongly correlated with elevation of the seed collection zone: provenances transferred from higher elevations grew more slowly than the local provenance (*i.e.* from the same elevation as the planting zone), but were less damaged in the laboratory freeze screening. An interesting result was that genetic differences between local and transferred provenances separated by a fixed difference in elevation increased with increasing elevation. Therefore, it appears that seed transfers of equal distances in elevation entail more risk at higher elevations. For example, if seeds from a provenance originating at 1200 m are transferred upward 250 m ($\Delta E = + 250$ m in Fig. 8.11), the expected result is a 10% increase in height growth and 7% increase in freeze injury, as compared to the local provenance. The same distance transfer of 250 m upward from a provenance originating at 1500 m is predicted to result in an increase in height growth of 18% and an increase in freeze damage of 13%. Therefore, the reforestation forester must weigh the potential for increased growth against the risk of maladaptation to freezing temperatures. Data from older trials would be required to adequately evaluate the long-term implications of these provenance differences observed at the seedling stage.

Before using any seed transfer guidelines, the reforestation forester must clearly understand how they were developed, in particular: (1) What were the objectives of the person who developed the guidelines (maintain local patterns of variation, ensure long-term natural adaptability or maximize yield); (2) What kind of data were used; and (3) What was the quantity and quality of the data? If the objectives of the developer do not match those of the user, or the quality of data is less than desirable, then the reforestation forester must rely on judgment and on additional information from the same or related species.

When allocating provenances to planting sites within the native range, there are several logistical considerations: (1) To ensure a broad genetic representation of each provenance, seed should be collected from multiple stands at different locations within the provenance area and from several trees per stand; (2) GPS should be used to record the exact positions of stands and trees from which seed were collected; (3) Sometimes there are regulations or certification processes that must be followed; and (4) When a reforestation unit is planted, the seed source identity, including provenance, year of seed collection and batch number, should become part of the permanent record associated with the developing plantation. While these steps may seem like common sense, there are many instances where a single lapse in this chain of events has resulted in loss of the identity and genetic origin of the seed used for reforesting large tracts of land.

SUMMARY AND CONCLUSIONS

Tremendous information exists about genetic patterns of geographic variation existing within a wide range of tree species. Information about genetic differences among provenances cannot be obtained by measuring phenotypes in natural stands because genetic and environmental influences on phenotypic expression are confounded. Rather, three types of investigations are used: genetic marker studies, short-term seedling experiments and long-term provenance trials. All three types of studies have advantages and disadvantages, and data from the three types can be complementary. In particular, information from the three types of investigations can be used together to untangle the complex interplay of evolutionary forces that have caused the geographic patterns observed today.

While a few species exhibit little genetic differentiation among provenances, most tree species exhibit marked genetic variation among provenances. The large majority of this variation is adaptive in nature meaning that provenances have evolved through differential natural selection to the local edaphoclimatic conditions in different areas of the species' range. Clinal variation is the most common pattern of adaptive variation manifested by the continuous relationship between measured traits and environmental gradients (such as rainfall, elevation and latitude). Ecotypic patterns of adaptive variation can arise when environmental differences are large and discrete.

The large majority of wide-ranging tree species exhibit complex patterns of geographic variation with clines, ecotypes and other patterns occurring within the same species over broad and fine geographic scales (Boxes 8.4, 8.5, 8.6). These patterns reflect adaptation to both current and past edaphoclimatic conditions, vagaries of biogeological history resulting in genetic drift and discontinuities in the species' distribution slowing gene migration.

Given that genetic patterns can be so complex and variable among tree species, proper genetic studies are critical. Results from short-term seedling studies and genetic marker studies can be useful for understanding evolutionary forces, planning gene conservation programs (*Chapter 10*) and defining preliminary seed transfer guidelines. However, definitive seed transfer guidelines should be based on long-term data from field provenance tests established in the planting zone. When these data do not exist, lessons learned from previous experiments indicate that: (1) Planting the local provenance is the safest choice, but may not result in maximum growth or product yield; and (2) Transfer of seed from a seed collection zone with a slightly milder climate to a planting zone with a slightly harsher climate may result in increased growth above the local source with little risk of increased maladaptation.

If the goal is to maximize product yield and sufficient data from long-term provenance trials are available, then three options exist for seed transfer, depending on the results: (1) When no significant differences among provenances are exhibited in the planting zone, the reforestation forester can deploy any provenance to any planting site (Fig. 8.9c); (2) When strong provenance differences are found, but no provenance x site interaction exists, the best provenance(s) can be deployed to all planting sites (Fig. 8.9d); or (3) When strong provenance x site interactions are observed, seed transfer guidelines must match provenances to the specific types of planting sites in which they excel (Fig. 8.9e).

CHAPTER 9
EVOLUTIONARY GENETICS – *DIVERGENCE, SPECIATION AND HYBRIDIZATION*

Life on earth began approximately 3.6 billion years ago with the occurrence of the first living cells. It was not until the Devonian Period or Carboniferous Period that gymnosperms evolved and the first forest trees were present on earth (300-400 million years ago). The forces of evolution led to the generation of an unknown number of tree species during this period. Many of these are now extinct, yet many thousand exist on earth today. Foundations of the theory of evolution were developed by Charles Darwin in his 1859 publication, *On the Origin of Species* (Darwin, 1859). Darwin's theory of evolution and Mendel's laws of inheritance (*Chapter 2*) together form the fundamental tenets of the science of genetics. In *Chapter 5*, we describe the primary evolutionary forces (mutation, selection, migration and genetic drift) that lead to the formation of new species and determine the genetic composition of individual species. In this chapter, the goals are to understand: (1) Processes of speciation and hybridization; (2) Taxonomic and phylogenetic relationships among species; (3) Molecular mechanisms by which genomes evolve; and (4) How mutualisms between species can influence evolution in both (*i.e.* coevolution). Some excellent general references on plant evolution include texts by Stebbins (1950), Grant (1971), Briggs and Walters (1997), and Niklas (1997).

DIVERGENCE, SPECIATION AND HYBRIDIZATION

Forest trees are found on every continent on earth except Antarctica. **Plant geography** is the science of describing and understanding historical and present-day distributions of plant taxa, locations of their origins and migration histories. Numerous classification systems are used to describe geographical patterns of plant distribution, including those describing forest types. Climatic (primarily moisture) and soil conditions most often determine major forest classification types. Physiognomic (closed or open forest) and structural (evergreen *versus* deciduous or coniferous *versus* broad-leaved) features are also used in classification systems.

A very general classification of forests around the world has four major zones (Kummerly, 1973): (1) Tropical lowland forests; (2) Tropical monsoon, savanna and dry forests; (3) Subtropical and temperate-latitude forests; and (4) Northern-climate coniferous forests (Fig. 9.1). Following this broad classification, more highly refined classifications have been developed. For example, a classification of forest types in North America includes eight major types (Duffield, 1990; Fig. 9.2). This classification includes structural features such as coniferous *versus* broad-leaf.

Species diversity (Gurevitch *et al.*, 2002) within any classification unit further describes a forest type. Species diversity can be measured in terms of the number of different species (**species richness**) or by the proportions of different species (**species evenness**). In general, diversity within forest types is highest in equatorial and tropical regions and decreases towards the poles. There are many more angiosperm forest tree species than gymnosperm species

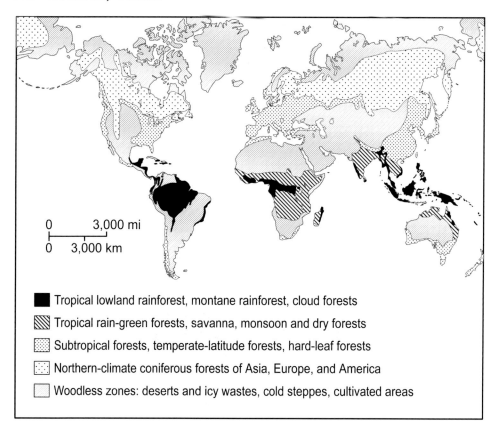

Fig 9.1. Geographic distribution of forest zones in the world (Kummerly, 1973).

(Table 9.1). For example, there are more than 700 species in the genus *Eucalyptus* alone, whereas there are only about 600 conifer species in total (Judd *et al.*, 1999). In this section, we discuss two important topics that determine species diversity: (1) Divergence and speciation; and (2) Hybridization and introgression. We begin, however, by defining a species.

Species Concepts

The definition of what constitutes a species is fundamental to the understanding of evolution. The classical definition of species is the **biological species concept (BSC).** Under the BSC a species is defined as a collection of individuals that interbreed and produce viable and fertile offspring, but cannot breed with members of another species (Niklas, 1997). This definition of species was likely inspired by thinking about animals, more so than about plants. Because plants are not free to move about like animals, all individuals of a species cannot freely inter-breed with one another. In plants, the situation also occurs that two groups are isolated from one another and have large morphological differences, but when brought into contact with one another they interbreed freely. There are so many exceptions to the strict BSC in plants that evolutionary plant biologists have developed other definitions for plant species.

Niklas (1997, p. 70-74) describes some alternative species definitions including: (1) Mate recognition concept; (2) Morphospecies concept; (3) Evolutionary species concept; and (4) Phylogenetic species concept. The mate recognition concept defines species mostly

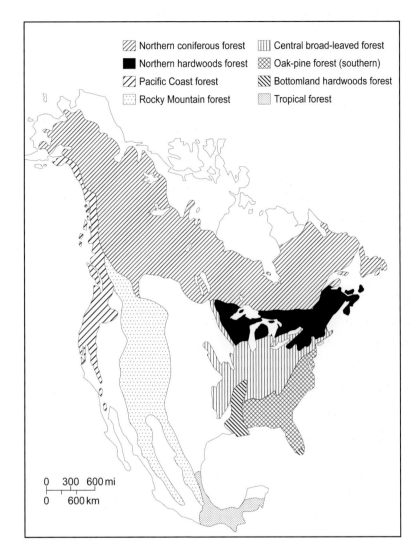

Fig. 9.2. Geographic distribution of forest regions in North America. (Reprinted from Duffield, 1999, with permission of John Wiley & Sons, Inc.)

on their ability to mate; thus, groups that can produce hybrids would be included as single species under this definition even if physical isolation prevents mating. The morphospecies concept gives priority to a species' unique and heritable phenotypic characteristics; thus, this definition includes species that reproduce asexually. Because BSC defines a species as a "collection of individuals that interbreed," asexual species do not fit under the BSC definition. The evolutionary species concept defines species based on derivation from a single evolutionary lineage. The phylogenetic species concept emphasizes fixed character states (see section on *Phylogenetics*), which may overestimate the number of species. There is also the issue of "splitting" and "lumping," which refers to the tendency of taxonomists to either over- or under-estimate, respectively, the true number of species. The main point is that no single definition of a species is adequate, especially for plants, and that species definition is not an exact science.

Table 9.1. Number of genera and species in a representative sample of families of forest trees.

Family	No. genera	No. species
Gymnosperms		
Araucariaceae	2	32
Cupressaceae	29	110-130
Pinaceae	10	220
Podocarpaceae	17	170
Taxaceae	5	20
Taxodiaceae	9	13
Angiosperms		
Betulaceae	6	157
Casuarinceae	1	70
Fagaceae	9	900
Juglanaceae	8	59
Myrtaceae	144	3100
Rosaceae	85	3000
Salicaceae	3	386
Ulmaceae	6	40

Mechanisms of Speciation

Regardless of how imprecise our definition of a species might be, it is clear that two key features in the development of new species (*i.e.* **speciation**) are genetic (and usually morphological) differentiation and the acquisition of reproductive isolation. Two general types of speciation are often recognized, allopatric speciation and sympatric speciation. In both types, the forces of evolution that cause genetic divergence among populations are natural selection and genetic drift, while gene flow retards divergence (*Chapter 5*).

Allopatric Speciation

Allopatric speciation, or gradual speciation, occurs over a long period of evolutionary time and is dependent largely on forces of natural selection to produce a new species from an existing species. Allopatric populations are physically separated and their ranges do not overlap. There are three major phases of allopatric speciation (Fig. 9.3). Initially, a subpopulation becomes spatially separated from the main population. Normal geological events such as building of continents, oceans, mountains, rivers, etc., can cause a subpopulation to become isolated. Next, the subpopulation becomes genetically differentiated from the main population through the processes of genetic drift and natural selection. Barriers to seed and pollen migration between the subpopulation and the main population ensure that genetic divergence is not overcome by gene flow. Finally, differentiation results in reproductive isolation such that it becomes impossible for the subpopulation and main population to ever interbreed again, hence forming two species. Allopatric speciation has probably been the most important type of speciation for the creation of species of forest trees. We return to this topic in the section on *Evolutionary History and Phylogeny* later in this chapter.

Sympatric Speciation

Sympatric speciation, or abrupt speciation, occurs in a relatively short period of evolutionary time and is usually associated with some major genetic change that causes reproductive isolation of some members of a population from the main population (Fig. 9.4). Sympatric populations have overlapping ranges. The types of major genetic changes that can lead to sympatric speciation are single gene mutation (**monogenic speciation**), doubling or loss of an entire chromosome (**aneuploidy**) or even multiplication of entire genomes (**polyploidy**). Polyploidy is most often responsible for abrupt speciation in higher plants and is common among angiosperm forest tree taxa, but rare in gymnosperm forest trees. We discuss polyploidy in more detail later in this chapter in the section on *Molecular Mechanisms of Genome Evolution*. Another potential mechanism of sympatric speciation (called **peripatric speciation**) is when genetic drift in a small geographically peripheral subpopulation causes major genetic modifications in its gene pool and subsequent rapid development of reproductive isolation from the main body of the species. This form of speciation may be more common in tree species than in other plants because of their potential for long distance seed dispersal, and therefore, the formation of small founder populations at the margins of a species range.

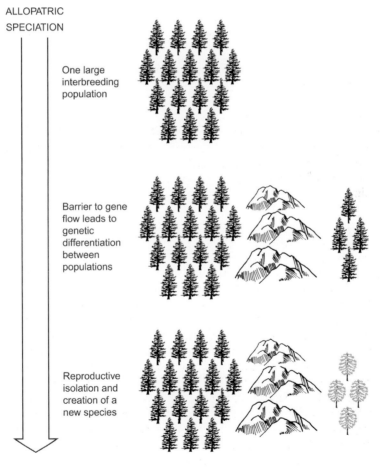

Fig. 9.3. Allopatric speciation occurs when a once large interbreeding population becomes partitioned into two or more spatially isolated groups (*e.g.* by the uplifting of mountains or rising sea level), and a barrier to gene flow develops. The subpopulations are then able to genetically diverge from one another until eventually they become reproductively isolated. At this time they become separate and unique species.

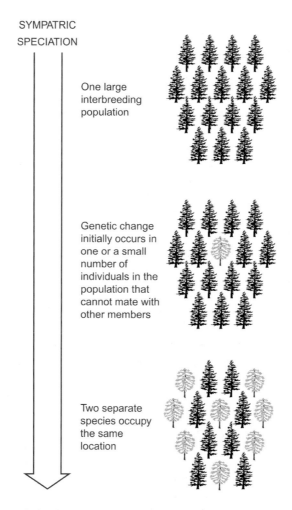

SYMPATRIC
SPECIATION

One large
interbreeding
population

Genetic change
initially occurs in
one or a small
number of
individuals in the
population that
cannot mate with
other members

Two separate
species occupy
the same
location

Fig. 9.4. Sympatric speciation is usually associated with some major genetic change that causes one or a small number of individuals in an interbreeding population to suddenly become reproductively isolated from the rest of the population. These individuals diverge genetically from the main population and eventually form a new species.

Hybridization and Introgression

Hybridization can be defined as "crossing between individuals belonging to separate populations which have different adaptive norms" (Stebbins, 1959). Although hybridization is generally meant to describe successful mating between individuals in different species, the term is also used to describe mating between different subspecies, ecotypes or even populations within a species. Hybridization occurs in nearly all plant groups and in many forest tree taxa (Wright, 1976). Traditionally, anatomical and morphological characteristics such as cone or leaf shape have been used to identify hybrids. More recently, molecular markers are used in studies of hybridization. Hybridization is generally viewed as a mechanism for reversal of evolutionary divergence between species; however, hybridization can also lead to the creation of new species (called **reticulate evolution;** see Box 9.1). In this case, two distinct species hybridize to form a hybrid population that eventually becomes reproductively isolated from the parent species and becomes its own unique species. In this chapter we are only concerned with the role of natural hybridization. In *Chapter 12*, the role of artificial hybrids in tree improvement and plantation forestry is discussed.

An interspecific F_1 hybrid results from the mating of two distinct species. In many cases such hybrids are not fertile and are therefore evolutionary dead-ends, unless they can reproduce vegetatively. Fertile hybrids have the opportunity to mate with other hybrids or with the parental species. A population that includes pure parental species and hybrids of varying degrees is called a **hybrid swarm**. When repeated backcrossing of hybrids with a parental species results in the transfer of genes from one parental species to another, it is called **introgressive hybridization** or simply **introgression.** Several examples of natural hybridization and introgression in *Pinus* and in angiosperm forest tree species are given in Boxes 9.1 and 9.2, respectively.

EVOLUTIONARY HISTORY AND PHYLOGENY

A series of fundamentally important events began about 650 million years ago (Myr) that led to the first multicellular, photosynthetic organisms appearing on earth (Table 9.2). Niklas (1997) lists the following seven successive evolutionary events over time: (1) Development of photosynthetic cells; (2) Development of organelles; (3) Development of multicellular structure; (4) Transition to terrestrial habit; (5) Development of vascular tissues; (6) Development of seeds; and (7) Development of flowers. Fossil evidence suggests that the first vascular land plants arose in the Ordovician Period (510 Myr), while it was not until the late Devonian Period (409 Myr) or early Carboniferous Period (363 Myr) that gymnosperms arose and the first trees were found on earth. In the following two sections we briefly trace the evolutionary history of trees through geologic time. We arbitrarily define early evolutionary history as the Paleozoic and Mesozoic Eras and recent evolutionary history as the Cenozoic Era.

Evolutionary History

Gymnosperms dominated early terrestrial forests. The first angiosperms did not evolve until the late Cretaceous, approximately 150 million years after the appearance of the first gymnosperms. Conifers were the dominant flora during the Mesozoic Era (225 Myr) and the greatest diversity of coniferous species existed at this time. Miller (1977, 1982) has studied the fossil record to confirm the hypothesis of Florin (1951) that modern families of conifers evolved from the Lebachiaceae that were found during the late Paleozoic via the Voltzlaceae that were found during the Mesozoic. The Lebachiaceae and Voltzlaceae are considered transitional conifers. The first modern conifer family to appear was the Podocarpaceae during the Triassic, followed by the Araucariaceae, Cupressaceae, Taxodiaceae and the Taxaceae during the late Triassic or early Jurassic (Fig. 9.5). The Cephalotaxaceae arose during the mid-Jurassic and the Pinaceae did not appear until the Cretaceous, although it is possible that they evolved much earlier (Fig. 9.5). The evolutionary history of Pinaceae is described in Box 9.3.

The majority of forest tree species, primarily angiosperms, evolved in the last 130 million years (Cenozoic Era). The Paleocene and Eocene epochs were hot and moist favoring the development of angiosperms. However, comparatively more is known about the evolution of *Pinus* during the Cenozoic (Box 9.3) than about angiosperm tree evolution.

Box 9.1. Hybridization and introgression in the genus *Pinus*.

The genus *Pinus* includes more than 100 species whose ranges are often sympatric. The genus is divided by taxonomists into two subgenera, four sections and 17 subsections (Price *et al.*, 1998). Natural hybridization occurs among species within subsections (Critchfield, 1975), although this is not surprising as the systematic classification is based partially on crossing data (Duffield, 1952). In contrast to hybrids of other plant groups, pine hybrids are often fertile; therefore, introgression and hybrid speciation are part of the evolutionary history of pines.

Natural hybrids are found among several pairs of species of the Australes section in the southeastern United States (Dorman, 1976). Important examples include *P. taeda* x *P. palustris* (Namkoong, 1966a) and *P. taeda* x *P. echinata* (Zobel, 1953). In the Ponderosae subsection, natural hybridization occurs between *P. ponderosa* and *P. jefferyi* in the western US (Conkle and Critchfield, 1988) and between *P. montezumae* and *P. hartwegii* in Mexico and Central America (Matos and Schaal, 2000).

One of the best studied hybrids in *Pinus* is the hybrid between *P. contorta* and *P. banksiana* (subsection Contortae) in western Canada. Both of these species have extensive ranges in North America (Fig. 1). *P. contorta* ranges from Baja California in Mexico to the Yukon, throughout much of the western US and Canada; *P. banksiana's* range extends from Nova Scotia to the Yukon in Canada and North-Central US. In central Alberta and southern Yukon the two species have large areas of sympatry.

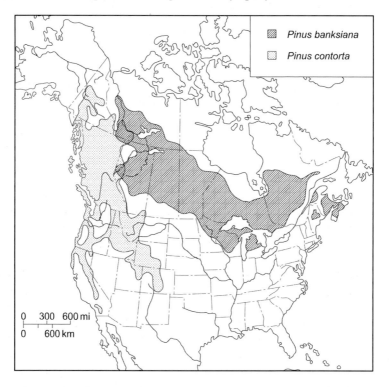

Fig. 1. Geographic distribution of *Pinus banksiana* and *Pinus contorta* in North America showing the zone of sympatry in western Canada.

(Box 9.1 continued on next page)

Box 9.1. Hybridization and introgression in the genus *Pinus*. (Continued from previous page)

   P. contorta x *P. banksiana* hybrids are known to occur in the zones of sympatry and direct evidence is provided through studies on morphological and biochemical characters (Moss, 1949; Mirov, 1956; Zavarin *et al.*, 1969; Pollack and Dancik, 1985). However, it is difficult to estimate the extent of introgression into allopatric populations or infer the evolutionary history from these studies (Critchfield, 1985). Wheeler and Guries (1987) concluded that introgression is ancient and extensive in the zone of sympatry based on data from 16 morphological (mostly seed and cone) characters and 35 allozyme loci. They calculated hybrid index measures of Anderson (1949) (Fig. 2).

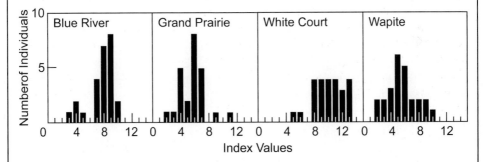

Fig. 2. Hybrid index histograms for trees sampled from four putative hybrid populations in the zone of sympatry between *Pinus contorta* and *P. banksiana*. Index values are aggregates of seven diagnostic traits. Pure *Pinus contorta* populations have index values that range from 0-3 while pure *Pinus banksiana* populations range from 8-14. The high frequency of intermediate index values in these populations support the conclusion that they have resulted from extensive hybridization and introgression between these species. (Reprinted from Wheeler and Guries, 1987, with permission from NRC Research Press, National Research Council of Canada)

   Wagner *et al.* (1987) conducted a study of cpDNA polymorphism in allopatric and sympatric populations of *P. contorta* and *P. banksiana* and found no evidence for introgression within allopatric populations but did within sympatric populations. These data suggest that contact between species is more recent and that introgression has yet to extend beyond the zones of sympatry.
   Evidence that new species can evolve from hybrids (following their stabilization and isolation from parental species; *i.e.* reticulate evolution) is reported for a number of plant groups (Niklas, 1997), including pines. Among the several species of *Pinus* found in China, *P. tabulaeformis* occurs in northern and central China and *P. yunnanensis* in southwestern China. It has been proposed that a third species, *P. densata*, is an ancient hybrid between *P. tabulaeformis* and *P. yunnanensis* that arose during the Tertiary (Mirov, 1967). *P. densata* occupies a higher elevation zone not occupied by either of the other two species. Wang *et al.* (1990) and Wang and Szmidt (1994) used isozyme and cpDNA markers to show that *P. densata* populations are intermediate in gene and haplotype frequencies, respectively. This appears to be a clear example of hybrid speciation in *Pinus.*

Box 9.2. Hybridization and introgression in angiosperm forest tree species.

Natural hybridization in angiosperm forest tree species is very common. Natural hybridization in the genus *Quercus* (oak) is so ubiquitous that it is difficult to apply the biological species concept to this genus. *Quercus* includes more than 450 species of trees and shrubs and the extensive literature on their hybrids is beyond the scope of this book (see reviews by Palmer, 1948; Rushton, 1993). Naturally occurring hybrid oaks are often found as isolated trees in areas of sympatry with both parental species. Because of the considerable amount of variation in morphological characters (*e.g.* leaf shape) within each species, it is difficult to assess the extent of introgression between species from morphological data and evaluate the importance of hybridization in the evolution of oaks.

Recent studies using chloroplast DNA markers provide new insight into these questions. Whittemore and Schaal (1991) studied introgression among five species of sympatric white oaks in the eastern US using maternally-inherited cpDNA polymorphisms. A cladogram generated from the cpDNA data showed that cpDNA haplotypes clustered by geographic region and not by species (*i.e.* they were more similar across species within the same region than among representatives of the same species in different regions; Fig. 1).

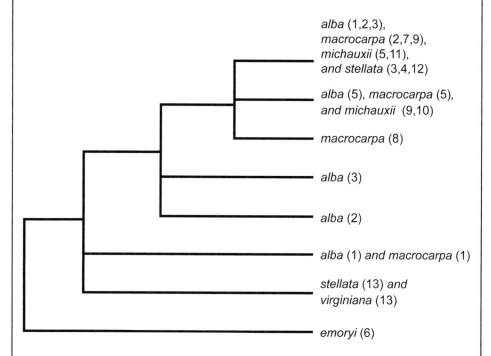

Fig. 1. Cladogram of eight different cpDNA genotypes found among five species of *Quercus* (*alba, macrocarpa, michauxii, stellata* and *virginiana*). *Q. emoryi* was included as an outgroup. Numbers in parentheses refer to sampling localities. (Reprinted by permission from Whittemore and Schaal, 1991; copyright 1991, National Academy of Sciences, USA)

(Box 9.2 continued on next page)

Box 9.2. Hybridization and introgression in angiosperm forest tree species. (Continued from previous page)

These data suggest that cpDNA genomes are freely exchanged among species, even though the individual species remain phenotypically distinct. Strong selection applied to the nuclear genome would help to maintain species identities in spite of considerable gene flow. Dumolin-Lepegue *et al.* (1999) conducted a similar study with four sympatric species of oaks in France. The extensive sampling and quantitative measures of gene flow used in this study further demonstrated high levels of contemporary gene flow and introgression, and that patterns of organelle DNA variation could not simply result from ancient hybridization events and species-similar patterns of post-glacial colonization.

The genus *Populus* includes 29 species, many of whose ranges are sympatric (Fig. 2). Natural hybridization is common among sympatric species both within and between different sections of the genus (Eckenwalder, 1984) (Fig. 3). For example, natural hybrids occur among species within and between the Aigeiros and Tacamahaca sections; whereas, species in the Populus section appear to be reproductively isolated from species in the other sections, even when their distributions are sympatric (Eckenwalder, 1984). Generally, hybrids are found in a narrow zone in the regions of sympatry. However more extensive hybrid swarms and introgression also occur in some hybrid complexes.

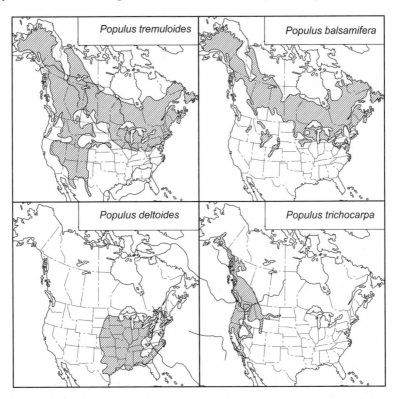

Fig 2. Natural ranges of four species of *Populus* (*tremuloides, balsamifera, deltoides, trichocarpa*). (Box 9.2 continued on next page)

Box 9.2. Hybridization and introgression in angiosperm forest tree species. (Continued from previous page)

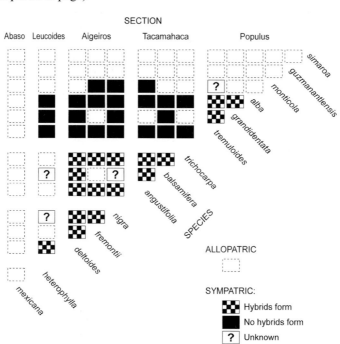

Fig. 3. Observed natural hybridization among species in the genus *Populus*. Natural hybridization is possible only in areas where species are sympatric. (Based on data from Eckenwalder, 1984)

Molecular markers have been used to study patterns of hybridization and introgression in *Populus*. Keim *et al.* (1989) used nuclear DNA RFLP markers to show that *P. fremontii* and *P. angustifolia* form F_1 hybrids in an elevational zone of sympatry in Utah (western US), but that within the hybrid swarm there was no evidence for mating among F_1s, and that hybrids only backcrossed to *P. angustifolia*. The implication of this unilateral introgression is that the hybrid swarm will not be self-perpetuating because hybrids only reproduce when crossed to *P. angustifolia*. *P. fremontii* is adapted to lower elevations and *P. angustifolia* to higher elevations. Nevertheless, *P. fremontii* may be gradually replaced by introgressed *P. angustifolia* if the *P. fremontii* genes in the introgressed species allow it to compete effectively with *P. fremontii* in its home environment.

The genus *Eucalyptus* includes more than 700 species, many of whose ranges are sympatric on the Australian continent (Pryor and Johnson, 1981). A survey of 528 species determined that 55% of these species potentially hybridize, although the observed number of natural hybrids is much less (Griffin *et al.*, 1988). Nevertheless, hybridization would seem to be a very important evolutionary factor in *Eucalyptus*. McKinnon *et al.* (2001) conducted an extensive study of cpDNA haplotype diversity within and between 12 species of the *Symphomyrtus* subgenus in Tasmania. Just as in the oaks, haplotype similarity corresponds more closely with the geographic proximity of individuals than with their species designation. In addition, species of the subgenus *Symphomyrtus* include possible examples of reticulate evolution.

Table 9.2. Geologic time scale. The first forest trees probably appeared on earth sometime during the late Devonian or early Carboniferous periods.

Eon	Era	Period	Epoch	Myr*
Phanerozoic	Cenozoic	Quaternary	Halocene	0.01
			Pleistocene	2.5
		Tertiary	Pliocene	7
			Miocene	26
		Paleogene	Oligocene	34
			Eocene	54
			Paleocene	65
	Mesozoic	Cretaceous	--	136
		Jurassic	--	190
		Triassic	--	225
	Paleozoic	Permian	--	290
		Carboniferous	--	363
		Devonian	--	409
		Silurian	--	439
		Ordovician	--	510
		Cambrian	--	570
Precambrian	Precambrian	--	--	4570

* Myr = million years ago.

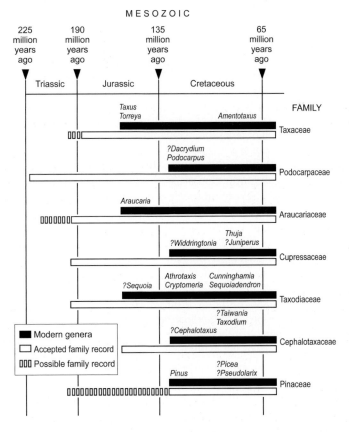

Fig. 9.5. Order of the first appearance of several families of conifers during the Mesozoic Era. Question marks (?) indicate that time of appearance is uncertain. (Reprinted by permission from Miller, 1977; copyright 1977, The New York Botanical Garden, Bronx, New York)

Phylogenetics

A **phylogeny** is a classification of a set of taxa based on their evolutionary history. The study of the evolutionary history of forest trees and their phylogenetic relationships are inextricably linked. It is beyond the scope of this text to describe the fundamental details of modern phylogenetic analyses, but several recent texts provide a comprehensive treatment on this topic (*e.g.* Judd *et al.*; 1999). We define some of the basic terminology and briefly describe some methods used to construct phylogenies in forest trees.

Traits must be measured on all the taxonomic units before a phylogenetic tree can be constructed. These traits are called **characters** and the observed values of the characters are called **character states.** Traditionally, morphological traits such as leaf shape (angiosperms) and cone characteristics (gymnosperms) have been used. Recently, biochemical or

---

Box 9.3. Evolutionary history of the Pinaceae.

Pinaceae first evolved during the Triassic and Jurassic and radiated during the Cretaceous (Miller, 1976) (Fig. 9.5). Analysis of cone fossils suggests that *Pinus* is the likely ancestral taxa and other genera of Pinaceae (now including *Abies, Cathaya, Cedrus, Keteleeria, Larix, Picea, Pinus, Pseudotsuga,* and *Tsuga*) radiated from *Pinus* during the Tertiary (Miller, 1976). Based on current understanding of plate tectonics, Millar (1993) proposed that the center of origin was in eastern North America and western Europe when both formed the single continent of Laurasia.

The first known species of *Pinus, Pinus belgica,* was found in fossil deposits from the Mesozoic. *Pinus* probably evolved from the extinct pinaceous genus *Pitystrobus,* which died out in the early Cenozoic (Fig. 1). The Cretaceous was a time of major radiation in *Pinus.* The *Strobus* and *Pinus* subgenera were formed in the first radiation, although it is not clear which subgenus came first. By the end of the Cretaceous, most of the sections and subsections had arisen. The Cretaceous was generally warm and dry, favoring the expansion of pines throughout mid-latitudes in the northern hemisphere.

The Cenozoic evolutionary history of *Pinus* is quite well understood (Axelrod, 1986; Critchfield, 1984; Millar, 1993, 1998; Millar and Kinloch, 1991). Climatic conditions were very tropical during the Paleocene and early Eocene that favored development and expansion of angiosperm species and pushed pines, and other conifers not adapted to tropical climates, into refugia at both high and low latitudes and warm-dry regions in mid-latitudes. This resulted in significant fragmentation of pines on a global scale. A secondary result of fragmentation was the development of secondary centers of diversity and radiation, such as in Mexico.

The end of the Eocene was marked by a significant drop in average global temperatures, leading to widespread extinction of tropical angiosperm species and the recolonization of mid-latitudes by pines from various refugia. Through the rest of the Tertiary, pines were widespread throughout the Northern Hemisphere and the radiation of the genus may have been largely complete.

The Quaternary history of the evolution of pines is characterized by rapid changes in species distributions and genetic structure of populations. The Pleistocene epoch was a time of expansion and contraction of glaciers that caused the extinction of some taxa and severe loss of genetic diversity of many others (Critchfield, 1984). Glaciation caused fragmentation and severe reduction in population sizes, which provides the opportunity for random genetic drift (*Chapter 5*) to genetically differentiate populations and reduce within population genetic diversity.

(Box 9.3 continued on next page)

Box 9.3. Evolutionary history of the Pinaceae. (Continued from previous page)

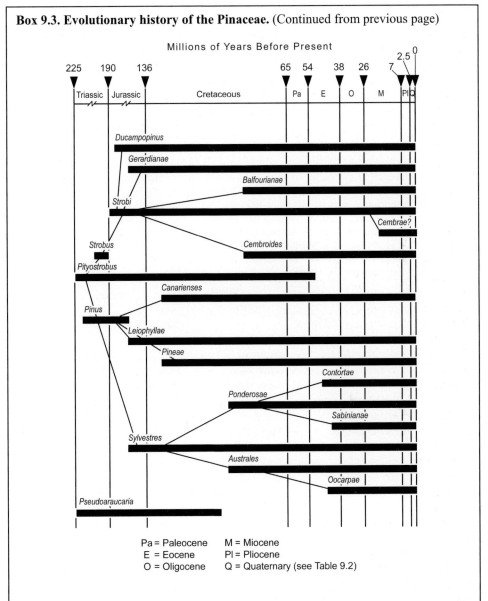

Fig. 1. Order of appearance of subgenera of the genus *Pinus* during the Mesozoic and Cenozoic eras. (Reprinted from Millar and Kinloch, 1991, with permission of the authors)

molecular characters such as isozymes, immunological assays and DNA sequences are more often employed. Measurements are made on all members of the study group (ingroup) and on one or two related taxa that are not part of the ingroup, which are called the **outgroup.** Inclusion of an outgroup is important as it allows for the phylogenetic tree to be **rooted.** Rooting of an evolutionary tree makes it possible to show the direction of evolutionary change. For example, sister taxa *Picea* and *Cathya* were chosen as outgroups in the *Pinus* phylogenetic tree constructed by Liston *et al.* (1999).

Phylogenetic trees based on different characters can often produce different sets of relationships. One reason that this might happen is because of **homoplasy**, which is the appearance of similar character states in unrelated organisms. It is easy to imagine how the same character could have evolved more than once in unrelated lineages. A group of individuals consisting of an ancestor and all of its descendents is called a **monophyletic group**, whereas a group consisting of two or more related groups (as a result of homoplasy) is called a **polyphyletic group**. A **clade** is a branch of a phylogenetic tree that includes a monophyletic group of taxa descending from a single common ancestor.

A variety of analytical methods have been developed for constructing phylogenetic trees and choosing the most likely relationships among different possibilities. The most widely used method is maximum parsimony, which aims to identify the tree requiring the fewest numbers of changes in character states. This approach is often used with morphological

Table 9.3. Taxonomic classification of species (111 total) in the genus *Pinus* as described by Price *et al.*, 1998. The taxa are arranged hierarchically into two subgenera (*Pinus* and *Strobus*), sections (2 within each subgenera), subsections (2-6 in each section), group (0-3 within each subsection), and species (1-19 within each group or subsection). Species within groups or subsections are the most closely related and decrease in relatedness across increasingly higher levels of classification.

Subgenus				***Pinus***		
Section				***Pinus***		
Subsection		***Pinus***		Subsection	***Canarienses***	
Species	*densata*		*pinaster*	Species	*canariensis*	
	densiflora		*resinosa*		*roxburghii*	
	heldreichii		*sylvestris*	Subsection	***Halepensis***	
	hwangshanensis		*tabuliformis*	Species	*brutia*	
	kesiya		*taiwanensis*		*halepensis*	
	luchuensis		*thunbergii*	Subsection	***Pineae***	
	massoniana		*tropicalis*	Species	*pinea*	
	merkusii		*uncinata*			
	mugo		*yunnanensis*			
	nigra					
Section			**New World Diploxylon Pines**			
Subsection	***Contortae***		Subsection	***Ponderosae***		
Species	*banksiana*		Species	*cooperi*		
	clausa			*durangensis*		
	contorta			*engelmannii*		
	virginiana			*jeffreyi*		
Subsection	***Australes***			*ponderosa*		
Species	*caribaea*			*washoensis*		
	cubensis			***Montezumae* Group**		
	echinata			*devoniana*		
	elliottii			*donnell-smithii*		
	glabra			*hartwegii*		
	occidentalis			*montezumae*		
	palustris			***Pseudostrobus* Group**		
	pungens			*douglasiana*		
	rigida			*maximinoi*		
	serotina			*nubicola*		
	taeda			*pseudostrobus*		

New World Diploxylon Pines (continued)			
Subsection	***Oocarpae***	Subsection	***Ponderosae*** (continued)
Species	***Oocarpa* Group**	Species	***Sabinianae* Group**
	greggii		*coulteri*
	jaliscana		*sabiniana*
	oocarpa		*torreyana*
	patula	Subsection	***Attentuatae***
	praetermissa	Species	*attenuata*
	pringlei		*muricata*
	tecunumanii		*radiata*
	***Teocote* Group**	Subsection	***Leiophyllae***
	herrerae	Species	*leiophylla*
	lawsonii		*lumholtzii*
	teocote		

Subgenus	***Strobus***		
Section	***Parrya Mayr***		
Subsection	***Balfourianae***	Subsection	***Cembroides***
Species	*aristata*	Species	*cembroides*
	balfouriana		*culminicola*
	longaeva		*discolor*
Subsection	***Krempfianae***		*edulis*
Species	*krempfii*		*johannis*
Subsection	***Rzedowskianae***		*juarezensis*
Species	*rzedowskii*		*maximartinezii*
Subsection	***Gerardianae***		*monophylla*
Species	*bungeana*		*nelsonii*
	gerardiana		*pinceana*
	squamata		*remota*

Section	***Strobus***			
Subsection	***Strobi***		Subsection	***Cembrae***
Species	*armandii*	*lambertiana*	Species	*albicaulis*
	ayacahuite	*monticola*		*cembra*
	bhutanica	*morrisonicola*		*koraiensis*
	chiapensis	*parviflora*		*pumila*
	dabeshanensis	*peuce*		*sibirica*
	dalatensis	*strobus*		
	fenzeliana	*wallichiana*		
	flexilis	*wangii*		

data. Other statistical approaches such as the minimum distance method and the maximum likelihood method are widely used with molecular data. Software programs such as PHYLIP (Felsenstein, 1989) and PAUP (Swofford, 1993) implement these methods.

The ease of obtaining large amounts of DNA sequence data through PCR and auto-mated sequencing technologies (*Chapter 4*) has created great interest in phylogenetic analysis in all types of organisms including trees. DNA sequences from the three genomes (chloroplast, mitochondrial and nuclear) have all been used. The choice of genome is de-pendent on the level of taxonomic classification desired. The chloroplast genome evolves more slowly and is thus best for higher level discrimination, *e.g.* species level and higher. Nuclear DNA sequences generally evolve more rapidly and are best suited for species level and lower discrimination. Phylogenetic relationships in the order Coniferales, the family Pinaceae, the genus *Pinus* (Table 9.3) and the genus *Populus* are described in Boxes 9.4-9.7, respectively.

Box 9.4. Phylogenetic relationships in the order Coniferales.

The order Coniferales (conifers) includes eight families, 63 genera and 500-600 species. Hart (1987) has conducted the most extensive study to determine the phylogenetic relationships among the conifer families (Fig. 1).

   He used 123 characters taken from herbarium and living specimens from all 63 known genera of conifers. He included several outgroups which confirmed the monophyly of conifers. Hart includes the Taxaceae as part of this monophyletic group whereas some taxonomists have excluded it. He also distinguishes the Sciadopityaceae as a separate family from the Taxodiaceae and Cupressaceae. The Taxodiaceae and Cupressaceae are closely related and form a single monophyletic group. Eckenwalder (1976) has proposed that the Taxodiaceae and Cupressaceae be merged. Tsumara *et al.* (1995) have also constructed a phylogeny of a subset of conifer families based on cpDNA RFLP data. Their data also support the inclusion of Sciadopityaceae and the monophyly of Taxodiaceae and Cupressaceae; however, their study did not include Araucariaceae or Podocarpaceae.

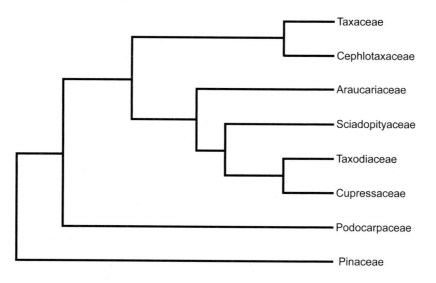

Fig. 1. Phylogenetic relationships among eight families of conifers. (Reprinted from Hart, 1987, with permission from the Arnold Arboretum of Harvard University)

Box 9.5. Phylogenetic relationships in the family Pinaceae.

The family Pinaceae includes ten genera: *Abies, Cathya, Cedrus, Keteleeria, Larix, Picea, Pinus, Pseudolarix, Pseudotsuga* and *Tsuga*. Hart's (1987) phylogenetic tree (Fig. 1a) is based largely on morphological characters.

   Contrast this tree with that from Price *et al.* (1987) (Fig. 1b) that was based on immunological characters. In the cases of *Tsuga, Picea* and *Pinus*, Price included two species from each genus. There is general agreement between these trees with the exception of the position of *Cedrus* and the position of the clade containing *Larix* and (Box 9.5 continued on next page)

Box 9.5. Phylogenetic relationships in the family Pinaceae. (Continued from previous page)

Pseudotsuga. Finally, Liston *et al.* (2003) have constructed a tree based on cpDNA rbcL gene sequences and nDNA and rDNA sequences (Fig. 1c) and also found a similar phylogeny except that *Cedrus* was placed at the basal position. These examples highlight the contrasting results among phylogenetic trees derived from single classes of characters and the need for synthetic analyses using multiple character types.

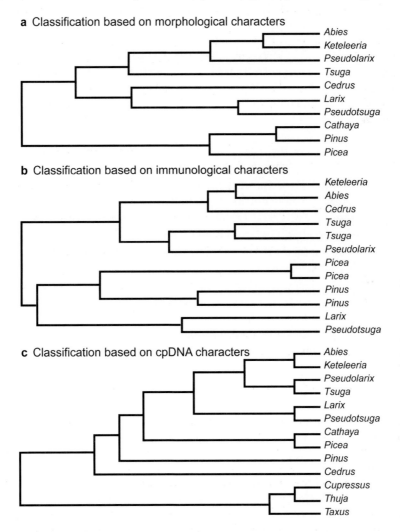

Fig. 1. Phylogenetic relationships among 9-10 genera of the family Pinaceae. (a) Phylogeny based largely on morphological characters (Reprinted from Hart, 1987, with permission from the Arnold Arboretum of Harvard University), (b) Phylogeny based on immunological characters (Reprinted from Price *et al.*, 1987, with permission from the American Society of Plant Taxonomists) and (c) Phylogeny based on DNA sequences (note, *Cupressus*, *Thuja* and *Taxus* comprise an outgroup in this cladogram; reprinted from Liston *et al.*, 2003, with permission from the International Society for Horticultural Science)

Box 9.6. Phylogenetic relationships in the genus *Pinus*.

Traditional classifications of *Pinus* are based on morphological traits, primarily foliar and reproductive (seed and cone) characters. The first two modern classifications of the genus *Pinus* were proposed by Shaw (1914) and by Pilger (1926). Both recognized two subgenera, *Strobus* (haploxylon or soft pine) and *Pinus* (diploxylon or hard pine). Numerous refinements of this classification have been made using crossability, cytogenetic and biochemical characters (reviewed by Price *et al.*, 1998). Duffield (1952) evaluated these classifications based on the results of crossability data and concluded that Shaw's classification agreed most closely to the crossing results. The classification of Little and Critchfield (1969) partitioned the genus into two subgenera, four sections and 14 subsections. The most recent classification proposed by Price *et al.* (1998) (Table 9.3) maintains the two subgenera and four sections. However, it now includes 17 subsections and many species in the subgenus *Pinus* have been reassigned to different subsections and sections.

It was not until the application of DNA-based methods that large numbers of species could be included in *Pinus* phylogenetic analyses. The first such studies were by Strauss and Doerksen (1990) and Govindaraju *et al.* (1992). They used cpDNA, mtDNA or nDNA RFLPs to construct phylogenies based on 19 and 30 species, respectively. Both studies generally confirm the traditional classifications based on morphological or anatomical characters. The most recent phylogenetic analysis of *Pinus* was conducted by Liston *et al.* (1999). This study included 47 species and was based on DNA sequence data from the rDNA internal transcribed spacer region (Fig. 1).

The molecular phylogenies are consistent with the traditional classifications for some parts of the genus (*e.g.* subgenus *Strobus*) and are inconsistent in other parts. This again underscores the need for constructing phylogenies based on several character types including data from the fossil record.

There have been numerous studies that attempt to clarify taxonomic and phylogenetic relationships within individual subsections of *Pinus* (Krupkin *et al.*, 1996; Wu *et al.*, 1999). An interesting example is that of North and Central American species of the *Oocarpae* and *Australes* subsections (Dvorak *et al.*, 2000). Species of the *Oocarpae* and *Australes* subsections of *Pinus* are of considerable economic importance. The evolutionary relationships among these species have been the subject of much debate (Little and Critchfield, 1969; Farjon and Styles, 1997; Price *et al.*, 1998). Earlier classifications combined California and Mesoamerican species into common groups. Dvorak *et al.* (2000) used RAPD markers to re-examine relationships of species in the *Oocarpae* and *Australes* subsections (Fig. 2). They found that species of the *Australes* (Clades 4 and 5) were more closely related to the Mesoamerican *Oocarpeae* (Clades 1-3) than California species of *Oocarpeae* (Clade 6) and hypothesized that the Mesoamerican *Oocarpeae* and the southeastern US *Australes* descended from a common ancestor, possibly *P. oocarpa*. They also proposed that the *Oocarpeae* ancestor migrated in two directions from a point in northern Mexico, one south further into Mexico and Central America (where the Mesoamerican Oocarpeae species evolved) and the other east into the southeastern US (where the Australes species evolved), and that these paths converged in Florida or the Caribbean.

(Box 9.6 continued on next page)

Box 9.6. Phylogenetic relationships in the genus *Pinus*. (Continued from previous page)

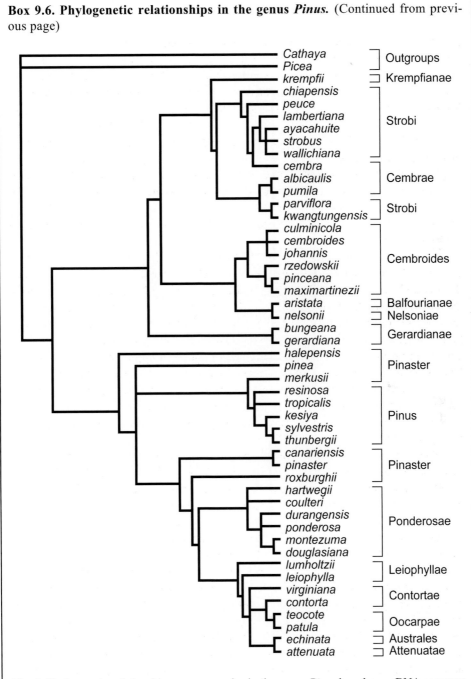

Fig. 1. Phylogenetic relationships among species in the genus *Pinus* based on cpDNA sequence variation in a 650 base-pair portion of the rDNA internal transcribed spacer region. Subsection names are given in the column to the right. (Reprinted from Liston *et al.*, 1999, with permission from Elsevier)

(Box 9.6 continued on next page)

Box 9.6. Phylogenetic relationships in the genus *Pinus*. (Continued from previous page)

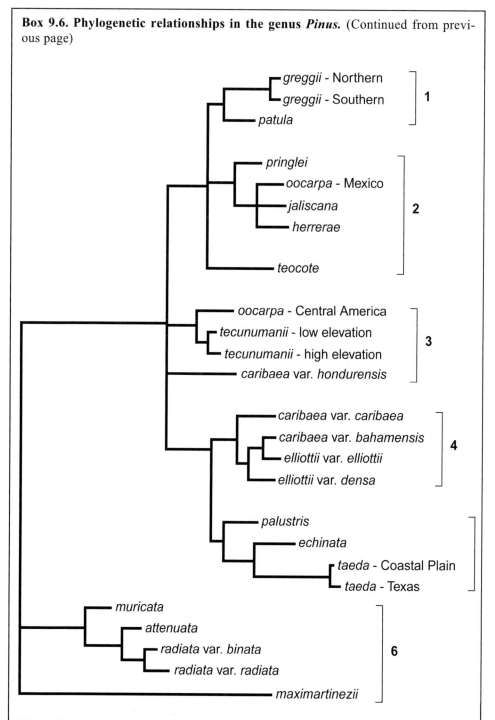

Fig. 2. Phylogenetic relationships among several Mexican *Pinus* species based on RAPD marker data. The six major clades are indicated by numbers 1-6. (Reprinted from Dvorak *et al.*, 2000, with permission of Springer Science and Business Media)

Box 9.7. Phylogenetic relationships in the genus *Populus*.

Eckenwalder (1996) recognized 29 distinct species in the genus *Populus* that fall into the six accepted sections, and conducted a phylogenetic analysis based on 76 morphological characters measured on all 29 species (Fig. 1). All sections form monophyletic groups except *Tacamahaca*.

The evolutionary position of sections, *Albaso* being basal, reflects the order of occurrence of species in the fossil record, providing support for the phylogeny at the section level. The relationship of species within sections is less well supported by fossil data.

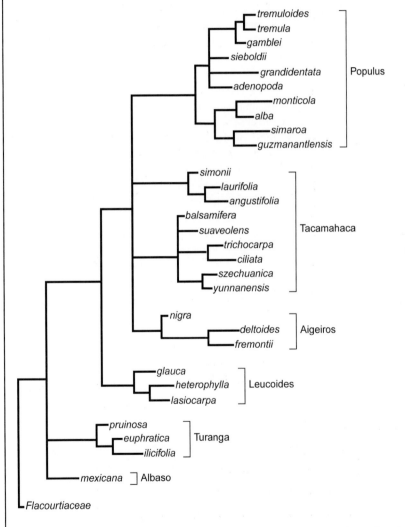

Fig. 1. Phylogenetic relationships among 29 species and six sections (named to the right of species) of the genus *Populus*. Flacourtiaceae was the outgroup for this analysis. (Reprinted from Eckenwalder, 1996, with permission from NRC Research Press, National Research Council of Canada)

MOLECULAR MECHANISMS OF GENOME EVOLUTION

In this section we briefly discuss some of the mechanisms that provide genetic variation upon which the forces of evolution operate. New variation can be generated at the level of individual nucleotides, genes, chromosomes or entire genomes. For a more thorough treatment of molecular evolution see texts by Li (1997) and Nei and Kumar (2000).

Mutation and Nucleotide Diversity

The concept of mutation as one of the forces of evolutionary change was introduced in *Chapter 5*. At the level of DNA sequences, three basic types of mutations are possible: substitutions, insertions and deletions (Fig. 9.6). Substitutions can either be **transitions** (purine to purine or pyrimidine to pyrimidine mutations) or **tranversions** (purine to pyrimidine or pyrimidine to purine mutations). Substitutions that do not result in changes of amino acid composition in the resulting polypeptide chain are called silent or **synonymous substitutions**, whereas those that lead to amino acid changes are called **nonsynonymous substitutions**. For example, due to the redundancy in the genetic code (Fig. 2.10), four different triplet codons code for the amino acid alanine; therefore, any of the four bases are possible at the third position without changing the code to another amino acid.

Although the mutation rate at an individual gene locus (μ) is generally thought to be about 10^{-5} to 10^{-6} per generation, there are very few empirically derived estimates of μ for forest trees. Dvornyk *et al.* (2002) estimated that the mutation rate in *Pinus sylvestris* is 0.15×10^{-9} per year for a phenylalanine ammonialyase gene and 0.05×10^{-9} for an alcohol dehydrogenase gene. These estimates are much lower than those reported for other plant species and are not consistent with estimates based on accumulated mutations per generation for chlorophyll deficiencies (*Chapter 7*).

Nucleotide diversity (ϕ) is a measure of the amount of DNA sequence polymorphism in a sample of individuals taken from a population and is a function of the mutation rate,

$$\phi = 4N_e\mu, \hspace{4cm} \text{Equation 9.1}$$

where N_e is the effective population size. An estimate of ϕ is given as,

$$\phi = p_s/\alpha 1 \hspace{4cm} \text{Equation 9.2}$$

where $p_s = S/n$ (S = number of segregating nucleotide sites and n = total number of nucleotides) and $\alpha 1 = 1 + 2^{-1} + 3^{-1} \ldots + (m + 1)^{-1}$ (m = number of sequences) (Nei and Kumar, 2000). This estimate of nucleotide diversity is independent of sample size.

In *Chapter 7* we presented statistics to estimate gene diversity (P, A and H_e). These measures are commonly used with genetic marker data (*Chapter 4*) and for most forest tree species, especially conifers, are generally higher than those for other plant species. However, the nucleotide diversity estimates that have been reported for pines are not as high as those for maize (Neale and Savolainen, 2004) (Table 9.4).

Gene Duplication and Gene Families

Another important way by which genomes evolve is by gene duplication. For example, a primitive species might just have one copy of a gene that codes for a single protein that is

A A C G T T C A C T G   Wild type

a) A A C G A T C A C T G   Substitution T ➔ A

b) A A C G C C T T C A C T G   Insertion of C C

c) A A C C A C T G   Deletion of G T T
      G T T

Fig. 9.6. DNA sequence mutations are of three basic types: (a) Substitutions, (b) Insertions and (c) Deletions.

used throughout the developmental life of an individual. A more highly evolved organism, however, might have multiple copies of that same gene that arose by duplication and mutated over evolutionary time that are slightly different in their function depending on the developmental state of the organism. An ancestral species might have a single copy of gene *A* (Fig. 9.7). Following speciation, both species 1 and species 2 have a single copy of gene *A*, denoted as *A* and *A'*, respectively. *A* and *A'* are called **orthologs** because they are related by descent. Subsequently, both *A* and *A'* undergo duplication to form genes *B* and *B'*, respectively. *B* and *B'* are called **paralogs**, because they are not directly related by descent, even though they may be very similar in functionally and at the DNA sequence level. It should be apparent that phylogenies need to be based on orthologs and that inclusion of paralogs will introduce error into estimation of evolutionary relationships, because of the possibility of making comparisons among genes that share sequence similarity but are not related by decent.

Most eukaryotic organisms undergo gene duplication to varying extents. In *Chapter 2*, we noted that conifers appear to have large numbers of pseudogenes, *e.g.* non-functional copies of duplicated genes. In addition, there are a few examples of conifer gene families that have more functional copies than their angiosperm counterparts (*i.e.* ADH gene family

Table 9.4. Estimates of nucleotide diversity (ϕ) for three conifer species. $\phi_{(average)}$ is an estimate from both gene-coding and non-coding regions. $\phi_{(non-synonymous)}$ and $\phi_{(silent)}$ estimate nucleotide diversity that is expressed and not-expressed, respectively, at the amino acid level (*i.e.* nucleotide substitutions that do ($\phi_{(non-synonymous)}$) and do not ($\phi_{(silent)}$) result in amino acid changes). Reprinted from Neale and Savolainen (2004) with permission from Elsevier.

Species	Number of genes (type)	ϕ (average)	ϕ (non-synonymous)	ϕ (silent)
Pinus taeda	19 (wood)	0.00407	0.00108	0.00658
Pinus taeda	28 (disease)	0.00489	0.00183	0.00588
Pinus taeda	16 (drought)	0.00460	0.00187	0.00700
Pinus sylvestris	1 (*pal1*)	--[a]	0.00030	0.00490
Pinus sylvestris	2 (phytochrome)	--[a]	0.00050	0.00300
Pseudotsuga menziesii	12 (cold related)	0.00853	0.00463	0.01256

[a]Not reported

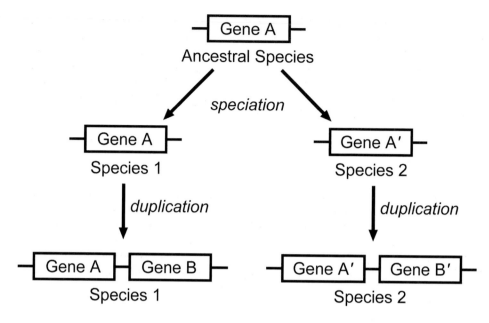

Fig. 9.7. Gene duplication. *A* and *A'* are orthologs because they are related by descent whereas *B* and *B'* are paralogs because they are not directly related by descent but rather are formed from duplication of individual genes in one species following speciation.

in *Pinus banksiana* (Perry and Furnier, 1996) and PAL in *Pinus banksiana* (Butland *et al.*, 1998)). In a recent analysis of genes involved in lignin and cell wall biosynthesis, however, Kirst *et al.* (2003) concluded that gene family sizes in *Pinus taeda* are not significantly greater than those in *Arabidopsis thaliana* (Table 9.5). Clearly, gene duplication is frequent in the evolution of forest tree genomes, but it is not yet clear if gene duplication is any more common in conifer genomes than in the evolution of any other plant taxa.

Polyploidy

The duplication or deletion of entire chromosomes (aneuploids) or entire genomes (polyploids) is found throughout higher plant taxa and is an important evolutionary mechanism

Table 9.5. Estimates of gene family sizes (*i.e.* number of functional gene copies) in *Pinus* and *Arabidopsis thaliana* (Sederoff, unpublished). Members of a gene family are functionally similar but may be separated on different chromosomes, and active in different cell types, tissues or developmental stages.

Gene family	*Arabidopsis thaliana*	*Pinus*
Cellulose synthase	10	6 to 10
Actin	5+	4 to 8
Alpha-tubulin	5	2 to 6
Beta-tubulin	9	3 to 8
cad/eli3 super family	8	2
CCR	10	2-8
C4H	1	2
4CL (1,2,3)	3	2-4

(Wendel, 2000). Two basic types of polyploidy are recognized: **autopolyploidy** (duplication of genomes derived from the same parental species) and **allopolyploidy** (duplication of genomes derived from different parental species through hybridization). Polyploidy is extremely rare in conifers (*Chapter 2*), but rather common in some groups of angiosperm forest tree species (Wright 1976). Forest tree genera with polyploid species include *Acacia*, *Alnus*, *Betula*, *Prunus* and *Salix* (see Table 19.1, pp. 406-409 in Wright, 1976). Polyploidy is certainly important in the evolution and speciation of these genera.

COEVOLUTION

Mutualisms, where two species each benefit from a biological interaction between species, occur throughout the natural world. If these mutualisms directly affect the evolution of the respective species, these species may have coevolved together. Futuyma (1998) defines **coevolution** as "reciprocal genetic change in interacting species owing to natural selection imposed by each on the other." In forest trees, two fairly well studied cases of mutualism and possibly coevolution are the pine-rust fungi symbiosis and the white pine-corvid symbiosis. Further reading on the subject of coevolution can be found in the text by Thompson (1994).

Pines and Rust Fungi

Rust fungi are a very large family (*Uredineles*) composed of more than 6000 species and 100 genera that date back to the Carboniferous. Ferns, gymnosperms and angiosperms are all hosts for various rusts. Classical theory says that rusts and their plant hosts have coevolved and that primitive rusts are associated with more primitive plants and likewise for more advanced types (Leppik, 1953, 1967; Millar and Kinloch, 1991). There are two genera of rusts associated with *Pinus*: *Cronartium* (11 species) and *Peridermium* (4 species). In a previous section, we discussed the evolutionary history and phylogenetic relationships in *Pinus* (Boxes 9.3 and 9.6). Hart (1988) estimated phylogenetic relationships among 30 genera of rusts including many that parasitize species of *Pinus*. He found that what appear to be more primitive rusts can occur on more recently evolved angiosperms, and more advanced rusts can occur on conifers and ferns that evolved much earlier. Hart concluded that co-speciation (coevolution) of rusts and plants may not be as important as generally thought.

White Pines and Corvids

A fascinating symbiosis occurs between the pines of the Strobus subgenus with wingless seeds and a group of birds of the Corvidae (corvids) (Tomback and Linhart, 1990). The most well studied of these symbioses is that between *Pinus albicaulis* (white bark pine) and *Nucifraga columbiana* (Clark's nutcracker; Fig. 7.5) (Lanner, 1980, 1982, 1996). Most pine seeds are winged and dispersed by wind. There are 20 pine species, however, that do not have wings (all but one are members of subgenus Strobus). Seeds of these pines are a source of food for corvid birds such as nutcrackers and jays. These birds harvest seed from ripening cones and cache seed in the soil to be eaten throughout the year. Because only a portion of the cashed seeds are eventually eaten, the birds effect dispersal and establishment of seedlings into new areas. The birds, on the other hand, have evolved special features to facilitate this task such as specialized beaks for opening cones and a sublingual pouch for carrying seed. It appears that winglessness in pine seeds and beak morphology in birds are co-evolving. These mutualisms probably date back to the Tertiary and it is likely they evolved by co-adaptation rather than by genetic drift.

SUMMARY AND CONCLUSIONS

Forest trees are found on all continents on earth except Antarctica. Forest types range from equatorial, tropical forests to boreal forests. Forest trees belong to both major plant groups, angiosperms and gymnosperms, although many more angiosperm species currently exist.

The classic definition of a species is called the biological species concept (BSC). It is defined as the collection individuals that interbreed and produce viable and fertile offspring, but cannot breed with members of another species. In both gymnosperm and angiosperm tree species, the BSC is not completely applicable since many species hybridize.

Two major forms of speciation are generally recognized: allopatric speciation and sympatric speciation. Allopatric speciation is a form of gradual speciation where subpopulations become physically separated and diverge due to the forces of genetic drift and natural selection. Sympatric speciation is a form of abrupt speciation where some type of major genetic change causes some members of a population to become reproductively isolated from the rest of the population.

Hybridization is the crossing between individuals belonging to separate populations. Hybridization can reverse the process of speciation, but it can also lead to the formation of new species. Introgression occurs when repeated backcrossing between the hybrids and parental species results in the transfer of genes between parental species.

Trees first evolved 300-400 million years ago during the Devonian Period or Carboniferous Periods. Conifers dominated forests during the Mesozoic Era and angiosperms did not appear until the late Cretaceous Period. Pines probably first appeared during the Cretaceous as well.

Phylogeny is a classification of a set of taxa based on their evolutionary history. Phylogenies are based on measures of characters such as morphological traits and DNA sequences. Different character states are more informative depending on taxonomic level of the phylogeny, *e.g.* family, genus or species.

Coevolution is defined as the reciprocal genetic change in interacting species owing to natural selection imposed by each on the other. Pines and rust fungi and pines and corvids are examples of coevolution in forest trees.

CHAPTER 10
GENE CONSERVATION – *IN SITU, EX SITU AND SAMPLING STRATEGIES*

Gene conservation, defined as the policy and management actions taken to assure the continued availability and existence of genetic variation (FAO, 2001), is an essential component of sustainable forestry. Genetic diversity is at the core of the adaptive value of species; it permits them to develop their unique adaptations and ensures their continued evolution under changing environments. Most forest tree species possess intrinsically-high levels of genetic variation (*Chapter 7*). As a result, provenances may diverge genetically in response to differing selection pressures and thrive in extremely dissimilar environments (*Chapter 8*). The within-population genetic variation that allows for responses to changes in local conditions is also necessary for long-term population viability. The ability to adapt may prove especially critical in the future, as species and populations encounter new stresses from habitat loss, introduced diseases and insects, pollution and climate change. In a changing world, the evolutionary potential of populations and species depends on whether they possess the genetic variation necessary to endure new conditions.

Conservation of genetic diversity has many different motivations, including ecological, economic, social and aesthetic ones. First, genetic variation within and between species enables forests to perform a wide variety of ecological functions that ultimately lead to forest productivity and resilience. In forest ecosystems, trees are often **keystone species**, with the survival of many other plant and animal species depending on them. Even when minor tree species are eradicated, a cascade of ecological interactions may destabilize a community and result in local extinction of additional species (Ehrlich and Ehrlich, 1981; Soulé, 1986; Terborgh, 1986). Therefore, conserving genetic diversity in tree species is essential to the long-term health of forest ecosystems.

Genetic variation in tree species also contributes economic benefits and sustains the capacity of forests to yield both timber and other products for human use. When tree species encounter new stresses, such as diseases, insects and climate change, the markets that depend on a supply of products are also affected. Because genetic variation enables tree species to evolve in response to new stresses (through either natural or artificial selection), it may prove critical for continued viability of forest industries. Genetic variation can also be tapped by tree breeders to meet the challenges of new product requirements, such as chemical derivatives, pulping and fiber characteristics, and wood of a specified density or grain (*Chapter 11*).

Finally, many people feel an ethical and aesthetic obligation to protect biological diversity, and, in particular, to safeguard endangered species from extinction caused by human exploitation or neglect (Ledig, 1988a). A diverse world is more interesting than a homogenous one, and each taxon, with its unique characteristics and life history, adds to the beauty of our planet. What's more, our knowledge of genetics is still very incomplete. The genetic diversity of a species may have future benefits that we cannot yet conceive of and that may never be realized unless we work to protect it.

For all of these reasons, gene conservation is a global imperative. So, this chapter begins with a discussion of the major threats to forest tree populations and their genetic di-

© CAB International 2007. *Forest Genetics* (T.W. White, W.T. Adams and D.B. Neale)

versity. Next, strategies and methods of gene conservation are described, including the preservation of genetic resources in intact natural forests in protected areas, managed forests, genetic tests, breeding arboreta and seed banks (or other forms of storage). The population sizes required and the appropriate number and location of gene conservation populations are discussed. Finally, the consequences of timber harvest and reforestation practices on genetic diversity are addressed. These topics are also explored in other books, including National Research Council (1991), Mátyás (1999), and Young *et al.* (2000).

THREATS TO GENETIC DIVERSITY

In today's world, the genetic integrity of many plant and animal species is imperiled, and forest trees are not immune to this threat. Over 7300 tree species globally are threatened with extinction, with many more subspecies, varieties, and races at risk (Table 10.1). No one knows exactly the current species' extinction rate, but many argue that it rivals the five mass extinctions of past geological history, including the one in which the dinosaurs went extinct (Wilson, 1992; Pimm *et al.,* 1995). Within taxa, populations are being eliminated along with all alleles that are unique to them. Even when populations are not lost entirely, they may go through extreme bottlenecks, resulting in the loss of some alleles, particularly those that were present in low frequency before the bottleneck (*Chapter 5*). If populations remain small for several generations after a bottleneck, random genetic drift will continue to erode their genetic variation and capacity for future adaptation. Another consequence of reduced population size is increased inbreeding. When inbreeding depression becomes great, forest productivity decreases, population viability declines, and the value of a population as a seed source for breeding and regeneration is diminished (*Chapters 5* and *7*). Additionally, the long-term evolutionary potential of many species is compromised when populations are lost or drastically reduced in size because of disruptions to their mating systems and patterns of gene flow (Young *et al.,* 2000).

Habitat Loss, Deforestation, and Fragmentation

Habitat loss and deforestation are the most pervasive threats to forest genetic resources (National Research Council, 1991). They are caused by urbanization, conversion of forests for agriculture and range land, overgrazing, overharvesting of fuel and industrial wood, natural disturbance, and poor management of production forests. Human population growth, poverty, and forest exploitation are often underlying factors, especially in developing countries. From 1990-2000 alone, forest area declined worldwide at an estimated rate of 9.4 million ha per year (FAO, 2001), with the greatest reductions occurring in the species-rich tropics and in developing countries. Habitat loss and deforestation can be especially damaging to genetic resources, because patterns of forest clearing generally follow topographic gradients with forests occupying lower-elevations and flatter ground more likely to be cut first. As the distribution of adaptive genetic variation in forest trees is often clinal and related to topography (*Chapter 8*), the rate of loss of genes controlling adaptive traits is higher than would be expected for genes controlling traits neutral to selection (which often vary little among populations in a region; *Chapter 7*). Logging can create a similar bias against adaptive genetic variation if the fastest growing, most pest-resistant trees are selectively harvested and less desirable individuals are left to supply seed for regeneration. This practice may lead to dysgenic selection (*Chapter 7*) (Ledig, 1992).

By fragmenting vegetation on the landscape, habitat loss and deforestation alter a wide variety of interrelated ecological and genetic processes (*Chapters 5, 7,* and *8*) (Nason *et al.*, 1997; Young and Boyle, 2000). The community composition of the remaining forest fragments is distorted, resulting in a myriad of changes in density, demography, and hence, genetic structure of tree species. Individual populations become more susceptible to demographic and environmental stochasticity (Lande, 1988b). The result is increased rates of local extinction and recolonization, with founder effects again producing genetic drift and inbreeding.

Fragmentation also alters patterns of gene flow, both directly by spatially separating individuals, and indirectly by impacting the abundance and behavior of the animals responsible for pollination and seed dispersal. When barriers to gene flow result, population differentiation increases, often with a concurrent decrease in reproductive success and increased inbreeding. In addition, because migration is often restricted, populations are less able to track environmental change by dispersing migrants to more hospitable environments. The effect of fragmentation on gene flow, however, is complex and cases exist where fragmentation actually resulted in increased pollen and seed movement (*Chapter 7*) (Nason and Hamrick, 1997; Aldrich and Hamrick, 1998).

Pathogens, Insects, Exotic Species and Movement of Genetic Material

Another major threat to genetic variation of forest tree species is posed by pathogens. Indigenous pathogens sometimes cause widespread death of their hosts, but more often, coevolution (*Chapter 9*) establishes a balance that enables both pathogen and host to survive. The balance results from a matching of genes for virulence in the pathogen with genes for resistance in the host (Burdon, 2001). In the absence of the pathogen, hosts expressing resistance alleles display decreased fitness, and this fitness cost contributes to the maintenance of the pathogen in the population (Burdon, 1987, 2001).

This balance between host and pathogen does not exist when pathogens are newly introduced and, consequently, exotic pathogens are more likely to cause epidemics that decimate populations and erode their genetic diversity (Byrne, 2000). For example, the exotic root disease *Phytophthora cinnamomi* has resulted in heavy mortality of *Eucalyptus marginata* in Australia and several other tree species in the same forest ecosystem (Shearer and Dillon, 1995). Even for exotic pathogens, however, alleles that confer resistance may be present in forest tree populations, often at low frequency (Box 10.1). Therefore, a justification for gene conservation strategies aiming to prevent the loss of (*i.e.* maintain) low frequency alleles is that among these alleles may be genes that confer protection against pathogens.

The accidental introduction of exotic insects into native populations can also impact population size and genetic diversity. For example, the introduction of the Chinese aldegid, *Adelges tsugae*, into eastern forests of the United States has had catastrophic effects on both *Tsuga carolina* and *Tsuga canadensis (*McClure *et al.*, 2001). Conservationists believe that *Tsuga carolina* may eventually be completely destroyed, and many populations of *Tsuga canadensis* in Virginia and North Carolina are highly threatened.

An additional hazard to genetic diversity of native species comes from the introduction of exotic tree species. Competition from exotics may displace native trees, especially in island floras where species evolved as colonists with relatively little competition. On San Cristobal in the Galapagos Islands, for example, the native evergreen forest has been nearly eliminated by the introduction of the two invasive tree species: guava (*Psidium guajava* L.) and quinine (*Cinchona succirubra* L.) (Schofield, 1989). Many of the forests in

Table 10.1. Examples of tree species threatened with extinction, their World Conservation Union (IUCN) red list category, and the major threats to their existence. Adapted from Oldfield et al., 1998.

Species (Family)	Range	IUCN category[a]	Major threats
Caesalpinia echinata (Fabaceae)	Brazil (9 states)	Endangered	Once widespread. Dramatic decline caused by exploitation as a dyewood dating from 1501-1920. More recently deforestation.
Cercocarpus traskiae (Rosaceae)	USA (California)	Critically endangered	Endemic to one gully on Santa Catalina Island. From 1897 to 1996, population declined from 40 adult trees to only six. Decline attributed to grazing and rooting by introduced herbivores and interspecific hybridization with the more abundant congener, *Cercocarpus betuloides* spp. *blanchea*.
Chamaecyparis lawsoniana (Cupressaceae)	USA (California, Oregon)	Vulnerable	Logged for valuable timber. Introduced pathogen, *Phytophthora lateralis*, kills mature trees and prevents regeneration.
Dahlgrenodendron natalense (Lauraceae)	South Africa (Eastern Cape, KwaZulu-Natal	Endangered	Rare palaeoendemic, with fewer than 200 individuals surviving in remnant forest patches. Rarely fruits, perhaps because of synchronized dichogamy, where male and female flowers are produced at different times. Isolation of remaining individuals provides little chance for pollination.
Dalbergia tsiandalana (Fabaceae)	Madagascar	Endangered	Endemic. Good quality rosewood is selectively logged.
Dendrosicyos socotrana (Cucurbitaceae)	Yemen (Socotra)	Vulnerable	Whole tree is cut and pulped for livestock fodder. One of the few Cucurbitaceae to grow to tree stature.
Magifera casturi (Anacardiaceae)	Indonesia (Kalimantan)	Extinct in the wild	Endemic in mango genus, with delicious fruit. Only known in cultivation.
Nesiota elliptica (Rhamnaceae)	St. Helena	Extinct in the wild	Originally known from one localized population. Last tree in the wild died in 1994. One cutting from it survives in Scotland. Monotypic genus.
Norhea hornei (Sapotaceae)	Seychelles	Vulnerable	Valuable timber species endemic to five islands. Main threat is introduced plants. Monotypic genus.
Northofagus allessandri (Fagaceae)	Chile (Maule)	Endangered	Endemic species that was once more widespread. Now occurs in eight scattered locations. Decline caused by deforestation and replacement with *Pinus radiata* plantations.
Ormocarpum dhofarense (Fabaceae)	Orman, Yemen	Vulnerable	Regional endemic. Wood cut for local use. Also threatened by grazing and livestock browsing.
Psidium dumetorum (Myrtaceae)	Jamaica	Extinct	The area containing the only known population was completed cleared.

Species (family)	Location	IUCN category	
Quercus dumosa (Fagaceae)	Mexico (Baja California), USA (California)	Endangered	Decline attributed to pollution, urbanization and industrialization.
Scalesia crockery (Asteraceae)	Ecuador (Galápagos)	Endangered	Formerly made up extensive woodland on slopes of two volcanoes. Decline caused by introduced herbivores and invasive weeds. Threat of volcanic eruption constant.
Shorea blumutensis (Dipterocarpaceae)	Indonesia (Sumatra), Malaysia (Peninsular Malaysia)	Critically endangered	Logged for valuable yellow meranti timber. Logging cycle shorter than the time required for tree to reach reproductive maturity.
Torreya tasifolia (Taxaceae)	USA (Florida, Georgia)	Critically endangered	Three pathogenic fungal species cause mortality before individuals reach reproductive maturity. Habitat loss and lower water table are contributing factors.
Warburgia salutaris (Cannellaceae)	Mozambique, South Africa, Swaziland, Zimbabwe	Endangered	Wide but scattered distribution. Exploited for use in traditional medicinal practices to treat head and chest ailments and to cure people who are 'bewitched'.
Widdringtonia cedarbergensis (Cupressaceae)	South Africa (Western Cape)	Endangered	Threatened by exploitation of wood and inappropriate fire management. Natural fires break out at intervals shorter than the species' maturation rate.
Wollemia noblis (Araucariaceae)	Australia (New South Wales)	Critically endangered	One population of fewer than 50 mature individuals discovered in 1994. Genus new but thought to be known from the Mesozoic fossil record.

a Definitions of IUCN categories. Extinct – a taxon is extinct when there is no reasonable doubt that the last individual has died. Critically endangered – a taxon is critically endangered when it is facing an extremely high risk of extinction in the wild in the immediate future. Endangered – a taxon is endangered when it is not critically endangered but is facing a very high risk of extinction in the wild in the near future. Vulnerable – a taxon is vulnerable when it is not critically endangered or endangered but is facing a high risk of extinction in the wild in the medium-term future.

Madagascar and the Pacific Islands face similar fates. Furthermore, exotic species may be capable of hybridizing with closely related native species and, through introgression (*Chapter 9*), overwhelm the indigenous genetic variation (Geburek, 1997; Carney *et al.*, 2000). For example, in Tasmania, Australia, introduced *Eucalyptus nitens* has been found to hybridize with the native *E. ovata* (Barbour *et al.*, 2002). Again, this phenomenon is most likely to harm rare or island species that lack well-developed reproductive isolating mechanisms.

Native forests in an area are sometimes replaced by exotic, domesticated species used in plantation forestry. As examples: (1) The most important plantation species in several European countries are indigenous to North America and not part of the native flora; (2) *Pinus radiata*, native to western North America is widely planted in Australia, Chile and New Zealand; and (3) *Eucalyptus grandis*, native to eastern Australia, has been planted in more than 20 countries in the southern hemisphere (*Chapter 12*). Importantly, these plantations of exotic species did not necessarily cause the direct displacement of native forests. In many cases, clearing of the native forests for other purposes occurred many decades before plantation establishment, and the plantations directly replaced grassland, savannahs, scrub or second growth forests.

Box 10.1. The importance of genetic variation for disease resistance.

The ability of a susceptible tree species to endure a pathogen's attack often depends on the existence of genetic variation that confers resistance. Disease resistance can be controlled by one or a few genes of major effect, as is the case with major gene resistance of *Pinus lambertiana* to white pine blister rust (*Cronartium ribicola*) (Kinloch *et al.*, 1970). More frequently, inheritance of resistance is quantitative, due to the cumulative action of a number of genes, each with a small effect (Burdon, 1987). For both mechanisms, resistance alleles for exotic pathogens commonly occur at low frequency and with an unpredictable distribution. They may be uniformly distributed, or concentrated either at the fringes of a species' range or in the center (Burdon, 2001). The low frequency and unpredictable distribution of valuable resistance alleles are important considerations in the development of gene conservation strategies.

For tree species severely impacted by disease, successful regeneration of natural populations may require artificial reforestation with disease-resistant seedling stock. This stock can be produced by tree improvement programs that first screen for resistance in trees after natural or artificial inoculation with the disease pathogen, and then produce seed from seed orchards incorporating the most resistant parents (*Chapter 11*). These programs typically aim to identify multiple forms and sources of resistance and diversity of resistance genes, so that resistance endures despite selection pressure in the pathogen for increased virulence. Examples of species where disease resistance breeding is being used as an integral component of gene conservation are: (1) *Eucalyptus marginata* attacked by *Phytophthora cinnamomi* (Stukely and Crane, 1994); (2) European and North American elms (*Ulmus* spp.) attacked by Dutch elm disease (*Ophistoma ulmi*) (Smalley and Guries, 1993); (3) North American white pines (*Pinus* spp.) attacked by white pine blister rust (Fig. 1) (Sniezko, 1996); and (4) North American southern pines attacked by fusiform rust (*Cronartium quercum* f. sp. *fusiforme*) (Cubbage *et al.*, 2000). The breeding and seed orchard populations in these programs are critical *ex situ* genetic resources for

(Box 10.1 continued on next page)

Box 10.1. The importance of genetic variation for disease resistance. (Continued from previous page)

these species. In addition to more traditional tree improvement methods, molecular tools such as marker-aided selection (*Chapter 19*) and genetic engineering (*Chapter 20*) are beginning to play a role in resistance breeding.

Fig. 1. Needle lesions (a) and stem cankers (b) on *Pinus monticola* caused by white pine blister rust. As part of a program screening for resistance, seedlings are inoculated with rust basideo-spores (c) and then screened for development of disease symptoms (*e.g.* seedling mortality) in nursery beds (dead foliage is light-colored in this black and white photograph (d). (Photographs courtesy of R. Sniezko, US Forest Service, Dorena Tree Improvement Center)

Extensive movement of seed within a species' range that sometimes occurs during re-forestation also has the potential to compromise the genetic variation and structure of natural populations. If replanting is done using genetically variable, local seed, the effects of reforestation on genetic diversity should be minor. Use of non-local, poorly adapted seed, however, may result in poorly stocked, unproductive forests that, nonetheless, disperse pollen and seed to surrounding, intact populations (Adams and Burczyk, 2000). The consequences of this contamination are not well understood, but could be serious if adaptation and genetic structure of natural populations are adversely impacted (Ledig, 1992; Adams and Burczyk, 2000). In many cases, non-local seed grows better than local seed and is, therefore, not poorly adapted. The theoretical reasons and practical implications for this are discussed in *Chapters 8* and *12*.

Pollution and Global Climate Change

In many parts of the world, pollution is another force affecting the genetic structure of forest tree species. Its impact is felt most strongly in heavily industrialized countries. Exceptionally damaging atmospheric pollutants include sulfur dioxide and ozone, while heavy metals and acid deposition are responsible for some of the worst soil degradation. The stress that these pollutants place on organisms may bring about heavy mortality. Because susceptibility varies by species, pollution often leads to changes in species' composition in the forest community (Geburek, 2000). Within species, genetic structure may be altered when selection acts on intraspecific genetic variation for tolerance that exists at the levels of provenance, family, and individual (or clone). The selection pressure exerted by pollutants can be strong, and given the presence of heritable genetic variation, tree species are expected to evolve in response (Karnosky, 1977; Scholz *et al.*, 1989).

In the next century, threats to genetic diversity may intensify, as global climate change is expected to proceed at a rapid and unprecedented rate. The exact extent and causes of climate change are controversial; natural climatic shifts, as well as atmospheric alterations brought about by the burning of fossil fuels and deforestation, are believed to be contributing factors (IPCC, 2001). While the average temperature of the earth is expected to rise, temperature and precipitation are likely to increase in some regions and decrease in others, with climatic fluctuations frequently becoming extreme. These changes could have a profound impact on all levels of biodiversity in forest ecosystems (Ledig and Kitzmiller, 1992; Rehfeldt, 2000).

Models based on climatic variables and present tree distributions predict massive displacement of tree species across latitudinal and elevational gradients (*e.g.* Thompson *et al.*, 1999). The actual migration of tree species may be limited by fragmentation of the landscape, long generation cycles, disruptions in the populations of animals responsible for seed dispersal, and decreased reproductive output of trees under stress. Species ill equipped for rapid migration may face extinction, but genetic variation helps to ensure that species that can migrate at sufficient rates will adapt to new conditions. In addition, genetic variation enables populations to endure a wider range of conditions in place than they presently experience, ensuring their survival if changes are not overly severe (Rehfeldt, 2000). This is another justification for conservation plans designed to maintain present day genetic diversity.

STRATEGIES TO CONSERVE GENETIC DIVERSITY

While threats to genetic variation continue to intensify in most parts of the world, the international community is increasingly committed to conserving biodiversity at all levels, including genetic variation within species (National Research Council, 1991; Kanowski, 2000). This commitment is evident in the many international and national programs that aim to conserve genetic diversity in forest trees (Box 10.2 and Table 10.2). Many tree species have been targeted for gene conservation activities including endangered species, keystone species and species with high economic value, much reduced distributions and abundances, and/or narrow or specialized habitat requirements (Eriksson *et al.*, 1993; Dvorak *et al.*, 1996; Namkoong, 1997). The species that will be most threatened in the future are those that are moderately to severely degraded, have little or no current economic value, and are not of great interest to conservation groups.

The gene conservation strategy that is most appropriate for each species depends on the size of its geographic range, population genetic structure and local attributes. For the majority of tree species, gene conservation is achieved by *in situ* methods, in which genetic variation is maintained in populations growing in their place of origin (National Research Council, 1991). *In situ* methods typically involve managing native forests as nature reserves, although secondary stands produced by local seed or remnant stands may also be used to protect genetic resources *in situ*. *In situ* conservation efforts have been successful in some countries, but have failed completely in others because of social and economic issues that make it difficult to protect designated reserves from further exploitation. *In situ* gene conservation is considered "dynamic" in that *in situ* gene conservation populations may continue to respond to normal evolutionary pressures (Eriksson *et al.*, 1993).

In contrast to *in situ* methods, *ex situ* gene conservation involves holding germplasm in cold storage (*i.e.* as seed, pollen, or vegetative tissue) or growing trees in plantations outside their native location. *Ex situ* methods are the primary means of gene conservation in agriculture (Kannenberg, 1983), but in forest trees, they are usually practiced only when natural populations are in jeopardy and cannot be adequately protected or when a species is actively being domesticated through a tree improvement program. While *ex situ* methods of gene conservation can be expensive, they are favored in some developing countries where *in situ* methods suffer from continuing threats to native populations in reserves or protected areas.

With both *in situ* or *ex situ* methods, primary considerations in the development of a gene conservation strategy revolve around: (1) Type and management of gene conservation populations or collections; (2) Population sizes needed to adequately conserve the species' genetic diversity; and (3) Number and location of protected or sampled populations.

In Situ Gene Conservation

Most *in situ* gene conservation populations reside in strict nature reserves and other types of protected areas (Box 10.3). When well-planned, protected areas serve many functions, ranging from conservation of ecosystems and individual species to maintaining a harmonious interaction of nature and culture. In some protected areas, management is mostly hands-off, whereas in others, habitats are actively manipulated.

Genetic diversity in forest tree species rarely influences decisions regarding selection and management of protected areas. Consequently, the extent to which the existing network of protected areas satisfies gene conservation objectives (in terms of maintaining appropriate population sizes and locations) for any given forest tree species varies considerably. Genetic resources for species and populations occupying scenic and other habitats where protected areas tend to be concentrated are more likely to be conserved *in situ* than those found in areas subjected to greater human impacts and afforded less protection.

Box 10.2. Examples of programs concerned with gene conservation of forest trees.

The following are brief descriptions of five gene conservation programs for forest trees in different parts of the world. Table 10.2 contains a more comprehensive list of programs that include gene conservation efforts and websites where additional information can be obtained.

CAMCORE: International Tree Conservation and Domestication, established in 1980 and based at North Carolina State University, is one of the oldest and most successful forest gene conservation programs. This cooperative is supported by government and industry members from 14 countries. Its mission is "to conserve, test, and improve forest species in the tropics and subtropics for the benefit of humankind." To accomplish this, CAMCORE collects genetic material from natural populations, establishes the material in provenance and progeny tests, and evaluates the tests to identify promising families for breeding. A strategy is then developed to achieve both breeding and conservation objectives. Conservation is *ex situ* and involves the designation and maintenance of genetic tests as long-term conservation plantings outside of the native range. Seed from collections in natural stands is kept in cold storage identified by mother tree. Seed collected from the *ex situ* genetic tests is now being reintroduced into the original donor countries. The organization attempts to include in its *ex situ* collections all alleles occurring at frequencies of 5% or greater in sampled, natural populations. To date, the cooperative has established programs to conserve genetic resources of 36 tree species.

CGIAR: The Consultative Group on International Agricultural Research is a large research organization created in 1971 with the goal of increasing agricultural productivity, protecting the environment, saving biodiversity, improving policies, and strengthening national research. CGIAR pursues these objectives through the diverse activities of 16 international research centers, three of which address forest gene conservation. A research program of the Center for International Forestry Research (CIFOR) is dedicated to biodiversity and genetic resource issues. The International Centre for Research in Agroforestry (ICRAF) has a program on domestication of agroforestry trees that focuses on genetic resources and improvement strategies, propagation and field testing. The Forest Genetic Resource Program of the International Plant Genetic Resource Institute (IPGRI) works to "ensure the continuous availability of these resources for present and future use, through *in situ* and *ex situ* measures that allow species' adaptation and evolution to changing environments."

DFSC: The DANIDA (Danish International Development Agency) Forest Seed Center, in operation since 1969, provides technical support to developing countries regarding several aspects of gene conservation, especially those related to seed procurement and storage. Its mission is "Upgrading the production, supply, conservation and correct use of physiologically sound and genetically well-adapted and improved planting material for tree planting activities in the region/country." The work is carried out in collaboration with national institutes, IGPRI, The Food and Agricultural Organization of the United Nations (FAO), and several other agencies. Examples of projects include the development of methods to produce and handle seed from tropical and subtropical species, including those with recalcitrant seed; *in situ* gene conservation and tree improvement of *Baikiaea plurijuga* in Zambia and *Tectona grandis* and *Pinus merkusii* in Thailand; assessment of *Azadirachta indica* international provenance trials; and evaluation of an international network of *ex situ* forest gene conservation stands of pines and eucalypts. (Box 10.2 continued on next page)

Box 10.2. Examples of programs concerned with gene conservation of forest trees. (Continued from previous page)

EUFORGEN: The European Forest Genetic Resources Programme, established in 1995, is a collaborative program among 30 European countries aimed at ensuring the effective conservation and sustainable use of forest genetic resources in Europe. It is coordinated by IGPRI and FAO and funded by the participating nations. EUFORGEN creates a forum for forest geneticists to analyze needs, exchange experiences and develop conservation plans for selected species, using both *in situ* and *ex situ* methods. The program also contributes to the development of conservation strategies for the ecosystems to which the species belong. The program is divided into five operating networks that focus on conifers, Mediterranean oaks, *Populus nigra*, temperate oaks and beech, and other hardwoods. Individual networks produce work plans that are carried out by participating countries.

FAO: FAO provides technical and scientific support to countries on all aspects of forest gene conservation. The organization's activities are planned by its Panel of Experts and involve extensive collaboration with other agencies, including DFSC, IGPRI, International Union of Forest Research Organizations (IUFRO), and the World Conservation Union (IUCN). FAO supports the exploration, conservation and testing of species and provenances across national boundaries. Ongoing programs include coordinating the evaluation of *ex situ* conservation stands in 12 developing countries and *in situ* conservation stands in three countries. Additionally, FAO produces numerous documents outlining strategies for gene conservation and publishes an annual bulletin, *Forest Genetic Resources*, in English, French and Spanish. It also maintains a database (REFORGEN) that is made available over the internet (Table 10.2), providing "reliable and up-to-date information on forest genetic resource activities for use in planning and decision making at the national, sub-regional, regional and international levels."

In situ gene conservation is more easily achieved for dominant forest trees, for which thousands of individuals may occur in even small protected areas, than for low-density species, including many outcrossing tropical trees, which require much larger areas to maintain genetic diversity and viable populations (National Research Council, 1991). In addition, species that form even-aged stands and depend on disturbance (*e.g.* aspens, many pines) may lose representation in protected areas over the long term when management precludes catastrophic disturbance events, such as fire, needed for their regeneration and growth. Management inputs such as prescribed fire or thinning are necessary to maintain these species.

When existing protected areas fail to adequately conserve the genetic resources of a forest tree species, management of supplementary *in situ* stands in a way consistent with achieving gene conservation objectives is desirable. Stands specifically designated for and managed to promote long-term gene conservation have been variously referred to as **gene resource management units** (Ledig, 1988b; Millar and Libby, 1991), **gene management zones** (Gulbaba *et al.*, 1998), and **genetic resource areas** (Tsai and Yuan, 1995). While such gene resource management units ideally contain representative samples of the genetic diversity of target populations, they are often designed to accomplish a variety of objectives simultaneously (Wilson, 1990). They may serve as seed production areas or as reference populations for breeding activities. They may also be harvested for timber, provided that regeneration is done naturally or is done using only genetically diverse, local seed, and hence does not compromise gene conservation objectives.

Table 10.2. Examples of programs with gene conservation efforts in forest trees.

Program	Region	Species involved	Webpage or References
CAMCORE: International Tree Conservation and Domestication	Central America and Mexico, South America, and southeast Asia	Conifers and some broadleaf species	http://www.camcore.org/
CIFOR: Center for International Forestry Research	Worldwide, emphasis is on tropical forests	Many	http://www.cifor.cgiar.org/
CGIAR: The Consultative Group on International Agricultural Research	Worldwide	All plants	http://www.cgiar.org
DFSC: The DANIDA Forest Seed Center	Developing countries	Tropical and subtropical forest trees	http://www.dfsc.dk/
EUFORGEN: The European Forest Genetic Resources Programme	Europe	Conifers, *Populus nigra*, oaks, beach, other hardwoods	http://www.ipgri.cgiar.org/networks/euforgen/euf_home.htm
FAO: The Food and Agricultural Organization of the United Nations	Worldwide	Many	http://www.fao.org/forestry/
ICRAF: International Centre for Research in Agroforestry	Worldwide	Agroforest species	http://www.icraf.cgiar.org/
INBAR: International Network for Bamboo and Rattan	22 countries with bamboo and rattan	Bamboo and rattan	http://www.inbar.int/
International Neem Network	Indian subcontinent, southeast Asia, Africa, Latin America, Caribbean	Neem (*Azadirachta indica*)	http://www.fao.org/forestry/site/5307/
IPGRI: International Plant Genetic Resource Institute	Worldwide	Many	http://www.ipgri.cgiar.org/
IUFRO: International Union of Forest Research Organizations	Worldwide	Many	http://www.iufro.org
Pacific Northwest Forest Tree Gene Conservation Group	USA (Oregon and Washington)	Conifers	Lipow *et al.*, 2001, Lipow *et al.*, 2003
Sprig: South Pacific Regional Initiative on Forest Genetic Resources	10 Pacific Island countries	Native and exotic species including sandalwood, mahogany and malili	http://www.ffp.csiro.au/tigr/atscmain/whatwedo/projects/sprig/
University of British Columbia Centre for Forest Tree Gene Conservation	Canada (British Columbia)	Forest trees and shrubs	http://genetics.forestry.ubc.ca/cfgc/

Box 10.3. **Types of protected areas for *in situ* gene conservation programs.**

The World Conservation Union (IUCN) defines a protected area as: "An area of land and/or sea especially dedicated to the protection and maintenance of biological diversity, and of natural and associated cultural resources, and managed through legal or other effective means" (IUCN, 1994). There are six IUCN management categories for protected areas designated according to the purpose of protection and the type of management they receive.

1. *Strict Nature Reserve/Wilderness Area* is a protected area managed mainly for science or wilderness protection.

1a. *Strict Nature Reserve* is a protected area managed mainly for science: An area of land and/or sea possessing some outstanding or representative ecosystems, geological or physiological features and/or species, available primarily for scientific research and/or environmental monitoring.

1b. *Wilderness Area* is a protected area managed mainly for wilderness protection: A large area of unmodified or slightly modified land, and/or sea, retaining its natural character and influence, without permanent or significant habitation, which is protected and managed so as to preserve its natural condition.

2. *National Park* is a protected area managed mainly for ecosystem protection and recreation: A natural area of land and/or sea, designated to: (a) Protect the ecological integrity of one or more ecosystems for present and future generations; (b) Exclude exploitation or occupation inimical to the purposes of designation of the area; and (c) Provide a foundation for spiritual, scientific, educational, recreational and visitor opportunities, all of which must be environmentally and culturally compatible.

3. *Natural Monument* is a protected area managed mainly for conservation of specific natural features: An area containing one, or more, specific natural or natural/cultural feature which is of outstanding or unique value because of its inherent rarity, representative or aesthetic qualities or cultural significance.

4. *Habitat/Species Management Area* is a protected area managed mainly for conservation through management intervention: An area of land and/or sea subject to active intervention for management purposes so as to ensure the maintenance of habitats and/or to meet the requirements of specific species.

5. *Protected Landscape/Seascape* is a protected area managed mainly for landscape/seascape conservation and recreation: An area of land, with coast and sea as appropriate, where the interaction of people and nature over time has produced an area of distinct character with significant aesthetic, ecological and/or cultural value, and often with high biological diversity.

6. *Managed Resource Protected Area* is a protected area managed mainly for the sustainable use of natural ecosystems: An area containing predominantly unmodified natural systems, managed to ensure long term protection and maintenance of biological diversity, while providing at the same time a sustainable flow of natural products and services to meet community needs.

Ex Situ Gene Conservation

Ex situ genetic resources include seed and pollen stores and trees in breeding populations, genetic tests, field conservation banks and breeding arboreta. Living trees protected in urban arboreta and botanical gardens also represent *ex situ* collections, but are generally of narrow genetic base. Other types of genetic archives, including tissue culture and DNA in genomic libraries, are rarely utilized for forest trees. They may become more prevalent as these technologies advance.

Seed and Pollen Stores

Seed storage is a widely practiced method of *ex situ* gene conservation. It is used both for threatened populations and for species in breeding programs. In the former case, seed stores may be the only protected genetic resource. Seed stores ideally consist of a **core collection** with a limited set of accessions chosen to represent the genetic spectrum of the whole collection and a **reserve collection** that may furnish genetic material for use in reforestation, breeding and evaluation of the resource (Frankel, 1986; Brown, 1989). Seed lots within populations may be bulked to minimize cost, or else stored separately by parent tree, thereby permitting the establishment of pedigreed breeding populations and the avoidance of inbreeding.

Seed storage has several advantages and disadvantages. The main advantage is that it usually demands a low level of initial investment and maintenance, especially in comparison to establishing living trees in plantations. Seed from many forest trees, chiefly temperate ones, remains viable for years or decades if held at subfreezing temperature and low humidity (National Research Council, 1991). Cryopreservation, in which seed is kept in liquid nitrogen at 196 C, can greatly extend viability over traditional cold storage methods, but is more costly than traditional methods. Seed from some taxa, however, cannot be stored for long periods, irrespective of storage conditions; these taxa, which include temperate oaks, eucalypts and many tropical species, are said to have **recalcitrant seed**.

For species that can be stored, periodic outplanting of seed stock and recollection of seed is required to "replenish" viability (time interval varies with storage conditions and storability of seed). Outplanting is expensive and labor intensive as trees are large and usually take several to many years to reach reproductive maturity. Other disadvantages of seed stores are that they require outplanting to identify adaptive genetic variation (*Chapter 8*), and they are "static" collections, unable to respond to normal evolutionary pressures. Seed stores may also be subject to genetic degradation as a result of heritable differences in tolerance to storage conditions (El-Kassaby, 1992; Chaisurisri *et al.*, 1993).

Pollen stores are generally not emphasized in forest tree gene conservation. Breeding programs sometimes store pollen for use in controlled crosses, but since the genetic variation in pollen is only expressed when a seed tree is pollinated, pollen stores are less useful for conservation purposes than seed stores. For species with recalcitrant seed, however, pollen stores may be the only way to archive a large amount of genetic material.

Genetic Tests and Breeding Arboreta for Gene Conservation

For forest trees undergoing domestication, valuable *ex situ* genetic resources reside in the hierarchy of populations that comprise the breeding program, including the breeding population (often grafted into breeding arboreta, also called clone banks) and its associated genetic tests (*Chapter 11*). Because breeding exerts strong directional selection, allele frequencies are altered and alleles may be lost from breeding populations during each cycle of recurrent selection. Nevertheless, if appropriate breeding designs are used (*Chapters 14 and 17*) (Namkoong, 1976; Eriksson *et al.*, 1993), moderately high levels of genetic variation can be retained in even relatively small breeding populations (in the order of 50 to 100 individuals). This level of genetic variation is sufficient to enable continued response (*i.e.* improvement) in selected traits over multiple generations of selection and breeding, without undue loss of genetic diversity (see *Population Sizes for Gene Conservation* below).

Genetic tests, including both progeny and provenance tests, have features that make them valuable for gene conservation purposes. Provenance tests provide information about geographic patterns of adaptive genetic variation (*Chapter 8*; also see *Number and Location of Populations for Gene Conservation* below), which can help in deciding the appropriate number and distribution of *in situ* gene conservation populations or *ex situ* collections. They can also direct delineation of breeding zones and seed transfer guidelines, which provide assurances that well adapted genotypes are deployed in artificial regeneration programs (*Chapters 8 and 12*).

Progeny tests, in which trees are identified by one or both parents, may be suitable for *ex situ* gene conservation, especially if they contain large numbers of families sampling appropriate native populations. Progeny tests have the added advantages that family identities are known and parents have desirable phenotypes. In addition, progeny are planted at multiple sites so that the families are exposed to a variety of environmental challenges for the life of the tests (often one to several decades). Therefore, if the species becomes confronted with new stresses, such as a disease epidemic or novel conditions brought on by climatic change, the tests are available for rapid screening. This can enhance the process of breeding for new traits. However, genetic tests have a finite lifespan, and one of the greatest needs in tree conservation efforts is development of advanced-generation strategies for recurrent cycles of genetic tests as *ex situ* gene conservation plantings.

A good example of the use of genetic tests for *ex situ* gene conservation is by CAMCORE (Box 10.2 and Table 10.2) which has collected seeds from the native range of more than 35 tropical tree species and planted them in provenance-progeny tests of member organizations. Efforts for a typical species include a sampling scheme designed to capture alleles present at frequencies of $p > 0.05$ which in this program usually means 1000 or so mother-tree collections from many native-population locations. Seedlings, with provenance and mother tree identified, are planted in genetic tests at 75 to 100 locations throughout the southern hemisphere. For example, in *Pinus tecunumanii*, there are 200,000 trees planted in 150 genetic tests and these genotypes represent 99% of the natural genetic diversity of the species.

Population Sizes for Gene Conservation

Some gene conservation strategies focus on conserving low-frequency alleles ($p < 0.05$), whereas others are more concerned with maintaining quantitative genetic variation (*e.g.* Brown and Hardner, 2000; Yanchuk, 2001). Low frequency alleles are more likely to be lost than alleles of medium or high frequency when establishing *in situ* gene conservation populations or when sampling native populations for *ex situ* collections. The capture of at least one copy of 95% of the genes occurring in a large target population at frequencies greater than 0.05 has been proposed as a gene conservation goal (*i.e.* benchmark conditions) for seed stores and other static *ex situ* collections (Marshall and Brown, 1975). Mathematical models suggest that this goal can be achieved with relatively small sample sizes in the range of 50-160 individuals, with the exact number required depending on model assumptions (Kang, 1979a; Namkoong *et al.*, 1988; Frankel *et al.*, 1995). These benchmark conditions are arbitrary and debatable (Marshall, 1990; Krusche and Geburek, 1991). Yet, decreasing the critical allele frequency below 0.05 or increasing the confidence level beyond 95% results in a sharp increase in the required sample sizes and hence in the effort and cost of sample collection and storage (Brown and Hardner, 2000).

Larger population sizes are needed if the goal is to capture and employ low-frequency alleles to "rescue" a target population or otherwise incorporate these alleles into a breeding program, because multiple copies are required. For a low-frequency allele to be useful for breeding, for example because it confers disease resistance, multiple copies are needed to avoid problems associated with monocultures and inbreeding (Yanchuk, 2001). In addition, a single copy of a recessive allele would not be detectable, as identification requires homozygous expression. Therefore, when sampling for beneficial and operationally important genes, it is much more difficult to capture recessive alleles than dominant ones. For example, very large samples (approximately 10,000) must be drawn in order to include, with 95% probability, 20 individuals homozygous for a recessive allele occurring at frequency 0.05 in the target population (Yanchuk, 2001). Only about 250 individuals, however, are needed to capture 20 individuals expressing a dominant allele, in either heterozygous or homozygous form, with the same frequency and probability. Such calculations show that conservation of usable low-frequency, dominant alleles is possible in moderately sized *in situ* gene conservation populations or *ex situ* collections (*e.g.* $N_e < 250$), as is conservation of recessive alleles of intermediate frequencies ($p > 0.25$). Because conservation of multiple, expressed copies of low-frequency, recessive alleles requires very large sample sizes, it is usually only feasible through the establishment of *in situ* populations in protected areas (Yanchuk, 2001).

An alternate gene conservation objective emphasizes maintaining the "normal" adaptive potential of an ancestral, pre-disturbance population by conserving a significant portion of its total genetic variation for quantitative traits (Eriksson *et al.*, 1993). This objective is directed towards creating conditions that permit naturally reproducing populations to respond to the range of selection pressures they might encounter in the future. Estimates of the effective population size required vary depending on the factors considered. Lande (1995), for example, recommended a target effective population size of 5000 after evaluating the balance between the creation of genetic variation by mutation and its loss by genetic drift. Lynch (1996) suggested an effective population size of 1000 based on his model, which included the impact of natural selection as well as genetic drift and mutation. This model predicts that once the effective population size exceeds 1000, the average genetic variance is controlled almost entirely by selection-mutation balance and is essentially independent of population size and genetic drift. Taken together, these and other analyses (*e.g.* Soulé, 1980; Franklin, 1980) point to an effective population size of one thousand to several thousand for the conservation, over many generations, of the normal adaptive potential of populations. Population sizes of several thousand also conserve multiple copies of most low-frequency alleles and therefore may be a sensible goal for many *in situ* gene conservation strategies.

Estimates of minimum effective population sizes for *in situ* gene conservation populations based on genetic considerations alone, however, must be interpreted with caution (Lande, 1988b, 1995). Demographic and environmental processes may pose a greater threat to small populations than genetic factors. *In situ* gene conservation populations must be large enough to endure population size fluctuations caused by, for example, fire, pest attacks, and climate extremes, and to permit the persistence of key elements in the ecosystem, including populations of animals important for pollination and seed dispersal. Unfortunately, untangling the relative risk of demographic, environmental, and genetic processes is extremely difficult, especially for long-lived trees (Millar and Westfall, 1992).

Number and Location of Populations for Gene Conservation

As genetic variation resides among, as well as within, populations (*Chapters 7* and *8*), gene conservation efforts must protect more than one population. Redundancy in protection is also needed to provide insurance against calamities such as catastrophic fire. The optimal number and location of populations needed for gene conservation through either *in situ* or *ex situ* means depends on the extent and pattern of genetic variation across the landscape, which differs greatly among species of forest trees (*Chapters 7* and *8*). Relevant information on population genetic structure can be obtained by allozymes or DNA marker studies, as described in Box 10.4, or by provenance and seed source tests (*Chapters 8* and *12*).

Knowledge of adaptive trait variation, obtained through provenance and seed source tests, is especially applicable to gene conservation assessments (Namkoong, 1997). Traits of adaptive importance include stem growth, environmental tolerances, and pest and disease resistance. Understanding how genetic variation for these traits is partitioned on the landscape is important since capturing a representative portion of genetic variation requires more conserved populations in species that concentrate genetic variation among, rather than within, populations. Analysis of adaptive trait variation can also expose whether a species has distinct ecotypes, each worthy of genetic conservation, and it may link the ecotypes to different environmental conditions encountered by the species across its range. Additionally, delineation of breeding and seed zones often reflects patterns of genetic variation for adaptive traits (*Chapters 8* and *12*). Given that breeding and seed zones are intended as guidelines for the limits of breeding populations and movement of germplasm during artificial regeneration, respectively, conservation of at least one representative *in situ* gene conservation population or *ex situ* collection within each breeding or seed zone is sensible (Millar and Westfall, 1992; Lipow *et al.*, 2003). Provenance and seed source tests can also highlight certain populations, such as those that display a high degree of resistance to disease or tolerance to climatic extremes, as having high fitness value and hence warranting conservation priority.

From an academic perspective, a disadvantage of using adaptive trait variation to guide conservation decisions is that provenance tests normally examine only a few traits, which may or may not reflect patterns of genetic variation for traits that will be needed in the future. Conversely, from a practical perspective, most governmental agencies, small landholders and local farmers, especially in developing countries, will only maintain conservation areas over the long term when there is an economic reason to do so. Results from provenance tests guide decision making about which populations should receive priority in conservation efforts.

Because it is doubtful that information on patterns of genetic variation will ever be complete enough, a number of geographical, ecological, and political criteria should also be considered when deciding on populations to target for gene conservation: (1) One or more populations should be from the center of the species' origin, where greatest genetic diversity is expected (Yanchuk and Lester, 1996), although there are exceptions (*Chapter 8*); (2) Populations growing on particularly unique environments should be conserved, since they are likely to contain unique adaptations; (3) Isolated or geographically peripheral populations may be especially important to a species' future evolution since founder effects, genetic drift, and/or extreme environmental conditions often contribute to their genetic uniqueness (Lesica and Allendorf, 1995); (4) Given that genetic variation is often correlated with environmental or ecological variation (*Chapters 7* and *8*), areas sharing climate, physiography, geology, soil, and/or plant and animal communities can be grouped into ecoregions for gene conservation (Bailey, 1989); (5) Remnant or threatened populations in areas most affected by changes in land use may deserve conservation (Morgenstern, 1996); and (6) If practical, *in situ* gene conservation populations should be located in areas expected to have long-term politically stability and where protection can be assured for the foreseeable future.

Box 10.4. Applications of molecular markers to gene conservation.

Molecular markers have many applications with respect to gene conservation of forest trees. At the broadest level, they can help delimit species and subspecies (Strauss *et al.*, 1992a). This is critical to assessing accurately the gene resource status of a taxon, and it may be a necessary step in the application of biodiversity legislation. For example, allozymes and random amplified polymorphic DNA (RAPD) markers were used to confirm that the controversial taxon, *Abies nebrodensis*, which is known from only one small population, is genetically distinct from the more widespread *Abies alba* (Vicario *et al.*, 1995). *Abies nebrodensis* is now listed as critically endangered by the IUCN (Oldfield *et al.*, 1998).

Within species, molecular markers (and particularly allozymes) have been widely employed to characterize patterns of genetic variation (*Chapters 7* and *8*). The patterns are often considered as baseline information when determining the appropriate number and distribution of conservation populations. An allozyme analysis of *Chamaecyparis lawsoniana*, for example, indicated that stands in the interior portions of the species range in California are more differentiated from each other than stands in coastal areas (Millar and Marshall, 1991). This led Millar and Marshall (1991) to suggest that, while a few large protected areas effectively conserve genetic diversity in coastal populations, a large number of *in situ* gene conservation populations, each requiring separate management, are needed in the interior.

The practice of developing gene conservation recommendations from genetic patterns of geographic variation based on allozyme or other molecular marker studies is common, yet also controversial. Molecular markers often fail to reflect patterns of differentiation in adaptive traits (Karhu *et al.*, 1996; Yang *et al.*, 1996; Reed and Frankham, 2001) (Box 5.6). Indeed, a summary analysis of 71 studies (involving both plants and animals) showed only a weak correlation between molecular marker diversity and quantitative trait variation, leading Reed and Frankham (2001) to conclude that, "Molecular measures alone cannot be relied upon to reflect the evolutionary potential of populations." This limitation must be kept in mind when attempting to derive conservation plans solely on the basis of molecular marker analysis.

Even though molecular markers are questionable surrogates for information on adaptive trait variation, they still have bearing on gene conservation strategies. Molecular markers enable characterization of a number of evolutionary forces that impact the maintenance of genetic diversity, including mating systems, gene flow, and genetic drift (*Chapters 5, 7* and *8*). Questions as to whether inbreeding increases in small populations or gene flow declines with fragmentation, for instance, can be best answered using molecular markers. Molecular markers can also reveal whether a small population is experiencing a recent bottleneck (Ledig *et al.*, 2000), and thus, is particularly susceptible to accelerated decline due to inbreeding depression, or whether it has been historically small (*i.e.* inbreds previously purged). In addition, studies based on molecular markers can help identify the locations of past refugia from glacial advances. Such refugia populations are expected to harbor high levels of genetic diversity, and thus, should be particularly targeted for gene conservation.

For many forest trees, at least some populations are protected *in situ* in existing networks of protected areas and gene resource management units or in *ex situ* collections. A common approach, called **gap analysis**, is used to assess the need for conservation of additional populations. Gap analysis proceeds by: (1) Developing an inventory of the extent and distribution of all existing conserved genetic resources in light of the species' population genetic structure, geography, and ecology; (2) Using geographic information systems to overlay species' distribution, protected areas, sampled populations and other information to find gaps in existing conservation efforts; and (3) Identifying new areas and populations that warrant additional protection and conservation to complement existing efforts (Box 10.5) (Dinerstein *et al.*, 1995; Yanchuk and Lester, 1996; Lipow *et al.*, 2003).

EFFECTS OF FOREST MANAGEMENT PRACTICES AND DOMESTICATION ON GENETIC DIVERSITY

Intensive forest management has the potential to greatly alter the genetic composition of tree species either positively or negatively. If sound genetic principles are adhered to, a wide range of management prescriptions can be employed without significantly eroding genetic diversity. In this section, we explore how various timber harvesting and reforestation practices impact levels of genetic diversity in managed forests.

Several studies investigating tree species in temperate forests have demonstrated that genetic variation is little impacted in stands that regenerate naturally following timber harvests, provided an adequate seed supply and appropriate conditions for stand establishment remain (Box 10.6) (Neale, 1985; Savolainen and Kärkkäinen, 1992; Adams *et al.*, 1998; El-Kassaby, 1999). Of the small number of tropical species examined, genetic diversity measures show no adverse changes after harvesting in some species, but in others, particularly those that normally occur at low density, some genetic variation is lost after logging (Ratnam *et al.*, 1999). For species where a severe reduction in diversity at marker loci is not apparent, changes in adaptive variation are also expected to be minimal, even when artificial selection has been applied. This is because adaptive traits such as growth and form typically display weak inheritance. Consequently, the mild selection that occurs when desirable genotypes are left to reseed a stand has only a minor effect on trait variance. This is why mild selection followed by natural regeneration, produces, at most, only modest genetic gain (*Chapter 7*). Regardless, high grading of the best genotypes, and leaving only the few poorest trees for regeneration, must be avoided, as it may decrease stand productivity and adaptation (*Chapter 7*).

Even if high levels of genetic diversity are maintained in a naturally regenerated stand, timber harvesting may have other genetic consequences. Low-frequency alleles may be lost during partial harvesting, especially if they have deleterious effects on tree growth or form and only the most vigorous individuals are left uncut following artificial selection (Box 10.6; Adams *et al.*, 1998; El-Kassaby, 1999). As the future importance of low-frequency alleles is unknown, a prudent strategy is to forego artificial selection, and therefore harvest randomly, in designated gene resource management units. Additionally, stands regenerating after a shelterwood and other partial harvest systems may produce a higher proportion of inbred seed due to the lowered density of mature (breeding) adult trees (*Chapter 7*). The impact of increased inbreeding at the seed production stage on genetic diversity is likely to be minor in the long-run, however, because weak, inbred progeny are at a competitive disadvantage in the early stages of stand establishment and usually don't survive to adulthood (*Chapters 5* and *7*).

Stands produced through artificial regeneration may also exhibit high levels of genetic diversity. The main factor determining the level of genetic diversity in a seed lot is the number of parent trees. Therefore, wild stand seed collections used for artificial reforestation should be made up of relatively even contributions of seed from at least 15-20 (preferably more) wild trees (Adams *et al.*, 1992b) if the objective is to create plantations with high levels of within-stand genetic diversity. As with natural regeneration, collecting wild seed from individuals displaying good characteristics has the potential to yield some genetic gain without significantly compromising genetic variance in adaptive traits (*Chapter 7*).

In order to achieve genetic gain through selection the number of genotypes in seed orchards is typically limited (ranging from as few as 10 to more than 100) (*Chapter 16*). Nevertheless, several comparisons of genetic diversity present among genotypes in first generation orchards relative to natural populations from the same breeding zones reveal that genetic (allozyme) diversity in seed orchards is little compromised by the selection of parents for desirable traits (Bergmann and Ruetz, 1991; Kjær, 1996; El-Kassaby, 1999; Godt *et al.*, 2001). As expected, however, seed orchard parents typically contain fewer low-frequency alleles compared to wild stands.

An equation is available to estimate how the total additive genetic variation (% V_A) in a first-generation seed orchard relates to the total additive genetic variation in the natural population from which the orchard clones are derived (Johnson, 1998):

$$\% \ V_A = (1 - (1/N)) * 100 \qquad\qquad \text{Equation 10.1}$$

where N = the number of clones in the orchard. This equation assumes unrelated clones undergoing random mating with no selfing and each clone contributing equally to the pollen and egg gametes comprising the seed crop. It predicts that a twenty-clone seed orchard possesses 95% of the total additive genetic variation found in the corresponding natural population.

Box 10.5. Gap analysis for evaluating genetic resources in protected areas.

Lipow *et al.* (2004) conducted a gap analysis to evaluate whether the existing network of legally defined protected areas (*e.g.* designated parks, wilderness areas, etc.) adequately conserve the genetic resources of conifer species in western Oregon and Washington. Gap analysis is a method of biodiversity planning in which a Geographic Information System (GIS) is used to intersect digital maps displaying protected areas with those showing species' distributions. The geographical area occupied by each species was subdivided into ecoregions (which represent areas with similar environmental characteristics) or seed zones (*Chapter 8*) in order to stratify their distributions into "populations," or units warranting gene conservation. The genetic resources within a particular ecoregion or seed zone were considered adequately protected *in situ* if there was at least one protected area in the region or zone containing 5000 or more mature trees of the species of interest.

While the gap analysis showed that genetic resources are adequately conserved in the existing network of protected areas for most of the investigated species in most ecoregions or seed zones, some species and areas appear to warrant additional gene conservation. For example, few *Pinus monticola* trees are present in protected areas in several ecoregions in the lowlands surrounding Washington's Puget Sound and adjacent portions of the Coast Range (Fig. 1). In these ecoregions, land conversion to (Box 10.5 continued on next page)

Box 10.5. Gap analysis for evaluating genetic resources in protected areas. (Continued from previous page)

plantation forestry, agriculture, and human development eliminated many wild populations of *P. monticola*, and white pine blister rust, an exotic pathogen, has caused widespread mortality and has seriously limited natural regeneration. *Ex situ* gene conservation involving seed collections from the unprotected populations in the area and breeding for blister rust resistance is recommended in concert with additional *in situ* protection.

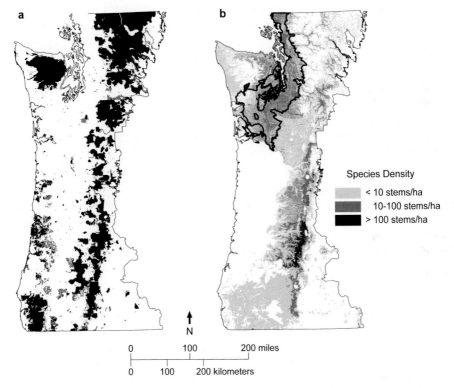

Fig. 1. Maps of the study region (*i.e.* western Oregon and Washington from the Pacific Ocean to the eastern foothills of the Cascade Mountains) showing: (a) Protected areas (in black); and (b) Distribution and estimated density of *Pinus monticola*, with the ecoregions where additional gene conservation measures are recommended outlined in black.

In reality, genetic variation in a seed orchard crop is reduced relative to its theoretical potential. Maximum genetic variation in a crop requires that the Hardy-Weinberg assumptions of random mating and panmictic equilibrium be achieved (*Chapter 5*). These assumptions are violated, however, when orchard clones (or families) differ in seed and pollen production (Fig. 10.1) (Muona and Harju, 1989; Roberds *et al.*, 1991; Nakamura and Wheeler, 1992) and when the phenology of reproduction is asynchronous among clones (thereby leading to nonrandom mating) (Roberds *et al.*, 1991; El-Kassaby, 1992, 1999). These effects can be countered by pollen management practices designed to equalize paternal contribution in crops and by collecting seed by clone and mixing the maternal families equally in seedlots (El-Kassaby, 1989).

Such procedures, however, incur added expense and may only marginally increase genetic diversity relative to the considerable variation present in mature orchard crops (*i.e.* when all or most clones are flowering heavily). Pollen contamination may increase the genetic diversity of the seed orchard crop, but is usually undesirable because it decreases genetic gain (*Chapter 16*). Seed should not be utilized from immature orchards (*i.e.* when few clones are flowering), not only because genetic diversity is reduced, but more importantly, because pollen contamination is likely to be considerable.

A variety of seed storage and seedling production practices used for both seed orchard and wild seed lots can lead to some losses of genetic variation in stocks used for artificial regeneration (Fig. 10.1). First, as seed lots age, they lose viability. If the rate of viability loss varies among genotypes, as is often the case, then genetic variation in a bulked lot will decline from collection to sowing (Chaisurisri *et al.*, 1993). Second, heritable, genotypic differences exist for seed size, so that the common practice of seed sizing during extraction and processing may erode genetic variation in a seed lot (Silen and Osterhaus, 1979; Friedman and Adams, 1982; Chaisurisri *et al.*, 1992). Third, germination parameters relating to dormancy, rate of germination and percent germination are also under genetic control (El-Kassaby, 1992; Edwards and El-Kassaby, 1995). Therefore, the common practices of multiple sowing followed by thinning in greenhouse containers to leave the largest

---

Box 10.6. Impact of harvesting on genetic variation in *Pseudotsuga menziesii* var *menziesii.*

Adams *et al.* (1998) investigated the impact of natural and artificial regeneration on the genetic composition of *P. menziesii* stands following three alternative harvesting methods, relative to uncut (control) stands. In group selection, all trees were harvested in 0.20 ha clearcut patches scattered throughout the stand, with about one-third of the stand removed. In the shelterwood method, 15-30 of the largest trees per hectare were left distributed relatively evenly over the entire area. In clearcuts, all trees were removed. Each harvest method and the control were replicated in three blocks. All harvested areas were replanted with seedling stocks (7 different types, total) derived from seeds collected in wild stands in the region. Natural regeneration, however, was adequate only in the three shelterwood replicates and one of the clearcut plots.

Allozymes were used to compare gene diversity of mature trees in control stands to that of mature trees in the group and shelterwood stands and to seedlings in the artificial and in natural regeneration (Fig. 1). Gene diversity was described by estimating the number alleles per locus (A), percentage of polymorphic loci (P), and expected heterozygosity (H_e) (*Chapter 5*). The fixation index (F_{is}) was also computed to assess the extent to which inbreeding may have occurred in each population (*Chapter 5*).

The results showed that harvesting followed by either natural or artificial regeneration produced offspring populations that differed little from adult populations in nearby uncut stands. When the smallest trees in the stand were cut to form shelterwoods, however, some rare, presumably deleterious alleles, were lost, resulting in slightly fewer alleles per locus among residual trees and natural regeneration than in uncut stands. For this reason, Adams *et al.* (1998) recommended that when regenerating stands in gene resource management units by the shelterwood method, parent trees of a range of sizes should be left to maximize allelic diversity in naturally regenerated offspring. Furthermore, average genetic diversity was found to be greater in the seedling stocks than in natural regeneration. Presumably, this is because the artificial seedling stocks were derived from seed collected from many wild stands, and therefore, sampled more diversity than found in individual populations.

(Box 10.6 continued on next page)

Box 10.6. Impact of harvesting on genetic variation in *Pseudotsuga menziesii* var *menziesii.*
(Continued from previous page)

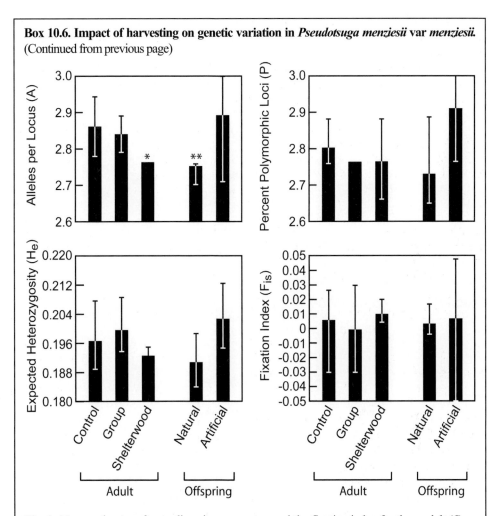

Fig. 1. Mean estimates of gene diversity parameters and the fixation index for three adult (Control, Group Selection (Group), Shelterwood) and two offspring (Natural and Artificial) population types of *Pseudotsuga menziesii* growing in the same vicinity. Brackets with each bar give the range over replicates (3 for each adult population type, 4 for natural offspring, and 7 for artificial offspring). Based on 17 allozyme loci for gene diversity statistics and 9-11 loci for the fixation index (n = 120 for each population replicate). * and ** mean significantly different from Control at $P < 0.10$ and $P < 0.05$, respectively. (Reproduced from Adams *et al.*, 1998, with permission of the Society of American Foresters)

germinant, and culling of smaller seedlings at the end of a nursery rotation, favor certain parents (Edwards and El-Kassaby, 1996). It is possible to avoid losing genetic variation at these various stages of seed storage and seedling production by storing and sowing seed lots by family. The seedlings could then be bulked prior to outplanting, if so desired. Losses of genetic diversity from mixed family sowing are not likely to be very large. Nevertheless, some organizations sow families separately in the nursery in order to enhance the efficiency of seedling production by tailoring nursery practices to family differences in optimal germination conditions, fertilization regimes, timing of lifting, etc.

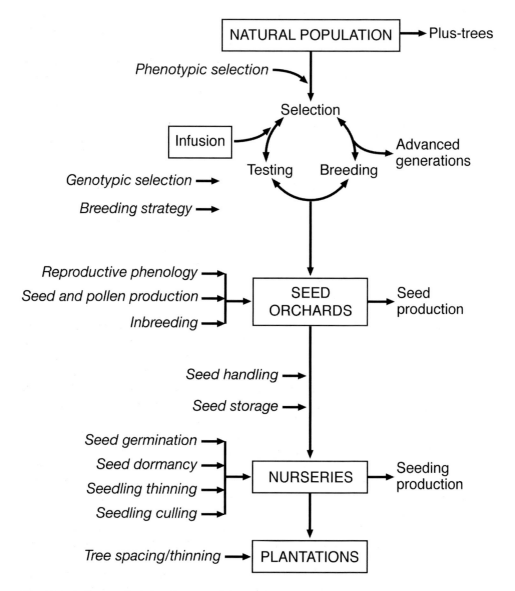

Fig. 10.1. A diagram depicting the steps of the tree improvement delivery system and the associated activities where genetic variation may be lost El-Kassaby, 2000, p 199). (Reproduced from Young, *et al.*, 2000, with permission of CSIRO Publishing)

SUMMARY AND CONCLUSIONS

More than 7300 tree species globally are threatened with extinction, with many more sub-species, varieties, and races at risk. Within species, some populations are being eliminated along with any alleles that may be unique to them. Genetic diversity is at the core of the adaptive value of species, permitting continued evolution under changing environments.

Yet, this genetic diversity is being impacted by natural and human causes. Therefore, gene conservation, including all policy and management actions to ensure continued availability and existence of genetic variation, is a global imperative.

Two methods of gene conservation are *in situ* and *ex situ,* and either or both of these can be used effectively in a gene conservation program. *In situ* methods involve establishing protected areas of forests such as wilderness areas, reserves and parks to maintain populations growing in their place of origin. *In situ* gene conservation has the advantage of protecting large numbers of species and trees of each species relatively inexpensively. When the protected areas are left unmanaged, natural selection and evolution proceed and species composition can change, sometimes with the loss of some species. Active management may be needed to maintain early successional species. *In situ* gene conservation methods often fail in countries where the reserved areas cannot be adequately protected from encroachment, destruction and poaching.

In *ex situ* gene conservation native populations of a species are sampled according to a plan, and the collected germplasm is stored in cold storage (*e.g.* seed or pollen), breeding populations (*e.g.* scion material from breeding populations grafted into clone banks or breeding arboreta), or genetic tests (*e.g.* seed planted in provenance-progeny tests). *Ex situ* methods can be expensive, but are favored when: (1) The species is being domesticated through a tree improvement program that can actively manage the gene conservation effort; or (2) *In situ* gene conservation reserves would be difficult to protect from threats.

Both *in situ* and *ex situ* gene conservation require strategies for sustaining effective conservation efforts in the long term. These strategies should consider: (1) Population sizes needed to maintain appropriate levels of genetic diversity; (2) The number and locations of protected *in situ* populations and sampled *ex situ* collections to properly sample the species' diversity; (3) Environmental, political and economic conditions that threaten or enhance the conservation efforts; and (4) Recurrent cycles of conservation efforts that ensure multi-generational programs are sustainable in the long term.

A relatively small number of individuals (say 50 trees) is needed to ensure continued response from selection and maintain high-frequency alleles in conserved populations. Conversely, large populations of several thousand are required to have a high probability that multiple copies of low-frequency alleles are maintained. Determining the number and geographical distribution of protected areas and sampled populations is greatly aided by understanding the genetic architecture of the species as estimated through studies with molecular markers or provenance-progeny tests. Since most tree species have substantial genetic variation among provenances, multiple populations, strategically located, must be protected and/or sampled to ensure adequate coverage.

Harvesting and other types of forest management can generally be practiced in forests without greatly impacting genetic diversity, provided adequate measures are in place for regeneration. Several studies have shown that most shelterwood and seed tree cuts do not greatly reduce genetic variation in the new stand compared to the parental stand, although some rare alleles may be absent in the new stand. It is important to leave good phenotypes to provide the seed for regeneration to avoid dysgenic selection. When clearcuts are re-planted with seedlings, the new plantations may have higher levels of genetic diversity than the preceding forests if the seed originates from many mother trees from several stands.

CHAPTER 11
TREE IMPROVEMENT PROGRAMS – *STRUCTURE, CONCEPTS AND IMPORTANCE*

Tree improvement programs are designed to develop genetically improved varieties used for reforestation and afforestation to increase the economic or social value of the planted forest. Tree improvement is an integral component of most plantation programs in the world, and involves repeated cycles of activities such as selection, intermating of selections and genetic testing. Beginning with the vast natural genetic variation that exists in an undomesticated tree species, a tree improvement program aims to change gene frequencies for a few key traits of that species. The ultimate goal may range widely including development of high yielding varieties for an industrial plantation program, stress-resistant varieties for reclamation of marginal sites, improved nitrogen-fixing trees for agroforestry systems or improved varieties for fuelwood or bioenergy production in a community reforestation program.

Variation in species biology, silviculture, product goals and economic considerations cause tree improvement programs to differ dramatically in both design and intensity. These differences, coupled with the large number of activities conducted as part of any program, may make it difficult to understand the reasons for any one activity and how all operations fit together to achieve the goals of the program. Nevertheless, there are basic concepts and activities that are common to most programs. It is the aim of this chapter to describe these common activities and to summarize the scope, structure, genetic progress and economic importance of tree improvement programs in the world. Here we stress general concepts; the details of how each activity is implemented and the many options available for conducting tree improvement programs are addressed subsequently in *Chapters 12-17*.

SCOPE AND STRUCTURE OF TREE IMPROVEMENT PROGRAMS

Tree improvement involves application of forest genetic principles along with good silviculture to develop high yielding, healthy and sustainable plantation forests. The overall impact of tree improvement programs on a global scale is to increase yields and value of plantations to meet the rising demand for forest products, while reducing the need to meet this demand from natural forests (*Chapter 1*). Each tree improvement program aims to develop genetically improved varieties in an economically efficient manner by maximizing genetic gain per unit time at the lowest possible cost.

Large-scale tree improvement programs began in earnest in the 1950s. Zobel and Talbert (1984, p 5) cite 23 papers from 14 countries published in 1950s advocating or describing tree improvement programs. Today, tree improvement is so widespread in the world that a listing of programs would take many pages and would include all countries with substantial plantation programs and all tree species that are planted in any quantity. Many countries in Africa, Asia, Australasia, Europe, North America and South America

© CAB International 2007. *Forest Genetics* (T.W. White, W.T. Adams and D.B. Neale)

have several distinct programs for the many important species planted. Therefore, it is apparent that tree improvement is a normal part of the majority of large plantation programs. In addition, there is considerable interest in the development of improved varieties for agroforestry systems, public-sponsored community planting programs and other non-industrial uses (Burley, 1980; Brewbaker and Sun, 1996; Kanowski, 1996; Simmons, 1996).

In most tree improvement programs, selection, breeding and testing activities are conducted by personnel in a separate department (referred to as the research department, tree improvement department or forest genetics department). Once a new variety is developed, it is usually the responsibility of the reforestation department (variously called department of operations, silviculture or plantations) to ensure that the plantations are established properly and well managed. It is critical that the two departments work well together for tree improvement to be successfully implemented. Poor plantation silviculture can mask the gains made from selection and breeding, while poor genetic stock reduces the value of intensive forest management. Therefore, good tree improvement and good silviculture interact synergistically to produce high yielding and healthy plantations.

The choice of the appropriate species to use in a plantation program is the first important genetic decision (*Chapter 12*). After one or more species are identified, the tree improvement program utilizes the natural genetic variation existing within each species to repackage it into desirable individuals that are eventually outplanted as improved plantations. It can take many years for each cycle of selection, testing and intermating depending on species biology and rotation length. In fact, few professions have a time horizon as long as that for tree improvement where personnel plan ahead for several rotations of plantations.

The long time horizon means that tree improvement programs need to be especially well organized and structured to maximize genetic gain, while minimizing the time for each cycle of breeding. With this in mind, there are essential aspects to the structure and efficient operation of a successful tree improvement program (Box 11.1). A single organization may have programs for multiple species bred for different site types and for different product objectives. Therefore, proper organization of these programs with the key elements described in Box 11.1 is essential for their efficient implementation.

A successful tree improvement program can be conducted by a large private company, by a government agency or by a cooperative organization composed of several private companies and public agencies. The cooperative type of structure deserves special mention because of its uniqueness. Tree improvement programs are expensive and difficult to undertake financially for smaller organizations. Yet by pooling resources, even smaller organizations participate in cooperatives that exist in the USA, Australia, New Zealand, Chile, Argentina and other countries. Although these cooperatives are composed of member companies that are competitors in every other segment of their businesses, they pool their efforts and work cooperatively in the area of tree improvement. Pioneers such as Dr Bruce Zobel recognized in the 1950s that, if properly managed, a **tree improvement cooperative** is an effective way to increase gains, decrease costs per member and share access to specialists funded by all members (Zobel and Talbert, 1984). Most cooperatives work together in selection, intermating, genetic testing and research activities, while operational mass propagation of commercial varieties and plantation management activities are conducted separately by the individual members.

Box 11.1. Essential features of tree improvement programs.

1. Clear projections of program objectives, breeding objectives, organizational capabilities and reforestation demand are needed to ensure that the tree improvement program is meeting the needs of the organization. These need periodic review.

2. Sound knowledge of the silviculture, biology and genetics of the species is required to implement a successful tree improvement program. For example, the following can have huge impacts on the design and intensity of the program: (1) For silviculture, the range of site types to be planted and different cultural treatments planned; (2) For species biology, knowledge of reproductive biology and ease of grafting, rooting and control pollination; and (3) For genetics, estimates of key genetic parameters for important traits such as heritabilities, genetic correlations and genotype x environment interaction (*Chapter 6*).

3. A sound breeding strategy is needed to smoothly guide the implementation of the program in both the near term (5 years) and longer term (15 years). A well-documented strategy facilitates budget preparation, permits annual work plan development and formally documents all assumptions.

4. Well-trained, dedicated personnel and a stable budget are needed for efficient implementation of the strategy. Trained foresters and biologists that have experience working with the species are critical to program success. An unstable budget that is too high in one year and too low the next can severely disrupt tree improvement programs. Even though the forest products business is cyclical by nature, tree improvement is a long-term commitment that needs to be buffered from these economic cycles.

5. Efficient propagation of high-quality plants for reforestation and excellent plantation silviculture is essential for realization of the genetic gains made by tree improvement. After the selection, intermating and testing phases, each new variety must be propagated on an operational scale to meet the forestation needs of the landowner or government organization. No social or economic gains are realized until these plantations, established with the genetically improved varieties, are harvested. Depending on species biology, mass propagation may produce seedlings, rooted cuttings or tissue culture plantlets for reforestation.

6. The maintenance of a broad genetic base is an important function of tree improvement programs. The future may bring new products, climates, soils, diseases and technologies that require different varieties for maximum genetic gain. The flexibility to form these new varieties relies on the maintenance of sufficient genetic diversity within the tree improvement program. This is a gene conservation function served by nearly all tree improvement programs.

7. Lastly, tree improvement programs benefit from supportive research. Research priorities differ from program to program, and may include both basic and applied research aimed at developing new technologies that increase gains or decrease costs associated with tree improvement.

THE BREEDING CYCLE OF FOREST TREE IMPROVEMENT PROGRAMS

The selection, intermating and other activities of most genetic improvement programs of cross-breeding plants and animals can be summarized using a conceptual model called the **breeding cycle**. Its description here (adapted from White, 1987b) stresses the potential set of activities and population types that can occur during a given cycle of improvement (Fig. 11.1). Programs differ widely in how these activities are implemented, in the size of the various types of populations and in program intensity; however, the rationale for each activity is readily understood in terms of the breeding cycle.

   Genetic improvement programs often employ several different types of populations that serve different functions. For example, there may be a larger population containing thousands of genotypes to maintain a broad genetic base and a population containing very few highly selected genotypes to produce the operational variety for reforestation. These population types are useful conceptual constructs because each different type of population is formed by an activity that is part of the improvement program. For example, the selected population is formed by selecting superior individuals from the base population (Fig. 11.1). Sometimes different conceptual population types are not physically distinct from each other; that is, a given physical population of trees planted in one location may serve several functions in the breeding cycle. This flexibility of forming one, two or several physically distinct population types during each cycle of improvement leads to the diversity in breeding strategies. In this chapter, we explain each population type as if it is physically distinct to stress the concepts underlying each type. In *Chapter 17*, we discuss many examples of how different breeding strategies combine population types and allow a single population of trees to serve several functions.

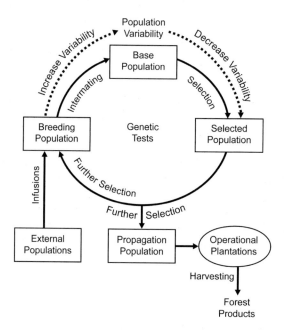

Fig. 11.1. The breeding cycle of forest tree improvement programs. Each of the core population types shown in the inner circle (base, selected and breeding populations) are formed once per cycle of improvement in the sequence shown, while the other population types may or may not be formed. (Reprinted from White, 1987, with permission of Springer Science and Business Media)

The core activities of the breeding cycle are selection and intermating (shown on the inner circle of Fig. 11.1). These are applied sequentially during each cycle of improvement; selections are made to form the selected population, then some or all of these selections form the breeding population and are intermated (cross bred) to induce recombination of alleles. Progeny derived from the intermatings are planted to form the next cycle's base population, and selection is again employed to choose superior individuals. If done effectively, the second cycle of selection results in additional genetic improvement, because superior first-generation parents are intermated and many of the second-cycle selections contain favorable alleles from these superior parents. Therefore, the breeding cycle is completed once per cycle of selection and intermating. The cycle is recurrent in the sense that a subsequent cycle begins after the previous has finished. Each cycle produces more genetic gain. The number of years between selection events in two successive cycles is called the **cycle or generation interval**. In the following sections, we describe the activities and population types that form one cycle of tree improvement.

While the terms cycle and generation are often used interchangeably, it is sometimes useful to distinguish between them. **Cycle** refers to the stage of a tree improvement program so that a program completing its third cycle has undergone three complete cycles of selection, intermating and genetic testing. Although **generation** is often used in the same sense as the term cycle, it also is used as a genetic term to refer to the generation of a specific selection. For example, in the first cycle of a tree improvement program, there are usually only first-generation selections. These are intermated and the selections made from among the offspring of first-generation parents are called second-generation selections. A program in the third cycle of improvement can involve first-, second- and third-generation selections if excellent selections from previous cycles are retained (*e.g.* through grafting) and reused for breeding in subsequent cycles. This gives rise to the concept of **overlapping generations** present in any given cycle of improvement and to rolling-front strategies in which all activities (such as selection and intermating) are conducted on a continuous basis annually, rather than in discrete phases (Borralho and Dutkowski, 1996; Hodge, 1997).

For convenience, we discuss the steps and activities of the breeding cycle as discrete phases in this chapter, and then address in *Chapter 17* how these steps can be packaged into effective breeding strategies. To illustrate the concepts described, the first cycles of two contrasting tree improvement programs are summarized in Boxes 11.2 and 11.3. The program for *Gmelina arborea* in Costa Rica (Box 11.2) is a low-intensity program for an exotic, tropical angiosperm planted on a small scale (Hamilton *et al.*, 1998). Conversely, the program for *Pinus taeda* in the southeastern USA (Box 11.3) is a high-intensity program for a native, temperate conifer planted on a large scale (Weir and Zobel, 1975; Li *et al.*, 1999). Through these and other examples included in the description of the breeding cycle, we hope both to reinforce the concepts and to convey the richness and diversity among programs of different species.

Base Population

The **base population** of a given cycle of improvement consists of all available trees that could be selected if desired (Zobel and Talbert, 1984). It is the population of trees that will be improved upon through selection and intermating, which is also called the **foundation population**. The base population is very large consisting of many thousands of genetically distinct individuals.

Box 11.2. First-cycle tree improvement program for *Gmelina arborea* in Costa Rica.

1. *Goal and scope* are to develop improved varieties of *G. arborea* for farmers and other small landholders on the peninsula of Nicoya in Costa Rica. This is a low-intensity program because farmers plant only 50 to 300 ha of plantations annually in this region, which is the breeding zone (Hamilton *et al.*, 1998).

2. *First-cycle base population* consisted of stands in the breeding zone of the local land race of *G. arborea*, which is an exotic in Costa Rica. *G. arborea* is native to south Asia (India, Vietnam and Myanmar, formerly Burma) and was first introduced into Costa Rica from exotic plantations in west Africa and Belize. The provenances in Asia, where seeds were collected to plant the exotic stands in Africa and Belize, were largely unknown.

3. *First-cycle selected population* contained 70 trees selected in the base population for rapid growth and stem straightness.

4. *First-cycle breeding and testing* consisted of 50 to 70 open-pollinated (OP) families (one family derived from the seed collected from each selection producing viable seed) planted in randomized, replicated tests on three site locations within the breeding zone. No control pollination was employed during the first cycle. The tests were measured at age 3 years (20% of rotation age of 15 years).

5. *First-cycle propagation population* was comprised of the three genetic tests that were converted into seedling seed orchards by thinning to remove inferior families and trees. Seed collected from these seed orchards formed the first-cycle improved varieties used for operational reforestation in the breeding zone (*i.e.* the peninsula of Nicoya).

6. *Advanced-generation tree improvement* has not begun due to the recent advent of the program, but the breeding strategy must be cost-effective to produce a positive economic return for this small-scale reforestation effort.

7. *Infusions* are likely in advanced-generation programs because 70 selections per breeding zone are not enough to produce substantial genetic gains from a recurrent breeding program. Possibilities include: (1) Sharing top first-cycle selections with other tree improvement programs within Costa Rica; (2) Obtaining proven selections from breeding programs of *G. arborea* in other countries; and (3) Returning to the native range in south Asia to make new selections in natural forests.

At the initiation of a program for a native species, the base population consists of all trees available for selection growing in natural stands and possibly plantations within the defined breeding zone (*e.g. P. taeda* in Box 11.3). The **breeding zone** is the set of environments for which an improved variety is being developed. Typically, this is a distinct geographical area, especially in the first generation of a tree improvement program. First-cycle base populations of exotic (*i.e.* non-native) species often include trees available for selection in: (1) The native range of the species; (2) Local land races occurring in the country of the program (*e.g. G. arborea* in Box 11.2) or other countries; and (3) Genetic test plantations in the country of the program or other countries.

Determining breeding zone boundaries is a critical decision in tree improvement programs, because each breeding zone has a separate improvement program with its own distinct base, selected, breeding and production populations (*Chapter 12*). Consider two examples of native species in the USA: *Pseudotsuga menziesii* and *Pinus elliottii*. The Northwest Tree Improvement Cooperative established 72 first-cycle breeding zones for *P. menziesii* in Oregon and Washington (Lipow *et al.*, 2003). These breeding zones were quite small, ranging in size from 12,000 to 50,000 ha of timberlands, with the first-cycle base population of each zone consisting of the trees growing in those timberlands. Therefore, there were 72 distinct improvement programs, each with its own breeding cycle, different set of population types, and improved variety being developed.

In contrast, the program for *P. elliottii* in the southeastern USA defined a single breeding zone consisting of the entire natural range of the species (approximately 4,000,000 ha of timberlands; White *et al.*, 1993). A single program with its base, selected and breeding populations is being conducted for this zone. The disparity in size and number of breeding zones between these two native species reflects both the more uniform climate in the southeastern USA and differences in breeding philosophy.

Advanced-generation base populations (*i.e.* after the first complete turn of the breeding cycle) consist of genetically improved trees formed by intermating members of the breeding population and planting their offspring in genetic test plantations. All trees in these genetic test plantations are available to be selected for advanced-generation selected populations.

Selected Population

Each cycle of the breeding cycle begins with selection of superior individuals from the base population. The selected individuals form the **selected population** which, for most programs, contains between 100 and 1000 selected trees for a single breeding zone (*e.g.* contrast the selected populations of the two programs in Boxes 11.2 and 11.3). In the first cycle of improvement, trees are selected from the base population solely by their superior phenotypic appearance (called **mass selection**, Fig. 13.1). However, advanced-generation base populations consist of pedigreed populations, and selection effectiveness is increased by using all information available about a candidate's progeny, relatives and ancestors. Although there are many different types of selection used at different stages and cycles of program development (*e.g.* mass, family, combined, indirect, and tandem; *Chapters 13-15*), all types have a single objective, which is to increase the frequency of favorable alleles at loci influencing the selected traits.

Genetic gain in a given trait is achieved only if the selected population has a higher frequency of favorable alleles than the base population. Genetic gain varies from trait to trait, and is greater if the selection is very intensive (only the very best individuals are selected) and the trait is under strong genetic control (*i.e.* has a high heritability). The optimum age to make selections, the appropriate traits to include in a program and the best methods of selection are all important topics in forest genetics research and in the development of a breeding strategy for any species (*Chapter 13*).

The selected population in each cycle always contains fewer individuals (say a few hundred) than the base population (many thousands). Allele frequencies differ between the two populations both by intent and by chance. The breeder selects superior individuals and therefore, intentionally alters allele frequencies for the few traits included in the selection criteria. In addition, allele frequencies for all traits (not just those included in the selection criteria) may change by random chance due to sampling (choosing a subset of trees from a larger base population). In fact, some very rare alleles present in the base population can be lost from the selected population (Kang, 1979a,b). However, when hundreds of individuals are included in the selected population, loss or large random changes in allele frequencies are expected to be extremely limited.

The superior trees chosen to form the selected population are always labeled with clear identification to distinguish them from other trees of the base population growing in natural stands, plantations or genetic tests. Often the selected trees scattered in several locations are vegetatively propagated through rooting or grafting and planted into a single location called a **clone bank.** This has the advantage of bringing all selections together in a well-protected location in which breeding can be conducted economically (since some or all of the trees in the selected population will form the breeding population).

Box 11.3. First-cycle tree improvement program for *Pinus taeda* in the southeastern USA.

1. *Goal and scope* are to develop improved varieties of *P. taeda* for industrial forest products companies and non-industrial landowners in nine states east of the Mississippi River in the southeastern USA. This high-intensity program, which began in the 1950s, is a cooperative effort by approximately 20 private companies and state agencies, all of which participate in the selection, breeding and testing activities under the guidance of cooperative staff from North Carolina State University (Weir and Zobel, 1975; Talbert, 1979; Li *et al.*, 2000). Combined, these organizations annually plant more than 350,000 ha of plantations (Li *et al.*, 1999).

2. *First-cycle base population* consisted of natural stands *P. taeda* divided into more than 30 local breeding zones. Each breeding zone was the local operating area of one or a few members of the cooperative. Many members were involved in distinct programs in multiple zones.

3. *First-cycle selected population* in each breeding zone contained varying numbers of selections made in the natural stands growing in that zone. The mass selection for superior phenotypes was very intensive searching for those trees with fast growth, straight stems, freedom from disease and appropriate branch, crown and wood characteristics. Each cooperator made 25 to 35 first-cycle selections in one or more breeding zones; this totaled 1050 selections for the entire cooperative across all breeding zones.

4. *First-cycle propagation population* was of two types in each breeding zone: (1) Original clonal seed orchards in the 1960s where each cooperator grafted their own 25 to 35 selections almost immediately after selection; and (2) 1.5-generation clonal seed orchards established in the 1970s with top tested clones (see step 5) that were exchanged among cooperating organizations. Seed collected from these seed orchards formed the first-cycle improved varieties used for operational reforestation.

5. *First-cycle breeding and testing* consisted of a control-pollinated (CP) incomplete factorial mating design called a tester system (*Chapter 14*), where each cooperator divided their say 30 selections into two groups with approximately five selections being used as pollen parents to mate with the other 25 selections used as female parents. This created a maximum of 125 CP families per cooperator that were established in randomized, replicated genetic tests on up to six sites (two sites planted in each of three years). The tests were measured at ages 4, 8 and some at 12 years (approximately 17%, 33% and 50% of rotation age of 24 years).

(Box 11.3 continued on next page)

Box 11.3. First-cycle tree improvement program for *Pinus taeda* in the southeastern USA.
(Continued from previous page)

6.   *Advanced-generation tree improvement* is now in its third cycle of breeding and enjoys a rich history of varying types of mating and testing designs that have evolved to take advantage of new technologies and to meet the changing needs of cooperators. Of particular note is the evolution of breeding zones which went from more than 30 first-cycle zones to eight testing zones in the second cycle to reflect knowledge about genotype x environment interaction gained from genetic testing. Now in the third cycle, the cooperative is using four main breeding zones with overlapping breeding populations where local breeding programs share selections in common with nearby programs.

7.   *Infusions* have been incorporated twice in the program: (1) More than 3000 selections from unimproved plantations were made to expand the first-cycle genetic base; and (2) Excellent selections known to perform consistently well over a wide geographic range have been used in multiple breeding zones.

Breeding Population

Some or all of the individuals in the selected population are included in that cycle's **breeding population** and are intermated (cross bred) to regenerate genetic variability through recombination of alleles during sexual reproduction. Many different mating designs are used to intermate the members of the breeding population (*Chapter 14*), and offspring from these intermatings are planted in genetic tests that form the next cycle's base population (Fig. 11.1 and examples in Boxes 11.2 and 11.3). This completes one cycle of the core activities of the breeding cycle, and the next cycle begins with new selections being made from these genetic tests (*i.e.* from the new base population).

When two superior parents are mated, not all their offspring are superior. There is as much variation among individuals of a given family of trees as there is among brothers and sisters in human families. The randomness of recombination during sexual reproduction (*Chapter 3*) ensures that some offspring in a family receive more favorable alleles from their parents than others. Therefore, intermating results in a large amount of genetic variation both among and within the families planted in the genetic tests that form the new base population. The breeder capitalizes on this variation by selecting superior individuals from superior families to form the next cycle's selected population.

If all members of the selected population are used as parents in the intermating schemes to create the progeny for the next cycle's base population, then the selected and breeding populations are identical. In this situation, intermating can begin soon after selection. However, sometimes further selection (called **two-stage selection**) allows the genetic quality of the breeding population to be increased above that of the selected population through exclusion of inferior individuals. Therefore, in the first stage the breeder might form a selected population of 1000 trees based on growth rate, and then in a second stage retain only the top 300 of those 1000 trees in the breeding population based on additional traits (*e.g.* wood quality, disease resistance and cold hardiness).

Both control-pollinated mating designs that form full-sib families (Box 11.3) and open-pollinated mating designs that control only the female half of the pedigree (Box 11.2) are used in different programs to effect the intermating of the trees in the breeding population; in fact, in some programs both types of matings are used. This allows tremendous flexibility in the development of a breeding strategy (*Chapters 14* and *17*). The important concept here is that intermating is a critical activity of any improvement program to create new genetic combinations in the progeny that form the next base population. Selection of superior trees from among these newly created progeny is the basis for making continuous genetic progress from recurrent cycles of selection and intermating.

Propagation Population

In each cycle of improvement, the **propagation population** (sometimes called the **production population** or **deployment population**) consists of some or all of the members of the selected population. Its function is to produce a sufficient quantity of genetically improved plants to meet the annual needs of the operational forestation program (Fig. 11.1). Collectively, these plants are sometimes called a **genetically improved variety**, and the activity of mass propagation and planting of an improved variety is called **deployment**. The primary realized benefit from most tree improvement programs is increased yield and health of plantations established with improved varieties.

The value of separating the propagation population from the main core of the breeding cycle (*i.e.* base, selected and breeding populations) is that the core activities focus on maintaining a broad genetic base and achieving genetic gains over many generations of improvement. On the other hand, the purpose of the propagation population is deployment of a variety that maximizes genetic gain in operational plantations of a given cycle of improvement.

Seed orchards, the most common type of propagation population (Boxes 11.2 and 11.3), are often formed by grafting the very best members of the selected population onto rootstock at a single location that is managed intensively for the production of seed (Box 16.1). The genetically improved seed formed by open pollination among the grafted trees is then grown by the nursery manager and the plants used for forestation. There are many other types of propagation systems and the best choice depends on many factors (*Chapter 16*).

It is common to include only the very best selections in the propagation population. For example, the selected population might contain several hundred genotypes with the propagation population consisting of the best 20 to 50 genotypes. Although this increases the genetic gain expected from the operational variety that is planted, it also reduces its genetic diversity. It also is common to continually upgrade the genetic quality of the propagation population even during a single cycle of improvement. As information becomes available from genetic tests, genetically inferior selections can be removed from the propagation population, while superior selections not originally included can be added. This means that the improved variety being planted may change in its genetic composition and its expected gain within a single cycle of improvement.

Most tree improvement programs limit selection to only a few (2-6) key traits because of the difficulty of making gains for many traits simultaneously. Other traits that are uncorrelated with the traits of interest are affected very little. This, coupled with the fact that most programs are only one or two generations removed from undomesticated natural populations, means that tree improvement programs have not dramatically altered the genetic structure of the species. In the future, after several cycles of breeding and with applications of new biotechnologies such as marker-assisted selection and genetic engineering (*Chapters 19* and *20*), the genetic structure of selected and propagation populations may begin to diverge substantially from the genetic structure of the species in nature. However, today's improved varieties of forest trees still bear strong resemblance to the undomesticated species, and this is quite unlike many agronomic crops.

Infusions from External Populations

Some years after the initial selections are made, many tree improvement programs infuse new selections into the breeding population that were not part of the original selected or breeding populations. These **infusions** can be aimed at improving a specific trait or at generally broadening the genetic diversity existing in the program. Some examples include: (1) Making new selections in native stands and/or plantations to broaden the genetic base, *e.g.* the *P. taeda* program in the southeastern USA made 3300 new selections (Box 11.3); (2) Making new selections to increase the frequency of favorable alleles for a specific trait, *e.g.* the *P. elliottii* program in the southeastern USA selected nearly 500 disease-free trees in stands highly infected with fusiform rust (*Cronartium quercuum* f. sp. *fusiforme*) to increase the frequency of resistance alleles in the breeding population (White *et al.*, 1993); (3) Obtaining proven selections from other breeding zones of the same species, *e.g.* Weyerhaeuser Company uses *P. taeda* selections from North Carolina in the breeding population developing improved varieties for their Arkansas timberlands some 2000 km miles away (Lambeth *et al.*, 1984); (4) Obtaining proven selections from other organizations working with the same species, *e.g.* breeding programs for *Eucalyptus grandis* in many different countries sometimes exchange material; and (5) Making intra- or inter-specific hybrids to increase genetic diversity or improve a specific trait, *e.g.* the improvement program of Mondi Forests in South Africa hybridizes *E. grandis* with other species of eucalypts to develop breeding populations for different types of lands (colder, more subtropical, etc.) for which *E. grandis* is not well adapted as a pure species.

In all cases of infusions, the new material should be evaluated for all traits through genetic tests to ensure that gains are not made in some traits with inadvertent losses in others. New infusions may be intermated in the breeding population for one or more generations prior to inclusion in the propagation population that produces the operational variety. This allows maintaining a broad genetic diversity in the breeding population without sacrificing gains in the operational plantations.

Genetic Testing

Genetic tests are central to all tree improvement programs and are established with pedigreed, well-labeled offspring or clonal plantlets produced by any of the population types in the breeding cycle. The tests are usually planted in field locations on forest sites (Fig. 14.11), but also may occur in nurseries, greenhouses and growth rooms. Usually a single series of tests has several objectives, and depending on the major objective may be called a progeny test, base population, yield trial or research experiment.

The concept of a common garden test was introduced in *Chapters 1* and *6*; many selections or their progeny are tested against each other in randomized, replicated experiments planted in one to several locations. The idea of common gardens is to grow genotypes in replicated environments so that the genetic effects on phenotypes can be isolated from confounding environmental effects.

Tree improvement programs rely on genetic testing to: (1) Evaluate relative genetic quality of selections made in any cycle of selection; (2) Estimate genetic parameters such as heritabilities, genetic correlations and genotype x environment interactions for key traits; (3) Provide a base population of new genotypes from which to make the next cycle of selection; and (4) Quantify or demonstrate genetic gains made by the program. These objectives along with appropriate mating and field designs are discussed in *Chapter 14*.

GENETIC GAINS AND ECONOMIC VALUE OF TREE IMPROVEMENT PROGRAMS

Genetic Gain Concepts and Types of Gains Estimates

It is important to quantify past and/or potential genetic gains at various stages of tree improvement programs to: (1) Choose between alternative mating and experimental designs for genetic tests; (2) Evaluate different breeding strategies; (3) Decide which traits to emphasize; (4) Justify program effectiveness (*e.g.* by comparing gains and economic returns from improved and unimproved plantations); and (5) Develop harvest schedules and estimate wood flows from a plantation estate consisting of many improved varieties of varying levels of improvement.

In all tree improvement programs, the goal is to increase frequencies of favorable alleles influencing selected traits. However, for polygenic traits it is impossible to measure the change in allele frequencies, and in fact, the frequency change from one cycle of selection is small at each of the many loci influencing any given trait. For this reason, the effectiveness of tree improvement programs, called **genetic gain** or **genetic progress**, is measured by the change in the population means for each trait, quantified as the increase in the average genetic value (or breeding value) between the two populations.

In *Chapter 6*, two different types of genetic values were defined: total genotypic value, G, which is also called clonal value; and breeding value, A, which is the portion of the clonal value passed on to offspring produced by sexual reproduction. Unfortunately, the symbol ΔG is used to represent genetic gains made in both:

$$\Delta G = \overline{G}_2 - \overline{G}_1 \hspace{5cm} \text{Equation 11.1}$$

or

$$\Delta G = \overline{A}_2 - \overline{A}_1 \hspace{5cm} \text{Equation 11.2}$$

where Equation 11.1 expresses genetic gain for any trait in terms of mean genotypic values of two groups or populations of trees and Equation 11.2 expresses genetic gain as the difference in mean breeding values between the two groups of trees. Equation 11.1 is more appropriate for comparing selected and propagation populations for programs that plant tested clones, while Equation 11.2 is appropriate for most other types of genetic gains comparisons (*i.e.* where population types are sexually propagated) (*Chapters 13* and *16*). The two groups or populations being compared can be: (1) Two populations in the breeding cycle during the same cycle of improvement (*e.g.* comparing the base and selected population in the same cycle measures gain from selection); (2) The same population in two successive cycles of improvement (*e.g.* comparing the first- and second-cycle propagation populations measures the gain in the commercial variety from an additional cycle of improvement); and (3) Any two groups of trees in which there is interest in comparing the difference in their mean genetic values.

Genetic gain estimates are commonly developed for different stages of tree improvement programs and for different population types. In any given cycle of improvement, genetic progress is expected to be lowest for the base population, higher for the selected population and highest for the propagation population (Fig. 11.2). The base population with a larger number of trees and more genetic diversity necessarily has a lower average genetic value than the smaller selected population formed by choosing superior individuals from that base population. The diversity is even smaller and genetic gain even greater for the propagation population, which contains the few very best individuals used to produce plants for forestation.

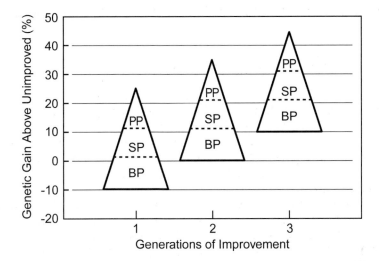

Fig. 11.2. Schematic diagram showing genetic gains for three population types in each of three different cycles of improvement. The entire triangle represents the base population (BP) for each generation. The next smallest triangle represents the portion of BP that is the selected population (SP), while the upper triangle is the propagation population (PP). Genetic gains are expressed above a starting point of 0% for the unimproved species (*i.e.* BP in generation 1). The area of each population type inside the triangle is proportional to its size (number of individuals) and diversity, while its height on the y axis expresses the mean genetic value for that population type.

This inverse relationship (more gain associated with less genetic diversity) is the primary reason for using different population types with different functions. Although the larger selected population has less genetic gain in any cycle of improvement, it retains sufficient genetic diversity to ensure excellent genetic gains for many cycles of improvement. Conversely, the propagation population aims to maximize gains in the operational plantations during the current cycle; however, it is re-formed with new genotypes every cycle so does not require as much genetic diversity. Note that the propagation population is schematically located outside of the breeding cycle proper in Fig. 11.1, which indicates that it is a "dead-end" population whose genotypes do not directly contribute to genetic gains in future generations.

Genetic gains are continuous and cumulative in the sense that each cycle builds upon the gains made in the previous cycle (Fig. 11.2). When a tree improvement program enters its third cycle, the gains expected for all population types benefit from the previous two cycles of selection, intermating and testing. Still in that third cycle, there is a hierarchy of population types that differ in their mean genetic value and genetic diversity. Therefore, there could be many different types of genetic gain estimates relevant to any tree improvement program (comparisons of different population types in each of several cycles); this is further complicated because gains can be estimated separately for each trait being improved. Another common problem is when gain estimates come from young genetic tests, while true interest is in gain achieved at rotation age. Assumptions must be made to translate gains estimated from the younger age to an older age, such as the age-age genetic correlation discussed in *Chapter 6*.

There are two methods of estimating genetic gains: (1) **Predicted gains** calculated from formulae derived from quantitative genetics theory (*Chapters 13* and *16*); and (2) **Realized gains** obtained from experiments that plant improved varieties next to unimproved varieties (or with varieties having varying levels of improvement) in randomized, replicated yield trials on several sites. Predicted gains (type 1) are always forward-looking expectations of gain that could be achieved from use of a new variety or from a type of selection; there are thousands of predicted gains in the literature. Realized gains (type 2) are retrospective, because results are obtained several years after planting the experiment. By then, the tested varieties are often no longer in use having been replaced by newer, even better varieties. Realized gains occur less frequently in the literature, because of the time and cost required to obtain reliable results; however, these types of experiments are critical, because they provide direct empirical evidence of the gains made from a tree improvement program (Eldridge, 1982; Lowerts, 1986, 1987; Shelbourne *et al.*, 1989; La-Farge, 1993; Carson *et al.*, 1999; Cotterill, 2001).

When either predicted or realized genetic gains are estimated for young ages, the estimates are sometimes incorporated into existing growth and yield computer models to estimate rotation-age yields and the ultimate impacts of genetic gains in terms of production of forest products. **Simulated gains** are a third type of genetic gains estimate, obtained by incorporating varying assumptions about improved and unimproved varieties into the computer model. The gains are estimated by comparing the simulated yields from computer runs for different levels of improvement as specified by these assumptions (Talbert *et al.*, 1985; Buford and Burkhart, 1987; Hodge *et al.*, 1989; Carson *et al.*, 1999).

These various types of gains estimates generated for different population types and for different cycles of improvement make it a challenge to compare the relative gains that may be expected from distinct varieties produced even by a single tree improvement program. Consider the plight of a non-industrial landowner wanting to purchase improved plants for reforestation and having many options available. An effective labeling method that clearly specifies the relative level of improvement expected in different traits for each available variety would be the most useful method of comparison. Although such labeling systems are rare, one has been developed for *Pinus radiata* in New Zealand in which GF (growth and form) ratings are assigned to each available variety (Shelbourne *et al.*, 1989; Carson *et al.*, 1999); GF 1 is an unimproved variety and higher GF numbers indicate more improvement (Fig. 11.3).

Genetic Gains Achieved for Different Traits

While early tree-breeding pioneers in the 1950s relied on their intuition and knowledge of the success of breeding programs with agronomic crops and farm animals to justify their efforts, we now have solid evidence that tree improvement programs of many species have produced improved varieties that deliver substantial genetic gains. In this section, we briefly highlight some gains made for different traits in the first generation of various tree improvement programs.

Most tree improvement programs expend considerable effort improving growth rate because of its high economic value. Therefore, even with low to moderate heritabilities ($h^2 = 0.10 - 0.30$; *Chapter 6*), appreciable gains have been made in the first generation due to the intensity of the efforts; reported gains in growth rate of 5-25% above the unimproved native species or local land race are common for both conifer and angiosperm species (Pederick and Griffin, 1977; Eldridge, 1982; Talbert *et al.*, 1985; Nikles, 1986; Shelbourne *et al.*, 1989; Rehfeldt *et al.*, 1991; McKeand and Svensson, 1997; Carson *et al.*, 1999). Even higher gains have been reported for first-generation programs planting tested clones (Franklin, 1989; Lambeth and Lopez, 1994).

Appreciable gains have also been made for stem straightness (Goddard and Stickland, 1964; Campbell, 1965; Ehrenberg, 1970; Talbert *et al.*, 1985; LaFarge, 1993; Nikles,

Fig. 11.3. Eight-year-old *Pinus radiata* planted in a realized gain experiment in the Kaingoroa Forest of New Zealand showing two different varieties developed by the Forest Research Institute (FRI). In this well-replicated experiment at age 17 years: (a) GF2, an unimproved variety, averaged 30.7 m in height, 50% merchantable stems and a mean annual increment (MAI) of 19.9 $m^3ha^{-1}yr^{-1}$; and (b) GF22, an improved variety, averaged 32.1 m in height, 80% merchantable stems and 24.9 MAI. (Photos courtesy of J. Barron, New Zealand Forest Research Institute)

1986) and branch quality traits (Shelbourne *et al.*, 1989). Zobel and Talbert (1984) assert that strong selection intensity for straight trees in the first generation can often produce excellent results that may allow selection for this trait to be relaxed in subsequent cycles of improvement. The high heritability associated with some wood quality traits such as wood density implies that genetic gains are certainly possible for these traits and gains have been reported (Jett and Talbert, 1982; Shelbourne *et al.*, 1986).

Genetic gains have also been reported for resistance to various fungal diseases in a range of tree species, and selection of disease-free trees in highly infected stands can be a particularly effective first step in the development of resistant varieties (Bjorkman, 1964; Zobel and Talbert, 1984; Shelbourne *et al.*, 1986, p 71; Hodge *et al.*, 1990; Wu and Ying, 1997). However, it should not be construed that resistant varieties provide complete immunity to a disease or that genetic variation in resistance in host tree species will be easily found for all fungal diseases that might be encountered. Dutch elm disease (*Ophiostoma ulmi*) and chestnut blight (*Cryphonectria parasitica*) are only two examples of fungal diseases that have injured or killed literally millions of trees in the USA with little evidence of resistance being detected. Perhaps it is not coincidental that these pathogens were new encounters for the host tree species in the sense that the pathogen and trees had not co-evolved over many tree generations (*Chapter 9*). Therefore, new pathogens on exotic tree species and exotic pathogens on trees in their native range (two different types of new encounters) may be particularly troublesome (*Chapter 10*).

There are literally thousands of other examples of genetic gain made for different types of traits, and some examples include rooting ability of stem segments, graft compatibility, drought resistance, cold hardiness, monoterpene content and seed production. In concept, nearly all traits exhibit genetic variation (even if quite low), and can be improved if enough effort is expended in selection, breeding and testing.

It is often desirable to express genetic gains in a form that integrates across all selection traits and is easily translated into value or total product gain. While not always possible, a particularly good example is the Celbi program for *Eucalyptus globulus* in Portugal (Cotterill, 2001). The first cycle of tree improvement produced a genetic improvement of 60% above unimproved varieties in bleached Kraft pulp production per hectare per year. This genetic gain value includes the gains made from selection, testing and breeding of three distinct traits: tree growth rate, wood specific gravity and pulp yield.

Economic Analysis of Tree Improvement Programs

Soon after large-scale tree improvement programs began in the 1950s, corporate managers began to ask if the considerable expenditures required represented a wise business decision. Several early studies outlined appropriate economic methods of discounted cash flow analysis that should be used to properly evaluate tree improvement relative to other possible investments (Davis, 1967; Carlisle and Teich, 1978). Using these methods, analyses of many specific tree improvement programs demonstrated favorable returns on investments under reasonable genetic and economic assumptions (Danbury, 1971; Dutrow and Row, 1976; Porterfield and Ledig, 1977; Reilly and Nikles, 1977; Ledig and Porterfield, 1982; Fins and Moore, 1984; Talbert *et al.*, 1985; Thomson *et al.*, 1989; McKenney *et al.*, 1989; Hamilton *et al.*, 1998). As a result of these findings, tree improvement has become an accepted silvicultural investment in most large plantation programs.

Given that the economic value of tree improvement had been generally established, recent analyses have focused not on justification of the entire program, but rather on use of economic tools to evaluate specific decisions such as whether it is better to use clones or seedlings, or when a seed orchard should be replaced (McKeand and Bridgwater, 1986; Thomson *et al.*, 1987; McKenney *et al.*, 1989; Williams and de Steiguer, 1990; Balocchi, 1996). A wealth of economic analyses has resulted in several conclusions that apply generally to programs across a wide range of species and can help guide program development and implementation (Box 11.4).

Box 11.4. General conclusions from economic analyses of tree improvement programs.

1. A low cost of capital (sometimes called the discount rate or compounding interest rate) always increases the economic favorability of a tree improvement program relative to shorter-term investments. Tree improvement is a long-term investment with economic benefits not realized until many years after program initiation. A high cost of capital means a high time value of money and implies that investors are less willing to wait for longer-term benefits.

2. Positive real rates of inflation of prices of wood products relative to other goods and services increases the economic value of tree improvement. In this scenario, genetic gains in yield or product quality are further magnified by future prices for wood products that are higher than today's prices even after adjustment for inflation.

3. Plantation programs that have higher reforestation rates (more hectares planted annually) benefit more from tree improvement. Although the costs of other silvicultural treatments like site preparation and fertilization are incurred on every hectare treated, most costs of tree improvement are associated with selection, breeding and testing, which do not directly depend on the number of hectares planted. Therefore, as the reforestation load increases, there is little increase in the cost of the program, yet a huge increase in the benefits expected. In other words, larger plantation programs benefit from spreading the cost over more planted hectares.

4. Increasing genetic gains or decreasing program costs directly increase the economic return and desirability of a tree improvement program. This is one reason why cooperative efforts, where many organizations pool their resources, have been so effective and stable through time.

5. Tree improvement is a more favorable investment on more productive timberlands. This is because the benefits are generally higher (more yield increase is expected) and are realized sooner (due to shorter rotation ages) than on less productive land.

6. Gains from tree improvement are cumulative across generations, and thus appear more economically desirable in analyses that consider several plantation rotations. Unlike most silvicultural treatments, which must be repeated with each rotation, the benefits from previous cycles of tree improvement accumulate at no additional cost in subsequent generations. Therefore, benefits compound in multiple cycles of improvement, and multiple rotations of plantations must be considered to properly realize this characteristic.

7. During a given cycle of improvement, any option that increases expected gain at a relatively low cost can have a huge positive economic impact. For example, once the costs of genetic testing have been incurred (and are considered sunk costs), the additional costs of roguing a seed orchard or establishing a 1.5 generation seed orchard (*Chapter 16*) nearly always produce an excellent economic return.

SUMMARY AND CONCLUSIONS

Large-scale tree improvement programs began in earnest in the 1950s, and today, they exist for all tree species that are planted in any quantity. In addition, there is considerable interest in tree improvement for agroforestry systems, for public-sponsored community planting programs and for other non-industrial uses. The following ingredients are critical to a successful tree improvement program: (1) Clear program and product objectives; (2) Sound knowledge of the species biology, silviculture and genetics; (3) Sound breeding strategies implemented by well-trained personnel who are supported by a stable budget; (4) Efficient mass propagation, nursery and plantation management systems to optimize yield and product quality; (5) Maintenance of a broad genetic base; and (6) Presence of a supportive research program.

The activities of tree improvement programs are conveniently summarized using a conceptual model called the breeding cycle, which depicts the potential set of activities and population types that can occur during a given cycle of improvement (Fig. 11.1). The core activities of the breeding cycle are selection and intermating; selections are made from the base population to form the selected population, then some or all of these selections (*i.e.* the breeding population) are intermated to induce recombination of alleles. Progeny formed from the intermatings are planted to form the next cycle's base population, and selection is again employed to choose superior individuals. The breeding cycle is completed once per generation or cycle of selection and intermating. The number of years between selection events in two successive cycles is called the generation interval.

The propagation population for each cycle of improvement generally consists of the very best members of the selected population whose function is to produce a sufficient quantity of genetically improved plants to meet the annual needs of the operational forestation program. These trees are collectively called a genetically improved variety, and the activity of mass propagation and planting of an improved variety is called deployment. The primary realized benefit from most tree improvement programs is the increased yield, health and product quality of the plantations established with an improved variety. The value of separating the propagation population from the main core of the breeding cycle (base, selected and breeding populations) is that the core activities focus on achieving genetic gains and maintaining a broad genetic base over many cycles of improvement, while the propagation population focuses on deployment of a variety that maximizes genetic gain in a given cycle.

Genetic gain estimates are commonly developed for different stages of tree improvement programs and for different population types. In any given cycle, genetic progress is expected to be lowest for the base population, higher for the selected population and highest for the propagation population (Fig. 11.2). The base population, with the largest number of trees and most genetic diversity, necessarily has a lower average mean genetic value than the smaller selected population formed by choosing superior individuals from that base population. Diversity is smallest and genetic gain highest for the propagation population, which includes the few very best individuals of that cycle and is used to produce plants for forestation. Genetic gains are continuous and cumulative in the sense that each generation builds upon the gains made the previous generation (Fig. 11.2).

Economic analyses utilizing discounted cash flow methods have demonstrated tree improvement to be a wise business decision with favorable economic returns for a large number of tree species under a wide variety of economic scenarios. Important factors contributing to a large positive return on investment are: (1) Favorable economic assumptions (low cost of capital and positive real rate of increase of forest products prices); (2) Large land base over which to spread the costs of the program; (3) Productive land that produces good yield in a short rotation; and (4) Innovations that increase expected genetic gains or reduce program costs.

CHAPTER 12
BASE POPULATIONS – *SPECIES, HYBRIDS, SEED SOURCES AND BREEDING ZONES*

Patterns of natural genetic differentiation among species and provenances within species were covered in *Chapters 9* and *8*, respectively. In this chapter, similar concepts are addressed with regards to two important taxa-level genetic decisions made in plantation programs: (1) Choosing appropriate species and seed sources to use for operational reforestation; and (2) Defining the founding sources upon which to base tree improvement efforts. At the beginning of any plantation program, defining the base population means determining which taxa to plant operationally. Subsequently, when tree improvement efforts begin, defining the base population means deciding which species and seed sources to utilize for selection. Once defined, trees in the base population are available for inclusion in the program, while trees in other sources are excluded from consideration (*Chapter 11*).

Choosing appropriate species and sources is the single most important genetic decision in a plantation program. Zobel and Talbert (1984, p 76) state very appropriately that, "The largest, cheapest and fastest gains in most forest tree improvement programs can be made by assuring the use of the proper species and seed sources within the species." The wrong choice of species or seed sources for reforestation can lead to loss in productivity or, at worst, complete plantation failure. Similarly, the use of less-than-optimal taxa in the base population of a tree improvement program can result in development of operational varieties that are poorly adapted to the edaphoclimatic conditions of the planting zone, do not grow well and/or do not produce the desired products. The overall goal both for operational reforestation and for tree improvement programs is to choose species and seed sources that are well adapted to the planting conditions and produce maximum quantities of the desired products in the least amount of time.

This chapter begins with a brief description of the many types of taxa that are available (species, hybrids and seed sources), focusing on their attributes for plantation programs. Second, we discuss how to choose species, hybrids and seed sources for reforestation including the process of field-testing a range of candidates. *Chapter 8* presents a related discussion of choosing provenances for reforestation within a species' native range, while this chapter focuses on exotic planting locations. There is considerable overlap in the concepts, and the reader may wish to refer to both chapters. Finally, this chapter addresses the definition of base populations for first-generation tree improvement programs for a range of situations and species. Advanced-generation base populations are discussed in *Chapter 17*.

TYPES OF TAXA AND THEIR ATTRIBUTES FOR PLANTATIONS

Species and Interspecific Hybrids

For some plantation programs, choice of species and even seed source may be predetermined or straightforward, for example, when an exotic species has been growing successfully in a region for several decades. In other programs, the forester may be given the task of planting several thousand hectares in a short period of time with little knowledge of which species is best for the situation. In a review of operational plantation programs in the tropics, Pancel (1993) found that all 42 projects started their forestation activities prior to having any field data upon which to base the choice of species planted, and 60% received information during the span of the project that resulted in changing to different species. Further, a review of the CABI literature database from 1987-1997 revealed nearly 800 articles dealing with field trials designed to determine appropriate species for various forestation projects. These articles spanned the entire world and a wide range of species, climates and product objectives. This is a very rich field of research that is much too complex to be given full treatment here. Rather, we briefly highlight different types of species and hybrid taxa and refer the reader to other more detailed treatments (Zobel and Talbert, 1984; Savill and Evans, 1986; Zobel *et al.*, 1987; Evans, 1992a; Hattemer and Melchior, 1993).

Native Versus Exotic Species

When species choice is not predetermined by biological, social or political reasons, one important consideration is whether to use native or **exotic** species (where an exotic species is any species planted outside of its natural range). Native species are the predominant choice for plantations in Canada and the USA, while both native and exotic species are grown throughout Europe. In the southern hemisphere, the majority of plantations are established with exotic species. There are many considerations contributing to these differences among countries and regions.

If available, the use of native species offers some advantages (Zobel *et al.*, 1987; Evans, 1992a): (1) If a local provenance is used, then adaptation to the climate and soils of the planting zone is ensured (*Chapter 8*); (2) Acceptable growth can be expected from adapted provenances, although growth may or may not be outstanding; and (3) Environmental concerns are minimized and social acceptance is maximized. There are many examples in the world of large, successful plantation programs based on native species: *Pseudotsuga menziesii* in the western USA and Canada; *Pinus taeda* and *P. elliottii* in the southern USA; *P. sylvestris* and *Picea abies* in Europe; and *Tectona grandis* in India. When available, first consideration should go to a native species unless field data from long-term field trials indicate that an exotic species is more desirable.

Often, especially in the southern hemisphere, exotic species are preferred because: (1) The native flora does not contain a species whose biology and products are sufficiently well-known to permit its immediate use in plantation culture; (2) There are many well-known exotic species available for a variety of climates, soils and product goals and often there is one that will grow faster and produce higher quality products than the native species; (3) Seed is readily available from many of these plantation exotic species; (4) Wood quality may be superior to native species for the desired products; and/or (5) Exotic species may not be susceptible to pest damage, at least for some period of time (Zobel *et al.*, 1987; Evans, 1992a; Kanowski *et al.*, 1992; Pancel, 1993; Boyle *et al.*, 1997).

Since there are many excellent examples of successful exotic plantations, it would be impossible to list them all. In the northern hemisphere, *Picea sitchensis*, *Pinus contorta*, *Pseudotsuga menziesii* and *Eucalyptus globulus* are all planted on a large scale in different parts of Europe (Savill and Evans, 1986; Morgenstern, 1996; Tibbits *et al.*, 1997), and *P. taeda* and *P. elliottii* are planted as exotics in China (Bridgwater *et al.*, 1997). In the southern hemisphere, species of *Pinus* (native only to the northern hemisphere) are widely used because they grow well and have long fibers desired for certain paper products, while several species of *Eucalyptus* (mainly native to Australia) with rapid growth rates are used for production of both industrial wood and fuelwood. As further examples, Box 12.1 summarizes the plantation programs for the two most widely planted exotic tree species in the world: an angiosperm, *Eucalyptus grandis* and a conifer, *Pinus radiata*.

As we have seen, many important native and exotic species produce excellent plantations and high quality products. The decision between species is sometimes easy and sometimes not. When the choice is not clear, the process outlined later in this chapter for choosing and field-testing candidate species is required to make the best decision.

Interspecific Hybrids

By far the largest fraction (well over 95%) of the approximately 150 million hectares of forest plantations in the world is planted with single, pure species. However, interspecific hybrids have long been known in nature (*Chapter 9*) and are becoming increasingly more important as commercial taxa in plantation programs (Zobel and Talbert, 1984; Namkoong and Kang, 1990; Nikles, 1992, 2000; Khurana and Khosla, 1998; de Assis, 2000; Retief and Clarke, 2000; Potts and Dungey, 2001). While interspecific hybrids initially must be formed by controlled pollination of trees in two different species, it is the ease of vegetative propagation (such as through rooted cuttings) of excellent hybrid clones that has contributed to the operational importance of hybrids in plantation programs of many genera including *Populus* and *Eucalyptus* (Box 12.2). While **F₁ hybrids** (formed by crossing trees of two different pure species) are the most common type of hybrids commercially utilized today, there is increasing interest in other types of hybrids: **F₂ hybrids** (formed by crossing F₁ trees together), **backcrosses** of F₁ hybrids to one or both pure species, **three-way** and **four-way** hybrids involving three or four pure species and other hybrid types (Nikles, 1992; de Assis, 2000; Verryn, 2000).

Two terms, **hybrid vigor** and **heterosis,** are used interchangeably to describe hybrid superiority over parental species. To make it even more confusing, either term may be used to define hybrid performance relative to average of the parents (*i.e.* the mid-parent value) or to the performance of the better parent. For example, two traits, bole volume and straightness, were measured on hybrids between *Pinus caribaea* var. *hondurensis* (PCH) and *Pinus elliottii* var. *elliottii* (PEE; Fig. 12.1). Focusing only on the performance of the F₁ hybrid for illustration, the hybrid showed better-parent hybrid vigor (or alternatively, better-parent heterosis) for volume since the volume of the hybrid was 31 dm³, while that of the better pure species (PCH) was 28 dm³. For the other trait, straightness, the F₁ exhibited mid-parent hybrid vigor (but not better-parent hybrid vigor) since the straightness of the hybrid (5.3 on a 1-8 scale) was statistically above the mid-parent value (5.1 = (5.6 + 4.6)/2), and closer to the better, straighter parent (PEE). In some cases, hybrids have negative hybrid vigor for one or more traits meaning that the hybrid is closer to or worse than the poorer-performing parental species (*e.g.* the F₂ hybrid for straightness in Fig. 12.1).

The underlying genetic and physiological mechanisms contributing to hybrid vigor are still poorly understood (Falconer and Mackay, 1996; Li *et al.*, 1998; Cooper and Merrill, 2000; Kinghorn, 2000) (*Chapter 5*). Further, the existence and degree of hybrid vigor sometime depend on edaphoclimatic conditions of the planting environment (Potts and Dungey, 2001). In this case, taxa x environment interaction is important with one or both pure species performing best in some environments and the hybrid being superior in other conditions. We still have much to learn about mechanisms conferring hybrid superiority, but new techniques of molecular genetics may help in this regard (*Chapter 18*).

Box 12.1. Plantation programs for two widely planted exotic tree species: *Eucalyptus grandis* and *Pinus radiata*.

1. Native to the eastern coast of Australia, *Eucalyptus grandis* is the most widely used exotic species of forest trees with nearly 11 million hectares of plantations in several countries in South America, Africa, China and elsewhere (Eldridge *et al.*, 1994; Wright, 1997) (Fig. 1a). Brazil and South Africa have the largest plantation estates of *E. grandis* where the plantations form the basis of vibrant industries. Growth rates are very fast with annual height growth of 3-7m and mean annual increments (MAI) for volume ranging from 30-70 $m^3ha^{-1}yr^{-1}$, depending on site quality. Most plantations are grown on relatively short rotations of 6-10 years, and the wood is mostly used for pulp (although some programs are aiming for higher-valued solid or reconstituted wood products). *E grandis* is somewhat specific in its site requirements and is susceptible to some fungal diseases; thus, interspecific hybrids with other species of *Eucalyptus* (*e.g. E. urophylla, E. nitens* and *E. camaldulensis*) are sometimes used to increase ecological amplitude and disease resistance.

Fig. 1. Plantations of the two most widely exotic tree species in the southern hemisphere: (a) Three-and-a-half-year-old stand of *Eucalyptus grandis* being grown on a 6-year pulpwood rotation by Smurfit Cartón de Colombia; and (b) Eleven-year-old stand of *Pinus radiata* after multiple thinnings and prunings being grown on a 25-year rotation for high-value solid wood products by Forestal Mininco in Chile. (Photos by T. White)

(Box 2.1 continued on next page)

Box 12.1. Plantation programs for two widely planted exotic tree species: *Eucalyptus grandis* **and** *Pinus radiata.* (Continued from previous page)

2. *Pinus radiata* has an extremely small native range of less than 7000 ha in coastal California, USA, and two islands off the coast of Mexico. Many trees in the native range grow relatively slowly and have a wind-swept form; thus, no one ever would have predicted its wide success as an exotic. Yet, plantations of *P. radiata* now occupy nearly 4 million hectares (principally in Chile, New Zealand and Australia) making it the most widely planted exotic conifer (Balocchi, 1997) (Fig. 1b). It has proved to be an extremely flexible species that performs well over a wide range of climatic (*e.g.* 500-2500 mm of annual precipitation in Australia) and soil conditions producing high quality solid wood and paper products (Balocchi, 1997). Annual growth rates range from 1-3m in height and 15-30 $m^3ha^{-1}yr^{-1}$ in MAI. When grown for pulpwood, plantations may be planted at dense initial spacings and managed on short rotations (less than 20 years) without thinning or pruning. For higher-valued solid wood products, plantations may be established at wider spacings and pruned and thinned multiple times during a rotation of 25-30 years (Fig. 1b).

(Fig. 1 continued from previous page)

Even without knowing the mechanisms, hybrids have been important for increasing product yield or quality in various crops (Cooper and Merrill, 2000), animals (Kinghorn, 2000) and trees (Box 12.2). In forestry, interspecific hybrids may be preferred over use of a single species when, compared to the pure species, the hybrid: (1) Grows faster (*e.g.* Fig. 12.1); (2) Extends the ecological amplitude of one of the pure species allowing its use in edaphoclimatic conditions to which it is poorly adapted (*e.g.* Fig. 12.2); (3) Confers disease or pest resistance; (4) Produces a suite of more desirable products (*e.g.* Fig. 12.3); and/or (5) Facilitates plantation culture or tree improvement (such as increasing rootability of stem cuttings).

The discussion above and the examples presented in Box 12.2 are not meant to indicate that hybrid vigor always occurs or that hybrids are appropriate taxa for all types of plantation programs. However, hybrids are already commercially important for some programs, and their importance will likely grow as cost-effective methods of vegetative propagation are developed for recalcitrant species. Depending on the underlying mechanisms conferring hybrid superiority, product objectives, edaphoclimatic conditions in the planting zone and silvicultural practices, any of several hybrid types (*e.g.* F_1 hybrids, backcrosses, 3-way hybrids) or the pure species may be most appropriate. Well-designed field trials are required to discriminate among the many taxa options.

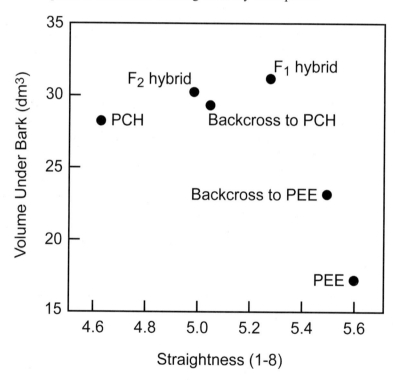

Fig. 12.1. Mean six-year performance for bole volume under bark (dm^3) and straightness score (1-8 scale with 8 being most straight) for pure species of *Pinus caribaea* var. *hondurensis* (PCH) and *Pinus elliottii* var. *elliottii* (PEE) and various hybrids across four sites in southern to central Queensland, Australia. The F_1 hybrid was formed by crossing various trees from both pure species, the F_2 was formed by intercrossing various F_1 trees together and the backcrosses were formed by crossing F_1 trees with PEE or PCH trees. (Reprinted from Powell and Nikles, 1996, with permission of the authors)

Box 12.2. Three examples of plantation programs utilizing interspecific hybrids.

1. Plantation and breeding programs for species of *Populus* in several countries of Europe and North America have long used interspecific hybrids (Zobel *et al.*, 1987, p 303; Stettler *et al.*, 1988; Bjorkman and Gullberg, 1996; Morgenstern, 1996, p 100). The rapid growth and good performance of interspecific hybrids of *Populus* contribute to their success in plantation culture. Also, the ease of producing interspecific hybrids through control pollination (*i.e.* **crossability**) and the ease of their vegetative propagation facilitate their breeding and operational deployment in tree improvement programs.

2. Another example stems from the pioneering work of Dr. Garth Nikles in Australia who began creating and testing the hybrid between *Pinus caribaea* var. *hondurensis* (PCH) and *P. elliottii* var. *elliottii* (PEE) in the 1960s (Nikles, 1992, 2000; Dieters and Nikles, 1997). Both pure species are important plantation species in their own right; however, there are some instances in which the hybrid performs better than both species (Fig. 12.1). PCH grows faster in mild subtropical climates, but it has lower wood density and is more sensitive to damage by cold, high water, tip moth and wind than PEE. The hybrid combines favorable features of both species into a single entity (growth and branch quality from PCH combined with wood density and resistance to damaging agents from PEE). It is planted on a commercial scale in Australia in "hybrid" environments in which it is superior to both species (*i.e.* shows hybrid vigor), and is being tested in other countries in South America and Africa.

3. Finally, hybrids with *Eucalyptus grandis* are important in countries such as Brazil, China, Colombia, Congo, South Africa and Venezuela (Denison and Kietzka, 1992; Endo and Lambeth, 1992; Nikles, 1992; Ferreira and Santos, 1997; Wright, 1997; de Assis, 2000; Retief and Clarke, 2000; Verryn, 2000). As mentioned previously (Box 12.1), *E. grandis* is the most widely planted exotic as a pure species. Its hybrids with various species, especially *E. urophylla*, but also *E. camaldulensis*, *E. tereticornis,* and *E. nitens*, are becoming increasingly important for enhancing yields on some types of sites, for increasing adaptation to drier and colder sites and for improving disease resistance (*e.g.* Fig. 12.2). In addition, the hybrids can have higher wood density and pulp yields (Fig. 12.3).

Subspecies, Varieties, Provenances and Land Races

There are several types of seed sources available within a given species that could be chosen for operational plantations or to form the base population of a tree improvement program. To briefly review terminology originally defined in *Chapter 8*, recall that provenance refers to a seed collection zone in the original geographic area *within the native range* of a species. Seed source is a more general term used only to denote the location where the seed was obtained, and may refer to a source of seed outside the native range such as *P. taeda* seed (native to the southeastern USA) obtained from an exotic plantation program in South Africa. In this case, the seed source (South Africa) is not the provenance and the original provenance in the native range may or may not be known. Seed source also may refer to a genetically improved variety obtained from a tree improvement program, and may refer to a single family or bulk collection with any level of genetic improvement.

Both provenance and seed source refer only to the area of collection, and do not necessarily imply genetic differences. When common garden studies have demonstrated genetic differentiation among provenances, they are called races and racial variation refers broadly to genetically determined population differentiation developed within a species (*Chapter 8*). Racial variation can be clinal if it is continuous and associated with environmental gradients such as elevation, latitude, etc; or may be ecotypic if provenances are more discrete in their differences. Both types can occur within a given species. A local land race (see below) is a geographic race of a species that develops in an exotic planting zone.

Varieties and Subspecies

Taxonomists use the terms variety and subspecies interchangeably to denote within-species subdivisions such that two different varieties or subspecies within the same species have different morphological characteristics, occupy distinct ranges and are given different Latin names (*Chapter 8*). It is important to remember that subspecies and varietal designations within a species are assigned by taxonomists usually based on observations in natural stands (*i.e.* in the absence of common garden experiments to assess genetic differentiation). The magnitude of genetic differentiation requires field trials in the planting zone for verification.

Generally, large genetic differences can be expected among varieties or subspecies (*e.g. Pinus contorta* and *P. caribaea*, described in *Chapter 8*); therefore, these taxa need to be considered as separate entities for the purposes of testing and choosing seed sources for plantation programs. In fact, the differences among varieties and subspecies may be known to be so great from previous work that only one variety or subspecies is even considered for a particular planting zone. Further, genetic differentiation within each variety or subspecies can also be great, necessitating consideration and field testing of many provenances or sources within each subdivision.

Provenances from the Native Range

Chapter 8 reviews the extensive information existing about provenance variation within species. Briefly, racial variation is common in wide ranging species that span diverse climatic regions. Further, this racial variation is almost always expressed in plantations established in both the native range and in exotic locations, although the magnitude and patterns of the variation may change depending upon the edaphoclimatic characteristics of the planting zone. The implication of strong racial variation when choosing the best provenance for a plantation program is that field-testing many provenances may be necessary to adequately evaluate a species.

Determining the most appropriate provenances to consider and field-test can be difficult. This is further complicated when a species exhibits clinal patterns of genetic differentiation because discrete boundaries between provenances are difficult to determine. In provenance studies discussed in *Chapter 8*, the main objective was to characterize racial variation across the entire range or a portion of the range of a species. When the objective is to find the appropriate seed source for a plantation program, it may not be necessary to sample the entire range. Information from existing field studies combined with biological judgment may indicate that certain parts of the natural range should be more intensively sampled, while other parts are unlikely to result in sources well suited to the planting zone.

For plantation programs in the native range of a species, it is advisable to use local provenances, with seed collected as near to the planting zone as possible, unless data from

Fig. 12.2. In South Africa, Mondi Forests plants clones of pure *Eucalyptus grandis* on a wide range of moist sites in temperate planting zones; however, hybrids show superiority and are used to extend the plantation range: (a) Progeny test growing in hot, dry sites in the sub-tropical area of Zululand showing row plots of clones of *Eucalyptus grandis* (right) and the superior F_1 interspecific hybrid between *E. grandis* and *E. camaldulensis* (left); and (b) Progeny test growing on a cold, high-elevation site showing row plots of clones of *Eucalyptus grandis* (left) and the superior F_1 interspecific hybrid between *E. grandis* and *E. nitens*. (Photos by T. White)

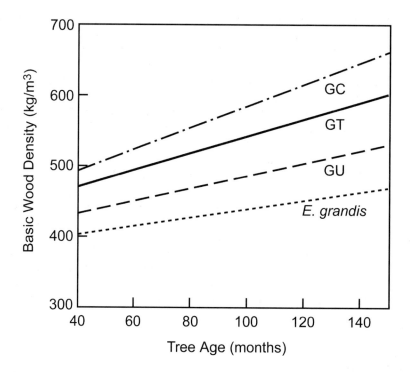

Fig. 12.3. Average wood density as a function of plantation age for *Eucalyptus grandis* and three of its F₁ hybrids in South Africa: GU = *E. grandis* x *E. urophylla*; GT = *E. grandis* x *E. tereticornis*; and GC = *E. grandis* x *E. camaldulensis*. While the wood densities of the other pure species are not shown, the hybrids are generally intermediate and near the mid-parent value of their two parental pure species. (Reproduced from Arbuthnot, 2002, with permission of the author)

long-term field trials indicate otherwise (*Chapter 8*). Use of a local seed source generally ensures that a well-adapted seed source is planted, although it may be possible to find non-local provenances that express superior vegetative vigor (growth) in the plantation environment. These concepts are discussed extensively in *Chapter 8*.

Local Land Races

A **local land race** forms when a species is introduced into an exotic planting environment and adapts through natural, and sometimes artificial, selection to the edaphoclimatic conditions of the new planting zone (*Chapter 8*). A local land race is a logical source of seed for an exotic plantation program if seed is readily available and the land race is observed to be doing well within the planting zone or in environmental conditions similar to it. Table 3.4 in Zobel *et al.* (1987) lists more than 20 studies of conifer and hardwood species in which seed collected from the local land race was equal or superior to imported seed from other sources. They conclude, "In most, but not all, instances of the numerous species listed in the table, the local selections from the exotic plantations were better than new provenances brought from the indigenous range." In our experience, however, there are numerous instances when the local land race is inferior to imported seed from other sources (*e.g.* Fig. 12.4). Therefore, it is important to consider the biological factors causing a well or poorly adapted local land race.

The following factors and sequence of events lead to a well-adapted local land race: (1) The seed introduced into the country comes from a large number of trees, thus providing much genetic diversity; (2) The original seed is from appropriate provenances that perform well in the planting zone; (3) There is strong selection intensity in the new exotic environment; (4) The land race has two or more generations to evolve and adapt; and (5) The land race occupies a large enough area in the exotic country to avoid extreme bottlenecks (genetic drift) and hence, reduction and non-adaptive shifts in genetic diversity. It is not often that all five of these factors are present; yet, local land races can still be a suitable starting point from which to initiate an operational plantation program.

The two most important conditions preventing the development of a well-adapted land race are limited genetic diversity and an inappropriate original seed source. If seed for the original introduction was collected from only a few trees, the resulting exotic plantations may have only limited genetic capacity to evolve adaptations (through natural selection) to the local environments, and mating among close relatives could lead to inbreeding depression in their offspring (*Chapters 5* and *8*; Box 12.3). If seed for the original introduction of a species comes from the wrong provenance, then new introductions from provenances better matched to conditions in the exotic planting zone may outperform local land races. An example is *Pinus taeda* in northeastern Argentina introduced from the southeastern USA (Baez and White, 1997). The original introductions were from unknown provenances,

Fig. 12.4. Two-year-old plantation of *Pinus taeda* growing in northern Argentina (Corrientes province) showing the slower-growing plantation that originated from seed collected from local plantations (*i.e.* from the local land race) and the faster-growing plantation established with seed collected from Marion County, Florida, USA in the native range of *P. taeda*. (Photo by T. White)

but likely from northerly, temperate parts of the native range. New collections from southerly, subtropical provenances (especially Florida) consistently grow faster than the local land race in field tests in Argentina (Figs. 12.4 and 14.2a). Plantation yields could be increased by importing seed from the appropriate provenances in the native range instead of relying on the local land race.

Even though the local land race may often be appropriate for the operational forestation in a new plantation program, it is seldom the best choice as the only source of material for tree improvement efforts of an exotic species. While the plantation pro-

---

Box 12.3. Species and seed source refinement trials: *Acacia* in Malaysia.

The *Acacias* are becoming increasingly important plantation species in Southeast Asia, South America, Africa and the Middle East with nearly 2 million hectares planted in 1996 and double that area expected by 2010 (Matheson and Harwood, 1997). They are grown for a range of products from industrial wood used for pulp and lumber to animal fodder. Their edible pods can provide a source of food and their nitrogen-fixing ability makes *Acacia* desirable for infertile and degraded soils and for agroforestry systems. A specific series of field trials of four seed sources employing bulked seedlots of *A. auriculiformis* and two sources of bulked seedlots of *A. mangium* was established in Sabah, Malaysia on two sites with very different soils. These two species are native to parts of Australia, Papua New Guinea (PNG) and Indonesia and are grown as exotics in Malaysia. Both species were originally introduced into Malaysia with seed collections from very few trees in the natural range and these introductions formed local land races (Evans, 1992a). The local land race of each species is one of the seed sources for the trials with the other sources coming from seed collections of provenances within the native range in PNG.

Species	Seed Source	Four-year Height (m)	
		Site 1	Site 2
A. auriculiformis	Local land race	9.87	11.51
	Balamuk, PNG	12.10	13.99
	Lowka, PNG	14.91	13.07
	Bula, PNG	11.59	12.27
A. mangium	Local land race	11.48	9.56
	Unknown, PNG	15.61	13.04

Assuming the means in this table are precisely estimated from a well-designed and well-replicated experiment on each site (so that the differences are real), several informative inferences can be made: (1) There are large differences in growth rate both among species and among sources within each species; (2) Species x site interaction indicates that *A. auriculiformis* grows better on site 2, while *A. mangium* is better on site 1; (3) There is little provenance x site interaction which concurs with other seed source refinement trials (Matheson and Harwood, 1997); and (4) The local land race is the poorest performing source in both species, likely due to inbreeding depression owing to the fact that the original introductions of both species came from few trees. Of course, final decisions must also consider other traits such as stem straightness, forking and wood quality because these species and sources are known to differ widely in these characteristics; however, this trial illustrates the large differences that can occur among seed sources and emphasizes the need to test on multiple sites.

gram can change seed sources as new information and options become available, the base population should sustain tree improvement efforts for many cycles of selection, breeding and testing. The first-cycle base population of a tree improvement program needs to have high levels of genetic diversity from appropriate provenances suitable to the planting zone. For most tree improvement programs, this means that the first-cycle base population should be composed of the local land race plus other materials, such as seed collections in the native range.

Other Sources of Seed within Species

For many plantation programs of both native and exotic species, other sources of seed exist that are neither from the native range nor a local land race in the country of the planting zone. Many species have been introduced into numerous countries in which there are active plantation programs and even formal tree improvement programs. Exotic plantations in another country may be an excellent source of seed for a new plantation program if: (1) The original introductions into that country had a sufficient genetic base and were from appropriate provenances; (2) Edaphoclimatic conditions in the planting zone are similar to those of the exotic plantation sites in the source country; and (3) Tree improvement programs in the source country have made substantial genetic gains.

   As just one of many examples, *Eucalyptus grandis* is planted widely in several countries (Box 12.1). Nearly all have supporting tree improvement programs, and some have created interspecific hybrids with other species as mentioned previously (Box 12.2). These improved varieties can be excellent seed sources both for reforestation and tree improvement efforts in new regions. For instance in south Florida USA, *E. grandis* is periodically exposed to colder conditions than in its natural range. Both natural selection and artificial selection by the tree improvement program in south Florida for cold hardiness and growth have resulted in a land race that is well adapted to colder growing conditions (Meskimen *et al.*, 1987; Rockwood *et al.*, 1993). Selected families from this program have done especially well compared to other sources in Uruguay where frosts and cold weather occur on low-lying sites (personal communication, A. Rod Griffin, 2000, Cooperative Research Center, University of Tasmania, Hobart, Australia.). In contrast, the same Florida-source families are inferior in growth to locally developed genetic varieties in Colombia, which has a much milder, tropical climate where frosts are not a problem (personal communication, Luis F. Osorio, 2001, Cartón de Colombia, Cali, Colombia). As with all decisions about movement of species and seed sources, it is important to consider the edaphoclimatic zones of the donor and receptor countries.

CHOOSING SPECIES, HYBRIDS AND SEED SOURCES FOR PLANTATION FORESTRY

Most plantation programs consist of several edaphoclimatic zones of varying sizes that differ to a greater or lesser extent in elevation, climate, soils and topography. The issue of defining which species and seed sources should be planted in which zones is critical. In this section, we focus on choosing appropriate species, hybrids and seed sources when the goal of the plantations is to maximize product yield in the shortest period of time. The discussion applies to choosing taxa and seed sources for community and social forestry (*e.g.* maximizing fuelwood production), for agroforestry projects, and for industrial plantations. *Chapter 8* addresses choosing appropriate provenances and developing seed transfer guidelines within a species' native range. Here the discussion is broadened to apply to plantation programs of both native and exotic species and to programs in which the most appropriate species or taxa have not been defined.

These taxa-level decisions usually involve balancing risk and reward, *i.e.* finding the species and sources with optimum production potential (maximizing the reward) that are well enough adapted to have a high probability of excellent survival and growth for the entire rotation (minimizing the risk of maladaptation). The acceptable level of risk is relative to the adaptation of the recommended sources to survive and grow in plantation conditions. This "plantation" fitness does not involve reproductive fitness (unless seed or fruit are desired products), but does involve fitness to special silvicultural conditions of the managed plantations (see *Local versus Non-local Provenances, Chapter 8*). To adequately assess both production potential (reward) and adaptation (risk), there is no substitute for good data from long-term field trials.

With these concepts in mind, a general process (Fig. 12.5) for determining the most suitable species, hybrids and seed sources for a plantation program is to: (1) Specify the product objectives, for example, pulpwood, solid wood products, fuelwood, animal fodder; (2) Determine the range of edaphoclimatic regions within the intended planting zone, and the soils, climates, and pests that occur in each region; (3) Specify the planned intensity of management of the plantations; (4) Prepare a list of candidate species, hybrids and

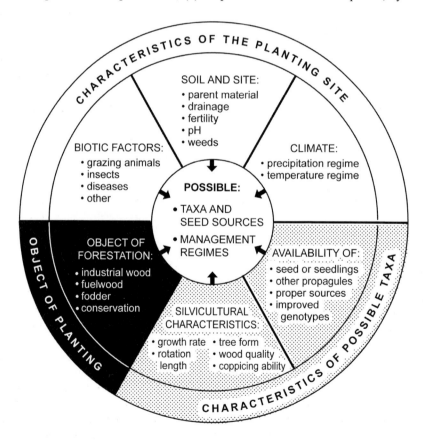

Fig. 12.5. Schematic diagram that can be used as a checklist of factors to consider when identifying potential species, hybrids and seed sources for a plantation program. (Reprinted from Savill and Evans, 1986, with permission of Oxford University Press)

seed sources that seem appropriate based on the product objectives, edaphoclimatic considerations and planned management intensity; (5) Define the range of seed sources for the candidate species and determine if seed is available; and (6) Conduct a series of field trials (common garden tests) to determine which are the best (Savill and Evans, 1986; Evans, 1992a). The first five of these factors are covered in the next section, followed by a discussion of field trial procedures. The final section in this topic area deals with how to use the information obtained from field trials to choose the optimum species, hybrids and seed sources for a plantation program.

Identifying Candidate Species, Hybrids and Seed Sources for Plantation Forestry

When the choice of taxa (*i.e.* species, hybrids and seed sources) is not obvious, a logical first step is to employ a decision-making process to identify a list of potential candidate taxa (Fig. 12.5). Depending on existing information and edaphoclimatic variability in the planting zone, the list may include only a few taxa (*e.g.* well-known seed sources in a few species) or may contain 50 or more candidate taxa. Once created, the list may have two immediate uses: (1) To identify the taxa to include in field trials in the planting zone (see next section); and (2) To choose the one or two most-likely candidates to begin the plantation program. At the same time as the most likely one or two candidate taxa are being planted operationally, field trials of many taxa may be established to subsequently confirm or refine the original decision.

The first step in the decision-making process to identify candidate taxa is to define the product objectives (*e.g.* pulpwood, solid wood products, fuelwood or fodder in Fig. 12.5). Clearly, different species are likely to be required for different product objectives. Conifers are preferred for some solid wood products and for paper products requiring long fibers, while some hardwoods may be better for fine writing papers. Multipurpose tree species exist, such as *Leucaena* spp., which can fix nitrogen and provide fuelwood and other products. The product objectives must be considered when choosing candidate species.

Second, the edaphoclimatic regions of the planting zone need to be identified and characterized (Fig. 12.5). If edaphoclimatic regions vary markedly in elevation, soils, climate or pests, then perhaps two or more different taxa will best optimize yields, and field trials will be necessary in all major zones. The third step is to determine the general philosophy of silviculture that will be applied. Will there be intensive management with good site preparation, fertilization, thinning, etc. or will the plantations be established and then left on their own to develop? Detailed silvicultural prescriptions will vary according to the species chosen, and this can influence the choice. For example, in the southeastern USA, all other things being equal, *Pinus taeda* responds better to more intensive management, while *Pinus elliottii* may be a better choice on some sites if less intensive management is planned (Colbert *et al.*, 1990).

The fourth task is to prepare a list of candidate species, hybrids and their available seed sources that are known to produce the desired products and are thought to be adapted to the intended planting zone. The objective is to match the biological tolerances and preferences of all available taxa with the characteristics of the planting zone. The process of choosing candidate species often begins by determining which species have been successful and not successful in nearby plantations, arboreta, botanical gardens, woodlots or city streets. It is surprising how much information can be obtained by these observations and by asking nearby landowners. In addition, it is important to consider which species have been successful and not successful in other parts of the world with similar climates and soils.

The process of matching species to climates and soils is aided by use of computer programs such as WORLD (Booth, 1990; Booth and Pryor, 1991) and INSPIRE (Webb *et al.*, 1984) and by lists of species whose products and biological tolerances are known (Webb *et al.*, 1984; Pancel, 1993). While there are many aids, there is no substitute for experience in this process because it can become quite complicated in some areas of the world; it is appropriate to seek guidance from silviculturalists who have experience with the taxa in similar regions.

Next, the importance of geographical variation within each candidate species and the availability of seed from appropriate seed sources need to be considered. If named subspecies or varieties exist within any of the species, these must be identified and considered as distinct taxa for subsequent field trials. Species with large natural ranges, or those that have exhibited distinct provenance variation should be represented by multiple sources in the field trials. After defining the likely importance of geographic variability within each candidate species, potential sources of available seed need to be located (Fig. 12.5). These potential sources include collections within the native range or local land races or may involve purchase of seed from plantation or tree improvement programs in other countries. The final list of candidate species, hybrids and sources for field-testing is comprised of all sources for which seed can be obtained. In this regard, there are seed dealers or government organizations in many countries that collect, label and distribute seed.

Multiphase Field Trials for Testing Species, Hybrids and Seed Sources

After the candidate taxa are identified and seed obtained, there are three potential phases of field testing to narrow down and eventually determine the optimum taxon for each edaphoclimatic region in the planting zone (Zobel *et al.*, 1987; Evans, 1992a): (1) *Phase 1,* species elimination trials to reduce the number of candidate species, hybrids and sources from as many as 50 to fewer than 5; (2) *Phase 2,* seed source refinement trials to more intensively assess adaptation and product yield for the most promising candidate taxa; and (3) *Phase 3,* large scale yield trials to quantify yields and product quality of the top sources in order to make yield projections and plan harvest scheduling. These phases sometimes proceed chronologically in sequence, waiting for the results of Phase 1 to begin Phase 2, etc. However, other times an entire phase is omitted (such as when there are few candidate taxa so that Phase 1 is omitted) or two phases proceed simultaneously (*e.g.* while many taxa are established during Phase 2 refinement trials, only a few of the most likely taxa are tested in a separate simultaneous series of Phase 3 yield trials).

Phase 1: Species Elimination Trials

Species elimination trials are short-term field tests designed to reject species obviously not suited to the planting conditions and to identify those that seem adapted (Fig. 12.6). There may be as many as 50 taxa (seed sources of different species and hybrids) each planted in rectangular plots (Fig. 14.9c) of say 25, 36 or 49 trees in each of two or more replications on a variety of sites. There should be at least one test site located in each of the major edaphoclimatic regions identified in the planting zone, and more than one test site in larger regions. Several taxa often fail completely or grow poorly within the first few years, and this makes the tests of little utility, long term, because differential competition exists between neighboring plots. For this reason and because the primary objective is to assess early adaptability, the field experimental design is not critical; often a randomized complete block with a few blocks (even as few as two blocks at each site) is used, but more sophisticated incomplete block designs are available (Williams *et al.*, 2002).

Sometimes unequal numbers of seed and plants from different taxa make it difficult to implement a balanced design for species elimination trials. The important thing is to place plots of all taxa in a variety of soils and climates within the planting zone and then to observe performance for approximately 25-50% of the intended rotation length. Since the goal is to eliminate poorly adapted taxa and define those that are well-adapted, traits for assessment include growth rate, survival, pest damage and hardiness to stresses such as cold (Fig. 12.6), heat, drought, flooding, etc. Normally, species elimination trials lead to identification of fewer than five species or hybrids (and perhaps a few seed sources within each those taxa) that are most appropriate for the planting zone.

Phase 2: Seed Source Refinement Trials

The second phase of field trials, variously called **seed source refinement trials, provenance trials** or **plot performance trials**, is designed to further test the most promising taxa and especially to characterize genetic variation within those taxa in order to find the best sources for the planting zone. Usually only the most promising few species or hybrids are tested in a given edaphoclimatic region with tests planted on three to five sites in each region. The focus is on longer-term survival, adaptation and growth so the duration is usually longer than for Phase 1 trials (approximately 50% of the intended rotation length). Seed source refinement trials are similar to provenance trials discussed in *Chapter 8* for native species and provenances, except that sources of seed outside of the species' natural range may be included, such as from local land races, and from tree improvement programs

Fig. 12.6. Species trial of *Eucalyptus globulus* (left) and *E. nitens* (right) growing in the foothills of the Andes Mountains in Chile. *E. globulus* grows well on sites in the milder, coastal region of the country, but suffers from cold damage on sites, such as this one, in colder regions. (Photo by T. White).

in other countries that plant the species as an exotic. There may be several different levels of genetic variation represented in Phase 2 trials with one to a few species or hybrids, varying numbers of sources within each species, and few to many open-pollinated families within each source.

Whenever possible for seed source refinement trials, it is best to maintain the mother tree identity of each seedlot and the location of each mother tree within each seed source. In this case, the tests may be called **mother tree tests**, and there may be from 2-50 mother trees from a given seed source depending on the assumed patterns of racial variation and sampling design. Normally open-pollinated seed is collected, and each mother tree should be of good form and vigor and free of obvious defects and diseases. Maintaining mother tree identity throughout testing has several advantages: (1) Relative importance of seed source to family-within-seed-source genetic variation can be determined for each trait measured; (2) Clinal patterns of geographic variation within seed sources, if they exist, can be quantified (*i.e.* if a large number of mother trees of known location are tested) (*Chapter 8*); and (3) The Phase 2 trials can be subsequently used as part of the base population for tree improvement efforts by making outstanding selections from top-performing families in appropriate sources. When it is not feasible to maintain mother tree identity, seed should come from at least 20 different mother trees to adequately represent each seed source. In this case, all seed from all mother trees in a source is combined together such that each seed source is represented by a single, bulked lot (see example in Box 12.3).

Experimental designs of seed source refinement trials vary widely depending on the numbers and types of genetic entries (see *Long-term Provenance Trials, Chapter 8*). Here, we describe only two designs illustrating extreme situations: (1) A few, discrete taxa (*i.e.* species, hybrids or seed sources) being tested; and (2) Many sources within a single species that shows clinal patterns of variation in its native range. In the first case, when only a few taxa are being tested and when taxa differences may be quite large, trees of the same taxon should be planted together in square or rectangular plots (Fig. 14.9c). This permits competition to occur among trees of the same taxon as would occur operationally in a plantation program. If competition between taxa becomes important, the outer row of each taxa plot becomes a buffer between plots, and only the remaining trees are measured. An example of a Phase 2 seed source refinement trial consisting of six total seed sources from two *Acacia* species is summarized in Box 12.3.

For species with clinal patterns of geographic variation (*Chapter 8*), it is difficult to form a few discrete seed sources, and it is usually more appropriate to collect seed from mother trees selected along one or more environmental gradients that are relevant to the planting zone (*e.g.* elevation, latitude, precipitation). Hundreds of mother trees may be required to represent a species. The experimental design is much more like those discussed in *Chapter 8* in which there are no discrete provenance or taxa plots, but rather open-pollinated families from the mother trees that are completely randomized within each block. Other times, there may be a few discrete taxa planted in rectangular plots with open-pollinated families nested and randomized within each plot. Often single-tree plots (one tree representing each family in each block, *Chapter 14*) is the optimal plot design to ensure maximum statistical precision.

While genetic composition and experimental design of seed source refinement trials vary widely, the common goal is to identify a few top potential taxa for operational forestation. Even early results from Phase 2 trials are often used to make preliminary decisions about which species and seed sources are most appropriate. By mid-rotation, the performance results should facilitate eliminating all but a very few taxa from further consideration.

Phase 3: Large-scale Yield Trials

The third and final phase of species and seed source trials, called **large-scale yield trials**, aims to evaluate long-term adaptability and growth, to quantify product yields and often to examine response to different silvicultural treatments. These trials normally include only those taxa that are being or will be planted in the operational plantation program, and the genetic composition of each taxon tested should mimic the composition of the taxon planted operationally. Therefore, these yield trials may include taxa planted as single clones, as single families or as bulk mixtures of many different families depending on what is deployed operationally (*Chapter 16*).

Trees of the same taxon are planted in rectangular plots (as in Fig. 14.9c) that are large enough to simulate operational stand conditions for an entire rotation, since the tests must be grown for this long to assess product yields. Because of the larger plots and hence large replication sizes, very few taxa can be compared in conventional randomized complete block designs (usually fewer than eight) and only a few more than this if incomplete blocks are employed (Williams *et al.*, 2002). Sometimes these trials are not conducted in replicated experimental designs, but rather permanent taxa plots are located within operational plantations and measured as part of the operational inventory. Either way, the longer duration of Phase 3 tests makes it important to establish them as soon as possible, even before the final species or taxa are known for certain. Sometimes the justification for an entire plantation program can depend on the outcome of these yield trials.

Using Available Information to Make Taxa Decisions for Plantation Forestry

While the optimum situation is to have data from long-term field trials, many times deployment decisions must be made before these data are available. In this case, data from field trials do not exist, are poor, or are from very young trials (also refer to Fig. 8.9a,b and the related discussion in *Chapter 8*). In this situation, the decision about which taxa to plant operationally must be based on other types of available information (see *Identifying Candidate Species, Hybrids and Seed Sources for Plantation Forestry* in this chapter), such as: (1) Rules of thumb based on studies of other species or taxa (*e.g.* move sources from a slightly milder climate of seed collection zone in the native range to a slightly harsher planting zone to achieve increased growth, *Chapter 8*); (2) Existing seed transfer guidelines for candidate species in other countries or planting zones; (3) Results from long-term field studies, short-term seedling studies and/or genetic marker studies of candidate taxa in other countries or planting zones; (4) Advice from silviculturalists who have experience with candidate taxa; and (5) Computer programs (*e.g.* WORLD and INSPIRE mentioned earlier in this chapter) that match species to climates and soils based on their known biological tolerances. Often, several types of information are combined together to make initial decisions about which taxa to plant operationally, and then these decisions are refined as data become available from field trials established in the planting zone.

When long-term data are available from well-designed, well-replicated field trials covering the varying edaphoclimatic regions in the planting zone, then the choice of taxa depends on results and inferences from these long-term trials. There are three possibilities. First, if there are no taxa differences in performance found in any of the edaphoclimatic regions of the planting zone (Fig. 8.9c), then taxa differences are not important and the reforestation forester may choose any of the tested taxa. While this scenario is not likely for widely varying species or hybrids, it is possible that several seed sources within a species might perform similarly and very well across all sites in the planting zone. If so, then any and all of these seed sources would be appropriate for operational use.

Second, if there are large taxa differences in performance, but no important taxa x location interaction across the edaphoclimatic regions (Fig. 8.9d), then the same one or a few taxa are most appropriate for all planting sites. This situation is the most common and is likely when the soils and climates do not vary markedly within the planting zone. It is logical in this case to plant the best source(s) across the entire zone. An example is New Zealand where *Pinus radiata* is the best performing species on a majority of sites, and the same provenances of *P. radiata* (Cambria and Monterey) are preferred over all sites.

Third, when taxa dramatically change performance rankings across planting sites (important taxa x location interaction, Fig. 8.9e), then it is necessary to define multiple planting regions (*i.e.* deployment zones; see next section) and to match taxa to planting regions. Each taxa deployment zone utilizes the taxon that performed best. Examples are multiple species planted by Mondi Forests in South Africa (Box 12.4) and *Pinus taeda* planted by Weyerhaeuser Company in Arkansas, USA (*Chapter 8*; Example 4 in Box 12.5).

These considerations and examples illustrate the complexities of choosing which species, hybrids and seed sources to plant operationally. The key is to have as much data as possible from field studies established in the planting zone. Without these data, the

---

Box 12.4. Taxa deployment zones, breeding zones and base populations for Mondi Forests in South Africa.

Mondi Forests manages a plantation estate of more than 400 000 ha in South Africa with a wide range of edaphoclimatic zones ranging from subtropical climates in Zululand to cold, temperate climates at high elevations and in the Eastern Cape (personal communication, Neville Denison and Eric Kietzka, 1999, Mondi Forests, Pietermaritzburg, South Africa). The company also makes a wide range of products. Therefore, the company plants several taxa deployed in the following manner: (1) *Eucalyptus grandis* is planted widely as a pure species on many of the temperate sites; (2) Interspecific hybrids, *E. grandis* x *E. urophylla* (GU) and *E. grandis* x *E. camaldulensis* (GC) are utilized in the warmer more subtropical regions; (3) The interspecific hybrid, *E. grandis* x *E. nitens* (GN), and the pure species *E. macarthurii* are deployed on colder sites at high elevations; and (4) *Pinus patula* is planted in many zones for some specific product objectives.

The company maintains tree improvement programs for both pine and eucalypts. Mondi defined a single, large breeding zone for each of the different taxa (*i.e.* five total breeding zones one for each of pure *E. grandis,* the GN, GU and GC hybrids and *Pinus patula*) corresponding to all site types planted with that taxon. There is a single-tree improvement program (with a single base, selected and breeding population) for each taxon. For pure *E. grandis,* a very broad genetic base population is maintained with selections made from the local land race and from several natural provenances in Australia. For the hybrid breeding programs, the *E. grandis* portion of the base population is weighted more heavily to the provenances and seed sources known to be better adapted to that edaphoclimatic zone (*e.g.* sources known to perform better in colder climates for the GN breeding zone and more sub-tropical seed sources of *E. grandis* for the GU and GC hybrids). The second part of the base population for each hybrid program utilized a variety of material of the second species available in South Africa. This example illustrates that definition of breeding zones and base populations is complicated in large programs spanning many edaphoclimatic regions and that the decisions can involve many species, hybrids and seed sources.

plantation program will never know how much yield or product quality has been sacrificed through inadvertent choice of the wrong taxon.

DEFINING BASE POPULATIONS FOR TREE IMPROVEMENT PROGRAMS

The **base population** of a given cycle of improvement consists of all available trees that could be selected if desired (Zobel and Talbert, 1984, p 417). Each base population is the foundation population for its own tree improvement program (with its distinct selected, breeding and propagation populations, *Chapter 11*). Before considering the biological, genetic and economic factors affecting the definition of base populations, there are several terms that need clarifying. The base population is specified for a particular **breeding unit** (also called a **breeding zone**), which is defined by a set of edaphoclimatic conditions (*e.g.* a given breeding unit may consist of all planting sites of certain soil types and site qualities lying within specified elevational and precipitation ranges). Each breeding unit is served by its own separate improvement program and distinct base, selected, breeding and propagation populations. A **multi-source base population** means that trees are selected and intermated from more than one seed source. For example, a base population for a given breeding unit might consist of several native provenances, local land races in one or more countries and improved genetic material from other tree improvement programs. Selections made in all of these sources would be used for a single breeding unit. A **multi-species base population** exists for all programs breeding interspecific hybrids because trees of more than one species are selected and included in the program (*e.g.* Box 12.4).

Two final terms are recruitment zone and deployment zone. With multi-source base populations, it is necessary to consider which sources to include and this gives rise to the term recruitment zone. The edaphoclimatic regions containing all seed sources included in the base population define the **recruitment zone**. In other words, all locations where selections are made form the recruitment zone, and this may include many regions and countries. When one large breeding unit is defined and breeding is for wide adaptability for that unit, it is often useful to create separate deployment zones within the breeding zone. A **deployment zone** is a smaller edaphoclimatic region within the breeding unit that has its own propagation population. The selection, breeding and testing (the central core activities in Fig. 11.1) are conducted as a single program for the entire plantation program (*i.e.* a single breeding unit), but in any given cycle of breeding, there could be multiple propagation populations (one for each deployment zone). For example, for two deployment zones there might be two different seed orchards–one containing selections more suited to drier sites and one containing selections more suited to moist sites.

There are two distinct issues to consider when defining base populations for a tree improvement program: (1) The number of distinct breeding units appropriate for a given plantation region; and (2) The composition of the base population for each breeding unit. These issues are discussed, in turn, in the following sections and some examples are given in Box 12.5.

Number and Size of Breeding Units

Each breeding unit should be as large as possible subject to the limitations imposed by adaptability. For a given plantation region (*i.e.* planting zone), larger breeding units (*i.e.* spanning large areas) mean fewer breeding units and hence fewer separate tree improvement programs. This has two main advantages: (1) Fewer, larger programs are simpler and more cost-effective to manage, due to overall lower costs of selection, breeding and testing for the entire region; and (2) More trees can be included in each base population (since there are fewer base populations) increasing genetic gain due both to the wider genetic diversity of the base population and the higher selection intensity that is possible. These advantages are tempered by the fact that all sources have limits of adaptability (*e.g.* the poor adaptability of North Carolina provenances of *Pinus taeda* on dry sites in Arkansas; *Chapter 8*, Box 12.5), and inclusion of poorly adapted sources in a base population can reduce overall suitability of genetically improved varieties developed by the breeding program. As stated by Namkoong *et al.* (1988, p 53), the question is, "how wide can any single [base] population be bred before substantial loss of local adaptability?"

Box 12.5. Examples of base populations, breeding units, recruitment zones and deployment zones.

Example 1: Many breeding units and local base populations. The Northwest Tree Improvement Cooperative program for *Pseudotsuga menziesii* in Oregon and Washington originated with 72 breeding units (Silen and Wheat, 1979; Lipow *et al.*, 2003). These 72 breeding units were defined based on the steep environmental gradients in the region and on the known adaptation of natural provenances to these gradients from short-term seedling trials (Campbell and Sorensen, 1978; Campbell, 1986). The base population in each breeding unit was composed of natural stands of the local provenance (*i.e.* a single-source base population for each unit using only local selections). Recruitment zone and deployment zone were identical for each breeding unit and were defined by the edaphoclimatic conditions of the unit.

Subsequently in the second cycle of tree improvement, many of the original breeding units were merged into fewer, larger units (Randall, 1996; Lipow *et al.*, 2003) because: (1) Subsequent long-term field trials have indicated less provenance x environment interaction than perhaps expected (White and Ching, 1985; Stonecypher *et al.*, 1996); and (2) The large number of initial breeding units decreased the economic feasibility of the program. Breeding programs for several species in Canada and Europe have also defined multiple breeding units to match edaphoclimatic zones (Morgenstern, 1996, p 141).

Example 2: Single breeding unit due to small seed source differences. Forestal Monteaguila plants *E. globulus* on approximately 25,000 ha in three different edaphoclimatic zones in Chile (coast, central valley and Andes foothills). The company has field tested 300 open-pollinated families collected from selections made in the local land race and 350 open-pollinated families collected from the natural range in Australia (Vergara and Griffin, 1997).

Results indicate: (1) No average differences between the two types of populations (local land race and natural range); (2) Small mean differences among provenances from the native range (this is consistent with data from field tests in other countries, Box 8.6); and (3) Little, if any, family x zone interaction across the three edaphoclimatic planting regions. Based on this information, the breeding strategy calls for a single breeding unit defined as the entire plantation estate with the multiple-source base population to include all sources of material tested (local land race and provenances from the native range; Vergara and Griffin, 1997). The recruitment zone includes all sources in the base population. The company may define separate deployment zones (and hence separate commercial propagation populations) for each of the major regions

(Box 12.5 continued on next page)

Box 12.5. Examples of base populations, breeding units, recruitment zones and deployment zones. (Continued from previous page)

(coast, central valley and Andes foothills) if some selections perform outstandingly in only a single region. Other examples of a single large breeding unit defined for a large plantation program are: (1) *Pinus elliottii* in its natural range (White *et al.*, 1993); and (2) *Pinus radiata* in both Australia (White *et al.*, 1999) and New Zealand (Shelbourne *et al.*, 1986).

    Example 3: Single breeding unit with large seed source differences. A cooperative breeding organization (called CIEF) began a breeding program in 1984 for *Pinus taeda* in northeastern Argentina (Baez and White, 1997). The edaphoclimatic conditions are fairly uniform in this region, so the decision was made to treat the region as a single breeding zone. *P. taeda* has a wide natural range in the southeastern USA and exhibits strong provenance differences when planted in both native (Wells, 1983; Lantz and Kraus, 1987) and exotic environments (Zobel *et al.*, 1987). In most subtropical climates, including Argentina, southern provenances (*e.g.* from Florida) grow more rapidly (Bridgwater *et al.*, 1997).

    For the first cycle of tree improvement, CIEF defined a multiple-source base population: (1) Trees growing in the southern provenances of the natural range (this meant new selections and seed collections from the USA); (2) Trees growing from southern provenances in field tests in Argentina; and (3) Trees growing in plantations of the local land race in Argentina. In total more than 600 selections were made from this base population to form the first-generation selected population. More selections were made from provenances known to perform better in Argentina (*e.g.* Florida), but the goal was to create a selected population of broad genetic diversity that will be bred for performance and adaptability in the Argentine environment. Since loblolly pine is an exotic in Argentina, this broad genetic base also serves an in-country gene conservation function (*Chapter 10*).

    Example 4: Predictable interactions and multiple breeding units. As discussed in *Chapter 8*, Weyerhaeuser Company learned in a series of long-term field experiments that *Pinus taeda* seed collected from North Carolina (NC) outperforms the local provenance on moist sites in the Company's planting zone in Arkansas (AR), while local sources perform better on dry sites (an important provenance x site interaction) (Lambeth *et al.*, 1984). Based on this information, Weyerhaeuser defined two breeding units (or zones) for the first cycle of tree improvement in Arkansas: moist sites and dry sites. The first-cycle base population for the dry-site breeding unit was composed only of trees from the AR provenance (a single-source base population), while the first-cycle base population for the moist site breeding zone was composed of only NC material. Subsequently during the first-cycle of field trials, families from many other provenances (*e.g.* from Alabama and Mississippi) were tested on moist sites in Arkansas. Selections from excellent families that performed well in these tests will be included in advanced cycles of improvement for the moist-site breeding unit (personal communication, Clements Lambeth, 1999, Weyerhaeuser Company, Hot Springs, Arkansas).

    The number and size of breeding units depend, to a large extent, on genotype x environment (g x e) interaction (*Chapter 6*). If seed sources change rank dramatically across different edaphoclimatic conditions in the planting zone, then this must be due to their differential response to environmental gradients. That is, specific seed sources are better suited to some environments, but not to others. One option then is to define several breeding units that correspond to the major edaphoclimatic conditions, and to compose each base population with the sources best suited to that breeding unit (see Box 12.4 and Examples 3 and 4 in Box 12.5). Even if seed sources have similar rankings across edaphoclimatic conditions, families or clones within sources may change rank in an important and predictable way across the conditions. In this case, there still may be need for more than one breeding unit.

If g x e interaction is only moderate or is unpredictable (*i.e.* not clearly associated with climatic gradients, soil differences or other identifiable patterns across the plantation region), it may be better to define a single breeding unit. Then, there is a single base population composed of the best selections bred and tested across the entire plantation region. In this situation with a single large breeding unit, there are two options: (1) If part of the g x e is predictable, it may be possible to define multiple deployment zones and operationally plant offspring (or clones) of selections in deployment zones in which they have performed well; and (2) If some selections perform well across many sites, then they can be deployed broadly (*e.g.* these selections might be included in a single seed orchard utilized for the entire plantation region).

The latter option gives rise to the notion of breeding for wide adaptability; selections that are both superior and stable across sites are bred together as members of a single base population with the goal of obtaining genetically improved varieties that are widely adapted to the edaphoclimatic and silvicultural conditions of the plantations. Even with moderate amounts of g x e interaction, this is the option favored by most tree improvement programs due to consideration of both economics (lower costs associated with a single breeding unit and single tree improvement program) and genetic gains (higher anticipated gains from a single large base population composed of many sources with wide genetic diversity). For example, the country-wide tree improvement cooperatives breeding *Pinus radiata* in both Australia (White *et al.*, 1999) and New Zealand (Shelbourne *et al.*, 1986) defined a single breeding unit consisting of all *P. radiata* plantation sites in each country (spanning a range in annual precipitation of 500mm to more than 2000mm in Australia). There is moderate family x environment interaction in both countries (Matheson and Raymond, 1984b; Johnson and Burdon, 1990); however, this was not thought large enough to warrant multiple breeding units. Rather, a single base population is being bred for wide adaptability in each country.

When large, predictable g x e exists, it is not simple to decide between the alternatives of several breeding units or a single breeding unit with several deployment zones. The choice depends on genetic gain expectations, future projections of changes in technology and climates, economic considerations and breeding philosophy. One thing is certain, however, that if data from long-term seed source refinement trials indicate that seed sources are stable in performance across all edaphoclimatic zones in the plantation region and there is little evidence of large family x location interactions, then there is no need for more than one breeding unit (assuming a separate gene conservation program exists to preserve extant patterns of genetic diversity among sources). Remember that there can still be large genetic differences among seed sources without interaction (in fact this is common, *Chapter 8*). It is the presence of large rank changes implying differential adaptation of different sources or families to the environmental gradients that argues for multiple breeding units.

Composition of Base Populations

The base population for each breeding unit should be composed of all seed sources having significant potential to contribute favorable alleles to improved varieties being developed for the plantation conditions of that breeding unit. Therefore, appropriate composition of the base population depends on knowledge of which seed sources are best suited to the edaphoclimatic conditions of the breeding unit. For example, for native species spanning large environmental gradients, there may be many small breeding units (if there is large g x e); however, each of the many base populations could be composed of local and non-local provenances known to perform well in that breeding unit (this statement assumes a separate gene conservation program exists to preserve natural genetic diversity among provenances).

In many instances, this concept of including all promising sources in the base population of a given breeding unit leads to a multisource base population of wide genetic diversity (even multiple species to form hybrids). In a sense, this might be called the 'big bang' theory of tree breeding–bring together many diverse sources in a single base population and then let the processes of selection, breeding and testing in the tree improvement program develop improved varieties most suited to the edaphoclimatic and silvicultural conditions in the breeding unit. In practice, it is possible to make selections from each source in proportion to its relative performance in the breeding unit. This forms a base population in which each seed source is weighted according to its presumed or measured potential in the breeding unit. For example, the *Pinus taeda* tree improvement program in Argentina (Example 3 in Box 12.5) made more selections from the more promising southerly provenances in the native range with fewer from the local land race.

A final concept is that of nondisjunct, overlapping base populations in which recruitment zones are shared by distinct breeding units. This means that excellent sources of selections can be included in the base population of more than one breeding unit. For example, for its third-cycle breeding strategy, the *Pinus taeda* breeding cooperative in the southeastern USA chose to define small, local breeding units and to recruit selections for each unit from that unit and from neighboring units as well (McKeand and Bridgwater, 1998). The main point is that the composition of the base population for each breeding unit can consist of diverse sources (all presumed to have a sufficient level of adaptation), and these sources may be blended in various proportions.

SUMMARY AND CONCLUSIONS

Three broad issues discussed in this chapter are: (1) Choosing an appropriate species or hybrid taxon for a plantation program; (2) Determining the best seed sources of that species to plant in the forestation program; and (3) Defining breeding units and base populations for tree improvement. All three issues involve consideration of similar factors: (1) The importance of taxa differences expressed in field trials in the plantation region: (2) The magnitude of genotype x environment (g x e) interactions in which taxa, families or clones change rankings across different edaphoclimatic conditions of the plantation region; and (3) The appropriate level of risk to assume in order to achieve genetic gains in growth rate or product quality. Throughout this chapter, we assume that there also is a gene conservation program separate from the mainline forestation and tree improvement programs; hence, the risk alluded to is the risk of either planting maladapted taxa (in the case of the forestation program) or breeding genetically improved varieties that are poorly adapted to planting environments and silvicultural treatments.

In general, the only genetic factor that argues for use of multiple operational taxa in the plantation region or for multiple breeding units for the tree improvement program is the presence of large and predictable g x e interactions exhibited by key performance traits in long-term field trials. In the case of the decision about operational taxa, if the same taxon excels in all edaphoclimatic conditions, then the only reasons to deploy multiple taxa would be for product diversity or some other extenuating circumstances. Only large taxa x location interactions provide genetic rationale for operational use of multiple taxa within a planting zone, and these interactions must be predictable in the sense that the edaphocli-matic niche for each taxon can be defined from results of field trials.

Similarly, only when large, predictable interactions exist are there genetic reasons to employ multiple breeding units. If the same sources, families and clones perform well across all test sites in the plantation region (no g x e), then a single breeding unit is appropriate for the entire region (unless the program chooses to also achieve gene conservation goals by maintaining multiple units). When a single breeding unit is defined, multiple sources of genetic material can be included in the base population such that the tree improvement efforts develop improved varieties that are widely adapted to the silvicultural, climatic and soil conditions of the plantation region.

CHAPTER 13
PHENOTYPIC MASS SELECTION – *GENETIC GAIN, CHOICE OF TRAITS AND INDIRECT RESPONSE*

After defining the base population for a given breeding unit (*Chapter 12*), all trees in that base population are candidates for selection and therefore, for inclusion in the tree improvement program for that generation of breeding. The selected individuals form the selected population, which normally contains between 100-1000 selected trees for a single breeding unit; all other candidate trees in the base population are excluded from further consideration in that generation. Some or all selections forming the selected population are subsequently used for two distinct purposes (Fig. 11.1): (1) Inclusion in the propagation population that produces trees for operational forestation; and (2) Inclusion in the breeding population and hence in the long-term breeding program. The goal for both purposes is to make genetic gain as quickly as possible by selecting individuals that, on average, are superior for a limited number of important traits. In addition, it is important to maintain an adequate genetic base for long-term breeding and gene conservation.

Issues that arise during the selection process are: (1) Which traits and the number of traits to select for; (2) Relative emphasis that should be placed on each trait; (3) Appropriate age at which to make selections; (4) Number of selections to include in the selected population; (5) Best method to choose superior candidates as selections; and (6) Amount of genetic gain expected for each trait. This chapter addresses these issues for initial selections made in first-generation tree improvement programs in which trees are selected solely by their superior phenotypic appearance (called mass selection). In advanced generations, selection effectiveness is increased by using all information available from candidates' progeny, relatives and ancestors growing in genetic tests; these topics are covered in *Chapters 15* and *17*.

GENERAL CONCEPTS AND THEIR APPLICATION TO MASS SELECTION

The Process of Selection

Before addressing selection in first-generation programs, we discuss some general concepts that apply more widely. Many different types of selection are used at different stages and generations of program development (*e.g.* mass, parental, family, combined, indirect, tandem), but all have the same objective: to increase the frequency of favorable alleles at loci influencing the selected traits. Only if the selected population has a higher frequency of favorable alleles than the base population is genetic gain achieved in a given trait. Selection does not create new alleles, but rather aims to locate and retain individuals that already possess superior combinations of alleles.

The task of selecting superior genotypes is made difficult because we cannot measure genotypes directly; rather, we measure phenotypes of individuals and their clones, off-spring, relatives or ancestors that may be growing in genetic tests. These measurements are analyzed and used to rank candidates. The best candidates are then retained in the selected population. Therefore, for all selection situations, the basic approach to making selections is the same and is described in Box 13.1.

Box 13.1. General process of selection.

1. *Define which traits to measure.* A maximum of three to five high-priority traits and perhaps a few lower priority traits are identified. High-priority traits are those that are highly correlated with the breeding objective, have high economic importance and are highly heritable (see section on *Selection Methods for Multiple Traits*).

2. *Measure traits on phenotypes of trees growing in the base population.* In first-generation tree improvement programs, traits are measured on candidate trees growing in base populations consisting of natural forests or unpedigreed plantations, while in advanced generations, base populations consist of genetic tests containing clones, full-sib, half-sib and/or open-pollinated families (*Chapter 14*).

3. *Analyze the data.* Best linear unbiased prediction (BLUP, *Chapter 15*) is used to predict the underlying genetic value for each trait for all candidates. For mass selection, each tree's phenotypic measurement is the only data used to predict that tree's genetic value for each trait. In advanced generations, measurement data are often combined across test locations and different types of relatives.

4. *Calculate a selection index value for each tree.* Based on economic and genetic considerations, the BLUP-predicted genetic values for different traits of a given candidate are aggregated into a single predicted worth for that candidate (this predicted genetic worth is sometimes called the selection criterion or selection index, see section on *Index Selection*).

5. *Rank and select the top candidates.* Trees with the highest predicted genetic worth are selected and form the selected population. The number of related selections (*e.g.* the number of selections from a single family) is usually limited to maintain an adequate genetic base and minimize potential inbreeding in the future.

6. *Calculate expected genetic gain for each trait.* For each trait, the difference between the mean predicted genetic value of the selected population and mean genetic value of the base population is the predicted genetic gain from selection (*e.g.* Equations 13.2 and 13.3 for mass selection).

7. *Label and preserve all selected trees.* Each selected tree is clearly labeled and its exact location noted using Global Position Systems (GPS). Arrangements are made for protection and future use of all selected trees in the long-term breeding program and operational propagation program.

The selected population in each cycle of improvement always contains fewer individuals (100-1000) than the base population (many thousands to millions). Allele frequencies differ between the two populations both by intent and by chance. The breeder selects superior individuals and therefore, intentionally alters allele frequencies for the few traits included in the selection criterion. In addition, allele frequencies for all traits change by random chance due to sampling (choosing a subset of trees from a larger base population); in fact, some very rare alleles present in the base population may be lost from the selected population (Kang, 1979a,b). However, random change in allele frequencies and loss of rare alleles are expected to be minor when hundreds of individuals are included in the selected population (*Chapters 10* and *17*).

Different stages of selection may be applied in a single generation of improvement and each stage uses the process of selection outline in Box 13.1. As an example of three stages, suppose: (1) 500 individuals are included in the selected population based on superiority of their predicted genetic worth aggregated across four measured traits; (2) The top 250 of these 500 are included in that generation's breeding population after measurement of a fifth trait that would have been too costly to measure on all trees in the base population (*e.g.* wood density); and (3) The top 50 of the 500 are selected to form the propagation population whose function is to produce plants for operational forestation. Compared to the breeding population, expected genetic gains are larger (due to the more intensive selection of the top 50) and the genetic diversity is smaller for the propagation population (see Fig. 11.2). The three stages of selection in this example all attempt to maximize the genetic worth of a particular population of trees (selected, breeding and propagation populations, respectively) subject to the constraint of maintaining an appropriate genetic base in that population.

Mass Selection in First-generation Tree Improvement Programs

Mass selection (also called **phenotypic selection**) means selection of individuals based solely on their phenotypic appearance without any knowledge of performance of their clones, ancestors, offspring or other relatives. Normally, this is the only type of selection possible at the initiation of a tree improvement program, because no pedigree information is available about trees in the base population, which consists of natural stands or plantations. Therefore, breeders choose selections based on phenotypic characteristics measured or observed on individual trees. For example to select for rapid bole volume growth, the breeder chooses trees that are both taller and larger in diameter at a particular age (Fig. 13.1). Volume growth, stem straightness, crown form and freedom from disease and defect are easily observable on many candidate trees and are commonly included as traits in first-generation programs.

Once selected, it is important to begin genetic testing of all selections to verify genetic superiority. Because mass selection is based solely on phenotype, environmental influences cause some trees to appear superior, when in fact they are genetically inferior. Therefore, some selections will prove genetically superior based on excellent performance of their offspring planted across many test locations, while others will not (*Chapter 14*). Genetically superior selections are retained and emphasized in the program, while inferior selections are eliminated or de-emphasized.

Zobel and Talbert (1984, p 145) present the following terms to clearly distinguish the stages and chronosequence of mass selection: (1) **Candidate tree** is one considered for selection, and may be measured intensively or simply observed ocularly; (2) **Plus tree or select tree** is selected for inclusion in the selected population based on superior phenotypic

Fig. 13.1. Mass selections of: (a) *Pinus taeda* selected by Weyerhaeuser Company in a natural stand growing in Alachua County, Florida (75-years-old, 38 m tall and 72 cm DBH); and (b) *Eucalyptus grandis* selected by the Instituto Nacional de Tecnología Agropecuaria (INTA) (federal research group) in a plantation of the local land race growing in Corrientes, Argentina (27-years-old, 52 m tall and 59 cm DBH). (Photos courtesy of G. Powell, University of Florida and J. A. Lopez, INTA, respectively)

characteristics, but has not yet been tested for genetic worth; and (3) **Elite tree or proven select tree** has proven itself genetically superior by excellent performance of its clonal or seedling offspring in genetic tests. Elite trees are the winners from the mass selection program and should be used in the propagation population to ensure that plantations are established with material of the highest genetic quality possible in that generation.

Methods of Mass Selection

Several methods of mass selection are used in natural stands and plantations, and the choice of the method depends on factors such as species biology, costs of selection, stand history and heritabilities of important traits. Once the traits to be measured are defined (Step 1 in Box 13.1), the breeder must decide how and where to assess candidate trees in order to find good selections. For example, the base population may consist of literally millions of trees growing in natural stands scattered over thousands of square kilometers. Since it would be impossible to measure all these trees, the first step is to identify desirable stands in which to concentrate selection (Zobel and Talbert, 1984, p 152): (1) Stands should be well distributed across the appropriate portion of the species' native or exotic range to ensure wide genetic diversity in the selected population; (2) Stands should be natural or have been established with an appropriate seed source; (3) For growth traits, selections should come from stands growing on above average sites compared to those in the planting zone; (4) For stress-related traits like disease resistance and cold tolerance, it is best to locate stands that have been significantly challenged by the stress to select those trees that appear stress resistant; (5) Stands should be as pure as possible in composition of the selected species; (6) Stands should be as even-aged as possible and range from one-half to near rotation age; (7) Stands should be on uniform sites with little history of disturbance, and especially be free from high grading that already harvested the better trees; and (8) Very small stands should be avoided to minimize problems with founder effects and inbreeding (*Chapters 5* and *7*). Note that criteria 1, 2 and 8 ensure wide genetic diversity from appropriate seed sources, while the remainder aim to increase genetic gain from mass selection.

After desirable stands are chosen, the next step is to determine the appropriate method of mass selection, and we outline three such methods. When stands have many of the desirable characteristics listed above, the **comparison tree method** is generally preferred and has been widely applied in even-aged, pure-species stands all over the world (Brown and Goddard, 1961; Ledig, 1974; Zobel and Talbert, 1984, p 153). Not all trees in each stand are measured; rather, breeders visually select a candidate tree by walking through the stand inspecting all dominant trees. When an excellent candidate is located, it is marked and four to six comparison trees are located nearby. The comparison trees should be dominant trees of the same age growing on similar microsites as the candidate tree. Then, the candidate and its comparison trees are measured intensively for all traits. Next, a single composite score is calculated for each tree using a weighting system that assigns relative importance to each trait measured (this is the predicted genetic worth or selection criterion for each tree). In the past, the composite score was often calculated with a grading sheet that assigned intuitive weights to each measured trait (Zobel and Talbert, 1984, p 158); however, the recommended procedure today uses genetic and/or economic information, when available, to combine measurements into a selection index (see the section on *Index Selection*).

The candidate is selected as a plus tree only if its composite score is superior to the mean of the comparison trees by a prespecified amount. When the candidate tree is not sufficiently superior to its comparisons, the search continues in that stand for other candidate trees. When a new candidate is found, new comparison trees are located and the process repeated. Rarely, one of the comparison trees has the highest predicted genetic worth and is selected in place of the original candidate.

The comparison tree method aims to reduce environmental noise and therefore, increase gain from selection by comparing the candidate tree to trees of the same age nearby. That is, the mean of the comparison trees is intended to be a measure of site potential, and plus trees must show superiority to other trees growing on the same site. A certain minimum distance (such as 500-1000 m) is required between plus trees selected in the same stand to avoid relatedness among selections. A maximum number of selections is set for each stand depending upon its expanse (larger stands being permitted more selections).

In some situations, it is not possible to find suitable comparison trees such as when stands are uneven aged, contain multiple species, or are very heterogeneous in terms of topography, tree density or microsites. Examples of such cases are mixed hardwood forests in both temperate and tropical climates. In these situations, two other methods of selection are preferred. The **regression or baseline method**, also called individual-tree selection in older literature (Brown and Goddard, 1961; Ledig, 1974), aims to adjust tree scores for differences in ages, competition and/or environmental gradients by fitting multiple regression functions that express tree performance (the dependent variable) as a function of one or more of these independent variables. Unlike the comparison tree method, which compares trees only within stands, the multiple regression is fit across all measured trees in all stands, and all measured trees are candidates for selection. The fitted multiple regression provides a predicted value (called the baseline) for each tree measured. A candidate is selected when its actual performance for one or more traits exceeds its baseline prediction by a certain amount. That is, a tree is selected when its measured performance is better than that predicted based on its age, competition and/or other environmental parameters included in the regression model.

As an example of the regression method, suppose that bole volume is measured on each of 10 dominant trees in 80 different uneven-aged stands (800 total trees) and at the same time, the tree's age and stand site index are measured. If the 800 measured volumes are called y_i, then the multiple regression function (also called the baseline) is $\hat{y}_i = f(age, site index)$ where $\hat{y}_i$ is the predicted volume for tree i calculated from the function. A tree is selected if its deviation from the baseline (*i.e.* $y_i - \hat{y}_i$) exceeds some prespecified value. Note that approximately half (400) of the trees have positive deviations and half negative.

The final method of mass selection, called the **subjective method**, is used when the breeder either believes that environmental noise is too large to be overcome by regression methods or believes that visual inspection will be effective at locating superior trees. With this system, the breeder drives or walks through designated stands selecting trees based mainly on visual assessment. There are no comparison trees and few, if any, formal measurements. The selection can be quite intensive when many candidate trees are inspected for each plus tree selected. Very intensive subjective selection can result in substantial genetic gains for traits with moderate heritabilities if the breeder is effective at visually assessing the aggregate genetic worth of the trees. When the selection is not intensive (such as when roadside selections are made by visually selecting superior phenotypes through the windshield of the car while driving through stands), the subjective system is sometimes called the **mother tree method.** As the name indicates, the selected tree immediately serves as a mother tree for subsequent progeny testing. In this case, little gain is expected from initial selection, and emphasis is on using test results to subsequently make gains by eliminating selections that prove inferior.

Whatever method of mass selection is used, the breeder must exercise extreme caution and diligence to properly label each plus tree and record all measurements and observations. The identification assigned stays with that tree for many years, and should be clearly marked on the tree and in the records. Global Positioning Systems (GPS) can be used to record the exact location of each selected tree. An effective, reliable data management system is required to ensure integrity of all information for many years.

Often selected trees scattered in several stands in the base population are vegetatively propagated through rooting or grafting onto rootstock located at a single site called a **clone bank** or **breeding arboretum**. The clone bank serves the following functions: (1) Preserves all genotypes in the selected population in a well-protected location for gene conservation purposes; and (2) Brings all selections together in a convenient location in which breeding can be economically conducted (since some or all of the trees in the selected population form the breeding population).

PREDICTING GENETIC GAIN FROM MASS SELECTION

Three methods of estimating genetic gain (predicted, realized and simulated) were introduced in *Chapter 11*, and all methods aim to quantify effectiveness and genetic progress made from selection. Here we discuss only predicted genetic gain, which is calculated using formula derived from the theory of quantitative genetics. This approach not only produces quantitative estimates of genetic gain from selection (useful for program justification and economic analysis), but also provides considerable intuitive insight into gains expected from different types of selection (useful for program planning and breeding strategy development).

To conceptualize predicted genetic gain, consider a hypothetical base population from which the tallest 5% of the trees are chosen to form the selected population (Fig. 13.2a). There are three population means relevant to genetic gain: (1) The original base population of parents which has a certain phenotypic mean ($\mu_p = 25$ m for the hypothetical population shown at the top of Fig. 13.2a); (2) A population of trees formed by taking a large sample of offspring at random from the parents in the base population (this progeny population is not shown in Fig. 13.2, but conceptually would have the same mean as the parental base population if the offspring are measured at the same age and planted in the same set of environments as the parents); and (3) The population formed by allowing only selected parents to intermate at random and then planting their offspring in the same set of environments ($\mu_o = 27$ m in the lower part of Fig. 13.2a). Gain from selection (ΔG) is the population mean of the offspring from the selected population (number 3 above) minus the population mean from either of the other two populations (number 1 or 2 above): $\Delta G = \mu_o - \mu_p = 27$ m $- 25$ m $= 2$ m for the example in Fig. 13.2a.

A distinct and yet equivalent expression is that genetic gain for each trait is the mean genetic value of the selected parents minus the mean genetic value of the entire base population, where for the purposes of this discussion 'genetic values' can be either clonal values or breeding values depending, respectively, on whether offspring are produced vegetatively or sexually (*Chapter 6*). Only genetic values are passed onto offspring, because environmental influences affecting phenotypes are newly formed each generation. Therefore, if selected trees are used as parents to form a new generation of offspring, those offspring are better than offspring taken at random from base-population parents only if the mean genetic value of the selected population is superior to that of the base population. In order to predict the change in the population mean between two generations as shown in Fig. 13.2, we predict the mean genetic value of selected parents and compare it to the mean genetic value of all parents in the base population. This is done separately for each trait measured.

Equations for Predicting Genetic Gain

The challenge for quantitative genetic theory is to predict the difference in mean genetic (clonal or breeding) values of the selected parents and the entire base population (*i.e.* genetic gain) from the difference in their mean phenotypic values which can be readily measured. This is approached as a linear regression problem and is:

$$\Delta G = b(\bar{y}_s - \mu_p)$$

$$= bS$$

Equation 13.1

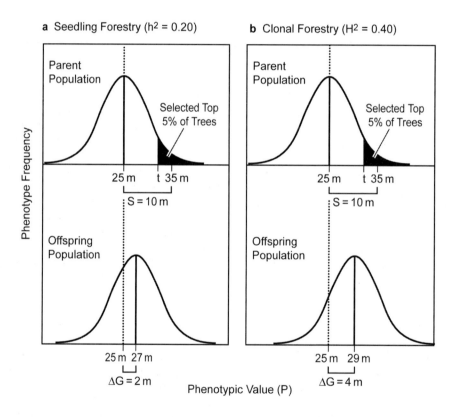

Fig. 13.2. Predicted gain (ΔG) from selection for stem height in a hypothetical parental base population (also shown in Figs. 6.1 and 6.2) that has mean $\mu_p = 25$ m and phenotypic variance $\sigma^2_p = 25$ m^2. The top 5% of the base population trees are selected and their offspring are planted in the same environments as the parents and grown to age 25 years. All selected trees have heights taller than the truncation point, t, and have a mean of 35 m such that the selection differential, S = 35 m – 25 m = 10 m. In (a), seedling offspring are planted, so the gain is calculated using the narrow sense heritability (0.20) and Equation 13.3. In (b), clonal offspring are planted, so the gain is calculated using the broad sense heritability (0.40) and Equation 13.2. Predicted gain is higher for clonal forestry than seedling forestry (4 m compared to 2 m).

where ΔG is predicted genetic gain, b is a regression coefficient, $\bar{y}_s$ is the phenotypic mean of selected trees (35 m in Fig. 13.2), μ_p is the phenotypic mean of the entire base population (25 m in Fig. 13.2), and S is the **selection differential**, $\bar{y}_s - \mu_p$ (10 m in Fig. 13.2). This equation applies to several types of selection (such as parental and family selection), but is explained here only for mass selection (see Falconer and Mackay, 1996, for a derivation of Equation 13.1 and for more advanced applications other than mass selection).

The regression coefficient needed is one that predicts genetic values from phenotypic values and this was shown in *Chapter 6* to be the heritability (Fig. 6.3). The general prediction Equation 13.1, then takes on two possible forms for mass selection, depending upon whether clonal values (ΔG_C) or breeding values (ΔG_A) are being predicted:

$$\Delta G_C = \bar{G}_s - \bar{G}_p = H^2 S \qquad\qquad \text{Equation 13.2}$$
and
$$\Delta G_A = \bar{A}_s - \bar{A}_p = h^2 S \qquad\qquad \text{Equation 13.3}$$

where for any trait measured, Equation 13.2 expresses genetic gain as the difference in mean clonal values ($\bar{G}$) between the selected population (subscript s above) and base parental population (subscript p) as a function of the broad-sense heritability, H^2, and selection differential, S, and, 13.3 expresses genetic gain as the difference in mean breeding values ($\bar{A}$) between the two populations of trees using the narrow sense heritability, h^2. Use of these equations is demonstrated with numerical examples in Box 13.2, and the two equations are compared in Box 13.3 (which is a comparison of clonal forestry *versus* seedling forestry).

The selection differential is conceptually the same for both equations and is the difference between the phenotypic mean of the selected individuals and the phenotypic mean of the base population (Equation 13.1, Fig. 13.2). The selection differential has units of the trait measured (*e.g.* meters in Fig. 13.2). Since heritabilities are unitless (*Chapter 6*), the gain predicted by Equations 13.2 and 13.3 has units of the measured trait. Predicted gain is converted into a percentage value by dividing by the mean of the base population from which the selections were made (Box 13.2). The selection differential is sometimes called the **reach** since it is the maximum upper limit to genetic gain. Heritabilities are always less than or equal to one, so gain predicted from either equation is always less than or equal to the reach (indirect selection is an exception discussed later in this chapter).

Equations 13.2 and 13.3 are appropriate for truncation selection (as depicted in Fig. 13.2) and for selections distributed through the population. That is, selections do not have to be the very best; some selections may be even below the mean of the base population. The equations above still apply as is demonstrated in Box 13.2 and developed further in *Chapter 15*. Equations 13.2 and 13.3 also do not depend on the phenotypic distribution of measurements being normally distributed. The equations provide best linear predictions of genetic gain regardless of the underlying distribution, meaning that the predictions have the minimum error variance among all possible linear combinations of the phenotypic measurements (White and Hodge, 1989). When the phenotypic measurements for a given trait are normally distributed, then the predictions are the best predictors, meaning that out of all possible linear and non-linear combinations of the phenotypic measurements, predictions from these equations have minimum error variance and maximize gain from selection.

Equations 13.2 and 13.3 apply to a specific trait and specific environment. Hence, if more than one trait is selected, heritabilities and selection differentials are needed for each trait to estimate gain for that trait. Further, the assumption is that the offspring are planted into similar environments as the parents or that genotype x environment interaction (*Chapter 6*) is negligible. The section on *Indirect Selection* addresses this further. To apply either equation, an estimated heritability must be used because the true heritability is never known. Heritability estimates can be quite imprecise and therefore, most predicted genetic gains are rough estimates.

Box 13.2. Numerical example of predicted genetic gain calculations.

Fig. 13.2 shows a frequency distribution for tree height at 25 years for a hypothetical base population with the following parameters: $\mu_p = 25$ m, $\sigma^2_P = 25$ m^2, $\sigma^2_A = 5$ m^2, $\sigma^2_I = 5$ m^2 and $\sigma^2_E = 15$ m^2 where μ_P is the population mean of tree heights, σ^2_P is total phenotypic variance, σ^2_A is additive variance, σ^2_I is non-additive variance and σ^2_E is variance due to micro-environmental effects. Broad and narrow sense heritabilities are calculated from Equations 6.11 and 6.12 as $H^2 = 0.40$ and $h^2 = 0.20$, respectively.

Equations 13.2 and 13.3 can be used to predict the underlying clonal value, G_i, and breeding value, A_i, of an individual tree, as well as for groups of selected trees (see below). For example, for a single tree from our hypothetical base population whose phenotypic measurement is 31 m, we find:

$G_i = \Delta G_C = H^2 S = 0.40$ x (31 m – 25 m) = 2.4 m
$A_i = \Delta G_A = h^2 S = 0.20$ x (31 m – 25 m) = 1.2 m

where the predicted clonal and breeding values are both expressed as deviations above the base population mean. This tree whose phenotype is 31 m, or 6 m above the mean of the base population, is predicted to have a clonal value and breeding value that are 2.4 m and 1.2 m above average. A tree 15 m tall is predicted to have clonal and breeding values of –4 m and –2 m, respectively, indicating that both are below average.

As originally defined in this chapter, these equations also apply to any group of selected trees. To see this, the table below shows predicted genetic values (G and A) for five possible selections made in the base population.

Tree	Phenotypic Measurement (P_i)	Phenotype Minus Mean ($P_i - \mu_p$)	Predicted Clonal Value (G_i)	Predicted Breeding Value (A_i)
1	31	6	2.4	1.2
2	33	8	3.2	1.6
3	35	10	4.0	2.0
4	37	12	4.8	2.4
5	39	14	5.6	2.8
Mean	35	10	4.0	2.0

Predicted genetic gains for these selections can be calculated in two equivalent ways (see *Chapter 15*): (1) Calculate the predicted clonal or breeding value for each selection and then average them (these averages are 4 m and 2 m, respectively, as shown in the last two columns of the table); or (2) First calculate the selection differential (S = 10 m) as the mean of the phenotypic measurements (35 m) minus the population mean (25 m) and then apply Equations 13.2 and 13.3:

(Box 13.2 continued on next page)

Box 13.2. Numerical example of predicted genetic gain calculations. (Continued from previous page)

$$\Delta G_C = H^2 S = 0.40 \text{ x } (35 \text{ m} - 25 \text{ m}) = 4 \text{ m}$$
$$\Delta G_A = h^2 S = 0.20 \text{ x } (35 \text{ m} - 25 \text{ m}) = 2 \text{ m}$$

Therefore, the values of 4 m and 2 m apply to any single tree whose phenotypic measurement is 35 m and also are the mean predicted clonal value and mean predicted breeding value for any group of selected trees that have S = 10 m (*e.g.* those pictured in Fig. 13.2).

When these selections are used to produce operational plantations, the predicted values are predictions of genetic gain. When clones are propagated from the selections, the plantations are predicted to be 4 m taller than plantations formed from clones taken at random from trees in the base population (Fig. 13.2b and Box 13.3). When plantations are established with seed produced by random mating among the selections, the predicted gain is 2 m above what would be obtained if the plantations were established with seed collected at random from the base population (Fig. 13.2a). Expressed as percentage values these gains are 4/25 x 100 = 16% and 2/25 x 100 = 8% above the mean of the base population, respectively.

If genetic gain and the selection differential from a previous generation of selection are known, the inverses of the two prediction equations are used to estimate what is called **realized heritability**. For example for Equation 13.3, $h^2_R = \Delta G_A / S$. Realized heritability can only be calculated retrospectively after the progeny from the selected trees are grown and the actual gains are estimated. Sometimes realized heritability is used to provide an independent validation of the heritability for a trait as estimated by other methods.

Selection Intensity

Equations 13.2 and 13.3 are often modified slightly by substituting for the selection differential, S, using the equality:

$$i = S/\sigma_P \qquad \qquad \text{Equation 13.4}$$

where i is a unitless value called the **selection intensity**, S is the selection differential in the units of the measured trait and σ_P is the phenotypic standard deviation (square root of the phenotypic variance for the trait measured), which also has the units of measure. Hence, selection intensity is the selection differential expressed in phenotypic standard deviation units. Using the example of Fig. 13.2, the selection intensity is $(35 \text{ m} - 25 \text{ m})/\sqrt{25 \text{ m}^2} = 10 \text{ m}/5 \text{ m} = 2$ meaning that the selected population shown in the figure has a phenotypic mean that is two standard deviations above the phenotypic mean of the base population. When $i\sigma_P$ is substituted for S in Equations 13.2 and 13.3, the following two commonly used equations result:

$$\Delta G_C = H^2 S = iH^2 \sigma_P \qquad \qquad \text{Equation 13.5}$$
and
$$\Delta G_A = h^2 S = ih^2 \sigma_P \qquad \qquad \text{Equation 13.6}$$

where all terms have been previously defined. These equations can be used to predict genetic gain from mass selection for a particular trait if estimates exist for the three parameters for that trait. For the example in Box 13.2, $\Delta G_C = iH^2\sigma_P = 2 \times 0.4 \times 5$ m $= 4$ m and $\Delta G_A = ih^2\sigma_P = 2 \times 0.2 \times 5$ m $= 2$ m, as obtained before. Selection intensity and heritability are unitless, while the phenotypic standard deviation has units of the measured trait; this results in gain predicted in the units of measure.

Box 13.3. Comparison of Equations 13.2 and 13.3 — clonal forestry *versus* seedling forestry.

Even though the details of propagation populations are not described until *Chapter 16*, we can preview the desirability of achieving clonal forestry by comparing genetic gain from two alternatives based on mass selection followed by immediate operational deployment of either clones or seedlings from the selected population.

Clonal forestry. If operational plantations are planted with clones propagated vegetatively from the selected trees without sexual recombination (clonal forestry in *Chapter 16*), then the entire clonal value of each phenotypic selection is expressed in those plantations, and Equation 13.2 is appropriate for predicting genetic gain from mass selection.

Seedling forestry. On the other hand, if operational plantations are planted with seedlings produced by random mating among the selections (such as from seed orchards, *Chapter 16*), then only the breeding value portion of each phenotypic value is passed onto offspring in those plantations. Equation 13.3 then is appropriate for predicting genetic gains from mass selection as they will be realized in operational plantations.

Comparison of genetic gain. For the same set of selections, predicted genetic gain is always larger for clonal forestry as opposed to seedling forestry, and is the ratio of the broad and narrow sense heritabilities: $\Delta G_C/\Delta G_A = H^2S/h^2S = H^2/h^2$ since the selection differentials are equal if we are considering the same set of selections. This is illustrated by comparing Figs. 13.2a, b where $\Delta G_C/\Delta G_A = H^2/h^2 = 4$ m/2 m $= 2$. The additional gain from clonal forestry comes from the fact that vegetative propagation captures the entire clonal value, G, of each selected tree, while seedling forestry captures only the breeding value, A. While desirable in theory, it is often difficult to achieve clonal forestry because of difficulties in vegetative propagation of selection-age trees (*Chapter 16*).

Genetic gains from breeding. Previous comments pertain to genetic gains in operational plantations, but it is also important to predict long-term gains that accumulate in breeding populations from several cycles of a genetic improvement program. Equation 13.3 is appropriate for calculating gains achieved from recurrent mass selection in a long-term improvement program involving the core activities of the breeding cycle (selection and intermating, Fig. 11.1). During the intermating phase, sexual recombination occurs and only breeding values are passed onto offspring. Therefore, the extra gains achieved by clonal forestry are realized in operational plantations established in any given generation of improvement, but are not accumulated in the breeding population. Cumulative gain from breeding depends on sustained improvement of the mean breeding value of the selected and breeding populations.

Sometimes it is useful to predict gain from selection before actually making the selections. For example, this may be important for choosing how much effort to put into the selection process and to evaluate different selection strategies. Before making selections neither the selection differential nor selection intensity are known (since no trees have been measured), but we can still make predictions when four conditions are met (Falconer and Mackay, 1996): (1) The phenotypic values in the base population follow a bell-shaped curve called a normal distribution (as shown in Fig. 13.2); (2) Truncation selection is employed meaning that only phenotypes above a certain truncation value, t, will be selected; (3) The proportion of trees selected is known; and (4) Estimates exist of the heritability and phenotypic standard deviation for the base population.

When these four conditions are met, there are tables (Becker, 1975; Falconer and Mackay, 1996) and graphs (Fig. 13.3) in which selection intensity is expressed as a function of the proportion of individuals selected. When the top 5% of the population is selected (the shaded portion in Fig. 13.2), the selection intensity is approximately 2 (actually $i = 2.06$ when the top 5% is selected from an infinite-size base population). As reference points, when the following top fractions of the population are selected 0.50, 0.25, 0.10, 0.01, 0.001, and

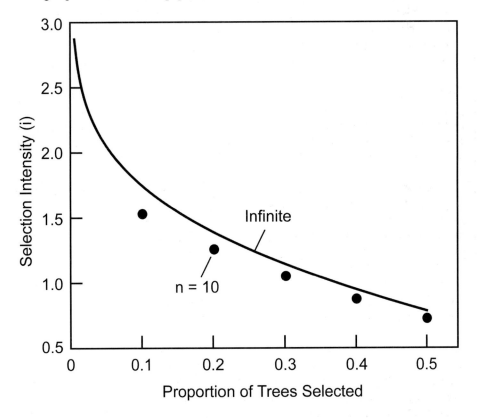

Fig. 13.3. Selection intensity, i, plotted as a function of the fraction of the population selected. The selection intensity is unitless and is expressed as standard deviations above the population mean. The selection intensity varies slightly with the size of the base population from which selections are drawn, and is shown for two population sizes: infinite and $n = 10$. Example: if the best tree is selected from a population of size 10, the proportion selected is 0.10 and $i = 1.54$, while for the same proportion selected from an infinite size population, $i = 1.76$. In practice, tables for small base populations are needed for precise predictions when $n < 50$. (Plotted from tables in Becker, 1975)

0.0001, the corresponding selection intensities are 0.80, 1.27, 1.76 and 2.66, 3.37 and 3.96. Therefore, when the proportion of the population selected is the best 0.0001 (*i.e.* the best tree selected out of every 10,000 measured), the mean of the selected population is nearly 4 standard deviations above the mean of the entire base population.

For most polygenic traits, the frequency distribution of phenotypic measurements in the base population is approximately normal. Further, deviations from normality generally have relatively minor effects on the predicted gain from Equations 13.5 and 13.6. Therefore, these two equations are commonly used to predict genetic gain from mass selection for a variety of traits with continuous distributions. For binomial traits measured as 0 or 1 (*e.g.* presence or absence of a disease), adjustments must be made to the predictions obtained from these equations (Dempster and Lerner, 1950; Lopes *et al.*, 2000).

Factors Affecting Genetic Gain from Mass Selection

Equations 13.5 and 13.6 provide intuitive insight into potential ways to increase gain from mass selection. Since gain is directly proportional to all three parameters on the right hand side of the equations, doubling any of the three factors doubles expected gain. Here we examine each of the three parameters in turn to consider what factors breeders can manipulate to increase gain from mass selection.

Selection Intensity

Selection intensity is the factor under most control by breeders. More intensive selection results in more genetic gain. In other words, the smaller the fraction of the population selected, the greater the selection intensity and the greater the expected gain. However, a point of diminishing returns occurs after a certain amount of effort is expended to measure candidate trees looking for the best ones to select (Fig. 13.4). Consider the following values of i as a function of the number of candidates measured: i is approximately 1, 2 and 4 when one tree is selected from 2.62, 20 and 10,000 trees measured (the first two values are from Fig. 13.4, the third is from tables in Becker, 1975). Therefore, if the goal for the selected population of a tree improvement program is to make 100 selections, it would be necessary to measure 262, 2000 and 1,000,000 trees to achieve selection intensities of 1, 2 and 4. It is much easier to double gain from a selection intensity of 1 to 2 (requiring a ratio of 2000/262 = 7.6 times as many trees measured) than to double gain again from intensities of 2 to 4 (requiring a ratio of 1,000,000/2000 = 500 times as many trees measured).

If costs of making selections are mainly associated with costs of measuring candidate trees, then the diminishing returns can be interpreted as increasing the cost per selection; costs rise rapidly for relatively small increases in expected gain as the curve in Fig. 13.4 flattens. It still may be justifiable to expend the extra costs associated with more intensive selection (since each increase in intensity is expected to increase gain); however, a complete economic analysis is required to make this decision.

When many traits are selected at the same time, the selection intensity is reduced for each trait considered individually. Suppose the breeder decides to select the best tree from every 100 candidates measured. For a single trait, the selection intensity is 2.66 (Fig. 13.4). If instead, the best tree is selected on the basis of two uncorrelated traits, then the selected tree is expected, on average, to be in only the top ten for each of the traits, which means i = 1.76. This is demonstrated by the product law of probability (Bayes Theorem) which states that if 1 out of 10 (*i.e.* 0.10) trees exceeds the truncation value for the first trait and the same fraction (*i.e.* 10%) exceeds the truncation value for a second, independent, trait, then the probability of a single tree exceeding the truncation values for both traits is the product of the individual probabilities (*i.e.* 0.10 x 0.10 = 0.01). This means that selection intensity has dropped from 2.66 for a single trait to 1.76 for each trait when there are two traits (assuming equal intensities). Therefore, predicted genetic gain is 50% more for the first trait when it is the only trait selected: $\Delta G_1/\Delta G_2 = i_1 h^2 \sigma_P / i_2 h^2 \sigma_P = i_1/i_2 = 2.66/1.76 = 1.51$.

As the number of traits increases, the selection intensity for each trait continues to decrease for a given number of candidates measured. In the example of one tree selected for each 100 measured, the inclusion of six uncorrelated traits means that the selected tree is only in the top half of the distribution for each of the six traits (*i.e.* using the product law of probability: $0.46^6 = 0.01$ or $\sqrt[6]{0.01} = 0.46$). This drops the selection intensity to $i = 0.87$ meaning that gain is three times more for any trait when it is the only trait selected compared to when it is included in a suite of six uncorrelated traits ($2.66/0.87 = 3.05$). This exercise demonstrates that to maximize genetic gains for the most important traits, tree improvement programs need to minimize the number of traits selected. Normally, no more than three to five high-priority traits should be emphasized during selection (see the section on *Defining the Breeding Objective*).

Heritability and Phenotypic Standard Deviation

Breeders have less influence over heritability than selection intensity. However, heritability (broad or narrow sense) increases as either the environmental error variance in the denominator decreases or the genetic variance in the numerator increases (see Equations 6.11 and 6.12). Many of the recommendations given previously in the section on *Methods of Mass Selection* aim to reduce the variance due to environmental factors. These include making selections in pure-species, even-aged plantations, and on uniform sites whenever possible.

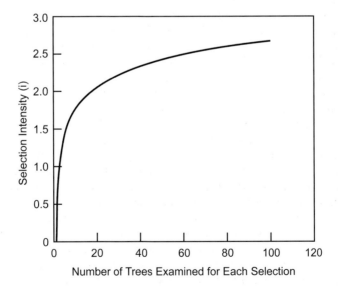

Fig. 13.4. Selection intensity, i, plotted as a function of the number of trees measured for each tree selected for an infinite population. Note that as the number of trees measured rises, there is a decreasing rise in the selection intensity. (Plotted from tables in Cotterill and Dean, 1990)

Increasing the numerator of the observed heritability is possible when a trait is differentially expressed in different environments. All pest and stress resistance traits require an environment in which stress is present in sufficient quantity to allow genetic differences in resistance to be expressed. For example, cold hardiness is not expressed unless there is a cold event that allows differential reaction of the trees in the base population. In addition to stress resistance traits, some other traits are better expressed in certain environments. It is often said that selections for growth rate should be made in stands with above-average site quality on the basis that genetic differences among trees are more clearly expressed when trees are growing more rapidly. A final example is stem straightness for which variation in some species is more clearly revealed on sites that have certain soils or growth characteristics. In all of these cases, the idea is to increase heritability and hence genetic gain by making selections in environments in which genetic differences among trees are well expressed.

More gain also is expected for traits with a larger phenotypic standard deviation and other things being equal. Low phenotypic variability means that there is little hope of finding trees substantially larger than the mean of the base population. When two traits have the same selection intensities and heritabilities, more gain is expected from the trait that is more variable in the population. Also, when distinct traits have differing heritabilities, focusing on traits with higher heritability leads to higher genetic gain (all other things being equal).

INDIRECT MASS SELECTION

Definition and Uses of Indirect Selection

The previous section dealt with direct mass selection. **Indirect mass selection** occurs when individuals are selected on the basis of one trait (called the measured trait) and gain is predicted for a second trait (called the **target trait**). The prediction equations (derived in Falconer and Mackay, 1996) require estimates of genetic correlations (*Chapter 6*) between the selected trait (y) and the target trait (t) and are:

$$\Delta G_{C,t} = i_y r_{G,ty} H_y H_t \sigma_{P,t} \qquad\qquad \text{Equation 13.7}$$
and
$$\Delta G_{A,t} = i_y r_{A,ty} h_y h_t \sigma_{P,t} \qquad\qquad \text{Equation 13.8}$$

where $\Delta G_{C,t}$ and $\Delta G_{A,t}$ are the genetic gains in trait t from indirect selection on trait y depending on whether clonal or breeding values, respectively, are being predicted; i_y is selection intensity on the measured trait, y; $r_{G,ty}$ is the genetic correlation between true clonal values for traits y and t; H_y and H_t are the square roots of the broad sense heritabilities for the measured and target traits; $\sigma_{P,t}$ is the phenotypic standard deviation for the target trait; $r_{A,ty}$ is the genetic correlation between true breeding values for traits y and t; and h_y and h_t are the square roots of the narrow sense heritabilities for the measured and target traits. The correlation in both equations must be a genetic correlation not a phenotypic correlation (see *Chapter 6* for distinction and further discussion of types of genetic correlations). Numerical application of Equation 13.8 is demonstrated in Box 13.4.

The first four terms in both equations are unitless, and the final term, $\sigma_{P,t}$, has units of the target trait; hence, predicted gains have units of the target trait. Therefore, the equations take the data on the measured trait and rescale these data so that the gains predicted are in the appropriate scale of the target trait. For example, if selection is based on measurements of tree height and the target trait is bole volume, then predicted gains have units of volume, m^3.

Indirect selection is extremely common in tree improvement programs, and the following examples illustrate the diversity of applications: (1) Selection for bole volume or height at early ages in the field when true interest is harvest-age volume (Lambeth *et al.*, 1983a; McKeand, 1988; Burdon, 1989; Magnussen, 1989; White and Hodge, 1992; Balocchi *et al.*, 1994); (2) Early selection based on a variety of young seedling traits expressed in greenhouse, growth room or other artificial environments when the target trait is field performance at harvest age (*e.g.* in *Pinus taeda* alone: Lambeth, 1983; Bridgwater *et al.*, 1985; Talbert and Lambeth, 1986; Williams, 1988; Li *et al.*, 1989; Lowe and van Buijtenen, 1989); (3) Screening of seedlings after artificial inoculation in the greenhouse to predict disease resistance in the field (Anderson and Powers, 1985; Redmond and Anderson, 1986; Oak *et al.*, 1987; de Souza *et al.*, 1992); (4) Visual scoring of damage or measurement of cellular electrolyte leakage after freezing of leaf tissue in the laboratory to predict cold hardiness in field conditions (Tibbits *et al.*, 1991; Raymond *et al.*, 1992; Aitken and Adams, 1996); (5) Depth of penetration of a spring-loaded pin into the outer wood of a tree (using a hand-held instrument called a **pilodyn**) to predict whole-tree wood density (Cown, 1978; Taylor, 1981; Sprague *et al.*, 1983; Watt *et al.*, 1996); and (6) Selection for the trait of interest, such as bole volume, in one field environment for deployment to plantations to be

---

Box 13.4. Numerical example of predicting genetic gain from indirect selection.

For this example, the target trait is stem height at 25 years, the same trait described for the hypothetical population in Box 13.2 and Fig. 13.2. Equation 13.8 is used to predict additive genetic gain at 25 years from mass selection on juvenile height at 8 years (Equation 13.7 for predicting gains in clonal values is not illustrated). Assumed genetic parameters for both traits are:

Target trait (t = height at 25 years): $\mu_t = 25$ m, $\sigma^2_{P,t} = 25$ m^2, $\sigma^2_{A,t} = 5$ m^2, $\sigma^2_{I,t} = 5$ m^2 and $\sigma^2_{E,t} = 15$ m^2
Measured trait (*y* = height at 8 years): $\mu_y = 12$ m, $\sigma^2_{P,y} = 10$ m^2, $\sigma^2_{A,y} = 2$ m^2, $\sigma^2_{I,y} = 2$ m^2 and $\sigma^2_{E,y} = 6$ m^2

where terms are defined in Box 13.2. In the absence of a genetic correlation estimated from real data, use of Equation 6.13 yields $r_{A,ty} = 0.67$. Note that the narrow-sense heritability for 8-year height is assumed the same as that for 25-year height ($h^2_8 = h^2_{25} = 0.20$). The selection intensities at both ages are also assumed to be identical: $i_8 = i_{25} = 2$.

Two genetic gains are compared: (1) Gain in 25-year height made from direct selection at age 25 calculated from Equation 13.6 as $\Delta G_{A,25,25} = 2 \times 0.2 \times \sqrt{25} = 2$ m or 8% of the 25-year mean height of the base population; and (2) Gain in 25-year height made from indirect selection at age 8 calculated from Equation 13.8 as $\Delta G_{A,8,25} = i_8 r_{A,25,8} h_8 h_{25} \sigma_{P,25} = 2 \times 0.67 \times \sqrt{0.2} \times \sqrt{0.2} \times \sqrt{25} = 1.34$ m which is 5.36% of the 25 m mean at age 25.

Although heritabilities of height at ages 8 and 25 were the same, predicted gain in 25-year height of offspring from parents selected at age 8 is less than predicted genetic gain if selection had been delayed until age 25. This is because there is only a partial overlap in the genetic control of height at the two ages (*i.e.* the genetic correlation between height at ages 8 and 25 was assumed to be only moderate, 0.67).

established in distinct field environments with different edaphoclimatic conditions than the selection environment.

The type of genetic correlation required depends on the type of indirect selection. For the first example above, the juvenile-mature genetic correlation is required to specify the correlation between two ages for the same trait (Box 13.4, *Chapter 6*). For examples 3-5, the genetic correlation is between two traits (*e.g.* a greenhouse symptom and disease resistance in the field for example 3). For example 6, the type B genetic correlation specifies the correlation between the same trait as expressed in two different environments (*Chapter 6*).

Inadvertent selection is a type of indirect selection that occurs when the breeder unknowingly affects a second trait by selection on the first trait. For example, if there is an unfavorable genetic correlation between stem growth rate and cold hardiness, the breeder might inadvertently reduce cold hardiness in the selected population by selecting for increased growth rate. In this case, the measured trait (y in Equations 13.7 and 13.8) is growth rate, while the target trait, t, is cold hardiness. To avoid unplanned effects on breeding populations, the breeder is wise to estimate genetic correlations between measured traits and important other traits that will not be routinely measured in the breeding program. The breeder can then take steps to ensure that inadvertent selection will not cause unwanted genetic changes in traits not measured, or at least minimize unfavorable responses.

Comparison of Indirect and Direct Selection

To compare relative effectiveness of direct selection and indirect selection, divide Equation 13.8 for indirect selection by Equation 13.6 for direct selection (or divide Equation 13.7 by 13.5):

$$\Delta G_{A,t}/\Delta G_A = (\text{Gain, Indirect Selection})/(\text{Gain, Direct Selection})$$
$$= (i_y r_{A,ty} h_y h_t \sigma_{P,t})/(i_t h^2_t \sigma_{P,t})$$
$$= r_{A,ty}(i_y/i_t)(h_y/h_t) \qquad \text{Equation 13.9}$$

where all terms have been previously defined and the subscript t has been added to the formula for direct selection to emphasize that all measurements, values and predictions apply to the target trait. Note that all terms are unitless, and that the ratio can be larger than one under certain conditions (*i.e.*, indirect selection produces more genetic gain than direct selection).

We consider each component of Equation 13.9 to understand intuitively those situations favoring indirect selection. The genetic correlation ranges between 1 and -1, and determines the amount of genetic information the measured trait provides about the target trait. When the genetic correlation is near one, there is a high degree of pleiotropy, meaning that many common gene loci affect both traits; hence, indirect selection on the measured trait changes allele frequencies in many of those common loci to produce gains in both traits. An example is DBH and bole volume, which have a large positive genetic correlation in all species. Selection of trees with large DBH produces genetic gain in both DBH and volume.

When the genetic correlation is near zero, few, if any, gene loci affect both traits and selection on the measured trait has little influence on allele frequencies at loci controlling the target trait. When a negative, unfavorable correlation exists between two traits, selection for improvement in one trait will result in a negative response in the other trait. For example, in species where stem growth rate and wood density are negatively correlated, selection of faster growing trees will inadvertently lead to lower wood density in the offspring of the selections (negative gains as calculated by Equations 13.8 or 13.9). Therefore, the magnitude of gains or losses in the target trait depends directly on the magnitude and sign of the genetic correlation.

The second component of Equation 13.9 is the ratio of selection intensities on the measured and target traits: (i_y/i_t). A value larger than one is not uncommon and is possible when more intensive selection can be applied on the measured trait. This occurs when the measured trait is more rapid or less expensive to measure than the real trait of interest. An example is the hand-held pilodyn used as a rapid method to assess wood density. The ability to measure many more trees increases the selection intensity for the proxy variable (*e.g.* pilodyn pin penetration) above that possible for the more difficult-to-measure target trait (*e.g.* whole-tree wood density as assessed from wood samples in a laboratory).

The final component of Equation 13.9 is the ratio of the square root of the heritabilities of the traits: (h_y/h_t). Sometimes a correlated trait might have a higher heritability than the target trait making it more favorable for selection. Two examples are: (1) Much more uniform sites with lower environmental variance are found in one region and selections are made there for use in target sites located in a different planting region; and (2) Traits measured in greenhouse or growth room environments have high heritabilities due to uniform experimental conditions and are used as proxy traits when the target trait is field performance.

When all three components of Equation 13.9 are favorable, indirect selection can produce substantially more gain than direct selection. Sometimes, indirect selection may be a viable alternative even when less gain is achieved. This commonly occurs for early selection when time saving is considered. Gain from early, indirect selection for stem growth in a given cycle of tree improvement is often less than that possible from direct selection for stem volume at rotation age. However, gain per unit of time (gain per year) is often greater with early selection, which permits a new cycle of improvement to start sooner and hence allows more cycles to be completed. To determine the optimum age for selection, breeders employ equations like Equation 13.8 and 13.9, but also factor in costs and the time value of money; ages that are 25%-50% of the harvest age are commonly cited as optimal (McKeand, 1988; Magnussen, 1989; White and Hodge, 1992; Balocchi *et al.*, 1994).

SELECTION METHODS FOR MULTIPLE TRAITS

Defining the Breeding Objective

All tree improvement programs aim to improve more than one trait at the same time, and there are two distinct types of traits to consider: target traits (discussed here) and measured traits, also called selection traits (discussed in the following three sections). The first step is to explicitly define the breeding objective, which is the combination of target traits that the breeder desires to improve (Borralho *et al.*, 1993; Woolaston and Jarvis, 1995; Bourdon, 1997, p 281). The **breeding objective**, written for a particular tree as T_i, is sometimes called the aggregate breeding value or economic genetic value (van Vleck *et al.*, 1987), and is defined as follows:

$$T_i = w_1 A_{1i} + w_2 A_{2i} + ... + w_m A_{mi} \qquad\qquad \text{Equation 13.10}$$

where w_1, w_2,...w_m are the economic (or social) weights for each of the m target traits included in the breeding objective and A_{1i}, A_{2i},....A_{mi} are the breeding values of tree i for each of the m traits.

The breeding objective is symbolized as H by some authors, but we prefer T because in every sense the breeding objective is the ultimate target that the breeder is trying to improve. In fact T can be thought of as a new aggregate breeding value that weights the breeding values of the original traits according to their relative importance. When economic gain is the goal of the tree improvement program, T is best expressed in units of dollars and measures the breeding value of a tree in monetary units. Traits with no economic importance have a weight of zero and are dropped from the breeding objective. All the important economic traits should be included, but the number of these high-priority traits should be five or less for reasons previously discussed in *Selection Intensity*.

Traits are included in the breeding objective only because of their importance and may or may not be measured on trees in the field. For example, harvest-age stem volume and pulp yield may be two traits in T, while juvenile height, pilodyn-assessed wood density and straightness may be the measured traits upon which selection is based. Therefore, this explicit definition of a breeding objective clearly separates the target aggregate genetic worth (*i.e.* the goal of improvement) from the measured selection criteria used in an attempt to reach that goal.

Forest tree improvement programs have not been as diligent as animal improvement programs in the explicit specification of breeding objectives (Woolaston and Jarvis, 1995) due in part to: (1) Difficulty of determining precise economic weights; (2) The complex and long-term nature of tree improvement; and (3) Emphasis on program implementation. A general idea of the overall goal is often formulated in tree improvement programs as a list of high-priority traits. Nevertheless, when the breeding objective is not explicitly defined, there still is an undefined, implicit set of weights for the traits in the program; however, this *de facto* set of weights is unlikely to achieve the desired results.

In summary, creation of a breeding objective with formal economic weights: (1) Exposes weaknesses in knowledge associated with the relative importance of different traits (and therefore allows these deficiencies to be addressed); (2) Introduces consistency to selection decisions; and (3) Provides a powerful tool for evaluating which traits should be measured and how they should be incorporated into the final selection criteria (Woolaston and Jarvis, 1995).

In a study to develop a breeding objective for a tree improvement program of *Pseudotsuga menziesii,* a multiple regression equation with three independent variables (stem volume, branch diameter and wood density) effectively predicted the value of trees sawn for solid-wood products (Aubry *et al.*, 1998; Box 13.5). The partial regression coefficient on each independent variable in the equation is its economic weight. Another excellent example of a breeding objective in tree improvement was developed for pulp production of *Eucalyptus globulus* in Portugal (Borralho *et al.*, 1993). One of the many important outcomes of this study was the clear economic importance of measuring wood traits in order to make selection decisions even though these traits had long been thought too costly to measure.

The two preceding examples notwithstanding, lack of reliable estimates of economic weights has been the main barrier to complete development of breeding objectives and use of index selection in tree improvement programs (Bridgwater *et al.*, 1983; Zobel and Talbert, 1984; Cotterill and Jackson, 1985). Since tree improvement programs are breeding improved varieties for future markets, the economic weights should reflect future product demand and pricing structures. With wood products markets being cyclical and new mill technologies always evolving, divining future values can be quite a challenge. Sometimes it is useful to develop several sets of economic weights corresponding to several future economic scenarios. If one set of weights is stable (*i.e.* robust) across several differing scenarios, then this set is chosen as appropriate to use in the breeding objective.

In the future, more tree improvement programs should explicitly define the breeding objective to maximize genetic gain from the program.

Choosing Which Traits to Measure

Breeders choose which traits to measure in an attempt to maximize genetic progress per unit time in the breeding objective, T. The most common mistake is including too many measured traits in a tree improvement program. As demonstrated in the section on *Selection Intensity,* inclusion of an additional trait reduces the gain on traits already in the program (assuming equal effort). Therefore, it is critical to limit the number of traits to as few as possible and to prioritize traits into groups of importance such as high, medium and low. There should no more than three to five high-priority traits that receive maximum emphasis, with one or two traits in each of the medium and low categories.

Desirable characteristics of high-priority traits include: (1) Strong correlation with the breeding objective, T; (2) High heritability; (3) Low cost and high precision of measurement; (4) Reliable assessment at an early age; and (5) No unfavorable genetic correlations with other high-priority traits. Seldom do all factors coincide for all traits of possible interest, and in most tree improvement programs economic importance (*i.e.* strong correlation with T) is the most important consideration. This is why stem growth rate is the highest priority trait measured in programs of many species even though it has a modest heritability (*Chapter 6*) and is not easily assessed at young ages.

Disease resistance also is important in many programs due to economic importance, good gain potential, ability to be assessed at early ages and lack of unfavorable correlation with growth rate. As a final example, although wood density is quite important economically for some products and has a high heritability, it is costly to measure and may have relatively limited phenotypic variability (thus reducing potential gain). Therefore, it is emphasized in some programs, but not others.

Lower priority traits are best used as ancillary information to guide selection decisions about trees that are borderline based on the measurements of higher priority traits. The important concept is to focus on measurements of high-priority traits and not permit lower priority traits to play a large role in selection decisions. An example from some programs is the relatively high emphasis placed on crown form when its economic importance and heritability are poorly understood. This only detracts from the high-priority traits and reduces expected genetic progress in the breeding objective, such as when an excellent candidate tree based on other traits is not selected due to poor crown form.

Index Selection

The best method for multiple-trait selection incorporates high-priority traits into a single **selection index**, I:

$$I_i = b_a(y_{ai} - \bar{y}_a) + b_b(y_{bi} - \bar{y}_b) + ... + b_p(y_{pi} - \bar{y}_p) \qquad \text{Equation 13.11}$$

where I_i is the calculated selection index for tree i; b_a, b_b...b_p are the numerical weights applied to the phenotypic measurements of each of the p traits measured on each tree; y_{ai}, y_{bi},...y_{pi} are the phenotypic measurements for tree i for traits a, b,...p; and $\bar{y}_a$, $\bar{y}_b$ ... $\bar{y}_p$ are the phenotypic means of the base population for each of the p traits. Note that $y_{ai} - \bar{y}_a$, $y_{bi} - \bar{y}_b$, etc. are the selection differentials expressing a tree's measured value above or below the mean of the base population for each trait. The letters a, b,...p are used instead of numbers 1, 2,...m to indicate that the measured traits are not necessarily those in the breeding objective of Equation 13.10.

We only briefly highlight the concepts of index selection. Readers are referred to the following sources for more in depth treatments, for derivation and theory (Henderson, 1963; White and Hodge, 1989; van Vleck, 1993; Bourdon, 1997), and for applications in forestry (Shelbourne and Low, 1980; Bridgwater *et al.*, 1983a; Christophe and Birot, 1983;

Box 13.5. Development of a breeding objective and economic weights for *Pseudotsuga menziesii.*

Background. A tree improvement program in western Oregon (OR) and Washington (WA), USA, is breeding *P. menziesii* for solid wood products, and three traits are known to influence yield and value of the lumber: stem volume, wood density and knot size. To increase value of the boards sawn from a tree, the goal is to increase volume per tree while maintaining high wood density and limiting knot size. To estimate reliable economic weights for these traits in a breeding objective, Aubry *et al.* (1998) conducted the product recovery study summarized below.

Field and laboratory methods. (1) 164 trees from 11 intensively managed stands in western OR and WA spanning a range of site qualities, growing conditions and ages (36-66 years) were measured for bole height and DBH, and diameter of the largest branch in the lowest log; (2) Increment cores (9 mm cores at breast height) were extracted from 92 trees and taken to the lab to determine wood density by x-ray densitometry; (3) Each tree in the field was felled, bucked into mill-length logs (4-8 m) and total tree volume calculated as the sum of the inside-bark volumes of all logs; (4) The logs from each tree were taken to a state-of-the-art mill and sawn into boards of standard dimensions; (5) All boards were graded by machine stress rating to assign a dollar value to each board; and (6) The dollar values from all boards sawn from all logs from a given tree were summed to estimate the tree's total value.

Data analysis. Multiple linear regression techniques were used to predict whole-tree value, obtained from step 6 above, as a function of several potential independent, regressor variables including height, DBH, branch diameter, bole volume, wood density and all of their two-way interactions. The final equation had an adjusted multiple coefficient of determination of $r^2 = 0.89$ and was:

$$T_i = -14.557 + 0.058(VOL) - 6.669(BD) + 0.065(WD)$$

where T_i is the predicted total value for a tree, -14.557 is the y intercept, VOL is the tree's bole volume in dm^3, BD is the diameter of the largest branch in the lowest log in cm, WD is wood density at breast height in kg m^{-3}, and the numbers preceding the three independent variables are partial regression coefficients.

Interpretation. The breeding objective, T_i, has units of dollars and hence, serves to integrate the relative importance of the three measured traits in terms of their effect on total product value from a tree. The three partial regression coefficients in the equation above can be used directly as economic weights for the three traits in the breeding objective. Each coefficient estimates the amount that the dollar value of a tree changes when the value of that trait changes by one unit and the other two traits are held constant. (Box 13.5 continued on next page)

Box 13.5. Development of a breeding objective and economic weights for *Pseudotsuga menziesii*. (Continued from previous page)

For example, 0.058 has units of \$ dm^{-3} and estimates that the tree's value increases \$0.058 for each 1 dm^3 increase in its volume (when BD and WD are held constant). The negative sign on the coefficient for BD indicates that increasing branch diameter decreases tree value. The coefficients in this equation depend on the scales of the measured traits. Therefore, it is not possible to look at the relative sizes of the coefficients and deduce their relative economic importance. Sometimes, standardized coefficients are developed to be scale independent.

It is important to note that this equation specifies only the breeding objective and not which traits to measure in the field to achieve this objective. For example, these data were taken on trees of ages 36-66 years using very intensive lab methods to obtain a precise measure of wood density. For selection purposes, it might be desirable to measure younger trees and use the pilodyn to rapidly estimate wood density. The exact traits to measure for selection depend on heritabilities, genetic correlations and other parameters (see section on *Index Selection*); however, all traits measured for selection should have strong genetic correlations with the traits in the breeding objective.

Dean *et al.*, 1983; Burdon, 1989; Cotterill and Dean, 1990; Borralho *et al.*, 1992b, 1993; Aubry *et al.*, 1998). Selection indices also are used to combine data from different relatives such as parents and offspring and this application is discussed in *Chapter 15*. Here we focus on mass selection.

After measurements are taken on trees in the base population, genetic theory is used to estimate the weights in Equation 13.11, and the equation is calculated as a single number for each tree that aggregates the data from all traits into the I_i for that tree. An index with two traits might be as simple as $I_i = 0.3 \times HT_i + 2.4 \times PILO_i$ where HT_i and $PILO_i$ are the height and pilodyn measurements used to calculate a single index for each tree measured. The trees are then ranked based on the calculated indices and trees having the highest values of I_i are selected (Fig. 13.5). Note that a tree with a low measurement on one trait can still be selected if its measurements for other traits are high enough (for example, trees A, B, X and W in Fig. 13.5 all have low measurements for one of the two traits). Also, note that when two traits are negatively correlated, it is difficult to find trees outstanding for both traits (Fig. 13.5b), and such trees are called **correlation breakers**.

The coefficients in Equation 13.11 are estimated from theoretical relationships to maximize the correlation between the calculated index, I_i, and the breeding objective, T_i (Henderson, 1963; White and Hodge, 1989; Borralho *et al.*, 1993). This means that if 1000 trees are measured, the 1000 indices calculated (one for each tree's measurements) have the highest possible expected correlation with the 1000 true, yet unknown, aggregate genetic worth of those trees. No other set of weights produces a higher correlation. When the traits are normally distributed, use of the selection index produces maximum genetic gain in the breeding objective from selection based on the measured traits.

To effectively aggregate the measurements of all traits into a single index value, the weights b_a, b_b,...b_p must incorporate economic information (since the weights maximize the correlation with T which is the economic target of the breeding) and genetic information (*i.e.* heritabilities of the measured traits, their correlations among themselves and their correlations with the traits in the breeding objective T_i). In general, higher weights are assigned to traits that are highly correlated with the breeding objective, have high heritabilities, and have the least unfavorable correlations with other traits.

One criticism of selection indices is that the weights are very difficult to estimate precisely due to lack of economic information, unstable future conditions, nonlinearity of economic relationships and imprecise estimates of the genetic parameters. To minimize the number of correlations and other genetic parameters that need to be estimated, it is important to include only a few high-priority traits in a selection index. Precise estimates of parameters for those traits should then be obtained from genetic tests with a large number of genetic entries (*i.e.* at least 100 families or clones). Under these conditions, selection indices are an extremely effective method of making selections.

When economic information is not readily available, other methods for assigning the index weights in Equation 13.11 can be based on achieving desired gains in individual traits or maximizing gain in one trait while limiting changes in others (Cotterill and Dean, 1990, p 37). These Monte Carlo methods, explained fully with an example in *Chapter 15*, involve trying many sets of weights (*i.e.* many different groups of randomly or systematically generated b values in Equation 13.11), and then choosing that particular set of weights that results in the combination of genetic gains over all traits that is deemed most desirable.

Independent Culling, Tandem Selection and Two-stage Selection

Independent culling is another method of multiple-trait selection in which truncation levels for selection are set for each trait separately, and trees are selected only if they exceed these minimum standards for all traits measured (Fig. 13.5c, d). Therefore, a truly exceptional tree for one trait is not selected if it fails to exceed minimum standards for all other traits (*e.g.* trees W and A in Fig. 13.5c, d, respectively). Selection using independent culling levels has intuitive appeal in breeding programs of plants and animals, because it is straightforward to apply in practice (Bourdon, 1997). Once the culling levels are set, the breeder simply chooses individuals that exceed the minimum standards for all traits. The difficulty is determining what those standard levels (truncation thresholds) should be for each trait measured; there is usually little theory used in this regard, and most often the culling levels are set intuitively. This relies heavily on the experience of the breeder and the breeder's understanding of all genetic and economic factors pertaining to both measured traits and to target traits in the breeding objective.

Selection using independent culling levels is most appropriately applied in combination with a selection index. High-priority traits are included in the selection index, and independent culling levels are set for lower priority traits. Normally, it is best to set relatively low minimum standards for these lower priority traits so that a large fraction of the trees selected on the basis of the index are also acceptable based on culling levels. For example, suppose that the measured traits are height (as a proxy for bole volume), pilodyn (as a proxy for wood density) and stem straightness. Suppose that for an ultimate breeding objective of pulp yield, the economic importance of height and pilodyn are high and well estimated, while that for stem straightness is unknown and suspected small. An appropriate selection strategy would be to first calculate a selection index for each tree combining the first two traits (height and pilodyn), and then to select that tree if it both has a large index value and exceeds the culling level established for straightness. By setting a low culling level for straightness (*i.e.* only trees that are unacceptably crooked are culled), the breeder achieves two desired results: (1) Trees that are unacceptably crooked are not selected so some minor gains in straightness are expected; and (2) More importantly, very few trees outstanding for the index including the traits of high economic importance are eliminated based on their straightness score.

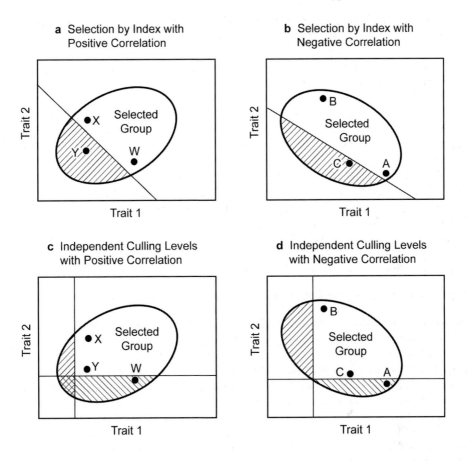

Fig. 13.5. Schematic plots of two hypothetical traits with either positive correlation (a and c) or negative correlation (b and d). Measurements on the base population fall within the oval, and the non-selected portion of the population is hatched. Measurements of three representative trees are shown in each situation (trees A, B and C for the negatively-correlated traits, and trees W, X and Y for the positively-correlated traits). Selection by index (a and b) weights the data from both traits and forms a single number (index value, I) for each tree. The downward-sloping line represents the index value (*i.e.* the same I on all points of the line) below which trees are culled and not included in the selected group. Independent culling (c and d) considers each trait separately, so that a poor measurement on either trait leads to culling; hence, trees W and A are not selected due to poor values for trait 2. (Reprinted from van Vleck *et al.*, 1987, with permission from W.H. Freeman and Company/Worth Publishers)

Another multiple trait selection method, called **tandem selection**, involves selection for only a single trait in each cycle of breeding and testing. Once the first trait has been improved to sufficient levels through several cycles of selection, testing and breeding, attention changes to the second trait, which is then the focus of selection, breeding and testing for several cycles. While used in crops with short generation intervals, tandem selection is not used in tree improvement programs due to the long cycle of breeding and testing that would require centuries to improve a few traits. However, tandem selection might inadvertently occur if a new trait becomes important in the future after several cycles of breeding focused on other traits. For further discussion of tandem selection, see van Vleck *et al.* (1987, p 338) and Bourdon (1997, p 278).

Two-stage selection is related to tandem selection in the sense that selection is first applied to one trait and then another; however, both traits are selected for in the same cycle of breeding and testing (Namkoong, 1970; Talbert, 1985; Lowe and van Buijtenen, 1989). Common application of two-stage selection in tree improvement proceeds as follows: (1) Early selection of trees for some trait such as growth rate in a field, greenhouse or growth room environment (first stage of selection); (2) Subsequent measurement of only the selected group for another trait such as mature growth rate or wood density; and (3) Selection based on the second trait from among the group previously selected for the first trait (second stage of selection). Two-stage selection also may be important in the future when the first stage of selection uses marker-assisted selection based on DNA markers (*Chapter 19*), and the second stage is based on field performance of those trees previously selected. Unlike other methods previously described, the second trait is measured only on the group of trees selected based on the first trait. For this reason, expected genetic gains depend heavily on the order of selection (*i.e.* which trait is first) and on the correlation between the two traits.

SUMMARY AND CONCLUSIONS

A general approach to making selections is: (1) Formulate a breeding objective, T, of target traits weighted by their economic or social importance; (2) Choose which traits to measure based on their ability to make rapid gains in T; (3) Measure phenotypes of trees growing in the base population; (4) Analyze the data and use genetic theory to predict the underlying genetic value for each trait for all candidates; (5) Based on economic and genetic considerations, aggregate the predicted genetic values for the different traits of a given candidate into a single selection index for that candidate; (6) Retain in the selected population those trees with the highest calculated indices; (7) Calculate expected genetic gain for each trait as a difference between the mean genetic value of the selected population and the mean genetic value of the base population; and (8) Label all selections (*i.e.* members of the selected population) and make arrangements for their protection and future use in the long-term breeding and operational propagation programs.

Mass selection, common at the inception of tree improvement programs, chooses trees solely on the basis of their phenotypes without additional information about performance of ancestors, relatives or progeny. In direct mass selection, the measured trait and target trait are the same, while indirect mass selection is based on a measured proxy trait attempting to make gain in the target trait. More genetic gain is expected from direct selection when a trait has a high heritability and when the breeder applies high selection intensity. Indirect selection, which is extremely common in tree improvement, is effective when the measured trait has a strong genetic correlation with the target trait, is highly heritable, lends itself to intensive selection, and can be easily assessed at young ages.

One of the most important selection decisions is deciding which traits to include in tree improvement programs, and it is very important to distinguish between target traits included in the breeding objective, T, and measured traits upon which selection decisions are made. Target traits are the ultimate breeding goal and are chosen for their overall economic or social importance. Measured traits are chosen for their ability to achieve maximum gains per unit time in the breeding objective and hence to make gain in the target traits. In tree improvement, target traits are nearly always harvest-age traits since these are the ultimate breeding goals, while measured traits are usually proxy traits measured earlier in the rotation. It is important to base selection decisions upon a minimum number of measured traits and to only include those that meaningfully contribute to making genetic gain in the target breeding objective. Three to five high-priority traits are best identified and included in a selection index. The process of forming the selection index, I, aims to maximize gain in T, and is effective at determining which measured traits should be included in the index. Lower priority traits can be used as ancillary information to help guide decisions about trees that are marginal based on the selection index. The important concept is to focus only on key traits in the index and not permit lower priority traits to play a large role in selection decisions.

CHAPTER 14
GENETIC TESTING – *MATING DESIGNS, FIELD DESIGNS AND TEST IMPLEMENTATION*

In earlier chapters, we described genetic testing for the purpose of examining patterns of geographic variation (*Chapter 8*) and for evaluating choices among species, provenances and seed sources (*Chapter 12*). In this chapter, we primarily focus on genetic testing that accompanies breeding efforts within individual tree improvement programs. Genetic tests are central to most tree improvement efforts and are established with seedlings or clonal plantlets from parents of any of the five population types in the breeding cycle (Fig. 14.1). The tests are usually planted in field locations on forest sites, but may also occur in farm fields, nurseries, greenhouses and growth rooms. These common-garden experiments (*Chapter 6*) allow separation of genetic from environmental effects that are confounded in the phenotypes of parents. Usually a single series of tests has several objectives, but depending on its primary objective may be variously called a progeny test, base population, realized gain trial, research experiment, clonal trial or demonstration planting.

When there are many objectives for a single test series, it is extremely important to first delineate and assign priorities to the various objectives and then to choose mating and field designs that maximize attainment of the most important ones. Therefore, this chapter begins with a section on the functions and objectives of various types of genetic tests and why these are important in forest tree improvement. This is followed by two major sections describing mating and field designs, respectively, commonly used in genetic tests of tree species. Each section discusses which designs are most appropriate for meeting different objectives. The final major section, *Test Implementation*, addresses important considerations in the establishment, maintenance and measurement of genetic tests. Other reviews of these topics are: Libby (1973), McKinley (1983), Zobel and Talbert (1984, Chapter 8), van Buijtenen and Bridgwater (1986), Bridgwater (1992), Loo-Dinkins (1992) and White (1996).

TYPES, OBJECTIVES AND FUNCTIONS OF GENETIC TESTS

Genetic tests can be established either with seedlings (Fig. 14.2a) or vegetative propagules, such as rooted cuttings (Fig. 14.2b) or tissue culture plantlets. Many of the comments in this chapter apply to both types of tests (seedling tests and clonal tests), and, for the purposes of this discussion, we use the generic term **genetic entries** to refer to the seedlots, families or clones that are being planted and compared in a series of genetic tests. Special attributes of clonal trials compared to tests established with seedlings are noted where appropriate.

Whether planted with seedlings or vegetative propagules, genetic tests are established to advance tree improvement programs and do this by enhancing steps in the breeding cycle. In this regard, functions of genetic tests fall into four broad categories associated with

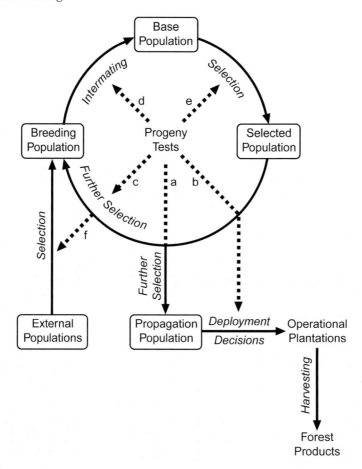

Fig. 14.1. Schematic diagram of the breeding cycle showing many ways progeny test information (dashed lines) is used to increase genetic gain. Predicted breeding values (*i.e.* rankings) of tested parents are valuable for: (a) Increasing genetic quality of the propagation population; (b) Maximizing gain from deployment by matching genotypes to appropriate soils and climates; (c) Upgrading genetic quality of the breeding population prior to initiation of intermating; (d) Optimizing mating designs through mating of superior or complementary selections; (e) Increasing gain from the next generation's selection by providing parental information for selection index development; and (f) Increasing the genetic quality of selections infused into the breeding population from external populations through progeny testing prior to infusion. Each of the five population types that typically occur in the breeding cycle are labeled with a box. Activities involved in the various stages of tree improvement (including the breeding cycle) are written along side the solid arrows.

activities and population types in the breeding cycle (White, 1987b): (1) *Defining Genetic Architecture* to understand the quantitative genetic makeup of a population; (2) *Progeny Testing* to evaluate the genetic worth of specific parents within a population; (3) *Establishing an Advanced-generation Base Population* from which to select the best trees for the next generation's selected population; and (4) *Quantifying Realized Gains* to estimate the progress made in successive steps or cycles of the program. We describe each of these functions separately, even though a single series of genetic tests often aims to accomplish more than one function.

Fig. 14.2. Examples of genetic tests illustrating differences among genetic entries: (a) *Pinus taeda* seed source and family test (established with seedlings) of Perez Companc at 2.5 years in northern Argentina showing row plots of a bulk seed lot of the local land race (shortest on left), a bulk seed lot from the Marion Co. Florida source (middle), and the best open-pollinated family from Marion County (right); and (b) *Eucalyptus grandis* clonal test (established with rooted cuttings) of Cartón de Colombia at 11 years in Colombia showing 6-tree row plots of a fast-growing clone (on right) and slow-growing clone (on left); (the first tree in each row is a border tree and not part of the clonal row plot). Note that row-plot tests, such as these, are appropriate for demonstration, but single-tree plots are preferred for statistical precision. (Photos by T. White)

Defining Genetic Architecture

An important function of genetic tests is to define the **genetic architecture** (*i.e.* the amount of variation in traits of interest, their genetic control, and genetic interrelationships among traits) of one or more populations in the breeding cycle. For this function, interest centers on parameter estimates at a population level, not on information about specific trees nor their selection. Types of genetic information needed (with example parameters in parentheses; see *Chapter 6* for definitions) include: (1) Importance and patterns of provenance and seed source differences; (2) Amounts of phenotypic (σ^2_P) and genetic (σ^2_G or σ^2_A) variation for individual measured traits; (3) Estimates of heritabilities (h^2 or H^2) for each trait; (4) Magnitude of genotype x environment interaction within the planting zone (r_B); (5) Age-age genetic correlations, also called juvenile-mature correlations, for each measured trait ($r_{A,age1,age2}$); and (6) Genetic correlations between pairs of traits ($r_{A,trait1,trait2}$). Specific features of clonal tests for defining genetic architecture are summarized in Box 14.1.

This population-level information from genetic tests is used during design and implementation of tree improvement programs to: (1) Determine appropriate size and location of breeding zones (*Chapter 12*); (2) Decide which traits to measure and their priorities in the selection process (*Chapter 13*); (3) Decide optimum selection ages and selection procedures (*Chapter 13*); (4) Choose optimal mating and field designs for genetic tests (this chapter); and (5) Calculate selection indices and predict breeding values of individual parents and their offspring (*Chapters 6, 13* and *15*). Integration of these different types of information into a long-term breeding strategy is the subject of *Chapter 17*. The effectiveness of any particular breeding strategy and its efficient implementation depend on precise estimates of many different genetic parameters, and such estimates require properly-designed genetic tests.

Box 14.1. Functions and features of clonal genetic tests.

The distinctive characteristic of clonal genetic tests is that each genotype is replicated as multiple ramets of the same clone. When these ramets are planted in different macro- and micro-environments, clonal genetic tests can be very efficient at separating environmental from genetic effects and, therefore, can be used successfully to meet each of the four functions of genetic testing (outlined below). However, care must be taken to minimize c-effects (*i.e.* effects common to ramets of a clone caused by similar treatment in the propagation system) that can inflate variation among clones and appear as genetic differences (*Chapter 16*).

1. *Genetic Architecture:* When unrelated clones are planted in genetic tests, true differences among clones represent differences among their total genetic values (assuming no c-effects). With data from these tests, broad-sense heritability (H^2 in Equation 6.11; see estimates for forest trees in Tables 6.1 and 6.2) and several types of genetic correlations based on total genetic values can be estimated. When properly structured populations of clones from full-sib families are grown in well-designed genetic tests, it is possible to partition total genetic variances and covariances into their components (additive, dominance and epistatic, and sometimes c-effects; *Chapter 6*) (Burdon and Shelbourne, 1974; Foster and Shaw, 1988; Bentzer *et al.,* 1989; Paul *et al.,* 1997).

(Box 14.1 continued on next page)

Box 14.1. Functions and features of clonal genetic tests. (Continued from previous page)

2. *Clonal Testing:* Genetic tests to rank clones are analogous to progeny tests of seedlings to rank parents. The clonal rankings can be used to select top clones for use in the breeding population or for use in the propagation population. When un-related clones are tested (*e.g.* at the initiation of a tree improvement program when one clone is generated from each of the many unrelated plus trees chosen by mass selection), the clones are ranked for their total genetic value, G_i in Equation 6.3 (*e.g.* Ikemori, 1990; Osorio, 1999). When many parents are bred to create many full-sib families of clones, then properly-designed genetic tests result in several different rankings, all of which are useful for different purposes. For example, assume that each of the 50 parents in a breeding population are crossed to four others creating 200 full-sib (FS) families and 5 offspring in each family are replicated by rooted cuttings to produce a total of 1000 clones. A well-executed test based on this sampling scheme would result in: (a) 50 predicted parental breeding values, based on additive genetic effects, useful for selecting top parents for breeding or operational seed production; (b) Rankings of 200 FS families, each based on the average breeding value of the two parents and their specific combining ability (Equation 6.9), useful for choosing the top FS families for operational planting (see *Family Forestry, Chapter 16*); (c) Predicted total genetic values (*i.e.* clonal values) of 1000 clones useful for selecting the top clones for operational planting; and (d) Predicted breeding values for 1000 clones, based only on additive genetic effects, useful for choosing which clones to use in the breeding program (since only additive genetic effects are passed on to seedling offspring in the breeding program).

3. *Cloning Base Populations:* Genetic gains from forward selection of the top individuals from within known-superior families to use in the breeding program can be enhanced by creating clonal replicates of the seedlings of each family (Shelbourne, 1991; 1992). Instead of existing as a single seedling growing in one microsite, as occurs when the base population is planted with seedlings, each genotype created by breeding is cloned and the ramets are planted in several environments. Then, clonal averages are used to rank the genotypes to choose the best individuals from top families to include in the next-generation's selected population (see *Cloning the Base Population, Chapter 17*).

4. *Realized Gains:* When clones are used to quantify realized gains, the genetic tests most often compare operational clones established in rectangular plots. That is, each plot contains many ramets of a single clone. Some plots can also contain seedlings to compare, for example, realized gains of seedlings from seed orchard seed with operational clones. Experimental design, implementation and analysis of clonal tests of realized gain are similar to those of seedlings.

Progeny Testing

The goal of **progeny tests** is to estimate the relative genetic values of parents based on performance of their offspring. Assume a first-generation selected population consists of 300 plus-trees. Since the trees were selected solely on their phenotypic appearance, progeny tests are the best way to evaluate the relative genetic worth of the selections. The tests might consist of 300 open-pollinated families (one from each parent) planted in randomized, replicated tests across a range of field locations in the planting zone. Plus trees whose progeny perform consistently well must have superior breeding values and are favored for inclusion in the breeding and propagation populations (Fig. 14.2). Therefore, the focus of progeny testing is ranking a set of individual parents; this is unlike the focus of genetic architecture, where estimating population-level parameters is the objective.

The rankings obtained from progeny tests are valuable at several stages of the breeding cycle (Fig. 14.1, Box 14.2), and more precise rankings mean more genetic gain at each of these stages. High precision means that parents are ranked very nearly according to their true breeding values with little influence from environmental sources of variation. Parental selection at each of these stages is called **backward selection** meaning that data from progeny are used to rank and re-select top parents, as opposed to selection of individual progeny trees within families (called **forward selection**).

Establishing Advanced-generation Base Populations

The third major function of genetic tests is to provide a base population from which to make advanced-generation selections. The intermating of superior parents of the breeding population results in a resurgence of genetic variability due to sexual recombination. After completion of matings, the offspring are planted as identified families in genetic tests. These tests form the next generation's base population, and the best trees from the best families are selected at an appropriate age to form the next generation's selected population.

The goal of the two functions previously mentioned (defining genetic architecture and progeny testing) was to provide information (population-level parameter estimates and parental rankings, respectively), not genetic material. For the base population function, the focus is on providing genetic material from which to make forward selections. Two desirable features of a base population are: (1) Ample genetic diversity which implies creation of many different families; and (2) Maximum number of unrelated families (with no parents in common) which sets the upper limit on the number of unrelated forward selections that can be made to form the next generation's selected population (*Chapter 17*).

There are two components to the forward selection process: choosing the best families in which to make selections and then selecting the best individuals within those superior families. Few, if any, selections are made from less-superior families, as judged by the predicted breeding values of their parents, and more selections are made from superior families. In some tree improvement programs, the same series of genetic tests is used to obtain both parental rankings and form the base population for forward selection. Other programs separate the two functions (see *Complementary Mating Designs* later in this chapter).

Quantifying Realized Gains

Estimating progress from a tree improvement program involves field tests of material of different levels of genetic improvement: improved *versus* unimproved, first-generation *versus* second-generation propagation population, commercial clones versus unimproved seedlings, etc. Like genetic architecture and progeny testing, this function focuses on information, namely the comparison of means of two or more genetic entries which can be various populations from the breeding cycle or improved varieties. Usually, the mating and field designs of **realized gains tests** (also called **yield trials**) attempt to simulate genetic and stand-level growing conditions of operational plantations so that the estimates of gain are appropriate for subsequent harvest scheduling and economic analysis. Therefore, the features of these tests are: (1) The genetic composition of tested materials often approximates operational varieties (or breeds); (2) It is often wise to include unimproved seedlots or unimproved clones to have a baseline comparison against which to compare

---

Box 14.2. Uses of parental rankings from progeny tests to increase genetic gains by backward selection.

1. *Increasing gain from the propagation population* (Fig. 14.1a). Low-ranking selections can be excluded from the current propagation population (*e.g.* by roguing them from seed orchards or not collecting their seed for operational forestation). Also new mid-generation propagation populations (called 1.5-generation orchards) can be established using only the highest-ranking backward selections (*Chapter 16*).

2. *Maximizing gain from deployment by matching genotypes to sites* (Fig. 14.1b). In the southeastern USA, seed from parents resistant to fusiform rust (based on progeny test data) are deployed to operational plantations located in high rust-hazard areas, while fast-growing, but less resistant families are deployed to plantations of high site quality and low rust hazard. If genotype x environment interaction exists, parents whose progeny are known to perform well in certain environments can be deployed to those environments.

3. *Upgrading genetic quality of the breeding population* (Fig. 14.1c). If intermating of the breeding population is delayed until progeny test data are available, low-ranking members of the selected population may be excluded from the breeding population, and hence their alleles are excluded from subsequent base populations. This increases gain in all subsequent cycles of the breeding cycle, but has two potential disadvantages: (1) Genetic diversity is reduced if a large number of selections are eliminated from the breeding population; and (2) Gain per unit time is reduced if many years are required to obtain reliable progeny test information.

4. *Optimizing mating designs through intermating of superior or complementary selections* (Fig. 14.1d). There are at least three ways that parental rankings can be used in mating designs to enhance genetic gains: (1) Using superior parents in more crosses so that more families in the subsequent base population carry their alleles (Lindgren, 1986); (2) Intermating superior parents with other superior parents (positive assortative mating) to create families that receive excellent alleles from both parents (Cotterill, 1984; Foster, 1986; Mahalovich and Bridgwater, 1989; Bridgwater, 1992); and (3) Intermating parents superior for one trait (*e.g.* growth rate) with those superior for another (*e.g.* disease resistance) to create offspring combining both traits (called **complementary breeding**).

5. *Increasing gain from the next generation's selection* (Fig. 14.1e). When selections are made from advanced-generation base populations, information about parental genetic worth (progeny test data) and individual tree performance are combined to select excellent trees from excellent families. Genetic gain is increased above that expected from simple mass selection of trees from the base population by using two levels of information: parental breeding values and measurement of individual tree phenotypes (*Chapter 15*).

progress (Fig. 11.3 and 14.2a); (3) Large plots of each genetic entry are required to simulate competitive conditions as they exist in operational plantations; and (4) Tests are of long duration (half to full rotation length) to quantify gains in harvest yield and product quality. Realized gains trials are also used for modeling stand growth, diameter distributions and bole taper and volume of genetically improved stock. This is best accomplished if the trials are measured periodically until near rotation age.

By their very nature, tests of realized gain are retrospective in the sense that results apply to varieties in operational use when the tests were established. These varieties have likely been replaced by newer, better varieties when results become available. Nevertheless, yield trials serve a valuable function by validating gains predicted from quantitative genetics theory (*Chapters 13* and *15*) and providing gains estimates on a per-unit-area basis.

MATING DESIGNS

The experimental design of genetic tests has two components: mating design and field design. The mating design specifies exactly how parents are intermated to create the offspring that are planted. Common mating designs are described below in three categories: (1) **Incomplete-pedigree designs**, in which at most one parent is known for each progeny planted, (2) **Complete-pedigree designs**, in which both parents are identified for each tree planted; and (3) Disconnected and complementary mating designs, in which variations and combinations of incomplete- and complete-pedigree mating designs are employed to more effectively satisfy multiple objectives. After the description of each design, its strengths and weaknesses are described in terms of meeting the previously described objectives of genetic tests. Most mating designs are effective for estimating realized gains so efficacy for this function is not always explicitly mentioned.

Incomplete-pedigree Mating Designs

No Pedigree Control (Single Bulk Collection)

When parents are allowed to intermate with no control on pedigree and a single bulk lot of seed is collected from the entire population, then there is no knowledge of parental identity. Neither the female nor male parent can be identified for any progeny trees planted in any resulting genetic tests. An example is where seed is collected and bulked without regard to parental identity from all producing ramets in a clonal seed orchard containing 30 selections. Alternatively, seed collections might be made from each of the 30 clones and then equal amounts of seed from each parent combined into a single bulk lot. This assures equal representation of each parent on the female side, but all pedigree information is lost in both cases.

Bulk collections are most useful for comparing means of different populations such as in realized gains trials. For example, suppose there are 200 selections in a selected population and that the top 20 of those selections are being used in the propagation population to produce seed for operational plantations. Two genetic entries (a bulk collection from each of the two populations) could be used effectively to compare genetic gain between the two populations. For this function it is not important to know the parentage of each tree planted, only which population it came from. A major advantage is that only two genetic entries are required to achieve the stated objective.

Bulk collections have been used in annual crops to form base populations in schemes of simple recurrent mass selection (*Chapter 17*) (Allard, 1960). In this breeding method, members of the breeding population intermate haphazardly; seed is bulked into a single lot and used to plant a base population from which advanced-generation selections are made solely on the basis of phenotype (with no knowledge of the parentage of the trees selected). This is repeated each generation: mass selection, random mating among selections and planting of the unpedigreed base population. This breeding method is simple, inexpensive and effective over the long term for traits with moderate to high heritabilities. In fact, the mass selection breeding method is essentially that used when positive selection is practiced in silvicultural systems employing natural regeneration, such as seed tree and shelterwood systems (*Chapter 7*). However, to our knowledge, it has never been used in intensive forest tree improvement programs because only marginal gains are expected each generation, and because of concern about increasing, uncontrolled relatedness among advanced-generation selections that could lead to inbreeding depression (*Chapter 17*).

Open-pollinated Mating Designs

In **open-pollinated (OP) designs**, seed is collected from the parents in the population and kept separate by parent for planting as OP families. For each tree planted, the female parent is known, but the male parent is unknown. When there are 200 plus-trees in the selected population, a test of all parents would involve planting 200 genetic entries (*i.e.* 200 open-pollinated families). Each progeny tree is labeled in the field with the identifying number of its female parent.

The success of OP mating designs for all four functions (genetic architecture, progeny testing, base population and realized gains) depends to a greater or lesser extent on how closely OP families approximate half-sib (HS) families. In a true HS family, the female parent mates at random with all male parents in the population. This never occurs perfectly for OP families, and, therefore, OP families can differ from HS families for a variety of reasons (Namkoong, 1966b; Squillace, 1974; Sorensen and White, 1988; White, 1996). Probably the most severe problem occurs when a high frequency of self-pollinated progeny suffering from inbreeding depression are unknowingly planted along with non-selfs in the same OP family. Family performance of these families is unfairly disadvantaged compared to families with fewer selfs and this affects both population-level parameter estimates (*e.g.* inflating heritability estimates) and parental rankings (OP families with more selfs fall in the rankings due to inbreeding depression). This problem may be most prevalent in insect-pollinated species when OP seed is collected from widely scattered trees (Hodge *et al.*, 1996) and in wind-pollinated species when OP seed is collected from isolated individuals (*Chapter 7*).

There are many instances when OP families can be used very effectively for defining genetic architecture and for progeny testing (Burdon and van Buijtenen, 1990; van Buijtenen and Burdon, 1990; Huber *et al.,* 1992). Recall that a parent's breeding value is defined as twice the true mean of its HS progeny (Equation 6.7) and that the additive variance (numerator of narrow-sense heritability) is the variance among true breeding values. Therefore, when OP families approximate HS families they are extremely effective for both predicting breeding values of the parents and estimating several types of genetic parameters involving additive variances and covariances (such as h^2, r_B, $r_{A,age1,age2}$, $r_{A,trait1,trait2}$). More than 100 OP families are required to obtain precise estimates of these parameters. OP designs, however, do not provide estimates of non-additive types of variance or covariance.

Use of OP mating designs for creating a base population from which to make selections can be very effective in first-generation tree improvement programs when OP seed is collected from mass-selected trees located in widely-scattered stands of the base population. In this case, selections made from different OP families are assumed to be unrelated if the parents of the families originated in different stands. On the other hand, use of OP base populations in advanced generations has been criticized because two selections from different OP families growing in the same genetic test could conceivably have the same male parent. In addition to leading to inbreeding in subsequent base populations, intermating of these selections in a propagation population (such as if both selections were grafted into the same seed orchard) could result in inbreeding depression and hence reduced gain in operational plantations. Nevertheless, a number of tree breeders have recommended using OP mating designs for creating advanced-generation base populations based on theoretical gains calculations (Cotterill, 1986), empirical evidence of substantial genetic gains (Franklin, 1986; Rockwood *et al.*, 1989) and logistical ease coupled with low cost (Griffin, 1982). This is discussed further in *Chapter 17*.

Polycross (Pollen Mix) Designs

In **polycross designs**, also called **pollen mix** (PM) or **polymix** designs, artificial **controlled pollination** is used to pollinate each female parent with a mixture of pollen from a number of male parents. Pollination bags are placed around the female flowers before receptivity to isolate them from other sources of pollen. Pollen, extracted from a number of male parents in the population and mixed together into a single mixture, is injected into the pollination bags one or more times during receptivity. As with OP designs, there is one family for each parent being tested (*i.e.* 200 PM families for 200 parents in a selected population), and progeny trees from each PM family are labeled with the number of their female parent.

PM designs have advantages over OP designs because trees in a PM family better approximate a half-sib family. When a large number of males are used in the mix (preferably 25 to 50 different male parents unrelated to the mother and each other), there is no selfing and differential fertilization by only a few male parents is minimized. Further, when the same males are used to pollinate all female parents, the common male parentage eliminates potential bias in comparison of rankings of female parents. In fact, for estimating many genetic parameters (h^2, r_B, $r_{A,age1,age2}$, $r_{A,trait1,trait2}$) and predicting parental breeding values, PM designs are as good or better than any other mating design discussed in this chapter (Burdon and van Buijtenen, 1990; van Buijtenen and Burdon, 1990; Huber *et al.*, 1992). Like OP designs: (1) More than 100 PM families are needed for precise estimates of population-level parameters; and (2) PM families cannot provide estimates of non-additive types of genetic variances (nor of clonal genetic values) unless clones are formed within each PM family.

The main disadvantage of PM compared to OP designs is increased cost associated with controlled pollination (pollen extraction, visiting each female tree multiple times to bag, pollinate, remove bag, etc.). Still, for many species the additional costs are slight, and PM designs are less expensive than most complete pedigree designs, because fewer families are required for a given number of parents.

Use of PM designs for creating a base population suffers from the same limitation as OP designs: only half the pedigree is known for each selection made from PM families. For this reason PM families have not been used historically for this purpose in intensive forest tree improvement programs desiring to maintain complete pedigree control. However, creative use of modified PM designs (Burdon and Shelbourne, 1971) or their use in combination with systems to manage inbreeding (such as sublining, *Chapter 17*) or molecular markers (*Chapter 19*; Lambeth *et al.,* 2001) may warrant their future use. The logistical ease, low cost and effectiveness for estimating genetic parameters and predicting parental breeding values certainly make PM designs attractive under some circumstances.

Complete Pedigree (Full-sib Family) Mating Designs

Mating designs in this section employ full-sib (FS) families and hence maintain complete pedigrees of all progeny planted. These designs share the following characteristics: (1) Artificial, controlled pollination is used to create the seedlings; (2) Identification of progeny trees in each family requires naming both parents and, by convention, the female is written first (*e.g.* A x B indicates a full-sib family with A as the female and B as the male parent); (3) The maximum number of unrelated forward selections is one-half the number of parents if parents are unrelated; (4) This maximum number of unrelated selections is only possible for certain types of FS designs; (5) Prediction of parental breeding values uses data from all connected FS families (White and Hodge, 1989) (*Chapters 6* and *15*) and four or more FS families (*i.e.* the common parent mated to four unrelated parents) are normally needed for each parent to precisely predict its breeding value; (6) A large number of parents (> 100 when possible) must be used to precisely estimate heritabilities; (7) Designs with multiple FS families per parent can estimate specific combining ability (SCA) of each FS family (Equation 6.9) and also the dominance variance due to SCA effects; and (8) FS designs both with multiple families per parent and with clones within each family are ideal for exploring all types of additive and non-additive genetic variances (Foster and Shaw, 1988) and for ranking clones (Box 14.1), assuming that c-effects are unimportant (*Chapter 16*).

Some FS designs can be very costly and logistically difficult to implement due to the large number of controlled crosses. The number of FS families produced for each design below is illustrated for N = 20 and N = 200 parents. This can be compared with 200 OP and 200 PM families required for testing N = 200 parents in incomplete pedigree designs. Even large tree improvement programs limit the number of total crosses for a given breeding zone, and several hundred families are approaching the feasible maximum even for species in which control pollinations are most easily accomplished.

Single-pair Matings

Single-pair matings, the simplest and least expensive FS design, involve crossing each parent with one other parent in the population (Fig. 14.3). Parents are first divided into two groups (females and males), and then pairs are formed such that each parent is used only once. There are 10 crosses (*i.e.* 10 FS families) required for N = 20 parents and 100 crosses for 200 parents. Of all FS designs, single-pair mating (SPM) produces the maximum number of unrelated families with the fewest number of crosses. For example, if 200 unrelated parents are crossed in a SPM design, all 100 families are unrelated. If one selection is made from each of these families, there will be 100 unrelated selections, which is the maximum of any full-sib design.

Single pair mating designs provide precise rankings of the FS families tested when the families are planted in appropriate field designs; however, the three components that contribute to the performance of each FS family (*i.e.* the breeding values of the two parents and SCA, Equation 6.9) cannot be separated (since each parent is only mated once). This characteristic of SPM designs means they are not useful for progeny testing (*i.e.* for ranking the parents used to create the FS families), because it is impossible to determine which of the two parents of any FS family contributed more to its performance. However, single-pair matings some-times are used for: (1) Testing FS families for possible deployment in operational plantations (see *Family Forestry* in *Chapter 16*); (2) Defining genetic architecture when effects due to SCA, are assumed small; and (3) Creating base populations for small tree improvement pro-grams that cannot afford to produce and plant multiple FS families per parent. Although effi-cient in terms of maximizing the number of unrelated families with the fewest crosses, the po-tential to make genetic gain from selection in base populations generated by SPM suffers from the lack of multiple crosses per parent, because the chance of two excellent parents being mated together increases with the number of crosses per parent. For this reason, single-pair matings are not widely used in intensive tree improvement programs.

Factorial Mating Designs

For **factorial mating designs**, parents are divided into two groups and all possible crosses are made. Two of the many types of factorials are: (1) A square factorial in which the parents are divided into two equal groups (Fig. 14.4a); and (2) A tester design in which 4-5 male parents (called the testers) are crossed with the remainder of the parents used as females (Fig. 14.4b). For a large number of parents, many more crosses are required for a square factorial compared to a tester design (*e.g.* 10,000 crosses *versus* 975 for N = 200 parents).

All factorial designs are useful for genetic architecture studies if enough parents are used. Two estimates of additive variance are obtained, one from the variability among means of female parents and one from variability among means of male parents. The more precise estimate comes from the group with more parents (*e.g.* the female group in the tester system shown).

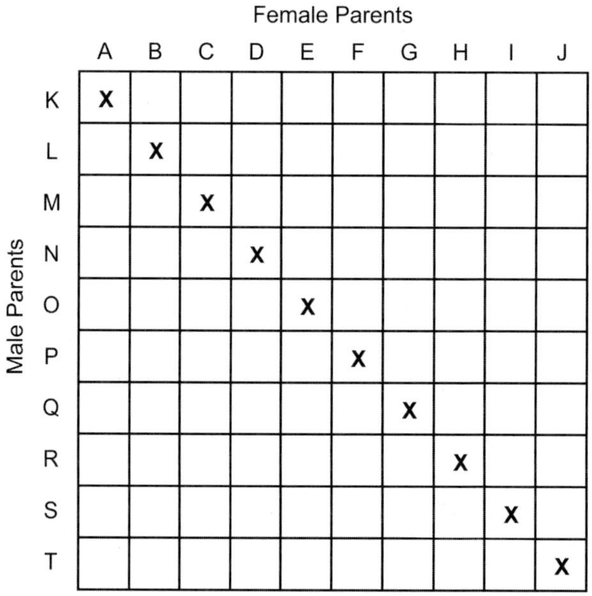

Female Parents

	A	B	C	D	E	F	G	H	I	J
K	X									
L		X								
M			X							
N				X						
O					X					
P						X				
Q							X			
R								X		
S									X	
T										X

Male Parents

Fig. 14.3. Single-pair mating design in which each parent is mated to only one other. Shown for N = 20 parents, the parents are first divided into two equal groups (A-J and K-T in this example) and then pairs are formed starting at the top using each parent only once to produce half the number of FS families as parents (10 FS families here). For N = 200 parents, 100 total crosses are made and 100 unrelated families are possible.

Factorials can also be used for progeny testing. Tester designs are especially useful for ranking large numbers of parents (*i.e.* progeny testing) using relatively few crosses. For example, more than 200 new selections of *Pinus radiata* were crossed with 5 female testers in New Zealand creating nearly 1000 FS families planted in progeny tests to rank the selections (Jayawickrama *et al.,* 1997). For all factorial designs, the male parents and female parents are not genetically connected, because each set of parents is crossed with a different set of parents from the other sex. For this reason, the rankings of the male parents cannot be directly compared (without assumptions) to those of female parents. Factorials are especially well-suited for dioecious species because males and females naturally divide into two groups.

a Square Factorial

Female Parents

	A	B	C	D	E	F	G	H	I	J
K	X	X	X	X	X	X	X	X	X	X
L	X	X	X	X	X	X	X	X	X	X
M	X	X	X	X	X	X	X	X	X	X
N	X	X	X	X	X	X	X	X	X	X
O	X	X	X	X	X	X	X	X	X	X
P	X	X	X	X	X	X	X	X	X	X
Q	X	X	X	X	X	X	X	X	X	X
R	X	X	X	X	X	X	X	X	X	X
S	X	X	X	X	X	X	X	X	X	X
T	X	X	X	X	X	X	X	X	X	X

(Male Parents = rows K–T)

b Tester Design

Female Parents

	A	B	C	D	E	F	G	H	I	J	K	L	M	N	O
P	X	X	X	X	X	X	X	X	X	X	X	X	X	X	X
Q	X	X	X	X	X	X	X	X	X	X	X	X	X	X	X
R	X	X	X	X	X	X	X	X	X	X	X	X	X	X	X
S	X	X	X	X	X	X	X	X	X	X	X	X	X	X	X
T	X	X	X	X	X	X	X	X	X	X	X	X	X	X	X

(Male Parents = rows P–T)

Fig. 14.4. Factorial mating designs in which parents are divided into two groups (females and males) followed by making all possible combinations of crosses: (a) Square factorial in which parents are divided into two groups of equal numbers; and (b) Tester design in which five male parents are designated as testers and all other parents are used as females. For N = 20 parents (A-T), there are 100 (10 x 10) and 75 (15 x 5) crosses for the square and tester designs, respectively. For N = 200 parents there are 10,000 (100 x 100) and 975 (195 x 5) crosses, respectively.

Factorials are also used to create base populations, but careful attention must be paid to the maximum number of unrelated families produced (because this sets an upper limit to the maximum number of unrelated selections possible for the next generation). Only square factorials create the maximum number of unrelated families, and therefore maximize possible genetic diversity in the selected population. Tester designs are especially poor for creating base populations because the maximum number of unrelated selections is limited to the number of testers (*e.g.* 5 in Fig. 14.4b).

Nested (Hierarchical) Mating Designs

For **nested mating designs** (also called **hierarchical mating designs**), the parents are divided into two groups: a common sex that is mated with only one parent of the opposite sex and a rare sex that is mated to more than one other parent; each parent of the rare sex is mated to a different set of parents of the common sex (Fig. 14.5). In forestry it is simpler to collect pollen from many trees than to pollinate many trees, so normally males are the common sex (in Fig. 14.5 there are 16 different males and only 4 different females). The number of crosses for each parent of the rare sex is called the size of the nest and usually ranges from two to eight (4-tree nests are shown in Fig. 14.5). The only advantage of nested designs is the fewer crosses required to intermate many parents compared to other FS designs (*e.g.* for 4-tree nests, there are 16 crosses with $N = 20$ as in Fig. 14.5 and 160 crosses for $N = 200$ parents).

Nested designs are not used widely in forest tree improvement programs, because normally another mating design is more desirable for one or more of the functions. For genetic architecture, smaller nest sizes are preferred (two is optimal) to increase the number of parents of the rare sex; this is because the estimates of additive variance (and hence heritabilities) come from the variability among parental means of the rare sex (*e.g.* among the four females in Fig. 14.5). The variability among parental means of the common sex contains additive and non-additive variance. For progeny testing, nested designs are not preferred because: (1) Predicted parental breeding values for the common sex are imprecise because each parent is involved in only one FS family; and (2) Predicted breeding values for parents of the rare sex are imprecise since each parent is crossed with a different set of parents of the common sex (Fig. 14.5).

Nested designs are also of limited utility for creating base populations since: (1) Expected genetic gain from forward selection is reduced since each of the common parents is mated only once and thus the probability of being mated with an outstanding parent is small; and (2) Genetic diversity is less than other designs since the number of unrelated families is not maximized. In fact the number of unrelated families is limited to the number of parents of the rare sex and depends on the nest size. For $N = 20$ and 200 unrelated parents in 4-tree nests, there are 4 and 40 possible unrelated families compared to maxima of 10 and 100 from some FS designs. The number of unrelated families for nested designs is maximized with 2-tree nests, and for $N = 200$ there are 67 unrelated families which is still fewer than the 100 maximum. The combined disadvantages of nested designs make them of limited use in forest tree improvement.

Diallel Mating Designs

A **full diallel** mating design is the most comprehensive and most costly mating design possible, because each parent is mated with all others including itself (Fig. 14.6). The parents

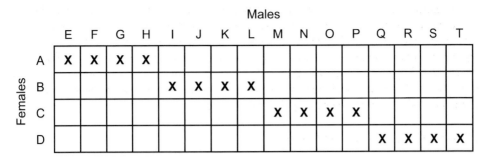

Fig. 14.5. Nested or hierarchical mating design in which each of the parents of one sex is mated to different sets of parents of the other sex. With N = 20 parents (A-T) and 4-tree nests, there are 16 crosses, and a maximum of 4 unrelated families. For N = 200 parents mated in 4-tree nests, there are 160 crosses (40 females x 4 different males per female) and 40 possible unrelated families.

are not divided into two groups as in factorial and nested mating designs; rather, each parent is used as both a female and a male parent. The crosses can be sub-divided into three types: (1) Selfs (on the diagonal) formed by putting pollen from an individual on its own female flowers; (2) Crosses created in the upper half of the table (cells labeled U in Fig. 14.6); and (3) Crosses in the lower half of the table (L) which are reciprocal crosses to those in the upper half (*e.g.* A x B in the lower half and B x A in the upper half). There are more crosses required than any other FS design (N^2 crosses required for N parents, *e.g.* 400 and 40,000 crosses for N = 20 and 200, respectively). This makes the full diallel mating design infeasible for even moderate size populations.

A common modification of the full diallel, called the modified **half diallel**, eliminates self and reciprocal crosses to reduce the number crosses required by more than half (producing only crosses labeled U or L in Fig. 14.6, but not both). For N parents there are (N^2 − N)/2 = N(N − 1)/2 crosses in the half diallel, so for N = 20 and 200 this means 190 and 19,900 crosses, respectively. This is still many more crosses than other FS designs. For this reason, half diallels are more commonly made in disconnected sets of 5 or 6 parents (see *Disconnected Mating Designs*). When half diallels are used, breeders usually assume *a priori* that reciprocal effects are not important (*i.e.* that the performance of a parent's progeny is not influenced by whether it is used as a female or male parent). For this reason it is common to make the crosses in the most convenient direction and/or to make a cross both directions and pool seed from the reciprocal crosses for planting as a single FS family (*e.g.* pooling A x B and B x A and planting as a single entry).

Another type of diallel is called a **partial diallel** in which only some of the crosses in the upper portion of a half diallel are made. There are many variants of partial diallels that make more or fewer crosses. A particularly useful partial-diallel design is the **circular mating design** which is formed by making crosses that fall along particular off-diagonals (Fig. 14.7). Each and every parent is mated in the same number of crosses, and yet the number of crosses is greatly reduced (compare Figs. 14.6 and 14.7). There is great flexibility in which off-diagonals are specified and thus in how many crosses are required for a given number of parents. More crosses per parent mean more precision for estimating population-level genetic parameters and for predicting breeding values, but also mean more work is required. Four or five crosses per parent (*e.g.* five in Fig. 14.7) seems optimal for most purposes assuming that all crosses attempted are successful. For N parents, the maximum number of unrelated families, 1/2 N, is created among the N crosses along any single off-diagonal (such as those labeled 1 in Fig. 14.7)

If the amount of effort required and logistical problems of planting so many FS families are not considered, full and half diallels are the most effective mating designs for all testing functions. Unfortunately, full and half diallels are not feasible for even moderate numbers of parents (even above 20 for most species). On the other hand, the partial diallels known as circular mating designs are probably the very best single mating design for accomplishing all functions of genetic tests. That is, if a single mating design must be chosen (in place of complementary mating designs described later), a circular partial diallel with say five crosses per parent: (1) Allows intermating of many parents all connected together in a single design (which facilitates direct comparisons among the parents); (2) Creates a manageable number of FS families to plant; (3) Provides for precise estimates of both genetic parameters and parental breeding values; and (4) Creates a base population that provides a maximum number of unrelated families and the potential for good gain from selection.

Female Parents

	A	B	C	D	E	F	G	H	I	J	K	L	M	N	O	P	Q	R	S	T
A	S	U	U	U	U	U	U	U	U	U	U	U	U	U	U	U	U	U	U	U
B	L	S	U	U	U	U	U	U	U	U	U	U	U	U	U	U	U	U	U	U
C	L	L	S	U	U	U	U	U	U	U	U	U	U	U	U	U	U	U	U	U
D	L	L	L	S	U	U	U	U	U	U	U	U	U	U	U	U	U	U	U	U
E	L	L	L	L	S	U	U	U	U	U	U	U	U	U	U	U	U	U	U	U
F	L	L	L	L	L	S	U	U	U	U	U	U	U	U	U	U	U	U	U	U
G	L	L	L	L	L	L	S	U	U	U	U	U	U	U	U	U	U	U	U	U
H	L	L	L	L	L	L	L	S	U	U	U	U	U	U	U	U	U	U	U	U
I	L	L	L	L	L	L	L	L	S	U	U	U	U	U	U	U	U	U	U	U
J	L	L	L	L	L	L	L	L	L	S	U	U	U	U	U	U	U	U	U	U
K	L	L	L	L	L	L	L	L	L	L	S	U	U	U	U	U	U	U	U	U
L	L	L	L	L	L	L	L	L	L	L	L	S	U	U	U	U	U	U	U	U
M	L	L	L	L	L	L	L	L	L	L	L	L	S	U	U	U	U	U	U	U
N	L	L	L	L	L	L	L	L	L	L	L	L	L	S	U	U	U	U	U	U
O	L	L	L	L	L	L	L	L	L	L	L	L	L	L	S	U	U	U	U	U
P	L	L	L	L	L	L	L	L	L	L	L	L	L	L	L	S	U	U	U	U
Q	L	L	L	L	L	L	L	L	L	L	L	L	L	L	L	L	S	U	U	U
R	L	L	L	L	L	L	L	L	L	L	L	L	L	L	L	L	L	S	U	U
S	L	L	L	L	L	L	L	L	L	L	L	L	L	L	L	L	L	L	S	U
T	L	L	L	L	L	L	L	L	L	L	L	L	L	L	L	L	L	L	L	S

(Male Parents — row labels at left)

Fig. 14.6. Diallel mating design in which each of 20 parents (A-T) is mated with all others. A full diallel includes all crosses shown (U = upper half, L = lower half and S = self) and each cross in the upper half has a corresponding reciprocal cross in the lower half (*e.g.* A x B and B x A are reciprocals). A half diallel includes only those crosses in the upper or lower half. For N = 20 parents there are 400 and 190 crosses for the full and half diallels, respectively; this is more than any other design. For N = 200, there are 40,000 and 19,900 for the full and half diallels. Both half and full diallels produce the maximum number of unrelated families (10 and 100 for N = 20 and 200).

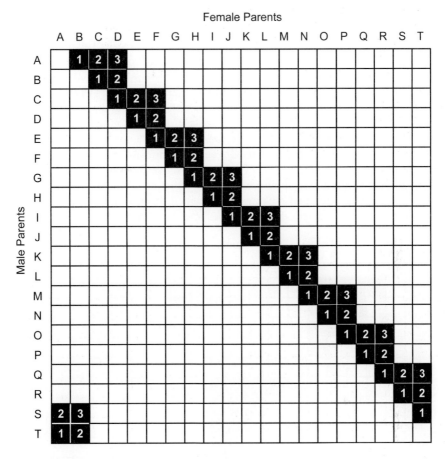

Fig. 14.7. Systematic partial diallel, called a circular mating design, in which crosses are made along specified off-diagonals, and each parent is crossed a certain number of times. For each off-diagonal completely filled with crosses, there are N crosses and each parent is crossed with two others. For N = 20 parents shown above, there are 20 crosses specified on each of the off-diagonals labeled 1 and 2. When every other cross is made along an off-diagonal (3), each parent is crossed with one other. For the three off-diagonals specified above (labeled 1, 2, and 3), there are a total of 50 crosses for 20 parents, and each parent is crossed with five others. For 200 parents, the same design (2-1/2 off-diagonals) creates 500 FS families, and each parent is crossed with five others.

Variations of Classical Mating Designs

Modern tree improvement programs often employ variants of the previously described mating designs to: (1) Reduce the number of crosses required for a large breeding population (say N = 300); (2) Accomplish multiple objectives simultaneously; (3) Facilitate mating superior selections more frequently than others; (4) Spread crossing and testing costs over several years; and (5) Manage inbreeding in the breeding population.

Disconnected Mating Designs

Disconnected mating designs divide parents into several subsets such that matings are made among parents within the same subset or group, but not among those in different groups. All mating designs previously discussed are possible with disconnected sets of parents, since once parents are divided into groups, the mating design is implemented in the normal fashion among parents within a group. Disconnected half diallels (Fig. 14.8) have been the most popular in tree breeding programs, but disconnected factorials and other designs have also been used.

Disconnected mating designs are used for two primary reasons: (1) To reduce the total number of crosses required, and hence the costs of genetic testing; and (2) To manage inbreeding by prohibiting intermating among parents in different groups (in this sense the groups are called sublines as described in *Chapter 17*). Disconnected mating designs are very effective for defining genetic architecture and creating base populations when enough parents and FS families are formed. They are less effective for progeny testing if the goal is to rank all parents in the breeding population together on a single scale. Since parents are divided into groups, parental rankings are obtained for the parents within each group and groups are not linked genetically (hence the name disconnected).

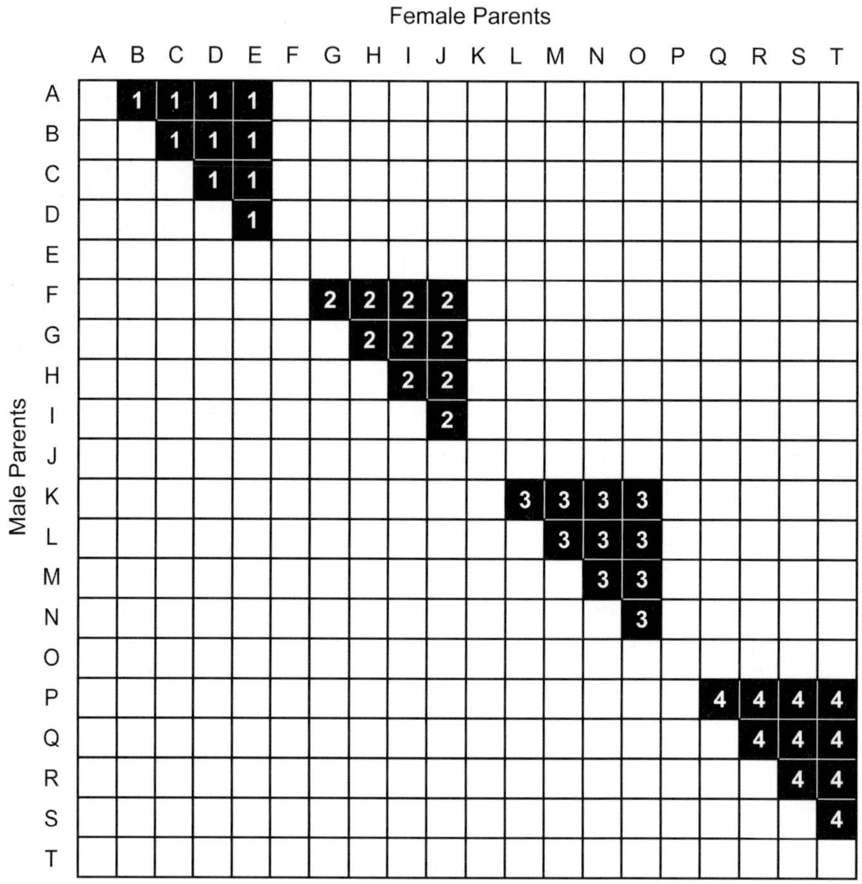

Female Parents

Male Parents	A	B	C	D	E	F	G	H	I	J	K	L	M	N	O	P	Q	R	S	T
A		1	1	1	1															
B			1	1	1															
C				1	1															
D					1															
E																				
F							2	2	2	2										
G								2	2	2										
H									2	2										
I										2										
J																				
K												3	3	3	3					
L													3	3	3					
M														3	3					
N															3					
O																				
P																	4	4	4	4
Q																		4	4	4
R																			4	4
S																				4
T																				

Fig. 14.8. Four disconnected half diallels (labeled 1, 2, 3 and 4) in which the 20 parents are subdivided into four disconnected groups of five parents each. A half-diallel consisting of 10 crosses is made for each of the four groups for a total of 50 crosses (note that this is the same number as for the circular mating design in Fig. 14.7). For N = 200 parents this design produces 500 FS families. For both N = 20 and N = 200, each parent is crossed with four others and the maximum number of unrelated families is created.

With interconnected full-sib designs, such as the circular mating design, there is genetic linkage among all parents and this linkage causes a covariance structure such that all pairs of parents have a non-zero covariance. Even if the 2 parents in the pair were never mated together directly, they still have a non-zero covariance induced by thread of connectedness provided by the design (*e.g.* crosses A x B, B x C and C x D connect A with C and D). In this way, all crosses provide information about all parents in a connected design. Therefore, disconnected designs are less efficient for ranking parents.

Complementary Mating Designs

Complementary mating designs (CMD) use more than one mating design for intermating the same population of parents to better achieve multiple objectives of genetic tests (Burdon and Shelbourne, 1971; van Buijtenen, 1976). The most common example is use of polymix (PM) crosses for progeny testing and defining genetic architecture in combination with one of many FS designs (usually with disconnected sets of parents) for creating a base population (van Buijtenen and Lowe, 1979; McKeand and Bridgwater, 1992; White *et al.,* 1993). With this scheme, each parent is used in both mating designs and planted in two types of genetic tests. PM crosses are planted in field designs best suited to ranking the parents for backward selection and estimating genetic parameters (single tree plots and randomized with many replications), while FS families are planted in larger plots to maximize gain from forward selection of superior trees within families (see *Field Designs*). The latter family selection plots should be duplicated for insurance against catastrophic loss, but are not replicated nor randomized for statistical purposes. Forward selection then proceeds as follows: (1) Measure and analyze the PM tests to predict parental breeding values and thereby choose the best FS families in which to make selections (based on mid-parent values); and (2) Visit the unreplicated family selection plots of only the chosen families to locate the best individual-tree selections.

The goal of CMDs is to maximize efficiency for distinct objectives by choosing mating designs that complement each other and planting these in the best possible complementary field designs. By definition, CMDs should be maximally efficient for multiple objectives, but they involve additional costs and effort of implementing two separate sets of genetic tests. CMDs, as outlined here, have been criticized because: (1) The crossing effort used to create the PM crosses could be used to create more FS families from which forward selections could have been made; and (2) There are no direct rankings of the FS families (PM tests are employed to rank parents and mid-parental values are used to rank FS families). These disadvantages of CMDs are countered by the efficient use of PM families to indirectly rank the parents and the flexibility of FS mating and field designs that allow unbalanced matings, use of superior parents in more crosses, unreplicated FS family plots and variable plot size for families producing different quantities of FS seed. The pros and cons require more research. However, for programs aiming to plant FS families in operational plantations, it is important to create and rank as many FS families as possible (*i.e.* to find the best ones; see *Family Forestry* in *Chapter 16*). This is a serious drawback to CMDs that do not result in direct rankings of FS families, of which only a few are created.

Rolling-front Mating Designs

All previous mating designs assume discrete generations of selection, intermating and genetic testing in which: (1) Crossing is conducted as rapidly as possible over a few years after formation of the breeding population; and (2) Genetic tests are established in one or a few major efforts as soon as possible. Conversely, **rolling-front mating designs** make a subset of planned crosses among members of the breeding population each year to spread the workload of breeding and testing over several years (Borralho, 1995; Borralho and Dutkowski, 1996; Araujo *et al.,* 1997). This mating design is flexible; each year parents are mated according to several criteria considering: (1) Availability of flowers; (2) Breeding values of flowering parents (higher ranked parents being crossed more frequently); and (3) Number of times a parent was crossed in previous years.

In addition, genetic tests are planted each year with the FS families for which seed is available. These tests are unbalanced in the sense that all crosses are not planted in a given test. Nevertheless, the various tests planted across all years are all interconnected in such a way that permits analysis by best linear unbiased prediction (BLUP, *Chapter 15*). The connectedness can be accomplished by planting a set of common families in all tests or by more complicated statistical means that ensure that parents in each test are interconnected directly or indirectly with all others (*i.e.* as described above for the circular mating design).

For some programs, rolling-front mating designs can effectively satisfy all three functions (genetic architecture, progeny testing, base population) if enough crosses are made per parent and the FS families are planted in appropriately interconnected genetic tests. Research is still needed to elucidate the relative effectiveness of the rolling-front designs compared to the more traditional FS designs described previously.

FIELD DESIGNS

The field design of genetic tests specifies the arrangement of progeny seedlings (or ramets of clones) in field locations including: (1) Plot conformation (shape and number of seedlings or ramets per plot); (2) Statistical design at each location (choice of design, type of blocking and number of replications); (3) Site selection (number and location of field sites); and (4) Inclusion of other seedlings or clones as checklots, borders and fillers. Specification of an optimal field design is complex, requiring consideration of statistical factors (outlined in Box 14.3) along with genetic, logistical and economic factors. Most important is consideration of the objectives of the tests, and just as for mating designs, some field designs are better for achieving certain objectives than others. Therefore, our discussion below stresses how different designs serve different test objectives (summarized in Table 14.1). These topics are discussed here for genetic tests planted in field locations, but the principles apply, as well, to genetic tests in greenhouses, growth rooms, nurseries and other situations.

Plot Conformation

A plot is the smallest area within a field experiment that contains trees from the same genetic entry (*e.g.* seedlings from the same family or ramets from the same clone). Normally there is one plot of each genetic entry in each complete block. Plot conformation includes plot shape (Fig. 14.9) and plot size (number of trees per plot). There are four types of plot shapes used in genetic tests: (1) **Rectangular plots** have many trees from the same genetic entry planted in a square or rectangular arrangement (Fig. 14.9c); (2) **Row plots** (also called line plots) contain two or more trees of the same entry planted in a row (Fig. 14.9b); (3) With **single-tree plots** (STP) each tree is a plot, so each genetic entry is represented by a single tree within each complete block (Fig. 14.9a); and (4) **Noncontiguous plots** (NCP) have multiple trees from the same family or clone randomized within each block rather

---

Box 14.3. Statistical factors impacting field designs of genetic tests.

1. *Unbiasedness-Reducing directional inaccuracy*: **Randomization** of families or clones to locations within blocks ensures that all genetic entries have equal opportunity to express genetic potential, and eliminates potential bias. When one genetic entry is unduly favored or handicapped in expression of its performance (*e.g.* if family *A* is always planted in the most fertile areas in every block), comparisons among entries are biased by this extraneous source of variation (*e.g.* family *A* appears better for non-genetic reasons).

2. *Precision-Obtaining estimates with small confidence intervals:* Whether the objective is estimating heritability or ranking parents, high **statistical precision** is desirable (meaning low error variance and small confidence intervals around the estimates). High precision is obtained through choice of proper experimental design (*e.g.* randomized complete block (RCB) or incomplete block (IBD) designs), increasing the number of **replications**, using border rows and fillers to reduce environmental noise, ensuring that all phases of the experiment are done carefully to minimize experimental error and conducting a proper analysis of the data (*Chapter 15*).

3. *Representation-Broadening the sphere of inference:* To ensure that the estimates apply to the intended population of planting environments, the field locations should sample multiple soils, climates, management conditions and planting years representative of the planting zone. Broadening the sphere of emphasis may also include growing the planting stock in multiple nurseries and establishing tests in multiple years.

4. *Error Control-Increasing precision of specific comparisons:* Field designs that increase precision of some or all comparisons among genetic entries include: (a) Block designs (*e.g.* RCBs and IBDs) that divide the field locations into homogeneous blocks to increase precision of comparisons among genetic entries; (b) Grouping of families into sets (*e.g.* planting families from the same provenance together within each block) to produce higher precision for comparisons of entries within the same set and lower precision for between-set comparisons; and (c) Split-plot designs (Cochran and Cox, 1957) to increase precision of comparisons among sub-plot treatments (usually the genetic entries), while sacrificing precision of comparing means among whole-plot treatments (usually silvicultural treatments).

---

than being planted in rows or rectangular plots. A NCP arrangement is derived from Fig. 14.9b by randomizing the five trees of each genetic entry within each of the four blocks; there are still four complete blocks, but the location of the five trees from each genetic entry sample five different microsites within each block.

Small plots, particularly STPs, always mean higher statistical precision (*i.e.* better population parameter estimates and better rankings of parents, families and clones) for two reasons: (1) There are more replications for a given level of effort, so each entry samples more of the microsite variability (*e.g.* in Fig. 14.9 with the same number of trees planted,

Table 14.1. Summary of field design requirements for genetic tests with different objectives.

Design Feature	Objective of Test			
	Genetic architecture: parameter estimates	Progeny testing: backward selection	Base population: forward selection[a]	Yield trials: realized gains
Randomization?	Yes	Yes	No	Yes
Replication?	Yes	Yes	No[b]	Yes
Number of entries[c]	Many	Many	Many	Few
Optimal plot type	STP, NCP[d]	STP, NCP[d]	Rectangular	Rectangular
Include controls?	No	No[e]	No	Yes
Include fillers	Yes	Yes	No	No

[a] If parental rankings are obtained from separate progeny tests in a complementary mating design, then the function of the base population is to maximize gain from forward selection of the best individuals from known-superior families. Hence, large rectangular plots with 49 to 100 trees per family are best for forward selection. If a single design is used both for progeny testing and forward selection, then the progeny test design is favored to best meet both objectives.

[b] With existing parental or family ranking from progeny tests, replication is not needed, statistically-speaking, to maximize gain from within-family selection. However, duplication of family plots at two test sites is prudent insurance against unexpected loss.

[c] The number of genetic entries (seedlots, families or clones) depends on size of program, but large numbers (> 100) are necessary to effectively meet the first three objectives.

[d] STP = single tree plot; NCP = non-contiguous plot.

[e] Strictly speaking, controls are not needed if genetically-connected mating designs are employed. Controls are required, however, to link disconnected sets of parents.

STPs result in 20 replications, row plots have 4 replications and rectangular plots have 1 replication); and (2) Block size is smaller for a given number of entries, so blocks are more homogeneous (block sizes are 0.005 ha, 0.025 ha and 0.1 ha for STPs, row plots and rectangular plots in Fig. 14.9). For this reason STPs are statistically optimal for estimating genetic parameters from genetic architecture tests and for ranking parents, families and clones from progeny and clonal tests (Table 14.1) (Lambeth *et al.*, 1983b Loo-Dinkins and Tauer, 1987; Loo-Dinkins *et al.*, 1990; Burdon, 1990; Loo-Dinkins, 1992; White and Hodge, 1992). This plot type facilitates testing large numbers of families or clones in many replications. Because every tree occupies a different plot, greater vigilance and expense are typically required in monumenting, labeling and maintaining data integrity in tests with STPs or NCPs (compared to row or rectangular plots); however, the statistical advantages of STPs outweigh these disadvantages.

There has been some debate over the relative advantages of STPs *versus* NCPs. As with STPs, each tree samples a different microsite in NCP designs. Therefore, for genetic tests in which thinning is not planned, the two are very similar in statistical precision. In fact, a STP design with 20 complete blocks can be analyzed as a NCP design with 5 blocks and 4-tree noncontiguous family plots by combining data from blocks 1-4, 5-8, 9-12, 13-16 and 17-20. When no thinnings of the genetic tests are planned, we favor STPs, because they readily accommodate more complex statistical designs such as incomplete block designs described later.

a RCB Design with 20 Blocks Planted in Single Tree Plots

C	B	A	G	G	H	A	C	A	C	E	G	C	D	E	G	A	B	C	E
F	D	E	H	B	E	F	D	H	D	B	F	A	F	B	H	D	F	G	H
H	G	F	E	B	C	D	F	A	F	G	H	D	C	B	A	H	F	D	B
D	C	B	A	G	H	A	E	E	B	C	D	H	G	F	E	G	E	C	A
B	D	F	G	E	F	A	B	C	B	E	H	G	A	B	H	F	E	G	A
A	C	E	H	G	H	C	D	G	F	A	D	C	E	F	D	B	D	C	H
E	F	G	H	G	H	A	C	F	A	C	E	D	C	H	G	D	C	B	A
B	C	D	A	F	E	D	B	D	B	G	H	A	E	F	B	H	G	F	E

b RCB Design with Four Blocks and Five-tree Row Plots

C	A	F	E	E	A	G	C	H	D	G	C	D	H	A	F
C	A	F	E	E	A	G	C	H	D	G	C	D	H	A	F
C	A	F	E	E	A	G	C	H	D	G	C	D	H	A	F
C	A	F	E	E	A	G	C	H	D	G	C	D	H	A	F
C	A	F	E	E	A	G	C	H	D	G	C	D	H	A	F
B	G	D	H	F	B	H	D	F	B	E	A	C	G	E	B
B	G	D	H	F	B	H	D	F	B	E	A	C	G	E	B
B	G	D	H	F	B	H	D	F	B	E	A	C	G	E	B
B	G	D	H	F	B	H	D	F	B	E	A	C	G	E	B
B	G	D	H	F	B	H	D	F	B	E	A	C	G	E	B

c RCB Design with One Block Planted in 20-tree Rectangular Plots

B	B	B	B	B	G	G	G	G	G	E	E	E	E	E	A	A	A	A	A
B	B	B	B	B	G	G	G	G	G	E	E	E	E	E	A	A	A	A	A
B	B	B	B	B	G	G	G	G	G	E	E	E	E	E	A	A	A	A	A
B	B	B	B	B	G	G	G	G	G	E	E	E	E	E	A	A	A	A	A
D	D	D	D	D	F	F	F	F	F	H	H	H	H	H	C	C	C	C	C
D	D	D	D	D	F	F	F	F	F	H	H	H	H	H	C	C	C	C	C
D	D	D	D	D	F	F	F	F	F	H	H	H	H	H	C	C	C	C	C
D	D	D	D	D	F	F	F	F	F	H	H	H	H	H	C	C	C	C	C

Fig. 14.9. Three different plot conformations for a Randomized Complete Block (RCB) experiment with 20 trees from each of 8 genetic entries (A, B, ... H), planted at a spacing of 2.5 m x 2.5 m for a total area of 0.10 ha for the 160 trees (not including borders or filler trees): (a) 20 blocks in single tree plots (block size = 0.005 ha); (b) 4 blocks in 5-tree row plots (block size = 0.025); (c) 1 block in 20-tree rectangular plots (block size = 0.10 ha). The genetic entries could be 8 provenances, families, clones or other seedlots. Internal lines indicate divisions between complete blocks.

When one or more thinnings are planned, NCPs can be planted in an **interlocking block design** which maintains balanced representation of all genetic entries in each sub-block after thinning, assuming no mortality (Fig. 14.10). Interlocking sub-blocks can be designed for one or more planned thinnings and can be planted to maintain equal distance among all trees after each thinning (Libby and Cockerham, 1980). Each thinning necessarily reduces the number of trees measured per genetic entry and, therefore, reduces the statistical precision of all estimates. Therefore, it is common to make a major assessment of all tests just prior to each thinning.

Row plots are not optimal for any of the four functions of genetic tests (Table 14.1), but were utilized in the past because of: (1) Convenience of laying out and monumenting row plots instead of STPs or NCPs; and (2) Concerns about analysis of data with missing cells due to mortality in STP designs. With modern-day methods of mapping, measuring and analyzing data from genetic tests, these concerns no longer argue in favor of row-plot designs. If row plots must be used, the number of trees per plot should be minimized (two to four) to minimize block size and allow for many blocks.

Rectangular plots (Fig. 14.9c) are preferred to row plots, STPs and NCPs for realized gains tests of harvest yield and, in certain circumstances, for forward selection from the base population (Table 14.1). Rectangular plots are favored in realized gains tests because they provide an environment that simulates stand conditions in operational plantations, and, therefore, produce unbiased estimates of yield on a per-unit-area basis (Foster, 1992).

a Before Thinning

1	2	3	1	2	3	1	2	3	1	2	3
2	3	1	2	3	1	2	3	1	2	3	1
3	1	2	3	1	2	3	1	2	3	1	2
1	2	3	1	2	3	1	2	3	1	2	3
2	3	1	2	3	1	2	3	1	2	3	1
3	1	2	3	1	2	3	1	2	3	1	2
1	2	3	1	2	3	1	2	3	1	2	3
2	3	1	2	3	1	2	3	1	2	3	1
3	1	2	3	1	2	3	1	2	3	1	2

b After First Thinning

1	2		1	2		1	2		1	2	
2		1	2		1	2		1	2		1
	1	2		1	2		1	2		1	2
1	2		1	2		1	2		1	2	
2		1	2		1	2		1	2		1
	1	2		1	2		1	2		1	2
1	2		1	2		1	2		1	2	
2		1	2		1	2		1	2		1
	1	2		1	2		1	2		1	2

Fig. 14.10. Example of one complete block from an RCB design with three interlocking sub-blocks (numbered 1, 2 and 3) before thinning (a) and after the first thinning, when trees in sub-block 3 are removed (b). A second thinning would remove sub-block 2. This example accommodates 36 genetic entries (there are 36 planting positions labeled with each number) and one seedling from each genetic entry is randomly assigned to a position in each of the interlocking sub-blocks. Therefore, before thinning, there are three seedlings planted for each genetic entry in each complete block (one in each of the three interlocking sub-blocks). Each thinning removes one tree from each entry. If equal distance is desired between trees after thinning, geometric designs with staggered rows are shown in Libby and Cockerham (1980).

Most progeny tests and base population genetic tests are measured at an early age (say 25 to 50% of harvest age), yet product yield and quality from realized gains tests are best judged at older ages. Inter-tree competition intensifies as trees age, and genotypes that start slowly may be disadvantaged. Therefore, means from small plots (row plots, STPs and NCPs) may be biased with results favoring genotypes that are strong competitors, especially at early ages. Large square or rectangular plots of 50 to 100 trees per plot avoid these biases since inter-tree competition is primarily among trees of the same genetic entry (*e.g.* same bulk mix, family or clone). The disadvantage of rectangular plots is that replication sizes are large, so the number of different genetic entries must be held to a minimum (normally less than ten different entries) and the number of replications is limited.

Rectangular plots are also sometimes preferred for forward selection from the base population. For example, when complementary mating designs are used, parental rankings are commonly obtained from polymix families planted in STPs (van Buijtenen and Lowe, 1979; McKeand and Bridgwater, 1992; White *et al.*, 1993). Then, the function of the base population is to maximize gain from forward selection of the best trees from superior families. This is enhanced by planting many trees (say 60 to 100) of the same family in a large rectangular plot so that all trees can be compared in a uniform environment. The parental rankings from the randomized, replicated STP tests determine which FS families are superior and, therefore, from which rectangular plots more forward selections should be made. When one design must satisfy both the progeny testing (backward selection) and base population (forward selection) functions, STPs and NCPs are preferred over rectangular plots.

Statistical Design (Field Layout) at Each Location

Assuming the number of genetic entries is predetermined, specifying the field layout at each location includes deciding on: (1) An appropriate statistical design; and (2) The number of blocks or replications. Here we consider the two most common statistical designs, randomized complete block designs and incomplete block designs. Choosing between these two hinges on desired block size which in turn depends on the number of genetic entries and planting density. Each block should be small enough to maximize uniformity among planting sites within the same block. Then, all genetic entries planted in the same block sample similar microenvironments and differences among entries are primarily due to genetic differences. As a general rule, block sizes of less than 0.1ha are appropriate for most forest situations (Matheson, 1989), but larger blocks may be acceptable for very uniform farm fields and smaller blocks needed for heterogeneous mountain environments. With common planting densities ranging from 1000 to 2000 trees per hectare, 100 to 200 trees can be planted in a single block of 0.1 ha.

Block Configurations

Randomized complete block designs (RCB) have been by far the most commonly used statistical design for genetic tests in forestry. Suppose that 150 different PM families are planted in a RCB design with single tree plots at a spacing of 2 m x 3 m (1667 trees per ha). Each block contains 150 seedlings (one from each PM family) randomly assigned to the planting locations in that block for a block size of 0.09 ha (2 x 3 x 150/10,000). RCB designs are simple to layout in the field, are straightforward to analyze and provide unbiased, precise estimates of genetic parameters and breeding values (when block sizes are small). These attributes explain the popularity of RCB designs.

Sometimes, it is not possible to fit trees from all genetic entries into a single complete block and still maintain each block below the desired size. Depending upon planting density, this occurs for STP designs when the number of genetic entries exceeds 100 to 200 or even smaller depending on site homogeneity. For rectangular plots, the problem occurs with far fewer genetic entries. For example, each block occupies 0.36 ha with only 6 genetic entries planted in rectangular plots of 100 trees at 2 m x 3 m spacing (2 x 3 x 100 x 6/10,000), and this is more than triple the size of a desirable block.

One option when block size is large is to group genetic entries into logical sets and then to plant the genetic entries within a set in contiguous plots within each complete block. For example, families might be grouped into sets according to their seed source. If there are, say, 20 families from each of 30 sources (a total of 600 families), then there are 30 sets of 20 families per set. Each complete block is sub-divided into 30 smaller units and there are two stages to the randomization: (1) The 30 sets are randomly assigned to 1 of the 30 units within each block in the field; and (2) One seedling from each of the 20 families from each set is randomly assigned to a location within that set's unit. This is still a RCB design (with each complete block containing 600 trees as 1 per family); however, families within the same set always occur in the same contiguous unit of 20 trees within each block. This means there are 2 levels of precision: families within the same set are compared more precisely than families in different sets. This **sets-within-reps** design was popular in the past and is still useful when the desire is to obtain more precise rankings within a set and less precise rankings of entries assigned to different sets.

An alternative to the above sets-within-reps design is **incomplete block designs** (IBD) which can be used anytime block sizes are large (or anytime micro-environments within a block become heterogeneous in field, greenhouse, nursery or growth room situations). With IBDs, each complete block (sometimes called a resolvable replication) is sub-divided into smaller units called incomplete blocks or, simply, blocks (Cochran and Cox, 1957; Williams and Matheson, 1994; John and Williams 1995). Each incomplete block contains a subset of the genetic entries planted in a more uniform micro-environment than possible if all entries were planted in one complete block. The incomplete blocks add another blocking factor used to adjust entry means for micro-site differences. IBDs increase the precision for comparing genetic entries by reducing experimental error through removal of variation among incomplete blocks as well as block-to-block variation. IBDs have proven useful for genetic testing in many forestry situations (Williams and Matheson, 1994; Fu *et al.*, 1998, 1999a,b).

The alpha-lattice design (Patterson and Williams, 1976; Patterson *et al.*, 1978) is a particularly useful and flexible type of IBD that is widely used for genetic tests with large numbers of entries. It requires a special computer program, originally called ALPHA+ or ALPHAGEN (Williams and Talbot, 1993) to layout the design in the field; however, with a little practice the design and analysis are straightforward. The newest version of the program, called CycDesigN (Whitaker *et al.*, 2002), also facilitates utilization of other types of computer-generated designs including row-column designs and Latinized designs (Box 14.4). The point of all of these modern experimental designs is to reduce experimental error, and therefore increase the precision of the rankings of genetic entries, by removing sources of environmental noise from comparisons among genetic entries.

Number of Blocks

Factors affecting the appropriate number of complete blocks, also called resolvable replications, at any location include: (1) Which parameters are being estimated (*e.g.* heritability, genetic correlation, parental breeding values, or realized gains); (2) Desired level of precision of the estimates; (3) Type of plot (*e.g.* STP or row plot); (4) Type of experimental design (RCB or incomplete block); (5) Total number of planned site locations; and (6) Practical and logistical constraints on the size of each test and difficulty of establishing multiple test locations. A variety of statistical methods have been utilized to investigate how these factors interact to influence the optimum number of complete blocks per site (Bridgwater *et al.,* 1983b; Cotterill and James, 1984; Lindgren, 1985; White and Hodge, 1992; Byram *et al.,* 1997; Osorio, 1999). More blocks at a site are required when: (1) The parameters being estimated have high error variance; (2) A high degree of precision is required; (3) STPs are used instead of row-plots; and (4) Few site locations are planned because practical considerations require establishing more blocks at one site rather than establishing more sites.

Box 14.5 discusses the factors governing the number of blocks needed per site for both seedling and clonal tests, and the following are rules of thumb. For seedling genetic tests planted in STPs, the total number of seedlings from a given family required across all sites ranges from a low of 50 (for small programs willing to accept lower statistical precision, but having uniform sites and selecting for few traits with moderate to high heritabilities) to 150 (for larger programs demanding high precision and making selections for many traits of low to moderate heritability). These 50 to 150 seedlings per family should be spread over as many macro-environments (*i.e.* site locations) and micro-environments (*i.e.* blocks within sites) as logistically feasible. Fewer ramets are needed per clone in clonal tests. For initial screening trials to be followed by subsequent testing of promising clones, 3 to 6 ramets per clone per site planted in STPs with several sites (say 30 total ramets per clone across all sites) is sufficient.

Selection of Sites

Site selection involves determining the number of sites and their location. All sites should be: (1) As uniform as possible (especially within blocks) to ensure maximum possible precision of estimates by minimizing environmental noise; (2) Stable in ownership for the planned duration of the test; (3) Accessible to machinery and labor for easy establishment, maintenance and data collection; (4) Large enough to accommodate the entire test and all border rows, firebreaks and access roads; (5) Representative of the lands in the planting zone; and (6) Conducive to promoting genetic expression of important traits.

The majority of sites are normally chosen to be above average in site quality and scattered among the major planting zones in approximate proportion to the amount of reforestation planned in each zone (*e.g.* three coastal sites and two inland sites given more planned reforestation in the coast). Higher-quality sites are often more uniform, have higher probability of good survival and also foster more rapid growth to promote early expression of genetic differences in growth among the families or clones.

In addition to these higher quality sites, some tests may be located in high-hazard lands where reforestation is planned (very cold sites, disease-prone sites, etc.). If certain areas are prone to biotic or abiotic stresses and selection for stress resistance is planned, then one or more sites should be located in these high-hazard areas to elicit expression of genetic differences in resistance.

Determining the appropriate number of sites depends on practical and statistical considerations, including the number of blocks allocated to each site (Lindgren, 1985; White and Hodge, 1992; Byram *et al.,* 1997). Statistically, it is always best to have as many sites as possible with fewer blocks per site; this is especially important when genotype x environment (g x e) interaction is appreciable. Having many sites broadens the sphere of inference so that genetic rankings and parameter estimates apply well to all lands in the planting zone. However, practical limitations and costs associated with additional sites often necessitate a compromise.

Box 14.4. Computer-generated experimental designs for genetic tests.

The interactive software called CycDesigN (Whitaker *et al.,* 2002) produces random-ized field layouts for a variety of experimental designs. John and Williams (1995) pre-sent the statistical theory of these designs, while Williams *et al.* (2002) describe appli-cations for genetic tests of forest trees. The concepts are summarized here and illus-trated in Fig. 1.

Incomplete Blocks: An **incomplete block,** or simply, block, is a small, homogene-ous area of field, nursery or greenhouse genetic test that contains fewer than the total number of genetic entries. Using the Whitaker *et al.* notation, the example in Fig. 1 contains v = 24 genetic entries planted in r = 3 resolvable replications with s = 6 blocks per replication (18 total blocks), with each block containing k = 4 ge-netic entries.

Block	1	2	3	4	5	6
	23	18	24	19	20	12
Replication 1	11	17	10	21	6	4
	7	22	9	1	5	8
	14	16	3	13	2	15

Block	1	2	3	4	5	6
	19	15	4	8	14	16
Replication 2	9	2	6	18	3	10
	5	21	22	20	17	7
	12	24	23	11	1	13

Block	1	2	3	4	5	5
	13	4	12	7	24	21
Replication 3	3	10	2	15	18	11
	6	20	16	17	23	9
	8	1	14	5	19	22

Fig. 1. Schematic diagram of a randomized field layout for an incomplete block design produced by CycDesigN. (Reproduced from Whitaker *et al.,* 2002, with permission from the authors)

Concurrence: CycDesigN minimizes **concurrence** of pairs of genetic entries across replications. The goal is to obtain as many intrablock comparisons of each genetic en-try with all others. So, once an entry occurs in one block with a set of entries, it is less likely to occur in other blocks (in another replication) with any of those entries. In the example, the concurrence is exactly 1 for all pairs of genetic entries meaning that all entries occur in exactly one block with every other entry. For example, entries 23, 11, 7 and 14 are in block 1 of rep 1, but do not occur together in any other block; they are located separately in blocks 3, 4, 6 and 5 of rep 2 and blocks 5, 6, 4 and 3 of rep 3, respectively. For most combinations of numbers of replications, genetic entries and block size, it is not possible to have a single level of concurrence for all pairs of entries and some designs could have various numbers of pairs occurring together 1, 2 or 3 times in blocks in different replications.

(Box 14.4 continued on next page)

Box 14.4. Computer-generated experimental designs for genetic tests. (Continued from previous page)

Resolvable Designs: We recommend using only **resolvable designs** in which the blocks and genetic entries are grouped such that there are complete replications containing one plot of each genetic entry. This way, the experiment could be analyzed as an RCB by ignoring the incomplete blocks nested within each complete replication. In the example, there are r = 3 resolvable and complete replications (sometimes called complete blocks), each replication having 1 plot of each of the 24 entries.

Row-column Designs: It is possible to minimize concurrence in both directions (in rows and columns). This requires detailed knowledge of the field orientation of replications, and is not discussed further here, but these are excellent designs.

Latinized Designs: When the replications are contiguous in the field, a further restriction on randomization, called Latinization, can be employed to ensure that the r plots of a given genetic entry are spread across the site. In the example, 12 of the 24 entries occur once in long column 1 which extends across block 1 of reps 1, 2 and 3. The other 12 entries occur in long column 2. This is called a t-Latinized design with t = 2, since it requires 2 long columns to accommodate all entries. With this Latinization, each entry occurs exactly once in each pair of long columns (1 and 2, 3 and 4, and 5 and 6), and hence in each third of the field layout when the site is divided longwise. This might be useful in accounting for variation if the thirds have 3 separate irrigation lines or a clear gradient running from left to right across the field.

Smaller tree improvement programs operating in more uniform environments with minimal g x e interaction might opt for only two site locations. This necessitates more blocks per location, and may mean that precision is sacrificed in extrapolating results to the entire plantation estate. Larger programs desiring high precision and working across diverse edaphoclimatic zones require a larger number of sites. There might be as many as ten or more site locations for some types of genetic tests with some of these scattered throughout the diverse edaphoclimatic zones and others planted in high-hazard sites or sites known to elicit genetic differences.

Including Additional Trees (Borders, Fillers and Controls)

All genetic tests should be surrounded with two or more rows of **border trees** planted at the same spacing as trees within the test. Border trees are planted around the outside of the entire test if all blocks are contiguous. If some blocks are separated from others, then border trees are also planted around these blocks. The goal is to prevent edge effects by ensuring that all measured trees grow in similar competitive conditions. Normally, some genetic entries produce more seedlings than required for the tests, and these are planted as border trees.

Another type of border tree is used in long-term realized gains tests, or other genetic tests planted in rectangular plots, when large performance differences are expected among genetic entries (Loo-Dinkins, 1992). In this case each rectangular plot consists of two parts: (1) The inner measurement plot (*e.g.* 36 trees in a 6 x 6 tree plot); and (2) The border plot (*e.g.* two rows of trees surrounding the measurement plot) planted with trees of the same genetic entry as in the measurement plot. In this example, each total rectangular

---

Box 14.5. Determining the number of blocks per site for seedling and clonal genetic tests.

1. From a purely statistical standpoint, it is always better to spread trees of a given genetic entry across as many different micro- and macro-environments as possible. This argues for STPs with few blocks at each location and many locations (White and Hodge, 1992; Byram *et al.,* 1997; Osorio, 1999), maximizing the precision of family or clone means (and other estimated parameters) by averaging each entry's performance across a wide spectrum of environmental sources of variation. Potential amounts of genotype x environment interaction and costs of establishing additional sites must be considered jointly when balancing the number of blocks per site against the number of sites (Lindgren, 1985).

2. For genetic tests employing seedlings, planted in STPs and aimed at three key functions (genetic architecture, progeny testing and forward selection from the base population), the optimum number of blocks per site ranges from 10 to 30. In one study of STP designs with loblolly pine (Byram *et al.,* 1997), 30 blocks per site were warranted when only two test sites were planned (60 total trees per family across both sites), while 10 blocks per site provided the same precision of family means when four test sites were planned (40 trees per family across all sites). Similar results were found in other studies (Lindgren, 1985; White and Hodge, 1992). A general rule of thumb for STPs is to use 15 to 20 complete blocks per site when there are four or more site locations.

3. For seedling genetic tests planted in row plots, the optimal number of blocks per site is smaller, but more total trees per family are needed to achieve equal precision as STPs. For example, depending on error variances, equal precision on family means might be achieved with six blocks of six-tree row plots (36 trees per family at each site) or 20 blocks planted as STPs (20 trees per family). This simply re-states the increased statistical efficiency of STPs.

4. For clonal genetic tests, both for ranking clones and for estimating genetic parameters, fewer ramets per clone are required per site than mentioned above for seedlings per family. This is because ramets within a clone are the same genotype, so fewer are required for precise estimates of clonal means. As a rule of thumb, 3 to 10 complete blocks of STPs (3 to 10 ramets per clone at each site) planted in a sufficient number of sites (five or more locations) are required (Russell and Libby, 1986; Russell and Loo-Dinkins, 1993; Osorio, 1999). Within this range, fewer ramets per site are needed when the tests are for preliminary screening to be followed by further testing of promising clones. Conversely, more ramets per site and more sites are warranted to increase precision on clonal means if high ranking clones will immediately be deployed to operational plantations.

---

plot, sometimes called the gross plot, consists of 10 (2 border trees + 6 measurement trees + 2 border trees) x 10 = 100 total trees. The goal is to prevent inter-plot interference from influencing the trees in the inner measurement plot. Regardless of the type of entry being tested (*i.e.* different species, provenances, families or clones), some entries may grow much faster and survive better than others. As this happens, border trees of poorer entries may be hindered in their development by shading or other competitive effects, but inner trees are buffered from these influences. Therefore, measurements of the inner measurement plot should reflect how that genetic entry would perform in a pure stand.

Two types of **filler trees** are planted within the test to provide uniform competition to neighboring test trees: (1) Fillers planted before or during test establishment into anomalous planting spots (such as low spots or next to tree stumps); and (2) Fillers planted shortly after test establishment to replace dead trees. For the first type, filler trees are planted into locations within the test that are deemed so unusual that performance will likely be quite different than other locations within the block. If planted with a measure tree, the data could increase experimental error; so, non-measured filler trees are planted into these abnormal microsites. For example, consider a RCB design with STPs and 150 families. Each block contains 150 measure trees (one from each family), and it is possible to designate a few extra spots (say 10 spots) in each block for fillers. If 160 planting positions are marked in each block before planting (Fig. 14.11), then fillers can be planted in the worst ten planting spots before the remainder of the test trees. Normally, only 5-10% of the planting spots should be designated for pre-plant fillers to maintain the block size as small as possible. Some trees usually die shortly after test establishment (3 months to two years depending on growth rates). These empty spots can be filled with filler trees or other comparable planting stock to provide equivalent competitive conditions for test trees. Both types of fillers should be clearly labeled and their measurements are usually excluded from analyses.

Controls (also called **checklots** or references) are sometimes planted to: (1) Provide a reference or benchmark against which to measure genetic gains; and/or (2) Link the rankings of genetic entries tested in disconnected sets or different locations so that all entries can be compared on a single scale. The first use is very important and all tree improvement programs should develop a set of reference populations that represent different stages of genetic improvement. When included as controls in genetic tests, these reference populations provide benchmarks against which to measure progress. For example, a standard commercial collection of unimproved material and a bulk lot of unrogued first-generation seed orchard seed are two reference populations that could be included in advanced-generation genetic tests.

The second use of controls is more problematic. Controls are not needed in tests designed primarily to estimate genetic parameters, and their use increases the size of each block. For predicting breeding values, it is preferable to select a mating design that interconnects all parents genetically (such as the circular mating design). Sometimes controls must be used to connect disjoint tests containing families from different sets of parents; however, their statistical efficiency for this purpose needs more research. When controls are used to link data from disconnected sets of genetic entries, it is often necessary to replicate the controls more frequently within each block than the other genetic entries to increase the precision of scaling all sets to the controls.

TEST IMPLEMENTATION

Care and proper technique in the implementation of tests are just as important to the success of genetic testing as the mating and field designs. In fact, it is not uncommon that entire tests must be abandoned because of improper care or unforeseen disasters. By their very nature, genetic tests in forestry are long term (5 to 50 years), and are subject to many unfavorable

Fig. 14.11. Implementation of genetic tests of *Pinus elliottii* in Florida: (a) Planting of a polymix test showing intensive site preparation, white plastic pipes marking block corners, pin flags marking planting positions and seedlings in white containers laid out prior to planting according to the single-tree-plot randomization scheme; and (b) Measurement of a 1-year-old test showing continued weed control, good growth and use of an electronic data recorder. (Photos courtesy of G. Powell, University of Florida)

influences during their duration (changes in ownership, turnover in personnel, natural catastrophes, etc.). The overall goal of test implementation is to install, monument, maintain and measure tests in a manner that reduces experimental noise (to increase precision of estimates) and ensures long-term viability of the tests and integrity of the data. The importance of proper documentation at all stages cannot be over emphasized.

Breeding and Nursery Phases of Test Implementation

During the breeding phase, important aspects to ensure high-quality results are: (1) Proper labeling of all pollen and seed lots; (2) Careful cleaning of equipment between processing different pollen and seed lots; and (3) Proper storage and handling of pollen and seed to ensure maximum viability and vigor. Surprisingly high errors in parental composition of full-sib families resulting from control pollinations have been reported in some tree breeding programs (Adams *et al.,* 1988). These errors must be minimized or else the credibility of genetic testing could be severely compromised.

Randomization and replication of test entries are critical during the nursery phase for outdoor nurseries, greenhouses and other types of production facilities. All facilities have gradients or patchy variability associated with shade, irrigation systems, edges of benches or beds, etc. If genetic entries (*e.g.* provenances, families or clones) are not randomized and replicated during the nursery phase, some entries are likely to grow in more or less favorable locations than other entries, and these effects can bias both nursery and subsequent field results.

When seedlings are ready for planting, seedlings of obviously poor quality should be culled (not planted). Sometimes there is a desire to plant only the very best seedlings of each entry (*e.g.* to apply within-family selection in the nursery), but this can also cause bias. Rather, the culling system should mimic that used for operational plantations.

Site Preparation and Test Establishment

Generally, sites for genetic tests are prepared with more intensity and greater care than are sites for operational plantations (Fig. 14.11). The goals are to: (1) Increase site uniformity to reduce micro-environmental noise and therefore increase heritabilities; (2) Ensure good survival and hence equal inter-tree competition and minimal data loss; (3) Promote early expression of genetic differences by encouraging rapid early growth; and (4) Mimic future site preparation techniques that may be operational when test results become available. In particular, weed and grass competition can be severe and patchy causing high levels of environmental noise within blocks. Therefore, pre-planting techniques to reduce woody and non-woody competition are most important. In some places, insect and animal control (*e.g.* fencing) may be needed to reduce damage from pests. Setting specified target levels for survival, test quality and early growth rate can provide effective incentives to field personnel.

Also prior to planting, decisions are made about proper block placement and layout. Blocks are not necessarily square; their shape and orientation should be chosen to create maximum site uniformity within blocks, while allowing site variability among blocks. With steep environmental gradients, the long axis of each block should be perpendicular to the gradient. For example on hillsides, rectangular blocks are placed with the long axis on the contour so that all seedlings within a block are at nearly the same elevation (Fig. 8.11 in Zobel and Talbert, 1984).

If a small area within a block is distinctly different (*e.g.* a low area or area where site preparation was less effective), it can be included in the block, but planted with filler trees. If larger anomalous areas occur, they are best excluded from the experiment. Non-test seedlings are then planted in these areas, and blocks are oriented to avoid them entirely. This may mean some blocks are spatially separated from others, but each complete block should remain intact. Two rows of borders should be planted around all edges of blocks that are not contiguous.

When feasible (*i.e.* when a site has no major anomalous areas), it is desirable to plant all blocks contiguously and to record the row and column position of each tree. Then, the entire test appears as a grid and the x,y coordinates of each tree can be entered into a spreadsheet. This allows certain types of analyses, called near-neighbor or spatial analyses (*Chapter 15*), to fit environmental surfaces to the entry data or otherwise account for local microsite (neighborhood) effects. These analyses reduce experimental error and increase statistical precision; so, maintaining row-column coordinates of each tree is definitely worthwhile, when possible.

The appropriate planting density can be debatable, but generally trees should be planted at operational densities or slightly closer than operational densities. Higher densities (closer spacing among trees) mean smaller block sizes (and hence more uniform blocks) and promote earlier inter-tree competition (which is desirable or undesirable depending on operational silviculture). Plans for thinning also affect initial planting densities.

Proper monumentation and documentation are critical for long-term integrity of genetic tests. Genetic tests of forest trees endure for many years during which time personnel turnover, changes in landscape and many other factors can contribute to loss of entire tests, of identity of genetic entries in tests and of past data. Proper monumentation and documentation are described in Box 14.6 and include: (1) Making maps that identify both test location and tree location within each test; (2) Monumenting test corners and block corners within each test; (3) Clearly identifying trees with durable labels; (4) Documenting all activities from breeding through final measurement; and (5) Maintaining a proper database of all data and other records.

Test Maintenance and Measurement

Proper test maintenance is needed periodically to ensure high-quality results. This may include competition control, pest control, mowing to permit easy access, fertilization, removal of wild, non-test seedlings that may confuse measurements, and re-monumentation of block corners and tree labels. Given the high cost of breeding and initial establishment, proper aftercare is relatively inexpensive and essential.

The first assessment of most genetic tests occurs within the first year. Normally only survival and pest damage are recorded to allow for replacement with filler trees and establish a baseline of test quality. After the initial assessment, tests are normally measured either at regular intervals, when key decisions need to be made (*e.g.* roguing a seed orchard), just prior to thinning, or when a special circumstance arises (*e.g.* a frost or disease epidemic provides an opportunity to measure genetic differences in tolerance or resistance). Among many considerations for successful measurements, some suggestions are: (1) Use experienced personnel that are re-trained periodically; (2) Ensure that a given crew completes a complete block or replication entirely before changing tasks; (3) Sub-sample and re-measure a few percent of the trees in each test with a crew of supervisors as a standard quality-control practice; (4) Minimize the number of traits scored subjectively (*e.g.* visual assessment of stem straightness, branching habit, crown form) because these may greatly increase measurement time and costs; (5) Use an even number of categories for subjective

Box 14.6. Monumenting and documenting genetic tests.

All genetic tests of forest trees need to be clearly monumented and carefully documented to ensure long-term integrity of the data. Some guidelines are:

Maps at two or more scales: At higher scale, the exact location of the genetic test should be referenced to permanent landmarks (surveying benchmarks, major roads, property boundaries, etc.), not to temporary features of the landscape that change with time. At lower scale, the test map shows the planting location of each tree with its identification (family number or seed lot code). When possible, trees are planted on a grid system and the row and column identifications are included on the map. Block corners are clearly marked on the map. Equally important is to show the starting point and direction that measurements are taken; this corresponds to the order of the trees in the electronic data recorder or data sheet. The test map should be created at time of planting to ensure proper location of all trees, and then re-verified within a week or so by a separate crew. This map may be the only method of identifying trees if field labels are lost or destroyed.

Monumentation at test and block corners: Large, permanent stakes should be located at all test and block corners. These should be clearly labeled with information about test and block composition.

Tree identification labels: It is not necessary to label each tree, but trees at regular intervals (*e.g.* every tenth tree) should be labeled with semi-permanent tags. Tags should be checked and replaced as necessary throughout the life of the test. The identifying code used for each genetic entry should be as short as possible (few digits) and completely unique to that entry. A given entry should have the same code in all test locations and related families need to be identified so that data analyses recognize these relationships. Normally it is preferable to label families with their unique parental codes (*e.g.* A01 x B03), and not to re-code these in the field.

Documentation: At time of test establishment, a detailed Test Establishment Report is completed with three types of information: (1) Test information (*e.g.* test identification number, location, objectives, number of genetic entries, types of entries and controls, reference to related tests that have similar entries); (2) Site information (*e.g.* prior uses, soils, elevation, climate); (3) Treatments applied (*e.g.* site preparation techniques, herbicides, pesticides, time-of-planting fertilization); and (4) Conditions at planting (*e.g.* date of planting, personnel involved, weather conditions, seedling quality). Photographs at various stages provide another valuable form of documentation.

Database management: A complete database system should be maintained for the entire life of all genetic tests and should include all test information (*e.g.* exact nature and composition of all genetic entries and checklots, statistical design, measurement data, maintenance records and photographs). The system should consist of at least two components: (1) Computer records containing data and digital information; and (2) Office files containing correspondence and other printed information. Computer records should be replicated and stored in two or more locations to prevent their catastrophic loss by fire, floods, etc.

scores (*e.g.* four categories with scores of 1, 2, 3, and 4) to avoid over-utilization of the middle category; (6) Use height poles when tests are young and trees are shorter than eight to ten meters and later use precision electronic devices (*e.g.* hypsometers); (7) Take key data (such as block, family and plot identification along with tree survival and height) from the previous assessment into the field (this verifies previous measurements, reduces measurement time and eliminates errors in re-entry of key information); and (8) Use electronic data recorders making certain to download previous data prior to field measurements and use in-field error checking capabilities of the software (*e.g.* checking permissible values for all traits measured).

After each test is measured, the data are cleaned and edited (*Chapter 15*). This involves many steps such as checking the minimum and maximum value of each trait measured to ensure that all measurements fall within a reasonable range. Plotting heights against diameters for all trees is also extremely useful for identifying anomalous measurements. After editing, the data are stored in a data management system until needed for analysis. The data file for each test and measurement date should be given a unique file name and should include a header prior to the data that completely describes the measurements and the format. To guard against loss in a catastrophe, the file should be stored in a minimum of two locations in separate buildings.

SUMMARY AND CONCLUSIONS

Genetic tests have four main functions: defining genetic architecture, progeny testing, creating base populations and quantifying realized gains. For any single objective, mating and field designs can be developed to maximize efficiency. In smaller tree improvement programs, practical considerations often mean that a single series of tests serves several functions and trade-offs must be accepted. By clearly determining the most important test objectives, the trade-offs can be made to ensure maximum efficiency for the highest priority. Sometimes in larger programs, complementary mating and field designs are utilized in which two series of tests are established to better achieve maximum efficiency for each of the multiple objectives.

Mating Designs -- Polymix and open-pollinated mating designs (when OP families approximate half-sib families) are optimal or nearly optimal over a wide range of genetic and environmental conditions for: (1) Estimating genetic parameters involving additive genetic variances and covariances (heritability, genotype x environment interaction, and genetic correlations); and (2) Predicting parental breeding values and hence maximizing gain from backward selection. Many full-sib designs (*e.g.* factorials, diallels and less structured rolling front designs) can also be useful for genetic architecture and progeny testing; four or five crosses per parent are sufficient for most uses. For estimating genetic parameters, it is most important to have a large number of parents or clones (>100 when possible).

In full-sib designs, matings among parents can be arranged in disconnected sets or in connected designs and these result in parameter estimates with similar statistical precision. When predicting breeding values, it is best that all parents being ranked are in a single genetically-connected mating design. This obviates the need for controls to link disconnected sets. Systematic partial diallels, called circular mating designs, are particularly well-suited for predicting breeding values. Rolling front mating designs where connectedness is checked each year before determining crossing patterns are also suitable.

For forward selection of the best trees within families comprising the base population, many mating designs are suitable, but full-sib designs are needed if complete pedigree control is desired. If other test objectives are being achieved with a complementary mating design, the mating design to create the base population can be very flexible and creative (*Chapter 17*): (1) Using better parents in more crosses; (2) Incorporating positive assortative mating of top parents; (3) Intermating parents that are good for different traits (called complementary breeding); and (4) Crossing trees producing flowers at early ages to minimize the time for completing breeding activities.

Field Designs and Test Implementation -- Randomized complete blocks (RCB) and incomplete block designs (IBD) are the two most common statistical designs for genetic tests in forestry. IBDs, such as alpha-lattices and row-column designs, are warranted when intra-block variability among planting spots is relatively large (when the size of each complete block exceeds 0.1 ha, as a rule of thumb). Single tree plots (STPs) are statistically optimal for obtaining precise parameter estimates and family rankings. Non-contiguous plots planted in interlocking blocks, however, are preferred when thinnings are planned in order to maintain equal representation of all entries throughout the life of the test. Rectangular plots are desirable for realized gains tests and for forward selection of individuals from a given family.

For a given fixed number of total blocks, it is better, statistically, to plant more test locations with fewer blocks per location especially when genotype x environment interaction is important. This increase in statistical efficiency must be balanced against the increased costs of establishing and maintaining more test sites. Smaller tree improvement programs operating in uniform environments with minimal g x e interaction might opt for only two site locations. This will mean more blocks per location (*e.g.* 30 blocks planted in STPs), and may mean that precision is sacrificed when results are extrapolated to the entire plantation estate. Larger programs desiring high precision and working across diverse edaphoclimatic zones may plant ten locations with fewer blocks per location (*e.g.* 15 blocks in STPs). Some of these sites are scattered throughout the diverse edaphoclimatic zones and others are planted in high-hazard sites or sites known to elicit genetic differences. All sites should be uniform, accessible and stable in ownership for the life of the tests.

Care and proper technique in the implementation of genetic tests are just as important to their success as the mating and field designs. The overall goal of test implementation is to install, monument, document, maintain and measure tests in a manner that reduces experimental noise (to increase precision of estimates) and ensures long-term viability of the tests and integrity of the data.

CHAPTER 15
DATA ANALYSIS – *MIXED MODELS, VARIANCE COMPONENTS AND BREEDING VALUES*

Effective data analysis and interpretation are the final stages of a successful genetic testing program. After large sums of money are spent to plant, maintain and measure genetic tests, it is equally important to invest in proper data analysis. There are two primary objectives of analysis of genetic tests in tree improvement programs: (1) To estimate genetic parameters (such as heritabilities, genetic correlations, etc.) that are used in a variety of ways to guide decisions and develop strategies in tree improvement programs (see *Defining Genetic Architecture* in *Chapter 14*); and (2) To predict genetic values (such as breeding values and clonal values of individual trees) that are used to identify the best candidates for selection (see *Progeny Testing* in *Chapter 14*). For both objectives, high-quality data analysis is required to ensure that the tree improvement program produces maximum genetic gain per unit time.

Chapter 6 presents an introduction to the estimation and interpretation of variance components and genetic parameters from genetic tests. This chapter focuses on a modern approach to data analysis utilizing **mixed model (MM) methods** that include a suite of analytical procedures for estimating variance components, predicting underlying genetic values and more. MM methods were first developed in the early 1950s by Dr. C. R. Henderson (1949, 1950), and are now widely used for data analysis in animal breeding programs (Mrode, 1996). This analytical approach has been adopted more slowly in tree improvement, because of theoretical and computational complexities (White and Hodge, 1988, 1989). However, with advances in computer technology and increased recognition of the advantages of these techniques (Borralho, 1995; White, 1996), MM methods are becoming the methods of choice for most advanced tree improvement programs.

The theoretical development of MM methods of data analysis is mathematically complex and beyond the scope of this book (see Henderson, 1984; White and Hodge, 1989; Mrode, 1996). Rather, we focus on a conceptual explanation of these methods and their usefulness for analysis of data from genetic tests. The chapter begins with a discussion of preliminary steps important for any data analysis and a conceptual overview of linear statistical models. This is followed by concepts and applications of MM methods and selection indices, with a final section on spatial analysis aimed at reducing experimental error. Throughout the chapter, we refer to a case study of *Eucalyptus grandis* genetic tests growing in Argentina to illustrate some of the concepts, capabilities and applications of MM methods. This case study is introduced in Box 15.1. Other discussions of MM methods in quantitative genetics, in increasing order of mathematical complexity, are: Borralho (1995), Bourdon (1997), White and Hodge (1989), Mrode (1996), and Lynch and Walsh (1998).

PRELIMINARY STEPS PRIOR TO DATA ANALYSIS

Editing and Cleaning of Data

Prior to data analysis, the data must be thoroughly cleaned and edited. It is especially important to rid the data set of invalid data points. Some invalid data points show up as **outliers**,

---

Box 15.1. Case study of *Eucalyptus grandis* from Argentina: Introduction and objectives.

The data come from a eucalypt tree improvement program conducted by INTA (the federal research organization) in the Mesopotamia region of Argentina (Marcó and White, 2002). These data are from 6 progeny test sites of *E. grandis* established by INTA between 1982 and 1992 (Table 15.1). The 203 open-pollinated families in this series of trials came from 14 different seed sources with 12 of those being provenances from the natural range in Australia (labeled N01, N02 ... N12) and the remaining two from the local land race growing in Argentina (one labeled selected and the other unselected to indicate that the parent trees in the plantations were or were not intensively selected, respectively). The test sites had different experimental designs, but in all sites, families were completely randomized within blocks, *i.e.* not nested or grouped by seed source.

In all sites, each tree was measured for height, diameter at breast height (DBH) and stem form. Form was measured on a scale of one to four with one being a straight, well-formed tree and four indicating a crooked and possibly multi-stemmed tree. Height and DBH were used in a volume equation to estimate over-bark stem volume for each tree. Two traits were then analyzed: stem volume and form.

The data were unbalanced in many ways (Table 15.1): (1) Single tree plots were used on some sites and row plots on others; (2) The number of blocks ranged from 3 to 20; (3) The number of seed sources planted in any site ranged from 2 to 13; (4) The number of families per site ranged from 31 to 179; (5) The number of families representing a seed source varied from 8 to 27; (6) Measurement ages spanned 4 to 16 years across test sites; (7) Test size varied from 0.4 ha and 349 trees in test 6 to 3.2 ha and 3332 trees in test 1; (8) Survival varied from 62 to 95%; and (9) Test environments and soils varied markedly causing differential site productivity which, measured as mean annual increment (MAI), ranged from 25 to 51 $m^3ha^{-1}yr^{-1}$.

The challenge of the mixed model analysis (Box 15.5) was to combine all data from all tests together, properly weighting the data and accounting for the many different types of imbalance. In particular, the objectives were to: (1) Determine the importance of different seed sources and estimate their relative rankings (Box 15.8); (2) Estimate genetic parameters (such as heritabilities and genotype x environment interaction) (Box 15.9); (3) Predict breeding values for stem volume and form for all 203 parents and all living 11,217 trees (Box 15.10); and (4) Combine the breeding values for the two traits into a selection index facilitating both backward and forward selection (Box 15.12).

---

data points distinctly outside the normal range of measurements for that variable (Mosteller and Tukey, 1977), while others may be within the usual range of measurements. Invalid outliers are especially troublesome because they can profoundly affect estimates of family means for the families in which they occur. Their effect is even more pronounced on estimates of genetic variances, since variances are quadratic functions greatly influenced by points far from the mean. Another concern that can lead to deletion of data prior to data analysis is selfs or other trees resulting from close inbreeding in open-pollinated families. Some inbred trees can be very small and cause several problems in data analysis (Sorensen and White, 1988; Hardner and Potts, 1995; Hodge *et al.*, 1996). The challenge of data cleaning is to find and eliminate invalid data points without removing legitimate measurements.

Table 15.1. Summary of experimental designs and results from single-site analyses of six progeny test sites of *E. grandis* growing in Argentina (Marcó and White, 2002). Two traits were analyzed: stem volume in units of m³ and stem form on a scale of 1 to 4 with 1 being good. Neither variable was standardized nor transformed for these test-by-test analyses. See Box 15.1 for details.

	Test 1	Test 2	Test 3	Test 4	Test 5	Test 6
			Experimental design			
Design[a]	STP	STP	STP	STP	RP	RP
Sources (#)	13	13	13	13	2	2
Families (#)	179	120	164	169	31	31
Blocks (#)	20	17	20	20	5	3
Trees/plot (#)	1	1	1	1	9	5
			Summary information			
Age (yrs)	5	5	5	4	16	16
Survival (%)	95	91	62	81	78	77
Living trees (#)	3332	1794	1971	2690	1087	349
MAI (m³ha⁻¹yr⁻¹)[b]	25	28	35	51	31	36
			Parameter estimates for volume			
Mean (m³)	0.11	0.13	0.26	0.21	0.56	0.65
σ^2_P (m³)² [c]	0.18	0.43	1.20	0.93	6.72	8.66
h^2 [d]	0.25	0.36	0.24	0.24	0.05	0.11
			Parameter estimates for form			
Mean	2.41	3.04	2.28	2.43	2.88	3.02
σ^2_P [c]	0.43	0.67	0.43	0.53	1.02	0.88
h^2 [d]	0.11	0.35	0.20	0.31	0.21	0.41

[a] STP=randomized complete block with trees planted in single tree plots.
  RP= randomized complete block with trees planted in row plots.
[b] MAI=mean annual increment for stem volume accounting for survival.
[c] Total phenotypic variance.
[d] Narrow-sense heritability.

There are elegant statistical methods available for searching for suspect data; however, simpler methods combined with common sense and experience are equally effective (Box 15.2). When a questionable value is found for any variable, there are four options: (1) Set the value equal to missing for that variable, but retain the data from other variables for that tree; (2) Delete the entire record for that tree; (3) Use judgment to modify the suspect value to a more reasonable value; or (4) Return to the genetic test site in the field to check the measurements for that tree.

The appropriate method depends greatly on the situation; however, normally when the number of outliers is low, there is no reason to return to the field. Rather, a combination of the other three options is preferable. For example, if there was clearly a column shift when the measurement was recorded (*e.g.* a tree 10.0 m tall was recorded as 100 m), then the value can be reasonably modified and used in the analysis. Other outliers with less apparent explanations should be deleted. When there are a large number of outliers (more than 1% of the observations), the entire process of data collection should be reviewed and changed to increase quality control.

Sometimes, the data from one or more blocks are eliminated when they have been adversely affected by spurious factors causing low survival or extreme variability within the block. Examples are when some blocks are severely damaged by herbivores or are located in low areas prone to flooding. Elimination of one or more blocks is appropriate when their key indicators are distinctly different from those of other blocks at the test location (step 6 in Box 15.2). However, judgment must be exercised to ensure that only data severely degraded in quality are deleted.

Elimination of data from an entire test site may be warranted if the site has been degraded by poor establishment practices, poor maintenance, unsure labeling of families, fire, road construction or natural catastrophes. However, climatic or biotic events such as droughts, frosts, ice storms, floods and disease outbreaks can provide an opportunity to measure genetic resistance to stresses. These test sites may require separate analysis if other sites have not been similarly challenged.

As a brief example of data editing, the data from all six *E. grandis* test locations described in Table 15.1 were subjected to the cleaning steps outlined in Box 15.2. In particular, graphs of stem diameter against height were conducted for each block in each test site (a total of 85 graphs with an example in Fig. 15.1). These plots revealed a total of 26 outliers believed to be recording errors during data collection. This represents only 0.23% of the 11,501 measured trees and the data for these trees were eliminated prior to analysis. In addition the plots for each block were examined for extremely small trees (*i.e.* runts) believed due to selfing or other forms of inbreeding. A total of 258 runts were identified and their data eliminated (2.2% of 11,501). After cleaning, 11,217 trees remained for analysis.

Transformations and Standardization

Some types of variables are often transformed prior to analysis. For example, the arcsine and logistic **transformations** are commonly used for binomial data, while the log transformation is sometimes recommended for growth variables (Snedecor and Cochran, 1967; Anderson and McLean, 1974). However, it is important to remember that the theoretical and practical value of these transformations have been demonstrated for fixed effects models (described later) in which the goal is to compare treatment means. The value of transformations for MM analysis in which the goal is prediction of random breeding values (see *Linear Statistical Models* in this Chapter) is less certain.

For example, for binomial data the above recommended transformations cannot be made for individual trees, but rather must be applied to cell or plot means. Individual tree breeding values cannot be predicted from transformed plot means; so, the value of transforming binomial data is questionable. Further, there is some information to indicate that MM analysis of untransformed 0,1 binomial data may be suitable under some circumstances (Foulley and Im, 1989; Mantysaari *et al.*, 1991; Lopes *et al.*, 2000). For continuous variables, the log transformation produces biased means when back transformed to the original units. For these reasons, the value and rationale for all transformations should be carefully considered prior to use. Many times, it is suitable to use untransformed data in MM analysis.

One exception is a special transformation sometimes called **standardization** which should be used in many cases prior to MM analysis for three reasons (Hill, 1984; Visscher

Box 15.2. Identifying questionable data values in genetic tests.

1. *Review written information about each test location:* Before starting data editing, it is useful to review the file containing information about unusual problems or special factors occurring before or after establishment: poor seedling quality, problems with site preparation of some blocks, natural catastrophes, etc. This can highlight problem areas to look for during the cleaning procedure.

2. *Verify descriptor variables:* Each tree is identified by several descriptor variables such as treatment, site, block and family and these should be checked for each observation. There are two ways to check for erroneous family identifiers: (a) Use a computer program to calculate the number of observations for each family in the test and look for families with too many or too few trees (these may indicate recording errors); and (b) Merge a separate computer file containing the known family identities onto the data set and look for non-matches.

3. *Obtain means, minimums, maximums and frequency distributions for each continuous variable:* These simple statistics are valuable for finding obvious outliers (*e.g.* a value for height that is simply not biologically feasible). Normally, it is useful to calculate the statistics separately for each block in each test site (*i.e.* to pool data from all families in a given block and obtain one mean, minimum and maximum for each variable for that block).

4. *Check for non-permissible values of discrete variables:* Discrete variables have a limited number of viable values. For binomial variables, there are only two possibilities (*e.g.* 0 and 1 or yes and no), while for stem form or other subjectively scored traits, there may be four or six categories. A computer program can check each value of each discrete variable and print out data from trees with non-permissible values for subsequent verification.

5. *Plot pairs of continuous variables that have strong phenotypic correlations:* This is a very effective way to identify unreasonable values of one or both variables (Fig. 15.1). It is useful to plot the data from all trees in a given block on a single graph.

6. *Examine key block statistics for dissimilarities among blocks in a test:* Statistics calculated for key variables using all data from a given block can identify those blocks that are distinctly different and candidates for elimination. Key variables include: mean survival, mean damage by herbivores or other damaging agents, and variation in stem growth. For the latter, the coefficient of variation (phenotypic standard deviation of all trees in a block divided by the block mean) can highlight blocks with extreme levels of variability.

7. *Compare measurements from different ages:* When a given test site has been measured at two or more ages, it is useful to compare measurements of each tree from consecutive ages and look for indications of recording errors: negative growth, unreasonably small or large changes in tree size, previously dead trees that have come to life, previously forked trees that are no longer forked and previously diseased trees that are no longer diseased.

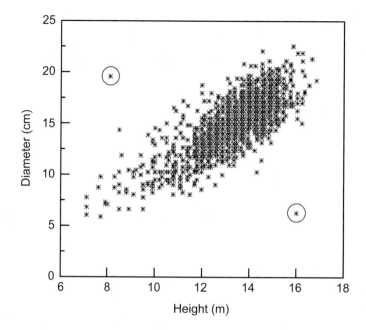

Fig. 15.1. Graph of stem diameter versus height from a single block of one of the 6 *E. grandis* test sites described in Box 15.1 and Table 15.1. Each point represents the data from one tree and all trees from the block are plotted. Note that the two circled points appear as outliers and, therefore, are candidates for subsequent modification or elimination.

et al., 1991; Jarvis *et al.*, 1995): (1) To homogenize variances that are pooled together across levels within a single effect in the linear model; (2) To eliminate statistically significant interaction for genotype x environment interaction due only to scale effects (as opposed to rank changes, *Chapter 6*); and (3) To facilitate straightforward, bias-free back transformations to predict genetic gains in the units of measure in various environments. With respect to the first point, the goal of many transformations is to homogenize within-block variances across blocks on a test location and across different test locations. Uniform variance at all levels (*e.g.* all blocks within all locations) within an effect (*e.g.* the block term in the linear model) is an underlying assumption of the analysis if these variances are pooled together into a single variance (*e.g.* the block variance). Since block variance is pooled across all sites, it is important to standardize across sites. Standardization at the block and location levels affects all of the variances estimated and eliminates scale-effect genotype x environment interaction.

An effective way to standardize at the block and location level is to standardize all tree measurements for each variable by: (1) Calculating the phenotypic standard deviation for each block (the square root of the variance calculated using all the data in the block for that trait); and (2) Dividing each tree's measurement in a given block by the standard deviation for that block. Sometimes this transformation is applied at the level of each test location instead of the block level (Jarvis *et al.*, 1995).

For stem volume in the *E. grandis* case study (Table 15.1), there were large differences among the six test sites in estimated total phenotypic variance (σ^2_P), ranging 50-fold from 0.18 to 8.66 (m³)², and these differences were positively associated with the large mean stem volume differences across sites. The standardization proceeded by dividing each tree's volume measurement by the phenotypic standard deviation calculated for the block containing that tree. The standardized volume for each tree was the variable used in all analyses that combined data across test sites. The phenotypic variance for stem form (Table 15.1), ranged only 2.5-fold among test sites (0.43 to 1.02) and this variance was not related to the site mean for stem form. Therefore, all subsequent analyses for form used the raw, unstandardized data values.

Note that after standardization, the phenotypic variance and standard deviation for each block (and hence each test site) are both one: Var (y/σ_y) = Var$(y)/\sigma^2_y$ = σ^2_y /σ^2_y = 1 (White and Hodge, 1989, p 46). This ensures that the scale of the measurements is identical across blocks and test sites and rids the data of interactions (such as genotype x environment interactions) due to scale effects. However, since all measurements within a block are divided by the same number, the relative comparisons among genotype and treatment means are not altered.

When breeding values are predicted using standardized values, the predicted values are in units of phenotypic standard deviations. Therefore, in order to present breeding values in absolute terms it is necessary to re-scale the predicted breeding values back to the original units of measure by multiplying them by an appropriate phenotypic standard deviation (*e.g.* by the pooled standard deviation across all test sites, or by the standard deviation of a single test site when performance in a specific target environment is of most interest). Another approach instead of re-scaling back to the original units of measure is to express the breeding values in percentages above or below the standardized mean for the variable. This yields the same result as first re-scaling and then expressing the breeding values as percentages above or below the unstandardized mean.

Exploratory Data Analysis

Mixed model (MM) analysis normally combines all data across all test sites and establishment years for a given trait and the analysis itself is a "black box" in the sense that the analyst often has little knowledge of the millions of arithmetic computations. Therefore, preliminary analyses to understand the data and guide formulation of the MM analysis are extremely important. These can include graphical analysis, age trends, simple phenotypic correlations among variables, test-by-test analyses and paired-test site analyses.

Particularly important is the estimation of heritabilities for each trait on a site-by-site basis and of type B genetic correlations (Box 6.6) for all pairs of test sites. The heritabilities indicate whether some test sites have drastically different variance structures and should be treated as separate traits in the MM analysis (*i.e.* multivariate mixed model as described later). The genetic correlations estimate the importance of genotype x environment interaction which guides both program-level strategic decisions and formulation of the linear models for the MM analysis. For example, if family x site interactions are large, it may be prudent to group test sites by environmental similarity and conduct data analyses separately for each group. This may also imply the need for multiple breeding units (*Chapter 12*) and/or targeted deployment of specific families to specific sites (*Chapter 16*).

For the *E. grandis* case study (Box 15.1), site-by-site estimation of variance components and heritabilities proceeded as described for the open-pollinated example of *Pinus taeda* in *Chapter 6* (Table 6.2). These analyses were conducted for each test site and variable (unstandardized stem volume and form) using the Mixed Procedure of the SAS software system (SAS Institute, 1996). The linear model included two fixed effects (blocks and seed sources) and two or three random effects (family within seed source, block x family interaction and within-plot error): the within-plot error term was necessary only in tests with row plots (RP). Single-site heritabilities (h^2) were estimated from Equation 6.17.

Single-site heritabilities are upwardly biased because the variance due to family x site interaction is confounded with the family variance in the numerator (*Chapter 6*); however, these values are useful for comparing genetic expression of a given trait in different test environments. It is common to find that estimated heritabilities range markedly across test sites and this was the case in these data: 0.05 to 0.36 for stem volume and 0.11 to 0.41 for form (Table 15.1). These values can vary both because of differential genetic expression in different test environments (affecting the numerator of h^2) or more commonly because of differences in environmental heterogeneity within test environments (affecting estimates of error variances ($\hat{\sigma}^2_{fb}$ and $\hat{\sigma}^2_e$) in the denominator of Equation 6.17). For these data, there was no indication that any test site had severely degraded data, and the family differences were nearly always highly significant. Therefore, all sites were included in subsequent MM analyses.

The final stage of exploratory analysis prior to the combined MM analysis was to estimate type B genetic correlations, r_B, (*Chapter 6*) for each variable (standardized volume and form) for each of the possible 15 pairs of test sites (6 test sites paired in all possible combinations). For each pair of test sites, the type B genetic correlation for each variable was estimated from Equation 6.19 as $\hat{r}_B = \hat{\sigma}^2_f / (\hat{\sigma}^2_f + \hat{\sigma}^2_{fs})$ where $\hat{\sigma}^2_f$ is the estimated variance component due to differences among families and $\hat{\sigma}^2_{fs}$ is the estimated variance component due to family by test site interaction. Remember that $\hat{r}_B$ values near one indicate little family by test interaction (stable family rankings) for a given pair of test sites, while values near zero are indicative of large rank changes in families between test sites.

For both stem volume and form, $\hat{r}_B$ ranged from 0 to 1 (data not shown). Extreme values occurred most frequently when a variance component estimate from one test site in the pair was zero. All estimates at the extremes occurred when one of the two test sites in the pair was test 5 or test 6. These test sites had only 31 families (Table 15.1), reinforcing the difficulty of obtaining good estimates of genetic parameters with few families. Averaged over all 15 pairs of test sites, the mean type B genetic correlation was 0.54 for stem volume and 0.68 for form, suggesting that family rankings for form are less influenced by environmental differences among test sites than family rankings for volume.

We examined the 15 correlations for each variable carefully for patterns that might lead to logical groupings of test sites. In particular for these data, we hypothesized that: (1) Test sites of similar productivity (MAI in Table 15.1) might have higher correlations than pairs of dissimilar levels of productivity; (2) Test sites in more similar latitudes might exhibit less family x test interaction; and (3) Test sites of similar age might logically group together in expression of interaction. No clear pattern of any kind was found for either stem volume or form, leading us to adopt a univariate mixed model for the combined MM analysis that treats data from all test sites as a single trait (described in the next section).

LINEAR STATISTICAL MODELS

Linear statistical models form the basis of data analysis in quantitative genetics and the first step in any analysis is to define one or more appropriate models for the data being analyzed. Several simple linear models were introduced in *Chapter 6* equating phenotypic observations to linear functions of underlying genetic and environmental causes (*e.g.* Equation 6.3 where $P_i = \mu + G_i + E_i$). Unfortunately, for all except the simplest cases, these models are not appropriate for analysis of real data from genetic tests, because they are conceptual models parameterized in terms of underlying **causal effects**.

Linear statistical models for data analysis are specified with **observable effects** that can be estimated from the data. The exact form of the model varies widely depending upon the mating and field designs, and in fact, different models can be written for the same data set (compare examples in Boxes 15.3, 15.4 and 15.6). Yet, all models have the same basic structure in which the phenotypic observation on each tree is equated to a linear function of fixed and random main effects and their interactions:

$$\begin{array}{ccccc} \text{Phenotypic} = \text{fixed} + \text{random} + \text{interaction} + \text{residual} \\ \text{value} \quad \text{effects} \quad \text{effects} \quad \text{effects} \quad \text{error} \end{array} \qquad \text{Equation 15.1}$$

A factor is considered a **fixed effect** in the model if interest centers on the specific levels of the factor included in the experiment. These levels are considered constants that have no variance nor covariances with other factors. For example, if there are three types of fertilizers applied in an experiment, then this factor, fertilizer, is considered a fixed effect since inferences are made only about the specific three types included in the experiment. In this book, fixed effects, such as the population mean, test effects and block effects, are designated with Greek letters such as μ, α and β (see Boxes 15.3, 15.4 and 15.6).

There are three types of factors treated as fixed effects in MM analysis: (1) Treatment effects (such as fertilizers, irrigation levels, planting density and other silvicultural treatments); (2) Environmental effects associated with the experimental design (such as location, year and block effects); and (3) Fixed genetic effects that are sometimes called group effects (such as provenance, seed source and generation of selection). At times, group effects may appropriately be considered random instead of fixed, but we do not address these situations in this book.

A factor is considered a **random effect** (*i.e.* random variable with variance and covariances) if the levels included in the experiment are a sample of a larger population and the inferences of the investigation are aimed at the entire population or at the future performance of the levels of the effect. For example, when 50 half-sib families are planted in genetic tests and used to estimate heritability, the investigator estimates a heritability that applies to the larger population from which the 50 families are a sample. Further, we may be interested in the future performance of the 50 families. Therefore, the family effect in the linear model is considered as a random variable. We write random effects in linear models as lower case Latin letters such "f" for family effects and "a" for breeding values in Boxes 15.3, 15.4 and 15.6. Genetic effects such as families, breeding values and clonal values are always considered random in MM methods of analysis.

When both fixed and random effects are included in a linear statistical model, the model is called a mixed linear model or more often a **mixed model**. Interactions between random and fixed effects are considered as random (see Boxes 15.3, 15.4 and 15.6 for examples). For further discussion of fixed effects, random effects and their specification in linear models, see Henderson (1984), Searle (1987), and Searle *et al.* (1992).

Before describing mixed models in further detail, it is important to distinguish between true parameters and their estimates obtained from experiments. The true value of any population parameter or effect in any linear model is never known, but rather is estimated from experimental data. From now on, we will use a hat to differentiate an estimate from its true value, for example: (1) $\hat{\alpha}_3 = -5$ from Box 15.3 is an estimate of α_3, the true effect of test location 3; and (2) $\hat{h}^2$ is an estimate of h^2, the true heritability. In MM analysis, the word **estimation** is used when referring to a fixed effect and **prediction** used for a random effect (Searle, 1974; Henderson, 1984). This is mainly semantics, but we adhere to this convention saying, for example, that $\hat{\alpha}_3$ is an estimate of α_3, while saying that breeding values and other random effects are predicted.

Box 15.3. Parental linear model for half-sib families planted at several locations.

Suppose that open-pollinated families (assumed here for simplicity to be half-sib families) are planted at several field locations, and at each location the experimental design is a randomized complete block with single tree plots (*e.g.* test sites 1 to 4 in Table 15.1, but ignoring source differences). A linear model for each tree's phenotypic measurement for a given trait (y_{ijk}) is:

$$y_{ijk} = \mu + \alpha_i + \beta_{ij} + f_k + fs_{ik} + e_{ijk} \qquad \text{Equation 15.2}$$

where

μ = fixed population mean of all trees averaged across all locations and blocks;
α_i = fixed effect of the ith test site environment, $i = 1, 2, ... t$;
β_{ij} = fixed effect of the jth block within the ith test site, $j = 1, 2, ... b_i$;
f_k = random effect of the kth half-sib family, $k = 1, 2, ... s$, $E(f_k) = 0$, $Var(f_k) = \sigma^2_f$;
fs_{ik} = random interaction of the kth family and the ith test site, $E(fs_{ik}) = 0$, $Var(fs_{ik}) = \sigma^2_{fs}$;
e_{ijk} = random residual error (*i.e.* block x family interaction) of a tree in family k, block j, test site i, $E(e_{ijk}) = 0$, $Var(e_{ijk}) = \sigma^2_e$;

and the covariances between all pairs of factors are assumed zero.

The fixed effects, μ, α and β, are associated with the population mean, test locations and blocks within test locations, respectively. There are t α_i effects, one for each of the t test locations (α_1 α_2 ... α_t), and each α_i is expressed as a deviation from the population mean, μ. As an example, if test location 3 has a mean that is 5 units below the population mean, then $\hat{\alpha}_3 = -5$. There can be a different number of blocks at each location (*e.g.* $b_1 = 4$ and $b_2 = 10$ for 4 and 10 blocks at locations 1 and 2, respectively). Each of these blocks has its own fixed effect, β_{ij}, associated with the average environmental influence of that block on the phenotypic measurement. Each block effect is expressed as a deviation above or below the test site mean.

 The three random effects in the model, f_k, fs_{ik} and e_{ijk}, are each assumed to have a mean of zero (indicated by the expected value operator, E, in the model above) and a specific variance (indicated by the variance operator, Var). For example, there are s random f_k effects, one for each of the s half-sib families being tested. These family effects are deviations centered around a mean of zero with a variance of σ^2_f. If there are large genetic differences among the families being tested, then σ^2_f is large. Genotype x environment interaction is quantified by σ^2_{fs} and this variance is large if the relative performance (rankings) of families changes substantially in the different test locations. All remaining genetic and micro-environmental sources of variation are combined together in the residual error term and each tree measured has a different e_{ijk}. Typically the residual variance, σ^2_e, is much larger than the other two variances.

 The linear model above is called a **parental model,** because for each family, f_k is the general combining ability of parent k and equivalently is half the breeding value of parent k (Equations 6.7 and 6.8). Therefore, the model is set up to rank the s parents being tested which is an important objective of genetic tests (*Chapter 14*).
(Box 15.3 continued on next page)

Box 15.3. Parental linear model for half-sib families planted at several locations.
(Continued from previous page)

In addition, the model is appropriate for estimating additive genetic variance, σ^2_A, and heritability, h^2, of the trait for the population from which the tested parents are a sample. The variance among half-sib family effects, σ^2_f, is equal to one quarter of the additive variance (Falconer and MacKay, 1996, p157). Further, the variance of the individual tree measurements, σ^2_y, is equal to the phenotypic variance. In fact, the variance among the phenotypic measurements, σ^2_P, is always estimated as the sum of the variance components of the random effects in the model since, by definition, fixed effects have no variance. Therefore, the narrow sense heritability is estimated from the observable variance components in the model as follows (which can be compared with Equation 6.12 which is presented in terms of causal components of variance):

$$\hat{h}^2 = \hat{\sigma}^2_A / \hat{\sigma}^2_p = (4 * \hat{\sigma}^2_f)/(\hat{\sigma}^2_y) = (4 * \hat{\sigma}^2_f)/(\hat{\sigma}^2_f + \hat{\sigma}^2_{fs} + \hat{\sigma}^2_e)\ \ \text{Equation 15.3}$$

Parental *versus* Individual Tree Models

Historically, quantitative genetics data in forestry and crop breeding were analyzed with **parental models** in which genetic effects in the model are specified in terms of parental influences on the families being tested (Shelbourne, 1969; Namkoong, 1979; Hallauer and Miranda, 1981; White and Hodge, 1989). Models written for half-sib and full-sib families have family effects in the model such as those in Equations 6.8 and 6.9. These family effects relate back to parental influences passed on to the offspring measured in the genetic tests. For example, a parent passes on half of its breeding value to offspring. This information is used both to predict parental breeding values and to estimate genetic variances (Box 15.3).

Parental models have been extremely useful in forestry and currently account for nearly all reports in the quantitative genetics literature of forest trees. However as normally applied, these models rely on a series of assumptions about the parents: (1) Parents are unrelated and non-inbred; (2) Parents are an unselected random sample from the population; (3) Parents must be from the same generation or cycle of selection; (4) Parents are randomly intermated to form the families tested (*i.e.* no **assortative mating** of best with best); and (5) All parents are mated in a single mating design (*i.e.* data can not be combined from several different designs). In addition, there are assumptions about regular diploid inheritance (Falconer and Mackay, 1996).

As tree improvement programs advance, most of the above assumptions are violated through use of intensively selected, related and possibly inbred parents from different generations intermated in a variety of mating designs. MM methods facilitate analysis of these types of data, and produce unbiased predictions of breeding values and estimates of variance components as long as the models are specified correctly and all data are included in the analysis (Kennedy and Sorensen, 1988).

All parental models are limited in one regard: there are no predicted breeding values for the trees measured, only the parents of those trees. This limitation led to the development of so-called animal models by animal breeders, which are called **individual tree models** for forestry applications (Borralho, 1995). With individual tree models, the breeding value of each measured tree is specified as a random effect in the model (Box 15.4). This greatly increases the size and computational complexity of the analysis, since there are many more breeding values to predict (one for each tree in every test site instead of only the parents of those trees). However, with modern computers and software, these analyses are possible even for very large data sets. See Box 15.5 for an example of the individual tree model adopted for the case study of *E. grandis* from Argentina.

Individual tree models can be used to predict breeding values of parents and ancestors simultaneously with those of offspring, even though parental and ancestral effects are not explicitly specified in the model. This is accomplished through use of a pedigree file and the **additive genetic relationship matrix** (Lynch and Walsh, 1998, p 750) in which additive genetic relationships are specified for all pairs of trees, parents and ancestors. Trees in the same family are siblings and have a certain genetic relationship with each other and with their parents and other ancestors. Trees in unrelated families have no relationship if the parents are unrelated. By specifying these relationships, the data from all measured trees is brought to bear on the prediction of the breeding values of ancestors, parents and offspring. This means that all breeding values (for ancestors, parents and offspring) are predicted on the same scale and are developed as a single set of rankings. This greatly facilitates choosing selections regardless of their generation as demonstrated later (*Selection Indices: Combining Information across Relatives and Traits*).

The relationship matrix also accommodates data from offspring of selected, related and inbred parents from different generations intermated in a variety of mating designs. Data from clonal genetic tests can also be analyzed with individual tree mixed models. Further, data from quantitative traits and quantitative trait loci (QTL in *Chapter 18*) can be combined together (Hofer and Kennedy, 1993). For all of these reasons, individual tree mixed models will undoubtedly become more common in future analyses of data from forestry genetic tests. However, many of these applications are not simple and require further research to determine the pitfalls and best methods to address specific problems.

Multivariate Linear Models

Sometimes a **multivariate linear model** is more appropriate. This is most obvious when the measurements are actually made on different traits (*e.g.* stem volume and straightness), but multivariate models are also useful when the same trait measured in different test sites is treated as distinct variables in the mixed model (Jarvis *et al.*, 1995). This application of multivariate linear models is most appropriate when there are indications that genetic control of a measured trait differs across environments: (1) Heritabilities are very different among test locations; (2) Test sites are measured at different ages; or (3) Genotype x environment interaction occurs at varying levels between different pairs of test sites (more interaction between some pairs than others).

The essence of this latter application of multivariate mixed linear models is that data from each test location are treated as completely different variables (even when a single trait is measured at all locations). For example, in Box 15.6, there are three test sites and three variables. This means that there is a different variance structure (*e.g.* different heritability) estimated for each test location. Then, when the data from all trees on all sites are combined in one simultaneous MM analysis, breeding values are predicted for all traits (*i.e.* all test locations) using all data. That is, the precision and accuracy of the breeding values predicted for each test location are increased because of incorporation of the data from the other locations. Once again, this happens because of the genetic relationships between the variables that are specified in the model (genetic correlations in Equation 15.6) and the relationships among trees specified in the relationship matrix. For example, tree diameter in one test site contributes information about breeding value of height for a sibling on another site.

Box 15.4. Individual tree model for half-sib families planted at several test locations.

For the same half-sib genetic tests in Box 15.3, an individual tree model is:

$$y_{ijk} = \mu + \alpha_i + \beta_{ij} + a_{ijk} + fs_{ik} + e_{ijk} \qquad \text{Equation 15.4}$$

where

μ, α_i, β_{ij}, fs_{ik} and e_{ijk} are identical in meaning to the same terms in Equation 15.2; and

a_{ijk} = random breeding value of tree ijk, $E(a_{ijk}) = 0$, $Var(a_{ijk}) = \sigma^2_a$.

The important difference between the models in Equations 15.2 and 15.4 is the specification of the breeding value. In Equation 15.2 there are s family effects ($s = 100$ if there are 100 families being tested) and the model is suitable for predicting breeding values for the s parents being tested. In Equation 15.4, there are many more genetic effects specified, one breeding value for each tree measured, and these breeding values are for the offspring generation, not the parental generation. With MM analysis, parental breeding values are also predicted from this tree-level model through the pedigree file and relationship matrix (not shown here) which specify the genetic relationships among trees and their parents.

To estimate heritability from the model, recall from *Chapter 6* that the additive genetic variance, σ^2_A, is the variance among true breeding values in the population. Since a_{ijk} in the model in Equation 15.4 is the breeding value of the tree measured, the variance among those breeding values estimated from this experiment, $\hat{\sigma}^2_a$, is an estimate of σ^2_A. As always, the phenotypic variance is estimated as the sum of the variance components of the random effects in the model. Therefore:

$$\hat{h}^2 = \hat{\sigma}^2_A / \hat{\sigma}^2_P = \hat{\sigma}^2_a / \hat{\sigma}^2_y = \hat{\sigma}^2_a / (\hat{\sigma}^2_a + \hat{\sigma}^2_{fs} + \hat{\sigma}^2_e) \qquad \text{Equation 15.5}$$

Sometimes edaphoclimatic data combined with results from exploratory data analysis indicate logical groupings for subsets of the test sites. In this case, the data for any trait within a group of test sites may be treated as a single variable, while data from sites from different subsets are considered as different variables in the analysis. This approach is appropriate when there is high genotype x environment interaction among test sites in different subsets of sites, but less interaction among sites within the same subset.

Multivariate individual mixed models are the most flexible and powerful for genetic test analysis; yet, they are also the most difficult to specify, compute and interpret. It is only in recent years that computer software has been written to analyze large data sets (Boldman *et al.*, 1993; Mrode, 1996). In the future, these models may become the most popular for: (1) Analysis of quantitative genetics data from field tests; (2) Analysis of genetic marker and QTL data (*Chapter 19*); and (3) The combination of data from quantitative traits with those from QTL and markers (Hofer and Kennedy, 1993).

Box 15.5. Case study of *E. grandis* from Argentina: Linear model and mixed model statistical methods.

The mixed model analyses combined data from all six test sites described in Box 15.1 to estimate fixed effects and variance components and to predict breeding values for each variable (standardized stem volume and untransformed stem form). The univariate individual tree model for each variable was similar to Equation 15.4 in Box 15.4. The model contained fixed effects for seed sources, test sites, seed source x test site interaction and blocks within test sites, and it contained random effects for tree breeding values (a_{ijk}), family x test site interaction (fs_{ik}) and the residual term (e_{ijk}). For convenience, we did not include a plot term in the model, even though two of the test sites had row-plot designs. An argument could be made for using a multivariate linear model treating the data from each test site as a separate variable (Box 15.6), since heritability estimates varied markedly among test sites; however, we opted for the simpler univariate model.

    The computer program MTDFREML (Boldman *et al.*, 1993) was used for two MM analyses (one for each variable). This program requires that a pedigree file be developed. In this case, this meant specifying the female parent for each measured tree, while the male parent was unknown and specified as zero for all trees. MTDFREML uses the restricted maximum likelihood (REML) method to estimate variance components and to predict breeding values for all parents and their offspring through best linear unbiased prediction (BLUP). All REML procedures are iterative and require a computer algorithm to find estimates of the components that maximize the value of the likelihood function given the data. MTDFREML uses the Simplex algorithm (Nelder and Mead, 1965) which can sometimes converge to a local maximum instead of a global maximum (meaning that the variance component estimates may not be optimal).

    To ensure obtaining the best estimates possible, the authors of MTDFREML recommend: (1) Starting with good prior estimates of all components that are entered into the program as the starting point for iterations (we obtained prior estimates from SAS Proc Mixed; SAS Institute, 1996); and (2) Re-running MTDFREML several times with different prior estimates to ensure consistent convergence to the same estimated values (we re-ran the program four to eight times with different prior estimates and slightly different linear models). MTDFREML never failed to converge and for all starting points converged after between 8 and 50 iterations with nearly identical estimates of fixed effects and predictions of random effects.

    Three useful types of results were obtained from the MM analysis for each variable: (1) Estimates of the fixed effects for seed sources, test sites and blocks (summarized in Box 15.8); (2) Estimates of variance components used to estimate heritabilities and type B genetic correlations (Box 15.9); and (3) Breeding value predictions for 11,420 candidates (203 parents + 11,217 measured offspring) (Box 15.10).

CONCEPTS AND APPLICATIONS OF MIXED MODEL METHODS

The main distinction between classical linear models previously used to analyze data from genetic tests and those used in MM analysis is that breeding values and clonal values are treated as random variables in MM analysis. While this may seem like a trivial distinction, the former assumption that these factors are fixed effects leads to the derivation of methods of analysis called **ordinary least squares (OLS)**. These classical methods are still taught in statistics classes where treatment means are compared with mean separation techniques such as least significant differences, Duncan's, Tukey's or Scheffe's methods (Neter and Wasserman, 1974).

When genetic effects are assumed random in the linear model, the derivations of all analytical methods proceed differently, because the factors have variances and also covariances with other random variables in the model. In place of OLS, this assumption leads to three different methods for predicting breeding values and clonal values (and hence for ranking candidates for selection): Selection index (SI), best linear prediction (BLP) and

---

Box 15.6. Multivariate individual tree linear model for half-sib families planted at several locations.

When phenotypic measurements from each test location are treated as distinct traits, a multivariate linear model is appropriate. For illustration, half-sib data from three sites are considered and the measurements on the same trait on the three sites are labeled x_{ij}, y_{ij}, and z_{ij} to indicate that these are assumed to be different variables. At the individual tree level, the model for three test locations is:

$$x_{ij} = \mu_x + \beta_{x,i} + a_{x,ij} + e_{x,ij}$$
$$y_{ij} = \mu_y + \beta_{y,i} + a_{y,ij} + e_{y,ij}$$
$$z_{ij} = \mu_z + \beta_{z,i} + a_{z,ij} + e_{z,ij}$$

Equation 15.6

where

μ_x, μ_y and μ_z are the population means for each of the three variables;
$\beta_{x,i}$, $\beta_{y,i}$ and $\beta_{z,i}$ are fixed block effects for each of the three variables;
$a_{x,ij}$, $a_{y,ij}$ and $a_{z,ij}$ are random breeding values for each tree measured with
$\quad E(a_{x,ij}) = E(a_{y,ij}) = E(a_{z,ij}) = 0$; $Var(a_{x,ij}) = \sigma^2_{x,a}, Var(a_{y,ij}) = \sigma^2_{y,a}, Var(a_{z,ij}) = \sigma^2_{z,a}$;
$e_{x,ij}$, $e_{y,ij}$ and $e_{z,ij}$ are the residual variances for each variable with
$\quad E(e_{x,ij}) = E(e_{y,ij}) = E(e_{z,ij}) = 0$; $Var(e_{x,ij}) = \sigma^2_{x,e}, Var(e_{y,ij}) = \sigma^2_{y,e}, Var(e_{z,ij}) = \sigma^2_{z,e}$; and
genetic correlations among the breeding values of the three variables of:
$\quad Corr(a_{x,ij},a_{y,ij}) = r_{A,xy}$, $Corr(a_{x,ij},a_{z,ij}) = r_{A,xz}$ and $Corr(a_{y,ij},a_{z,ij}) = r_{A,yz}$.

The model at each test location contains fewer terms than Equation 15.4, because the test location effects and their interactions are removed. Instead in this model, three different population means are estimated, one for each site. The breeding values are also modeled as different variables in different test locations; however, the MM analysis uses all data from all sites and utilizes the genetic correlations among the breeding values in the different tests. This requires variance-covariance matrices and relationship matrices that are not shown.

There are three heritabilities to estimate, one for each trait, and the formulas are:

$$\hat{h}^2_x = \hat{\sigma}^2_{x,a} /(\hat{\sigma}^2_{x,a} + \hat{\sigma}^2_{x,e})$$
$$\hat{h}^2_y = \hat{\sigma}^2_{y,a} /(\hat{\sigma}^2_{y,a} + \hat{\sigma}^2_{y,e})$$
$$\hat{h}^2_z = \hat{\sigma}^2_{z,a} /(\hat{\sigma}^2_{z,a} + \hat{\sigma}^2_{z,e})$$

Equation 15.7

best linear unbiased prediction (BLUP). The first two are special cases of BLUP; so, here we focus on BLUP. However, many concepts described below also apply to SI and BLP. White and Hodge (1989) present detailed comparisons among SI, BLP and BLUP, and the capabilities of BLUP compared to OLS are summarized in Box 15.7. The next major section, *Selection Indices: Combining Data Across Relatives and Traits,* describes combining information from multiple traits after a BLUP analysis of each trait.

When BLUP methods are used for prediction of random genetic values, generalized least squares (GLS) is used for estimation of fixed effects in the linear model. Usually, GLS and BLUP are found together in computer software packages that also estimate variance components for the random effects in the linear model by a method called Restricted Maximum Likelihood (REML). Each of these three methods, GLS, REML and BLUP, are briefly described in the following sections, and together they make up **mixed model analysis (MM)**. Most software packages currently available for MM analysis were written by animal breeders, for example: ASREML (Gilmour *et al.*, 1997); DFREML (Meyer, 1985, 1989); PEST (Groeneveld *et al.,* 1990); and MTDFREML (Boldman *et al.*, 1993). However, there are some newer packages aimed specifically at MM analysis of forest genetic data: GAREML (Huber, 1993) and TREEPLAN (Kerr *et al.*, 2001).

Prior to the availability of appropriate computer software, classical OLS methods of data analysis were used in tree improvement programs to rank parents, families and clones. These methods give very similar rankings as MM methods when a number of assumptions are met: (1) Data are balanced (*i.e.* all genotypes are present with the same number of trees in all blocks at all test sites and there are equal numbers of blocks per site); (2) Parents are unselected, unrelated and randomly mated; and (3) All data come from a single age and mating design.

MM methods have distinct advantages over OLS methods when the above assumptions are not met (Box 15.7), which is likely to be the case in many tree improvement programs as they move into advanced generations of breeding. For example, in advanced generations, there can be advantages to preferentially mating more superior selections more frequently and to using unstructured mating designs (*Chapter 17*). In addition, advanced-generation programs will benefit from combining data across multiple generations of genetic tests of varying mating and field designs. These needs, coupled with the availability of computer software, imply that MM methods of analysis will be the method of choice for tree improvement programs in the future as it is now for animal breeding programs (Mrode, 1996).

Effective use of MM methods does not depend on balanced data nor structured mating designs. However, MM methods require that the various test sites being analyzed together are connected genetically. **Connectedness** among test sites means that some families, parents or other relatives occur in more than one site in a pattern that interlinks all sites (Foulley *et al.*, 1992). This connectedness establishes a genetic linkage among sites that results in a covariance structure such that all pairs of parents have a non-zero covariance. For example, crosses A x B, B x C and C x D connect A with C and D even though there are no crosses between A x C nor A x D. Once again, it is the assumption that genetic effects are random variables that gives rise to the notion of connectedness. When breeding values are random variables, knowledge of a family's performance in one site provides information about its performance and the performance of connected crosses in other sites of the same or different mating design. This provides a statistical linkage among the test sites. Proper connectedness among sites eliminates the need for checklots or controls to effectively predict breeding values (*Chapter 14*).

Box 15.7. Capabilities of mixed model analyses not possible with ordinary least squares.

1. *Data from several mating designs:* Data from half-sib, diallels, factorials and unstructured mating designs can all be analyzed in a single combined analysis if genetic entries are connected among the different tests (*Chapter 14*).

2. *Spanning several generations:* Breeding values of parents, offspring, siblings and other relatives from several generations can be predicted in a single combined analysis given proper connectedness of the data. Breeding values and rankings of all candidates are generated on a single scale for convenient comparison and selection of candidates with the highest genetic merit regardless of generation.

3. *Related parents and non-random mating:* Related parents may be used in the same mating design and may be mated assortatively (*e.g.* mating best with best). Better parents may be mated more frequently.

4. *Effects of selection:* Effects of previous selection or culling are accounted for to produce unbiased breeding value predictions if all data from all generations are included in the analysis (Kennedy and Sorensen, 1988).

5. *Traits not in the linear model*: Breeding value predictions can be obtained for target traits not measured if estimated correlations between measured and target traits are available (*e.g.* can predict breeding values for mature bole volume from measurements of juvenile tests).

6. *Genetic gain predictions:* BLUP-predicted breeding or clonal values can be used directly to estimate genetic gains for a variety of commercial deployment and breeding options.

7. *Multivariate linear models:* Data on a single trait from several test sites can be treated as different traits in order to account for different heritabilities on different sites and/or complex patterns of genotype x environment interactions (Box 15.6). Also, data from several distinct traits (*e.g.* height, diameter and volume) can be analyzed simultaneously with multivariate linear models.

8. *Major genes and minor genes:* MM methods can be used to screen for major gene effects in phenotypic data and can incorporate both major gene effects (such as QTLs, *Chapter 18*) and breeding values due to polygenic effects into the same model. (Kennedy *et al.,* 1992; Hofer and Kennedy, 1993; Kinghorn *et al.,* 1993; Cameron, 1997, p 150).

Estimation of Fixed Effects

In the previous section on linear models, fixed effects were divided into three types associated with treatments, environmental influences and fixed genetic groups. In MM analysis of data from genetic tests, accurate estimation of these fixed effects is important for two reasons. First, there may be interest in the fixed effects themselves, *e.g.* are fertilizer treatments or seed sources significantly different? Second, accurate estimates of fixed effects are required for unbiased prediction of breeding values. That is, the estimated fixed effects are used to properly adjust predicted breeding values so that they are not biased by treatments, design factors or other spurious "nuisance" effects. As an example, consider the common case when some families are unequally represented across blocks, sites or treatments. Without proper estimation and adjustment for these fixed effects, families with more trees in faster-growing blocks, better sites or superior treatments "appear" more genetically superior for growth than they really are.

MM analysis uses an elegant method of estimating fixed effects called **generalized least squares, GLS**. The difference between OLS and GLS estimates once again relates to the assumption in GLS that breeding values and their interactions with fixed effects are random variables. In GLS the knowledge of a family's or clone's performance in all blocks, treatments and tests is used to increase the accuracy of estimation for the fixed effects. For example, suppose that one test location contains predominantly good families just due to their availability at time of planting. OLS does not account for the fact that these families are generally superior at all locations, and, therefore, overestimates the environmental effect for the test location of interest. Basically, the better performance of the superior families is confused for a higher quality site. Or in other words, the estimate of $\hat{\alpha}_i$ from Equation 15.2 is upwardly biased for that test location. Unlike OLS, GLS uses the knowledge of these families' superior performance at other sites to produce an unbiased estimate of the test effect.

The estimates of fixed effects from GLS are called **best linear unbiased estimates (BLUE)**, because among all other possible linear functions of the data that are unbiased, the GLS estimates have the smallest error variance (*i.e.* they are best in this sense of minimum error variance). Therefore, the GLS estimates are both unbiased and as precise as the data will allow. There are no other linear functions of the data that produce unbiased estimates of higher precision (*i.e.* of lower error variance). For these optimality properties of GLS to hold (that is, for GLS estimates to be BLUE), the variance and covariance components associated with the random effects must be known without error. In practice, these underlying variances and covariances are never known, but must be estimated from the data. Therefore, high quality estimates of the fixed effects require high quality estimates of the variance components and this is the subject of the next section.

In the case study of *E. grandis* data, the computer software program MTDFREML was used to calculate GLS estimates of all levels of the four fixed effects specified in the linear model (Box 15.8). The estimates of test site effects and blocks within test sites were not of any inherent interest, but were used to properly adjust breeding value predictions. Conversely, GLS estimates of seed source effects and seed source x test site interactions were of direct interest. Estimates of seed source performance were used to infer the relative importance of the various seed sources to future breeding efforts.

Estimation of Variance Components and Genetic Parameters

By definition, each random effect in a linear model has an associated variance component (*Chapter 6*). Some examples are shown in Boxes 15.3, 15.4 and 15.6. As mentioned previously, these variance components are associated with populations of genotypes and environments. Since we never measure all possible genotypes grown in all possible environments, the exact values of these components are never known, but rather must be estimated from the sample of genotypes and environments represented in genetic tests.

High quality estimates of variance and covariance components are needed for three distinct purposes: (1) Estimation of fixed effects, since the GLS procedure described above requires estimates of the variance components to solve for estimates of the fixed effects; (2) Estimation of various genetic parameters such as heritabilities and type B genetic correlations that provide insight into the genetic and environmental variance structure of the population (see examples in Equations 15.3, 15.5 and 15.7); and (3) Prediction of genetic values, since BLUP requires variance component estimates as part of the process. These 3 uses of variance components are illustrated in Boxes 15.8, 15.9 and 15.10, respectively, for the case study of *E. grandis* from Argentina.

ANOVA versus REML Estimates of Variance Components

Prior to the 1980s in all fields of quantitative genetics, the vast majority of variance components were estimated from analysis of variance (ANOVA) methods in which observed mean squares are equated to their expected values expressed as functions of the underlying variance components (Searle, 1987,Chapter 13; Searle *et al.,* 1992, Chapter 4). This ANOVA method has been widely used and expected mean squares for the structured mating designs described in *Chapter 14* are given in several books (*e.g.* Hallauer and Miranda, 1981) and automatically calculated by most computer packages designed for analysis of experimental data from structured experimental designs.

When certain conditions are met, ANOVA estimates of variance components have highly desirable properties. In fact, ANOVA estimates are the best unbiased quadratic estimates (meaning they have the minimum error variance among all possible quadratic functions of the data that produce unbiased estimates, Searle *et al.,* 1992) under the following conditions: (1) Data are balanced, meaning that there is 100% survival and all families are planted in equal numbers of blocks at all test sites; (2) Parents are unselected, non-related and come from the same generation; and (3) Parents are intermated in a single, structured mating design (such as a factorial or diallel) to produce a single type of collateral relatives (such as full-sib families).

When data are unbalanced, ANOVA methods may still be used to estimate variance components, but there is not a single set of variance component estimates produced that meets any known optimality properties. In fact, for unbalanced data, there are, theoretically, an infinite number of different types of ANOVA-based estimates obtainable from a single data set. Some of the more popular ANOVA estimation procedures for unbalanced data are called Henderson's methods 1, 2 and 3 and SAS types I, II, III and IV; the relative merits of these different methods depend on the exact type of imbalance and underlying variance structure (Freund and Little, 1981; Milliken and Johnson, 1984; Searle, 1987; Henderson, 1988).

More recently, a completely different method of estimating variance components called **Restricted Maximum Likelihood (REML)** has gained widespread use among quantitative geneticists (Patterson and Thompson, 1971; Searle, 1987; Searle *et al.,* 1992; Lynch and Walsh, 1998). Unlike the ANOVA approach, REML is an iterative process that searches for estimates of the variance components that maximize the likelihood that the observed data came from a given probability distribution. In almost all mixed model computer software packages, the probability distribution is assumed to be a multivariate normal distribution.

When the three conditions mentioned above for ANOVA are met (data are balanced, etc.), ANOVA and REML methods yield identical estimates for all variance components in the linear model. The one exception is when some of the true variance components are near zero; ANOVA methods can produce negative estimates, while most REML algorithms do not allow negative estimates (*i.e.* negative estimates are forced to zero) (Searle *et al.,* 1992; Huber *et al.,* 1994). Even when data are fairly unbalanced, REML and ANOVA estimates are quite similar, although REML generally produces slightly better estimates for most situations common to forest genetics (Huber *et al.,* 1994).

The real advantage of REML compared to ANOVA methods for estimating variance components is for analyzing more complex, multi-generational data sets from advanced-generation tree improvement programs. ANOVA methods cannot be used for these complex data sets; yet, REML methods can. The additive genetic relationship matrix is used to specify all possible genetic relationships in the data set, and REML takes these relationships into account to produce unbiased, precise estimates of variance components.

Box 15.8. Case study of *E. grandis* from Argentina: Estimation of fixed effects.

There were four fixed effects specified in the linear model (Box 15.5) for the *E. grandis* data described in Box 15.1: test sites, blocks within test sites, seed sources and seed source by test site interaction. For each trait (stem form and standardized volume), the mixed model analysis with computer software MTDFREML produced: (1) Generalized least square estimates of all levels of each of these four effects; and (2) Predicted breeding values for 203 parents and 11,217 trees that are properly adjusted for the test sites, blocks and seed sources in which they occur.

The first two effects (test sites and blocks within test sites) are nuisance effects in the sense that there is little interest in the means of the various test sites or blocks within sites. However, with data as unbalanced as these, it is imperative that estimates of seed source effects and breeding value predictions properly account for the fact that seed sources and families are not planted in all blocks and sites. Therefore, the breeding value prediction for a particular tree growing in a certain block of a given site is properly adjusted for the mean levels of that site and block.

Unlike test sites and blocks, there was special interest in knowing which seed sources were better. For both volume and form, the seed source effect was highly significant in the combined MM analysis, but seed source x test site interaction was not significant for either variable. Comparing the two seed sources from the local land race in Argentina (*i.e.* unselected and selected; Fig. 1), it is clear that the 16 selected parents produced faster-growing offspring of better stem form than the 16 unselected parents. All 32 parent trees were sampled from the same stands in the Entre Rios province of Argentina; the only difference was that 16 parents were selected as being phenotypically superior and 16 were not. These data indicate that mass selection in the local land race population was effective for both stem volume and form, and the realized gain (*i.e.* mean of selected trees minus mean of unselected trees) was approximately 8% for both traits.

Another interesting result was the relatively mediocre performance of the progeny of selected trees from the local land race compared to progeny of unselected trees from 12 provenances in the native range in Australia (*i.e.* N01-N12; Fig. 1). Means for about half of the native provenances were better than means for the selected trees for both traits. As discussed in *Chapter 12*, several factors influence whether local land races are superior or inferior to newly introduced material from the native range. In this case, perhaps the local land race originated from a less-than-optimal provenance in the native range. Whatever the reason, both the reforestation program and tree improvement program can benefit from introduction of new genes from native provenances.

(Box 15.8 continued on next page)

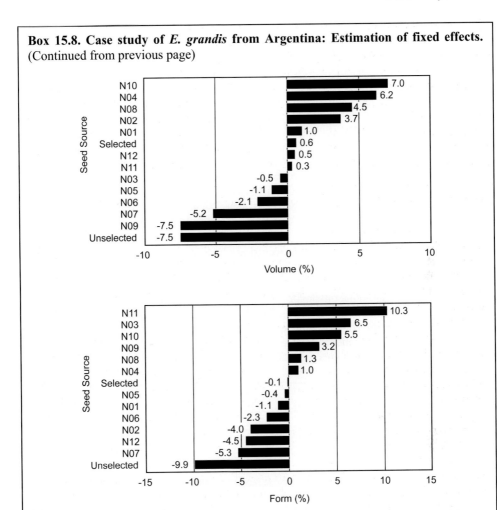

Box 15.8. Case study of *E. grandis* from Argentina: Estimation of fixed effects. (Continued from previous page)

Fig. 1. Seed source means for volume (top) and form (bottom) for *E. grandis* obtained from the mixed model analysis of data from all 6 test locations from the case study in Argentina (Boxes 15.1 and 15.5). Seed source effects were significantly different (P < 0.01) in the mixed model analyses of both traits. Seed source means are expressed as percentages above or below the average of all 11,217 trees. Positive values are good for both traits.

Importance of Large Data Sets for Estimating Variance Components

Large amounts of data (large numbers of families or clones tested in several environments) are required to obtain high-quality estimates of variance components (*i.e.* have small confidence intervals). The importance of large data sets for precise estimates cannot be over emphasized. The exact amount and type of data required, however, depends on the parameters being estimated and many other factors.

To illustrate this concept, consider the estimation of narrow sense heritability from completely balanced half-sib data when we assume that the true heritability is $h^2 = 0.10$, the type B genetic correlation between pairs of sites is 0.60 and 20 complete blocks are planted per test site in single tree plots. The magnitude of a 95% confidence interval on the estimated

Box 15.9. Case study of *E. grandis* from Argentina: Estimation of variance components.

There were 3 random effects in the linear model (Box 15.5) for the *E. grandis* data described in Box 15.1: tree breeding values (a_{ijk}), family x test site interaction (fs_{ik}) and the residual term (e_{ijk}). A pedigree file specified the relationships among the 203 parents and their 11,217 offspring, so that MTDFREML could properly account for all genetic relationships in the estimation of the three variance components for each trait: (1) Additive variance, $var(a_{ijk}) = \sigma^2_a$; (2) Variance due to site x family interactions, $var(fs_{ik}) = \sigma^2_{fs}$; and (3) Micro-site residual variance, $var(e_{ijk}) = \sigma^2_e$. These values sum to the total phenotypic variance (σ^2_P) as shown in Table 1 below for both traits. Note that the phenotypic variance for stem volume is near 1.0 since this variable was standardized prior to the MM analysis. For fixed effects, MTDFREML produces a GLS estimate for each level of each effect in the model (*e.g.* 14 estimated seed source means and 6 estimated test site means); however, there is only a single estimate for each random effect: the estimated population variance of that effect.

Table 1. Genetic parameter estimates for *E. grandis* obtained from the mixed model analysis of data from all 6 test locations for the case study in Argentina (Boxes 15.1 and 15.6). Note that stem volume was standardized before analysis, so the values are unitless (not in m^3). The values for stem form are in the units of measure (*i.e.* a scale of 1 to 4). All variance components are defined in Box 15.4. The type B genetic correlation, r_B, is defined in *Chapter 6*.

Trait	Mean	σ^2_a	σ^2_{fs}	σ^2_e	σ^2_P	h^2	r_B
Volume (Standardized)	2.30	0.140	0.034	0.815	0.989	0.142	0.58
Form (Scale 1-4)	2.55	0.097	0.011	0.455	0.563	0.172	0.74

The variance component estimates were used to estimate an individual tree heritability and type B genetic correlation for each trait using Equations 15.5 and 6.19, respectively. The heritability estimates indicate that stem volume and form are under similar genetic control (h^2 = 0.14 and 0.17 in Table 1), but that there is more family x test location interaction for stem volume than form (estimates of r_B are 0.58 and 0.74). Both of these results are consistent with the individual test site analyses and with the analyses of pairs of sites. To illustrate that the single site heritabilities are upwardly biased by variance due to family x test interaction, compare the estimates obtained from the combined MM analysis (0.14 and 0.17 for volume and form, respectively) with the average of the estimates from the test-by-test analyses (0.21 and 0.26 for volume and form obtained by averaging the six values for each trait from Table 15.1).

The parameter estimates indicate that substantial genetic progress can be expected from selection for either trait in a tree improvement program. There is more genotype x environment interaction for stem volume than commonly found for families of eucalypts; perhaps this is because the test locations span nearly 500 km in Argentina across a range of soils and rainfall patterns.

heritability under the above assumptions is plotted in Fig. 15.2 for increasing numbers of families when the total number of test sites is 2, 4 or 8. The 95% confidence interval is $\hat{h}^2$ ± 0.10 when 15 families are planted in 8 test sites, when 23 families are planted in 4 sites or when 55 families are planted in 2 sites. At this level of precision, the 95% confidence interval includes zero for an estimate of $h^2 = 0.10$ (the true value); so, the estimate would not be significantly different from zero at the 5% level of probability.

To obtain a 95% confidence interval equal to the true value of h^2 (*i.e.* $\hat{h}^2$ ± 0.05 so that the interval is 0.10), it would be necessary to plant approximately 55 half-sib families in 8 test sites or more than 90 families in 4 sites (Fig. 15.2). Even in these cases, the confidence interval is quite large compared to the size of the true heritability, reinforcing the need for large data sets. More than 100 families planted in several test sites would be required to reduce the confidence interval still further, and 1000 families would be even better.

The situation is only slightly different for clonal tests where stronger genetic control might be expected (*e.g.* true broad sense heritability of 0.30 and type B genetic correlation of 0.80). With 10 ramets of each clone planted in 10 complete blocks of single tree plots at each location, a 95% confidence interval of 0.10 (*i.e.* $\hat{H}^2$ ± 0.10) requires 80 clones planted in 8 test sites or nearly 100 clones in 4 sites.

Genetic correlations among two different traits are even more difficult to estimate than heritabilities of traits, and very large amounts of data are required for precise estimates. This is why hundreds of families or clones are often mentioned as the minimum needed for genetic tests aimed at parameter estimation.

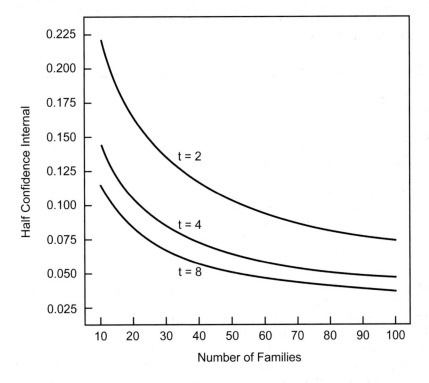

Fig. 15.2. Half or one side of the 95% confidence interval for narrow sense heritability estimates using Dickerson's method (1969) plotted as a function of the number of test locations (t = 2, 4 and 8) and the number of half-sib families planted at each location. It was assumed that: (1) True narrow-sense heritability $h^2 = 0.10$; (2) Type B genetic correlation between pairs of test sites is 0.60; and (3) Test design is single tree plots with 20 blocks per location.

Box 15.10. Case study of *E. grandis* from Argentina: Prediction of breeding values.

Two separate MM analyses, one for standardized stem volume and one for stem form, were conducted on the data described in Box 15.1 using the methods described in Box 15.5. The tree-level linear model (Equation 15.4) meant MTDFREML predicted a breeding value for each measured tree and also for each of the parents whose relationship to the offspring was specified in the pedigree file. Therefore, MTDFREML calculated breeding value predictions for volume and form for 11,420 candidates (203 parents plus 11,217 measured trees). Two subsequent steps after the MM analysis were: (1) Raw breeding value predictions from MTDFREML were converted to percentage values (as described below); and (2) Estimated seed source fixed effects (also expressed in percent, Box 15.8) were added back so that the final predicted breeding values reflect both seed source and within seed source contributions. After conversions, the final predictions were called volgain and formgain (Table 1).

Table 1. Summary statistics for 203 parental breeding value predictions and 11,217 tree-level breeding value predictions for *E. grandis* obtained from the mixed model analysis of data from all 6 test locations for the case study in Argentina (Boxes 15.1 and 15.5). Volgain and formgain are the breeding value predictions for volume and form, respectively, as symbolized by a_{ijk} in Equation 15.4. Index = 0.5*volgain + 0.5*formgain, where the units of all variables are percentages above the mean of 11,217 measured trees. See Boxes 15.11 and 15.12 for details.

Variable	Mean	Standard deviation	Minimum value	Maximum value
Summary of 203 parental breeding value predictions				
Volgain	0.00	14.4	-49.9	33.4
Formgain	0.00	11.7	-35.7	31.5
Index	0.00	10.3	-24.2	23.5
Summary of 11,217 tree-level breeding value predictions				
Volgain	0.00	10.3	-31.2	38.8
Formgain	0.00	8.8	-28.4	27.3
Index	0.00	7.7	-22.4	30.0

To re-express breeding values as a percentage above or below the mean of all 11,217 trees (baseline), different methods were used for the two traits. For form, this meant dividing estimates of seed source effects and predicted breeding values by the overall mean for all trees (2.55; in Table 1, Box 15.9) and then multiplying by 100. For convenience of presentation, form values were multiplied by -1 so that larger percentage values indicate better form (in the original scale low values were good). For volume, the estimates and predictions first had to be multiplied by the average of the phenotypic standard deviations used to standardize the data, and then the re-scaled values were converted to percentages around the overall mean. Therefore, all values reported are in percentages above or below the baseline.

(Box 15.10 continued on next page)

Box 15.10. Case study of *E. grandis* from Argentina: Prediction of breeding values. (Continued from previous page)

With BLUP, parental and tree-level breeding value predictions are properly shrunken or regressed to account for differential data quality among candidates. Differential shrinkage (as described in Box 15.11) happens in this data set because each of the 203 families has a different number of measured trees (varying from 11 to 121 depending on the number of sites and blocks the family is contained in). Parental breeding values are more regressed towards the mean of zero for families containing fewer trees, because the quantity of data is less and there is less ability to judge the underlying genetic merit of the parent using BLUP. Conversely, parental breeding value predictions are more spread out (larger variance among predictions) for families with more trees. There can still be good and poor parents from families with few trees, but the performances of their trees must be truly extreme since the weights applied to the tree-level deviations are smaller resulting in more regression towards zero.

All 11,420 breeding value predictions for each trait are on the same scale and can be put together into a single listing to make the best selections regardless of the generation (parent or offspring). This is discussed further in Box 15.12. In addition, the predicted breeding values are direct predictions of genetic gain. So, if the best parent for stem volume (with predicted breeding value of 33.4%, Table 1) is mated with the second best parent (with predicted breeding value of 32.8%), the offspring are predicted to grow 33.1% above the mean of the population of all tested trees (ignoring specific combining ability). Similarly, predicted gain from a seed orchard composed of 10 parental (*i.e.* backward selections) and 20 offspring (*i.e.* forward selections) is the unweighted average of the 30 selections' predicted breeding values (assuming equal seed and pollen production by all selections).

Prediction of Genetic Values

Prior to BLUP, a variety of analytical methods was used to rank parents, families and clones from genetic tests. All of these are variants of ordinary least squares (OLS) in which genetic values are considered fixed effects. All OLS methods proceed by: (1) Adjusting the data for fixed effects of blocks, treatments and test locations; (2) Perhaps transforming or standardizing the data; and (3) Averaging all adjusted, transformed data of a given family (or clone) to estimate the genetic value of that family (or clone). The average for a given family may include data from many blocks, treatments and test locations. Selection then proceeds by choosing the best families based on the rankings developed by aggregating all family averages into a single list.

One undesirable property of OLS methods is that there is a tendency to select candidates that are more poorly tested (*i.e.* that have less data or data of poorer quality). Animal breeders were the first to point this out. They noticed that when bulls first entered the artificial insemination program, there was a tendency for their OLS rankings to be extreme (very good or very poor) and the highest ranked bulls were often selected for further breeding. After some years in the program as more data accumulated, there was a tendency for the means of the extreme bulls to regress back towards the mean of all bulls. They no longer appeared so superior (or inferior for those starting at the other end of the rankings). So, there was less of a tendency to select the same bulls that had appeared so outstanding only a few years before when their rankings were based on less data.

This tendency to select a larger fraction of more poorly tested individuals, parents, families or clones also occurs in forestry when the rankings are based on OLS methods of data analysis. In essence, the tendency occurs because OLS means that are less-precisely estimated have a wider range and variance due to larger contributions of experimental and microsite error. Since there is more variation among less-precisely-estimated OLS means than among more-precisely estimated OLS means, a greater number of candidates appear superior from the imprecise group. For example, if there are two different groups of 100 half-sib families with group 1 planted in one test site and group 2 planted in five test sites, there are generally more higher ranking families chosen from group 1 (White and Hodge, 1989, p 54). In both groups, a high ranking family for growth can be superior due to its genetic superiority or due to its location on more favorable microsites (which is due to happenstance of randomization). In group 1 (tested on a single test site), microsite variability has more influence on OLS family means, causing those means to have a wider range and variance than OLS family means in group 2. Therefore, more families in group 1 will be above any truncation point chosen for selection leading to more families being selected from group 1 than group 2.

Animal breeders sought new approaches to data analysis that would not have this problem associated with OLS methods, and eventually developed **best linear unbiased prediction**, BLUP (Henderson, 1974, 1975). Genetic values predicted by BLUP have the lowest error variance and highest correlation with the true (but unknown) breeding values of any possible linear functions of the data that produce unbiased predictions (White and Hodge, 1989; Chapter 11). The details of BLUP are beyond the scope of this book, but BLUP and OLS are compared for mass selection in Box 15.11. As with REML and GLS described previously, BLUP can be used to analyze data from a wide variety of different situations, with related and selected parents mated in different mating designs, planted in different test designs, and measured for different traits. BLUP assigns weights to data based on the quality and quantity of that data as judged by the variances and covariances that are estimated by REML.

There are three important properties of all BLUP-predicted genetic values that make them different from OLS methods: (1) All phenotypic data are weighted in accordance with the genetic information they provide and this results in a larger "regression" or "shrinkage" of less reliable data (*e.g.* lower h^2 results in more shrinkage; see Box 15.11); (2) As a result of differential shrinkage for data of different quality, there is a tendency to select more winners from groups of candidates that are better tested; and (3) The predicted genetic values can be used directly for estimating genetic gain because they have been properly adjusted and shrunken (*i.e.* there is no need to apply further formulas such as those in Falconer and Mackay, 1996). These concepts with regards to BLUP-predicted breeding values are illustrated for the simple case of mass selection in Box 15.11 and are discussed in Box 15.10 for the case study of *E. grandis* data from Argentina.

The concept of regression or shrinkage is especially important. All phenotypic measurements are composed of underlying genetic effects and environmental noise and linear models are written to specify those various underlying effects. When environmental noise is high, there is necessarily less genetic information contained in a given phenotypic measurement. Also, when an observation is made on a distant relative (say a cousin of the candidate being ranked), it provides less information about the candidate than data from near relatives. BLUP considers all of these factors when assigning weights to data.

For example, if one group of parents has progeny planted in only a single test location and that test location is beset by poor survival and large environmental variation, then the phenotypic data is discounted greatly and given little weight (near zero). The BLUP-predicted breeding values are all regressed or shrunken back towards a population mean of zero (assuming the models are written correctly and variances estimated properly). If a second group of parents is represented by progeny planted in more locations with better survival and less environmental noise, the BLUP-predicted breeding values are less shrunken towards the mean of zero. In other words, there is a wider range and variance among the predictions for the second group of parents. Therefore, when making selections across both groups, there is a tendency to make more selections from the second, better-tested group since more candidates appear further above the mean (property 2 above). See Box 15.11 for further illustration of these points.

Implementation and Limitations of Mixed Model Analyses

In theory MM methods appear suitable for all analyses of complex quantitative genetics data in forestry; yet, the implementation of these methods is not always straightforward and there are some important limitations. While new software is being developed for forestry applications (GAREML, Huber,1993; TREEPLAN, Kerr *et al.*, 2001), most software packages currently available for MM analysis were written by animal breeders (see citations at the beginning of this section). These programs are large "black boxes" in which the inner workings are generally too complicated to understand in detail. Therefore, it is very easy to produce results that seem reasonable, but are completely erroneous. To use these MM programs effectively in forestry applications, the analyst must: (1) Install the software on the analyst's computer platform; (2) Obtain a manual to learn how to appropriately use the software; (3) Sometimes make modifications to alter default values and assumptions; and (4) Validate as many of the answers produced by the candidate software as possible against answers produced by a second, well-tested computer program. Each of these steps can be straightforward or difficult depending on the situation, but care at each step is critical for successful application.

Even when software is properly installed and well understood by the analyst, there are still some important issues to consider. First, the theoretical derivation of MM methods and their desirable statistical properties depend on the assumption that the variance and covariance components are known without error. In practice, this is never the case; rather, these components are estimated from the data (generally with a REML algorithm as described above). Therefore, all applications of MM are approximate in the sense that the assumptions are never fully met. This limitation of MM methods has two implications: (1) Extreme care is warranted when MM methods are applied to small data sets, because imprecise estimates of the variance and covariance components can lead to poor-quality genetic value predictions; and (2) The linear models should not be over specified with too many model effects, because it is more difficult to estimate variances and covariances for complex systems. Put another way, MM methods are most suitable for large data sets in which the linear models are well-specified to account for important sources of variation.

Another issue to consider with MM analyses is in the application of the results during selection. Software packages can predict breeding values (or clonal values) for every tree measured and some not measured. For example, if there are 100 trees in a half-sib family then there could be 101 predicted breeding values for each trait (one for each family member plus the parent of the family). During selection, there is a tendency to make many selections from the better families. In fact, if a constraint is not placed on the number of trees selected from each family, the genetic diversity of the population of trees selected can be severely reduced (Jarvis *et al.*, 1995). Therefore, the breeder using the breeding value predictions must decide on the appropriate amount of relatedness permitted, and this may change with the objective (*e.g.* selection for a propagation population *versus* a breeding population) and with the ranking of the family (*e.g.* permitting more relatives from a high ranking family than a low ranking one, *Chapter 17*).

Even though there are limitations and concerns with applications of MM analyses in quantitative genetics, these methods are growing in popularity and will almost assuredly be emphasized by large tree improvement programs in the future. As more forest geneticists become familiar with MM methods, the applications will become routine.

Box 15.11. Comparison of best linear unbiased prediction (BLUP) and ordinary least squares (OLS) for mass selection.

This box illustrates several general properties of BLUP for the special case of mass selection in which a single phenotypic measurement of a trait for each tree is used to rank many individual candidates and subsequently to select higher ranking candidates. The general properties are: (1) BLUP and OLS give different rankings when the genetic quality of the data available for different candidates varies; (2) BLUP-predicted breeding values are shrunken or regressed back towards the population mean breeding value of zero; (3) More selections come from candidates whose BLUP-predicted breeding values are based on higher quality data; and (4) BLUP-predicted breeding values can be used directly to estimate genetic gain from mass selection.

The BLUP and OLS equations for predicting the breeding value of each tree measured are:

$$\hat{a}_{i,blup} = h^2_i(y_i - \hat{\mu})$$

$$\hat{a}_{i,ols} = (y_i - \hat{\mu})$$

Equation 15.8

where $\hat{a}_{i,blup}$ and $\hat{a}_{i,ols}$ are the alternative breeding values calculated for the same tree (tree i) using BLUP and OLS, respectively; h^2_i is the heritability appropriate for tree i and may change for trees measured in different environments, and $y_i - \hat{\mu}$ is the phenotypic measurement adjusted for fixed effects that are assumed to include only the site mean estimated as $\hat{\mu}$.

Note that the OLS-predicted breeding value is simply the adjusted phenotypic value, while the BLUP-predicted breeding value is regressed towards zero when h^2 is less than one (Fig. 1). For example, a tree that is 10 units above the site mean has a BLUP-predicted breeding value of 10, 5 and zero when $h^2 = 1.0$, 0.5 and 0.0. Uniform stands with low levels of environmental noise and high levels of trait expression have higher heritabilities, and, therefore, the breeding values are less regressed for these stands than others. When $h^2 = 0$, BLUP-predicted breeding values are zero for all trees, because there is no genetic information contained in the phenotypic data. OLS does not account for this difference among stands where the trees are measured, nor are the OLS breeding values regressed or shrunken.

To illustrate the effect on selection, suppose that a group of 100 trees is measured in each of the three stands depicted in Fig. 1. Suppose we arbitrarily set a truncation point of 5 units meaning that any tree with a predicted breeding value of more than 5 is selected. Using rankings from OLS-predicted breeding values, an equal number of high ranking candidates are selected from all three stands (on average). Using rankings from BLUP-predicted breeding values, there are no trees with predicted breeding values greater than 5 in stand 3 (where $h^2 = 0.0$), so no trees are selected from that stand. Only the very best trees (with measurements more than 10 units above the site mean) are selected from site 2 (where $h^2 = 0.50$), and most of the selections come from site 1 with the highest heritability ($h^2 = 1.0$).

(Box 15.11 continued on next page)

Box 15.11. Comparison of best linear unbiased prediction (BLUP) and ordinary least squares (OLS) for mass selection. (Continued from previous page)

Finally, to illustrate that the BLUP-predicted breeding values are measures of genetic gain, consider Equation 13.3 for predicting additive genetic gain, ΔG, from mass selection:

$$\Delta G = h^2(\bar{y}_s - \hat{\mu}) = h^2 S$$

where $\bar{y}_s$ is the mean of the phenotypic measurements for the selected group of trees, $\bar{y}_s - \hat{\mu}$ is the selection differential, S, and all other terms are defined above. The only

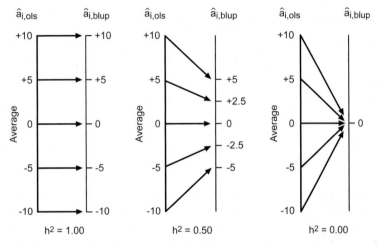

Fig. 1. Schematic diagram comparing OLS-predicted breeding values $(\hat{a}_{i,ols})$ and BLUP-predicted breeding values $(\hat{a}_{i,blup})$ for mass selection in three stands with heritabilities (h^2) for the measured trait equal to 1.0 (Stand 1), 0.5 (Stand 2) and 0.0 (Stand 3), respectively. (Reproduced from van Vleck *et al.,* 1987, with permission of W.H. Freeman and Company/Worth Publishers).

difference between Equations 13.3 and 15.8 is that the gains are predicted for a group of selected trees in the former and a single tree in the latter. It can be shown (Hodge and White, 1992b) that the average of the BLUP-predicted breeding values for any group of selected trees is the estimated genetic gain for the offspring from those trees. In other words, genetic gain can be calculated in two different, yet equivalent ways (Box 13.2): (1) Using Equation 13.3, first average the phenotypic measurements of the selected trees, and then adjust this average, $\bar{y}_s$, by the site mean to obtain the selection differential which is multiplied by the heritability; or (2) Average the BLUP-predicted breeding values of the selected trees.

SELECTION INDICES: COMBINING INFORMATION ACROSS RELATIVES AND TRAITS

Concepts of Selection Indices

Selection index, SI, is a classical approach to combining data from different traits and different types of relatives into a single index (Hazel, 1943). Candidates are ranked and subsequently selected based on their values of the index. Selection indices have been widely used in both plant sciences (see references in Baker, 1986) and forestry (Baradat, 1976; Namkoong, 1976; Burdon, 1979, 1982b, 1989; Bridgwater and Stonecypher, 1979; Shelbourne and Low, 1980, Cotterill *et al.,* 1983; Rehfeldt, 1985; Cotterill and Dean, 1990; Talbert, 1986; Land *et al.,* 1987; King *et al.,* 1988; White and Hodge, 1992).

An example of combining data across traits was introduced in *Chapter 13* when measurements of different traits on the same tree are combined into a single index for subsequent mass selection (Equation 13.11). An example, of combining data across relatives for a single trait is when data from a family's performance in several test locations are averaged and then combined with the phenotypic measurement of a given tree in that family to form an index predicting the breeding value of the tree (Hodge and White, 1992b). For example, if there are 50 trees measured for height from each of 100 families, then there are 5000 index values. There are two pieces of information used to calculate the index value for each tree: the family mean and the height measurement of the individual tree. Each tree's index value is calculated by first multiplying each piece of information by an appropriate coefficient (called an index weight), and then summing the weighted values. Each index value is a predicted breeding value. Once calculated for all trees, the predicted breeding values are then used to select the best trees from across all families in what is called **combined selection**.

Unfortunately, selection index theory assumes that all data are balanced meaning that there is the same quantity and quality of data for all candidates being ranked. This assumption means that a single set of weights is appropriate for all candidates. For the above example, there are two index weights (one for the family mean and one for the tree's measurement). These same two weights are used for calculating all 5000 index values; the data change from tree to tree, but the relative weights applied to the family mean and tree-level measurements stay the same. The assumption is that all families are tested equally (*i.e.* present in the same genetic tests with equal survival), and, therefore, one set of weights is appropriate.

As we have seen in the previous section, BLUP automatically combines data from all types of relatives into a predicted breeding value (or clonal value) for each candidate. In fact, BLUP calculates a separate set of index weights for each candidate depending upon the quality and quantity of data available for that candidate. That is why selection index is called a special case of BLUP: they give the same answers when data are balanced since a single set of index weights suffices for all candidates. When data are unbalanced, BLUP is the appropriate method for combining data across relatives to predict breeding values.

Calculating Selection Indices

When data are balanced and a single selection index is appropriate for all genetic entries, then classical methods can be used for calculating index weights (Hazel, 1943; Baker, 1986; White and Hodge, 1989, Chapters 9 and 10). We focus on a more general approach appropriate for balanced and unbalanced data: (1) Use BLUP to predict a breeding value for each measured trait for all candidates (*e.g.* if there are 3 measured traits for each of 1000 candidates, then there are 3000 predicted breeding values); (2) Determine the appropriate weight for each trait (*e.g.* if there are 3 traits, there would be 3 weights); and (3) Calculate a single index value for each candidate by summing the products of the weights times the predicted values (*e.g.* there would be 1000 index values combining the data from 3 traits). More formally:

$$I_i = w_1 \hat{g}_{i1} + w_2 \hat{g}_{i2} + w_3 \hat{g}_{i3}$$

Equation 15.9

where I_i is the index value calculated for genotype i, w_1 - w_3 are the three index weights if there are three traits being combined and $\hat{g}_{i1}$ - $\hat{g}_{i3}$ are the three BLUP-predicted breeding values for genotype i, one for each trait. This approach is illustrated in Box 15.12 for two traits for the *E. grandis* case study.

The validity of this multiple-step approach rests on a property that the BLUP of any linear combination of traits is equal to that linear combination of the BLUP-predicted values of the individual traits (proven in White and Hodge, 1989, p 165). So, for selection indices that are linear functions, the index values calculated as described in the previous paragraph are the best linear unbiased predictions of that linear combination of the traits. The analyst can first use BLUP to predict values for all genotypes on a trait-by-trait basis, and then combine the traits into an index value for each candidate using the appropriate weights. A multivariate approach combining all traits in a single BLUP analysis is certainly possible, but may be computationally difficult.

A difficult issue that arises with both SI and BLUP is determining the appropriate weight to place on each trait (step 2 in the sequence above). Two methods to choose appropriate weights are: (1) Based on relative economic importance of the traits; and (2) Based on desired gain produced for each trait (sometimes called the Monte Carlo method). For either approach, it is important to remember that the measured traits actually assessed in the genetic trials may not be the same as the target traits in the breeding objective (*Chapter 13*), and the weights should be chosen to either maximize economic or social gain in the breeding objective or to produce desired gain in the traits of the breeding objective.

Economic analyses have often been used to choose appropriate index weights in forestry (Bridgwater and Stonecypher, 1979; Busby, 1983; Cotterill and Jackson, 1985; Bridgwater and Squillace, 1986; Cotterill and Dean, 1990, p 33; Borralho *et al.,* 1992b, 1993; Cameron, 1997, p 90). This is a desirable method when the relative economic importance of the measured traits can be evaluated and when these relative weights are expected to be stable across the time period during which the selected genotypes are bred and deployed. Sensitivity analysis is useful to examine how much the composition of the selected group of candidates changes under varying economic scenarios.

The other approach to estimating index weights, called the desired gain method, tries many different sets of weights (*e.g.* many different sets of w_1 - w_3 if there are three traits), and then the breeder chooses that set of weights producing the most desirable gains in all traits (Cotterill and Dean, 1990, p 35). (see example in Box 15.12). This approach works well when traits are positively correlated to find a set of weights that ensures near-maximum gain in most or all traits. However, when traits have unfavorable correlations, the desired gain method does not directly incorporate economic or social information to optimize gain in the breeding objective, *per se*. Rather, the method relies on the judgment of the breeder or analyst.

Making Selections and Calculating Genetic Gain

All selection processes begin by ranking the available candidates based upon some estimate of their relative genetic worth, such as the calculated values of the selection index as described in the previous section. Then, the breeder uses the rankings to choose candidates subject to: (1) Constraints on relatedness; (2) Field verification of suitability; and (3) Any other constraints or considerations specified by the breeder. We illustrate the general process of making selections using BLUP-predicted breeding values and selection indices, with the case study of *E. grandis* data from Argentina (see Box 15.1 for description of data, Box 15.5 for methods of analysis and Box 15.12 for calculation of selection index values).

Box 15.12. Case study of *E. grandis* from Argentina: Calculation of selection index values.

In Argentina, the INTA tree improvement program (Box 15.1) breeds improved varieties for many products and for small and large landowners. Therefore, appropriate economic weights to place on stem volume and form are difficult to calculate and in fact change with intended product and landowner. For this reason, the desired gain method of developing an appropriate selection index was used. The steps involved in choosing index weights to later apply to the entire data set were as follows: (1) For computational simplicity and because most selections will be top individuals, a computer file of 1421 trees was formed containing the best seven trees from each family (203 families x 7 trees per family); (2) The Monte Carlo trial-and-error process created 25 different indices of the form $I = w_1*volgain + w_2*formgain$ in which the coefficients w_1 and w_2 were varied between 0 and 1 with $w_1 + w_2 = 1$ and volgain and formgain are the BLUP-predicted breeding values described in Box 15.10; (3) The top 150 out of 1421 trees (approximately top 10%) were selected based on the index values for each of the 25 different indices (a different set of 150 trees was obtained for each index tried because of the varying coefficients); (4) The genetic gains for volume and form were calculated for each of the 25 sets of selections by averaging the volgain and formgain values for the 150 trees in that set; and (5) These gains were expressed as a percentage of the maximum value possible for each trait if the 150 trees were selected based solely on the basis of volgain or formgain (Fig. 1).

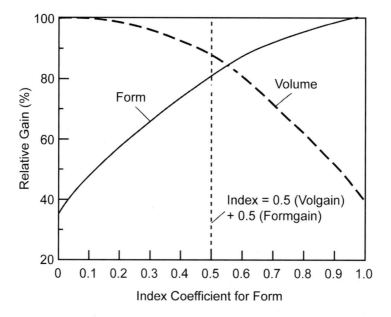

Fig. 1. Relative genetic gains estimated for stem volume and stem form when the top 10% of *E. grandis* trees are selected on the basis of a variety (25 total) of indexes of the form $I = w_1*volgain + w_2*formgain$, where volgain and formgain are the BLUP predicted breeding values for each trait and the index coefficients (w_1, w_2) were generated by a Monte Carlo trial-and-error process (with the restrictions that both w_1 and w_2 lie between 0 and 1, and $w_1 + w_2 = 1$).
(Box 15.12 continued on next page)

Box 15.12. Case study of *E. grandis* from Argentina: Calculation of selection index values. (Continued from previous page)

When $w_1 = 1$ and $w_2 = 0$, the index selects 150 individuals based solely on volgain. This produces the maximum genetic gain in stem volume possible for a 10% selection intensity, subject to the constraint that a maximum of 7 trees are selected per family. Conversely, when $w_1 = 0$ and $w_2 = 1$, maximum gain is achieved for stem form. Indices with intermediate values of w_1 and w_2 result in below-maximum gains in both traits, and the breeder then chooses the particular index that appears most suited to the breeding objectives.

We chose the index with w_1 and w_2 equal to 0.5 (*i.e.* I = 0.5*volgain + 0.5*formgain). This index achieved nearly 90% of the maximum possible gain in volume, yet still achieved nearly 80% of the maximum possible for form (Fig. 1). Sometimes it is advisable to choose two or more indices from the same curve. For instance, if there were two distinct products (pulp and solid wood), the index for pulp would have less weight on form and more on volume than a second index created for solid wood. Or, one index might be used to select a propagation population and another to choose members of a breeding population. For the INTA program of *E. grandis,* a single index was thought appropriate.

Once the 50:50 index was chosen, the numerical value of the index was calculated for all 11,420 candidates (203 parents + 11,217 measured offspring). All 11,420 candidates then have a single value used to judge their relative genetic worth. These index values were the basis for making selections, subject to constraints on relatedness and other decisions made by the breeder.

Using the selection index weights described in Box 15.12, a single index value combining the stem volume and form breeding values was calculated for each of the 11,420 *E. grandis* candidates (203 parents and 11,217 offspring across all test sites). After ranking all candidates, the next step was to determine the amount of relatedness to be permitted in the selected group. INTA decided to restrict the pool of candidates to the top 7 trees from any given family, based on their index values. All other trees were deleted from the candidate list, and this placed an upper limit on maximum relatedness. Based on this constraint, the top 30 candidates are listed in Table 15.2. Note that four of the candidates are parents (type = backward) and 26 are offspring (type = forward). Three of the parents (0033, 0038 and 0093) are from the native provenances in Australia (N06, N10 and N12), and are, therefore, not available for selection. They would need to be excluded from the final candidate list. Parent 0022 from the local land race in Argentina is a viable candidate assuming the mother tree was identified at time of seed collection and grafted into a clone bank.

The very best candidate is 20610033 indicating that this is tree number 1 growing in block 6 of test number 2 and is an offspring of parent 0033. From the list of the top 30 candidates, seven are related to this top candidate (six siblings with ranks 1, 3, 4, 11, 18 and 24, and their parent, rank 6). It is always the case that excellent families produce many top ranked candidates, and the breeder must decide how many related individuals to select. For a breeding population, it is advisable to permit several offspring of the very best families in the breeding population, since it is not known which of these trees are actually the very best genetically. That is, breeding value predictions are subject to error, so the selected trees need to be tested in the next generation of progeny tests. For a breeding population of 300 total selections, we suggest including 3 to 7 from each of the very top families, fewer from moderate families, and only 1 from marginal families. This increases genetic gain through inclusion of multiple candidates from top-ranked families, yet maintains a broad genetic base.

For a propagation population of 30 selections, the following restrictions on relatedness are suggested: (1) For clonal seed orchards, no more than 1 tree from a given family should be selected to avoid potential problems from inbreeding depression (although see *Chapter 16* where two relatives per family are included in some orchards); and (2) For a clonal propagation program we might allow 2 or 3 selections from the top 5 or 6 families, since inbreeding is not a concern when clones are propagated vegetatively.

Once the appropriate amount of relatedness is determined, the breeder begins at the top of the ranking of all available candidates and proceeds downward selecting candidates in order. In most cases, the breeder is indifferent to the generation (forward or backward), and makes selections based solely on the predicted breeding values. When the number of selected candidates in any family reaches the allowable number, no more candidates are considered from that family. Then, the candidates that are offspring (*i.e.* forward selections) are verified in the field to ensure that the trees are truly superior to their neighbors, and that they have other desirable characteristics (free from disease, etc.). It is important not to put much emphasis on secondary traits that are not in the index, because this can dramatically reduce the genetic gain expected for traits in the index (*Chapter 13*).

Another consideration is that different candidates can rank high for a selection index for different reasons. In the *E. grandis* example (Table 15.2), consider candidates 0022, 10210093 and 0038, all with similar index values. The first (0022) is outstanding for stem form (31.46%), but only above average for volume (15.58%). Conversely, 10210093 is only slightly above average for form (6.89%), but outstanding for volume (37.66%). The third candidate is above average for both traits, but not outstanding for either. While not a major problem for an index with two traits, indices combining more traits often identify candidates that are above average for all traits, but not outstanding for any one trait. For breeding populations it is important to select some trees for each trait that are truly outstanding for that trait, even if their index values are smaller than other candidates. This allows capture of superior alleles for all traits and facilitates mating designs that use complementary breeding among selections that are truly excellent for different sets of traits.

The process of selection and verification continues until the desired number of selections have been selected and field verified. Then, genetic gains can be estimated. For all types of selected populations (breeding populations, seed orchards and clonal programs), additive genetic gains can be estimated by averaging the BLUP-predicted breeding values of the candidates in the selected group. For a 30-clone seed orchard in our *E. grandis* example, the gain in stem volume above the first generation is calculated by averaging the volgain values (Box 15.10) of the 30 clones selected for the orchard. If the clones are represented by different numbers of ramets in the orchard, a weighted average can be calculated.

It is not necessary to use heritabilities or any complicated formulae to estimate genetic gain. The BLUP process has done this automatically, and all predicted breeding values are already expressed in the form of genetic gain. Therefore, simple averages suffice to predict additive genetic gain for any group of selections. Note that two types of predicted genetic values are required to estimate genetic gain for clonal programs: (1) BLUP-predicted breeding values, as described here, to estimate gain for clones selected for the long-term breeding effort (since only additive genetic effects are passed on to offspring, (*Chapter 6*); and (2) BLUP-predicted clonal values (incorporating additive, dominance and epistatic genetic effects) to estimate genetic gain for clones deployed to operational plantations (see *Clonal Forestry, Chapter 16*).

Table 15.2. Listing of the top 30 candidates out of 11,420 total candidates (203 open-pollinated parents + 11,217 trees) for *E. grandis* obtained from the mixed model analysis of data from all 6 test locations from the case study in Argentina (Box 15.1 and 15.5). Candidates were ranked on the basis of the selection index (Box 15.12) where index = 0.5*volgain + 0.5*formgain and volgain and formgain are the BLUP-predicted breeding values for stem volume and form expressed as a percentage above or below the mean of 11,217 measured trees (Box 15.11). Treeid completely specifies each candidate using eight digits for offspring (forward) and four digits for parents (backward). For forward selections, column 1 specifies the test site (1, 2, ..6), columns 2 and 3 indicate the block (01, 02, ... 20), column 4 is the tree number inside the block, and the last four digits designate the family. Parents are indicated by the same four digits as the last four columns of their offspring. Therefore, candidate 0033 is the parent of 20610033.

Rank	Treeid	Type	Seed source	Volgain (%)	Formgain (%)	Index (%)
1	20610033	forward	N10	38.79	21.25	30.02
2	0022	backward	select	15.58	31.46	23.52
3	21510033	forward	N10	26.58	19.50	23.04
4	31710033	forward	N10	29.76	15.25	22.51
5	10210093	forward	N12	37.66	6.89	22.27
6	0033	backward	N10	30.29	14.19	22.24
7	0038	backward	N06	20.96	23.49	22.23
8	21510182	forward	N11	18.85	25.17	22.01
9	50180022	forward	select	22.37	21.61	21.99
10	11410121	forward	N02	29.81	13.77	21.79
11	11510033	forward	N10	32.33	11.10	21.72
12	0093	backward	N12	30.03	13.37	21.70
13	21710028	forward	N10	21.26	21.59	21.43
14	21110038	forward	N06	22.53	20.26	21.39
15	20410028	forward	N10	21.21	21.49	21.35
16	40910022	forward	select	20.54	22.14	21.34
17	50460022	forward	select	17.81	24.82	21.31
18	30110033	forward	N10	32.57	9.68	21.12
19	41010145	forward	N08	28.14	13.81	20.98
20	31510093	forward	N12	29.66	12.28	20.97
21	21410129	forward	N02	27.74	13.90	20.82
22	21410032	forward	N10	23.43	17.95	20.69
23	50440022	forward	select	16.47	24.82	20.64
24	32010033	forward	N10	25.36	15.87	20.61
25	31010182	forward	N11	20.38	20.81	20.59
26	50560021	forward	select	26.00	15.00	20.50
27	20210093	forward	N12	30.88	10.00	20.44
28	20910169	forward	N09	21.91	18.94	20.42
29	60230022	forward	select	21.67	18.98	20.33
30	40810145	forward	N08	26.69	13.94	20.31

SPATIAL VARIATION AND SPATIAL ANALYSIS IN GENETIC TRIALS

We leave the topics of mixed model analysis and selection indices to address an important issue in all data analysis, including breeding value prediction. All field, greenhouse, nursery and growth room experiments in forestry are spatial in nature, in the sense that the trees are arrayed on a two-dimensional grid in space. Therefore, the trees' locations may be specified by either their x,y coordinates, row-column positions or latitude and longitude. The aim of the field of statistics called **spatial analysis** is to increase the power of detecting treatment and genetic differences in a measured trait through the incorporation of this spatial information directly in data analysis (Cressie, 1993; Littell *et al.*, 1996). In particular, in forest genetics trials, spatial analysis can improve the power of detecting genetic differences, as well as increase heritability estimates and gain from selection (Bongarten and Dowd, 1987; Magnussen, 1989, 1990, 1993; Fu *et al.*, 1999c; Silva *et al.*, 2001; Gezan, 2005).

All methods of spatial analysis require knowledge of the x,y or row-column coordinates, and it is this information that allows spatial analysis to model and adjust for any spatial variation that might exist on the site. A trial with n rows and m columns is similar to a spreadsheet with tree 1,1 in the upper left and tree n,m in the lower right. From this information, we know that: trees 1,1 and 1,2 are nearest neighbors in the same row; trees 1,1 and 2,2 are diagonal nearest neighbors; and trees 1,1 and 10,10 are 10 spaces apart along the diagonal. Knowing the spacing (say in meters) between rows and columns, then allows the exact distance among all pairs of trees to be calculated. Since the data sets for the case study of *E. grandis* from Argentina did not contain spatial coordinates, we are unable to illustrate spatial analysis methods with these data. In the future it will be important to maintain row-column coordinates for each tree in a field site to permit spatial analysis.

In the next section, we introduce some general concepts of spatial variation and discuss motivations for spatial analysis in genetic tests. Then, in the following section some of the many methods for incorporating spatial coordinates into data analysis are briefly outlined. Reviews and applications of spatial analysis in other disciplines include: ver Hoef and Cressie (1993) in plant ecology; Gilmour *et al.* (1997) in agronomy and horticulture; and Clark (1980) and Wakernagel and Schmitt (2001) in geology, oceanography and mining.

Concepts of Spatial Variation

The environment is heterogeneous in nearly all forest genetics experiments, and this is the main reason that heritability estimates for growth traits are generally small (Table 6.1). For example, for a typical broad sense heritability of $H^2 = 0.30$, 30% of the observed phenotypic variance is attributable to genetic differences among clones, while the large majority (70%) is due to environmental variation. Environmental variation that occurs in field experiments is conveniently divided into 3 components (Gilmour *et al.*, 1997): (1) **Global variation** (also called large-scale, gradient or trend variation) that occurs across the entire field and may be caused by gradients in, say, slope, drainage, wind exposure, soil depth, soil fertility, or other factors that change across the field (Fig. 15.3a); (2) **Local variation** (also called small-scale, patchy or stationary variation) that is specific to areas within the trials (Fig. 15.3b); and (3) **Extraneous variation** (also called non-spatial, nugget or unexplainable microsite variation) that is due to both non-spatial micro-environmental variation and experimental procedures such as variation in planting stock quality, planting procedures, labeling mistakes or measurement errors. All three types of environmental variation may occur in the same trial (Fig. 15.3c), although one may predominate more than the other two.

In a study of 66 progeny trials of *Pseudotsuga menziesii* in British Columbia assessed for growth rate, 90% of the sites had important amounts of spatial variation due to gradients, patches or both (Fu *et al.*, 1999c). Similarly, Silva *et al.* (2001) found important levels of spatial variation in 11 of 12 progeny and clonal trials of three conifers spanning Denmark, Portugal and Australia. In both studies, direct incorporation of the x,y coordinates into the data analysis significantly decreased experimental error and increased gain from selection on those sites with explainable spatial variation. In agronomy, Gilmour *et al.* (1997) state that they analyze more than 500 replicated variety trials annually in Australia, and are convinced of the usefulness of spatial analysis for improving accuracy and efficiency.

Conceptually, there are three distinct, though related, issues associated with the impacts of spatial variation on data analysis. First, if effects of global and local environmental variation are not removed from treatment means (*e.g.* means of genetic entries), then treatment comparisons are less precise. For example, if many of the replicates of genotype 1 happen to be randomly located in patches with favorable growing conditions, the mean of genotype 1 appears larger than it would if the patch effects were removed or adjusted for. Some authors call this a bias in the treatment means in the sense that the means can be differentially affected by environmental effects (*i.e.* some means affected upward and some downward depending upon where their replicates happen to lie). However, randomization ensures that treatment means are unbiased over a large number of realizations or runs of the experiment (this is, in fact the purpose of randomization, see Box 14.3); so, we prefer to view this issue as a lack of precision in treatment means. Either way, removing patch effects or global trends can improve the efficiency of comparing genetic entries and making selections.

A second issue deals with the environmental variance (written as σ^2_E in the causal Equations 6.10-6.12) which is conceptually similar to the residual variance or mean square error from an experiment (written as σ^2_e in linear statistical models of Equations 15.2, 15.4 and 15.6). Suppose, we partition this environmental variance into two components such that $\sigma^2_E = \sigma^2_S + \sigma^2_N$, where σ^2_S is the portion of the environmental variance due to spatial variation (both global and local effects) and σ^2_N is the non-spatial, extraneous portion of the environmental variability (called the nugget in geostatistics). Spatial analysis attempts to explain and remove some or all of σ^2_S from the residual variance, thereby reducing σ^2_E, and increasing both heritability estimates and gain from selection.

The third and final impact of spatial variation on data analysis is associated with the fact that both global and local sources of spatial variation cause measurements from neighboring trees to be more similar, *i.e.* more correlated, than measurements from trees planted some distance apart. This is called **spatial autocorrelation** and is an extension to two dimensions of the temporal autocorrelation dealt with in time series and repeated measures analyses (Littell *et al.*, 1996). As Sir Ronald Fisher, the father of modern statistics, said in the 1930s (cited by ver Hoef and Cressie, 1993), "After choosing the area [for an experiment], we usually have no guidance beyond the widely verifiable fact that patches in close proximity are more commonly alike, as judged by the yield of field crops, than those which are far apart." Spatial autocorrelation violates one of the key assumptions of most data analysis methods, that the experimental errors are uncorrelated and independent. Therefore, a key aspect of spatial analysis is incorporation of correlated errors (CE) in the residual covariance matrix (Cullis and Gleeson, 1991; Zimmerman and Harville, 1991) as discussed in the next section.

Fig. 15.3. Schematic diagrams illustrating spatial variation in environments (*e.g.* elevation, soil pH, level of pest infestation) across field sites with n = 50 rows and m = 50 columns for a total of 2500 tree locations indicated by the intersection of lines: a) A location with global variation caused by gradients across the site; b) A location with local variation caused by patches of different sizes; and c) A location with all three types of spatial (global, local and extraneous microsite) variation. Spatial patterns in stem growth and other measured tree characteristics (*e.g.* stem form and foliage frost damage are likely to reflect patterns in environmental variation. (Fig. courtesy of S. Gezan, University of Florida)

It is important to mention that proper experimental design (*Chapter 14*), especially blocking, helps reduce spatial variation and autocorrelation. So, choosing blocks wisely and using modern designs (such as incomplete block designs and Latinized row-column designs described in *Chapter 14*) are key to increasing experimental precision and power. However, blocks are rarely successful at completely eliminating spatial variation and auto-correlation. This is because blocks are usually not homogeneous entities, and they are discrete areas, whereas spatial variation and spatial autocorrelation are continuous. For example, it is possible to have gradients or patches that span parts of several blocks or are more severe on one side of a block. Also, measurements of trees on one side of a block can be more correlated with each other than measurements from trees on the opposite sides of the same block. Consider again the numerical example at the beginning of this section where a typical broad sense heritability value for growth traits is given as $H^2 = 0.30$ (Table 6.1). Variance among blocks has been removed from the residual environmental variance, and yet still the environmental variance dominates. Clearly, complete blocks alone are not sufficient. Therefore, proper experimental design and spatial analysis are complementary methods of maximizing experimental efficiency.

Methods of Spatial Analysis

Spatial analysis is not simple (Littell *et al.*, 1996), which may explain why it is not widely applied in analysis of data from forest genetics experiments. There is a myriad of available techniques which can be complex to understand and implement. Further, there is no single model or approach that is best for all data sets (Stroup *et al.*, 1994; Grondona *et al.*, 1996; Gilmour *et al.*, 1997; Gezan, 2005) meaning that the analyst usually attempts to fit several models, determines the best fit, and then eventually, after several iterations, settles on an appropriate approach. Finally, nearly all of the papers dealing with spatial analysis consider treatment effects as fixed effects and conduct the analyses on a site-by-site basis. In forestry, we are most interested in genetic effects being treated as random variables (BLUP) and nearly always wish to combine data across multiple site locations. For all of these reasons, it is impossible here to make broad recommendations about preferred methods of spatial analysis; rather, we summarize several methods and then close with some general observations.

   Available methods for spatial analysis can be broadly grouped into two categories: (1) Methods that adjust individual tree measurements for global or local environmental trends (*i.e.* gradients or patches) and, thereby, reduce the residual environmental variance (σ^2_E from the previous section) by eliminating or reducing variance related to spatial variation (*i.e.* by reducing σ^2_S); and (2) Methods that model the spatial autocorrelation that exists among neighboring plots for each trait and directly account for this autocorrelation in the error or residual covariance matrix of the analysis (these are called CE methods for correlated errors). These two approaches are complementary in the sense that most data sets benefit from application of both, simultaneously. However, some methods from category 1 can be applied as part of an ordinary least squares analysis (OLS) which assumes that residuals from neighboring observations are uncorrelated (*e.g.* Bongarten and Dowd, 1987, in forestry). On the other hand, all methods in category 2 require a more complex analysis (*e.g.* generalized least squares or mixed model analysis), requiring modeling the residual covariance matrix such that the relationships among error terms of all pairs of trees are specified.

Methods of Spatial Analysis that Adjust for Global and Local Variation

There are numerous methods that fall into this category. We describe three of them here to introduce the principles, but see Cressie (1993) for others including: median polishing; first- or second-order differencing (called Kriging in geostatistics); and moving averages. First, a popular method for removing global trends from the data is the fitting of polynomial functions or smoothing splines to the row (R_i) and column (C_j) coordinates (Bongarten and Dowd, 1987; Brownie *et al.*, 1993; Gilmour *et al.*, 1997; Brownie and Gumpertz, 1997; Gezan, 2005). The simplest method is to fit a quadratic equation to the individual tree measurements (y_{ij}) such as: $y_{ij} = b_1 + b_2 R_i + b_3 C_j + b_4 R_i C_j + b_5 R_i^2 + b_6 C_j^2$ where $b_1 - b_6$ are regression coefficients and i and j range from 1 to n and m, the number of rows and columns, respectively. After the model is fit, subsequent data analysis (including sites, blocks, treatment effects, genetic effects and their interactions) is conducted on the residuals from this model (r_{ij}) obtained as $r_{ij} = y_{ij} - \hat{y}_{ij}$ (where $\hat{y}_{ij}$ is the predicted value for the ij^{th} row-column position). Enhancements to this method include: (1) Fitting higher order polynomials, such as cubic; and (2) Fitting the polynomials using a method called smoothing splines which divides the rows and columns of the entire field into multiple segments, and then fits a separate cubic polynomial to each segment subject to the constraint that the first and second derivatives must be equal at join points (to ensure smooth transitions across segments). While the latter method uses more degrees of freedom (since there are numerous splines to fit), it is considerably more flexible for fitting complex surfaces.

A second method for adjusting individual tree measurements for global and local spatial variation, is to add rows and columns as effects in the linear model (Zimmerman and Harville, 1991; Grondona *et al.*, 1996; Brownie and Gumpertz, 1997; Gilmour *et al.*, 1997; Gezan, 2005). If there are n rows and m columns, there are n-1 and m-1 degrees of freedom accounted for by rows and columns in the model and each tree's observation is adjusted for the estimated average effects of the row and column in which it is located. For example, if row 3 and column 4 happen to have higher-than-average effects (*e.g.* more rapid growth than average), then the observation for the tree in position 3,4 is adjusted downward to reflect the fact it is in a favorable microsite. Two suggestions for using this method are: (1) When there are many genetic entries or large trials, fit rows and columns as nested effects within complete blocks, since this allows more flexibility by permitting different patterns of spatial variation within blocks; and (2) Fit rows and columns as random effects to allow for maximum efficiency (called recovery of interblock information; John and Williams, 1995).

A third method that has proved useful for removing local trends and reducing spatial autocorrelation is called near or nearest neighbor analysis (NNA). This method traces back to the late 1930s (Papadakis, 1937) and is still used in agronomy (Zimmerman and Harville, 1991; Ball *et al.*, 1993; Stroup *et al.*, 1994; Grondona *et al.*, 1996; Vollman *et al.*, 1996; Brownie and Gumpertz, 1997) and forestry (Magnussen, 1993). A typical NNA for each measured trait proceeds in the following steps: (1) Conduct a randomized complete block (RCB) analysis of the experiment and estimate residuals for each tree (adjusted for estimates of block and genetic entry); (2) For each tree, average the residuals of the nearest neighbors on either side in the same row [$r_{ij} = 0.5*(r_{i-1,j} + r_{i+1,j})$ where r_{ij} is the averaged row residual for tree i, j] and in the same column [$c_{ij} = 0.5*(c_{i,j-1} + c_{i,j+1})$]; (3) Run a new analysis that includes all previous factors from the RCB analysis (*e.g.* blocks and genetic entries), but also includes two new covariates, $b_1 r_{ij}$ and $b_2 c_{ij}$, where b_1 and b_2 are regression coefficients; and (4) Repeat, iteratively, steps 2 and 3 until there is little change in the mean square residual and rankings of genetic entries. If, on average, a tree resembles its neighbors due to spatial variation, then the regression coefficients will be positive (indicating higher observations for tree ij when the residuals of its neighbors are positive). In this case, observations of trees on favorable microsites are adjusted downward. Sometimes, this method is modified to include two or three neighbors in each row and column, instead of only the nearest neighbors.

Methods of Spatial Analysis that Model Spatial Autocorrelation

To discuss correlated error (CE) methods of spatial analysis (*i.e.* those in category 2 above), it is necessary to consider the residual or error covariance matrix, since all CE methods involve modeling the structure of this matrix (Grondona *et al.*, 1996; Littell *et al.*, 1996; Gilmour *et al.*, 1997; Cullis *et al.*, 1998b; Silva *et al.*, 2001; Gezan, 2005). Let **e** be an nm x 1 vector of residuals, e_{ij}, the last term in the linear model (*e.g.* Equations 15.2, 15.4 and 15.6). There is one residual for each of the n x m trees in the trial and it is convenient, though not essential, to order the observations in **e** to correspond to the field layout of the trial (*i.e.* residuals for all of row 1 in the trial, followed by those for row 2, etc.). Ordinary least squares analyses (OLS) assume that var(**e**) = **R** = $\mathbf{I}\sigma^2_e$ where **R** is an array or matrix with the variances of the residuals on the diagonal and the covariances among all pairs of residuals in the off diagonal positions, **I** is an array with 1's on the diagonal and 0's elsewhere and σ^2_e is the residual variance, as from Equation 15.2 (and similar to the conceptual σ^2_E above, although not computationally equal).

The implications of the above OLS assumption are: (1) All residuals have the same variance (*i.e.* σ^2_e); and (2) All possible pairs of residuals are uncorrelated. Of course, the second implication means that there is no spatial variation, otherwise measurements of trees in near proximity will be more correlated than those at a distance. If spatial correlation does exist, the OLS assumption results in too little weight being given to genotypic differences of neighbors and too much weight being given to genotype comparisons from distant trees. Intuitively, if neighboring trees have residuals that are correlated (*i.e.* similar), then the comparison of their measurements more precisely estimates their genotypic differences and should be weighted more heavily when estimating genotype means. This is the goal of CE methods of analysis.

CE methods, as implemented in generalized least squares or mixed model analyses, assume that spatial correlations exist; so, the structure of **R** is more complicated than assumed by OLS. It would be possible, at least in concept, to allow **R** to have a completely general structure such that the residuals among all pairs of trees could have different estimated correlations. However, this is never done because: (1) There would be too many parameters to estimate (one for every off-diagonal element of the **R** matrix); and (2) It would be difficult to estimate so many parameters and still produce an **R** matrix that is positive definite (*i.e.* that has an inverse as required for solution). Rather, the approach taken is to model the spatial correlations to estimate a relatively few parameters that still make every off-diagonal element of **R** non-zero (meaning a non-zero covariance between all pairs of residuals). There are many such models available (Littell *et al.*, 1996).

A common model for CE analysis, called the autoregressive (AR) approach, uses an exponential model for estimating all cells of **R**, and can be implemented as follows (Grondona *et al.*, 1996; Littell *et al.*, 1996; Gilmour *et al.*, 1997):

$$\text{Var}(e_{ij}) = \sigma^2_e = \sigma^2_S + \sigma^2_N \qquad \text{for all diagonal cells } i = j$$

and
$$\text{Cov}(e_{ij}, e_{i'j'}) = \sigma^2_S \rho_x^{dx} \rho_y^{dy} \qquad \text{for all off-diagonal cells}$$

where σ^2_S and σ^2_N are the spatial and non-spatial variance, as defined above, $\text{Cov}(e_{ij}, e_{i'j'})$ is the covariance for each and every pair of residuals, ρ_x and ρ_y are the spatial correlations between nearest neighbors in the row and column direction, respectively, and dx and dy are the metric distances between the two locations ij and i'j' whose residual is being estimated. Relative to analyses that do not account for spatial variation, there are only three additional parameters to estimate, σ^2_S, ρ_x, and ρ_y, because dx and dy are calculated directly from the row-column coordinates and σ^2_N can be derived by the difference from σ^2_e once σ^2_S is estimated. For example, assuming a 1 m x 1 m spacing, suppose ρ_x, and ρ_y are estimated as 0.9 and 0.7, then any two trees that are nearest diagonal neighbors (*e.g.* trees 1,1 and 2,2 and also trees 100,100 and 101,101) have residuals that are assumed to have a correlation of $\rho_x^{dx} \rho_y^{dy} = (0.9)^1(0.7)^1 = 0.63$ and a covariance of $0.63\sigma_S^2$. Trees that are 3 positions apart in the same row (*e.g.* 1,1 and 1,4) have residuals with correlation $0.9^3*0.7^0 = 0.73$, while trees 3 positions apart in the same column have residuals with an estimated correlation of $0.7^3 = 0.34$.

By accounting for the spatial correlation that exists, CE methods increase precision of rankings and gain from selection (*e.g.* Silva *et al.*, 2001; Gezan, 2005). There are two main modifications of this AR method, neither of which we advocate for forestry experiments. The first combines the two variances so that a single residual variance is estimated. This drops the nugget, non-spatial source of variation (σ_N^2) which seems to be important in forestry experiments (Fu *et al.*, 1999c; Silva *et al.*, 2001). The second modification, called **isotropy**, assumes that the spatial correlation is the same in both the row and column direction and estimates only a single spatial correlation (*i.e.* assumes $\rho_x = \rho_y$). We recommend the anisotropic approach, with a separate correlation for each direction, because of its flexibility. Finally, it is important to note that all CE methods assume **stationarity** which means that the variance and correlation structures are the same across the entire field (*e.g.* trees that are the same distance apart have the same correlation regardless of their location in the test site). Due to this assumption, it is usually important to reduce effects of gradients and patches using category 1 methods described above in conjunction with CE methods that model the correlation structure of **R**.

Spatial variation and its direct incorporation into data analysis of forestry field experiments seems very promising. However field, greenhouse and growth room experiments in forest genetics have different patterns of spatial variation, different treatment structures and contain different organisms than do experiments in agronomy and horticulture. Therefore, while we can learn much from research in these related disciplines, more research in forest genetics is needed to define the best methods for various types of experiments and various patterns of spatial variation.

SUMMARY AND CONCLUSIONS

The most important feature of all mixed linear models in quantitative genetics is that breeding values and clonal values are considered random variables, not fixed effects. This assumption leads to three analytical methods for genetic data: Selection index (SI), best linear prediction (BLP) and best linear unbiased prediction (BLUP). Both SI and BLP are special cases of BLUP. BLUP is often incorporated into computer packages that provide complete mixed model (MM) analysis of fixed and random effects in the linear model. MM analysis can be used for a single large data set or to combine data from many genetic tests with different mating designs, different field designs and/or different generations of breeding. Under certain circumstances, even with parents that are selected, related and non-randomly mated, BLUP still produces unbiased predictions of breeding values.

An important difference between BLUP-predicted breeding values and those predicted by ordinary least squares (OLS) is that BLUP-predicted breeding values are more regressed or shrunken back towards the mean when the data for those candidates are of lower quality or quantity. This means that with BLUP there is a tendency to choose as winners the better tested candidates, while the opposite is true for classical methods based on OLS. Another advantage of BLUP is that the predicted breeding values can be used directly to estimate genetic gain from selection. There are no textbook formulas appropriate for calculating gain when data are unbalanced, come from many tests of different designs or use highly selected, related parents. The better option for these situations is to first use BLUP to predict breeding values, and then estimate gains as weighted averages of the BLUP-predicted breeding values.

MM methods are most appropriate for large data sets from which the required variance and covariance components can be effectively estimated. It is important that an experienced person conducts the analysis, and that the results are validated against those from other computer programs to ensure that the programs were properly installed and implemented.

If breeding values are predicted for each trait separately (*i.e.* a separate BLUP analysis for each measured trait), then selection indices can be formed to combine data from all traits into a single index value for each candidate. Selection then proceeds by selecting the candidates with the largest index values subject to constraints on relatedness and any other considerations imposed by the breeder.

Important levels of spatial variation exist in most genetic trials in forestry. Incorporation of the x,y coordinates into mixed model analyses through spatial analysis has the potential to reduce experimental error and increase heritability estimates and gain from selection. However, spatial analysis is not currently used widely in analysis of forest genetic data, and further research is needed to determine optimum methods.

CHAPTER 16
DEPLOYMENT – *OPEN-POLLINATED VARIETIES, FULL-SIB FAMILIES AND CLONES*

The propagation population of most tree improvement programs is made up of the very best selections and is used to produce genetically improved plants for meeting annual operational forestation needs. The improved plants established in plantations are collectively called a breed or genetically improved variety, and the activity of mass propagating operational quantities of new varieties and planting them is called **deployment.**

Applied tree improvement is conveniently divided into two phases (Fig. 11.1): (1) Activities within the breeding cycle (selection, testing and breeding) aimed at development of new varieties; and (2) Formation of the propagation population and deployment of the newly developed variety. The propagation population is purposely placed outside of the breeding cycle because its objectives are distinct. The activities within the breeding cycle develop the improved varieties for a given cycle of improvement, while also maintaining a broad genetic base to provide for gene conservation and to ensure long-term gains from many cycles of tree improvement; therefore, the planning horizon is several generations of breeding and testing (*Chapter 17*).

The planning horizon for the propagation population is a single generation and the goal is to use the very best genotypes available to produce plants for reforestation that will maximize genetic gain in the operational plantations. A tree improvement program can have the best breeding program in the world, but social and economic benefits are realized only when plantations established with the improved varieties are harvested. In other words, the principal benefits (called realized gains) from any applied tree improvement program are measured as the difference in yield and product quality of the improved plantations compared to those of plantations established with unimproved material. Therefore, the formation and management of the propagation population are critical to the real success of any tree improvement program.

There are many types of propagation populations managed in different ways to produce the plants for operational forestation (*e.g.* seed orchards that produce improved seed or clonal gardens that produce rooted cuttings). All types of propagation populations, however, share the following common attributes: (1) The selected genotypes included in the propagation population are the best ones available; (2) These outstanding genotypes are relatively few in number (say 10 to 70 selections) compared to the number in the selected and breeding populations (100 to 1000); (3) This means higher genetic gain and lower genetic diversity in the propagation population compared to the selected population (Fig. 11.2); (4) As new or better data become available from genetic tests, there is almost always the opportunity to upgrade the genetic quality of the propagation population, and therefore to increase gains still further by eliminating some inferior genotypes (*e.g.* by roguing seed orchards) or by including newly identified superior genotypes; (5) Therefore, rather than producing a single improved variety in a given cycle of breeding, a continuum of varieties that change in composition and increase in genetic quality is generated as more genetic information becomes available; and (6) The management objectives for the propagation population are to produce mass quantities of high-quality plants at the lowest possible cost.

This chapter begins by describing interim means of obtaining seed for reforestation in the early phases of a tree improvement program before improved varieties are available. Then, the most common types of propagation populations (seed orchards, family forestry and clonal forestry) are described in some depth. The final major section of the chapter compares the various options in terms of their genetic diversity. It is not possible to describe all possible approaches to the design and management of propagation populations because these vary widely depending on program size and species biology. Therefore, we focus on common principles, and the reader is referred to other reviews on the biology and management of propagation populations: Faulkner (1975), Wright (1976), Zobel and Talbert (1984), and Eldridge *et al.* (1994). Detailed formulas for calculating genetic gains from different seed orchard and deployment options are in: Namkoong *et al.* (1966), Shelbourne (1969), Foster and Shaw (1988), Borralho (1992), Hodge and White (1992a), and Mullin and Park (1992).

INTERIM OPTIONS FOR MEETING IMMEDIATE SEED NEEDS

Organizations with extensive plantation programs need large quantities of seed for reforestation. If the tree improvement program begins at the same time as the plantation program, operational quantities of seed are needed in the first year, while the tree improvement program requires some years to produce genetically improved varieties for operational use. One initial method is to follow logging operations and harvest seed from the best phenotypes after they are felled. When this is not an option, many organizations purchase seed collected from one or more sources: natural stands, local land races and other tree improvement programs of the same species.

For most organizations an extremely high priority is to become self sufficient in seed production as soon as possible to reduce costs and reliance on outside sources of seed. This has two implications: (1) An interim source of seed is required to meet immediate seed needs (and two methods of doing this are described in this section); and (2) The organization should focus on rapid development of their own propagation population (*e.g.* seed orchard or clonal garden) to produce higher-quality improved plants for operational reforestation as soon as possible. When sufficient plants are produced by the formal propagation populations, the interim sources of seed are phased out.

Seed Production Areas

A **seed production area** (SPA) is a natural stand or plantation that is thinned by removing poorer phenotypes; the good trees are left to intermate and produce seed collected for operational forestation. Genetic gains from SPAs are minimal, especially for traits with low heritability, such as bole volume. A similar conclusion was drawn in *Chapter 7* for genetic improvement expected from natural regeneration in stands subjected to selective thinning. The combination of low heritability and low selection intensity (since a substantial fraction of the original trees is normally retained to meet seed needs) mean that a maximum of a few percent gain in volume is expected above seed collected at random from all trees in the unthinned stand. In addition, pollen often comes not only from the SPA, but also from unselected trees in surrounding stands. This reduces gain still further. Nevertheless, this few percent gain is achieved at very low cost. So, if the stand is from the appropriate seed source, establishment of a SPA is an excellent investment to provide a reliable, low-cost source of well adapted seed until an improved variety of higher-genetic quality becomes available.

SPAs are very popular in the early phases of plantation programs of many exotic species. Plantations of the local land race of the species are converted to SPAs to provide a reliable source of seed that is less expensive than importing seed from external sources (Fig. 16.1). For example, during the period 1980 to 1985 in Brazil, there were 51 certified SPAs of 13 different *Eucalyptus* species occupying nearly 1150 ha (Eldridge *et al.*, 1994, p 213).

Fig. 16.1. Seed production areas (SPAs) formed by conversion of plantations of the local land race of exotic species: (a) 12-year-old *Pinus taeda* in Argentina; and (b) 5-year-old *Eucalyptus nitens* in Chile. The best phenotypes are retained and other trees felled. The SPAs are then managed to promote seed production. (Photos by T. White)

Design and Establishment of SPAs

Natural stands or plantations chosen for conversion into a SPA should have as many of the following characteristics as possible: (1) Originate from an appropriate seed source known to be well adapted to the edaphoclimatic conditions of the planting zone; (2) Be of sufficient size and contain enough good phenotypes (at least two hectares with 50 to 150 good phenotypes per hectare) to ensure adequate pollination and sufficient seed production after poorer phenotypes are thinned; (3) Show clear evidence of past seed production and have sufficient crown surfaces to continue adequate seed production; (4) Be located on sites that are easily accessible during seed collection times; (5) Be located a sufficient distance from stands established with maladapted seed sources that might pollinate the stand chosen for conversion; and (6) Have the desirable characteristics mentioned in *Chapter 13* that enhance gain from mass selection (*i.e.* even-aged, well-stocked, with mostly a single species growing on a uniform site).

After one or more stands are chosen, they are converted into SPAs by marking the best dominant and co-dominant trees as leave trees and cutting the unmarked trees. The leave trees should be selected based on the same breeding objectives and selection criteria (*Chapter 13*) as defined for selection in the tree improvement program. This usually means selection of large, straight, and disease- and insect-free trees.

The thinning serves two purposes: (1) Achieving some genetic gain through selection; and (2) Opening the crowns to full sunlight that stimulates flower production in most tree species. To maximize gain, all poor trees should be removed regardless of spacing. In addition, if several good phenotypes occur in a clump, some good phenotypes may also need to be felled in order to open the crowns of the leave trees. Both purposes (genetic gain and crown opening) must be weighed when marking trees for thinning.

The cutting of the trees should be timed to: (1) Produce seed crops as soon as possible; (2) Ensure safety of the leave trees since insects and disease may be more of a problem in certain seasons of the year in some species; and (3) Minimize physical damage to soils. Logging residue should be removed from the site to improve physical access during seed harvests, reduce potential of pest infestations, and diminish wildfire hazard (Zobel and Talbert, 1984).

Management of SPAs and Seed Collection

After thinning, the SPA should be managed to maximize seed production and ease of seed collection; this may include removal of understory species to improve access, flower stimulation through fertilization and other chemical or mechanical means, and application of pesticides to control flower, seed and cone insects.

There are two different strategies for seed collection. Sometimes, it is logistically more feasible and less costly to establish a large SPA in one or more stands and then harvest the seed at maturity by cutting the trees in a portion of one or more areas. The seed is collected from the felled trees. This can be more cost effective than climbing or other seed-collection methods, but means that a portion of the SPA is destroyed each year that seed is harvested.

In the second strategy the total area of SPA is smaller because seed is collected repeatedly from the same trees across different years. Normally this means climbing the trees; however, bucket trucks, poles and mechanical tree shakers serve for some species. With tree shakers care must be taken not to damage tree crowns in any way that would impair seed production capacity in subsequent years. If seed supplies are sufficient, additional genetic gain can be made by collecting from only a few of the very best trees in the

SPA. In this case, all trees remaining after thinning serve as pollen parents, but seed is collected from a smaller number of the very best trees that are clearly marked for seed collection (Fig. 16.1).

Directed Seed Collections

After formation of the first-generation selected population (*Chapter 13*), seed can be harvested from some or all of the plus trees while they are still growing in the plantations or natural stands of the base population (called **directed seed collection,** DSC). An example from the past is *Pseudotsuga menziesii* in the Pacific Northwest of the USA. Plus trees were selected in natural stands in widely scattered locations within each breeding unit and established in seed orchards (*i.e.* as grafted clones or seedling offspring); however, the seed orchards required a number of years to produce operational quantities of seed (Silen, 1966; Silen and Wheat, 1979). In the interim, some organizations collected seed directly from the very best selections growing in the natural stands.

Theoretically, genetic gain from DSC is as half the gain from mass selection as calculated in Equations 13.3 and 13.6. These equations are those used for clonal seed orchards. Gain for DSC is half the gain for a clonal seed orchard (assuming equal selection intensities) because the seed collected from the scattered plus trees for operational forestation results from pollination by neighboring and distant trees (*Chapter 7*) that are unselected. The mean breeding value of these unselected pollen parents is expected to be zero (half the parents better than average and half worse, see *Chapter 6*). Genetic gain is the breeding value of the offspring (*i.e.* seed collected), and this is always calculated as the average of both parental breeding values. Since the mean breeding value of the male parents is zero, the genetic gain is half the mean breeding value of the female plus trees, and by definition this is half the genetic gain from mass selection.

The theoretical gain from DSC is reduced if: (1) Some plus trees are growing in stands of poorly adapted seed sources (this reduces the breeding value of the pollen below zero); or (2) There is inbreeding, either due to selfing or due to pollination of the plus trees by close relatives, such that the seedlings produced suffer inbreeding depression (*Chapter 5*).

Economically, DSC is a wise decision when selection intensity is high enough to produce appreciable genetic gains (say 5% in bole volume) and costs of seed collection are reasonable. The latter consideration means the plus trees need to be readily accessible.

In DSC, the plus trees serving as the source of operational seed often are growing in widely scattered stands of the base population. These stands must be protected from operational logging. In addition, the plus trees should be clearly labeled to facilitate easy identification. This is usually done with tree-marking paint and by attaching metal tags to the bark that have permanent selection numbers. Neighboring trees are felled to open the crowns of the plus trees, and fertilization is sometimes applied to promote flowering. Other treatments (such as protection against fruit and seed insects) may also be necessary to maximize seed yields.

SEED ORCHARDS

A **seed orchard** is a collection of selected clones or families established in one physical location and then managed to produce genetically improved seed for operational forestation. By locating the improved material in a single location, both male and female parents are selected and thus theoretical genetic gains are doubled compared to directed seed collection. Seed orchards are by far the most common type of propagation population, and nearly all tree improvement programs employ seed orchards at some point in the program to produce genetically improved seed for operational deployment. In most cases, mating among the trees in the orchard is uncontrolled (*i.e.* mating is by open pollination that is mediated by wind, insects or animals), and we assume this is the case in this section. Matings effected through controlled pollinations are addressed in the next major section on *Family Forestry*.

There are two types of seed orchards (described below): (1) **Clonal seed orchards** (CSO) for which vegetative propagation (grafting, rooting or tissue culture) is used to establish clones of selected trees in the seed orchard; and (2) **Seedling seed orchards** (SSO) established with open-pollinated or full-sib offspring (*i.e.* seedling families) from selected trees. In the 1960s, there was a strong debate about the advantages and disadvantages of the two types of seed orchards (Toda, 1964; Zobel and Talbert, 1984, p 178), but it is clear that both types are valuable in different situations. In fact, a single organization may use both SSOs and CSOs in a tree improvement program of a single species.

We first describe CSOs and SSOs separately emphasizing principles of design, establishment and management that differ between the two types of orchards. Then, in subsequent sub-sections, we discuss issues common to both orchard types such as size and cultural management.

Clonal Seed Orchards

Clonal seed orchards are the most common method for mass-producing genetically improved seed for operational forestation. CSOs are popular for both gymnosperm and angiosperm species, and in all parts of the world. For example, there are more than 2000 ha of CSOs of *Pinus taeda* and *P. elliottii* operated by approximately 35 different organizations in the southeastern United States (McKeand *et al.*, 2003). Collectively, these orchards produce 1.4 billion genetically improved seed each year, enough to reforest approximately 700,000 hectares annually which is 90% of the total annual conifer reforestation in the region.

CSOs are favored over SSOs when: (1) Vegetative material from the selected trees is easy to obtain and propagate by rooting, grafting or tissue culture; (2) The species flowers only at older ages when trees originate as seedlings, but flowers readily when material (*e.g.* branch shoots, stump sprouts) from older selections is vegetatively propagated; and (3) The orchard is used for the single purpose of mass producing genetically improved seed, and not for additional purposes such as genetic testing. Genetic gains are usually similar for both types of seed orchards; therefore, the above factors are more important in determining which orchard type is most appropriate.

A typical sequence of events for establishing a CSO is illustrated using *Pinus taeda* as an example in Box 16.1. Since CSOs are established by cloning the selections, the selection in the field is called the **ortet**, while the plants of the same genotype in the orchard are called **ramets**; together, the ramets and ortet of the same genotype form a **clone** (hence the name clonal seed orchard). Typically, a clonal seed orchard is established with between 20 and 60 clones from the best selections available. The number of ramets per clone is determined by the number of total trees required to meet seed demand for forestation. Often, 50 to 100 ramets per clone are planted in an orchard.

The phenotypic appearance of trees in the orchard is not important since the genetic values of the clones are determined by genetic tests, not by their performance in the orchard. Almost always, trees in the orchard have a wolf-tree, heavily-branched appearance due to the wide spacing. Many times, the trees become gnarled and malformed from repeated seed harvest. The only objective of the CSO is to produce improved seed, and all cultural practices are aimed at maximizing genetic quality and seed production at a minimum cost.

In the early stages of seed production, it is common that a few of the most precocious clones produce most of the pollen and/or seed; further, there may be more inbred seed and also more pollen contamination (see *Chapter 7* and *Pollen Dilution Zones and Pollen Enrichment Zones* in this Chapter) (Adams and Birkes, 1989; Erickson and Adams, 1990; Adams and Burzcyk, 2000). Therefore, it may be best to wait until the orchard is nearer full production to use the seed for operational forestation.

Genetic Designs for Clonal Seed Orchards

Designing a CSO entails two aspects just as for genetic testing: (1) Genetic design including the number and types of clones to be established; and (2) Field design specifying how the ramets of all clones are physically arranged on the ground (van Buijtenen, 1971; Giertych, 1975; Hatcher and Weir, 1981; Zobel and Talbert, 1984; Hodge and White, 1993; and Eldridge *et al.*, 1994). For the discussion below, we first focus on open-pollinated seed orchards of species that are not dioecious (*i.e.* for species that have either perfect flowers or are monoecious). Designs for dioecious species are then mentioned briefly.

The overall goal of the genetic design is to maximize genetic gain while maintaining an acceptable level of genetic diversity in the seed crop. The first step is to determine the appropriate number of clones. Too few clones may mean a narrow genetic base, increased pollen contamination (if the orchard is small) or increased selfing due to close proximity of ramets of the same clone. Too many clones make management of the orchard more complex and reduce expected genetic gain (because more clones necessarily mean that some are less superior and this reduces the selection intensity). This latter consideration is particularly important, because a subtle advantage of CSOs above both SSOs and directed seed collection is the ability to create 50 or 100 copies (*i.e.* ramets) of the very best genotypes.

As a rule of thumb, a relatively large number of clones (say 60) is needed for first-generation orchards established with clones of untested plus trees since: (1) Progeny testing will identify clones of inferior genetic quality that can be removed to increase genetic gain; and (2) Some clones will prove unsuitable for other reasons (*e.g.* graft incompatibility, lack of seed production or phenological disparity in time of flowering with other clones). Removing all ramets of half or even two-thirds of the initial clones based on progeny test rankings can substantially increase genetic gain.

At the other extreme, 20 clones is a minimum number of clones to establish in a CSO and is appropriate when: (1) The clones have been previously tested (*i.e.* their breeding values have been estimated from previous genetic tests); and (2) Differences among clones in **fecundity** (*i.e.* magnitude of pollen or seed production) and phenology of flowering are expected to be minor. If all clones have proven superior in genetic tests, then there is little opportunity to increase gain from roguing the seed orchard in the future; therefore, most if not all clones will remain in the orchard. Such an orchard is known as a **tested orchard** or sometimes a **half-generation orchard** (*e.g.* 1.5 generation in the first generation of improvement or 2.5 generation if second-generation selections are tested before being established in a CSO).

In advanced-generation CSOs, the opportunity exists to use clones from different generations (*i.e.* both backward selections (parents from a previous generation) and forward (offspring) selections) (*Chapter 14*). There may be both biological and genetic reasons to include backward selections since they are: (1) Physiologically older and in some species will flower sooner after establishment than grafts from younger trees (*e.g.* Parker *et al.*, 1998); and (2) Known to be superior because they have been tested in the previous generation (Hodge, 1997). With this type of an orchard, backward and forward selections can be strategically positioned on the ground (Hodge and White, 1993) such that the backward selections are relied on for the early years of seed production. Later as genetic testing provides information about the forward selections, only the best forward selections and best backward selections are retained.

Another desirable option for advanced-generation seed orchards is inclusion of more ramets of clones of higher genetic quality to increase expected genetic gain. Algorithms have been developed that maximize genetic gain for a constant level of genetic diversity by assigning ramet frequency as a function of the predicted breeding values of the orchard clones (Lindgren and Matheson, 1986; Lindgren *et al.*, 1989; Hodge and White, 1993). These algorithms work by increasing the number of ramets of better clones, while at the same time including more clones of lower breeding values to accumulate genetic diversity.

Box 16.1. Typical sequence of events in establishment of a clonal seed orchard (CSO).

1. Selections are made in the base population. In first-generation programs, this entails mass selection of plus trees growing in natural stands or plantations. In advanced-generation programs, the base population consists of pedigreed material growing in genetic tests, and index selection is used to identify the very best candidates. These may be backward or forward selections (*Chapter 14*) or a combination of both.

2. Vegetative shoots, called **scion** material, are obtained from each selection. Most commonly, branch tips (Fig. 1a) are collected, but sometimes the selections are felled and sprouts from the stumps are used to establish the orchard.

3. The scion material is propagated by vegetative methods (grafting, rooting or tissue culture) to produce the plants for establishing the seed orchard. Grafting is the most common means for establishing seed orchards (Fig. 1b), but any form of vegetative propagation ensures that all plants established in the orchard have the same genotype as the selection from which the scion was obtained (barring somatic mutations).

4. The vegetatively propagated plants are established in the orchard location at relatively wide spacings (*e.g.* 10 m x 10 m) compared to plantations, providing full sunlight to the crowns to promote flowering (Fig. 1c). Tighter spacing is used if thinning, rouging or mortality is expected. Each plant is identified by its clone number, and ramets of the same clone are planted at a distance from each other to reduce cross pollination between them (*i.e.* self pollination among ramets of the same clone). Typically, a CSO is established with between 20 and 60 clones and 50 to 100 ramets per clone.

5. After planting, the orchard is managed intensively to produce mass quantities of open-pollinated seed, and this seed is used for operational forestation. Cultural management of CSOs is completely different than that of plantations and genetic tests due to the very different objective: seed production.

6. Data from a series of progeny tests are used to predict breeding values and poor clones are felled and eliminated from the orchard in the process called roguing. This results in non-uniform spacing, but increases the genetic quality of the seed obtained from the orchard (Fig. 1d).

(Box 16.1 continued on next page)

Box 16.1. Typical sequence of events in establishment of a clonal seed orchard (CSO).
(Continued from previous page)

Fig. 1. Typical sequence of establishment of a grafted clonal seed orchard of *Pinus taeda* in the southeastern United States. (Photos courtesy of S. McKeand, North Carolina State University)

Field Designs for Clonal Seed Orchards

The **field design** specifies the location of all ramets, and for most orchards the most important consideration is to maintain adequate distance between ramets of the same clone and between other types of relatives. Planting spacing among ramets depends on species' biology and anticipated thinning of the orchard. Commonly, CSOs are established at spacings ranging from 3 m x 3 m for species that flower early and for orchards that will be heavily thinned to 10 m x 10 m for later flowering species planted at near-final anticipated spacing.

Inbreeding (selfing or other types of mating among close relatives) is highly undesirable in seed orchards since inbred seed nearly always means the seedlings used in operational plantations suffer from inbreeding depression. This reduces the expected gain from the orchard. Therefore, it is important to maximize the distance between relatives.

Many types of field designs have been employed for open-pollinated CSOs (see review by Giertych, 1975) and we mention three broad classes: (1) Modified randomized complete block designs; (2) Permutated neighborhood designs; and (3) Systematic layouts. In modified randomized block designs (Fig. 16.2a): (1) One ramet per clone is planted in each block of the seed orchard (similar to single tree plots in genetic tests, *Chapter 14*); and (2) The clones are randomly assigned to positions in each block (normally with a computer program), subject to the constraint that a minimum distance is maintained between related trees both within and between blocks. The distance constraint ensures separation between both ramets of the same clone and ramets of related clones in the same or adjacent blocks (this is important in advanced-generation orchards when some clones may be related as parent-offspring or siblings).

The permutated neighborhood design was originally developed (La Bastide, 1967) to provide a layout in which: (1) There is a double ring of trees (first-order and second-order neighbors) of other clones isolating each and every tree in the orchard from other ramets of the same clone; and (2) The number of occurrences of any combination of two clones as first-order neighbors is minimized. The original concept was modified by Bell and Fletcher (1978) to increase flexibility in the number of unrelated neighbors surrounding each tree, and their computer program, called COOL (for coordinated orchard layout) has been used successfully to establish first-generation clonal seed orchards of many species. The permutated neighborhood design was also modified to make it more suitable for designing advanced-generation seed orchards of *Pinus taeda* (Hatcher and Weir, 1981).

Systematic orchard designs utilize repeating blocks, rows or other patterns in which clone positions do not vary from block to block (Fig. 16.2b). Systematic designs are the easiest to layout, install and manage since once clone positions are assigned to the first block, the positions are maintained throughout all blocks. Although systematic designs were sometimes recommended for first-generation CSOs (Giertych, 1965; van Buijtenen, 1971), they were criticized on the grounds that repeated neighborhoods could lead to large, repeated holes in the orchard if a group of neighboring clones proved inferior and were therefore removed from all blocks in the orchard.

Systematic orchard designs, however, may be favored for advanced-generation seed orchards when: (1) Flowering phenology and fecundity of orchard clones are similar; (2) Related clones (such as parents and offspring or siblings) are included in the orchard; and (3) Some clones are utilized in higher frequency to maximize genetic gain (Hodge and White, 1993). For example, in Fig. 16.2b the upper case letters represent one set of clones (*e.g.* backward, parental selections) and the lower case letters represent a second group of clones related to the first group (*e.g.* forward, offspring selections). The design maintains maximum distance between ramets of the same clone (*e.g. A* from another *A* and *a* from *a*) and also between related clones (*e.g. A* and *a*). Two clones with the same upper and lower case letters (*e.g. A* and *a*) could be relatives (*e.g.* as parent and offspring or siblings) or they could be the same clone. The latter consideration means that the frequency of the best clones could be doubled in the orchard by assigning each of the top clones to all positions with the same upper and lower case letters, and this has been shown to result in additional genetic gains (Hodge and White, 1993). Systematic orchard designs can be used for any number of clones, and are easy to establish and manage.

Genetic and Field Designs for Clonal Seed Orchards of Dioecious Species

For open-pollinated CSOs of dioecious species, the male trees serve as pollinators and seed is collected from female trees. Selfing is not possible and inbreeding is not an issue as long as any related clones are the same sex. Thus, genetic and field designs have entirely different considerations than described above. At one extreme of genetic diversity, it is possible to maximize genetic gains through establishment of a two-clone orchard with the best clone of each sex. To increase genetic diversity, one or both clones could be changed periodically in different blocks of the orchard.

In one sense, it is wise to minimize the number of male trees in the orchard, because these trees occupy space without providing any seed. On the other hand, genetic diversity, as measured by effective population size (N_e in *Chapter 5*), is largely determined by the number of individuals in the lower frequency sex. So, the number of male genotypes and trees needed depends both on desired level of genetic diversity and pollination biology of the species. The number of male trees needs to be sufficient to ensure adequate seed production.

a Modified RCB Design   b Systematic Design

X	O	P	A	G	I	W	U
I	G	E	R	L	H	E	T
K	C	D	F	Q	O	R	P
J	Q	L	H	X	V	K	S
W	U	S	M	J	C	F	M
B	N	T	V	B	N	A	D
X	D	J	L	X	P	W	B
R	U	W	F	G	T	K	V
P	T	V	N	J	I	R	F
B	G	Q	M	D	C	N	S
I	K	O	S	E	H	L	O
H	E	C	A	U	Q	A	M

A	B	C	D	A	B	C	D
k	l	i	j	k	l	i	j
E	F	G	H	E	F	G	H
c	d	a	b	c	d	a	b
I	J	K	L	I	J	K	L
g	h	e	f	g	h	e	f
A	B	C	D	A	B	C	D
k	l	i	j	k	l	i	j
E	F	G	H	E	F	G	H
c	d	a	b	c	d	a	b
I	J	K	L	I	J	K	L
g	h	e	f	g	h	e	f

Fig. 16.2. Clonal seed orchard designs illustrated for 24 clones (designated by different letters) and 4 blocks of the orchard (a total of 96 trees): (a) Modified randomized complete block (RCB) design in which clones are completely randomized in each block subject to the constraint that at least three orchard positions are maintained as a minimum distance between ramets of the same clone; and (b) Systematic design with two sets of clones, 12 clones in set 1 identified by capital letters and 12 clones in set 2 identified by small letters. Clones in the same set are assumed unrelated, but any two clones with the same letter in different sets (*e.g.* A and a) may be related or may be the same clone.

The field design also depends on pollination biology, but usually the male trees (*i.e.* the pollinators) are systematically located throughout the orchard. One example is to have a "pollinator" in the center of every group of nine trees (3 rows x 3 columns) which means, there are eight female trees surrounding each male tree. The groups of nine trees can be repeated spatially to form the entire orchard and both the male clone in the center or any of the eight female clones surrounding it can be changed for some groups.

Genetic Management of Clonal Orchards and Seed Deployment Options

To **rogue** a seed orchard is the process of removing inferior genotypes based on rankings from progeny tests. After rouging, seed collected for operational forestation has improved genetic quality since only superior clones remain in the orchard. Most CSOs are rogued several times during their life span to continually increase expected genetic gain. The first roguing normally occurs when data from progeny tests are just becoming available and/or when crowns of orchard trees begin to compete thus reducing future seed production potential.

It is sometimes confusing that breeding values obtained from "seedling" progeny tests are used to rogue a "clonal" orchard. However, improved seed is the propagule used to establish the operational plantations, and the clones in the orchard pass on only their breeding values (not their total clonal value) to their seedling offspring. This is described in more detail in *Chapter 6.*

As data from progeny tests get better (*i.e.* data from more and older tests are used to rank the orchard clones), several options become available to increase genetic gains: (1) Continue to rogue the orchard by eliminating inferior clones; (2) Establish a new half-generation orchard or tested orchard containing only the very best clones; and (3) Collect seed from the very best subset of clones, while allowing others to serve as pollinators. All of these are useful options at different stages in a tree improvement program.

The first option is repeated roguing of an existing seed orchard. This is a low-cost option for increasing genetic gain because the only incremental cost is the felling of orchard trees. However, after repeated roguings: (1) A point of diminishing returns is reached in the sense that incremental genetic gain decreases from each subsequent roguing; (2) The number of clones decreases to a point below which there is concern about inadequate genetic diversity; and (3) The absolute number of trees and their density in the orchard may be too low to ensure adequate pollen production and/or seed supply.

At some point, rather than continually rouging an existing orchard, genetic gain can often be substantially increased by establishing a new CSO with the best tested clones. In most tree improvement programs, hundreds of selections form the selected population each generation (*Chapters 13* and *17*), while only a small subset of these selections (say 60) are grafted in the initial CSO. After progeny testing of the several hundred selections, many new ones are usually identified as superior to those grafted into the initial orchard. Thus, a new orchard is established containing only those clones known to be genetically superior based on progeny test results. The additional genetic gains from a tested orchard are substantial if high-quality data from progeny tests have identified outstanding selections not included in the initial orchard.

At any point in the life of an orchard, it is possible to collect seed from only the best subset of clones. After progeny test results provide predicted breeding values for all clones in the orchard, the worst clones are eliminated by roguing, while clones of intermediate breeding values are left as pollinators (but not used for seed collection). Then, seed is collected from only the very best clones. When considering both genetic diversity and genetic gain, this option is often superior to continued roguing that reduces the number of clones below 15 or 20 (Lindgren and El-Kassaby, 1989). One prerequisite is that the clones used for seed collection must produce enough seed to meet annual demands for operational forestation.

Extending this notion of seed collection from only some clones a little further, many organizations collect and store the seed by individual clone (rather than combining seed-lots together into a single orchard mix). Thus, if there are 50 ramets of any specific clone in a CSO, the seed from all 50 ramets is collected, combined together into a single lot identified by the clone number, grown separately in the nursery and deployed as an open-pollinated family in operational plantations. All seedlings in a single lot planted in a single plantation share the same maternal parent, but have many different paternal parents.

This option of deploying open-pollinated families leads to additional gains in several ways: (1) Nursery operations are more efficient if different families germinate at different rates and require different cultural management; (2) Genetic gain is increased by deploying families superior for different traits to specific sites (*e.g.* families resistant to a specific disease are deployed to sites where that disease is known to cause problems); and (3) Genotype x environment interaction for bole volume growth can be exploited by deploying specific families to the types of sites where they grow best.

Deployment of OP families has become the method of choice in the southeastern USA where approximately 60% of all plantations of *Pinus taeda* are established in blocks of single OP families and this number increases to 80% on industrial, company-owned lands (McKeand *et al.*, 2003). Different organizations vary widely in how many families are deployed per region from a minimum of 4 to more than 40 families. Average block size is 35 ha for a single OP family, but this varies markedly.

Seedling Seed Orchards

Seedling seed orchards are established with seedlings from open-pollinated or full-sib matings among the selected trees, and thus the families established in the orchard are off-spring of the selected trees. SSOs are favored over CSOs when: (1) Trees originating as seedlings flower at an early age and/or vegetative propagation is difficult (*e.g.* severe graft incompatibility in *Pseudotsuga menziesii* made SSOs popular until graft-compatible root-stock was developed, Copes, 1974; 1982); (2) Scion material from the selected trees is not available (*e.g.* many first-generation programs for exotic species can easily import seed from selected trees growing in the native range or from other tree improvement programs, while import of scion material is restricted); and (3) Several objectives are combined into a single location that is first managed as a genetic test and later converted into a production seed orchard (*e.g.* the low intensity *Eucalyptus grandis* breeding strategy in Florida, detailed in *Chapter 17*; Rockwood *et al.*, 1989).

A typical sequence of events in the establishment and management of a SSO is illustrated using *Eucalyptus grandis* as an example in Box 16.2. An important distinction between CSOs and SSOs is that many SSOs are established for multiple purposes. While CSOs are established almost exclusively for seed production, a particular SSO may serve as a genetic test for some years, then later be converted into a production seed orchard, and finally may be used as a base population for making selections for the subsequent generation (Wright, 1961; Rockwood and Kok, 1977; Byram and Lowe, 1985; Franklin, 1986; La Farge and Lewis, 1987; Barnes, 1995). Therefore, unlike CSOs which are designed and managed for a single purpose, most SSOs must be designed and managed to meet several objectives. This nearly always means compromises in the sense that there is no single design or management style to optimize all objectives. For example, intensive cultural practices needed to stimulate early and sustained flowering may compromise the genetic test results. Conversely, conditions needed to simulate plantation environments for a genetic test (*e.g.* tight spacing, typical soils, lower cultural intensity) may not be conducive to seed production.

Box 16.2. Typical sequence of events in establishment of a seedling seed orchard (SSO).

1. Selections are made in the base population just as for clonal seed orchards (Box 16.1) and/or selections are identified in tree improvement programs being conducted by other organizations. These may be backward or forward selections (*Chapter 14*) or a combination of both.

2. Open-pollinated (OP) seed is collected from each selection, and kept identified by mother tree (*i.e.* OP family). Less commonly, controlled pollination is conducted among the selections to produce seed of full-sib (FS) families. The seed is sown in a greenhouse or nursery to produce seedlings that are planted to form the SSO (Fig. 1). Typically, a SSO is established with between 50 and 300 families. There may be 15 to 50 seedlings per family, identified by their family and seedling numbers.

Fig.1. Aerial view of two generations of seedling seed orchards (SSO) of *Eucalyptus grandis* growing in southern Florida, USA. The third-generation SSO is the 8-year-old, more widely spaced planting in the foreground. The poor families and poor seedlings from within each family have been felled to leave only genetically superior trees, thereby converting this planting into a production seed orchard. The fourth-generation SSO is the 4-year-old, more densely spaced planting in the background. It contains more than 31,000 seedlings from 529 OP families, and is being managed as a progeny test. After the fourth-year data are collected, inferior trees will be cut based on their predicted breeding values to convert this planting into a widely spaced, production seed orchard. Then, seed will no longer be collected from the less-improved third-generation orchard in the foreground (*Chapter 17*). (Photo courtesy of D. Rockwood, University of Florida)

(Box 16.2 continued on next page)

Box 16.2. Typical sequence of events in establishment of a seedling seed orchard (SSO).
(Continued from previous page)

3. A SSO is established in a statistically sound, randomized and replicated design, and used as a genetic test for the first few to several years. More than one test location is often established (*i.e.* multiple SSOs with the same families) to assess family x environment interaction. Measurements are taken periodically to estimate genetic parameters (*Chapters 14* and *15*) and to predict breeding values of parents and offspring.

4. There are multiple objectives and phases for the SSO. During the "genetic test" phase, the SSO is managed like genetic tests and operational plantations. Further, the SSO is planted at initial densities similar to those of operational plantations.

5. At some point, one or more test locations are converted from a genetic test to a production seedling seed orchard. A breeding value for each and every tree is predicted based on the combined data of the family's performance across all locations and all replications within each location with the individual tree's phenotypic measurements (*Chapter 15*). Then, trees with inferior predicted breeding values (poor phenotypes in poor-performing families) are felled and rogued from the orchard. Seed is collected from the remaining trees and used for operational forestation.

6. At time of conversion, cultural management objectives change, and the SSO is managed intensively for seed production. Then, the SSO's value as a genetic test is reduced if not eliminated due to the roguing and altered cultural practices. Site locations not converted to SSOs can still serve as genetic tests.

Even with the compromises involved with SSOs, they have been used quite successfully with both angiosperm and conifer tree species (see references above): (1) During first-generation improvement programs of some exotic tree species when scion material is difficult to obtain for establishing CSOs; (2) To reduce costs of improving less-important species since a single SSO can serve for testing, selection and seed production; (3) To reduce costs of managing multiple populations of the same species in which each sub-population has a different breeding objective; and (4) To manage the main population of a breeding program at reduced costs so that more resources can be focused on fewer more elite genotypes. These issues deal with breeding strategies and are discussed more fully in *Chapter 17*.

Genetic and Field Designs for Seedling Seed Orchards

Both the genetic and field designs of SSOs are greatly influenced by the need to satisfy multiple objectives (*i.e.* genetic testing, seed production and perhaps advanced-generation selection). Both full-sib families and open-pollinated (OP) families have been used to establish SSOs, and the design criteria are similar for both. However, OP SSOs are by far more common because it is less expensive and quicker to obtain OP seed from selections than to make control-pollinated crosses among them.

For unbiased and precise estimates of both genetic parameters and breeding values, design criteria for genetic tests (*Chapter 14*) must be considered. This means randomized, replicated designs should be installed in multiple locations. Other factors mentioned in

Chapter 14 in connection with test implementation (such as surrounding the SSO with border rows to reduce edge effects and finding uniform blocks) must also be considered.

Several different designs have been employed successfully for SSOs (Byram and Lowe, 1985; Adams *et al.*, 1994; Eldridge *et al.*, 1994; Barnes, 1995). We recommend either randomized complete blocks (if each block size is maintained smaller than 0.1 ha) or incomplete block designs (such as alpha-lattice designs, Williams *et al.*, 2002) established with single tree plots. For example, in an SSO with 150 OP families, the design at each location might entail 25 complete blocks with one seedling from each family planted in each block. This means a total of 150 seedlings per block and 3750 seedlings in an entire test (not counting border rows). The SSO should be established in a minimum of two and preferably several locations in the planting zone. The number of blocks at each location is adjusted by seed needs (more blocks for larger programs requiring more seed) and number of locations (fewer blocks per location if there are more locations).

When multiple locations are used, sometimes only one or two are converted into production seed orchards. The others are used as genetic tests for their entire life; these may have fewer blocks and should be located on a variety of sites and soils throughout the planting zone. The sites meant for conversion to production seed orchards may have more blocks (to meet seed needs) and are located on sites better suited to seed production (see *Determining Seed Orchard Location and Size*, below). In fact the statistical designs may vary between these two types of plantings.

Genetic Management of Seedling Orchards and Seed Deployment Options

Roguing a SSO is essentially the process of forward selection as described in *Chapter 15*, and begins as soon as reliable data are available. First, breeding values are predicted for all trees using all information available. The family-level information comes from all replications at all locations of the SSO and this is combined with the individual-tree phenotypic measurements. After breeding values are predicted, a field crew visits the SSO location being converted to seed production and chooses the best candidates to retain in the orchard based on breeding values, field observations, and spacing considerations. As with CSOs, SSOs can be rogued multiple times during their lifespan as crowns close and genetic information continues to improve.

It is very important to reduce the chance of inbreeding among trees in the same family (*i.e.* among siblings), which could lead to inbreeding depression in the seed collected for operational forestation. If the SSO is planted in row plots or rectangular plots such that there are several trees of the same family that are neighbors, then it is important to leave only the best sibling per plot for seed production. Distances between siblings in adjacent blocks and among individuals from related families in the same and adjacent blocks should also be maximized.

A method of seed collection often used when large areas of SSO exist (such as when multiple locations have been established) is to fell the trees at time of seed collection. This is especially beneficial if the trees are very large and costs of seed collection are high. First, the SSO is rogued, and allowed to form seed by matings among the superior remaining genotypes. Then, a portion of the SSO is felled and seed collected. The amount felled each year depends on annual seed requirements.

As described for CSOs, seed can be collected from only a subset of the very best genotypes, while the larger population of pollinators (male parents) provides genetic diversity. It is also possible to collect seed by family, maintain family identity and deploy OP families (as described for CSOs). Both of these methods can increase genetic gain.

Considerations Common to both Clonal and Seedling Seed Orchards

Determining Seed Orchard Location and Size

For both types of seed orchard, physical location is critical to success and the chosen site should have as many of the following desirable characteristics as possible: (1) History of excellent seed production (in both native and exotic species some locations may not produce abundant quantities of seed and these must be avoided); (2) Near operating facilities to ensure adequate supplies of labor, water and fire protection; (3) Distant from contaminating sources of pollen, especially poorly adapted sources that could severely reduce genetic gain; (4) Flat or gently sloping terrain with soils that facilitate entry for management and seed harvest; (5) Stable ownership that ensures usage of the site for the life of the orchard; (6) Away from population centers and located in areas where insecticides and other management tools can be used; and (7) Good air drainage if frosts are a problem during the flowering season (since flowers are more sensitive to cold damage than vegetative tissues in most tree species).

The chosen site for a seed orchard does not necessarily have to be within the breeding unit or planting zone if seed production is the only goal; however, SSOs also serving as genetic tests should be located in the breeding unit on sites representative of those of the operational plantations. If seed production is the only goal, then the objective is to produce genetically improved seed in a cost-effective and logistically efficient manner. In theory, any site that optimizes this goal can be selected, and there are some examples of organizations establishing orchards some distance (even hundreds of kilometers) from the planting zone. Before this is done, it is important to know that the orchard site has a history of excellent seed production for the species, and that either the orchard will be isolated from contaminating pollen or that pollen contamination that does occur will not compromise the adaptability of seed for use in the planting zone.

Once a site is selected, orchard size is most often determined by the amount of annual forestation and hence the annual demand for seed. Another consideration is that the orchard must be large enough to ensure adequate pollen supply. In general, seed orchards vary from as small as 1 ha to larger than 100 ha. The process for determining orchard size based on operational demand for improved seed is: (1) Estimate the number of plantable seedlings produced annually by one hectare of seed orchard; (2) Divide the total annual requirement for seedlings by the number obtained in step 1; and (3) Make any adjustments to this figure deemed necessary based on other considerations (*e.g.* desire to sell seed to other organizations and need to completely meet the annual demand for seedlings before the orchard has reached complete reproductive maturity). An example application of this process is given in Box 16.3 for *Pinus elliottii*.

Cultural Management and Record Keeping for Seed Orchards

Appropriate cultural practices may include irrigation, fertilization, mowing and herbicides to control competition, pollarding (topping or pruning to control tree height and crown shape), sub-soiling to reduce soil compaction, chemical or mechanical treatments to induce flowering (such as use of paclobutrazol in angiosperms, Griffin, 1989, 1993), and pesticides to control insects that damage seed crops.

Cultural practices in seed orchards are very specific to the needs and biology of the species, and we do not review here the various practices used with different species. Zobel and Talbert (1984, *Chapter* 6) present a detailed review with general recommendations, and Jett (1986, 1987) reviews the range of practices used in CSOs of *Pinus taeda* and *P. elli-ottii*. All practices that increase seed production and decrease costs should be considered.

Box 16.3. Calculating the number of hectares required for a clonal seed orchard of *Pinus elliottii*.

Average seed yield from 56 different clonal seed orchards of *P. elliottii* located in the lower Coastal Plain of the southeastern United States (Powell and White, 1994) is approximately 34 kg ha^{-1}yr^{-1} for fully productive orchards in which pesticides are used to control cone and seed insects (Fig. 1). Assuming 25,000 seeds per kg and an 80% nursery yield (*i.e.* 80% of the seeds sown in the nursery become plantable seedlings), then one hectare of mature orchard produces 680,000 plantable seedlings each year (680,000 = 34 * 25,000 * 0.8). At an initial planting density in plantations of 1800 trees per hectare, one hectare of orchard can provide, on average, enough plantable seedlings for 375 ha of forestation annually (680,000/1800). To calculate total orchard size for an organization with 400,000 ha of timberlands, assume that the area harvested each year is 1/25 of the total based on a 25 year rotation. Then, the 16,000 ha of reforestation each year requires 42.7 ha of seed orchard; a smaller organization with 2000 ha of annual reforestation requires a seed orchard of 5.3 ha.

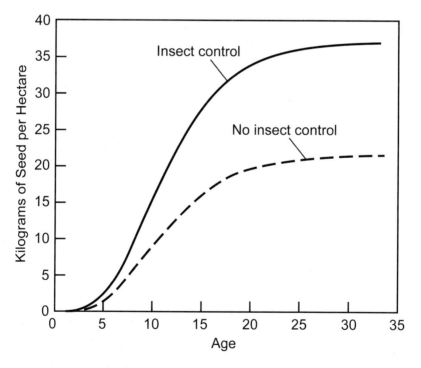

Fig. 1. Seed yields (kg ha^{-1}yr^{-1}) from seed orchards of *Pinus elliottii* managed with and without control of cone and seed insects. The values are predicted values from a data set containing seed yields from 56 different orchards. Yields of individual orchards vary widely around these average values. (Reproduced from Powell and White, 1994, with permission of Society of American Foresters)

(Box 16.3 continued on next page)

Box 16.3. Calculating the number of hectares required for a clonal seed orchard of *Pinus elliottii.* (Continued from previous page)

Factors inflating the required orchard size are: (1) The size should be doubled if no or little insect control is planned due to the dramatic reductions in seed yields caused by cone and seed insects (Fig. 1); (2) The size should be doubled if the desire is to meet all of the organization's seedling requirements when the orchard is only 10 years old, because the productivity per hectare of 34 kg ha^{-1}yr^{-1} is not reached until 20 years after orchard establishment; and (3) Orchard sites vary markedly in seed yields and many organizations establish larger orchards if uncertain about seed yields at the specific location.

One cultural practice that deserves special mention is the use of pesticides to control insects that damage flowers, cones and seeds. In many species, insects can cause significant losses of seed (Fig. 1, Box 16.3) and insect control is needed to realize full production potential. Sometimes several applications each season are required (*e.g.* Powell and White, 1994).

An extremely important aspect of seed orchard management is the maintenance of good records. Trees need to be clearly labeled and mapped. Most seed orchard managers record daily weather data, make notes about cultural practices, and keep detailed records of timing and quantity of seed production. These records are useful in many ways, and formal seed orchard monitoring systems have been developed (*e.g.* Bramlett and Godbee, 1982; Merkle *et al.*, 1982). In fact, many clonal seed orchards maintain floral phenology and seed production records for each clone. Clones differ markedly in timing of flowering and fruit or cone crop maturity. Sometimes, clones that are extremely early or late in flowering time are rogued from the orchard due to concerns about limited male parentage in their seed crops or selfing. In addition, seed yield and quality from individual clones are maximized if seeds are collected when fruits (or cones) are completely mature.

Pollen Dilution and Enrichment Zones

Pollen contamination, defined as influx of pollen from outside the orchard, reduces expected genetic gains of seed collected from seed orchards, because the selected material in the orchard is pollinated by unknown parents of lower genetic quality than the selected clones in the orchard. More formally, pollen contamination is designated *m*, for migrants (see *Chapter 5*), and is quantified as the proportion of orchard seeds resulting from fertilization by pollen coming from outside the orchard.

A variety of procedures is used to estimate pollen contamination in seed orchards, but those employing genetic markers are preferred (Smith and Adams, 1983; El-Kassaby and Ritland, 1986; Devlin and Ellstrand, 1990; Xie *et al.*, 1991; Adams *et al.*, 1992c; Stewart, 1994; Adams *et al.*, 1997). The most commonly used method, paternity exclusion, compares genotypes of the seed produced in the orchard and determines the fraction that could not have been fertilized by orchard trees. These seed must have been fertilized by male parents from outside the orchard. This information, along with knowledge of allele frequencies in surrounding stands, is used to estimate *m*. A simple, one-locus example of this method is illustrated in Box 5.5, but in practice the paternity exclusion method is more powerful when multilocus genotypes are determined for all seeds and genotypes of orchard clones and surrounding stands.

In a review of 21 investigations of pollen contamination in seed orchards of six coni-fer species in eight countries, estimates of pollen contamination ranged from a low of $m = 1\%$ for an eleven-year-old *Picea glauca* orchard in Canada to a high of 91% for an eight-year-old seed orchard of *Pseudotsuga menziesii* in Oregon, USA (Adams and Burczyk, 2000). These orchards spanned a range of ages, sizes and isolation (distance to nearest stands of the same species). The mean of all the orchards was $m = 45\%$, and the majority of orchards (3/4 of the orchards with estimates of m) had more than 33% pollen contami-nation. Therefore, it appears that conifer seed orchards are highly susceptible to pollen contamination, and crops where one-third or more of the seed are pollinated by trees out-side of the orchard may not be unusual.

Pollen contamination of 50% reduces expected genetic gain by 25% assuming the outside pollen sources are unimproved, average material. It is extremely important that the orchard is not located in an area surrounded by plantations of poorly adapted sources (*e.g.* established with poor provenances or seed sources). If poorly adapted sources provide the contaminating pollen, then genetic gains can be severely reduced.

Clearly, pollen contamination is an important issue in conifer seed orchards, and probably in many seed orchards of angiosperm species. In the past, it was common to rec-ommend that a **pollen dilution zone** of approximately 150 m be established surrounding a seed orchard. The idea was to reduce the amount of pollen contamination by partially iso-lating the orchard from surrounding stands of the same species. Adding a pollen dilution zone, however, increases the size of the physical site required for an orchard, and therefore the costs. For example, a 10 ha square orchard is 316 m on each side. A pollen dilution zone of 150 m increases this to 616 m on each side (316 + 150 + 150) which is a total size of 38 ha.

It now appears that pollen dilution zones of 150 m are not effective at reducing pollen contamination in conifer seed orchards, and isolation of 500 to 1000 m is necessary for at least some protection (Adams and Burczyk, 2000). Even these large distances may not be adequate, because there is evidence that large amounts of pollen can be dispersed to seed orchards from stands that are more than 50 km away (DiGiovanni *et al.*, 1996).

Conifers are wind pollinated, and this partly accounts for the ineffectiveness of pollen dilution zones. For insect- and animal-pollinated species, the potential effectiveness of pollen dilution zones depends on the travel distances of pollen vectors. For example, based on the work of van Wyk (1981) and Griffin (1989), Eldridge *et al.* (1994, p 225) state that a 100 to 200 m pollen dilution zone planted with a non-interbreeding species would be effective for many insect-pollinated species such as *Eucalyptus*.

Instead of pollen dilution zones, three other options appear viable. One option is to establish **pollen enrichment zones** in which plantations of the best genetically improved material of the same species are planted surrounding the orchard. Given the fact that some pollen contamination is likely, the idea is to increase the genetic quality of the contaminat-ing pollen. While effectiveness of pollen enrichment zones is not yet proven, they seem a logical alternative to pollen dilution zones, especially for insect-pollinated species in which insects certainly would visit surrounding stands.

A second option is to increase the pollen production within the orchard so as to lessen the importance of contaminating pollen. This could be accomplished by establishing a lar-ger orchard, since larger orchards produce more pollen. Another approach is to use cul-tural practices to stimulate within-orchard pollen production using a variety of pollen in-duction techniques (*e.g.* fertilization, hormone treatments, stem girdling and root pruning) (Bonnet-Masimbert and Webber, 1995).

A third option is to alter the floral phenology of orchard trees so that they flower earlier or later than other stands in the same region. One example is "bloom delay" which has proven effective in reducing pollen contamination in seed orchards of *Pseudotsuga menziesii* (El-Kassaby and Ritland, 1986; Wheeler and Jech, 1986). This technique uses overhead irrigation systems to sprinkle water on the orchard trees in winter and spring to retard flower development through evaporative cooling. The method is somewhat costly and its effectiveness may vary from year to year; nevertheless, "bloom delay" is an example of an innovative method for reducing pollen contamination.

FAMILY FORESTRY

Family forestry is the large-scale deployment of superior (*i.e.* tested) full-sib (FS) families in operational plantations (Box 16.4). While family forestry could conceivably be defined to include deployment of open-pollinated (OP) families, we restrict the definition here to FS families. Deployment of OP families was discussed in the previous section on clonal seed orchards.

Box 16.4. Typical sequence of events in family forestry.

1. Family forestry begins after genetic tests have identified either superior parents or superior full-sib (FS) families. The first step is to create 5 to 20 superior FS families by control pollination. Parental rankings (*i.e.* predicted parental breeding values) can be used to choose the parents for these control pollinations if data come from only polymix or open-pollinated genetic tests (*Chapter 14*). Otherwise, genetic tests of many FS families are used to identify those families that combine excellent parental breeding values and have positive specific combining ability (*Chapter 6*).

2. The goal is to create large numbers of individuals from these superior FS families, and to plant them in operational plantations (called operational deployment of FS families). For species whose flowering biology facilitates large scale production of operational quantities of seed through control pollination (CP), the FS families may be deployed directly as CP seed. With this option, large quantities of CP seed are produced by one of many large-scale control pollination methods (*e.g.* Fig. 16.3). The seed is then sown in containerized or bare root nurseries, and the CP seedlings are subsequently outplanted in plantations.

3. Often the production of large quantities of CP seed is quite expensive, and vegetative multiplication (VM) is used to clonally multiply the limited number of seedlings obtained from CP seed. In this option, the VM system produces plantlets that are planted operationally. Rooting of cuttings is a common VM system used to produce the operational plantlets of FS families for forestation, and the sequence of events is: (a) Limited quantities of CP seed are obtained for each FS family; (b) These seed are germinated and the seedlings are used to form hedges (Fig. 16.4); (c) The hedges are severely pruned and intensively managed to produce copious quantities of branch tips for rooting; and (d) The branch tips are stuck into containerized or bare root nurseries, allowed to root and when sufficiently developed are planted operationally. In most cases, rooting success and subsequent field performance of rooted cuttings diminish after hedges are a few years old. This means the hedges must be re-established periodically from additional CP seed.

Family forestry typically begins after a tree improvement program has been operating for several years when data from genetic tests are available to identify the top few parents (say between 10 and 30) to use in creating FS families for operational deployment. The key differences between the use of FS families for operational deployment as described here and their use for genetic testing in *Chapter 14* are: (1) Large quantities of seed are required for family forestry as opposed to small amounts needed for genetic testing; and (2) The genetic worth of the parents and families are already known in family forestry (and only top parents are chosen for use), while one objective of genetic testing is often to determine these genetic values.

Realized genetic gains from family forestry are nearly always substantially higher than those obtained by deploying seed from seed orchards. Theoretical advantages of family forestry compared to use of OP families or bulk mixtures of orchard seed are (Burdon, 1986; 1989): (1) The selection intensity can be very high since only the top few parents are used to create the FS families, while more parents are typically included in seed orchards to reduce the risk of inbreeding during open pollination; (2) Pollen contamination, which reduces genetic gain from seed orchards, is avoided in family forestry; (3) Inbreeding also can be completely avoided; (4) Additional gains through capture of specific combining ability (see Equation 6.9) are possible if large numbers of FS families have been screened in genetic tests; (5) Family forestry greatly facilitates operational deployment of interspecific hybrids between species which would not inter-pollinate reliably in an open-pollinated seed orchard (*e.g.* the deployment of the F_1 hybrid of *P. caribaea* var *hondurensis* x *P. elliottii* in Australia) (Nikles and Robinson, 1989); and (6) Family forestry is very flexible and facilitates development of specialized varieties targeted to changing markets or specific products, or adapted to special environmental conditions.

Family forestry is not a breeding strategy, but rather a deployment strategy. Therefore, the tree improvement program still requires selection, testing and breeding activities. Further, a given organization may use a complementary mix of deployment strategies. For example, an organization may have clonal seed orchards from which they deploy bulk mixtures of seed or OP families to lands of lower productivity, while using family forestry to capture additional genetic gains on higher-valued timberlands.

In family forestry, there are two common propagation options used to create sufficient plants for operational forestation: (1) The FS families may be planted as seedlings produced from control pollination (CP); and (2) Vegetative multiplication (VM) techniques (tissue culture, rooting or other methods) are used to clonally multiply limited quantities of individual seedlings produced from CP seed. Average genetic gains are expected to be the same for both options, so the choice between them is usually based on the relative ease and costs associated with control pollination compared to vegetative propagation. Each of these options is briefly described in the subsequent two sections.

Family Forestry Based on Control-pollinated (CP) Seedlings

In some species, control pollination is logistically feasible and comparatively straightforward, while vegetative propagation (even of juvenile material) is difficult on an operational scale. Technologies for VM are always evolving, but when formation of CP seed is less tedious and less costly than VM, then direct deployment of CP seedlings is the better option. With this option, CP seed of the superior FS families is grown in containerized or bare root nurseries and the CP seedlings are used to establish operational plantations.

When CP seedlings of superior FS families are planted operationally, costs of control pollination must be minimized for the program to be economically feasible. For example, a medium-sized organization reforesting 2000 ha per year requires approximately 4 million CP seed annually. This is quite an undertaking compared to the few thousand CP seed needed for genetic testing programs. As an example of the scale, the first operational quantity of CP seed was produced in New Zealand for *Pinus radiata* in 1986 with 3000 isolation bags, but by 1996 the number of bags used annually in the country totaled nearly 500,000 (Fig. 16.3) (Vincent, 1997). The number would have been larger if all operational planting relied on CP seed; rather, some of the seed was used to establish hedges for rooted cutting production as explained in the next section.

Owing to the scale of control pollination required for family forestry, much research has been conducted on efficient techniques for producing large quantities of CP seed. In general, these methods aim to increase the speed of pollination, even if some small amounts of pollen contamination occur. Compared to the ultra-careful methods of control pollination for research purposes, these more rapid techniques may employ more expedient pollen extraction and delivery methods, less costly bagging systems (or no bagging of flowers at all), and fewer applications of pollen to each flower. Examples include: (1) Supplemental mass pollination (Bridgwater and Bramlett, 1982; Bridgwater *et al.*, 1987); (2) Controlled mass pollination (Harbard *et al.*, 1999); (3) One stop pollination (Griffin, 1989); and (4) Liquid pollination (Sweet *et al.*, 1992). These methods have had varying levels of success, depending upon species.

Fig. 16.3. Meadow orchard of *Pinus radiata* in New Zealand in which thousands of pollination bags are used annually to create large quantities of control-pollinated (CP) seed of superior full-sib families. The CP seed of these families is sometimes planted operationally, but more commonly the CP seed forms the basis of a vegetative multiplication (VM) system in which rooted cuttings are planted operationally (see Fig. 16.4). (Photo courtesy of M. Carson, Forest Genetics, Rotorua, New Zealand)

When control pollination is done on an operational scale, the design of the breeding orchard used to create CP families is often drastically different than designs for open-pollinated (OP) seed orchards as previously described. In general, pollen contamination, inbreeding and panmixis do not influence design criteria as they do for OP seed orchards. Rather, ease and efficiency of pollination are the main factors (Shelbourne *et al.*, 1989). Therefore, clones are often grafted into rows or clonal blocks to facilitate pollen collection and control pollination. Sometimes, the trees are pruned severely (called pollarding) to maintain them as short as possible for convenient pollination. Examples include meadow orchards (Arnold, 1990; Sweet *et al.*, 1992) (Fig. 16.3), hedged artificially pollinated or-chards (HAPSOs, Butcher, 1988) and clonal row orchards (Bramlett and Bridgwater, 1987).

Family Forestry Using Plantlets from Vegetative Multiplication

The VM method of family forestry begins with the production of CP seed of superior FS families as described in the previous section, and relatively large amounts of CP seed are still required. Therefore, the remarks above about the need for cost-effective CP technology also apply to the VM method. Following formation of CP seed, vegetative propagation is used to clonally multiply individuals derived from the seed, so that each seed may eventually yield hundreds or thousands of plantlets for operational forestation. The goal is to use VM to expand the costly or limited number of CP seed available. The VM option is sometimes confused with clonal forestry which entails deployment of fewer tested clones (Box 16.5). The VM option of family forestry is particularly well-suited to those species that have delayed or inherently poor seed production (*e.g. Picea abies* and *Pinus patula*), but is useful for any species for which an efficient VM method exists (*e.g. P. radiata*) (Burdon, 1989).

The most appropriate VM method depends on species biology and any VM method that is cost effective may be used. The most common propagation method is rooted cuttings, and we focus the discussion here on family forestry implemented through large-scale rooted cuttings programs. Some examples of species with operational VM programs based on rooted cuttings are: *Eucalyptus globulus, Pinus radiata, P. patula*, and the F_1 hybrid of *P. caribaea* var *hondurensis* x *P. elliottii*.

Once CP seed is obtained, each seed is germinated to produce a seedling. After allowing growth for a brief period (*e.g.* several months to a year), the seedling top and/or branch tips are cut from each seedling. These scion pieces (also called cuttings) are placed into soil or rooting medium (with or without a rooting hormone), and given proper conditions to promote roots. Once the cuttings have rooted and grown for some period of time, the process is repeated as follows (called **serial propagation**): (1) New cuttings are harvested from the original rooted cuttings; and (2) These are rooted to form a second stage of rooted cuttings. There may be several stages of serial propagation depending on the species.

At some point, the rooted cuttings are pruned and intensively managed to form plants capable of producing large numbers of cuttings for operational production (Boxes 16.5 and 16.6); the plants begin to resemble "bonzai" plants due to the heavy pruning. These plants are variously called mother plants, stool bed plants and hedges (Fig. 16.4). An organiza-tion may need many hectares of hedges to produce enough cuttings to meet the annual demand for reforestation. The operational cuttings are produced by: (1) Taking cuttings from the hedges; (2) Placing them in soil or rooting medium in either a greenhouse or out-door nursery to promote rooting; and (3) Growing the rooted cuttings to a size and quality that performs well after planting. The process must be efficient to produce millions of rooted cuttings in a cost effective manner.

Multiplication rates can be as high as 10 to 50 fold per year (Vincent, 1997) meaning that a single seed can produce 10 to 50 cuttings in the first year and each of those can produce 10 to 50 in the second year. Therefore, a single seed can produce literally thousands of rooted cuttings within a few years. These rates mean that thousands of costly CP seed can be multiplied into millions of plantable rooted cuttings for operational forestation.

Box 16.5. Family forestry via vegetative multiplication (VM) compared to clonal forestry (CF).

All rooted cuttings originating from a single seed have the same genotype (barring somatic mutations), and are therefore members of the same clone. For this reason, the VM method of family forestry is sometimes confused with clonal forestry (CF). The distinction is that VM family forestry deploys juvenile, untested clones, while CF deploys tested clones (Burdon, 1989). In family forestry, predicted breeding values of parents (A values from *Chapter 6*) or performance of full-sib families in genetic tests are used to rank the families that are deployed operationally. In CF, the clones themselves are ranked on the basis of their total genetic value, sometimes called clonal values (G values from *Chapter 6*).

In VM family forestry, each seed within a family has a different genotype due to recombination during meiosis (*Chapter 3*), and these genotypes are untested. So, if 1000 CP seed from a given family are used to produce 1,000,000 rooted cuttings, there are 1000 different genotypes (*i.e.* 1000 different clones). If 20 different FS families are deployed, there are 20,000 clones (*i.e.* genotypes) established in plantations. This contrasts markedly with the 10 to 50 clones often used operationally in CF.

VM family forestry is used instead of CF for species in which it is not possible to clone individual trees on an operational scale after genetic test results identify the top clones. For example, suppose clonal genetic tests are established with the 20,000 clones just mentioned to identify the best clones to deploy operationally. In most tree species, the age of selection is older than five years; this is the time in CF programs when the top few clones (maybe 20) would be identified from the test results and then propagated operationally. Unfortunately by this time, the hedges used to produce the tested clones have aged physiologically (now 6 years old or more from seed), and in many species rooting ability and field performance have declined to the point where CF is not feasible or desirable (Box 16.6). In the VM method of family forestry, the influence of ageing on clones is avoided by re-mating the top parents to form CP seed and planting operationally juvenile (but untested) clones derived from this seed. This reinforces that in family forestry the parents or families are selected based on breeding values, while in CF the tested clones at selection age are chosen based on their total genetic value.

In VM family forestry, the clones from a given FS family are sometimes mixed prior to planting, and this means that each plantation contains many genotypes from a single FS family. Other times, single clones are planted and each plantation contains a single genotype. For the latter option, the genetic diversity within a stand is the same as for monoclonal blocks in CF; *i.e.* there is no genetic diversity within a stand. Perhaps this is one reason why the two options are sometimes confused. However, two important distinctions remain: (1) In VM family forestry, the clones are untested meaning that there are better and poorer genotypes within each family and all of these are deployed operationally; and (2) At a landscape level, genetic diversity for family forestry is always greater than for CF since thousands of different genotypes are deployed.

Box 16.6. Juvenility, ageing and hedge management for production of rooted cuttings.

Several terms, including maturation, cyclophysis, ontogenetic ageing and phase change, refer to morphological, anatomical and physiological changes that occur in woody plants with increasing age (Greenwood and Hutchinson, 1993). These changes are apparently associated with differential gene expression at different ages meaning that gene regulation plays a major role (*Chapter 2*). Changes with increasing age in most tree species include (Bonga, 1982; Zimmerman *et al.*, 1985; Greenwood, 1987; Greenwood and Hutchinson, 1993): (1) Reduced growth rates; (2) Reduced branchiness; (3) Onset of flowering and seed production; and (4) Decline in rooting ability; and (5) Decline in growth of cuttings obtained from older trees.

The latter two changes are especially important in the operational production of rooted cuttings. Vegetative propagation of juvenile material is relatively easy in most tree species, but rooting becomes increasingly difficult with increasing age. The rate of decline in rooting ability varies among species, but can occur as early as one year of age (Greenwood and Hutchinson, 1993; Talbert *et al.*, 1993). In addition, cuttings obtained from older trees often retain characteristics of older trees including slower growth. Therefore, in nearly all large-scale rooted-cutting programs of all conifers and most angiosperms, the cuttings are obtained from juvenile material. Talbert *et al.* (1993) present an excellent review of 25 organizations with large-scale rooted-cuttings programs.

Hedging (meaning severe shearing, toping and pruning) of donor plants is often used to retard maturation (ageing) and to sustain rooting ability (Fig. 16.4). In many species, hedged donor plants retain acceptable levels of rooting ability and subsequent field performance of rooted cuttings for some years longer than non-hedged trees (Menzies and Aimers-Halliday, 1997). Therefore, proper management of donor plants, including timing and severity of shearing and proper nutrition, is critical to the success of large-scale rooted-cutting programs. In most tree species, rooting ability and field performance of rooted cuttings decline as hedges age (even with proper management); so, the hedges must be replaced periodically. This is done by creating new CP seed from the superior full-sib families and producing new hedges.

Some programs use genetic tests planted with seedlings to select the parents or FS families that are subsequently deployed operationally. However, in the VM method of family forestry rooted cuttings, not seedlings, are planted in operational plantations. Presence of C-effects (Box 16.7) implies that some families respond differentially to the VM method. For example, some families that are excellent performers as seedlings may not root well or may age more rapidly as hedges leading to slower growth of cuttings from those families. Therefore, it is unlikely that a correlation of one exists between family performance tested as seedlings and rooted cuttings. For these reasons, strong consideration should be given to establishing the genetic tests used to choose operational FS families with the same propagation method and maturation states that are used operationally.

CLONAL FORESTRY

Clonal forestry, CF, (Box 16.8) refers to the large-scale deployment of relatively few (typically 10 to 50), known-superior clones that have proven their superiority in clonal tests (Burdon, 1989; Libby and Ahuja, 1993; Talbert *et al.*, 1993). Other characteristics of CF are the abilities to: (1) Deploy the same well known, reliable clones for many years, even several rotations; and (2) Adopt clone-specific management practices including matching clones to specific site types and optimizing silvicultural regimes for different clones. Sometimes the CF and the VM options are confused (Box 16.5), but the key to CF is the ability to propagate selected, individual clones on a large scale after test results are available.

Fig. 16.4. Different styles of hedges used to produce rooted cuttings of: (a) *Pinus radiata* in Chile (containerized); (b) *P. radiata* in Chile (bareroot); (c) *P. patula* in South Africa; and (d) The F$_1$ hybrid of *P. elliottii* x *P. caribaea* var *hondurensis* in Queensland, Australia. In all cases, it is imperative to keep the hedges very short by topping, and hedges are replaced after a few years due to declining rooting ability and subsequent tree growth. (Photos by T. White)

Box 16.7. C-effects and their impact on clonal forestry and VM family forestry.

The term **C-effects** includes several environmental, developmental and atypical genetic effects that are common to a group of relatives and cause the differences among groups of relatives to be larger than predicted by the classical genetic theory described in *Chapter 6* (Bonga, 1982; Burdon, 1989; Foster, 1993; Frampton and Foster, 1993). In the broadest sense, types of C-effects include: (1) **Topophysis**, positional effects associated with branches located in different parts of the crown such that vegetative propagules taken from the lower and central portions of the tree crown generally possess more juvenile characteristics than those from the upper or peripheral portions; (2) **Cyclophysis**, maturation effects as described in Box 16.6; (3) **Periphysis**, environmental effects that have caused pre-conditioning of tissue such that donor plants conditioned in different environments produce propagules with distinct characteristics; (4) **Maternal effects**, such that members of the same maternal family share common attributes (*e.g.* due to seed size differences among female parents); and (5) **Non-nuclear genetic effects**, such as caused by inheritance of mitochondrial and chloroplast DNA (*Chapter 3*).

The last two types of C-effects apply to families and tend to make families more different than predicted by classical genetic theory; we ignore these here. The first three types apply to vegetative propagules produced by either clonal forestry or VM family forestry, and their biological manifestations are: (1) Propagules, (*e.g.* rooted cuttings) from the same clone may perform differently depending on the environmental and maturation state of the donor plants or the crown position of shoots taken for cuttings; and (2) Different clones may respond differently to the propagation system, so that apparent genetic differences between clones or VM families may in part be due to C-effects.

There are two major impacts of C-effects on operational programs. First, differences due to C-effects are not necessarily the same if the clones or families are tested in different propagation systems or with donor plants from different states of maturation or environmental pre-conditioning. Rather, each set of conditions results in a new set of C-effects that differentially affect propagation and subsequent performance. For example, suppose that the hedges of a superior clone age more rapidly than those of most other clones. Then, rooted cuttings from that clone perform well from young hedges, but the clone's relative superiority in growth decreases when cuttings are taken from older hedges. The presence of C-effects means that clones or VM families change rank depending on how each clone or family responds to the altered set of conditions. Therefore, if C-effects are important, clones or VM families should be tested and ranked using propagation conditions and maturational states identical to those used in operational propagation for reforestation.

The second impact is that heritabilities and genetic gains are over estimated if C-effects are important, because clonal or family differences are inflated above those predicted by classical genetic theory. For example, if some clones respond better or poorer to the propagation system (*e.g.* root differently and perform differently), then these clones can appear superior or inferior for both genetic and C-effect reasons. This inflates the observed differences among clones above that caused by only genetic differences. In quantitative terms, the observed, estimated variance among clones (σ^2_{Clones}) is the sum of the total genetic variance (σ^2_G from *Chapter 6*) and the variance caused by C-effects ($\sigma^2_{C\text{-effects}}$): $\sigma^2_{Clones} = \sigma^2_G + \sigma^2_{C\text{-effects}}$. To estimate broad-sense heritability (Equation 6.11) and predict gain from clonal selection (Equation 13.5), we typically assume that $\sigma^2_{C\text{-effects}}$ is zero and use σ^2_{Clones} as an estimate of the total genetic variance. This results in an upward bias in both estimating broad sense heritability and predicting genetic gain if C-effects are important.

As with all deployment options discussed in this chapter, CF is not a breeding strategy. All programs employing CF should have an underlying breeding program of selection, breeding and testing to identify clones of ever-increasing genetic merit. Discussion of how breeding strategies may change when CF is the deployment option is presented in *Chapter 17*, and many breeding options can successfully result in long-term genetic gains in both the breeding population and deployment population. However, when CF is the deployment option, at least some of the genetic tests must be established with identified clones for the purpose of ranking clones and selecting which ones to propagate operationally.

Success of CF programs depends to a large extent on the biology of the species, and most operational clonal forestry programs are for species in which trees can be vegetatively propagated on a mass scale at selection age (Box 16.6). Examples include: (1) Several species in the genus *Eucalyptus,* especially *Eucalyptus grandis* and several of its interspecific hybrids (Box 16.8) in many tropical and sub-tropical countries (van Wyk, 1985b; Denison and Quaile, 1987; Campinhos and Ikemori, 1989; Lambeth *et al.*, 1989; Zobel, 1993; Duncan *et al.*, 2000); (2) Species and hybrids in the *Populus* and *Salix* genera in temperate regions of both the Northern and Southern Hemisphere (Zsuffa *et al.*, 1993); (3) *Cryptomeria japonica,* mainly in Japan (Ohba, 1993); and (4) *Picea abies* in Europe (Bentzer, 1993). There are other examples described in the second volume of *Clonal Forestry* (Ahuja and Libby, 1993). Further, there are still more applications of CF in agroforestry, Christmas trees and landscaping in urban settings (Kleinschmit *et al.*, 1993). Finally, the advent of operationally feasible tissue culture methods (*Chapter 20*) may facilitate CF in species that are difficult to propagate at selection age.

Advantages of Clonal Forestry

Advantages and desirable attributes of CF have been thoroughly discussed (Libby and Rauter, 1984; Carson, 1986; Burdon, 1989; Libby and Ahuja, 1993; Lindgren, 1993), and include the six points given in the previous section for family forestry. CF also has desirable attributes beyond family forestry that can be grouped into three broad categories dealing with increased genetic gain, better control over genetic diversity in operational plantations and enhanced operational logistics.

Advantages of CF relating to increased genetic gain include: (1) Better utilization of additive genetic variance, because differences among tested clones include all the additive variance, while that between full-sib and half-sib families includes only one-half and one-quarter of the additive genetic variance, respectively; (2) Complete capture of non-additive genetic variance since deployed clones retain their entire genetic value (G from *Chapter* 6); (3) Exploitation of genotype x environment interaction through site-specific deployment of tested clones to edaphoclimatic conditions in which they are known to perform well; and (4) Identification of "correlation breakers" for traits with unfavorable genetic correlations, meaning selection of outstanding clones for two traits even when the correlation implies that this would be difficult to reproduce in genetically diverse seedling offspring.

Lack of genetic diversity is often mentioned as a drawback of CF, and this is discussed further in the next section; however, in some ways, the forest manager is better able to manage genetic diversity in plantations established through CF than through other deployment methods: (1) Organizations desiring uniform plantations to reduce costs of harvest and produce more uniform products can establish monoclonal plantations in which a single clone (a single genotype) is deployed across many hectares (Fig. 16.5); and (2) Organizations preferring genetically diverse plantations can plant intimate mixtures of many

Box 16.8. Typical sequence of events in clonal forestry.

1. The first step is to establish clonal trials with large numbers of clones (hundreds or thousands). These clonal trials are established on several sites in the planting zone and each test site contains multiple ramets (typically 5 to 20) of each clone. These ramets are planted in randomized, replicated designs, usually in single tree plots (*Chapter 14*), to facilitate the precise ranking of clones at each site and across all sites.

2. When the clonal tests reach an appropriate selection age (as young as three years for fast-growing species of *Eucalyptus* and *Populus*, but older for slower growing species), measurements taken in the tests are used to predict clonal genetic values (*G* values from *Chapter 6*) for all of the clones. The best ones (typically 10 to 50) are used for operational deployment. If there is substantial clone x site interaction (*e.g.* Osorio, 1999), then specific clones are chosen for deployment to particular types of sites within the planting zone.

3. Next, the selected clones are propagated on a large scale, most commonly by formation of rooted cuttings or by tissue culture (*Chapter 20*), to produce plantlets for operational forestation (*e.g.* Fig. 1). These plantlets have the same genotypes as the selected clones; that is, there is no sexual reproduction and no recombination between the time the tested clones are selected and deployed. So, if 20 clones are selected in the genetic tests, then there are 20 clones deployed operationally.

Fig. 1. Typical events in clonal forestry of *Eucalyptus grandis*: (a) Selected trees are felled and juvenile stumps sprouts are formed; (b) These stump sprouts are rooted to form many hedge plants of the tested clone; (c) Cuttings are obtained from the hedges and stuck in rooting medium to produce cuttings that are planted operationally. (Photos by T. White)

(Box 16.8 continued on next page)

Box 16.8. Typical sequence of events in clonal forestry. (Continued from previous page)

Fig. 1. (Continued)

clones in every plantation, and these clones can be specifically chosen to be: (a) Complementary in the sense of exploiting different ecological niches; (b) Complementary in the sense of having different mechanisms for resistance to a particular pest; and/or (c) Genetically diverse in all measured traits to increase genetic diversity.

Lastly, several logistical benefits accrue to an organization that deploys a reasonable number (say fewer than 50) tested clones in each cycle of breeding. After several years of deployment, the organization comes to "know" each clone, and this can be formalized as a profile for each clone that specifies its nursery requirements (*e.g.* time and best conditions for rooting), field performance (*e.g.* performance in different edaphoclimatic conditions), and product quality (*e.g.* wood quality, tendency for splitting). This can lead to clone-specific management prescriptions in all phases (nursery, plantation and processing facility) with concomitant cost savings and gains in efficiency of operation.

Issues and Concerns about Clonal Forestry

The several concerns about CF can be categorized as technical, biological and social/ethical (Carson, 1986; Burdon, 1989; Kleinschmit *et al.*, 1993; Libby and Ahuja, 1993; Lindgren, 1993). The technical concerns are sometimes viewed as the most pressing and involve: (1) Difficulty in efficiently producing propagules from selection-age material thereby precluding use of CF in some species; (2) Maturation (Box 16.6) which can limit the useful operational life of tested clones; (3) Influences of C-effects which can both complicate clonal testing programs and confound gain predictions (Box 16.7); (4) Somatic mutations that can potentially cause within-clone genetic variation; and (5) The large number of initial clones that must be screened to identify 10 to 50 for operational deployment. Regarding the last concern, many programs of *Eucalyptus* (Brandao, 1984; Zobel, 1993; Lambeth and Lopez, 1994) screen thousands of initial selections in clonal trials to find the few that are most desirable in terms of both plantation performance and being amenable to the propagation system (meaning the selections must both sprout well and root well as cuttings).

By far the most important technical issue is maturation, which in some ways influences all five issues listed in the previous paragraph. There are many active research projects addressing maturation with two main strategies being investigated. One approach is to attempt to maintain copies of all clones in a juvenile state (through repeated hedging, serial propagation or storage of tissue cultures), while other ramets of the clones are tested in field trials. When the top clones are identified in the field, the copies kept in the juvenile state can then be operationally propagated. This implies maintaining hundreds or even thousands of clones in the juvenile state in a special facility for many years at a time. For example, a tissue culture method, called somatic embryogenesis (*Chapter 20*), is being used in some commercial species of pines to generate clones from seed tissue (Grossnickle *et al.*, 1996; Menzies and Aimers-Halliday, 1997; Bornman and Botha, 2000). Some ramets of each clone are held in tissue culture in liquid nitrogen storage, while others are planted in field tests used to rank the clones. Then, the cultures of selected clones are retrieved from cold storage and propagated operationally. The second option is to attempt to propagate the top clones at selection age using the ramets growing in field tests. In some species of *Eucalyptus*, ramets of the selected clones can be felled and the stumps form sprouts. These sprouts are juvenile and can be propagated through rooted cuttings (Box 16.8).

Biological concerns about CF relate to genetic diversity in both operational plantations and breeding populations. In operational plantations, the concern is that reduced within-stand diversity associated with planting a single clone may increase the probability of a major calamity, such as total devastation by disease. As mentioned above, CF permits substantial control on within-stand genetic diversity; however, many organizations do plant **monoclonal plantations** in which all trees are the same genotype.

There is no direct evidence in trees, but in crops, increased genetic diversity of plantings generally reduces problems associated with diseases; however, results vary widely across the range of experiments and situations reviewed (Lindgren, 1993). For example, mixtures of genotypes can sometimes be more susceptible than a single resistant genotype when unknown, susceptible genotypes in the mixture facilitate establishment of a disease epidemic. Also, if the pathogen has many host species (such as the root pathogen *Armillaria mellea* in Europe), then it seems unlikely that a range of genotypes of the same species (*e.g. Picea abies*) in the same stand would help to slow disease spread. Further, there are examples from both horticulture (such as some grape cultivars) and trees (*e.g.* the Lombardy popular) in which a single clone has been successfully used for several decades. In summation, it seems very important that organizations considering CF carefully evaluate the risks in their specific situation, and these risks seem higher for large expanses of single clones of tree species requiring long rotations. The optimal number of clones to plant is discussed in the next section.

Operational deployment of a relatively few clones also raises concerns about erosion of genetic diversity in the species as a whole. This could be a major issue if large portions of the natural range of a species are planted with few clones. This underscores the need to use CF wisely and to support CF with both breeding and gene conservation programs.

The social/ethical concerns about CF involve the general feeling that any managed forest represents an undesirable manipulation of nature and that CF is the most undesirable form of this manipulation. Our views, as expressed in *Chapter 1*, are that: (1) All types of forests are needed; and (2) Intensive management of a small fraction of the earth's forests to meet world demand for wood products would make available a larger fraction of natural forests for all other uses.

Operational Deployment of Clones

When tested clones are available for operational deployment, forest managers must decide how many clones to deploy and how those clones should be patterned on the landscape. For the latter question, there are two common options: (1) Few to many clones are mixed together for planting within the same stands (sometimes called intimate mixtures of clones); or (2) Single stands of trees (*e.g.* plantations of 20 ha) are planted with a single clone and different clones are deployed to surrounding stands thus forming a mosaic of **monoclonal blocks** across the landscape. Both the number of clones and their pattern of deployment have been the subject of many theoretical and empirical investigations, and we only provide a brief summary (see reviews by Carson, 1986; Libby, 1987; Foster, 1993; Lindgren, 1993).

The appropriate number of clones for operational deployment depends both on productivity (the reward or genetic gains associated with deployment of a very few top clones) and diversity (the risk associated with potential catastrophic loss if too few clones are deployed, see previous section). General conclusions are: (1) Deployment of very few clones increases expected genetic gains due to a higher selection intensity (*Chapter 6*); (2) Many theoretical studies indicate that there is little value, in terms of risk reduction, to mixtures above 30 clones and, in some situations, deployment of very few clones actually minimizes risk (Libby, 1982; Huehn, 1987, 1988; Huhn, 1992a,b; Roberds and Bishir, 1997); and (3) Mixtures of 7 to 30 clones appear as safe as mixtures of larger numbers of clones (Libby, 1987).

Taking all of these factors into consideration, we suggest that relatively few clones (7 to 20) are appropriate for deployment in any given cycle of breeding within a single breeding unit for short-rotation species planted in areas where there is no known risk of catastrophic loss. Conversely, larger numbers of clones (perhaps up to 50) seem appropriate for long-rotation species planted in areas where risks are high. Also, genetic gain is maximized for any specified level of genetic diversity by deploying more ramets of the very best clones (Lindgren *et al.*, 1989). This means planting more hectares of stands with the top few clones and then planting several additional clones on fewer hectares to enhance overall diversity.

In terms of the pattern of clonal deployment on the landscape, **intimate mixtures** of clones are suggested for species with long-rotations and planned thinnings (Burdon, 1982b). This permits poor-performing clones to be removed naturally or during thinnings without complete loss of a stand. Conversely, mosaics of monoclonal blocks of 2 to 25 ha are recommended for species with short rotations and no planned thinnings. The short time between planting and harvest reduces the risk associated with complete stand loss. In addition, the landowner obtains the previously mentioned benefits of increased stand uniformity (Fig. 16.5) and the ability to monitor specific clones for development of clonal profiles.

Finally, some countries have laws or regulations governing the marketing or deployment of clones. Two countries, Sweden and the Federal Republic of Germany, have established specific regulations for marketing, planting and distributing clones, while several other countries (including Denmark, Belgium, New Zealand and Canada) are considering either regulations or voluntary guidelines relating to deployment of clones. When implemented, these regulations must be followed as part of any CF program.

GENETIC DIVERSITY CONSIDERATIONS IN DEPLOYMENT OPTIONS

Genetic diversity among trees established in operational plantations depends on both the genetic diversity among selections in the propagation population and on the deployment method (*e.g.* seedlings from bulk orchard seed, family forestry or clonal forestry). The propagation population is a subset of the selected population (Fig. 11.1), and therefore genetic diversity in plantations also depends on the levels of genetic diversity maintained by the tree improvement program in the selected and base populations.

Both theoretical considerations (reviewed in *Chapter 17*) and empirical studies (Adams, 1981; Cheliak *et al.*, 1988; Bergmann and Ruetz, 1991; Chaisurisri and El-Kassaby, 1994; Williams *et al.*, 1995; El-Kassaby and Ritland, 1996; Williams and Hamrick, 1996; Stoehr and El-Kassaby, 1997) (*Chapter 10*) indicate that even modest numbers of selections in either breeding populations or seed orchards maintain most of the genetic diversity present in natural populations in the same breeding zone of the species for many generations. Certainly, tree improvement programs that maintain effective population sizes above 300 conserve the large majority of natural genetic diversity. Therefore, we begin by assuming that there are accompanying breeding and gene conservation programs maintaining appropriate levels of genetic diversity for the long term.

Comparison of various deployment options entails consideration of genetic diversity in the plantations established under each option. This diversity can be considered on three levels of scale: (1) Within individual trees; (2) Among trees within a stand; and (3) Among stands patterned on the landscape. At the first level of scale, we have already discussed (*Chapter 7*) that trees are among the most heterozygous organisms known in nature with

a

b

Fig. 16.5. Monoclonal plantations are more uniform than those from seedling offspring: (a) A single clone of *Eucalyptus grandis* on the right and a seedling plantation on the left from Cartón de Colombia; and (b) Harvesting of a single *E. grandis* clone by Mondi Forests in South Africa. (Photos by T. White)

high average levels of observed heterozygosity in most species. Therefore, operational deployment of a single clone (which is a single genotype barring somatic mutations) is not equivalent to similar deployment of an inbred crop species which may be highly homozygous. In trees, there are two alleles at a large number of loci and this provides some genetic diversity even within the same tree or same clone.

In theory, the presence of genetic diversity within trees (heterozygosity) may be valuable in helping them adapt to environments that vary over time due to changing weather patterns and ecological succession (called heterozygote superiority or heterosis, *Chapter 5*). The importance of heterozygote superiority to adaptation has been widely debated and few actual examples have been reported. The main importance of heterozygosity in trees and other outcrossing species may be in masking deleterious alleles that otherwise would be expressed in homozygous form (*Chapter 5*). Still, differences among alleles within a locus is one of the forms of genetic diversity in forest tree species, regardless of the deployment method, that is not present in most inbred crops.

At the second level of scale, different deployment options affect genetic diversity among trees in the same stand. We first consider plantations established from seed from seed orchards, because this deployment option has been most widely studied. In general, genetic marker studies comparing genetic diversity in natural stands compared with plantations established with seed from seed orchards indicate similar or even greater levels of genetic variation in plantations from seed orchards compared to their natural counterparts (Adams, 1981; Williams and Hamrick, 1996; Stoehr and El-Kassaby, 1997; Schmidtling and Hipkins, 1998) (*Chapter 10*). More specifically, the percentage of polymorphic loci, average number of alleles per locus and observed and expected heterozygosities are similar.

Genetic diversity can even be higher in plantations from seed orchard seed than within natural stands perhaps due to: (1) A higher degree of outcrossing in seed orchards if the natural stands contain neighborhood structures such that mating occurs among relatives growing in close proximity (*Chapter 7*); (2) Large diversity among seed orchard clones selected from many widely dispersed stands within the breeding unit; and (3) Pollen contamination that can mean infusion of genetic diversity from outside of the seed orchard.

At the third level of scale, when genetic diversity is measured across the landscape, replacement of all natural stands within a breeding zone using seed from a single seed orchard will likely result in lowered genetic diversity contained in the entire zone. In particular, some rare alleles present in the natural stands may not be present in the plantations established from seed orchard seed (*Chapter 10*). Consider the study of 17 allozyme loci in *Picea* sp. sampled in British Columbia (Stoehr and El-Kassaby, 1997). The total number of alleles summed across all loci ranged from 33 to 40 per natural stand (averaging 36), and totaled 46 across all stands sampled. In the only seed orchard plantation sampled, there were 39 alleles, more than all except one natural stand, but not as many as all stands combined.

The issue of genetic diversity of various deployment options can also be approached using the quantitative genetics theory developed in *Chapter 6*, and this is particularly useful since deployment options other than seed orchards have been less well studied with genetic markers. First based on the preceding discussion, we take plantations established from seed orchard seed as a baseline and assign them a value of 100% to indicate that most of the natural genetic variation within a breeding zone exists among trees within plantation stands (Table 16.1). Then, for any other deployment option, the expected amount of genetic variation existing among individuals within the same plantation can be expressed as a fraction of that in plantations from seed orchards. This relative amount of genetic variation does not depend on the heritability of the trait, but does depend on the ratio of additive to non-additive genetic variance (and here we assume that variance due to epistatic interactions is negligible).

With these assumptions, between 75 and 88% of the total genetic variation expected among seed orchard offspring, still remains, on average, among trees in a plantation when

Table 16.1. Theoretical genetic variation among trees within plantations established with different deployment options for three different levels of additive (V_A) vs. non-additive (V_{NA}) genetic variance. Plantations from bulk mixtures of seed from seed orchards are taken as the baseline at 100%, and other options expressed relative to this baseline. Genetic variance due to epistasis is assumed zero yielding the following expectations for genetic variance among trees within the same stand (V_w): $V_w = 0.75V_A + V_{NA}$ among trees in the same half-sib family; $V_w = 0.50V_A + 0.75V_{NA}$ among trees in the same full-sib family; and $V_w = 0$ among ramets of the same clone (barring somatic mutations). For example, if it is assumed that $V_{NA} = 1/2\ V_A$, then the total additive genetic variance among seed orchard offspring, $V_{w,seed\ orchard} = V_A + V_{NA} = V_A + 0.50V_A = 1.50V_A$, while the total additive variance within a single half-sib family, $V_{w,HS\ family} = 0.75V_A + V_{NA} = 0.75V_A + 0.50V_A = 1.25V_A$. Then, the ratio of genetic diversity in plantations from seed of a half-sib family to that of a seed orchard is $1.25V_A / 1.50V_A = 0.83$ as shown in the table.

Deployment option	V_w, Genetic variation within stands (%)		
	$V_{NA} = 0$	$V_{NA} = \frac{1}{2}\ V_A$	$V_{NA} = V_A$
Seed orchards	100	100	100
Single half-sib family	75	83	88
Single full-sib family	50	58	63
Single clone	0	0	0

those trees are from a single half-sib family. Similarly, more than half of the genetic variation still remains among trees when a single full-sib family is deployed to a plantation site. Therefore, deployment of individual families (either open-pollinated families from seed orchards or full-sib families from control-pollinated seed) is expected, on average, to retain the majority of the genetic diversity among trees within stands. This conclusion also holds for family forestry using vegetative multiplication to create large numbers of plantlets, if plantlets from different clones (*i.e.* different seed from the same family) are mixed in equal proportions prior to planting.

Clearly, when a stand contains only ramets of a single clone, there is no genetic variation remaining among trees. This occurs in family forestry or clonal forestry whenever monoclonal plantations are established. In fact, this genetic uniformity is one of the previously mentioned advantages of clonal forestry to some organizations (Fig. 16.5). If within-stand genetic diversity is desired, more than one clone must be planted in intimate mixtures in the same stand. The alternative is to plant mosaics of monoclonal plantations; then, within-stand genetic diversity is nil, but there can be considerable overall diversity if several genetically distinct clones are deployed across the landscape (see *Operational Deployment of Clones*).

SUMMARY AND CONCLUSIONS

Applied tree improvement is conveniently divided into two phases (Fig. 11.1): (1) Activities within the breeding cycle (selection, testing and breeding) aimed at development of new varieties; and (2) Formation of the propagation population and deployment of a newly developed variety. In contrast to the first set of activities, the planning horizon for the propagation population is only a single generation and the goal is to use the very best genotypes available to produce plants for reforestation that maximize genetic gain in operational plantations. The principle benefits (called realized gains) from any applied tree improvement program are measured as the difference in yield and product quality of the plantations established with improved varieties compared to plantations established with unimproved stock.

In any cycle of breeding and testing, an organization may use multiple types of propagation populations and deployment options. A seed orchard is a collection of selected clones or families established in one physical location and then managed to produce genetically improved seed for operational forestation. There are two types of seed orchards: (1) Clonal seed orchards (CSO) for which vegetative propagation (grafting, rooting or tissue culture) is used to establish clones of selected trees in the seed orchard (Box 16.1); and (2) Seedling seed orchards (SSO) established with seed (open-pollinated or full-sib) from selected trees (Box 16.2). Seed orchards are by far the most common type of propagation population, and nearly all tree improvement programs employ seed orchards at some point to produce genetically improved seed for operational deployment.

Family forestry (FF) is the large-scale deployment of superior (*i.e.* tested) full-sib (FS) families in operational plantations (Box 16.4). FF typically begins after a tree improvement program has been operating for several years. Data from genetic tests are employed to identify the top few parents (say between 10 and 30) that will be used to produce FS families for operational deployment. The FS families can be deployed as seedlings created by control pollination (CP), but more commonly vegetative multiplication, (VM; *e.g.* production of rooted cuttings) is used to multiply scarce CP seed to generate the number of plants needed for operational forestation.

Clonal forestry (CF) refers to the large-scale deployment of relatively few (typically 10 to 50) clones that have proven their superiority in clonal tests (Box 16.8). Sometimes CF and the VM option of family forestry are confused. The difference is that in CF individual clones are propagated on a mass scale for operational reforestation after testing has demonstrated the superior genetic value of those clones. In VM family forestry, parents or full-sib families are tested, but not the actual clones that are deployed.

Genetic gains are cumulative through cycles of selection, breeding and testing, and therefore, extra gain from any deployment option adds to the gain already captured in the breeding cycle for the plantations established with that option. The implications are: (1) Gains from deployment options include the gains made in the breeding cycle proper, therefore all deployment options benefit from a well designed and properly implemented breeding strategy (*Chapter 17*); (2) Gains from a deployment option apply to the plantations established with that improved variety, and the benefits are realized when those plantations are harvested; and (3) Incremental gains from a deployment option above those from the breeding cycle are not generally carried back into the breeding cycle (*e.g.* the extra benefits of CF apply only to the plantations established during that generation; those gains are not cumulative as are the gains from the central activities of the breeding cycle (Fig. 11.1)).

When an organization is deciding which of the many deployment options to employ in any given generation of breeding, there are many logistical, genetic and economic factors to consider. Some guiding principles are as follows:

- Generally speaking the expected genetic gains from CF are greater than those from FF which are greater than those from seed orchards (Matheson and Lindgren, 1985; Borralho, 1992; Mullin and Park, 1992). CF provides the highest gains for the reasons detailed in *Advantages of Clonal Forestry*. Similarly the additional gains from FF over seed orchards were described in *Family Forestry*.

- When tested clones have short operational lives due to early onset of maturation, clonal tests need to be established more frequently to identify new clones. This raises costs associated with CF, and reduces its overall returns compared to other options.
- If vegetative propagation is expensive and inefficient (*e.g.* a large proportion of trees do not root well), the costs of either CF or FF with vegetative multiplication, increase and genetic gains decrease. For example, suppose that only 25% of cuttings root on average, with some clones or families having extremely poor rooting (near zero) and others having acceptable rooting (say above 50%). In this situation, there will be a natural tendency to choose operational clones or families from among the group that root well. This adds another selection objective (rooting ability), and therefore reduces the selection intensity on the other traits. This means reduced genetic gain on other traits, because some outstanding clones or families may not root well.
- Both CF and FF eliminate pollen contamination as a source of reduction in genetic gains. This increases their attractiveness, especially for conifer species where pollen travels great distances and pollen dilution zones around seed orchards seem mostly ineffective.
- CF, FF or seedling seed orchards are favored over clonal seed orchards if there is a very long lag time (> 10 years) from time of grafting to time of seed production. This lag substantially delays the return on the investment, and becomes more important with higher real discount rates.
- If genetic diversity among trees within the same stand is important, then CF and FF with monoclonal plantations are not options. On the other hand, all other options maintain a substantial portion of the natural genetic diversity within plantations, and in some cases plantations from seed orchards may be more genetically diverse than their natural counterparts.

CHAPTER 17

ADVANCED-GENERATION BREEDING STRATEGIES – *BREEDING POPULATION SIZE, STRUCTURE AND MANAGEMENT*

Many tree improvement programs in the world are already in the second or third cycle of breeding, and gains from these advanced-generation efforts promise to be even greater than those from the first generation. Continued progress depends on a long-term breeding program, and the activities of this program are guided by a breeding strategy. A complete **breeding strategy** (sometimes called a tree improvement plan) is a detailed document that specifies the design, timing and implementation logistics of all components of tree improvement. This includes selection, testing, breeding, development of propagation populations, deployment of commercial varieties, infusion of new material, gene conservation efforts and research. Thus, the breeding strategy deals with all activities in the inner circle of the breeding cycle that aim to develop new varieties (Fig. 11.1), as well as with the propagation and deployment of those varieties.

The goals of tree breeding programs vary, but include: (1) Achieving near-optimal short-term genetic gains in a few traits of high economic or social importance; (2) Maintaining sufficient genetic diversity in the breeding population to ensure near-optimal long-term genetic gains in the same or different traits as markets, products, technologies and environments change in the future; (3) Ensuring sufficient flexibility in program design to facilitate change in direction and incorporation of new technologies (such as the biotechnologies discussed in *Chapters 18-20*); (4) Conserving genetic diversity in the species; and (5) Conducting all of these activities in a timely, cost effective way that yields appropriate economic and/or social returns.

The long time interval of a typical breeding cycle (several years to decades), coupled with the large size and costly nature of tree breeding, pose special problems for tree breeders compared to crop breeders. When a single cycle of breeding means many years and high costs, it is critical the breeding strategy achieves the objectives efficiently; yet, there are many uncertainties when planning decades into the future. Also, the best strategy depends on a myriad of biological, silvicultural, genetic, and management assumptions. Thus, developing breeding strategies is not an exact science, and we agree with Shelbourne *et al.* (1986) who state that "because intuition and subjective judgment play no small part in strategy development, it is rightly viewed as an art."

For these reasons no two breeding strategies are identical, nor would two forest geneticists likely devise the same strategy even for the exact same program given the same assumptions. In fact, it is impossible to present a "recipe" for breeding strategy development. Thus, we have structured this chapter such that the five major sections (*General Concepts of Advanced-generation Breeding Strategies, Breeding Population Size, Breeding Population Structure, Mating Designs for Advanced-generation Breeding,* and *Making Advanced-generation Selections*) describe many of the factors that should be considered when developing long-term breeding strategies. In each of these sections, we attempt to put these factors into the context of real-world breeding programs and provide citations to the literature so that the reader can refer to the rich diversity of breeding strategies that have been developed.

Throughout this chapter, we assume that appropriate breeding units and base populations have been delineated (*Chapter 12*), and that there is a separate breeding, testing and deployment program for each unit. Thus, we implicitly are discussing development of a breeding strategy for a single breeding unit, and multiple units could conceivably have different strategies. This is only to simplify the discussion, because strategies would not usually be developed in isolation for each unit. Other reviews of breeding strategy development in forest trees include: Kang (1979a), Zobel and Talbert (1984), Kang and Nienstaedt (1987), Namkoong *et al.* (1988), White (2001), and White (2004).

GENERAL CONCEPTS OF ADVANCED-GENERATION BREEDING STRATEGIES

Organization of a Breeding Strategy

While there are many ways to organize and present an advanced-generation breeding strategy, we prefer a complete and detailed plan (Box 17.1) that: (1) Justifies the need and sets overall goals for the program; (2) Reviews current knowledge about the species available in the literature and other sources; (3) Explicitly states all assumptions required to develop the strategy; (4) Catalogs all of the organization's previous efforts in tree improvement of that species; (5) Provides a detailed strategy for one entire breeding cycle; (6) Provides a timeline showing all activities needed to implement the proposed breeding strategy; (7) Speculates on the suitability of the strategy for the long term of several cycles of breeding given changing assumptions, new technologies, etc.; (8) Discusses plans for gene conservation of the species; and (9) Sets research priorities by identifying research projects that would greatly benefit the tree improvement program. In addition, other sections sometimes found in a breeding strategy include analyses of expected genetic gain, economic returns and effective population sizes in different generations.

Principles of Recurrent Selection

Most tree improvement programs span many cycles of breeding, selection, testing, and deployment, and recurrent selection is the general term that refers to repeated cycles of genetic improvement aimed at the gradual and cumulative improvement of a few traits in a population (Shelbourne, 1969; Namkoong *et al.,* 1988; *Chapter 3*). In its simplest form, called **simple recurrent selection**, each cycle involves: (1) Mass selection of individuals from the base population of generation k to form the selected population of that generation; and (2) Random mating of these phenotypically-superior trees (*i.e.* without pedigree control) to produce the offspring that become the base population of generation $k + 1$. In this form, the selected and breeding populations of the breeding cycle in Fig. 11.1 are identical and composed of the selected trees. Simple recurrent selection is the oldest form of recurrent selection, and was the method used more than 10,000 years ago by ancient farmers to improve their field crops (Briggs and Knowles, 1967) where seed from superior individuals was retained for next year's crop.

Simple recurrent selection is rarely used in breeding programs today, because it is less efficient at achieving genetic gains than other forms of recurrent selection that incorporate genetic testing and pedigree control. Nevertheless, most breeding programs of plants, animals and trees employ some form of recurrent selection, and these programs share many common features for quantitative traits assuming that large enough breeding populations are maintained (Allard, 1960, Chapter 23; Namkoong *et al.*, 1988, Chapter 3; Falconer and Mackay, 1996; Comstock, 1996, Chapter 12): (1) Genetic gain in the selected traits results from changes in gene frequencies of alleles at the loci controlling expression of those traits; (2) Success at achieving genetic gain depends on the amount of genetic variation in the original founding population, because recurrent selection programs with closed populations do not create new genetic variation (rather, existing variation is repackaged into individuals containing a higher frequency of the favorable alleles); (3) In the first several cycles (say 10 cycles), genetic variability for the selected traits is little changed from that in the initial founding population and genetic gain is similar in each of those cycles assuming similar selection intensities and breeding strategies; (4) It takes many cycles of recurrent selection (more than 50) to lead to a plateau after which selection is ineffective (due to fixation of favorable alleles or other causes); (5) Beginning with larger founding populations and infusing unrelated material into the population extend the number of cycles before a selection plateau is reached; and (6) The vast majority of the population's characteristics associated with traits uncorrelated with the selected traits is unaffected by the recurrent selection program.

Recurrent Selection for General Combining Ability

Breeding strategies of forest trees vary dramatically due to different selection, breeding, testing and deployment methods; however, nearly all strategies for population improvement of a single species (hybrids are discussed in the next section) are founded on a single type of recurrent selection, called **recurrent selection for general combining ability** (RS-GCA) (Shelbourne, 1969; Namkoong *et al.*, 1988, p 44). There are other methods, but they are rarely used in forest trees and not discussed here (Allard, 1960; Briggs and Knowles, 1967; Hallauer and Miranda, 1981; Bos and Caligari, 1995).

In RS-GCA, genetic testing follows selection and the selections are ranked based on their GCA values (*Chapter 6*) for the selected traits. Then, only the selections with the highest GCAs (or their offspring) are included in future cycles of breeding, and poor selections are excluded. Genetic testing greatly increases the genetic gain above that from mass selection (especially for traits with low heritabilities). This is especially important to maximize the genetic gain per unit time in forest trees with long generation cycles.

Many different mating designs (*Chapter 14*) can be used to estimate the GCA values of the selected parents, but unlike simple recurrent selection, there must be at least some control of pedigree. When open-pollination or pollen mixes are used to create the offspring for genetic tests, then the pedigree is maintained only on the female side. Conversely, full-sib designs maintain knowledge and control of both male and female parents. Both half- and full-pedigree designs are used successfully in recurrent selection programs of forest trees, as discussed later in this chapter. The common themes are that the selections in a given generation are tested by planting their offspring in multiple test sites within the breeding unit and that this information is used to favor those selections with the higher breeding values.

Genetic tests provide two other important advantages over simple recurrent selection based on mass selection: (1) Information on genetic parameters (h^2, g x e, correlations, etc.) is obtained and used in many ways to enhance the breeding program (*Chapters 6, 13, 15*); and (2) The offspring planted in the genetic tests sometimes serve as the next generation's base population meaning that the best individuals within the best families are selected for the next generation.

Box 17.1. Simplified structure of a breeding strategy or tree improvement plan.

Section I, Introduction. (a) Organization's overall objectives including products, markets, species mix; (b) Organization's breeding objectives with specific traits and their justification; and (c) Overview of key issues and main points of the breeding strategy.

Section II, Review of Species' Silviculture, Biology and Genetics. (a) Species' importance including distribution and edaphoclimatic requirements in both its natural and exotic ranges; (b) Products, uses and plantation estates summarized by country; (c) Summary of regeneration, silviculture and harvesting systems used in different countries (including rotation ages and mean annual increments); (d) Key elements of species' biology such as systematics (taxonomy and hybrids with related species), reproductive biology (flowering and pollen biology, mating systems, ease of control pollination, and ages and quantity of seed production), propagation biology (grafting, rooted cuttings, tissue culture) and sprouting and coppice ability; (e) Current knowledge of species' genetics including provenance and geographic variation, estimates of quantitative genetic parameters (*e.g.* heritabilities, g x e, age-age correlations); and (f) Summary of world genetic resources describing countries and organizations with breeding programs for this species along with the degree of availability of improved materials from each organization.

Section III, Assumptions, Premises and Constraints. (a) Major edaphoclimatic regions where the species is to be planted (soils, climate, and anticipated annual reforestation for each region); (b) Silvicultural assumptions for each edaphoclimatic region (site preparation, major treatments and rotation lengths); (c) Flowering biology (ages and quantities of flowers for breeding, difficulty and costs of pollen extraction and control pollinations, estimated number of control pollinations that can be conducted in one year and expected seed yields); (d) Propagation technology for all possible types of operational propagation populations (assumptions needed to plan sizes and logistics of seed orchard and rooted cutting production facilities); (e) Breeding objectives including final target traits of operational importance and the selection criteria for achieving these targets (see *Chapter 13)*; (f) Protocols for genetic tests including measurement techniques, timing of selections and reference populations to be used as controls and against which long-term gains will be measured; (g) Potential uses of hybrids and sources of infusions including unrelated germplasm from other programs; and (h) Other considerations (any special constraints or assumptions that will impact the breeding strategy).

Section IV, Current Genetic Resources and Past Breeding Programs. The advanced-generation breeding strategy is greatly affected by what has already transpired and by the current genetic resources that are available to build upon. So, this section details previous efforts by the organization: selections, genetic tests, breeding activities, seed orchards, etc. Any genetic resource that could be drawn upon in advanced generations should be summarized. This includes provenance trials, seed source tests, hybrid experiments and other research trials that are often a rich source of unrelated material for use as infusions to increase effective population sizes in advanced generation programs.

(Box 17.1 continued on next page)

Box 17.1 Simplified structure of a breeding strategy or tree improvement plan.
(Continued from previous page)

Section V, Advanced-Generation Breeding Strategy. This is the heart of the tree improvement plan and presents the details of one complete generation of the breeding cycle (Figure 11.1) including justification and rationale for each planned activity: (a) Base population (complete delineation of each breeding unit and definition of the trees, clones and other materials available for selection in each unit); (b) Selected population (how many selections, selection methodology and sources of all planned selections); (c) Breeding population and genetic tests including population size and structure (*e.g.* sublines, main, elite), mating design(s), field designs (with numbers of crosses, blocks, and sites along with detailed experimental designs), and measurement schedules; (d) Propagation populations and deployment plans specifying each proposed type of new operational variety such as seed orchards or rooted cutting facilities (with genetic designs and field designs); and (e) Plans for infusion of new material into the breeding population to maintain an adequate genetic base.

Section VI, Time Line for Implementation of Advanced-Generation Strategy. This is a detailed implementation plan showing the timing (by year and season of the year) for each activity for the entire cycle of breeding proposed in the previous section. This section is useful for planning yearly work schedules and budgeting.

Section VII, Looking into the Future. While it is not usually advisable to make detailed plans for more than a single cycle of breeding, it is imperative to evaluate the proposed strategy with a view to the longer term: (a) What would be the impact of new biotechnologies and could they be readily incorporated; (b) Is the strategy flexible enough to adjust to changes in any of the assumptions and premises from Section II; and (c) Are there adequate plans for gene conservation?

Section VIII, Research Needs. This section identifies and sets priorities for future research projects that could enhance genetic gains, reduce time to realization of those gains, increase program efficiency or lower costs. Research priorities can be established by evaluating each candidate project in terms of: (a) Probability of successfully completing the project (*i.e.* the likelihood of developing the new technology or otherwise solving the problem if the project is undertaken); (b) Impact on the program if successfully completed (*i.e.* the benefits of solving the problem); (c) Timing of the realization of the benefits; and (d) Costs of the candidate project.

For all of these reasons, RS-GCA is favored by forest tree breeders over simple recurrent selection, and all programs discussed in this chapter are based on RS-GCA (except breeding for hybrids). Even tree improvement programs involving clonal forestry for operational deployment (*Chapter 16*) normally utilize a long-term breeding strategy based on RS-GCA. In these programs, ranking of clones for operational use in the current generation's propagation population is based on the total genetic value (clonal value, G, in *Chapter 6*); however, the propagation population is outside of the main circle of breeding activities in Fig. 11.1. Each generation, activities in the main circle are conducted to recombine alleles from superior selections and produce better clones for subsequent generations. The genetic gains in the long-term breeding population are based on RS-GCA. This is because only additive genetic effects (*Chapter 6*) are passed on to sexually-produced offspring. Thus, breeding to achieve long-term gains is based on additive gene effects, and, in clonal forestry, each cycle of clonal testing is used to identify the best clones for operational use based on additive and non-additive genetic effects (*i.e.* based on total genetic value).

Recurrent Selection Methods for Breeding Interspecific Hybrids

Breeding for interspecific hybrids is becomingly increasingly popular in forest trees (see the collection of papers in Dungey *et al.*, 2000), and there are four potential mechanisms that make hybrids attractive (Namkoong and Kang, 1990; Nikles, 1992; Stettler *et al.*, 1996; Li and Wu, 2000; Verryn, 2000): (1) Hybrids can combine desirable traits from two or more species presumably through additive gene action (called complementation or combinatorial hybrids); (2) Hybridization may lead to heterosis or hybrid vigor due to non-additive gene action; (3) Hybrids may exhibit greater homeostasis (*i.e.* phenotypic stability) due to higher levels of heterozygosity; and (4) Hybrids may allow the planted range to be extended to sites where one or both parental species are marginal (perhaps through one or a combination of the above mechanisms). Regardless of the mechanism, production of hybrids brings alleles together that would rarely if ever occur together in nature and this sometimes produces a hybrid taxon better than either single species involved. While not absolutely necessary, efficient means of vegetative propagation greatly facilitate operational deployment of hybrids. Thus, there are hybrid breeding programs in several countries for species of *Eucalyptus, Populus* and *Salix*.

Breeding of hybrids is more complicated than that of a single species since the breeder begins with two or more initial founding populations (there are as many initial populations as there are species to be combined into the hybrid taxon). While the topic is too complex for complete review here (Namkoong *et al.*, 1988; Nikles, 1992; Li and Wyckoff, 1991, 1994; Bjorkman and Gullberg, 1996; Stettler *et al.*, 1996; Griffin *et al.*, 2000; Kerr *et al.*, 2004a,b), the broad array of strategies available fall into two general categories (Fig. 17.1): (1) **Multiple population breeding** in which each of the original species is maintained as a separate breeding population for many cycles of breeding, but hybridized each generation to produce F_1 hybrids for both testing and operational deployment; and (2) **Single population breeding** in which the original species are hybridized initially to form a single hybrid breeding population that is managed and improved recurrently as a single, synthetic taxon.

Multiple population breeding for hybrids implies that each species has it own breeding population and all of the associated activities of the breeding cycle (Fig. 17.1a). The species' identities are maintained intact by conducting essentially separate breeding programs. Each species is improved each cycle. During each cycle of breeding, selections from each species are also intermated to form F_1 hybrids that are planted in genetic tests. These tests are used to rank the parents in each species based on the quality of hybrids they produce (called general hybridizing ability, GHA, Nikles, 1992) as well as to identify specific hybrids for operational propagation. The hybrid crosses each cycle are a dead end in the sense that they are used only in that cycle of breeding; no hybrid selections are used in future breeding work. Rather, new selections are made within each of the parental species based on GHA values, and these selections are used to create new F_1 hybrids for testing in the next cycle. Thus, each parental population is being improved to produce increasingly better F_1 hybrids each generation.

In single population hybrid breeding, the goal is to develop a single synthetic taxon by combining all species into a hybrid population that is then managed as a single breeding population (Fig. 17.1b). The first step is to make crosses among many parents of two or more species to form many different F_1 hybrid families. In subsequent cycles, many types

a Multiple population breeding for hybrids

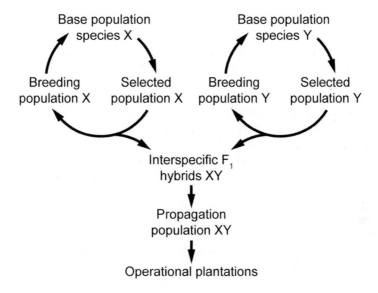

b Single population breeding for hybrids

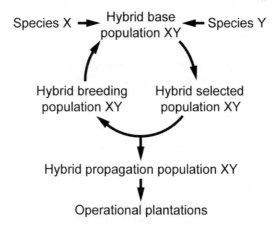

Fig. 17.1. Simplified schematic diagrams of hybrid breeding programs for creating interspecific hybrids between two pure species labeled X and Y: (a) Multiple population breeding program; and (b) Single population breeding program. In both cases each breeding cycle consists of creating each of the types of populations shown.

of crosses may be made (backcrosses to one or more of the original species, F_2, F_3, three-way crosses involving multiple species, etc.). If backcrossing is planned, then the pure species populations are maintained (*e.g.* in breeding arboreta), but the pure species may or may not be actively improved. The main distinguishing feature is that there is a single hybrid breeding population consisting of various types of crosses to which the selection, breeding and testing activities are applied. At one extreme, this permits introgression of alleles from one species into another (through repeated backcrossing) as may be desirable if only a single trait (such as rooting ability or cold hardiness) is desired from one species for introgression into an otherwise high-yielding species. Conversely, the crossing pattern could be highly complex. Each cycle a wide variety of types of crosses is tested and selections made in any of those that perform well. This leads to development of a completely new synthetic taxon combining alleles from all founding species in unknown frequencies.

There are advantages and disadvantages of multiple and single population breeding for hybrids and computer simulation models have been developed to compare options given some assumptions (Kerr *et al.*, 2004a). Multiple population breeding is more costly and sometimes involves longer generation intervals (since separate breeding populations are maintained for each species), but it may result in more genetic gains if those gains depend on heterosis observed in F_1 populations (Kerr *et al.*, 2004b). Multiple population breeding also improves the founding species separately, and these could conceivably be used as pure species or in different hybrid combinations if warranted in the future.

On the other hand, single population breeding of a synthetic hybrid taxon may be more efficient in many circumstances (assuming that viability and fecundity of the hybrids are not diminished due to drastic karyotypic differences between the species) (Kerr *et al.*, 2004b). Single population breeding conveniently allows more than two species to be combined, and facilitates development of a "designer" taxon especially adapted to the edaphoclimatic, silvicultural and product objectives of the organization. For reasons of flexibility in combining species in multiple ways, lower costs and potentially higher gains, we believe that more hybrid programs will opt for single population breeding in advanced generations.

Management of Genetic Diversity and Inbreeding

To illustrate the impact of multiple cycles of breeding on genetic diversity, consider that in each generation of breeding with full-sib families, the number of possible unrelated selections is reduced by half in the absence of any selection to eliminate inferior families. So, if a program begins with a breeding population of 400 unrelated selections, the maximum number of unrelated full-sib families that can be created is 200 (since two parents are required to create one full-sib family, see *Chapter 14*). After a single selection is made from each of these families, then all additional selections are related to a previous selection. After 5 cycles of breeding, the maximum number of unrelated selections in the breeding population with this example is 25 ($400 \rightarrow 200 \rightarrow 100 \rightarrow 50 \rightarrow 25$). If some families are eliminated by selection each generation, then the maximum number of unrelated selections falls even more rapidly. There can be thousands of selections in the fifth-generation breeding population; it is the number of unrelated selections that is quickly reduced.

This example illustrates that build up of relatedness inevitably occurs in recurrent breeding programs that employ closed breeding populations. The loss of genetic diversity due to increasing relatedness is quantified as a reduction of the effective population size, N_e (*Chapter 5*). So, even if a breeding population is maintained at census number of N = 400 selections throughout each of several cycles of breeding, N_e falls gradually each generation. The decrease in N_e is not nearly as precipitous as perhaps implied in the above illustration of unrelated selections and depends on several factors discussed later in this section. However, all closed breeding populations eventually become completely inbred (with an inbreeding coefficient of F = 1, *Chapter 5*) with no genetic diversity. To some extent this problem can be ameliorated by infusing new unrelated material into advanced-generation breeding populations (as described in *Chapter 11*). However, this only slows the progression, and in later generations it becomes difficult to find unrelated material to infuse into the breeding population that is of acceptable genetic quality.

The inexorable increase of relatedness in advanced-generation breeding populations coupled with the near-universal importance of inbreeding depression in forest trees (*Chapter 5*) means that management of inbreeding is a key issue in the development and implementation of all advanced-generation breeding programs. There are three issues associated with management of genetic diversity and inbreeding: (1) Activities that enhance near-term genetic gain in the first few cycles of improvement also tend to speed decrease in genetic diversity and hence may hamper continued genetic progress in later generations; (2) The problem of inbreeding depression means that inbreeding should be avoided in any trees planted in operational plantations; and (3) Use of directed inbreeding to rid the breeding population of unfavorable deleterious alleles is a positive aspect of inbreeding that can be incorporated into breeding strategies as long as the trees produced by the propagation population for operational plantations are outcrossed. We briefly introduce each of these issues in this section because they permeate all aspects of breeding strategy development including determination of breeding population size and structure as well as mating and field designs.

Most tree improvement programs, especially those associated with private industry, are justified based on economic or social returns produced by the first few cycles of improvement. This means that many activities are focused on producing large short-term genetic gains. Unfortunately, most of these activities (such as intensive selection that eliminates inferior families and putting more emphasis on superior material) also result in more rapid reduction in genetic diversity in the breeding population (King and Johnson, 1996; Lindgren *et al.*, 1996; McKeand and Bridgwater, 1998). Long-term genetic gains depend directly on the amount of genetic diversity maintained in the breeding population and are maximized when there is less intensive selection each cycle (Robertson, 1960). Thus, there is an apparent conflict between activities that maximize short-term gain and those that maximize long-term gain.

There is no single solution to the trade-off between rapid short-term gains and reduction in genetic diversity; rather, advanced-generation breeding strategies aim to achieve the best of both worlds (near-optimal short-term and long-term gains) through a combination of several methods described in more detail in later sections of this chapter: (1) Starting with large initial breeding population sizes; (2) Structuring the breeding population into subpopulations (*e.g.* elite, main and gene conservation populations) that are improved at different rates; and (3) Using restricted selection indices that limit the number of selections made from any single family.

The second issue raised above is the need to avoid inbred trees that are planted in operational plantations. If inbred offspring are planted, genetic gains from selection and breeding could be offset by loss of productivity due to inbreeding depression. Thus, it is axiomatic in forest tree improvement that related matings are avoided among selections in the propagation population; this means avoiding the use of related clones in the same seed orchard and avoiding related matings among top parents in family forestry (*Chapter 16*).

These deleterious effects of inbreeding in operational plantations can be avoided in two ways: (1) If full pedigree is maintained on all crosses in the breeding population, then coancestry can be managed and related matings avoided as long as possible; and (2) The breeding population can be subdivided into distinct sublines (Burdon and Namkoong, 1983) such that all breeding is conducted among selections in the same subline. Inbreeding builds up within each subline, but crosses for operational plantations are restricted to parents from different sublines (that are unrelated since all breeding has been within sublines).

The third issue mentioned above identifies a potentially beneficial use of inbreeding in recurrent selection programs. Inbreeding has long been used in crop breeding to purge breeding populations of deleterious alleles (by increasing the frequency of individuals expressing homozygous deleterious alleles and selectively removing them) and to increase genetic variance among inbred lines (Allard, 1960; Baker and Curnow, 1969; Hallauer and Miranda, 1981). Recognition that these positive attributes of inbreeding can mean increased genetic gain in the breeding population led Lindgren and Gregorius (1976) to recommend that tree breeders at least consider some form of directed inbreeding in advanced-generation tree improvement programs.

Some programs now employ intentional, directed inbreeding as part of the breeding strategy (Gullberg, 1993; White *et al.,* 1993; McKeand and Bridgwater, 1998; Kumar, 2004), and all of these programs use methods that ensure that inbreeding is confined to the breeding population and avoided in the propagation population. The exact methods and types of inbreeding employed are areas that need more research to maximize the benefits from directed inbreeding in forest tree improvement programs. Selfing of top selections leads to the most rapid increase in homozygosity, but has two potential disadvantages: (1) Loss of vigor and fecundity in selfed offspring making subsequent breeding more difficult; and (2) Overly rapid fixation of alleles at some loci before selection can act to fix favorable alleles. The latter means that some loci may become fixed for undesirable alleles meaning loss of the corresponding favorable alleles at those loci. Thus, some programs are employing various types and degrees of directed inbreeding as part of the breeding strategy, but it is a major focus in few, if any long-term tree improvement efforts.

Placing More Emphasis on Better Material

In advanced-generation breeding programs, each selection in the breeding population has a predicted breeding value based on its own performance and that of its relatives and ancestors in previous cycles of genetic tests (*Chapter 15*). Some selections have higher predicted genetic merit than others. At every stage in the program, the breeder should consider placing more emphasis on better selections if the objective is to maximize short-term genetic gains (*i.e.* from the first few generations of breeding). For programs aimed at maximizing economic returns, analyses have shown that the gains made in the first few cycles of breeding have an extraordinary impact on total returns from an infinite series of breeding cycles (*Chapter 11*). Thus for these programs, the first few cycles are critical, and short-term gains are paramount to overall long-term economic success.

Depending on the strategy and species' biology, placing more emphasis on better material could mean (Lindgren, 1986; Lindgren and Matheson, 1986; Lindgren *et al.,* 1989; Hodge and White, 1993; White *et al.,* 1993; Lstiburek *et al.,* 2004a): (1) Including progressively more relatives of increasingly better selections in the breeding population (*e.g.* eight or ten total relatives of various types for high ranking candidates and only one of low ranking candidates); (2) Stratifying the breeding population into tiers or hierarchies such that upper tiers contain better selections and are managed more intensively (see *Breeding Population Structure)*; (3) Including better selections in more crosses in mating designs aimed at producing the next generation's base population (from which the next generation's selections will be chosen); (4) Obtaining more precise breeding values on better selections by testing them (or their offspring) in more locations; (5) Including more ramets of the top clones in advanced-generation seed orchards; and (6) Deploying the best material (families or clones) to a larger area of operational plantations.

Placing more emphasis on better material in breeding programs exacerbates the problem of increasing relatedness discussed in the previous section, and may reduce long-term genetic gains (which depend on large effective population sizes). To counteract this disadvantage, lower ranking (but still desirable) selections can be maintained in the breeding population for purposes of gene conservation, providing flexibility and enhancing long-term gains. If these lower-ranking selections are bred and tested less intensively, then the associated costs are small. This is discussed further in *Mating Designs for Advanced-Generation Breeding*.

BREEDING POPULATION SIZE

Determination of breeding population size in advanced-generation programs is critical, because genetic diversity is the raw material for making genetic gains from breeding. Thus, achieving near-maximum genetic gain from a long-term breeding program depends in part on preservation and management of genetic diversity in the breeding population. Unfortunately, determining the appropriate size is quite subjective, because considerations based on population and quantitative genetic studies (reviewed below) provide only very broad guidelines (Kang, 1979a) and because the optimum starting size is a complex function of costs, desired genetic gain, genetic parameters (such as heritability) and other factors. This section (augmented from White, 1992): (1) Presents results from theoretical studies; (2) Considers factors that influence determination of population size; and (3) Summarizes breeding population sizes of some current advanced-generation breeding programs of forest trees.

It is first necessary to develop a quantitative measure of breeding population size so that populations of different sizes can be readily compared. Two terms were introduced in *Chapter 5* (see also Hallauer and Miranda, 1981, p 306; Namkoong *et al.,* 1988, p 60; Falconer and Mackay, 1996, p 59): (1) Census number, N, is the total number of selections retained in the breeding population in any given generation; and (2) Effective population size, N_e, is the theoretical number of selections when relatedness among members of the breeding population is considered. Another term, **status number,** N_s has been introduced more recently (Lindgren *et al.,* 1996; Gea *et al.,* 1997), and is defined as $N_s = 0.5/f$ where f is the average coancestry of the parents in the breeding population (Falconer and Mackay, 1996, p 85).

Status number is similar to effective population size in that both quantify the loss of genetic diversity that inevitably occurs in breeding programs with closed breeding populations. Breeding populations with related individuals have smaller effective sizes (measured by N_e or N_s) than their census number (N). Status number is preferable when several unrelated populations are considered (such as sublines described later). Both N_e and N_s can be interpreted as the size of a non-inbred population in which all selections are unrelated. Thus, if N_s is 150 and N is 300, the total number of selections in the breeding population is 300 (the census number), but the population is equivalent (in terms of its genetic diversity) to a population of 150 non-inbred, unrelated selections. The **relative effective size,** N_r, is defined as N_s/N, and in the above example, N_r is 0.5. Thus, due to build up of relatedness, this breeding population is functioning as a population half as large as its census number. In this chapter, we use the terms effective population size and status number interchangeably since they are similar conceptually.

Guidelines for Breeding Population Sizes from Theoretical Studies

Theoretical studies have examined the issue of appropriate size of the breeding population from a variety of perspectives. Assumptions often used in these studies are: (1) Additive gene action (no dominance or epistasis); (2) Phenotypic mass selection (no family and/or index selection); (3) One locus influencing a trait or many unlinked, equivalent and non-interacting loci (no epistasis); (4) Balanced mating systems (random mating among all members of the breeding population meaning no emphasis on better material); (5) Closed populations (no infusion of new material after any generation); (6) Identical mating designs and selection criteria for all generations; and (7) Selection for a single trait.

While these studies are based on simple models with somewhat unrealistic assumptions, they do provide broad guidelines for appropriate population sizes. The studies are grouped below according to different objectives: (1) Population size needed to maximize long-term genetic gain (obtained after many, often an infinite, number of generations of breeding); (2) Size needed to sustain appreciable short-term gain (for 5 to 10 generations); and (3) Size needed to maintain neutral alleles in the population to provide a broad genetic base for flexibility or gene conservation.

The ultimate, maximum cumulative gain from selection occurs after many generations of breeding and selection when all positive alleles at all influential gene loci are fixed (*i.e.* have a gene frequency of 1). Based on a mass selection model, Robertson (1960) showed that an infinitely large starting population is required to reach this ultimate selection limit, because otherwise some favorable alleles (especially those with low starting gene frequencies) are lost by chance along the way (*i.e.* due to genetic drift, *Chapter 6*). Also, the fraction of the maximum cumulative gain ultimately obtained is directly proportional to N_e; thus, from this perspective bigger is always better.

Based on a mass selection model with 100 unlinked, additive loci and $h^2 = 0.2$, Kang (1979a) examined the necessary initial population size needed to be 95% assured of fixing desirable alleles after an infinite number of generations of breeding and selection (Table 17.1). Larger population sizes are needed to ensure fixation of rare alleles and to ensure fixation at lower selection intensities. Very small population sizes are adequate for achieving fixation of common alleles or even those of intermediate frequencies (*e.g.* q = 0.5 requires N_e = 6 to 18). If the breeder is content with a high probability of fixing alleles with intermediate frequencies (q = 0.25 to 0.5), then initial effective population sizes of 10 to 40 are required. If the breeder desires to ultimately fix extremely rare alleles (q = 0.01), then initial effective population sizes of 300 to 1000 are required.

Baker and Curnow (1969) examined gain from mass selection based on a model with 150 unlinked, additive loci, a heritability of 0.2, and a selection intensity of 2.2 (Table 17.2). In terms of ultimate gain from many generations of selection, initial population sizes of 16 achieved approximately 50% (115/240) of the gain achieved by a breeding population of infinite size. Population sizes of 32 and 64 achieved approximately 75% and 92% of that achieved with an infinite population.

From these studies on ultimate gain from many generations of mass selection, it appears that initial effective population sizes of 20 to 60 are large enough to achieve 50 to 90% of the maximum cumulative response (*i.e.* attainable with an infinite population size) and will result in fixation of all but rare desirable alleles. The number of generations required to achieve half of this ultimate gain is approximately 1.4 N_e (Robertson, 1960; Bulmer, 1985, p 236); so for example, 28 generations of breeding are required (over 300 years for most tree species) to reach half of the ultimate gain for a breeding population with $N_e = 20$.

Gain after several generations of breeding (say 5 or 10, but not infinite as above) is also a useful criterion for determining appropriate population size. In a simulation study incorporating both family selection and positive assortative mating (*i.e.* preferentially mating best with best), a population of size 48 responded well to selection for 10 generations of selection (Mahalovich and Bridgwater, 1989). In the Baker and Curnow (1969) study (Table 17.2), an initial population of 32 resulted in 95% (16.8/17.6) and 91% as much expected gain as an infinitely large population after 5 and 10 generations, respectively. These gains are not necessarily achieved by any one population of a given size, but rather are expected gains when averaged over many populations of that size treated in the same fashion (Baker and Curnow, 1969; Nicholas, 1980).

Variance of response is greater among smaller populations due to genetic drift among the populations (Baker and Curnow, 1969). Nicholas (1980) used a mass selection model to examine the size needed to ensure that most of the gain predicted would actually be achieved. After five generations and for selection intensity of 1 and heritability of 0.2, his equation number 14 can be used to calculate that an initial effective population size of N_e = 50 is required to achieve 75% or more of the predicted response 90% of the time (*i.e.* 90% of the replicate populations treated in the same manner). A larger N_e = 328 is required to achieve at least 90% of the predicted response 90% of the time.

A final criterion for assessing appropriate population size is the need to maintain neutral alleles in a population throughout many generations of recurrent selection. Selection criteria may change over time due to changing target traits, environments, technologies, etc.; so what were once neutral alleles may become desirable in the future. Also, since gene conservation is an aim of most tree improvement programs, maintenance of neutral alleles is an important consideration. Kang (1979a) calculated that breeding population sizes of 20 and 50 would maintain uncommon neutral alleles (q = 0.2) in the population for 30 and 80 generations, respectively, with a probability of 0.95. For rare alleles (q = 0.01), sizes of 160 and 430 were needed to ensure (with probability of 95%) presence after 30 and 80 generations, respectively. Namkoong and Roberds (1982) suggested doubling these sizes to account for the possibility of linkage of neutral and undesirable alleles.

Table 17.1. Effective population size (N_e) required to be 95% assured of fixing a desirable allele after an infinite number of generations of mass selection for different levels of selection intensity, i (in standard deviation units, see *Chapter 6*), and different levels of initial frequency of the desirable allele, q. (Reproduced from Kang, 1979a, with permission of the Southern Forest Tree Improvement Committee)

Initial gene frequency (q)	Selection intensity (i)				
	2.67	2.06	1.76	1.27	0.80
0.01	281	364	426	590	937
0.05	56	73	85	118	187
0.10	28	36	43	59	94
0.25	11	15	17	24	38
0.50	6	7	8	12	18
0.75	3	4	5	6	10

Table 17.2. Total accumulated expected genetic gain (in simulation units) after 1, 5, 10 and infinite (∞) generations of mass selection for different values of N_e, the effective population size, and number of generations to achieve half of the ultimate gain (half-life). (Reproduced from Baker and Curnow, 1969, with permission of the Crop Science Society of America)

Size (N_e)	Number of generations				Half-life[a] (# gens)
	1	5	10	∞	
16	3.3	16.0	31.4	115	22
32	3.3	16.8	34.6	177	44
64	3.3	17.2	36.4	221	89
256	3.3	17.5	37.8	240	358
∞	3.3	17.6	38.0	240	--

[a] Calculated as 1.4 N_e (Robertson, 1960; Bulmer, 1985, p 236)

Further Considerations about Size of Breeding Populations

At least six factors not usually considered in the theoretical studies discussed above argue for starting with larger breeding population sizes than indicated by those studies. First, if there is substantial relatedness among selections in the initial advanced-generation breeding population, then effective population size (N_e) can be much less than the census number (N). Second, high selection intensities required to achieve large short-term gains reduce N_e, and may sacrifice long-term progress (Dempster, 1955; Robertson, 1960, 1961; Smith, 1969; James, 1972; Cockerham and Burrows, 1980; Askew and Burrows, 1983; Cotterill and Dean, 1990, Chapters 2 and 10). A selection intensity of i = 0.8 (equivalent to selection 1 out of 2 or 50% of the base population retained in the breeding population) is optimal for long-term progress (Dempster, 1955; Robertson, 1960, 1961; Cockerham and Burrows, 1980; Bulmer, 1985, p235), but this is clearly less intensive than most breeding programs.

Third, use of family or unrestricted index selection reduces N_e faster than the mass selection approach assumed in most theoretical studies (Robertson, 1960, 1961; Burdon and Shelbourne, 1971; Squillace, 1973; Askew and Burrows, 1983; Cotterill, 1984; Cotterill and Jackson, 1989; Cotterill and Dean, 1990, Chapter 10; Gea et al., 1997). Faster increase in relatedness results from the selection of multiple individuals from superior families and complete elimination of poor families. Fourth, placing more emphasis on better material (such as by crossing the better parents more often, Lindgren, 1986) reduces the effective population size faster than balanced mating designs (Kang and Namkoong, 1988; Mahalovich and Bridgwater, 1989; Lindgren et al., 1996; Gea et al., 1997). If plans are to achieve large early gains by these means, it may again be wise to start with a larger initial breeding population in order to ensure near maximum long-term gains.

Fifth, programs with many traits under selection (especially if some are unfavorably correlated), must either increase the size of the breeding population or settle for reduced gains on each trait. Other things equal, the gain achieved for any single trait declines as the number of traits in a program increases (*Chapter 13*). Finally, the structure of the breeding population affects the build up of relatedness, or conversely loss of genetic diversity (see *Structure of the Breeding Population*). In particular, structuring the breeding population into many sublines of fewer selections per subline is more effective at maintaining genetic diversity (higher N_e) than fewer sublines with more selections per subline (Lindgren et al., 1996; Gea et al., 1997; McKeand and Bridgwater, 1998). Of course, this comes at some sacrifice of short-term genetic gains.

Although each of the above considerations argues for starting with and maintaining a larger breeding population, infusion of unrelated material in later generations will increase effective size at the time of infusion (Zheng *et al.,* 1998). In this case, it may be appropriate to start with a smaller initial breeding population with plans to infuse material in the future.

As a final point of consideration about breeding population size, it is of interest to note the sizes being used by applied tree programs around the world (Table 17.3). While the list is not exhaustive, many programs have breeding populations with census numbers (N) on the order of 300 to 400. This size is large enough to support appreciable gains in a tree improvement program for many generations.

Table 17.3. Approximate census number (N) for breeding populations of some advanced-generation tree improvement programs. N is on a "per breeding unit" basis for programs with multiple breeding units.

Species	Program	N	Citation
Eucalyptus globulus	CELBI - Portugal	300	Cotterill *et al.,* 1989
	APM - Australia	300	Cameron *et al.,* 1989
Eucalyptus grandis	ARACRUZ - Brazil	400	Campinhos and Ikemori, 1989
Eucalyptus nitens	APM - Australia	300	Cameron *et al.,* 1989
Eucalyptus regnans	APM - Australia	300	Cameron *et al.,* 1989
	FRI - New Zealand	300	Cannon and Shelbourne, 1991
Eucalyptus urophylla	ARACRUZ - Brazil	400	Campinhos and Ikemori, 1989
Picea abies	Sweden	1000	Rosvall *et al.,* 1998
Picea glauca	Nova Scotia - CAN	450	Fowler, 1986
Picea mariana	New Brunswick	400	Fowler, 1987
Pinus banksiana	Lake States - USA	400	Kang, 1979a
	Manitoba - CAN	200	Klein, 1987
Pinus caribaea	QFS - Australia	250	Kanowski and Nikles, 1989
Pinus elliottii	CFGRP - USA 2nd	900	White *et al.,* 1993
	CFGRP - USA 3rd	360	White *et al.,* 2003
	WGFTIP - USA	800	Lowe and van Buijtenen, 1986
Pinus radiata	STBA - Australia	300	White *et al.,* 1999
	FRI - New Zealand	350	Shelbourne *et al.,* 1986
	NZRPBC - NZ	550	Jayawickrama and Carson, 2000
Pinus taeda	NCSU - USA	160	McKeand and Bridgwater, 1998
	WGFTIP - USA	800	Lowe and van Buijtenen, 1986
Populus tremuloides	Interior - CAN	150	Li, 1995
Pseudotsuga menziesii	BC - CAN	350	Heaman, 1986
Salix spp.	SLU - Sweden	200	Gullberg, 1993
Tsuga heterophylla	USA and CAN	150	King and Cartwright, 1995

Recommendations for Sizes of Breeding Populations

Taking all of the theoretical and practical considerations into account, the following conclusions and recommendations seem reasonable. First, breeding populations with effective sizes (N_e) of 20 to 40 will support selection and breeding programs with appreciable genetic gains for several generations. However, such small populations are subject to larger deviations of actual versus predicted progress and lead to more rapid build up of relatedness. Thus, populations of N_e = 20 to 40 are too small upon which to base entire long-term tree improvement programs, but are appropriate for elite populations being supported by a larger main population and for the number of selections in a subline or multiple population if there are many such groups comprising the entire breeding effort.

Second, breeding populations with a few hundred members, common in advanced-generation tree improvement programs (Table 17.3), are large enough to achieve near-maximum long-term gains even when several traits are selected and when intensive efforts are used to achieve large gains in the first few generations. Breeding population sizes of 300 to 400 also are consistent with guidelines and recommendations made by many other forest geneticists (Burdon and Shelbourne, 1971; Namkoong, 1984; Zobel and Talbert, 1984, p 420; Kang and Nienstaedt, 1987; Namkoong *et al.,* 1988, p 63; Lindgren *et al.,* 1997).

Third, quite large populations (1000 or so) may be required to ensure that rare alleles are maintained in the population for many generations. Thus, if a breeding population is also serving a gene conservation role or providing long-term flexibility, then these larger sizes seem warranted (*Chapter 10*).

BREEDING POPULATION STRUCTURE

Early discussions of advanced-generation breeding strategies focused on the use of a single, unstructured breeding population, where unstructured means that all members are chosen based on the same selection criteria and are considered available for breeding with all other members. More recently, reasons have surfaced for imposing structure on the breeding population, and there are several types of structure now used routinely (Fig. 17.2). For convenience in the following three subsections, we discuss each type separately. However, in real tree improvement programs, several different types of structure can be imposed simultaneously on a single breeding population. Therefore, the fourth subsection presents two examples of breeding population structures from intensive tree improvement programs to illustrate how various types of structures can be combined.

Structures that Promote Emphasis on Superior Material

In sheep breeding, the desire to focus maximum emphasis on top selections led to development of the nucleus breeding population structure (Jackson and Turner, 1972; James, 1977) which was subsequently adapted for use in forest tree improvement (Cameron *et al.,* 1988; Cotterill and Cameron, 1989; Cotterill, 1989; Cotterill *et al.,* 1989). The breeding population is subdivided into two groups (Fig. 17.2a): (1) The **nucleus** or **elite population** consisting of the highest ranking members of the breeding population (on the order of the 10% which is the top 30 selections if the breeding population has 300 selections); and (2) The larger **main population** containing the remaining say 90% of the members. More intense selection, breeding and testing on fewer, better selections in the elite population means more rapid genetic gains in the first few generations and the best of these can be propagated operationally. On the other hand, the purpose of the larger main population is to maintain genetic diversity to ensure long-term genetic gain, flexibility in the future, and genetic conservation. Relatives (such as two selections from the same family) may be allocated to either or both groups.

This subdivided structure of the breeding population has been employed in many tree improvement programs, and allows limited resources of a breeding program to be focused on the better genotypes in the elite population for more rapid progress in that segment (Cotterill and Cameron, 1989; Cotterill *et al.,* 1989; Mahalovich and Bridgwater, 1989; White *et al.,* 1993; McKeand and Bridgwater, 1998; White *et al.,* 1999; Jayawickrama and Carson, 2000; Lstiburek *et al.,* 2004a,b). This could mean: (1) Different mating designs (*e.g.* full-sib mating designs in the elite and polycross or open-pollinated mating in the main); (2) More crosses per member among elite selections; (3) Different genetic test designs and/or more extensive testing (and hence more precise rankings) for elite selections; (4) Directed inbreeding only in the elite; (5) Cloning of the progeny of elite crosses via vegetative propagation to improve the efficiency of within-family selection (see *Cloning the Base Population*); and (6) More rapid turnover of generations in the elite population to increase gain per unit time.

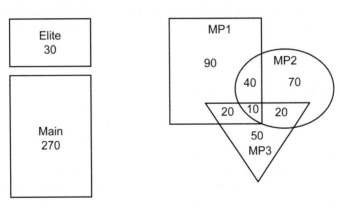

a Elite and main populations   **b Multiple populations**

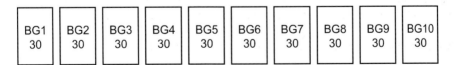

c Breeding groups or sublines

Fig. 17.2. Schematic representation of three types of structure that could be imposed on a hypothetical breeding population containing a census number of N = 300 selections: (a) Elite and main populations in which say 10% of the selections (30 out of 300) are included in the elite population to promote additional emphasis in breeding and testing; (b) Three multiple populations (MP1, MP2 and MP3) of different sizes and shapes to indicate that they are being bred for different purposes using three distinct sets of selection criteria (note that different numbers of selections are in common in two or three of the MPs); and (c) Ten breeding groups (BG1-BG10), synonymously called sublines, each containing 30 selections formed so that relatedness is completely contained within breeding groups.

Since selections in the elite population (or their offspring) form each generation's propagation population, increased emphasis on the elite population contributes directly to increasing realized genetic gain through operational deployment of improved propagules. That is, genetic gains in operational plantations are increased by focusing efforts on the best material used to form the propagation population.

If there is no interchange of selections between the elite and main populations for many cycles of breeding, two things can happen: (1) As a result of the increased efforts on the elite population, the mean breeding values of main and elite population begin to diverge with the elite becoming increasingly better than the main; and (2) Relatedness builds up in the smaller elite population reducing genetic diversity. Two-way gene flow between the two segments of the breeding population reduces the disparity and also maintains a larger effective population size in the elite population (Cotterill and Cameron, 1989; White *et al.*, 1999). Typically, 10 to 50% of the elite selections each generation could be the very best selections moved up from the main population, while 5 to 25% of the main members could be selections moved down from the elite population (Cotterill and Cameron, 1989; Cameron *et al.*, 1989; White, 1992).

The elite-main structure described above divides the breeding population into two discrete segments; however, the general concept of placing more emphasis on superior material does not necessarily lead to a structure with only two segments. In fact, in some situations (Lindgren, 1986; Lindgren and Matheson, 1986) it is better not to allocate selected individuals to discrete groups, but rather apply a continuous function that allocates progressively more emphasis (*e.g.* in breeding or testing) to superior material in linear proportion to increasing breeding value. However, this may be impractical to implement and may, in some instances, result in too rapid reduction in effective population size (Kang and Namkoong, 1988).

As an operational compromise, some programs have subdivided the selections of the breeding population into a hierarchy of three or four populations such that progressively better selections are allocated to groups at the top of the hierarchy. For example, a cooperative breeding program for *Pinus taeda* in the southeastern USA formed three subpopulations in each breeding unit (McKeand and Bridgwater, 1998): (1) Elite population with the top 40 selections; (2) Main population with 160 selections; and (3) Genetic diversity archives with 600 selections. Also in the southeastern USA (White *et al.*, 1993), the second-generation program for *P. elliottii* formed a four-tiered hierarchy of populations with progressively better selections and more intensive breeding and testing of the upper tiers (see *Examples of Breeding Population Structures*). In both of these cases, the lowest tier serves mainly for gene conservation and to sustain the genetic diversity in the breeding population required to achieve near-optimal long-term genetic gains. Many selections can be maintained in these "gene conservation" populations through low intensity breeding and testing efforts.

Multiple Populations

When the breeding population is subdivided for the purpose of applying different selection criteria to different segments, the subdivisions are called **multiple populations** (Fig. 17.2b) (Burdon and Namkoong, 1983; Namkoong *et al.*, 1988, p 71). For example, if there are three different product objectives (*e.g.* pulpwood, sawtimber and veneer) then selection in each of the three multiple populations will be for one of the objectives. Other instances for which use of multiple populations might be appropriate are (Namkoong, 1976; Burdon and Namkoong, 1983; Barnes *et al.*, 1984; Barnes, 1986; Namkoong *et al.*, 1988, p 72, 97;

Carson *et al.,* 1990; White, 1992; Jayawickrama and Carson, 2000): (1) There are many traits in the breeding program making it difficult to make gains for all traits in any single population, so each multiple population is bred for only a subset of traits; (2) There are unfavorable correlations between some traits, so the multiple populations are formed around groupings of compatible traits; (3) Economic weights for traits are uncertain and/or may change in the future, so multiple selection indices with different weights on the different traits are employed for different multiple populations as a hedge against changing conditions; and (4) There are different edaphoclimatic zones requiring distinct selection indices.

There may be few or many multiple populations (*e.g.* Barnes, 1986; Barnes and Mullin, 1989; Jayawickrama and Carson, 2000) and there is no specific control on relatedness between the multiple populations; such that, selections and their relatives may occur in more than one of the multiple populations. For example, if there is a negative correlation between growth and wood density and hence different populations for the two traits, there may still be "correlation breakers" that are good for both traits. These "correlation breakers" or their relatives may be assigned to both multiple populations as long as they meet the selection criteria for both populations (*e.g.* the 40 selections suitable for both MP1 and MP2 in Fig. 17.2b).

Sublines or Breeding Groups

As discussed previously, use of a single, unstructured breeding population leads to a buildup of relatedness among selections in the breeding population. When the breeding population is subdivided primarily to manage or control relatedness among members of the propagation population (and hence to avoid inbreeding depression in trees planted operationally), the subdivisions are called **breeding groups** or, synonymously, **sublines** (Fig. 17.2c) (van Buijtenen, 1976; Burdon *et al.,* 1977; van Buijtenen and Lowe, 1979; McKeand and Beineke, 1980; Burdon and Namkoong, 1983). Structuring the breeding population into breeding groups has been recommended for many tree improvement programs as an effective means of avoiding inbreeding depression in operational plantations (Kang, 1979a; McKeand and Beineke, 1980; Purnell and Kellison, 1983; Coggeshall and Beineke, 1986; Lowe and van Buijtenen, 1986; Carson *et al.,* 1990; White *et al.,* 1993; Baez and White, 1997; McKeand and Bridgwater, 1998; White *et al.,* 1999; Jayawickrama and Carson, 2000; White and Carson, 2004).

Implementing a breeding group structure involves initially assigning all members of the breeding population to a number of smaller breeding groups in such a way that all relatedness is contained within breeding groups. In this way, the different breeding groups are samples of the breeding population, and in the purest sense of the definition, all breeding groups are bred for the same objectives using the same selection criteria. Subsequently, all breeding is conducted among selections in the same breeding group, and each generation's selections are assigned to the same breeding group as were their progenitors. Over generations of breeding, relatedness builds up within each breeding group; however, members from different breeding groups are unrelated. Therefore, any mating between members from different breeding groups results in outcrossed progeny that do not suffer from inbreeding depression. Thus, these offspring are suitable for planting in operational plantations. So, by choosing selections from different breeding groups to form the propagation population, only outcrossed offspring are produced and planted operationally.

The appropriate number of breeding groups and the number of selections per breeding group depend both on the overall size of the breeding population and the type of production populations (*e.g.* wind-pollinated seed orchards *versus* family forestry in *Chapter 16*). Theoretically, more genetic gain is expected from use of one large compared to use of many small breeding populations (Madalena and Hill, 1972; Madalena and Robertson, 1975; Namkoong *et al.,* 1988, p 72; McKeand and Bridgwater, 1998). This would argue for employing the minimum number of breeding groups that are operationally feasible.

Compared to using fewer but larger breeding groups, having a large number of breeding groups each with fewer selections has the following implications (Baker and Curnow, 1969; Kang and Nienstaedt, 1987; Lindgren *et al.,* 1996; Gea *et al.,* 1997; McKeand and Bridgwater, 1998): (1) Less genetic gain for the first several generations of breeding; (2) More rapid build up of relatedness and inbreeding within each of the small groups; and (3) Maintenance of more genetic diversity in the population as a whole with most of the variability being distributed among groups.

Taking all of these factors into account, the minimum of two breeding groups is appropriate for programs employing family forestry (*Chapter 16*). In this case, half the breeding population is allocated to each breeding group, and all breeding and testing are conducted within each of the two breeding groups (see *Examples of Breeding Population Structures*). Then, the very best selections in each breeding group are used to form the propagation population, so that top selections from one breeding group are crossed with top selections from the other to produce the trees planted operationally. The full-sib seed can be planted directly in plantations or be used to form hedges to produce rooted cuttings planted operationally (*Chapter 16*). The important concept here is that, even after many generations, the full-sib families planted operationally are not inbred, because all breeding and testing is conducted within breeding groups and families used for deployment are made between groups.

A larger number of breeding groups is required for programs employing wind-pollinated seed orchards in order to ensure that a sufficient number of unrelated clones are available for orchard establishment after many generations of breeding. Most applied programs based on wind-pollinated seed orchards have opted for 10 to 30 breeding groups (van Buijtenen and Lowe, 1979; Kang, 1979a; McKeand and Beineke, 1980; Purnell and Kellison, 1983; White *et al.,* 1993; White *et al.,* 2003). However, a few have opted for a much larger number of groups with few selections per group (McKeand and Bridgwater, 1998). With 10 to 30 breeding groups, the number of selections within each breeding group often ranges between 20 to 40 (Burdon *et al.,* 1977; Namkoong, 1984; Kang, 1979a; van Buijtenen and Lowe, 1979; McKeand and Beineke, 1980; Coggeshall and Beineke, 1986; White *et al.,* 1993; White *et al.,* 2003). Considering the previous discussion of breeding population size, $N_e = 20$ to 40 allows good progress from breeding and selection within each breeding group for many generations. Also, these are often convenient sizes for logistics and management of breeding and testing.

Examples of Breeding Population Structures

As mentioned previously, many advanced-generation breeding programs employ two or all three types of structuring of the breeding population to accomplish all multiple objectives: place more emphasis on better material (*e.g.* elite and main populations); apply multiple sets of selection criteria (*i.e.* multiple populations); and manage inbreeding (*i.e.* breeding groups or sublines). This can make for structures that appear complicated until the purposes are described. Two examples of advanced-generation breeding population structures are described in Box 17.2 (*Pinus elliottii* in the US) and Box 17.3 (*P. radiata* in New Zealand).

MATING DESIGNS FOR ADVANCED-GENERATION BREEDING

Mating Designs used by advanced-generation tree improvement programs are described below and the activities associated with *Making Advanced-Generation Selections* are addressed in the final major section of this chapter. Definitions and details of implementing various mating and field designs are described in *Chapter 14*, and the principles of selection are presented in *Chapter 13*. In this chapter, we focus on issues and implications related to development and implementation of breeding strategies.

The primary constraint in the management of breeding populations of forest trees is the length of time required to complete an entire cycle of breeding, testing and selection which can be several years to decades depending upon the species. The need to achieve genetic gains as rapidly as possible to justify expenditures associated with tree improvement has led to the concept of maximizing genetic gain per unit time instead of per cycle of breeding. Strategies that produce the most gain per cycle, may not maximize gain per unit time if many additional years are required to achieve the gains.

The desire to maximize genetic gains per unit time has led to two major areas of research associated with the two primary means by which the time required to complete a cycle of improvement can be decreased: (1) **Early selection** which aims to make selections in field tests or artificial environments when trees are as young as possible (see *Optimum Selection Age*); and (2) **Accelerated breeding** whose goal is to shorten the time required to complete the breeding after selections are made (Box 17.4).

Once selections have been made and are flowering, some or all of those selections are included in the breeding population for that generation. These selections are interbred according to a mating design and the offspring planted in specific field designs (*Chapter 14*). Two primary functions of mating designs in advanced-generation breeding strategies are to: (1) Create the next generation's base population consisting of progeny produced by intermating trees in the current generation's breeding population (Fig. 11.1); and (2) Allow breeding value prediction for each of the selections in the breeding population and all of the progeny in the new base population to maximize genetic gain from making the next generation of selections.

In early years of tree improvement, many programs employed a single, balanced full-sib mating design for all breeding activities (the 6-parent, disconnected diallel described in *Chapter 14* was especially popular); however, more recently: (1) Some programs are employing open-pollinated management of some or all of the breeding population; (2) Complementary mating designs are sometimes used in which different mating and field designs are employed for different purposes (*e.g.* for predicting breeding values *versus* making selections); (3) Unbalanced mating designs are being advocated that place more emphasis on superior selections by crossing them more frequently; (4) Rolling front mating designs are gaining popularity to maximize gain per unit time by breeding selections as they become available rather than in discrete generations; and (5) Different mating and field designs are being used on distinct segments of the population (*e.g.* different designs for the elite compared to the main population).

All of these new developments are important and have a role in advanced-generation breeding strategies of some programs. Each program must evaluate the range of mating designs available in the context of the two primary functions of breeding mentioned above, the species' biology and the operational logistics of the program. Thus, the discussion of different mating designs below emphasizes the potential advantages and disadvantages of each design and their potential roles in advanced-generation breeding.

Box 17.2. Structured breeding population of *Pinus elliottii* in southern USA.

The Cooperative Forest Genetics Research Program (CFGRP) of *Pinus elliottii* in the southeastern USA formed a second-generation breeding population to meet three objectives (White *et al.,* 1993; White and Carson, 2004): (1) Maximize short-term genetic gain (first few cycles of breeding) in both breeding and propagation populations; (2) Provide for near-maximum long-term genetic progress; and (3) Maintain a broad genetic base for gene conservation and flexibility. To achieve objectives 2 and 3, the CFGRP formed a large breeding population with N = 936 and N_e = 625. While the CFGRP employed a slightly different population size and structure for the subsequent third-generation of breeding (White *et al.,* 2003), most of the concepts and objectives are similar to the second generation.

To achieve all three objectives in a cost-effective manner, the CFGRP subdivided the large breeding population into 4 hierarchical tiers or strata (Fig. 1). The top 60 selections compose an elite population managed very intensively to achieve rapid near-term genetic gains. The 936 selections in the main population (which includes the 60 elite selections that are in both the elite and main populations) were divided into three equal-sized strata (I = top, II = middle, and III = bottom) based on predicted breeding values. Breeding, selection and testing are progressively more intensive the higher the stratum. In fact, the selections in stratum III are mainly managed for maintenance of genetic diversity.

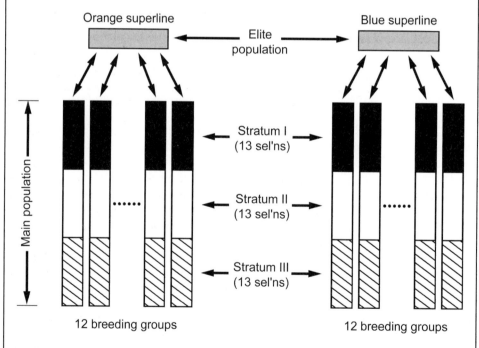

Fig. 1. Schematic diagram of the structured, second-cycle breeding population of *Pinus elliottii* being managed by the Cooperative Forest Genetics Research Program. (Reproduced from White *et al.*, 1993, with permission of J.D. Sauerländer's Verlag)

(Box 17.2 continued on next page)

Box 17.2. Structured breeding population of *Pinus elliottii* in southern USA. (Continued from previous page)

> To manage inbreeding in the propagation population, the 936 selections in the main population were subdivided into 24 breeding groups with 39 selections per breeding group. This relatively large number of breeding groups was employed because some members of CFGRP rely upon wind-pollinated clonal seed orchards as the primary type of propagation population. All relatedness and full-sib breeding in the main population were confined within breeding groups. Each breeding group had 13 clones in each of the three strata (top, middle and bottom third) and all breeding groups were managed and bred similarly with the same selection criteria (they were not multiple populations).
>
> Superimposed upon the breeding group structure were two superlines to serve for family and clonal forestry (*Chapter 16*). Twelve breeding groups were nested within each superline with the top segment in each superline (called the elite population) being comprised of the very best material from each of the 12 breeding groups. The elite population in each superline was managed quite intensively including crosses among selections in different breeding groups within that superline to mate the best with the best. CFGRP members employing family forestry can intermate the top members of the orange elite with those from the blue elite to create full-sib families for operational deployment. These deployment crosses are always free of inbreeding depression since the two superlines are unrelated.

Open-pollinated (OP) Management of the Breeding Population

With OP breeding: (1) OP families are planted in genetic tests to form the base population of generation *k*; (2) New selections comprising the best trees in the best families are made in these tests at an appropriate age; (3) OP seed is collected from these selections still growing in the genetic tests; (4) This OP seed is used to establish another set of genetic tests that are the base population of generation *k* + *1*, thus completing a complete cycle of breeding; and (5) After selection, one or more of the tests in generation *k* may be converted into seedling seed orchards.

OP breeding can be used as the only mating design to manage the entire breeding population or only a portion of the breeding population (such as for the main population in a breeding program in which the elite population is managed through a full-sib design requiring controlled pollinations). An example of an entire tree improvement program based on OP management and establishment of a single genetic planting each generation is the *E. grandis* program in south Florida (Box 17.5).

There are two aspects of implementing OP breeding strategies worth noting. First, as with all breeding designs, it is important to minimize the time between selection and seed collection. Since the OP seed is collected from selections made in genetic tests, it is sometimes useful once the selections are identified to fertilize them and to remove neighboring trees to increase sunlight to promote flowering and seed production. Partial girdling of selected trees is also used sometimes to promote earlier flowering.

Second, the genetic quality of the pollen cloud is important for maximizing genetic gain from OP breeding, and this has three aspects: (1) The genetic tests should be located far from unfavorable contaminating pollen sources (*Chapter 16*); (2) The genetic tests containing selections can be thinned prior to open pollination of the selections to remove non-selected trees (especially those with the lowest breeding values that would reduce

Box 17.3. Structured breeding population of *Pinus radiata* in New Zealand.

The New Zealand Radiata Pine Breeding Cooperative (NZRPBC) established a large breeding population of *Pinus radiata* (N = 550, N_e = 400) with similar objectives as those described for the CFGRP in Box 17.2 (short-term gains, long-term gains and genetic diversity). The structured breeding population (Fig. 1) combined features of all three structure types (Carson *et al.,* 1990; Jayawickrama and Carson, 2000; White and Carson, 2004). First, the entire population was divided in half into two breeding groups (called superlines) such that no relatedness existed between superlines. Each superline was then further divided into a larger main population, and several smaller elite breeding populations, called **breeds**. The functions of the main population are to maintain broad genetic diversity and identify excellent selections to use in subsequent breeding cycles in one or more breeds. There are also 1300 clones grafted in genetic diversity archives for gene conservation purposes (not shown in Fig. 1).

The breeds combine the multiple population and elite population concepts, because they are smaller populations bred very intensively (as are elite populations) and each breed employs different selection criteria for different end-use product objectives (*e.g.* a growth and form breed and a high wood density breed). Within a superline, the same selection is included in more than one breed if it meets the stringent and different selection criteria for those breeds (shown as overlap of breeds in Fig. 1). So the breeds within a superline do not serve as unrelated breeding groups. Rather, the superlines are unrelated and the same set of breeds is being developed in parallel in both superlines. This allows directed crossing to produce outcrossed, full-sib families for operational plantations for each breed (*i.e.* breed-specific family forestry). For example, even after several cycles of breeding, top members of the HWD breed from superline 1 can be crossed with top members of the HWD breed of superline 2, and these full-sib families are free of inbreeding depression and suitable for operational deployment. Alternatively, top selections from the HWD breed can be crossed to top selections in the growth and form breed of the other superline to create complementary families intermediate for the various traits.

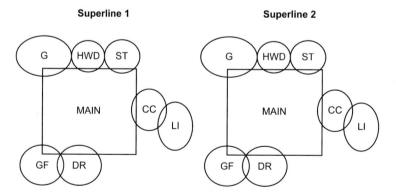

Fig. 1. Schematic diagram of the breeding population of *Pinus radiata* being managed by the New Zealand Radiata Pine Breeding Cooperative. Each superline is composed of a large main population with more than 100 selections and several elite populations that they call breeds (each with 10 to 30 selections) that are managed intensively for different selection objectives: CC = clear cuttings; DR = *Dothistroma* resistance; G = general; GF = Growth and form; HWD = high wood density; LI = Long internode; and ST = Structural timbers. (Reproduced from Jayawickrama and Carlson, 2000, with permission of J.D. Sauerländer's Verlag)

Box 17.4. Accelerated breeding techniques for promoting early flowering.

Some species of forest trees flower in only a year or so after being established through grafting or rooted cuttings into a breeding arboretum. More commonly there is a delay of at least a few years, and some species may not flower for more than a decade. This has led to much research to develop accelerated breeding techniques aimed at reducing the number of years between selection and breeding (Greenwood, 1983; Greenwood *et al.*, 1987). The various techniques outlined below are successful with different species, and several techniques are sometimes combined on the same species.

1. *Breeding arboretum management.* When selections are established in a breeding arboretum or clone bank located in a field or forest location, the goal is to promote full tree crowns and rapid early growth so that trees reach flowering size as soon as possible. Techniques employed include wide spacing (*e.g.* 5 m x 5 m or wider), fertilization, irrigation, and sub-soiling (Barnes and Bengtson, 1968; Ebell and McMullan, 1970; Greenwood, 1977; Gregory *et al.*, 1982; Wheeler and Bramlett, 1991).

2. *Greenhouse breeding arboreta.* For some species, selections are established in pots and after some time moved into a greenhouse to promote flowering (often in combination with other treatments). This technique has been employed successfully with several species including the southern pines of the USA (Greenwood *et al.*, 1979; Greenwood, 1983; Bower *et al.*, 1986).

3. *Hormones and chemicals.* Gibberellins sometimes promote flowering with varying degrees of success in conifers (Hall, 1988; Ho, 1988a,b; Almqvist and Ekberg, 2001). In angiosperms, paclobutrazol is often effective at promoting flowering one year after application (Griffin, 1993).

4. *Stress treatments.* Girdling, drought and other treatments that stress trees are sometimes successful at promoting early flowering (Cade and Hsin, 1977). With these techniques, the goal is to stress the tree to the point of flower production without killing it.

5. *Top grafting.* Grafting scion from advanced-generation selections onto branches located in the crowns of mature, flowering trees has successfully promoted early flowering in some pines (Bramlett and Burris, 1995; Bramlett, 1997; McKeand and Raley, 2000; Almqvist and Ekberg, 2001; Lott *et al.*, 2003).

---

genetic gains if they contribute to the pollen cloud); and (3) Siblings of selected trees should be removed from the genetic tests before pollination to avoid inbred seed. Instead of removal of relatives of selected trees, some programs design the genetic tests initially so that the distance between siblings is maximized; computer programs are available for the purpose of randomization with this constraint (Cannon and Shelbourne, 1993; Cannon and Low, 1994).

Recently, breeding strategies utilizing OP management of the breeding population have been proposed for some programs based on (see more detailed review by White, 1996): (1) Theoretical genetic gain calculations indicating favorable gains per unit time (compared to full-sib designs) due to rapid turnover of breeding cycles (Cotterill, 1986;

Box 17.5. Breeding of *Eucalyptus grandis* based on open pollination.

A tree improvement program for *E. grandis* began in south Florida in the 1960s (Meskimen, 1983; Franklin, 1986; Reddy and Rockwood, 1989; Rockwood *et al.*, 1989; White and Rockwood, 1993). The species is planted only on a very small scale in the region, so the tree improvement program necessarily adopted a low-intensity, low-cost breeding strategy. The program has completed more than four generations of breeding using only open-pollinated (OP) management of the breeding population, and has made substantial genetic gains for the three traits under selection: bole volume growth, frost resilience and coppice re-growth volume after hard freezes (Table 1).

Table 1. Realized gains from four generations of the *Eucalyptus grandis* tree improvement program in south Florida. Bole volume at 2.5 years is for seedling growth (Franklin, 1986), while the other two traits are for coppice growth after harvest (Reddy and Rockwood, 1989). The first-generation population was poorly adapted to the Florida climate and soils so that genetic gains in Generation 1 are small compared to later generations. Frost resilience was scored from 0 to 3 with 0 indicating no damage. Values in parentheses indicate percentage genetic gain above the first generation. (Reproduced from White and Rockwood, 1993, with permission of the authors)

Generation of breeding	Bole volume at 2.5 years (dm^3)	Coppice bole volume at 5 years (dm^3)	Frost resilience at 5 years
1	7.5	7.0	1.75
2	14.6 (95)	21.5 (207)	1.39 (20)
3	17.0 (127)	25.9 (270)	1.34 (23)
4	19.7 (163)	40.2 (474)	1.25 (29)

Key features of this breeding strategy (Fig. 1) are: (1) A single field location each generation which serves four functions to reduce costs (genetic test to rank families, base population in which selections are made, breeding arboretum to provide OP seed for the next generation and propagation population for collection of improved seed for operational deployment); (2) A single, unstructured breeding population managed with open-pollinated breeding to obviate the need for control pollinations; (3) A short generation time of approximately four years involving early selection at 2 years (25% of the 8-year rotation) followed by collection of OP seed for the next generation two years later; and (4) Substantial infusion of new material from external programs each generation to maintain a genetically-diverse breeding population.

Each generation, called generation *k*, begins with formation of a base population comprised of 500 OP families selected as follows: (1) 300 to 400 selections are made in the OP tests of the previous generation; and (2) 100 to 200 new selections (infusions) are obtained from external sources to maintain a broad genetic base and limit the buildup of inbreeding. This base population of 500 OP families is planted on a single site location in south Florida (year 0, generation k in Fig. 1). The 15-ha site contains 60 seedlings of each of the 500 OP families for a total 30,000 seedlings planted in a completely randomized design with single tree plots (pictured in Fig. 1, Box 16.2). (Box 17.5 continued on next page)

Box 17.5. Breeding of *Eucalyptus grandis* based on open pollination. (Continued from previous page)

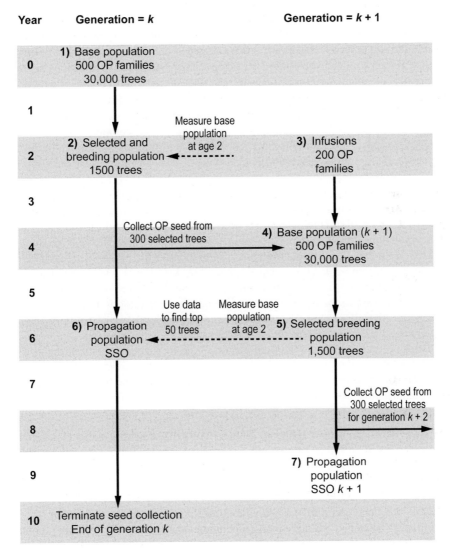

Fig. 1. Breeding strategy for two generations (labeled k and $k + 1$) of the *Eucalyptus grandis* program in south Florida.

The single planting first serves as the base population, and is next used as the only genetic test in generation k. At age 2 years (with mean height of 5m), the 30,000 trees are measured for several traits. The rankings of the 500 families and the individual performance of trees within families are used to select 1500 trees to form the selected population of generation k. The process retains at least one tree from each of 400 of the original 500 families, and up to 10 trees are retained from the top families.

(Box 17.5 continued on next page)

Box 17.5. Breeding of *Eucalyptus grandis* based on open pollination. (Continued from previous page)

The breeding population is formed by cutting and removing the 28,500 trees not in the selected population. Thus, the breeding and selected populations are the same and consist of 1500 trees selected at age 2 (year 2, generation k in Fig. 1). Open pollination is allowed to occur as the breeding method, and, at age 4, 300 new selections are made. OP seed is collected from these 300 trees (out of 1500 present) and combined with 200 new infusions to form the next generation's base population (generation *k + 1*). The 500 OP families are planted in a second field location (year 4, generation k + 1 in Fig. 1) adjacent to the single field planting of generation *k*. This completes a generation of selection, testing and breeding in four years. The *k + 1* generation proceeds in the same fashion as the previous generation and by year 8, two generations of breeding have been completed.

The field planting in generation k has one remaining function which is to serve as a seedling seed orchard for production of open-pollinated seed for deployment to operational plantations. Predicted breeding values from a second measurement of the base population (now 4 years old) combined with the data from the first measurement at age 2 of the *k + 1* base population are used to identify the best 50 seed orchard trees for operational seed collection (year 6, generation k in Fig. 1 and pictured in Fig. 1, Box 16.2). All 1500 trees are left to grow in the seedling seed orchard to maintain a broad genetic base in the pollen cloud and minimize inbreeding in the seed deployed to operational plantations. Seed is collected from this orchard from years 6 through 10, when the orchard from generation *k* + 1 begins commercial production.

Thus, in each generation the field planting lasts for 10 years, first serving as a base population and genetic test, then being converted into a breeding arboretum at age 4, and lastly into a seedling seed orchard at age 6. At any one time there are two adjacent field plantings that are in different stages and serving different functions.

Shelbourne, 1991; King and Johnson, 1996); (2) Realized genetic gains evidenced from four generations of an *Eucalyptus grandis* breeding program (Table 1, Box 17.5; Franklin, 1986; Rockwood *et al.,* 1989); and (3) Logistical ease of breeding when control pollinations are too costly or logistically infeasible (McKeand and Beineke, 1980; Purnell and Kellison, 1983; Griffin, 1982).

There are many programs worldwide utilizing OP breeding either for the entire breeding population or for the main population segment of the breeding population (Purnell and Kellison, 1983, in the USA; Cannon and Shelbourne, 1993, in New Zealand; Ampie and Ravensbeck, 1994, in Nicaragua; Osorio *et al.,* 1995, in Colombia; Barnes *et al.,* 1995, in Zimbabwe; Dvorak and Donahue, 1992, in the CAMCORE program; Baez and White, 1997, in Argentina).

The success of OP management of the breeding population depends at least in part on how closely OP families; approximate half-sib (HS) families. OP families can differ from HS families for several reasons (most importantly, when there are selfs or other inbred trees in OP families; see review by Sorensen and White, 1988), and these problems are likely worse in angiosperm species with animal- or insect-mediated pollination mechanisms (*e.g.* Hodge *et al.,* 1996). The array of potential violations of the assumption that OP families are true HS families is enhanced by variation among trees within the same species in the type and extent of violations. So for example, if there are

100 OP families they likely vary in the number of inbred trees and amount of inbreeding depression experienced.

These violations can reduce the precision and accuracy of the breeding values predicted for both parents and progeny generated by OP breeding (one of the two primary functions of mating designs introduced above), and result in reduced genetic gains from selection if predicted rankings differ from true rankings (Hodge *et al.,* 1996). While data analysis techniques may help ameliorate the problem (White, 1996), some programs have opted to use polymix breeding (*Chapter 14*) to avoid potential problems with OP families (Li, 1995; White *et al.,* 1999).

Another criticism of OP breeding is the loss of full knowledge of pedigree (since the paternal parent is not known for any selection). This raises concerns about avoiding inbreeding in the propagation population, since conceivably two selections producing trees for operational plantations could be related. This problem can be overcome by subdividing the breeding population into breeding groups.

While there are problems with OP breeding designs, the advantages outweigh these disadvantages for some programs and this explains the increasing popularity cited above. OP breeding designs seem especially appropriate for: (1) Minor species requiring low cost breeding strategies; (2) Tropical hardwood species that are difficult to control pollinate or for which flowering biology is poorly understood; (3) Programs managing several species that must minimize costs and efforts on all species; and (4) Complementary management of the main population when the elite population is managed with full-sib designs.

Full-sib (FS) and Complementary Mating Designs for Managing the Breeding Population

The steps for a complete cycle of breeding using full-sib mating designs are (refer to Fig. 11.1): (1) Selections in the breeding population of generation k are intermated using control pollination (CP) to form FS families; (2) These FS families are planted in genetic tests that form the base population of generation $k + 1$; (3) Selections of superior trees in superior families (based on performance in the FS tests) are made to form the selected and breeding populations of generation $k + 1$; and (4) These selections are intermated to create FS families that are planted to establish the base population of generation $k + 2$. FS breeding can be employed for the entire breeding population or for only the elite population when the main population is managed with OP or polymix breeding. Also, FS designs can be employed such that all breeding is conducted within breeding groups to manage inbreeding (*e.g.* McKeand and Bridgwater, 1998; Jayawickrama and Carson, 2000; White *et al.,* 2003).

FS breeding is sometimes preferable to OP breeding, because (McKeand and Bridgwater, 1998): (1) Full-control of pedigree facilitates management of co-ancestry and inbreeding; (2) Higher genetic gains per cycle can be achieved since there can be selection on both the maternal and paternal side of the pedigree; and (3) It is possible to identify FS families that have positive specific combining ability (SCA). Additional genetic gain is realized if FS families with positive SCA are deployed operationally as part of the propagation population. However, these additional gains are not captured in the breeding population, since only additive genetic effects associated with general combining ability (GCA) are accumulated in recurrent selection programs for GCA.

A key issue concerning implementation of FS breeding is the delay between making the selections in the genetic tests and conducting the control pollinations. For many reasons, it is usually not feasible to make the CP crosses on the selections growing in the genetic tests. So, the selections are often propagated vegetatively, through grafting or rooting, into some type of a breeding arboretum, clone bank or other breeding facility. Then, a variety of methods is used to stimulate flowering so that the control pollinations can be made as soon as possible (Box 17.4).

In the past, most tree improvement programs employed **balanced mating designs** in which all selections in the breeding population are crossed the same number of times with other selections; thus, all selections theoretically contribute the same number of progeny to the next generation's base population. A variety of balanced mating designs have been used, and all are appropriate if a sufficient number of unrelated families are created to ensure maintenance of adequate genetic diversity in the base population of progeny produced (*Chapter 14*). In fact under some assumptions, all balanced mating systems result in the same long-term gain for a fixed effective size of the breeding population (Kang and Namkoong, 1979).

With balanced FS mating designs, genetic gain is enhanced by making more crosses. This means that **random single pair matings** in which each selection is crossed only once result in less gain than designs in which each selection is mated more frequently. The reason is that after the FS genetic tests have been analyzed and breeding values predicted, there is a higher probability of finding outstanding FS families when more FS families are tested. Theoretical studies show that a point of diminishing returns is reached such that little extra gain is achieved if each selection is crossed more than 3 to 5 times (van Buijtenen and Burdon, 1990; King and Johnson, 1996; Gea *et al.*, 1997).

Unbalanced mating designs cross some parents more often than others so that favored selections contribute more progeny to the next generation's base population. Unbalanced mating designs that cross superior selections more frequently increase the probability of maintaining and concentrating superior alleles in future selections. This means more genetic gain is achieved from use of these unbalanced mating designs (van Buijtenen and Burdon, 1990; King and Johnson, 1996; Gea *et al.*, 1997; Lstiburek *et al.*, 2004b). As an example, a second-generation breeding strategy for *Pinus elliottii* called for selections in the top third of the breeding population (stratum I) to be crossed to eight other members, while those in the bottom third (stratum III) were crossed only once and only to a selection in stratum I (Box 17.2). As with most activities that increase genetic gain, using unbalanced mating designs also results in more rapid loss of genetic diversity in the breeding population. Thus, unbalanced mating systems generally lead to less ultimate long-term gain for a fixed N_e (Kang and Namkoong, 1980; Namkoong *et al.*, 1988). Programs employing these designs can maintain sufficient population diversity by starting with larger breeding populations or by establishing some populations managed less intensively for genetic diversity.

As described in *Chapter 14*, complementary mating designs are employed in some advanced-generation programs because it is difficult to find a single mating design that efficiently achieves all objectives. As one example, the second-generation program of *P. elliottii* mentioned above and described in Box 17.2 employed polymix designs for ranking selections in both the main and elite populations, and unbalanced FS designs to create progeny to establish a third-generation base population in which to make selections.

All FS designs can be implemented in either discrete cycles or **rolling front** designs (*Chapter 14*). When discrete cycles are used, all FS families planned for a given generation of breeding are planted in a single series of genetic tests. This means that the tests cannot be planted until seed from all FS families is available. For some species this can mean a delay of several years waiting for seed from selections that are slow to flower. Rolling front designs plant several series of genetic tests over many years using all families for which seed is available in a given year. Under some circumstances, rolling front designs can produce more genetic gain than using discrete generations (Borralho and Dutkowski, 1996, 1998; Araujo *et al.*, 1996).

When employing FS mating designs in either discrete generations or rolling front designs, it is often desirable to ensure genetic connectedness among the crosses made (*Chapter 15*). This connectedness facilitates precise prediction of breeding values of all parents through best linear unbiased prediction (BLUP, *Chapter 15*). The systematic partial diallel called a circular mating design (Fig. 14.7) is useful for this purpose. For example, if there are 30 parents in a breeding group and each parent is crossed with four others, there are 60 total crosses created. Additional crosses can be made among the better parents scattered throughout the diallel to create an unbalanced mating design, and this increases the precision of the BLUP-predicted breeding values for all 30 parents.

Regardless of which mating design is used (OP or one of the many FS mating designs), the offspring are planted into genetic tests in the field. Details of various field designs are presented in *Chapter 14*. Randomized complete blocks or incomplete block designs planted in single tree plots are recommended for all genetic tests aimed at ranking parents or families (*i.e.* for predicting parental breeding values for use in backward selection, see Box 14.1). Further, these field designs are robust and also serve well for making forward selections (*i.e.* for predicting breeding values of individual trees and making selections of superior trees growing in the genetic tests).

MAKING ADVANCED-GENERATION SELECTIONS

Within-family Selection

In the first generation or two of a tree improvement program, it is common to eliminate large numbers of inferior parents (or families) from contributing to future breeding populations based on poor performance of their progeny (or siblings) in genetic tests. As an oversimplified example, suppose a first-generation breeding program starts with 300 unrelated parents and eliminates the bottom half of the parents and their progeny based on their predicted breeding values. In this example, second-generation selections come from families in the top half of the ranking, and this among-family selection (also called **family selection**) results in substantial genetic gain.

As with most activities aimed at increasing short-term genetic gains, family selection also results in more rapid buildup of relatedness in the breeding population. In the above example, such intense selection among families could not continue for several generations without a dramatic reduction in the effective population size of the breeding population. Therefore, in advanced-generation breeding there is tendency to maintain representation from more families and increase the emphasis on **within-family selection** meaning the selection of the best individuals within a family.

The above discussion is oversimplified, because during the selection process in advanced-generation programs, selection indices are commonly used that combine data across relatives and traits (see *Selection Indices and Other Methods of Selection*). This leads to **combined selection** that weights family and individual performance appropriately to predict a breeding value for each parent and tree in genetic tests. Thus, we do not usually speak of family *versus* within-family selection. Still, the concept is valid that in advanced-generation breeding programs there is a tendency to increase emphasis on within-family selection instead of eliminating large numbers of inferior families, as a mechanism to slow the build up of relatedness in the breeding population (see *Balancing Genetic Gain and Genetic Diversity*).

For these reasons, it is instructive to understand the factors affecting genetic gain from within-family selection to judge how many individuals of a given family to establish in genetic tests or in unreplicated family plots. Fig. 17.3 shows the additive genetic gains expected from within-family selection within half-sib families for a hypothetical trait as a function of heritability (Fig. 17.3a) and number of trees per family (Fig. 17.3b). This figure is appropriate for both within-family selection from seedling populations, as discussed here, and for cloned breeding populations discussed next. The methods used to calculate these predicted genetic gains are described in Box 17.6.

a Selection of the top clone from c * n = 100 trees

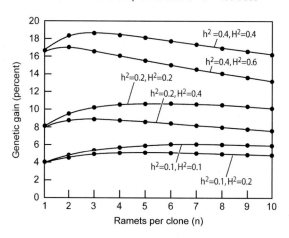

b Selection of the top 3 clones from various family sizes
($h^2 = 0.2$, $H^2 = 0.4$)

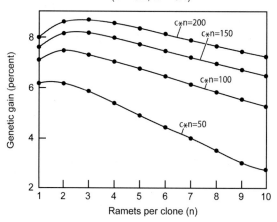

Fig. 17.3. Genetic gain (%) from within-family selection in half-sib families from a cloned breeding population: (a) Selection of the top clone from families having 100 total trees such that c*n = 100 where c is the number of clones per family and n is the number of ramets per clone; and (b) Selection of the top three clones from plot sizes ranging from c*n = 50 to c*n = 200. Note that when n = 1, there is only one ramet per clone, and the graphs show genetic gains from forward selection within an uncloned, seedling half-sib family with c individuals. The gains are for a hypothetical trait, and should be used only for comparative purposes. The trees could be planted in a single, unreplicated plot or scattered across replications and site locations. Details of assumptions are in Box 17.6.

Figure 17.3 shows only genetic gain from additive genetic effects (those effects associated with breeding values and general combining abilities, *Chapter 6*); these are the effects captured in breeding populations under recurrent selection for GCA. Non-additive genetic effects are captured each generation in some types of propagation populations (*e.g.* family forestry and clonal forestry, *Chapter 16*); so more gain is expected from selection of operational clones than is shown in Fig. 17.3. However, these additional gains from non-additive genetic effects are not accumulated in the breeding population in subsequent generations.

For test populations planted as seedlings (seedling population), each tree is a different genotype, and hence a different clone; thus, there is one ramet per clone ($n = 1$) and the reader should focus on $n = 1$ on the *x*-axis. First, gain from within-family selection is nearly directly proportional to the narrow sense heritability, h^2, as can be seen by noting that gain doubles from 4 to approximately 8% (for selection of the best seedling out of 100 per family) as h^2 doubles from 0.1 to 0.2. The same doubling in gain occurs from 8 to 16% as h^2 doubles from 0.2 to 0.4. This indicates that within-family selection is more effective for traits with higher heritabilities (such as wood density), than with lower heritabilities (such as growth traits).

Second, the amount of non-additive genetic variance does not affect expected gain from within-family selection from genetic tests established with seedlings. This is seen by noting that for given size of h^2 in Fig. 17.3a (0.1, 0.2 or 0.4), the size of the broad sense heritability does not influence genetic gain (for $n = 1$ which applies to seedling populations). This is not surprising since only additive genetic gain (Equations 11.2 and 13.6) associated with breeding values and average allele effects are accumulated in the breeding population.

Finally, the expected gain from within-family selection increases with the number of trees planted per family. However, the gains are not linearly related to the number of trees per family; rather, the amount by which gain increases diminishes with increasing numbers of trees per family. For example, when the top three trees are selected from 50 planted per family, the gain is 6.2% (Fig. 17.3b with $n = 1$ on the *x*-axis and c*n = 50). When the number of trees per family is doubled to 100, gain increases by 1% (from 6.2 to 7.2%). However, another doubling of family size (from 100 to 200), results in only an additional 0.7% gain (from 7.2 to 7.9%).

Taking all these factors into consideration, family sizes between 50 and 150 seem appropriate for balancing gains and costs. All seedlings in a family might be planted in a single, unreplicated plot if replicated OP or polymix trials are also being established to rank parents as part of a complementary mating design. Conversely, the total number of seedlings could be distributed across many replications and site locations. For smaller programs focusing on only one trait, 50 seedlings per family seem appropriate. When there are several traits involved, the selection intensity for each trait is reduced and more trees per family are warranted. Finally, programs placing more emphasis on better families can plant more trees per family of better families and fewer progeny of families formed by mating less superior selections in the breeding population.

Cloning the Base Population

The weak link in achieving gains from within-family selection in seedling populations as described in the previous section is that there is only one copy (*i.e.* one tree) of each seedling. Therefore, choosing the best seedling out of say 100 trees within a family is made difficult by the fact that the 100 seedlings are planted in 100 different microsites and selection is based solely on phenotypic appearance.

Genetic gains from within-family selection can be increased by **cloning the base population** which entails use of vegetative propagation (such as rooted cuttings) to make multiple copies of each seedling (Shelbourne, 1991; Rosvall *et al.*, 1998). Gain is increased, because each genotype (*i.e.* each cloned seedling from a family) is planted as multiple ramets on several microsites, and the clone mean upon which selection is based tends to average out the environmental noise associated with different microsites to provide a better estimate of the true underlying genotype of each seedling clone. Genetic gains from within-family selection from cloned base populations can be predicted using the methods described in Box 17.6.

The simplest field design for a cloned base population is to establish each family in its own unreplicated plot containing *c* clones per family (each different clone originates from a different seedling of that family) and *n* ramets per clone. The *n* ramets of each seedling

---

Box 17.6. Calculating genetic gains in the breeding population from within-family selection.

Calculating genetic gains from different types of within-family selection allows comparisons of options, such as seedling versus cloned base populations with various numbers of trees planted per family. Only additive genetic gains accrue to the breeding population (*Chapter 6*), and additive genetic gains from within-family selection are shown in Fig. 17.3 using the methods outlined in Shelbourne (1991) for hypothetical traits having different heritabilities and different amounts of additive and non-additive genetic variation. The gains shown are for selection from within half-sib (HS) families and C-effects are ignored. The trends are similar for selection within full-sib (FS) families; however, the gains for FS families are smaller (by approximately 1/3) than gains from HS families since there is more genetic variation among individuals within a HS family.

The equation used to generate the gains shown in Fig. 17.3 is:

$$\Delta G_{A,w} = i_w (0.75\sigma^2_A)/(0.75\sigma^2_A + \sigma^2_{NA} + (\sigma^2_E/n))^{1/2} \qquad \text{Equation 17.1}$$

where

$\Delta G_{A,w}$ is the additive genetic gain from selection within a HS family;
i_w is the within-family selection intensity in units of standard deviations (*Chapter 6*);
σ^2_A is the additive genetic variance ($0.75\sigma^2_A$ remains among trees in a HS family);
σ^2_{NA} is the non-additive genetic variance;
σ^2_E is the environmental variance;
n is the number of ramets of each seedling clone within a given HS family; and
$0.75\sigma^2_A + \sigma^2_{NA} + \sigma^2_E$ is the phenotypic variance among seedlings within a HS family.

Without loss of generality for any trait P, we can set the total phenotypic variance (Equation 6.10), to $\sigma^2_P = 10$ squared units ($\sigma^2_A + \sigma^2_{NA} + \sigma^2_E = 10$) and the trait mean to $\overline{P} = 15$ units. Then, within-family additive genetic gain in percentage units is $\%G_{A,w} = 100(\Delta G_{A,w}/\overline{P}) = 100(\Delta G_{A,w}/15)$.

Given the above assumptions, the broad and narrow sense heritabilities (H^2 and h^2) can be calculated for any combination of values assumed for σ^2_A and σ^2_{NA} using Equations 6.11 and 6.12, respectively. For example, if we assume $\sigma^2_A = 2$ and $\sigma^2_{NA} = 2$, then $H^2 = 0.4$ and $h^2 = 0.2$.

(Box 17.6 continued on next page)

Box 17.6. Calculating genetic gains in the breeding population from within-family selection. (Continued from previous page)

So, by setting σ^2_P and $\overline{P}$ to arbitrary values and then varying the assumed values for σ^2_A, σ^2_{NA} and n in Equation 17.1, we can simulate expected genetic gains for within-family selection for a range of conditions. These gains are for comparative purposes, and apply to situations in which all trees from a family are planted in a single, unreplicated family plot or are scattered across many replications and site locations.

Equation 17.1 can be used to calculate genetic gains from selection among seedlings within a family (see *Within-family Selection*) or among clones within a family if the base population has been cloned (see *Cloning the Base Population*). When n = 1, there is a single ramet of each clone, and this is the situation when seedlings are planted for each family. Then, Equation 17.1 (with n = 1) predicts genetic gain from selecting the best seedlings within a HS family. Let c be the number of clones per family. Then when n = 1, c is the number of seedlings per family (since each seedling is a different genotype). For example, set $\sigma^2_A = 2$, $\sigma^2_{NA} = 2$, n = 1 and c*n = 100, and assume a selection intensity of i = 2.508 equivalent to choosing the best seedling out of 100 seedlings planted for that family. Then, $\Delta G_{A,w} = i_w(0.75\sigma^2_A)/(0.75\sigma^2_A + \sigma^2_{NA} + (\sigma^2_E/n))^{1/2} = 2.508(0.75*2)/(0.75*2 + 2 + 6)^{1/2} = 1.221$ units. The gain in percentage units is $\%G_{A,w} = 100(\Delta G_{A,w}/15) = 100*1.221/15 = 8.1\%$ as in Fig. 17.3a with n = 1 and the assumed heritabilities.

The calculations proceed the same way if the base population is cloned, except that different values of n must be specified. It is useful to keep the product c*n a constant to reflect equal amounts of work. So, in Fig. 17.3a the plot size c*n = 100 total trees planted per family which could be created by 2 ramets of each of 50 clones (2 on the x axis), 5 ramets from each of 20 clones, or other combinations.

---

clone are planted in randomly assigned locations within the family plot if a completely randomized design is used. Alternatively, one ramet per clone is planted in each of *n* blocks for a randomized complete block design (RCB). As an example of an RCB design with *c* = 25 and *n* = 4, each family plot contains four complete blocks of 25 clones per block for a total of 100 trees. Note that if 300 FS or HS families are created as part of the base population, then there would be 300 family plots, each planted in an RCB design.

With this "family plot" approach to cloning the base population, each family is planted as a single non-replicated plot. Therefore, a separate series of tests (planted as seedlings or cuttings) is required to rank the families and parents (see complementary mating designs in *Chapter 14*). Another approach to cloning the base population is to plant a single type of genetic test, planted with rooted cuttings, in which all families are randomized and replicated over many blocks and sites. Each family is represented by several seedling clones and the ramets of each clone are distributed across the replications and site locations.

Cloning the base population provides additional genetic gain from within-family selection above that expected from within-family selection from seedling families (Fig. 17.3). For example, Fig. 17.3a shows genetic gains from plots containing a total number of 100 trees (*c**n = 100) of a given family. For traits with $h^2 = 0.1$, gain from selecting the best tree from a HS family is 4% for an uncloned, seedling population (*n* = 1 on the *x*-axis meaning *c* = 100 seedlings per family). The additional genetic gain from cloning the base population is 1 to 2% (bringing the total gain to 5 or 6%), which is a sizable increase of 25 to 50% on a base of 4%.

The relative impact of cloning the base population is more for traits with lower h^2. This is because traits with lower h^2 are more impacted by environmental variation that is averaged out by cloning. Numerically this can be seen by noting that: (1) Additional gains from cloning the base population remain approximately constant across different h^2 values at 1 to 3% above those for a seedling population; and (2) This additional gain is a much larger fraction of 4% (when $h^2 = 0.1$) than of 16% (when $h^2 = 0.4$).

For a fixed number of trees per family plot, planting more ramets per clone (higher n) means planting fewer clones (lower c). The optimum number of ramets per clone ranges between $n = 2$ and 7 and is higher for lower heritabilities because more copies of each clone are needed to average out the high amount of environmental noise.

More gain is achieved by planting more ramets and more clones of each family (*i.e.* by increasing the plot size for each family). However, the additional gains increase at a diminishing rate as $c*n$ increases (Fig. 17.3b). This is seen by noting that the curves in Fig. 17.3b are progressively closer together for each increase in plot size of 50 more trees. Taking all factors into consideration, $n = 3$ to 5 ramets per clone and c = 25 to 30 seedling clones seem appropriate for many situations.

Throughout this discussion, we have ignored C-effects (*Chapter 6*) that, if present, can reduce gains below those calculated in Fig. 17.3. For example, if rooted cuttings are used to clone the base population, differential C-effects could mean that some seedling clones would be selected because they root more easily or develop better adventitious root systems. If seed from wind-pollinated seed orchards is the primary method of commercial propagation, then these gains in the breeding population would not be realized in operational plantations.

Therefore, cloning the base population to increase gain from forward, within-family selection is most recommended for programs that employ family forestry or clonal forestry (*Chapter 16*) and establish the cloned base population using the same vegetative propagation methods as are used operationally. This provides the additional benefit of basing selection on performance of the operational type of propagule. For example, many plantation programs of *Pinus radiata* utilize rooted cuttings in operational plantations. Making within-family selections of the best clones based on clonal performance as rooted cuttings ensures that the selected trees are amenable to the propagation methods.

Selections from Overlapping Generations

Best linear unbiased prediction (BLUP) was described in *Chapter 15* as a data analysis technique that makes best use of data from genetic tests for predicting breeding values. An important advantage of BLUP in advanced-generation breeding programs is that all data from multiple generations of genetic tests can be combined into a single analysis (*e.g.* data from three cycles of testing with the third cycle tests consisting of grand-offspring from the first cycle). The BLUP-predicted breeding values for all selections in all cycles are: (1) Adjusted to account for the effects of the progressively increasing genetic mean in later cycles (Kennedy and Sorensen, 1988); and (2) Predicted on the same scale (as described for backward and forward selections in *Chapter 15; e.g.* Table 15.2).

Putting all predicted breeding values on the same scale leads naturally to the use of overlapping generations as is common in animal breeding (van Vleck *et al.*, 1987) and advocated for tree improvement programs (Lindgren, 1986; White *et al.*, 1993; Hodge, 1997; McKeand and Bridgwater, 1998). When only two generations are involved, the discussion simplifies to the relative advantages of making backward (*i.e.* parental) or forward (*i.e.* offspring) selections, and this issue has received considerable attention

(Hodge, 1985; Lindgren, 1986; Hodge and White, 1993; Hodge, 1997; Ruotsalainen and Lindgren, 1997).

Hodge (1997) illustrates the real-world impact of overlapping generations on selection by considering examples from two tree improvement programs. First, the cooperative breeding program for *Pinus elliottii* in the southeastern USA crossed 2200 parents to make 2100 full-sib families with over 170,000 progeny. Of the top 30 selections for bole volume, 14 were parents (backwards selections) and 16 were progeny (forward selections). Similarly in the *Eucalyptus globulus* program in Australia (Jarvis *et al.,* 1995), 52,000 open-pollinated offspring were planted from more than 500 parents. Of the best 30 trees, 15 were parental and 15 were offspring selections. Clearly, it is important to include the backward (parental) selections in both the breeding and propagation populations to capture additional genetic gains.

The situation becomes more complex to study theoretically when several generations are involved. However, using BLUP-predicted breeding values makes it straightforward for the breeder, because all breeding values are predicted on the same scale and can be directly compared. Then, it is normally best to disregard the "generation" of the candidates and choose them based solely on their predicted breeding values (subject to constraints on relatedness). Some original mass selections from the first cycle still might rank high enough (*e.g.* the well-known 7-56 in *Pinus taeda* and NZ55 in *P. radiata*) to warrant inclusion in second, third and fourth cycle breeding populations. Of course, some of their offspring and grand offspring would also warrant inclusion. This naturally leads to an advanced-generation breeding population composed of overlapping generations.

Selection Indices and Other Methods of Selection

Rarely are breeders interested in only a single trait, and selection indices incorporate data from multiple high priority traits to form a single index score for each candidate. The concepts were introduced for mass selection in *Chapter 13*, and illustrated for two traits combined across two generations of data in *Chapter 15*. The idea is that candidates are ranked on the basis of a single score that combines data across several traits, and then selections are made based on this single score (subject to constraints on relatedness as discussed in the next section).

When all data are balanced, classical methods of forming selection indices are appropriate (see references in *Chapter 13*). However, in most advanced-generation tree improvement programs the following process is preferred: (1) Use BLUP to predict breeding values for all candidates for each trait separately (*e.g.* if there are 10,000 candidates and 3 traits measured, then there are 30,000 predicted breeding values, one set for each trait); and (2) Apply appropriate weights for each trait to aggregate the predicted breeding values into a single index score for each candidate (there are 10,000 index scores in the previous example). This method is preferred, because BLUP accounts for unbalanced data and puts predicted breeding values for all candidates from all generations of breeding on the same scale.

Ideally, the index weights incorporate economic or other data that define the relative importance of the traits (Cotterill and Dean, 1990; Borralho *et al.,* 1992b). However, often this information is not available, and the Monte Carlo approach is used in which several sets of weights are applied in order to examine the impact of different weights on the predicted gains made in each trait. This method is illustrated in *Chapter 15*, and the idea is to find that particular set of weights that produces the desired amount of gain from selection in each of the composite traits making up the index. With the Monte Carlo approach, several different indices that assign different weights to the composite traits can be developed for varying purposes. For example, if the breeding population is subdivided into two multiple populations being bred for different product objectives (*e.g.* sawlogs and pulpwood), then the relative weights on bole volume, straightness and wood density could be quite different when making selections for each of the two populations. Only candidates that rank high enough for both indices would be included in both multiple populations.

One problem with using selection indices to form the breeding population is that as more traits are included, there is a tendency for the index to identify candidates that are slightly above average for all traits, but outstanding for none. This means it is possible to exclude some of the most outstanding candidates for any single trait from the breeding population. These "excluded" candidates clearly have outstanding alleles for at least one trait, and are therefore useful for breeding in a recurrent selection program. For this reason, some programs adopt the following strategy to select a breeding population: (1) Include only the two or three most important traits in a selection index and conduct the primary screening based on this index; (2) Conduct a second stage screening of all candidates and select those that are truly outstanding for each trait in the index (even if they fail to meet the minimum score for the index); and (3) Use other means of selection (such as low intensity independent culling) for the secondary traits not included in the index.

As an example of the final point about secondary traits, crown form is rarely important enough to be included in an index, but could be used as the final stage screening to exclude only those candidate trees that have extremely poor, unacceptable crown form. By excluding only the "unacceptable" candidates (as opposed to including only very good ones), the selection intensity on this secondary trait is reduced. As discussed in *Chapter 13*, it is impossible to make substantial gains on many traits simultaneously, so some mechanism is needed to reduce the emphasis on secondary traits whose overall impact on the desired breeding objective is less certain.

A final approach to selection that is more commonly used in crops, but sometimes used in forest trees, is to define an **ideotype** (Box 17.7), which is a conceptual model that explicitly describes the phenotypic characteristics of a plant hypothesized to lead to greater yield (Donald, 1968; Dickmann, 1985; Dickmann and Keathley, 1996; Martin *et al.*, 2000). It is sometimes argued that the commonly used indirect selection for individual-tree growth in relatively young genetic tests planted in single tree plots may not maximize gain in the true target trait which is improved stand yield at rotation age. If underlying biological mechanisms are well understood, an ideotype can be composed of morphological, physiological and even molecular traits that produce greater gain in stand yield than selection for juvenile growth of individual trees.

The ideotype approach to selection is a form of indirect selection (*Chapter 13*) that produces more genetic gain than the more conventional approach only if a few key underlying traits are found that have higher heritabilities and stronger genetic correlations with the target trait than the conventional traits being measured. Martin *et al.* (2000) argue that the ideotype approach may become more tractable as physiological and biochemical mechanisms controlling growth and yield become better understood and more easily measured. Functional genomics (*Chapter 18*) may lead to these discoveries and make selection based on ideotypes more tractable.

Balancing Genetic Gain and Genetic Diversity

Unrestricted use of predicted breeding values or selection indices leads to concentrating selections in a few of the top families and unacceptable loss of genetic diversity. The impact of unrestricted selection is greatest for traits with low h^2 values (such as 0.1 and 0.2

commonly found for important traits in tree species), because it is for these traits that family performance from progeny testing is most useful for determining genetic worth (Robertson, 1961; Askew and Burrows, 1983; Cotterill and Dean, 1990, p 8). For traits with low h^2, individual tree performance is given little weight in predicting that tree's breeding value since most of the variability from tree to tree is due to environmental noise. Thus, once a family is predicted to have a high genetic value, many of its trees also have high predicted values (*i.e.* the family's value dominates the predicted values of the trees in that family). This leads to selection of many trees from the superior families when restrictions on relatedness are not enforced.

As an example, consider the *Eucalyptus globulus* program in Australia whose objective was to select a second-generation breeding population consisting of 600 total selections; candidates for selection were from the first-generation program consisting of 52,000 open-pollinated (OP) offspring from more than 500 parents (Jarvis *et al.,* 1995). Unrestricted selection would have resulted in selection of 600 trees from only 63 of the original 500 OP families, and this was deemed an unacceptable reduction in effective population size. By limiting the number of selections from a given family to five trees, the 600 selections came from 181 of the original 500 families (a threefold increase in the number of families represented).

The appropriate balance between genetic gain and relatedness depends on many factors (Lindgren *et al.*, 1997; Lindgren and Mullin, 1998; Ruotsalainen and Lindgren, 2001), but the following conclusions are universal: (1) Some algorithm should be developed to limit the number of related selections and this may be a simple intuitive rule (*e.g.* no more than five selections per family as in the *E. globulus* program) or a more complex computer-based method; and (2) The algorithm should consider both the breeding value of the candidate and the effect that selecting that candidate would have on genetic diversity of the breeding population.

The term "group merit" has been used (Wei and Lindgren, 1995; Lindgren and Mullin, 1998; Danusevicius and Lindgren, 2002; Lstiburek *et al.*, 2004a) to combine the concepts of genetic gain and effect on relatedness. A candidate has a higher group merit (and hence is more likely to be selected) if it has a high breeding value and would have small negative impact on population relatedness. This means that, in general, candidates with extremely high breeding values are selected even if they are related to other selections already in the breeding population.

Lower in the rankings of breeding values, the group merit for each candidate is dominated by its contribution to overall population diversity, and less affected by its breeding value since there are many candidates with similar, moderate breeding values. In this way, the top selections may be quite related and have a negative impact on the effective population size of the breeding population, but this is offset by less superior selections that come from many different families.

An example of applying the concept of group merit to selection of a breeding population without using a sophisticated computer algorithm is the second-generation *Pinus elliottii* program described in Box 17.2. Within the main population, stratum I (the top third based on predicted breeding value) had the highest mean breeding value (volume gain is 28% in Table 17.4), but contributed least to genetic diversity of the breeding population ($N_e = 174$ and relative effective size of only $N_r = 0.56$). Selection intensity was very strong in stratum I and several relatives of the top parents were included. This was done purposely to achieve rapid near-term genetic gains. In lower strata, more emphasis was placed on obtaining unrelated selections to increase genetic diversity. Thus, in the bottom third of the main population (stratum III in Table 17.4), volume gain was only 6%, but $N_e = 288$ and $N_r = 0.93$.

Box 17.7. Ideotype concepts for forest trees.

Three main categories of ideotypes have been described (Donald and Hamblin, 1976; Cannell, 1978; Martin *et al.,* 2001): (1) An **isolation ideotype** grows well in the absence of competition, *i.e.* open-grown conditions, and tends to have a large, long crown and spreading root system; (2) The **competition ideotype** is similar to the isolation ideotype except it is a stronger competitor, rapidly exploiting site resources and aggressively expanding its own crown and roots to the detriment of its neighbors; and (3) A **crop ideotype** efficiently utilizes available resources, while not competing strongly with neighboring trees. The crop ideotype could have efficient nutrient utilization and narrow, more vertical crowns that interfere less with neighboring crowns. The crop ideotype is predicted to produce the greatest yield per unit area in intensively-managed monocultures.

    The conventional process of selection for individual tree growth in tests with few trees per plot (such as single tree plots) could have three disadvantages in terms of making genetic gain in rotation-age stand yield: (1) Selection for growth in young progeny tests prior to crown closure could bias in favor of isolation ideotypes; (2) Selection in the same tests after crown closure may bias toward competition ideotypes; and (3) Weak competitors (crop ideotypes) are perhaps overlooked. That is, the crop ideotype predicted to excel if planted in pure stands of similar genotypes is at a competitive disadvantage in a progeny test planted with single tree plots of each different genotype.

    Constructing ideotypes composed of key underlying traits related to stand yield at maturity could conceivably increase genetic gain from selection if the underlying mechanisms controlling stand growth and yield were understood. Then, those traits producing crop ideotypes could be selected. This is a large challenge for forest biologists, because of the difficulty of measuring underlying traits and understanding their impact on final stand yield. Nevertheless, new methods that combine plant physiology with molecular biology and functional genomics (*Chapter* 18) make this a promising area of research (Martin *et al.,* 2001).

Optimum Selection Age

Nearly all tree improvement programs make selection decisions prior to rotation age to capture genetic gains sooner and therefore increase gains per unit time. This early selection is a form of indirect selection (*Chapter 13*), because the measured trait upon which selection is based (early performance) is not the target trait (performance at maturity). As shown in Equations 13.7 and 13.8, the efficacy of early selection depends, in part, on the heritability of the measured trait and its genetic correlation with the target trait. Studies to determine optimum selection age are based on equations similar to Equation 13.9 in which these factors are integrated to determine the potential genetic gains from making selections at different ages.

    When genetic gain per cycle is calculated, gains are often higher when selection is made at rotation age (the genetic correlation is 1.0, because the measured trait is the target trait). However, gains per cycle do not incorporate the benefits of earlier deployment or earlier breeding when selections are made at younger ages. Thus, studies of optimum selection age are commonly based on gains per unit time (Lambeth *et al.,* 1983b) or incorporate the time value of money in a manner analogous to present value analyses and

Table 17.4. Census number (N), effective population size (N_e), relative effective number (N_r = N/N_e) and average genetic gain in bole volume (% gain above unimproved material at 20 years from White *et al.*, 1988) for the second-generation breeding population of *Pinus elliottii* in the southeastern USA described in Box 17.2. The breeding population with 936 selections is subdivided into an elite population and main population. The main population is further subdivided into three equal-sized strata based on predicted breeding values of the selections. The 60 elite selections are also included in Stratum 1 of the main population. (Reproduced from White *et al.*, 1993, with permission from J.D. Sauerländer's Verlag)

Subdivision	N	N_e	N_r	Bole volume gain (%)
Elite	60	47	0.78	35
Stratum 1	312	174	0.56	28
Stratum II	312	266	0.85	17
Stratum III	312	288	0.93	6
Total	936	625	0.67	17

discounted cash flow in economics (McKeand, 1988; Balocchi, 1990; White and Hodge, 1992; Osorio, 1999). These latter studies base the decision about optimum selection age on the **discounted selection efficiency** which incorporates both genetic factors and the time value of money.

When selection decisions are based on discounted selection efficiency, selection at later ages is penalized to various degrees depending on the assumed interest or discount rate. When higher discount rates are used, there is an implicitly high time value of money, and this leads to younger optimum selection ages. Very low discount rates mean that the breeder is less concerned about capturing early genetic gains, because the extra profits from early selection would accrue relatively low interest. Low interest rates lead to older selection ages.

The following are recommendations about optimum selection ages: (1) Traits associated with adaptation during early years in the rotation (such as damage or survival due to frost, drought or early-rotation diseases) can be measured and selected at the young ages they are normally expressed; (2) Optimum selection ages for stem growth, wood quality and other production-oriented traits are commonly 25 to 50% of the rotation age (McKeand, 1988; Magnussen, 1989; White and Hodge, 1992; Balocchi *et al.*, 1994; Osorio, 1999); (3) For growth traits, optimum selection age is nearer 50% of the rotation age for short-rotation species, *e.g.* three years for *E. grandis* which has a rotation age of six years in Colombia (Osorio, 1999) and nearer 25% of the rotation age for longer-rotation species, *e.g.* 10 years for *Populus tremuloides* with a 40-year rotation (Li, 1995); and (4) The optimum selection age for making forward selections may be a few years older than for making backward selections (Balocchi *et al.*, 1994).

SUMMARY AND CONCLUSIONS

Breeding strategies of forest trees vary dramatically due to different selection, breeding, testing and deployment methods; however, nearly all strategies for population improvement of a single species are founded on a single type of recurrent selection, called recurrent selection for general combining ability (RS-GCA). In RS-GCA, genetic testing follows selection and the selections are ranked based on their GCA values for the selected traits. Then, the selections with the highest GCAs (or their offspring) are included in future cycles of breeding, and poor selections are excluded. Over several cycles of breeding and testing, the breeding population gradually accumulates alleles with positive additive effects on the few traits being selected.

There are three issues associated with management of genetic diversity and inbreeding in advanced-cycle breeding programs of forest trees: (1) Activities (such as the use of intensive selection) which increase near-term genetic gain in the first few cycles of improvement accelerate the decrease in genetic diversity and hence may reduce potential for making near-optimal long-term genetic progress (say after 15 cycles of improvement); (2) The near-universal problem of inbreeding depression in forest trees means that inbreeding should be avoided in trees planted in operational plantations; and (3) Use of directed inbreeding to rid the breeding population of unfavorable deleterious alleles is a potentially positive aspect of inbreeding that could be incorporated into breeding strategies as long as the trees produced by the propagation population for operational plantations are outcrossed.

For programs aimed at maximizing economic returns, analyses have shown that the first few cycles are critical, and short-term gains are paramount to overall long-term economic success. These programs should aim to place more emphasis on better material at all stages of the breeding efforts by: (1) Including more relatives of better selections in the breeding population; (2) Stratifying the breeding population into tiers or hierarchies such that upper tiers contain better selections and are managed more intensively; (3) Employing unbalanced mating designs that include better selections in more crosses; and (4) Obtaining more precise breeding values on better selections by testing them in more locations. Programs employing these tactics should start with larger breeding populations and use other methods to offset the reduction in genetic diversity in the breeding population that these activities cause.

Breeding Population Size and Structure – Considering both theoretical studies and practical implications, the following conclusions seem reasonable about the size of a breeding population. First, breeding populations with effective sizes (N_e) of 20 to 40 will support selection and breeding programs with appreciable genetic gains for several cycles of breeding. Populations with N_e between 20 and 40 seem appropriate for elite populations being supported by a larger main population and for the number of selections in a subline or multiple population, if there are many such groups comprising the entire breeding population in a breeding unit.

Second, breeding populations with a few hundred members, common in advanced-generation tree improvement programs (Table 17.3), are large enough to achieve near-maximum long-term gains even when there are several traits and when intensive efforts are used to achieve larger gains in the first few generations. Third, quite large populations (1000 or so) may be required to ensure that rare alleles are maintained in the population for many generations. Thus, if a breeding population is also serving a gene conservation role or providing long-term flexibility, then these larger sizes seem warranted.

Today, most tree improvement programs do not manage the breeding population as a single unified population in which all selections are treated equally. Rather, the breeding population is structured (*i.e.* subdivided) for one of three purposes: (1) Elite populations and hierarchical strata are formed on the basis of predicted breeding values to facilitate placing more emphasis on superior material; (2) Multiple breeding groups (also called sublines) are created that allocate related selections to and constrain breeding within the breeding groups to manage inbreeding in production populations; and (3) Multiple populations are formed by applying different selection criteria to different segments of the population being developed for different environments, product objectives, etc. In practice, tree improvement programs may employ more than one, or even all three types of structures in the same breeding population.

Breeding Population Formation and Management – The desire to maximize genetic gains per unit time has led to two major areas of research: (1) Early selection which aims to make selections in field tests or artificial environments when trees are as young as possible; and (2) Accelerated breeding whose goal is to shorten the time required to complete the breeding after selections are made.

Two primary functions of mating designs in advanced-generation breeding strategies are: (1) To create the next generation's base population consisting of progeny produced by intermating trees in the current generation's breeding population (Fig. 11.1); and (2) To facilitate breeding value prediction for each of the selections in the breeding population and all of their progeny in the new base population to maximize genetic gain from the next generation selections.

In early years of tree improvement, many programs employed a single, balanced full-sib mating design for all breeding activities; however, more recently: (1) Some programs are employing open-pollinated management of some or all of the breeding population; (2) Complementary mating designs are becoming more common; (3) Unbalanced mating designs are being advocated that place more emphasis on superior selections by crossing them more frequently; and (4) Rolling front mating designs are gaining popularity to maximize gain per unit time by breeding selections as they become available.

In terms of field designs, randomized complete blocks or incomplete block designs planted in single tree plots are recommended for all genetic tests aimed at ranking parents or families (*i.e.* for predicting parental breeding values for use in backward selection, see Box 14.2). Further, these designs are robust and also serve well for making forward selections (*i.e.* for predicting breeding values of individual trees and make selections of superior trees growing in the genetic tests). Forward selections can involve the use of large unreplicated family plots when complementary mating designs are employed.

Regardless of the mating and field designs used for management of the breeding population, there are at least three important issues to consider when making selections from these tests to form the next cycle's breeding population. First, the use of best linear unbiased prediction combines data from different designs and different generations to put all predictions on a single scale. This leads naturally to the use of overlapping generations in which superior selections from several previous cycles of breeding are included in each cycle's breeding population.

Second, selection indices should be limited to only a few of the most important traits, and unrestricted use of selection indices reduces population diversity too severely by making too many selections from superior families. Thus, normally it is important to use selection indices with care, and to check carefully to make sure that excellent selections are being made and that an adequate effective population size is being maintained.

Finally, early selection is warranted in most cases, because waiting until rotation age to make selections reduces genetic gain per unit time. Based on genetic, logistic and economic considerations, optimum selection ages are often between 25% and 50% of the rotation age. Usually, the optima are closer to 50% for fast-growing short-rotation species, and closer to 25% for longer-rotation species.

Final Comments – There is probably no such thing as a perfect breeding strategy. Rather, each organization adopts a strategy appropriate to its set of biological, economical and logistical conditions. The three most important aspects to developing a breeding strategy are: (1) Be creative and consider all possible options to find those most suited to the organization's specific set of forecasted conditions; (2) Make sure the strategy can be implemented as planned and keep it as simple as possible, because there is no reason to have the most elegant strategy if it is impossible to carry it out; and (3) Ensure that the strategy is flexible to changes in assumptions, developments of new technologies, and other changes in the future. For most species, the operational plantations planted as the product of this cycle's breeding strategy are many years (more often decades) in the future. Thus, breeders must be futuristic as well as practical and logical; this is the challenge of developing a good tree breeding strategy.

CHAPTER 18
GENOMICS – *DISCOVERY AND FUNCTIONAL ANALYSIS OF GENES*

Genomics is the study of the structure, location, function, regulation and interaction of large numbers of genes simultaneously. Genomics is an extension of some of the traditional sub-disciplines of genetics, such as transmission and molecular genetics, to the entire genome. Genomics is a science that is made possible and is dependent upon technologies that allow rapid analysis of hundreds or thousands of genes and often in many individuals at once, such as by automated DNA sequencing. Whereas traditional molecular genetics enabled the discovery and functional analysis of one or a small number of genes in single experiments, genomics not only seeks to understand the function of large numbers of genes in parallel, but also seeks to understand the complex interactions among genes. Genomic sciences should ultimately lead to deeper insight into the relationship between an organism's total genetic composition and its complete phenotype, including a mechanistic understanding of gene action, pleiotropy, GxE interaction and epistasis.

Three major sub-areas of genomics are generally recognized: structural genomics, functional genomics and comparative genomics. **Structural genomics** seeks to identify all the genes in the genome and determine their locations on chromosomes. This is usually achieved by DNA sequencing of individual genes or of entire genomes and by genetic or physical mapping. **Functional genomics** seeks to determine the function of genes and how genes determine the phenotype. A variety of new techniques and experimental approaches have been developed to help understand how genes and genomes function. **Comparative genomics** seeks to understand either the structure or function of genes by making comparisons across taxa. DNA sequencing and genetic mapping techniques are most often used for comparative genomic analyses.

Genomic science technologies have largely been developed through the Human Genome Project. The pharmaceutical and biomedical industries have made significant investments in genomic science research for the discovery of new drugs and treatment of diseases. These same technologies are now being applied to agricultural plants, animals and forest trees. The large investments being made by private corporations lead to patenting of genomic information and control of intellectual property. So although genomic sciences promise to achieve a deeper understanding of genomes and the function of genes, this knowledge may not be free for all to use; the products developed through genomics may be owned by individual corporations.

In this chapter, we discuss how structural, functional and comparative genomic sciences are being applied to forest trees. In *Chapters 19* and *20*, we illustrate how knowledge of the structure and function of genes and genomes can be applied to develop varieties of trees for specific end-product needs. Other recent reviews of genomic sciences in forestry include Sederoff (1999), Krutovskii and Neale (2003), and Kumar and Fladung (2004).

STRUCTURAL GENOMICS

The genomes of forest trees, like all eukaryotes, are organized into chromosomes, each of which contains a large number of genes (*Chapter 2*). The total number of genes in the genome is not yet known for any tree species. Estimates from a few animal and plant species have been made as a result of complete genome sequencing (Table 18.1), and from these data it is expected that tree species have in the range of 30,000 genes. The central task of structural genomics is to build a catalog of all genes in the genome, sometimes called **gene discovery**, and determine their locations on chromosomes by genetic and physical mapping.

Gene Discovery

Several different approaches are used in gene discovery. The most direct approach is to determine the DNA sequence of the entire genome, which includes DNA sequences for both gene-encoding and non-encoding portions of the genome. The individual genes can be identified from the total DNA sequence using sophisticated computational algorithms that take into account many factors related to the known structure of genes such as open reading frames, intron-exon splice sites, initiation and termination codons and much more (*Chapter 2*). This approach has been used for a variety of organisms, many of which have small genomes that make it experimentally and economically feasible, and it has also been used for the human genome because of its importance (IHGSC, 2001; Venter *et al.*, 2001). The complete genome sequence of two model plant species, *Arabidopsis thaliana* (TAGI, 2000) and rice (Goff *et al.*, 2002; Yu *et al.*, 2002), were the first to be determined and several more are in progress. Full genome sequencing of the *Populus* genome was started in 2002 and a first draft sequence was released in September 2004.

A complementary approach to complete genome sequencing, called **expressed sequence tags (ESTs**, *Chapter 4*), has been applied to scores of organisms, including several forest tree species. EST sequencing is based on identifying only the DNA sequences that code for genes that are expressed. The basic approach of EST sequencing is to construct cDNA libraries (*Chapter 4*) for the many genes expressed in one or more tissue-types and determine the DNA sequences of short segments of a large number of cDNA clones from the libraries (Fig. 18.1). The DNA sequences are most often obtained from the 5' end of the cDNA because there is generally greater sequence conservation at this end of genes. This raw sequence information is submitted to gene databases and compared to all other sequences in the database to identify matches (Fig. 18.2). If there are matches to gene sequences whose functions have already been determined, then the likely identity of these ESTs will have been made without further investigation.

Table 18.1. Estimated number of genes in several organisms based on complete genome sequencing.

Species	Number of genes
Saccharomyces cerevisiae (yeast)	6,034
Caenorhabditis elegans (nematode)	19,099
Drosophila melanogaster (fruit fly)	13,061
Mus musculus (mouse)	~30,000
Homo sapiens (human)	~40,000
Arabidopsis thaliana (flowering plant)	25,498
Oryza sativa (rice)	59,855

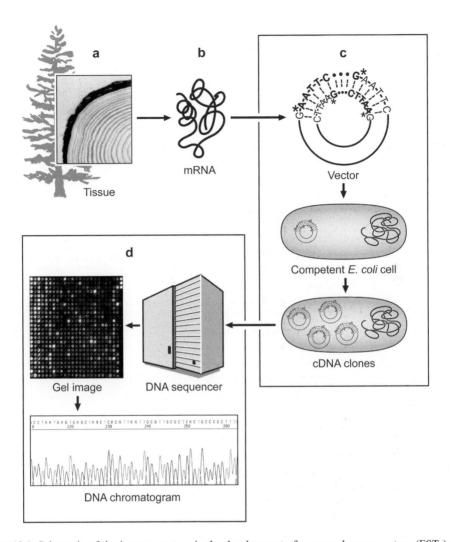

Fig. 18.1. Schematic of the important steps in the development of expressed sequence tags (ESTs): a) Tissue is harvested from the tree, such as xylem in this example; b) Messenger RNA (mRNA) (see Fig. 2.9) is isolated from the xylem tissue; c) The mRNA is converted to complementary DNA (cDNA) (see Box 4.2); and d) DNA sequences are determined from cDNA clones using automated instrumentation.

EST sequencing was first applied to identify genes expressed in the human brain (Adams *et al.*, 1991) and has since been applied to scores of plant species. Some of the first EST sequencing projects in forest trees focused on genes expressed in wood forming tissues (Allona *et al.*, 1998; Sterky *et al.*, 1998; Whetten *et al.*, 2001; Kirst *et al.*, 2003). These projects discovered many genes involved in lignin and cellulose biosynthesis and other components of cell walls. Genes for basic metabolism in woody plant cells have also been identified (Fig. 18.3). EST sequencing projects are in progress for many forest tree species and for a wide variety of tissue types (*e.g.* roots, leaves/needles, stems, meristems and reproductive structures) (Table 18.2). As of 2005, more than a million sequences from forest tree species were found in Genbank (see *Bioinformatics and Databases*), and this number is growing rapidly.

a

```
SOURCE      Pinus taeda (loblolly pine)
  ORGANISM  Pinus taeda
            Eukaryota; Viridiplantae; Streptophyta; Embryophyta; Tracheophyta;
            Spermatophyta; Coniferopsida; Coniferales; Pinaceae; Pinus; Pinus.
REFERENCE   1  (bases 1 to 1406)
  AUTHORS   MacKay,J.J., Liu,W., Whetten,R., Sederoff,R.R. and O'Malley,D.M.
  TITLE     Genetic analysis of cinnamyl alcohol dehydrogenase in loblolly
            pine: single gene inheritance, molecular characterization and
            evolution
  JOURNAL   Mol. Gen. Genet. 247 (5), 537-545 (1995)
  MEDLINE   95327049
   PUBMED   7603432
REFERENCE   2  (bases 1 to 1406)
  AUTHORS   MacKay,J.J.
  TITLE     Direct Submission
  JOURNAL   Submitted (29-SEP-1994) John J MacKay, Forestry, North Carolina
            State University, Room, 6113 Jordan Hall, Raleigh, North Carolina,
            27695-8008, USA
```

b

```
BASE COUNT      404 a    258 c    371 g    373 t
ORIGIN
        1 atagcttcct tgccatctgc aaggcaatac agtacaagag ccagacgatc gaatcctgtg
       61 aagtggttct gaagtgatgg gaagcttgga atctgaaaaa actgttacag gatatgcagc
      121 tcgggactcc agtggccact tgtcccctta cacttacaat ctcagaaaga aaggacctga
      181 ggatgtaatt gtaaaggtca tttactgcgg aatctgccac tctgatttag ttcaaatgcg
      241 taatgaaatg ggcatgtctc attacccaat ggtccctggg catgaagtgg tggggattgt
      301 aacagagatt ggtagcgagg tgaagaagtt caaagtggga gagcatgtag gggttggttg
      361 cattgttggg tcctgtcgca gttcggtaac tgcaatcag agcatgaac aatactgcag
      421 caagaggatt tggacctaca atgatgtgaa ccatgacggc accectactc agggaggatt
      481 tgcaagcagt atggtggttg atcagatgtt tgtggttcga atcccggaga atcttcctct
      541 ggaacaagca gcccctctgt tatgtgcagg ggttacagtt ttcagcccaa tgaagcattt
      601 cgccatgaca gagcccggga agaaatgtgg gattttgggt ttaggaggcg tggggcactt
      661 gggtgtcaag attgccaaag cctttggact tcacgtgacg gttatcagtt cgtctgataa
      721 aaagaaagaa gaagccatgg aagtcctcgg cgccgatgct tatcttgtta gcaaggatac
      781 tgaaaaagatg atggaagcag cagagagcct agattacata atggacacca ttccagttgc
      841 tcatcctctg gaaccatatc ttgcccttct gaagacaaat ggaaagctag tgatgctggg
```

c

```
>gi|12482543|gb|BG039862.1|BG039862   NXSI_104_H07_F NXSI (Nsf Xylem Side wood
Inclined) Pinus taeda cDNA
            clone NXSI_104_H07 5' similar to Arabidopsis thaliana
            sequence At3g19450 putative cinnamyl alcohol
            dehydrogenase 2 see
            http://mips.gsf.de/proj/thal/db/index.html.
            Length = 558

 Score = 1084 bits (547), Expect = 0.0
 Identities = 555/558 (99%)
 Strand = Plus / Plus

                                   ↓
Query: 303 cagagattggtagcgaggtgaagaagttcaaagtgggagagcatgtaggggttggttgca 362
           ||||||||||||||||||||||||||| |||||||||||||||||||||||||||||||||
Sbjct: 1   cagagattggtagcgaggtgaagaaattcaaagtgggagagcatgtaggggttggttgca 60

Query: 363 ttgttgggtcctgtcgcagttgcggtaactgcaatcagagcatggaacaatactgcagca 422
           ||||||||||||||||||||||||||||||||||||||||||||||||||||||||||||
Sbjct: 61  ttgttgggtcctgtcgcagttgcggtaactgcaatcagagcatggaacaatactgcagca 120

Query: 423 agaggatttggacctacaatgatgtgaaccatgacggcacccctactcagggaggatttg 482
           ||||||||||||||||||||||||||||||||||||||||||||||||||||||||||||
Sbjct: 121 agaggatttggacctacaatgatgtgaaccatgacggcacccctactcagggaggatttg 180

Query: 483 caagcagtatggtggttgatcagatgtttgtggttcgaatcccggagaatcttcctctg 542
           |||||||||||||||||||||||||||||||||||||||||||||||||||||||||||
Sbjct: 181 caagcagtatggtggttgatcagatgtttgtggttcgaatcccggagaatcttcctctg 240

Query: 543 aacaagcagcccctctgttatgtgcaggggttacagttttcagcccaatgaagcatttcg 602
           ||||||||||||||||||||||||||||||||||||||||||||||||||||||||||||
Sbjct: 241 aacaagcagcccctctgttatgtgcaggggttacagttttcagcccaatgaagcatttcg 300

Query: 603 ccatgacagagcccgggaagaaatgtgggattttgggtttaggaggcgtggggcacttg 662
           |||||||||||||||||||||||||||||||||||||||||||||||||||||| |||
Sbjct: 301 ccatgacagagcccgggaagaaatgtgggattttgggtttaggaggcgtggggcacatgg 360
                                                                     ↑
```

Fig. 18.2. A report from the DNA sequence database Genbank for the cinnamyl alcohol dehydrogenase (*cad*) gene sequence from *Pinus taeda*: a) Sequence annotation showing accession number, gene name, species, authors and journal reference; b) Nucleotide sequence of the *cad* gene from *P. taeda*; and c) The *cad* nucleotide sequence from *P. taeda* was compared to all sequences in the database and aligned with the most similar sequence found. In this case, a nucleotide sequence from *Arabidopsis thaliana* (Sbjct) was completely identical to a portion of the *P. taeda* sequence (Query) except for two nucleotide differences (indicated by upward and downward arrows at *P. taeda* positions 328 and 659, respectively).

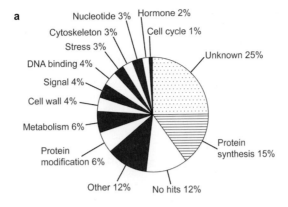

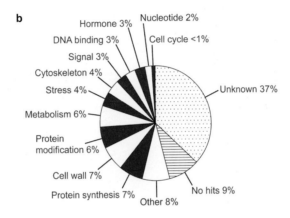

Fig. 18.3. Functional classification of: a) 4809 ESTs from *Populus* cambium; and b) 833 ESTs from *Populus* xylem. Note the large proportion of ESTs that could not be assigned to a functional class. "Unknown" indicates a match to a gene of unknown function in the database and "No hits" indicates that no match was found. (Reproduced with permission from Sterky *et al.*, 1998; copyright 1998, National Academy of Sciences, USA)

Genetic Mapping

Genetic mapping is central to genomic sciences because it provides an organizational framework to the genome. In *Chapter 4*, the many types of genetic markers available for use in forest trees are described. The value of genetic markers is enhanced if their location in the genome is known, *i.e.* their map position on a specific chromosome. Genetic markers can be positioned onto two kinds of maps: physical maps and genetic maps. **Physical maps** provide the exact location of genes or genetic markers on chromosomes. Techniques such as radio-labeled or fluorescent *in situ* hybridization are used for physical mapping (Fig. 2.7). Ribosomal RNA genes and a few highly repeated DNA sequences have been physically mapped in forest trees using such techniques (Brown *et al.*, 1993; Brown and Carlson, 1997; Doudrick *et al.,* 1995). Current technologies, however, are not sensitive enough with the large genomes of conifers to map molecular markers that are based on relatively small segments of DNA, such as cDNAs. Large fragments (~100 kb) of cloned DNA, such as bacterial artificial chromosomes (BACs) (see section on *Positional Cloning of QTLs* for a description of BACs), would be ideal for physical mapping in conifers.

Table 18.2. The number of ESTs found in the Genbank database for a sample of tree species.

Pinus taeda	173,680
Populus tremula x *Populus tremuloides*	65,981
Picea glauca	55,108
Populus balsamifera subsp. *trichocarpa*	54,660
Populus balsamifera subsp. *trichocarpa* x *Populus deltoides*	33,134
Populus tremula	31,288
Pinus pinaster	18,254
Populus deltoides	14,645
Populus balsamifera subsp. *trichocarpa* x *Populus nigra*	14,281
Populus euphratica	13,903
Populus tremuloides	12,813
Picea engelmannii x *Picea sitchensis*	12,127
Picea sitchensis	12,065
Populus x *canescens*	10,446
Populus euramericana	10,157
Pseudotsuga menziesii var. *menziesii*	6721
Cryptomeria japonica	6589
Cycas rumphii	5952
Juglans regia	5025
Tamarix androssowii	4756
Welwitschia mirabilis	3732
Betula pendula	2545
Gnetum gnemon	2128
Camellia sinensis	1989
Ginkgo biloba	1953
Pinus sylvestris	1663
Eucalyptus grandis	1574
Eucalyptus tereticornis	1131
Populus tomentiglandulosa	1127

The alternative approach is to develop **genetic maps** by segregation and linkage analysis (*Chapter 3*). Genetic maps identify the relative distance between two markers based on the number of recombination events between the markers. That is, physical maps show the locations of genes in base-pair distance units from each other while genetic maps show the relative locations of genes in terms of recombination units. The relationship between a physical map and a genetic map is not direct, because the amount of recombination between any two equidistant markers can vary significantly throughout the genome. Genetic linkage mapping is discussed in this chapter because this has been the most widely employed mapping technique in forest trees.

There are three major steps to constructing a genetic map: (1) Selecting an appropriate mapping population; (2) Obtaining genetic marker segregation data from the mapping population; and (3) Applying linkage analysis to place the segregating loci in relative positions and distances from one another. Genetic markers were addressed previously (*Chapter 4*); so, here we limit the discussion to mapping populations and linkage analysis. Genetic maps have been constructed for a large number of forest tree species and the types of genetic markers, mapping populations, and methods of linkage analysis all vary among mapping projects (Cervera *et al.*, 2000; http://dendrome.ucdavis.edu/).

Genetic Mapping Populations

Genetic maps are based on segregation and linkage analysis performed on some type of segregating or mapping population. In *Chapter 3*, the concept of genetic linkage was illustrated using a simple example of two linked allozyme loci in *Pinus rigida* (Box 3.4). In that case the mapping population was a sample of segregating haploid megagametophytes in open-pollinated seed from a single mother tree. The two general types of mapping populations most commonly used in forest trees are haploid conifer megagametophytes and full-sib pedigrees.

Megagametophyte mapping populations were first used with allozyme markers to estimate linkages among allozyme loci (*Chapter 3;* Guries *et al.,* 1978; Rudin and Ekberg, 1978; Conkle, 1981b; Adams and Joly, 1980). In the last 20 years, megagametophyte mapping populations have been used to estimate linkages among genes coding allozymes in dozens of conifer species. More recently, megagametophyte mapping populations have been employed to estimate linkages and construct genetic maps based on RAPD and AFLP markers (*Chapter 4*) (*e.g.* Tulsieram *et al.*, 1992; Nelson *et al.*, 1993; Binelli and Bucci, 1994; Remington *et al.*, 1999). Clearly, the megagametophyte mapping population system will continue to be used for genetic mapping of molecular markers in conifers.

Segregating populations can also be generated by controlled crossing. These populations are essential for linkage mapping in angiosperms (which lack haploid megagametophytes) and also can serve as an alternative to megagametophyte mapping in conifers. When inbreeding depression is not a concern, such as in many crop plants that are naturally predominately self-pollinated, ideal mapping populations can be generated by first creating highly inbred lines that are homozygous at most, or all, marker loci. By crossing two different inbred lines, F_1 progeny are produced that are heterozygous at many loci. Backcrossing one of the F_1 individuals to one of the parental inbred lines produces a segregating population for linkage mapping. Backcrosses are ideal, because all recombination occurs in the gametes of one parent (the heterozygous F_1), and the linkage phase of all doubly-heterozygous combinations of loci is known (*i.e.* coupling ABab or repulsion AbaB). The segregation data obtained with a backcross mapping population is similar to that derived from segregating megagametophytes, except that linkage phase is not known with megagametophyte data.

In outcrossing plants, like most trees, inbred lines cannot be produced because the close inbreeding required to create these lines results in weak, infertile individuals (*Chapter 5*). Any two potential parent trees chosen from a natural (not inbred) population are likely to be heterozygous at many loci (*Chapter 7*) and if crossed, the progeny will segregate at these loci. Linkage analysis, however, is complicated because the linkage phase of pairs of markers is not known and mating types (Table 18.3) of markers can vary. Different mating types are possible because for any marker, just one or both of the parent trees can be heterozygous. This makes the estimation of linkage more complicated than in the simple inbred backcross case, because differences in mating types produce different two-locus ratios in the progeny. Still, these types of mapping populations involving crosses among heterozygous parents are widely used in forest trees out of necessity.

Linkage Analysis and Map Construction

Once marker segregation data are obtained from a mapping population, it is possible to estimate linkages among the markers and construct a genetic map. Estimation of the recombination fraction (r), or linkage, between a pair of markers was described in *Chapter 3*.

Table 18.3. Informative full-sib mating types for genetic mapping.

Maternal genotype	Paternal genotype	Mating type
A_1A_2	A_1A_1	Maternally informative
A_1A_1	A_1A_2	Paternally informative
A_1A_2	A_1A_2	Intercross
A_1A_2	A_1A_3	Fully informative
A_1A_2	A_3A_4	Fully informative

If the number of markers analyzed is very small (<10), as is the case with some allozyme marker data sets, it is possible to estimate all pairwise linkages and order markers manually. Some loci may appear to be closely linked (r is small), or located very far apart on the same chromosome or on different chromosomes (r approaches 0.5). However, as the number of markers increases, it becomes tedious to estimate all linkages manually and extremely difficult to manually determine marker order on linkage groups. Rather, specialized genetic mapping software is used; two commonly used programs are Mapmaker (Lander *et al.*, 1987) and JoinMap (Stam, 1993).

All genetic mapping software follows a similar approach for constructing maps. The first step is to group all the genetic markers into linkage groups. This is done by first calculating all the pair-wise linkage distances (r values) among all markers in the data set. The programs use this matrix of linkage estimates to determine which sets of markers are likely linked to one another and should be assigned to a common linkage group. Once the linkage groups are determined, markers are then ordered within each group. The statistical approach to ordering varies among programs and the extent to which the user can control the ordering analysis also varies. While in concept a linkage group is equivalent to a chromosome, the number of linkage groups often exceeds the number of chromosomes due to inadequate marker coverage or large regions with suppressed recombination (*e.g.* inversions) that prevent detection of linkage between regions on the same chromosome.

The Mapmaker software is most often used for megametophyte mapping population data, because this data type is very similar to the backcross and F_2 data types for which the software was designed. Genetic mapping in full-sib pedigree mapping populations in trees is significantly more complicated. The marker genotypes of the progeny result from independent segregations in both the maternal and paternal parent. Given this situation, two mapping strategies can be employed: (1) Construct individual genetic maps for each of the parents; or (2) Construct a single "sex-averaged" map using data from both parents. Groover *et al.* (1994) used the JoinMap program to create individual parent tree maps in *Pinus taeda*. Sex-averaged maps have been constructed for *Pinus radiata* (Devey *et al.*, 1996), *Eucalyptus nitens* (Byrne *et al.*, 1995) and *Pseudotsuga menziesii* (Jermstad *et al.*, 1998) using JoinMap. Sewell *et al.* (1999) used Mapmaker and JoinMap to create individual parent tree maps for the four parents of two unrelated, full-sib pedigree mapping populations in *Pinus taeda*. JoinMap was used to merge these four maps into a consensus map (Fig. 18.4).

Gene Mapping by Bulked Segregant Analysis

Once a genetic linkage map has been constructed using genetic markers, it can be used to map the position of genes of special interest. A gene controlling a qualitatively inherited trait can be mapped simply by scoring the Mendelian segregations between this gene and marker genes of known location, adding these data to the full marker data set and reconstructing the genetic map. In many cases however, the gene controlling the qualitative trait

might only be segregating in a population for which a genetic map has not been cons
In this situation, a shortcut approach to identifying markers linked just to the qualitati
gene can be used. This method is called **bulked segregant analysis (BSA)**.

BSA was initially developed by Michelmore *et al.* (1991) for mapping disease ι ۔۔s-
tance genes in lettuce. This method is especially amenable to dominant molecular markers
such as RAPDs and AFLPs (*Chapter 4*), although it can also be used with codominant
markers. The basic principle of BSA relies on very strong linkage (*Chapter 3*) between the
qualitative trait gene and one or more genetic markers. To begin, two pooled DNA samples are created by combining DNA from individuals sharing each of the alternative alleles
controlling the qualitative trait. For example, in a backcross for a dominant disease resistance gene, *Rr* x *rr*, two genotypes segregate in the progeny, *Rr* and *rr*. DNA from a small
number (10-20) of *Rr* progeny is combined to form one pooled DNA sample and likewise
a small number of *rr* progeny DNA samples are combined to form the other pooled DNA
sample. In the next step, these two pooled-DNA samples are assayed for a large number of
genetic markers such as RAPDs or AFLPs. The dominant markers tightly linked to the
gene controlling the qualitative trait are detected by being present in one DNA pool and
absent in the other DNA pool (Fig. 18.5). Markers that are not tightly linked to the gene of
interest are either present in both pools or absent in both DNA pools.

BSA was first used in trees by Devey *et al.* (1995) to map a gene for resistance to
white pine blister rust (*Cronartium ribicola*) in *Pinus lambertiana*. BSA has been used to
map genes for resistance to black leaf spot (*Stegophora ulmea*) in *Ulmus parvifolis* (Benet
et al., 1995), to *Melampsora larici-populina* in *Populus* hybrids (Cervera *et al.*, 1996; Villar *et al.*, 1996), and to fusiform rust (*Cronartium quercuum*) in *Pinus taeda* (Wilcox *et
al.*, 1996). BSA has also been used to map the pendula gene controlling the narrow crown
phenotype in *Picea abies* (Lehner *et al.*, 1995).

FUNCTIONAL GENOMICS

Gene discovery and structural genomics can provide a wealth of information about the
types and numbers of genes encoded in the genome, but they provide little understanding
of the function of all these genes. Functional genomics seeks to understand the function of
all genes in the genome using techniques that often allow study of hundreds or thousands
of genes in parallel. Gene function can be assessed at the biochemical, cellular, developmental and adaptive level. Functional genomic experimental methods are developing rapidly; we discuss just a few techniques that are used in forest trees.

Comparative Sequencing

The simplest way to predict the biochemical function of a gene is to determine its DNA
sequence and compare it to DNA sequences of genes of known function in databases. This
activity is a routine component of gene discovery, as was discussed earlier. For example, if
an EST from *Pinus* matches the DNA sequence for an alcohol dehydrogenase (ADH) gene
from corn, then the predicted function of the EST is as an ADH gene. This is not absolute
proof; that can only be determined through biochemical assays. The limitation of using
EST database comparison to assign biochemical function to gene sequences is that only a
proportion of the genes can be identified in this way. For example, only 55% of the ESTs
identified in a *Pinus taeda* study could be assigned function (Allona *et al.*, 1998) and only
39% in a *Populus* study (Sterky *et al.*, 1998).

Once a large number of ESTs from different tissue types and/or developmental states are obtained and functions putatively assigned based on database comparison, it is possible to ask additional questions relating to gene function. For example, Whetten *et al.* (2001) obtained 22,233 ESTs from several wood-forming tissues in *Pinus taeda*. They found quantitative differences in the abundance of mRNAs involved in lignin and cellulose biosynthesis between compression wood samples and normal wood samples. Comparisons of gene content and gene expression across species can also be made. Kirst *et al.* (2003) identified gene homologs from the angiosperm *Arabidopsis thaliana* in a sample of 20,377 genes from *P. taeda*. These results suggest that genes may be highly conserved in seed plants since angiosperms and gymnosperms diverged from one another more than 300 million years ago.

Gene Expression Analysis

Traditionally, levels of transcription are determined from Northern blot analysis (*Chapter 2*) where separate assays are performed to assess mRNA abundance for each gene. PCR-based techniques have also been devised to measure mRNA abundance, but like Northern blot analysis, only a small number of genes can be analyzed at once. A new technique, called **DNA microarray analysis**, makes it possible to study differential gene expression for thousands of genes at once (Schena *et al.*, 1995; Schenk *et al.*, 2000).

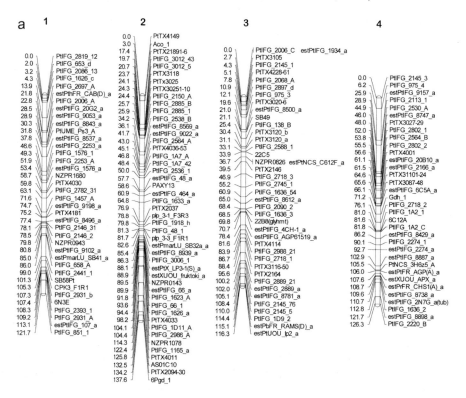

Fig. 18.4. *Pinus taeda* genetic linkage map showing the positions of genetic markers on each of the 12 chromosomes of *Pinus taeda*: a) Linkage groups 1-4; b) Linkage groups 5-8; and c) Linkage groups 9-12. Each linkage group likely corresponds to one of the 12 chromosomes in this species. (Based on data from Sewell *et al.*, 1999)

(Fig. 18.4 continued on next page)

Fig. 18.4. (Continued from previous page)

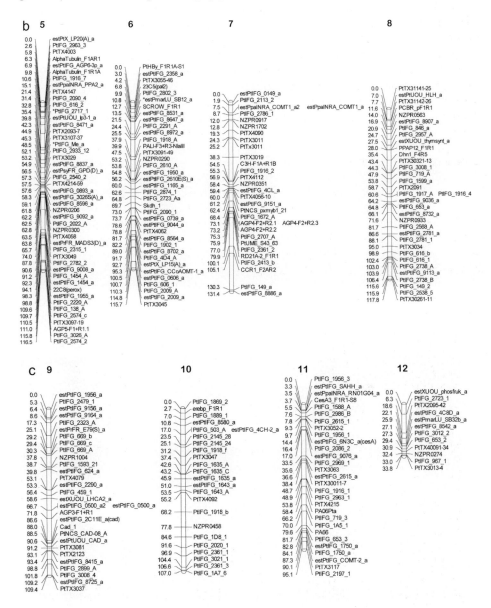

Small amounts of DNA from 1000 or more ESTs are first arrayed onto glass slides (Fig. 18.6). RNA is then isolated from two samples for which differences in gene expression will be measured. For example, one might wish to know which genes are expressed in xylem tissue compared to phloem tissue or in trees resistant versus susceptible to a particular disease. The RNA isolated from each of these samples is labeled with a different fluorescent dye (*e.g.* one with red, the other with green) and the two samples are hybridized to the same glass slide containing the ESTs spotted in a grid pattern. Specialized equipment is then used to measure the amount of fluorescence at each position on the grid. The software then combines results from both fluorescent scans into a single display. EST spots showing

a

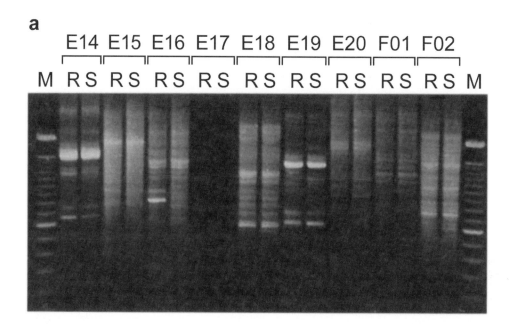

b

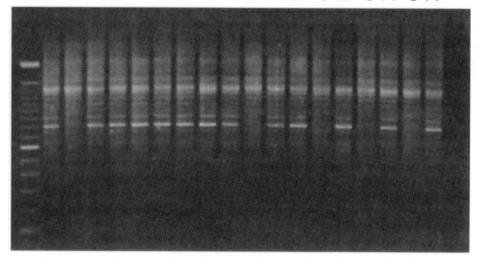

Fig. 18.5. Bulked segregant analysis to identify markers linked to a dominant resistant gene to white pine blister rust (*Cronartium ribicola*) in *Pinus lambertiana*: a) Pooled DNA samples from resistant (R) and susceptible (S) seedlings were assayed with several RAPD markers (E14, E15, etc). RAPD marker E16 clearly shows the presence of a band in the R pool that is absent in the S pool. This indicates that the E16 marker may be linked to the resistance gene; and b) Assay of the DNA samples from individual genotypes comprising the R and S pools confirms that the E16 RAPD marker is linked to the resistance gene. (Reproduced with permission from Devey *et al.*, 1995; copyright 1995, National Academy of Sciences, USA)

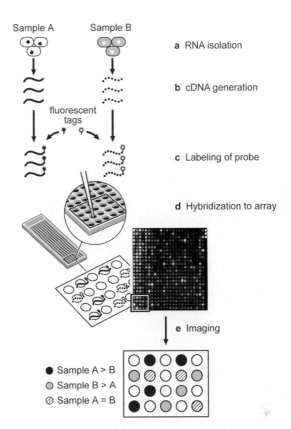

Fig. 18.6. Gene expression analysis using DNA microarrays: a) mRNA samples A and B are prepared from individuals subjected to different treatments, of different genotype, from different developmental stages, different tissue types, etc.; b) cDNA is synthesized from mRNA; c) cDNA is labeled with fluorescent dyes, red (dark) in sample A and green (light) in sample B; d) labeled cDNAs are hybridized to arrays containing hundreds or thousands of known ESTs in specified locations; and e) the amount of fluorescence at each position in the array is evaluated to determine gene expression differences between the samples for that gene and this is repeated for all locations (*i.e.* genes on the array). Open circles represent positions where no hybridization occurred.

the color of one of the sample types (red *vs.* green) reveals a gene more abundantly expressed in one tissue versus the other tissue. A spot showing intermediate color (yellow) represents a gene whose level of expression is roughly equivalent in both samples.

By quantifying differential gene expression for large numbers of genes in parallel, it becomes possible to identify coordinated patterns of gene expression and regulatory networks. Microarray analysis is a very powerful tool for understanding patterns of differential gene expression between tissues, during development, in response to stresses, and among genotypes. The first application of DNA microarray analysis in a forest tree was by Hertzberg *et al.* (2001) who measured differences in the expression of 2995 genes among different developmental stages of xylem formation in a *Populus* hybrid. DNA microarray analysis has also been used to study patterns of gene expression in xylem tissues over a seasonal cycle (Egertsdotter *et al.*, 2004). In addition, patterns of gene expression during embryogenesis (van Zyl *et al.*, 2002; Stasolla *et al.*, 2003) and during drought stress (Heath *et al.*, 2002) have been investigated. Clearly, the application of this technology has just begun and our understanding of the coordinated patterns of gene expression in trees will develop rapidly in the future.

Forward and Reverse Genetic Approaches

There are two general approaches to understanding the function of a gene: forward genetics and reverse genetics (Fig. 18.7). **Forward genetics** begins with a well-characterized phenotype, such as a tree resistant to a disease, and then works toward identifying the gene(s) responsible for the phenotype. **Reverse genetics** begins with a gene, for example a protein kinase, and works toward determining which phenotype(s) it determines. Forward genetics approaches such as T-DNA tagging, transposon tagging and gene or enhancer traps require inserting foreign DNA into the host tree genome (*i.e.* gene transformation, see *Chapter 20*). These methods alter the expression of the target gene in some manner such that it then reveals a relationship with a specific phenotype that has been altered by the insertion of the foreign gene. Genetic mapping approaches such as quantitative trait locus (QTL) mapping and association mapping, discussed later in this chapter, are also forward genetic approaches and are often used because gene transfer is not required.

Reverse genetic approaches such as gene silencing by RNA interference (RNAi) or anti-sense RNA (see *Chapter 20*) are methods whereby foreign DNA of some kind is introduced into the host genome and the expression of individual genes is in some way disrupted. In some cases, the disruption of a gene can cause a visible mutation in the plant. Some of these reverse genetic approaches can be applied on a large scale thus enabling the assignment of many genes to specific functions. These approaches, however, require the ability to genetically transform the host plant (*Chapter 20*), so they are likely to be used in just a few tree species, such as in *Populus*, which can easily be transformed. Reverse or forward genetic approaches requiring transformation are less likely to be used in conifers in the near future because of the difficulties with transforming conifers.

Quantitative Trait Locus (QTL) Mapping

One of the most common applications of genetic maps in forestry is for **QTL** mapping. As the name implies, the goal of QTL mapping is to identify the chromosomal regions within which one or more genes reside that effect the quantitative trait. In practice, chromosomal regions identified in QTL mapping experiments rarely include just a single gene but many other genes not related to the trait. Sax (1923) developed the theoretical basis for QTL mapping and the method was first empirically demonstrated by mapping bristle number genes in *Drosophila* (Thoday, 1961). It was not until the advent of molecular genetic maps in the late 1980s that QTL mapping was applied to a variety of plant and animal species.

There are four basic components to all QTL mapping experiments: (1) Segregating or mapping population(s); (2) Phenotypic measurements of the quantitative trait(s) for all members of the mapping population; (3) Genetic marker data; and (4) Statistical analysis for mapping QTLs and estimating the magnitude of their influence on phenotype (Box 18.1). Quantitative trait phenotypes (*e.g.* tree size) and marker genotypes (*e.g.* BB, Bb and bb) are scored on all members of a mapping population and then a statistical analysis is performed to associate phenotype with marker genotype which provides evidence for the existence of a QTL. For the example shown in Box 18.1, the B locus appears to be associated with tree size because BB homozygotes are large trees, Bb heterozygotes are immediate size trees and bb homozygotes are small trees. It is easy to see how the effects of alternate alleles at the B locus on tree size phenotype can be estimated.

Each of the four components of QTL mapping is discussed below in the context of QTL mapping in forest trees. The fundamental difference between QTL mapping in agronomic crops, such as corn, soybean and tomato, *versus* forest trees, is that inbred lines generally do not exist for trees. This difference affects the types of mapping populations, choice of genetic marker systems and statistical methods that can be used for detecting QTLs.

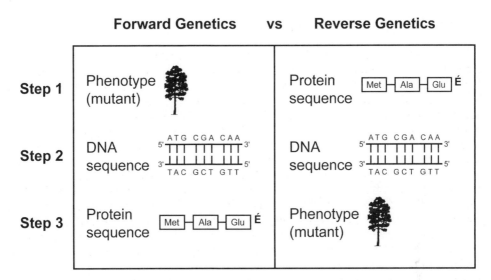

Fig. 18.7. Forward and reverse genetic approaches to understanding the function of genes. The forward genetic approach begins with phenotype and works toward identifying the gene that determines the phenotype. Reverse genetics begins with a gene or protein and works toward understanding its effect on the phenotype.

Mapping Populations for QTL Estimation

A variety of population types are used in QTL mapping of forest trees. The most important consideration in choosing a mapping population is to maximize segregation of the QTLs. Due to the long generation times in many forest tree species, mapping populations are often selected from among existing populations, such as those used in breeding programs. Mapping populations can be derived from: (1) Inter- or intra-specific crosses; (2) Outbred or inbred pedigrees; (3) Single- or multi-generation pedigrees; (4) Full-sib, half-sib or open-pollinated families; and (5) Combinations of all these family structures.

The most commonly used QTL mapping population types are: (1) The three-generation outbred pedigree; (2) The open-pollinated family; and (3) The two-generation "pseudotestcross." Groover *et al.* (1994) first used the three-generation outbred pedigree to map QTLs for wood specific gravity in *Pinus taeda*. This mapping population type most resembles the F_2 used in inbred crops (Fig. 18.8). Two pairs of grandparents (first generation) are chosen: one grandparent of each pair being from the opposite end of the phenotypic distribution for the quantitative trait within the population from which they were selected. This helps to ensure that the two grandparents within a pair will differ genetically at the QTL locus. A single F_1 individual of intermediate phenotypic value is chosen from each of these matings (second generation) and mated; their offspring (third generation) form the segregating mapping population.

Several variations on the three-generation outbred type have been used, such as the three-generation interspecific *Populus* hybrid mapping populations (Bradshaw and Stettler, 1995; Frewen *et al.*, 2000; Howe *et al.*, 2000). In general, a three-generation pedigree, combined with highly informative genetic markers, can track up to four different QTL alleles potentially segregating at any one QTL locus; however, three-generation pedigrees with large families do not exist for most forest tree species and can take many years to develop.

Box 18.1. Quantitative trait locus (QTL) mapping in forest trees.

Step 1. Mate two parent trees at the extremes of the phenotypic distribution of the trait (*e.g.* a large tree and a small tree) that are also different for a large number of genetic markers (*e.g.* AA versus aa) (Fig. 1). Mate an F_1 progeny that is intermediate for the phenotype and heterozygous at marker loci with a similar F_1 (F_1'). This can be either a full-sib mating or more often a cross between two unrelated F_1's. The progeny resulting from the F_1 x F_1' mating segregate for both the phenotype and the genetic markers.

Step 2. Each of the progeny is genotyped at all marker loci. Many different marker types can be used (*Chapter 4*). In the example the B locus marker is codominant so both homozygotes (BB and bb) and the heterozygote (Bb) can be scored.

Step 3. A statistical test is then performed to test for differences in mean phenotypic values among genotypic classes. In this example, BB genotypes are associated with large trees, Bb heterozygotes with intermediate size trees and bb homozygotes with small trees. The inference that can be drawn is that a QTL for tree size resides on the chromosome somewhere near the B genetic marker. This analysis is performed for all markers on all chromosomes to discover QTLs controlling the phenotype.

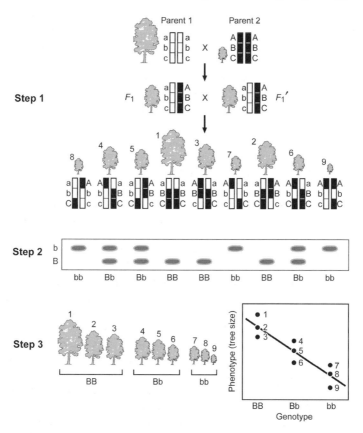

Fig. 1. A schematic description of the basic approach to quantitative trait loci (QTL) mapping in an outbreeding forest tree.

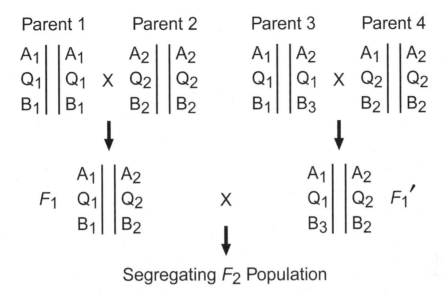

Segregating F_2 Population

Fig. 18.8. Example of a three-generation pedigree that might be used for QTL mapping in a forest tree. In the ideal situation, as in the pedigree shown, the markers (A and B) and QTL (Q) loci are homozygous for alternative alleles in the P_1 and P_2 parents. Thus the F_1 parents are completely heterozygous and the linkage phase of marker and QTL alleles is known. However, more complex configurations are possible in trees, as shown in the cross between P_3 and P_4, where there may be more than two alleles at any locus and parent trees are not necessarily homozygous at all marker loci. The linkage phase of marker and QTL alleles in the F_1', parents are not known with certainty. QTL mapping with these pedigree structures is possible but is more complicated than the standard F_1 and backcross pedigrees used with inbred crops.

Alternatives to multi-generation pedigrees are open-pollinated or half-sib families and two-generation full-sib families. In conifers, a population of seed from an individual tree can be used for QTL mapping, where the megagametophytes are employed for genetic marker analysis of segregation in gametes from the maternal parent. Most often, dominant markers (*e.g.* RAPDs and AFLPs) have been used with this approach. Phenotypes are measured on diploid offspring of open-pollinated or half-sib progeny. Because genetic marker data are obtained from megagametophytes, which represent only meioses in the maternal parent, there is no genetic marker information from the paternal parent of the progeny, and QTLs contributed from the pollen parents cannot be estimated. This approach has less power to estimate QTLs, but can be easily applied to any tree where megagametophytes can be assayed in offspring, and does not require crossing.

In angiosperm forest trees such as *Populus* and *Eucalyptus*, the haploid megagametophyte system is not possible; however, a similar approach, called the pseudotestcross, has been used (Grattapaglia *et al.* 1995). Phenotyping and marker genotyping are performed on the progeny of a full-sib cross. However, dominant markers in the backcross configuration only (*e.g.* Aa x aa or aa x Aa) are used to track QTL segregation in the parents. Therefore, QTLs segregating in each of the parents are estimated independently and it is not possible to simultaneously estimate QTLs that are segregating in both parents, which in crosses made between forest trees is often the case. This approach can be applied to any full-sib cross, but again has low power.

Phenotypic Measurements on Quantitative Traits

The success of a QTL mapping experiment is in part dependent upon the genetic control of the quantitative trait. The first consideration often made is to estimate the number of genes controlling a trait. Theoretical methods have been developed for estimating the number of genes controlling a complex trait, but in practice these were not used in forest trees prior to QTL mapping experiments. The question is often whether a trait is purely polygenic (controlled by many genes of small effect) or is oligogenic (controlled by a smaller number of genes of larger effect). Intuitively, QTLs for a trait under oligogenic control should be easier to detect than for a trait under polygenic control. In practice, QTL mapping experiments must be performed without this knowledge and draw inference of genetic control following the outcome of the experiment. As will be described later, QTL mapping experiments have contributed significantly to the understanding of the genetic control of complex traits in forest trees.

Factors that affect power and precision of QTL estimation are much the same as those that affect quantitative genetic parameter estimation (*Chapter 15*). Minimizing environmental variation and therefore increasing heritability, increases the power to detect QTLs. Clonal testing of the progeny in the mapping population is the most efficient way to minimize the effect of environmental variation and obtain better estimates of the phenotypic value because data on multiple randomized copies (ramets) of a given genotype are averaged together (Bradshaw and Foster, 1992).

The quantitative traits chosen for QTL mapping in forest trees are much the same as those involved in tree improvement because of the interest in marker-aided breeding (*Chapter 19*). Stem growth, wood quality, hardiness, disease resistance, and reproductive traits have been included in QTL mapping experiments (Sewell and Neale, 2000). QTL mapping might also help to understand the genetic architecture of non-commercial traits and provide insights into genome evolution and speciation.

Genetic Markers and Maps

The types of genetic markers available for use in forest trees were described in detail in *Chapter 4*. Allozymes, RFLPs, RAPDs, AFLPs, SSRs, and ESTPs are all useful for QTL mapping, although some are better suited than others for application with different mapping population types. This is due to the extent of **marker informativeness** of different marker types (Table 18.3). Informativeness is a general term, which describes how completely a genetic marker "marks" the segregating variation in the region of the genome where it is located. For example, codominant markers (*e.g.* RFLPs, SSRs and ESTPs) are more informative than dominant markers (*e.g.* RAPDs and AFLPs) because the heterozygote (Aa) can be distinguished from the dominant homozygote (AA). However, there is no loss of information using dominant markers in conifer haploid megagametophyte mapping populations, because genotyping is performed on gametes (A or a) and not on zygotes (AA, Aa or aa).

Genetic marker data need not be organized into the form of a genetic map to perform QTL detection (see single-factor approach below). However, the goal of most QTL experiments is to both detect the presence of QTLs and determine their location in the genome. Thus, genetic maps are usually constructed from marker data prior to QTL mapping. One type of QTL detection approach, called interval mapping (see discussion of interval mapping below) requires that markers be mapped.

Statistical Methods for Detecting QTLs

In theory, the basic statistical approach for detecting QTLs is simple: all methods test for a relationship between the quantitative trait values and the marker genotypic classes in the mapping population. In practice, however, statistical methods and computational procedures can be quite complex, with two general approaches distinguished: (1) The single-factor approach; and (2) The multi-factor or interval approach.

The single-factor approach uses analysis of variance or regression analysis to test for differences in the quantitative trait means among marker genotypic classes (Edwards *et al.,* 1987). For example, does the height value mean of individuals in the AA genotype class differ from that of individuals in the aa genotypic class. Analyses are performed one marker at a time. A statistically significant association (*i.e.* significant F-value) is considered evidence that the QTL has a map location somewhere near the genetic marker. The single-factor approach requires only basic statistics and can be performed using standard statistical analysis software packages. The major limitation of this approach is that the exact position of the QTL, relative to the genetic marker, cannot be easily determined. This limits the power of detection and precision in the estimation of the magnitude of effects of the QTL on the phenotype (Lander and Botstein, 1989).

The multi-factor or interval approach was developed to overcome the limitations of the single-factor approach (Lander and Botstein, 1989). QTLs are detected using the information in pairs of markers that flank a segment of a chromosome. The statistical estimators are complex, often using maximum likelihood procedures and computationally demanding solutions. The interval mapping software Mapmaker/QTL has been widely used in forest trees, even though it was written for use with inbred species (Lincoln *et al.,* 1993). The interval mapping method of Knott *et al.* (1997) was designed specifically for use in outbred forest trees.

QTL Discovery in Forest Trees

QTL mapping research in forest trees has provided new insight into the genetic control of complex traits by: (1) Estimating the numbers and magnitude of effects of QTLs; (2) Revealing the genomic locations of QTLs; and (3) Describing environmental and developmental patterns of QTL expression. Knowledge of such factors can, in some cases, be used directly in tree improvement through marker-aided breeding (*Chapter 19*) or indirectly through influencing the choice of breeding designs.

QTLs have been identified for a variety of traits and in a variety of species and were reviewed in a paper by Sewell and Neale (2000). Recent examples of QTL studies in some representative forest tree genera include: *Pinus* (Hurme *et al.,* 2000; Sewell *et al.,* 2002; Brown *et al.,* 2003; Devey *et al.,* 2004); *Pseudotsuga* (Jermstad *et al.,* 2003); *Populus* (Frewen *et al.,* 2000); *Eucalyptus* (Thamarus *et al.,* 2004; Kirst *et al.,* 2005); *Quercus* (Saintagne *et al.,* 2004; Scotti-Saintagne *et al.,* 2004); and *Fagus* (Scalfi *et al.,* 2004). In general the number of QTLs detected and the relative magnitude of their effects vary little among traits (Sewell and Neale, 2000). With a few exceptions, individual QTLs for growth and development, wood quality, adaptability and reproduction each account for no more than 5-10% of the phenotypic variance in their respective traits and the number of QTLs detected per trait is generally less than ten.

These results suggest that the polygenic mode of inheritance is most likely for quantitative traits of commercial interest in forest trees. QTLs fitting the oligogenic model might eventually be detected in specific crosses involving trees carrying major genes, but it seems safe to conclude that few major QTLs will be detected such that a single QTL, alone, determines a large proportion of the variation in a quantitative trait.

Determination of the chromosomal location of QTLs in genomes will be important to enable use of transgenic technology in tree improvement (*Chapter 20*). Eventually it may become possible to genetically engineer the promoter regions of genes coding for QTLs or even replace endogenous genes with engineered genes (*Chapter 20*). These types of genetic manipulations, however, require precise knowledge of the location of genes to be modified. The molecular basis of pleiotropy may also become apparent as QTLs for many different traits are localized on genetic maps.

QTLs, just as the quantitative traits themselves, can vary in their expression developmentally and when trees are grown in different environments. However, QTLs that have stable expression across time and space will be most valuable for marker-aided breeding applications. Developmental stability of QTL expression has been shown for growth traits in *Pinus pinaster* (Plomion and Durel, 1996) and wood quality traits in *Eucalyptus* (Verhaegen *et al.*, 1997) and *Pinus taeda* (Sewell *et al.*, 2000; Brown *et al.*, 2003), but not for growth traits in *Pinus taeda* (Kaya *et al.*, 1999) or *Populus* (Bradshaw and Stettler, 1995). Likewise, QTL by environment interactions have been shown for wood density in *Pinus taeda* (Groover *et al.*, 1994) and for bud flush in *Pseudotsuga menziesii* (Jermstad *et al.*, 2001, 2003). Clearly, patterns of QTL expression are complex and experiments need to be repeated across environments and years to fully understand how specific QTLs determine quantitative trait phenotypes.

Positional Cloning of QTLs

Once a QTL or simple Mendelian trait has been precisely mapped, it then may be possible to clone the underlying gene based solely on the knowledge of its genetic map position. Once cloned, its DNA sequence and ultimately its biochemical function can be determined. This is known as **positional or map-based cloning**. The first step is to identify a large piece of cloned genomic DNA that contains two mapped genetic markers which flank, on opposite sides, the QTL or gene coding a simple Mendelian trait. Large-insert DNA libraries from the host genome are constructed in yeast artificial chromosomes (YACs) or in bacterial artificial chromosomes (BACs) (Fig. 18.9).

In the past, genetic maps were not sufficiently dense with genetic markers such that it could be expected that the genetic markers flanking the qualitative trait gene would reside on the same YAC or BAC clone. A technique called "chromosome walking" was employed to identify the target gene. A physical map was constructed in the region including the QTL gene by ordering the overlapping YAC or BAC clones based on genetic markers or DNA sequences contained within the clones. Eventually a single YAC or BAC clone could be identified that contained the target QTL and its DNA sequence could be determined. Several QTLs may be encoded on the YAC or BAC clone, so it is yet another step to identify the true target gene. With the development of very dense genetic maps the "walking" step can now often be eliminated, if it is likely that the flanking markers reside on the same YAC or BAC. This approach is called "chromosome landing" (Tanksley *et al.*, 1995) and has been used to clone a number of disease resistance genes in crops. A final step that is required to confirm positional cloning of the gene is to insert the cloned gene into a host genome that does not have the phenotype, for example, inserting a cloned disease resistance gene into a susceptible host. Strong constitutive or inducible promoters are usually attached to the target gene to ensure that it is expressed once integrated into the new host genome. This confirmation test is called a complementation test.

The feasibility of positional cloning of genes is highly dependent on the size of the genome. Many genes have been cloned in this way from species with small genomes such as *Arabidopsis* and rice, but the task is much more difficult and expensive for species with

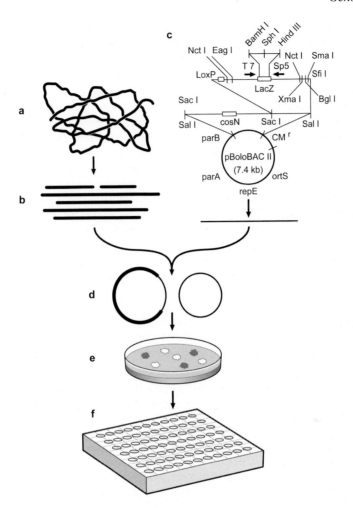

Fig. 18.9. Bacterial artificial chromosome cloning: a) High molecular weight DNA is isolated; b) This DNA is cut into large fragments using a restriction enzyme; c) The bacterial chromosome vector is prepared to accept large insert DNA; d) The DNA fragments are inserted into the bacterial chromosome; e) The bacterial colonies containing the cloned DNA are plated out onto agar plates; and f) DNA is isolated from the bacterial chromosomes and put into microtiter plates.

large genomes. No genes have yet been positionally cloned from forest trees. However, several efforts in *Populus* are underway (Stirling *et al.*, 2001; Zhang *et al.*, 2001). Positional cloning of genes from conifers is a daunting task and success using this approach may have to await the development of new technologies.

Association Genetics

QTL mapping studies are very useful for estimating the number, size of effects and approximate location in the genome of genes controlling complex traits. As we discussed in the last section, however, it is a very difficult task to identify the exact gene underlying the QTL. Another approach, called **association mapping**, can be used to more precisely determine the exact genes controlling complex traits, and therefore ultimately identify mutations responsible for phenotypic differences among individuals for a quantitative trait. Each functional variant (*i.e.* mutation) is referred to as a **quantitative trait nucleotide (QTN)**.

The fundamental difference between QTL mapping and association genetics is that the former relies on genetic linkage (*Chapter 3*) following a generation or two of crossing, whereas the latter takes advantage of population-level linkage disequilibrium (LD) (*Chapter 5)* between genetic markers and QTNs following many generations in a large inter-mating population (Fig. 18.10). The association mapping approach has been developed for identifying genes controlling complex traits in humans (Risch, 2000; Cardon and Bell, 2001; Weiss and Clark, 2002) and has only recently been applied to plants (Remington *et al.*, 2001; Rafalski, 2002; Neale and Savolainen, 2004).

There are two basic approaches to association genetics: (1) Genome scan; and (2) Candidate gene. In the genome scan approach, genetic markers are populated throughout the genome so that the entire genome can be searched for QTNs. In the candidate gene approach, genetic markers are used only within individual genes that are thought to be involved in determining the phenotype. The genome scan approach is thus more exhaustive in its search but is also more expensive to conduct.

Association genetics requires four components similar to those necessary for QTL mapping: (1) A mapping population; (2) Phenotypic measurements of the quantitative trait(s) for all members of the mapping population; (3) Genetic marker data for all members of the mapping population; and (4) A statistical method for associating genotype with phenotype. Issues related to phenotyping are the same as with QTL mapping and need no further discussion; however, there are many differences for the other three components.

Association Mapping Populations

An association population is generally constructed by sampling individuals from a large random mating population. However, some association statistical tests such as the transmission disequilibrium test (TDT) require family structure (Lynch and Walsh, 1998). The informational content of an association genetics population depends on the amount of population-level LD between genetic markers and QTNs. The amount of LD in turn depends on population history, such as the occurrence and degree of past bottlenecks (*Chapter 5*) and subsequent recombination. For example, a population that has undergone a recent bottleneck and/or has a very low rate of recombination would have high LD relative to a large random mating population with high recombination. Most, but not all, natural forest tree populations are expected to fall into the latter category. However, artificially constructed populations, such as breeding populations, might be expected to have much higher LD. The greater the LD, the more likely that associations between genetic markers and QTNs will be detected. This is advantageous for breeding applications (*Chapter 19*); however, if the goal is to exactly pinpoint the QTN, somewhat less LD would be required. This is because with high LD more than one genetic marker could be in complete LD with the QTN making it impossible to ascertain which polymorphism is truly the QTN.

Simulation studies show that association mapping populations should include at least 500 individuals (Long and Langley, 1999). In addition, measures should be taken to increase precision of evaluating phenotypes, including use of appropriate field designs (*Chapter 14*) and clonal replication of individuals (*Chapter 17*).

Single Nucleotide Polymorphisms

Association genetics is based on LD over short physical distances along chromosomes and therefore requires detection of polymorphisms over these short distances. Most of the

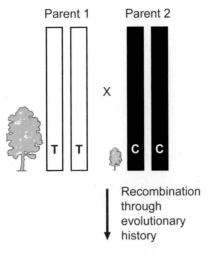

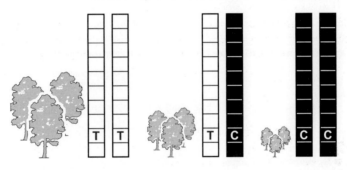

Fig. 18.10. Association genetics of complex traits in forest trees. Association genetics is similar to QTL mapping (Box 18.1) except that chromosomes being mapped are more highly recombined than those of QTL mapping populations due to the accumulation of recombination events over evolutionary time. This enables greater resolution of genes controlling complex traits. In this example, a T/C single nucleotide polymorphism (SNP) is associated with tree size; therefore, this region of the chromosome is presumed to include a gene that in part determines tree size.

genetic marker types described in *Chapter 4* are not sensitive enough for this purpose. Therefore, markers derived from **single nucleotide polymorphisms (SNPs)** are primarily utilized in association genetics. SNPs are nucleotide differences (*i.e.* polymorphisms) occurring at specific locations in the genome that have arisen through mutations, for example, T and C in Fig. 18.10. Methods based on DNA sequencing are used to discover SNPs. In *Pinus taeda*, it is estimated that a SNP occurs once every 60 base-pairs (bp) (Brown *et al.*, 2004). In contrast, SNPs are found approximately 1/1000 bp in humans; therefore SNPs are clearly abundant in some forest trees and their discovery is straight-forward. SNPs are usually bi-allelic, although very rarely (<1%) tri- or tetra-allelic SNPs are found. This is because the likelihood of a second mutation occurring at the same nucleotide position in a population is exceedingly small.

SNPs can be discovered either by multiple alignment and comparison of ESTs in electronic databases (electronic SNPs) or by *de novo* sequencing of a sample of individuals taken from a population. Haplotype-based association tests sometimes have more power to detect associations with phenotypes; therefore, it is desirable to either infer or determine directly the haplotypes of alleles in the population. A **haplotype** is defined as the distinct combination of SNPs on a single DNA fragment. In conifers, this can be easily accomplished by using haploid megagametophyte tissue. For example, there were approximately 16 different haplotypes among a sample of 31 loblolly pine megagametophyte DNA samples for the AGP6 gene (Fig. 18.11). These SNPs and haplotypes were determined by DNA sequencing.

Once SNPs have been discovered in a small sample, SNP genotypes of all members of the association population must be determined. In forest trees, where nucleotide diversity is generally quite high, it is necessary to prioritize SNPs for genotyping in large association populations because it would be cost-prohibitive to type all discovered SNPs. Priorities can be assigned based on the haplotype structure, functional potential of the SNP (*e.g.* synonymous or nonsynonymous) or tests of selective neutrality (*e.g.* whether a SNP is under selection or selectively neutral). There are many different ways to genotype specific SNPs, each method with its own specialized chemistries and instrumentation (see review by Syvanen, 2001), but all require high throughput technologies to enable genotyping hundreds or thousands of trees at hundreds or thousands of SNPs. SNP genotype methods are evolving rapidly and there is intense competition among commercial laboratories to lower assay costs and increase throughput.

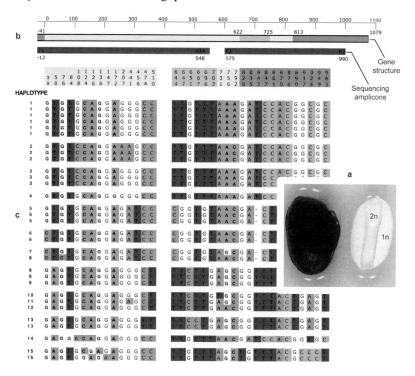

Fig. 18.11. An example of direct determination of haplotypes in conifers using 1n seed megagametophytes: a) Megagametophyes are excised from the seed and DNA is isolated from the megagametophyte tissue; b) Gene structure of the AGP6 gene from *Pinus taeda* -- positions 5′ to position − 41 make up the 5′ flanking region, − 41 to 622 is the 1st exon, 622 to 725 is an intron, 725 to 813 is the 2nd exon and from 1080 to 3′ is the 3′ flanking region. Two fragments (amplicons) were selected for DNA sequencing − 12 to 548 and 575 to 990); and c) Results of sequencing 31 megagametophyte DNA samples for both amplicons. Only nucleotide positions that were polymorphic among megagametophytes are shown. These are called single nucleotide polymorphisms (SNPs). In total, 16 haplotypes are revealed.

Association Tests

A statistical test for an association between a genetic marker (genotype) and a quantitative trait (phenotype) has similar properties to QTL detection. Members of the association mapping population are classified based on their genotype and standard statistical analyses (ANOVA or regression) can be performed to test for differences in the phenotypic means among genotypic classes. SNP genotypic classes can be assigned based on individual haploid or diploid SNPs or based on a multi-SNP haplotypes.

Association Mapping in Forest Trees

Association mapping experiments in forest trees have only recently been implemented. Gonzalez-Martinez *et al.* (2006) reported results from a candidate gene approach to search for associations between genes involved in lignin and cellulose synthesis and wood property traits in *Pinus taeda*. SNP discovery was performed in a sample of 32 megagametophytes (Brown *et al.*, 2004). Linkage disequilibrium between SNPs within genes was detected but the magnitude of LD decreased rapidly with distances between base pairs (essentially non-detectable at distances beyond 2000 base pairs) (Fig. 18.12). An association population of 425 *Pinus taeda* clones was evaluated for several wood property traits including wood specific gravity, microfibril angle and percent lignin and cellulose. Associations between SNP genotypes in a few of the lignin biosynthetic pathway candidate genes and wood property phenotypes were found. This preliminary study demonstrates that it is possible to use an association mapping approach to identify individual genes controlling complex traits in forest trees. This approach enables application of marker-assisted breeding for both within-family selection and between-family selection (*Chapter 19*).

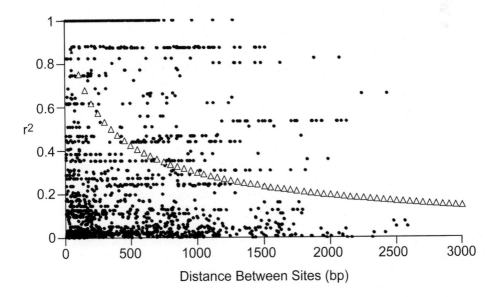

Fig. 18.12. Plot of the measure of linkage disequilibrium, r^2, versus distance in base pairs among polymorphic SNPs in 19 *Pinus taeda* genes. The fitted curve shows that linkage disequilibrium decays rather rapidly in *Pinus taeda*. (Reproduced with permission from Brown *et al.*, 2004; copyright 2004, National Academy of Sciences, USA)

COMPARATIVE GENOMICS

Comparative genomics has developed into an experimental science of comparing genomes among species. Comparisons are most often made at the level of DNA sequences and genetic maps. There are several plant and animal species which are rich in gene discovery and genetic map information, such as human, mouse, *Arabidopsis*, rice and poplar. These model organisms have: (1) Very dense genetic maps; (2) Thousands of ESTs sequenced and mapped; and (3) Genomes that are completely sequenced. For a number of plant model species, every gene and its genetic map position will soon be known. The function of all genes in model species is also well on the way to being understood.

For the vast majority of species, including tree species, much less genomic information will exist in the near future. However, the information from model species can be directly "accessed" if genetic maps can be compared. This situation is already true in several major species groups, most notably in the grasses (Bennetzen and Freeling, 1993). The genetic maps of rice, corn, sorghum, barley, wheat, rye, millet, and sugarcane can be directly compared due to the discovery of synteny among the genomes of these related species. **Synteny** means the conservation of gene order on linkage groups or chromosomes following speciation and evolution. Although chromosome number and ploidy levels can vary among related species, large chromosomal segments with genes in the same order can be identified. Candidates for qualitative trait genes known only by phenotype or genes underlying QTLs in non-model species can be identified by comparing to map positions in model species. This approach of identifying genes from related species based on their similar location in the genome can facilitate the "chromosome landing" method described in a previous section (see *Positional Cloning of QTLs*).

Comparative mapping between distantly related plant groups should be possible in the future, for example between *Populus* or *Eucalyptus* and *Arabidopsis*, but for the moment comparative mapping in trees is limited to comparisons between species within a genus or family. Comparative maps have been constructed in the Pinaceae and Fagaceae families and for the genus *Eucalyptus*.

Comparative mapping requires **orthologous genetic markers** to be mapped in each of the species being compared. Orthologs are genes that have descended from a common ancestral locus, whereas paralogs are loci that have originated by gene duplications within an individual species (Fig. 9.7). Most of the anonymous marker types (RAPD, AFLP, SSR) cannot be used for comparative mapping because loci are not orthologous across species. However, SSR markers have been used for comparative mapping in *Eucalyptus* (Marques *et al.*, 2002) and Fagaceae (Barreneche *et al.*, 2003).

Genetic markers based on genic DNA sequences, such as RFLPs and ESTPs, are most useful for comparative mapping. Because RFLPs are assayed by Southern hybridization, both orthologs and paralogs are revealed, thus RFLPs can be used for comparative mapping as long as orthologs are identified. Devey *et al.* (1999) used RFLP loci from both *Pinus taeda* and *Pinus radiata* to construct comparative maps between these species. However, because orthologs and paralogs are not easily distinguished in RFLP markers and because they are difficult to apply, RFLPs are unlikely to be used widely for comparative mapping. ESTP markers have many positive attributes needed for comparative mapping (Temesgen *et al.*, 2001); ESTPs among species are usually orthologous and only occasionally paralogous. ESTP markers from *Pinus taeda* have been used to construct comparative maps between *Pinus taeda* and three other important species in the genus: *Pinus elliottii* (Brown *et al.*, 2001), *Pinus pinaster* (Chagne *et al.*, 2003) and *Pinus sylvestris* (Komulainen *et al.*, 2003). It is now possible to treat *Pinus* as a single genetic system and perform comparative genomic analyses among species within the genus (Fig. 18.13). A comparative map between *Pinus taeda* and *Pseudotsuga menziesii* has also been constructed (Krutovsky *et al.*, 2004), thus extending comparative mapping between genera within the Pinaceae.

BIOINFORMATICS AND DATABASES

The high-throughput technologies of genomic sciences enable the collection of overwhelming amounts of data. **Bioinformatics** combines aspects of biotechnology, statistics, information science, and computational biology to devise new ways to analyze and extract knowledge from the vast amount of genomic data. The primary types of genomic data are: (1) DNA sequences; (2) Genetic mapping data; and (3) Data resulting from functional analyses, *e.g.* from DNA microarray experiments. Aside from bioinformatic methods to analyze genomic data, databases must be developed, curated and made accessible to other researchers. Some of the more commonly used bioinformatic tools and database structures used in forest tree genomics are briefly described in this section.

The National Center for Biotechnology Information (NCBI) is the primary site in the United States for DNA sequence databases and DNA sequence analysis tools. The primary database is called GenBank. NCBI also provides on-line access to the Basic Local Alignment Search Tool (BLAST) programs, which are the primary tools used to search the data bases and identify matches among sequences. All of these resources are free and publicly available through the World-Wide-Web (http://www.ncbi.nim.nih.gov).

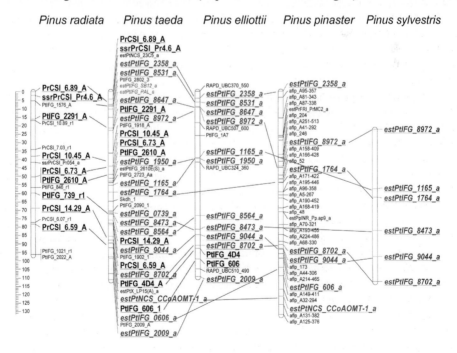

Fig. 18.13. Comparative genetic mapping in the genus *Pinus* showing a single homologous chromosome for *P. radiata, P. taeda, P. elliottii, P. pinaster* and *P. sylvestris*. Genetic markers shown in larger font are the orthologous genetic markers used to establish syntenic relationships between species.

The basic principles of linkage analysis and genetic mapping were described in *Chapter 4* and earlier in this chapter. Descriptions of linkage and QTL mapping methods can be found in Liu (1998). Rockefeller University hosts a web site of genetic analysis software, including linkage and QTL mapping programs (http://linkage.rockefeller.edu/soft/list.html). A number of software packages have been developed which provide programs for both linkage mapping and QTL analysis. Two such suites of programs are the Mapmaker/EXP (Lander *et al.*, 1987) and Mapmaker/QTL programs (Lincoln *et al.*, 1993) and the JoinMap (Stam and van Ooijen, 1995) and MapQTL (van Ooijen, 2004) programs. These and other genetic mapping programs have been used extensively with forest trees, although none are designed to optimally analyze forest tree data.

The primary repository of conifer genomic data is the TreeGenes Database that is maintained by the Dendrome Project at the Institute of Forest Genetics, Davis, California (http://dendrome.ucdavis.edu). TreeGenes contains a variety of data-types and is an object-oriented database that allows complex queries and searches. Through the use of databases and bioinformatic tools it is possible to perform experiments *in silico* and begin to understand all the complex relationships among genes and how they work together to determine the phenotype.

SUMMARY AND CONCLUSIONS

Genomics is the study of the structure, location, function, regulation and interaction of large numbers of genes simultaneously. Genomics is an extension of some of the traditional sub-disciplines of genetics, such as transmission and molecular genetics, to the entire genome. Genomics is a science that is made possible and is dependent upon technologies that allow rapid analysis of hundreds or thousands of genes and often in many individuals at once, such as by automated DNA sequencing.

The discovery and cataloging of all genes in the genome is an integral component of genomics. One approach is to determine the DNA sequence of the entire genome and infer genes from the DNA sequence. This approach has been applied to *Populus*, but is currently not feasible in conifers because of their large genome sizes. An alternative approach is to determine the DNA sequences for just the gene-coding regions. This can be accomplished by sequencing cDNA derived from mRNA of genes being expressed at the time of the experiment. These sequences are called expressed sequence tags (ESTs). ESTs are submitted to databases and compared to all other sequences in the database to see if they match genes whose function has already been determined. EST databases of tens of thousands of ESTs have been developed for *Pinus*, *Populus*, and *Eucalyptus*.

The construction of genetic linkage maps is another integral component of genomics. Genetic maps show the position of genes relative to one another on chromosomes and are valuable for understanding genome organization and evolution. Maps are extremely useful tools for identifying genes controlling interesting phenotypes. Qualitatively inherited traits, such as disease resistance genes, can be located on maps and then cloned based on knowledge of their map position. The map positions of the individual genes controlling quantitatively inherited traits, called QTLs, can also be determined from analyses using genetic maps. QTLs for a variety of growth, wood quality, and other economic traits have been identified. Knowledge of the number and size of effects of QTLs controlling a quantitative trait can assist tree breeders.

Cataloging and mapping all the genes is, however, only an initial step in genomics. The ultimate goal is to understand the function of all genes and their interactions. Techniques such as microarray analysis are used to study the expression patterns of genes. Eventually all genes in all biochemical pathways will be known, as well as how these genes and gene products interact. Functional genomic studies also seek to determine the relationship between the vast amount of allelic diversity in genes and the array of different phenotypes found in populations.

Comparative genomic analysis is possible in a few forest tree genera (*Pinus* and *Eucalyptus*) following the development of comparative genetic maps based on orthologous genetic markers. Complete genome sequencing in *Populus* will greatly enhance comparative genomic analysis in these taxa.

CHAPTER 19
MARKER-ASSISTED SELECTION AND BREEDING
– *INDIRECT SELECTION, DIRECT SELECTION AND BREEDING APPLICATIONS*

The principles of genetic improvement of forest trees were introduced in *Chapter 11* through a conceptual model called the breeding cycle. Each phase of the cycle is described in detail in *Chapters 12-17* in terms of classical tree improvement methods of selection, testing, breeding and deployment. In addition, the concepts of molecular markers (*Chapter 4*) and genomic technologies (*Chapter 18*) have been described for studying natural populations of forest trees (*Chapters 7, 8, 9* and *10*) and for understanding the inheritance of complex traits (*Chapter 18*). In this chapter, we describe how molecular markers and genomic technologies can be used in applied tree breeding programs.

While applications of genomic technologies and molecular markers are just beginning in tree improvement, these new technologies have the potential to enhance every phase of the breeding cycle (Fig. 11.1) by increasing genetic gains, shortening the generation interval, and eventually leading to genetic improvement based directly on the genotype instead of traditional approaches based solely on the phenotype. Potential applications fall into two broad categories: marker-assisted selection and marker-assisted breeding. **Marker-assisted selection (MAS)** is the selection of trees with desirable traits based on their molecular genotype. MAS can be used alone or in combination with the classical methods of selection discussed in *Chapters 13, 15* and *17*. **Marker-assisted breeding (MAB)** includes many applications of molecular markers to enhance mating designs, genetic testing programs, deployment strategies and overall quality control in tree improvement programs.

The chapter begins with a discussion of the *Concepts of Marker-assisted Selection* followed by two major sections describing different types of MAS: *Indirect Selection Based on Markers Linked to QTLs* and *Direct Selection Based on Genes Coding for Target Traits*. The final major section in this chapter, *Marker-assisted Breeding*, describes various other applications of genetic markers in breeding programs. While some of these applications have existed for many years, new methods will continue to develop and become integral parts of tree improvement as knowledge and technologies evolve.

There are many review papers on marker-assisted selection and breeding, although most focus on crops or livestock (Kearsey and Farquar, 1998; Kumar, 1999; Young, 1999; Dekkers and Hospital, 2002; Koebner and Summers, 2002; Morgante and Salamini, 2003; Barone, 2004; Francia *et al.*, 2005). A comprehensive review of marker-assisted selection and breeding in forest trees has not been written; thus, readers are referred to the primary literature cited in this chapter. In addition, marker-assisted selection has only been applied in privately owned forest companies, not in public breeding programs as in some agricultural crops; so, access to details from relevant case studies is not available.

CONCEPTS OF MARKER-ASSISTED SELECTION (MAS)

Definitions and Concepts Related to MAS

The goal of any tree improvement program is to increase the mean genotypic value of the breeding and production populations. Traits of long-standing interest to tree breeders include growth and bole volume, wood properties and resistance to disease. While significant genetic gains in these traits have been achieved using classical methods of selection (*Chapters 13, 15* and *17*), marker-assisted selection has the potential to enhance gains and shorten the generation interval.

One type of MAS is described in *Indirect Selection Based on Markers Linked to QTLs*. In this method, pedigreed mapping populations are used to identify associations (*i.e.* linkages) between molecular markers and quantitative trait loci (QTLs, *Chapter18*) such that selection for specific alleles at the marker locus results in increased frequency of favorable alleles at the QTL.

All forms of indirect selection involve selection for one trait to make improvement in a different trait, called the target trait (*Chapter 13*). Examples of classical indirect selection not involving molecular markers include: (1) Selection on seedling height to improve rotation-age bole volume (Lambeth, 1980); and (2) Selection on pilodyn penetration of trees in the field to improve total bole wood density (Sprague *et al.*, 1983; Watt *et al.*, 1996). The efficiency of indirect phenotypic selection compared to direct phenotypic selection on the target trait itself depends on the heritabilities of the two traits (indirect and target) and their genetic correlation (Equations 13.7 and 13.8). Indirect MAS is an extension of classical indirect selection, because selection is based on genetic markers, the indirect traits, that are correlated to target phenotypic traits. In indirect MAS, the correlation between the marker and the target trait normally results from the marker being linked to a region of the chromosome containing a gene affecting the target trait (*i.e.* a QTL, *Chapter 18*).

A second and developing, form of MAS is to select directly on the individual alleles at one or more loci affecting polygenic traits (see *Direct Selection on Genes Coding for Target Traits*). This form of marker-assisted selection requires knowledge, at the molecular level, of some or all of the genes controlling the target trait and can be viewed as direct selection, rather than indirect selection, since selection is for specific, favorable alleles at those loci. In this case, the markers are causal polymorphisms directly affecting the polygenic target trait (*e.g.* quantitative trait nucleotide, QTN, or insertion/deletions, *Chapter 18*). To be successful, direct molecular selection would likely need to be applied to many, if not all, of the individual gene loci that code for polygenic target traits. Indirect selection based on genetic markers is dependent on QTL mapping methods described in *Chapter 18*, whereas direct selection on genes coding for target traits is dependent on association genetics, also described in *Chapter 18*.

All forms of MAS can be applied separately or in conjunction with classical methods of selection (mass, family, within-family, combined and index selection) and can be utilized to make selections for selected, breeding and/or production populations. Two possible ways to utilize marker and phenotypic information together in selection programs are two-stage selection (*Chapter 13*) and combined index selection (*Chapters 13* and *15*). In two-stage selection, a first round of selections is made at a very young age based solely on marker genotypes measured on all potential candidates for selection. Those candidates having desirable alleles at the marker loci are chosen and only the selected individuals are established in field progeny tests. Later, when the tests are old enough, a second round of selection is made based on phenotypic measurements. The two stages of selection can be based on the same or different traits (*e.g.* selection for molecular markers related to disease resistance in the first round followed by selection for growth and wood density measured in the field tests).

Alternatively, marker data at one or more loci can be included in a combined mixed model analysis (*Chapter 15*) with phenotypic measurements made on the same and/or different individuals (*e.g.* Hofer and Kennedy, 1993). Then, selection is based on the predicted genetic worths of the individuals combining both marker and phenotypic data. For example, if two SNP loci are known to affect bole volume growth in a tree species, then trees growing in genetic tests could be measured for volume as well as genotyped at both SNP loci. All these data can theoretically be combined to predict aggregate genetic values.

Benefits, Limitations and Challenges of MAS

There are many benefits to marker-assisted selection as a supplement, or even as an alternative, to phenotypic selection. The primary advantages include: (1) Early selection that can potentially decrease the breeding cycle time; (2) Decreasing costs by reducing expensive progeny test establishment, maintenance, and measurement; (3) Increasing selection intensity (i, *Chapter 6*) because more individuals can possibly be evaluated in the laboratory using markers versus phenotypic selection in field tests; and (4) Increasing the relative efficiency of selection on low heritability traits (Lande and Thompson, 1990). Advantages 1-3 have been highlighted repeatedly in discussions of marker-assisted selection in a variety of crop, livestock and forest tree species. A symposium was held in 1991 in Gatlinburg, Tennessee, USA to discuss the potential of marker-assisted breeding in forest trees and the proceedings were published in 1992 in the *Canadian Journal of Forest Research* (Volume 22, Number 7). Case studies illustrating the advantages of marker-assisted selection were reported for several traits including wood properties (Williams and Neale, 1992), disease resistance (Bernatzky and Mulcahy, 1992; Nance *et al.*, 1992) and abiotic stresses (Tauer *et al.*, 1992). In addition, a paper was included on the advantages of clonal propagation of progeny for more precise phenotypic evaluation (Bradshaw and Foster, 1992). Although these papers were among the first to evaluate the potential of marker-assisted selection in forest trees, none directly addressed the relative efficiency of marker-assisted selection versus phenotypic mass selection. This issue was first addressed by Strauss *et al.* (1992b).

Despite the many potential advantages of marker-assisted selection, there are a number of significant hurdles to its effective application (Neale and Williams, 1991; Strauss *et al.*, 1992b). Many of these hurdles apply equally to the effectiveness of marker-assisted selection in any crop, livestock or forest tree species. Foremost, is the genetic architecture of the trait to be improved by indirect or direct marker-assisted selection. If the target trait is truly polygenic and under the control of hundreds or even thousands of gene loci, each with very small effect on the phenotype (Fisher's infinitesimal model, Fisher (1930)), it might be intractable to detect all such genes and identify associated markers.

Beavis (1995) showed through simulation that the numbers of QTLs controlling a polygenic trait are usually underestimated and the size of their effects on the phenotype (either proportion of phenotypic or genetic variance explained) is almost always overestimated in QTL detection experiments with modest numbers of progeny tested (~100 progeny). Ranges in the number of QTLs identified and the sizes of their effects vary considerably among published QTL detection studies in forest trees (Sewell and Neale, 2000). In general, however, it can be concluded that nearly all traits of economic interest in forest trees are controlled by many genes, and in most cases, individual genes have only a small effect on each trait. A few notable exceptions are genetic markers associated with disease resistance genes (Devey *et al.*, 1995; Wilcox *et al.*, 1996) that segregate as single Mende-lian characters and thus explain all the variation (*i.e.* susceptible or resistant). In all other cases, the challenge becomes one of detecting all the individual gene loci controlling the target trait. We return to a discussion of the feasibility of QTL identification when we con-trast the indirect selection approach based on markers linked to QTLs versus the direct selection approach based on specific alleles controlling phenotypes.

A second factor limiting the effectiveness of marker-assisted selection is variable de-tection of the QTL in different environments or genetic backgrounds leading to QTL by environment and QTL by genetic background interactions, respectively (Neale and Williams, 1991; Strauss *et al.*, 1992b). An example of the first interaction is when a QTL detected in one soil or climate is not detected in other soil or climatic conditions. The second type of interaction means that QTL detection is influenced by the specific mapping population used (*e.g.* a QTL may be detected in one family, but not another). QTL detection studies in forest trees have rarely been repeated across environments or across genetic backgrounds, but where they have (Brown *et al.*, 2003; Jermstad *et al.*, 2003), interactions have been found.

Clearly, if there are important QTL by environment interactions, then selection must take this into account just as phenotypic selection must account for g x e interactions. Nev-ertheless, this can be overcome if the investment is made in QTL detection across many environments. Then, just as with phenotypic selection, the breeder either selects for QTLs detected in most environments to develop stable genotypes, or alternatively, matches genotypes to sites by selecting on QTLs detected in specific types of planting environments.

The greatest challenge in the application of marker-assisted selection in forest trees is the expected low level of linkage disequilibrium (*Chapter 5*) between genetic markers and QTLs (*Chapter 18*, Neale and Williams, 1991; Strauss *et al.*, 1992b). The implication of low levels of linkage disequilibrium in natural and breeding populations of forest trees is that marker by QTL linkages might be in coupling phase in one genotype and repulsion phase in another genotype (Fig. 19.1). Therefore, linkage phase detected in one genotype (or segregating family) cannot be assumed in other genotypes (or families), because a spe-cific marker allele might be linked to the favorable QTL allele in one family but to the unfavorable QTL allele in other families. For this reason it has been concluded that marker-assisted selection in forest trees can only be applied to selection within the same families in which the QTL detection was performed (Strauss *et al.*, 1992b). We return to the topic of the disadvantage that linkage equilibrium may or may not cause when we con-trast the two general approaches to marker-assisted selection.

INDIRECT SELECTION BASED ON MARKERS LINKED TO QTLS

Indirect selection based on markers linked to QTLs was the first marker-assisted selection approach in forest trees imagined and has been the topic of several papers (O'Malley and McKeand, 1994; Kerr *et al.*, 1996; Kerr and Goddard, 1997; Johnson *et al.*, 2000; Kumar and Garrick, 2001; Wilcox *et al.*, 2001; Wu, 2002). Many different scenarios might be developed to apply indirect selection based on markers linked to QTLs; however, within-family selection (forward selection, *Chapter 17*) is the only application considered by most authors. Family or parental selection (backward selection) has been regarded as impracti-cal due to the general lack of strong linkage disequilibrium between markers and QTLs in tree populations (Fig. 19.1). For example, selection on the A_1 and B_1 marker alleles in Tree 2 would not lead to selection for the desired large tree phenotype (Fig. 19.1).

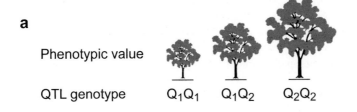

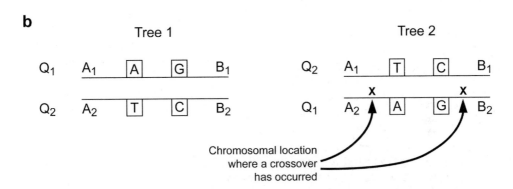

Fig. 19.1. Why linkage disequilibrium in forest tree populations makes it difficult to apply loosely linked flanking markers for family selection but is possible with very tightly linked single nucleotide polymorphisms (SNP). a) Assume there is a QTL, with two alleles in the population Q_1 and Q_2. The Q_1 allele contributes to small tree size and the Q_2 allele contributes to large tree size. b) The tree size QTL is mapped in Tree 1 using segregating genetic markers flanking the QTL. In Tree 1, the A_2 and B_2 marker alleles are in coupling linkage phase with the desirable QTL allele, Q_2. Progeny of Tree 1 carrying the A_2 and B_2 marker alleles would be selected in a marker-based breeding program. However, if marker alleles and QTL are in linkage equilibrium in the breeding population there will be many trees like Tree 2 where the A_2 and B_2 marker alleles are in coupling linkage phase with the undesirable QTL, Q_1. Selection of progeny from Tree 2 based on the A_2 and B_2 marker alleles would result in the selection of trees with small phenotypes (undesired phenotype). A solution to the limitation imposed by linkage equilibrium between flanking marker alleles and QTL alleles is to identify very tightly linked genetic markers where recombination has never, or rarely, occurred in the evolutionary history of the species. The A/T and G/C single nucleotide polymorphisms (SNPs) are examples of tightly linked genetic markers that could be used for selection in the progeny of both parent Tree 1 and Tree 2 (*i.e.* TC remains associated with Q_2 regardless of crossovers between the more loosely-linked QTL and marker (A,B) loci).

This limitation could be largely overcome if marker by QTL phase relationships are determined in all families in a breeding program (*e.g.* by constructing individual genetic maps for all selections in a breeding program); however, this idealized notion has not been attempted in tree breeding programs due to the typically large number of selections and the prohibitive cost of determining all phase relationships. Furthermore, it has been argued that family heritabilities are generally quite high and little additional gain in backward selection might be expected using markers beyond what can be achieved by phenotypic selection alone (*Chapter 15*, Johnson *et al.*, 2000).

The concept of relative efficiency (RE) of genetic gain from MAS compared to phenotypic mass selection was first introduced by Lande and Thompson (1990). A mathematical expression for relative efficiency is:

$$RE = [p/h^2 + (1-p)^2/(1-h^2p)]^{1/2}$$ Equation 19.1

where p is the proportion of additive genetic variance associated with the marker and h^2 is the heritability of the trait. This relationship shows RE increases as p increases and h^2 decreases. Using simulation, they compared the efficiency of phenotypic selection alone compared with phenotypic selection combined with indirect marker selection. The general conclusion resulting from this study was that combined phenotypic and indirect marker selection was superior for low heritability traits.

Marker-assisted selection studies in forest trees have all incorporated the general concept of relative efficiency and have all reached the same conclusion as above: that is, indirect marker selection is most beneficial for low heritability traits (O'Malley and McKeand, 1994; Kerr *et al.*, 1996; Kerr and Goddard, 1997; Johnson *et al.*, 2000; Kumar and Garrick, 2001; Wilcox *et al.*, 2001; Wu, 2002). Two studies also included an economic component where the cost of marker genotyping was included (Johnson *et al.*, 2000; Wilcox *et al.*, 2001); however, these are based on assumptions of marker costs that reflect technologies at the time of study. Marker genotyping technologies are developing rapidly and costs continually decline; so, conclusions reached in these studies may not be relevant in the future.

Wu (2002) performed a series of five different simulations that nicely illustrate the concept of RE and are described in the following sections.

Marker-assisted Early Selection (MAES) *versus* Mature Phenotypic Selection

For the case of mass selection, genetic gain per year under mature phenotypic selection for a target trait is:

$$R_M = i\,h\,\sigma_A/T_c$$ Equation 19.2

where R_M is ΔG_A on a per year basis, ΔG_A is the additive genetic gain in the mature trait (*Chapter 6*), i is the selection intensity on the mature trait, h is the square root of the heritability of the mature trait, σ_A is the standard deviation of the additive genetic variance of the mature trait and T_c is the number of years to complete the normal breeding cycle. This equation follows directly from Equation 13.6 by substituting $h = \sigma_A/\sigma_P$, canceling terms and dividing by the generation interval, T_c.

In contrast, the additive genetic gain per year from MAES is:

$$R_{MAES} = i_y\,r_{MA}\,\sigma_A/T_E$$ Equation 19.3

where i_y is the selection intensity at the time of marker-assisted selection, r_{MA} is the genetic correlation between the marker information and the additive genetic value of the mature trait and T_E is the number of years to complete a cycle of MAES mass selection. This equation can be derived from Equation 13.8 by assuming that the marker information is not influenced by environmental error and therefore has a perfect heritability of 1 (*i.e.* h_y in Equation 13.8 equals 1) and then substituting, canceling and dividing as described above for Equation 19.2.

The relative efficiency of MAES compared to mature phenotypic selection is:

$$RE_1 = (R_{MAES}/R_M) = (p/h^2)^{1/2}\,(T_C/T_E)$$ Equation 19.4

where p is the proportion of additive genetic variation accounted for by the markers. This equation assumes that $i = i_y$ meaning that the selection intensities are equal on the marker and phenotypic target trait. This relationship shows that RE increases when the proportion of additive variation accounted for by the markers increases and the heritability of the mature trait phenotype decreases. This is the same conclusion reached earlier by Lande and Thompson (1990) and many other authors. In practical terms, it means that MAES might be considered for low heritability traits such as growth and that markers need to be found that account for a large proportion of the additive variation in the target trait.

Marker-assisted Early Selection (MAES) *versus* Early Phenotypic Selection

A second scenario considered by Wu (2002) was whether MAES could be more efficient than phenotypic early mass selection for a mature target trait. This compares two forms of indirect selection at the same age, one based on genetic markers and the other based on the early phenotypic measurement. Since the ages of selection are assumed equal, $T_C = T_E$, differences in breeding cycle length need not be considered. The RE of MAES versus early phenotypic selection is (Wu, 2002):

$$RE_2 = [p/(h^2_X * r^2_A)]^{1/2} \qquad \text{Equation 19.5}$$

where p was defined in Equation 19.1, h^2_X is the heritability of the early phenotypic trait and r_A is the genetic correlation between the early and mature traits. It can be seen again that MAES might be preferred over early phenotypic selection if p is high and/or the heritability of the early trait is low. Also, if the genetic correlation between the early versus mature trait is low, then the relative efficiency of MAES improves.

Combined Phenotypic and Marker-assisted Early Index Selection *versus* Early Phenotypic Selection

Intuitively, it seems apparent that combined phenotypic and marker early selection would always have a higher RE than early phenotypic selection alone, assuming the markers account for at least some of the additive variation in the target trait. Wu (2002) showed this in a rather complicated mathematical relationship:

$$RE_3 = [(1-2p^{1/2} + p/(h^2_X * r^2_A))/(1-h^2_X * r^2_A)]^{1/2} \qquad \text{Equation 19.6}$$

where all terms are defined above and this again applies only to the case of mass selection.

As before, RE of combined phenotypic and marker index selection increases as p increases. Furthermore, RE also increases if heritability of the early trait decreases and the genetic correlation between the early trait and the mature trait decreases. A combined phenotype and marker index selection approach was the original concept envisioned by Lande and Thompson (1990).

Marker-assisted Selection and Combined Within-family and Family Selection *versus* Combined Within-family and Family Selection Alone

As we described earlier, marker-assisted indirect selection in forest trees is limited in its application due to linkage equilibrium conditions in forest trees. Wu (2002) then considered whether marker-assisted within-family selection, combined with normal family and within-family phenotypic selection could have higher RE than combined family and within-family phenotypic selection alone. Wu (2002) derived a very complicated mathematical expression for RE under this scenario that is not presented here (see Wu, 2002, p 265).

Wu (2002) used the results of two simulations to model the RE of combined within-family marker-assisted selection and phenotypic family and within-family selection assuming full-sib (Fig. 19.2) and half-sib (Fig. 19.3) families. These simulations show trends similar to all other scenarios, *i.e.* RE increases as p increases and as target trait heritability decreases. The proportion of variation accounted for by the markers, p, is clearly the more important factor. It is not until p approaches a value of 0.5 that RE reaches a value (say 1.2-1.5) where it might be economic to employ markers.

Although Wu (2002) did not include an economic component to his analyses, it seems clear from studies by Johnson *et al.* (2000) and Wilcox *et al.* (2001) that marker costs must be quite low and/or investments spread out over very large plantation acreages to justify marker-assisted selection approaches with REs in the range of 1.2-1.5. These results probably explain why indirect selection based on markers linked to QTLs has not been applied in a significant way to forest tree breeding. It was not until high throughput genomic technologies began to be applied to forest trees (*Chapter 18*) that alternative approaches to marker-assisted selection could be envisioned. One such approach is described in the next section.

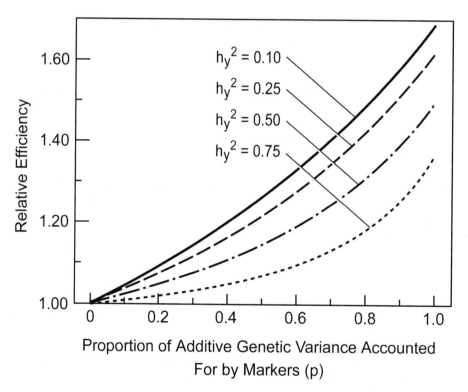

Fig. 19.2. Relative efficiency of within-family marker-assisted early selection and combined family and within-family index selection *versus* combined family and within-family index selection alone, when full-sib families are assumed and four different trait individual heritabilities are modeled. (Reproduced from Wu, 2002, with permission of J.D. Sauerländer's Verlag)

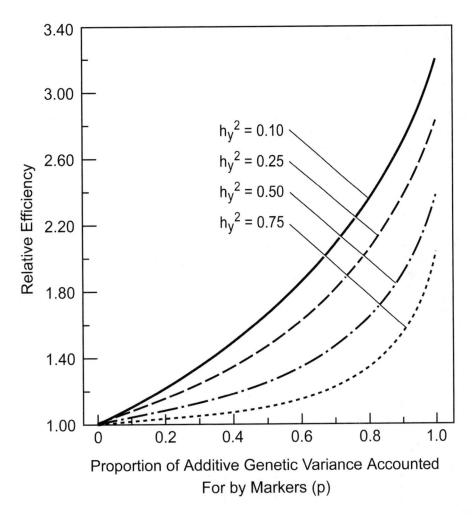

Fig. 19.3. Relative efficiency of within-family marker-assisted early selection and combined family and within-family index selection *versus* combined family and within-family index selection alone, when half-sib families are assumed and four different trait individual heritabilities are modeled. (Reproduced from Wu, 2002, with permission of J.D. Sauerländer's Verlag)

DIRECT SELECTION BASED ON GENES CODING FOR TARGET TRAITS

In this section, we discuss approaches for applying genetic marker selection that involve specific alleles coding for the desirable phenotype. We have chosen to call this marker selection approach "direct selection" rather than "indirect selection" to highlight a fundamental difference that relates to the knowledge at the molecular level of allelic differences affecting phenotype and the ability to select directly at the DNA level for favorable alleles. In the previous section, indirect selection was based on markers (often anonymous markers) linked to QTLs to increase gain in a phenotypic trait, and is clearly a form of indirect selection. However, this distinction between indirect and direct selection may be somewhat artificial and others may choose to refer to the approaches described in this section as indirect selection as well, because technically selection for DNA alleles is still a form of indirect selection on the molecular phenotype.

The use of association genetics to dissect complex, polygenic traits into their individual gene components was described in *Chapter 18*. Two approaches can be employed, the genome scan approach or the candidate gene approach (*Chapter 18*), although the candidate gene approach will more likely be used in forest trees, particularly in conifers (Neale and Savolainen, 2004). This is because linkage disequilibrium is very low in natural populations of trees and it would be cost prohibitive and inefficient to saturate the genome with markers (genome scan approach) when the vast proportion of the genome does not code for genes. The candidate gene approach is likely to be more efficient and cost effective for discovering genetic markers associated with phenotypes and furthermore, such markers are likely to be found at or very near the causative mutation within the gene that encodes the allelic variation leading to phenotypic variation (Fig. 19.4).

A comprehensive and successful candidate-gene-based association genetics study would yield several important outcomes including: (1) An estimate of the number of individual gene loci controlling the quantitative trait; (2) An estimate of the proportion of phenotypic variation explained by each locus; (3) The identity and putative function (*e.g.*

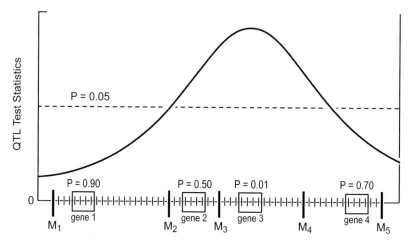

Fig. 19.4. Comparison of the relative power of the candidate-gene-based association genetics approach compared to the QTL approach for identifying individual genes controlling phenotypes in populations where linkage disequilibrium decays rapidly. In the QTL approach, the vertical axis is the test statistic for the presence of the QTL affecting the phenotype and the horizontal axis is the chromosome or linkage group with the positions of genetic markers (large vertical bars). The solid curve represents the likelihood function testing for the presence of a QTL linked to the marker loci (M) in a mapping experiment using a segregating population of modest size. Positions on the likelihood curve above the P = 0.05 significance threshold are significant for the presence of a QTL. Thus, the region bounded by markers M_2 and M_4 harbors a QTL. In the association genetics approach, SNPs (small vertical bars in the figure) within four different, previously identified candidate genes are tested for association with the phenotype and only SNPs within gene 3 were significant (*i.e.* allelic differences in candidate loci 1, 2 and 4 were not significantly associated with phenotypic differences). SNPs not within candidate gene loci were not tested for association with phenotype to reduce the overall size and cost of the experiment. This hypothetical example shows how the QTL approach would not determine whether gene 2 or gene 3 was associated with the phenotype whereas the candidate-gene-based association approach excludes gene 2 and implicates gene 3 as being associated with the phenotype. The favorable SNPs (*i.e.* alleles) in gene 3 can now be used for direct marker-assisted selection for both within family and among family selection because SNPs within gene 3 are unlikely to recombine with the causal polymorphism in gene 3.

metabolic role) of each gene locus based on the knowledge gained from its previous selection as a candidate; (4) Allelic variation within candidate gene loci found in the population including the effect of each allele on phenotype; (5) SNP allele and haplotype frequencies in the population; (6) Mechanisms of gene action at each locus (additive, dominant) and estimates of the effects of allelic substitution within candidate genes on phenotypic expression of the complex trait; and (7) Genetic markers (SNPs) that are either the causative mutation (*i.e.* the quantitative trait nucleotide, QTN) or are in complete or nearly complete linkage disequilibrium with the QTN.

Outcomes 1 and 2 are also common to the QTL mapping approach to marker-assisted indirect selection, but items 3-7 are unique to the association genetics approach to direct candidate gene selection. Selection can now be applied directly to increase the frequency of desirable alleles in the population. It may ultimately become possible to estimate positive epistatic interactions among alleles at candidate gene loci and also capture that component of non-additive genetic variation using a marker-assisted approach.

Research has only begun to determine if direct selection on candidate gene alleles can be applied to forest trees, but a number of potential advantages of this approach in forest trees compared to highly domesticated and inbred crops are discussed in Neale and Savolainen (2004). Several studies have shown that allelic variation in candidate genes can easily be detected and associations with phenotypes can be found (Brown *et al.,* 2004). The size of the effect of a single gene locus or a specific allele on phenotypic expression appears to be small for some traits examined, supporting earlier QTL studies and Fisher's infinitesimal model. Although the exact number of genes controlling traits of interest is not yet known, it remains possible that some traits are controlled by so many genes, all with small effects, that even a direct marker-assisted approach would not improve breeding efficiency over classical methods of phenotypic selection described in *Chapters 13, 15* and *17* (Bernardo, 2001; Gupta *et al.*, 2005).

MARKER-ASSISTED BREEDING

In addition to marker-assisted selection described in the previous sections of this chapter, there are many other existing and potential applications of molecular markers in breeding programs of forest trees. These applications can occur in any and all of the phases of the breeding cycle including breeding and mating designs, genetic testing, deployment, propagation and evaluating infusions. The goals of these applications are to increase genetic gains, reduce costs, shorten the generation interval or improve quality control.

Some applications of molecular markers in forest tree improvement programs have been in existence for many years, and have been mentioned in previous chapters or other reviews (Adams, 1983; Wheeler and Jech, 1992; Friedman and Neale, 1993; O'Malley and Whetten, 1997). Other new applications are emerging or will emerge as technologies evolve and costs of marker analysis decline. The goals of this section are to highlight some current applications of markers in breeding programs and speculate about potential uses in the future.

Quality Control in Tree Improvement Programs

There are several stages during tree improvement programs when anticipated genetic gains are lost if genetic identities are not maintained and/or documented: (1) Creating full-sib families through control pollination among members of the breeding population; (2) Testing in which literally thousands of families, clones and trees are planted and labeled; (3)

Making hundreds of selections from thousands of individuals in a base population; (4) Establishing a propagation population (such as with the top 30 clones out of thousands); (5) Exchanging selections with other organizations or countries such as for infusions; and (6) Patenting or registering clones or selections to protect property rights. In each of these activities, quality control on genetic identity is critical throughout the chain of custody involving the breeding cycle, deployment and registration. If genetic identities are lost or confused, then genetic gains and even property rights can be lost.

While proper labeling and care during all aspects of the tree improvement program are essential, **genetic fingerprinting** is also an important tool that should be used to enhance quality control. One example is verification of controlled crosses that are created as part of the breeding or deployment phase in family forestry (*Chapter 16*). Using isozyme genetic markers, Adams *et al.* (1988) showed that errors in controlled crosses do occur and can easily be detected. They found that there were different types of identification errors (wrong female parent, wrong male parent and multiple male parents) and that the incidence varied greatly among the tree improvement organizations making the crosses. Needless to say, a misidentified full-sib family being used for operational deployment on thousands of hectares could lead to substantial losses in anticipated gains.

While isozymes have been used successfully for genetic fingerprinting, their ability to discriminate individual genotypes is limited due to: (1) The relatively small number of loci available (usually 20 or so); (2) The relatively low number of alleles per locus (2 to 6); and (3) Most loci being dominated by 1 or 2 alleles at very high frequencies.

Newer molecular markers have much more power to detect genetic differences. SSR markers (*Chapter 4*) are probably the most widely used markers for fingerprinting applications because: (1) SSRs are most always codominant; (2) There can be hundreds of SSR loci available; and (3) Many loci are highly polymorphic with 5 to 30 alleles per locus in a range of intermediate frequencies. These characteristics make SSR markers extremely powerful for detecting genetic differences. For example, Kirst *et al.* (2005) genotyped each of the 192 unrelated selections in a breeding population of *Eucalyptus grandis* in Brazil. Using only 6 highly informative SSR loci (having a total of 119 alleles), the probability of finding two identical genotypes in the population was 2×10^{-9} or nearly zero. Therefore, they could confidently fingerprint each of the selections and distinguish among them all. Similarly, only three highly variable SSR loci (with a total of 68 alleles) were needed to uniquely identify the 51 clones in a *Pseudotsuga menziesii* seed orchard block (Slavov *et al.*, 2004).

Breeding and Mating Designs

Molecular markers can also be used to enhance mating designs (*Chapter 14*) in advanced-generation breeding programs (*Chapter 17*). One example is the **polymix breeding with paternity analysis** (PMX/WPA) proposed by Lambeth *et al.* (2001). The polymix breeding design has the advantage that more pollen parents can be tested per unit of investment (*e.g.* pollen from 50 male parents can be mixed for use in a single control pollination with each female). However, this mating design has been avoided by programs wishing to maintain full pedigree control in the breeding population, because the male side of the pedigree is not known as it is in full-sib breeding. In the PMX/WPA approach, the distinct genotypes of all possible males are identified prior to polymix breeding using highly polymorphic molecular markers such as SSRs. Following breeding, it is then possible to determine the male parentage of all progeny selected and to make selections of known degrees of relatedness.

Molecular markers can also be used to enhance full-sib mating designs in at least three ways: (1) Complementary breeding; (2) Gene pyramiding; and (3) Diversity index breeding. While in their infancy in tree improvement programs, these three applications have the potential to increase genetic gains by using molecular marker genotypes to make more informed choices of which parents to intermate. As a hypothetical example of **complementary breeding**, suppose that there are three potential parents to breed together (P_1, P_2 and P_3) and that the molecular genotypes are known at two unlinked molecular loci affecting wood density: P_1 is AAbb; P_2 is aaBB; and P_3 is AAbb. All other things being equal and assuming that dominant alleles increase wood density, it would make sense to cross P_1 X P_2 and P_2 X P_3 but not to make the cross P_1 x P_3. The genotypes of P_1 and P_3 are complementary to that of P_2 (since they are superior at different loci), but a cross of P_1 x P_3 would not yield superior offspring at both loci because the genotypes are not complementary. This concept could also be extended to complementary breeding involving multiple traits.

The idea of complementary breeding can be applied to **gene stacking** or **gene pyramiding** as has been done for disease resistance genes in crop breeding that employ classical methods of breeding and selection. Using molecular markers, suppose there are 10 resistance genes for a fungal disease, that trees can be genotyped at all 10 of these loci, and that most resistant trees have the favorable (resistant) allele at only one or two of the 10 loci. Breeding should favor crosses among parents superior at distinct resistance loci such that over time, individuals could be selected that "stack" together resistance alleles from multiple loci.

Finally, consider the likelihood that most QTLs or QTNs have relatively small effects on polygenic target traits (*Chapter 18*) and that there are multiple target traits in most breeding programs (*e.g.* volume growth, disease resistance and wood density). If true, then it would be difficult to apply complementary breeding in the manual way described above due to the sheer number of loci involved. In this case, the idea of **diversity index breeding** (number 3 above) involves the calculation of pairwise diversity indices among all possible pairs of parents based on their molecular genotypes at many (even hundreds of) loci influencing multiple target traits. Then, pairs of selected parents could be chosen for mating to maximize their pairwise genetic diversity to enhance the probability that the two parents in each cross have genotypes that complement each other.

To our knowledge, diversity index breeding has not been attempted in tree improvement programs, and a diversity index is not the only criterion to consider when designing a breeding scheme. For example, it is also important to consider the predicted breeding values of all parents involved and favor those with higher breeding values (*Chapter 17*). So, at a minimum, this means taking into account parental breeding values and pairwise diversity indices for all traits when making decisions about mate selection in a breeding program. In animal breeding, similar reasoning has led to the concept of breeding for total progeny merit (Hayes and Miller, 2000). Clearly, this is an area of tree improvement that needs more research.

While diversity index breeding has not yet been applied in advanced-generation tree improvement programs, a potentially useful index based on molecular markers, the **Pairwise Genetic Diversity (PGD)**, has been proposed and used in animal breeding and tree improvement for other purposes (Bowcock *et al.*, 1994; Ciampolini *et al.*, 1995; Tambasco-Talhari *et al.*, 2005; Kirst *et al.*, 2005). For any two parents, the PGD is calculated as:

$$D_m = 1 - (\textstyle\sum_r \text{shared alleles})/2r \qquad\qquad \text{Equation 19.7}$$

where D_m is PGD that ranges from 0 to 1, $\sum_r$ indicates a summation over all marker loci of the number of alleles in common to the two parents, and r is the number of marker loci. D_m = 1 when two parents share no alleles in common and D_m = 0 when they share all alleles in common. A hypothetical example of calculating and employing PGD is presented in Box 19.1.

As a real example of its use, Kirst *et al.* (2005) calculated 18,336 values (192x191/2) of PGD from all possible pairs of 192 unrelated selections in a breeding population of *Eucalyptus grandis* in Brazil based on genotypes at 6 SSR loci containing 119 different alleles. They found that D_m ranged from 0.33 to 1 and that 97% of the parental pairs had D_m > 0.6. The large average genetic distance (0.857) among the 192 selections in this population likely reflects the fact that they were unrelated, and could imply that D_m should only be a minor consideration in a breeding scheme.

PGD is different than Nei's (1987) genetic distance (D) introduced in *Chapter 8*. In particular, Nei's genetic distance is based on differences in estimated allele frequencies between pairs of populations. As such, Nei's genetic distance cannot be used to quantify differences between two individuals whether they are in the same or different populations; rather, Nei's distance quantifies genetic distances at the population level.

PGD, on the other hand, is calculated directly from molecular genotype data between pairs of individuals, not from population statistics. Therefore, PGD can quantify genetic distances among pairs of individuals in the same or in different populations and these data have direct use as described above and in the later section on *Hybrid Breeding*. Then, if desired, PGD values can also be averaged across all pairs of individuals within the same population and across all pairs of individuals from different populations to examine genetic architecture in an analogous manner as Nei's genetic distance (*e.g.* Tambasco-Talhari *et al.*, 2005, among and within breeds of cattle).

Propagation Populations and Deployment

There are numerous applications of molecular markers in propagation and deployment populations including: (1) Ensuring varietal protection of registered, patented or deployed material; (2) Quantifying and maintaining genetic diversity within one and across several generations; (3) Quantifying seed orchard efficiencies and pollen contamination levels; and (4) Increasing genetic gains and capturing genotype by environment interaction. Ensuring varietal protection is a fingerprinting application (explained above in *Quality Control in Tree Improvement Programs*) in which improved clones, families or varieties are fingerprinted to ensure that they can be protected, marketed and deployed securely and repeatedly.

For numbers 2 and 3 above, isozymes and increasingly microsatellite/SSR markers have been used to monitor pollination patterns and contamination in seed orchards (Wheeler and Jech, 1992; Adams *et al.*, 1996; Grattapaglia *et al.*, 2004; Slavov *et al.*, 2005). These methods can be used to quantify panmixia (*i.e.* whether all orchard clones are contributing in nearly equal amounts to the pollen and seed pools). Box 19.2 presents a study in which several changes in cultural practices and genetic composition of a seed orchard were adopted after microsatellite markers were employed to genotype offspring from the orchard growing in operational plantations (Grattapaglia *et al.*, 2004).

For number 4 above, consider a hypothetical example in which a tree improvement program experiences important g x e interaction for volume growth, but chooses to establish a single large breeding unit and breed for genotypes that are stable over all planting zones. Even though the presence of g x e implies a sacrifice of gain through use of a single breeding population, some of this loss could conceivably be captured in each generation's propagation through application of molecular markers. If there are marker loci reliably expressed in some planting zones, but not others, then it would be possible to select genotypes for deployment based on favorable molecular alleles expressed in distinct planting environments.

Finally, for number 4 above, consider the possibility of using molecular markers to rogue a seed orchard by genotyping the progeny growing in operational plantations established with seed from that orchard (Box 19.2, Grattaglia *et al.*, 2004). These investigators

---

Box 19.1. A hypothetical example of pairwise genetic distance (PGD)

1. *Genotyping the individuals.* In this simple example, we assume: (1) Only four loci are genotyped that affect the trait(s) of interest; (2) There are only two alleles at each locus with the dominant allele being favorable; (3) Each parent is homozygous at two of the four loci affecting the trait; and (4) Each locus has an equal effect on the trait with no epistasis. That latter two assumptions imply all four hypothetical individuals below have the same breeding value:

   Individual 1: AABBccdd
   Individual 2: AABBccdd
   Individual 3: aabbCCDD
   Individual 4: AAbbCCdd

2. *Calculating PGD values.* There are eight possible alleles across four loci ($r = 4$) in any diploid organism and this is the maximum number that any pair of individuals can share. Individuals 1 and 2 have all 8 alleles in common, and using Equation 19.7, $D_m = 1 - (8/2r) = 1 - (8/8) = 0$ meaning that PGD is zero since they are the same genotype at these four loci. Similarly for individuals 1 and 4, there are 4 alleles in common (A, A, d and d) and $D_m = 1 - (4/8) = 0.5$. The table of D_m values is:

Individual	2	3	4
1	0	1	0.5
2		1	0.5
3			0.5

3. *Utilizing PGD values in breeding.* Since the breeding values of all 4 individuals are assumed equal, this criterion does not need to be considered. When breeding for a deployment population of full-sib families for family forestry, the matings 1 x 3 and 2 x 3 (with D_m values = 1) would produce progeny (AaBbCcDd) that maximize deployment value given complete dominance. When breeding for a multiple-generation recurrent selection program based on general combining ability (*Chapter 17*), there is less reason to mate individuals 1 and 2 than other pairwise combinations that are more genetically distant and bring together superior alleles at different loci into the same individuals.

4. *Potential future applications.* This is an oversimplified example that does not begin to address many of the issues of using D_m values in combination with other criteria to optimize advanced-generation mating designs for breeding and deployment populations. However, we believe this approach has merit and warrants future research.

demonstrated the feasibility of using microsatellite markers to genotype plantation progeny (*in lieu* of establishing progeny tests) to identify those clones in the seed orchard that were contributing a higher fraction of desirable progeny in the plantations. These clones could then be favored for seed collection. While this study was done with a small number of orchard clones, the authors state that "sufficient power of discrimination can easily be attained to resolve maternity and paternity of offspring derived from orchards of several tens of unrelated parents." The authors do not advocate eliminating progeny tests, but have shown the potential of using molecular markers to genotype progeny growing in operational plantations for the purposes of retrospective (*i.e.* backward) selection to either rogue an orchard or utilize superior parents in a breeding program.

Box 19.2. Applications of molecular markers in a *Eucalyptus* seed orchard.

Grattapaglia *et al.* (2004) used microsatellite markers in several ways to validate the concept of and improve efficiency of a hybrid seed orchard of *Eucalyptus* in Aracruz, state of Espiritu Santo, Brazil. The results are briefly highlighted below.

1. *Seed orchard.* In a novel design, a single clone of *E. grandis* that was putatively self sterile was surrounded by six pollinator clones of *E. urophylla*, and this hexagonal pattern was repeated across the landscape to establish a 7.5 ha seed orchard. The idea was for the *E. urophylla* clones to serve as pollinators such that F_1 hybrid seed (called *E. urograndis*) would be produced from the single *E. grandis* clone. A 300 m buffer of native forest surrounded the orchard to provide isolation from nearby eucalypt plantations.

2. *Progeny samples for genotyping.* Rather than measuring phenotypes of progeny growing in randomized, replicated genetic tests, the investigators used microsatellite markers to genotype six-year-old (rotation-age) progeny growing in operational plantations established with orchard seed. Their samples included: (1) Selected individuals whose circumference at breast height was one standard deviation above the plantation mean (n = 144 trees); (2) A control sample of n = 72 random, non-selected individuals; and (3) Very slow-growing runts (n = 10).

3. *Genotyping and parentage analysis.* Total DNA was extracted from leaf tissue of each sampled six-year-old progeny tree. Out of a total of 47 microsatellite loci, a battery of 14 marker loci were selected to be informative and used throughout the study. For each tree, parentage analysis included: maternity analysis to determine if the mother was the single *E. grandis* clone in the seed orchard; and paternity analysis to determine whether the pollen parent was one of the six *E. urophylla* clones in the orchard and if so, exactly which one.

4. *Results.* The maternity analysis revealed that 8.3% of the plantation progeny were not offspring from the single *E. grandis* clone from which all seed had presumably been collected. This meant that quality control could be enhanced during seed collection, seed processing and/or nursery production. Paternity analysis showed:

   • Two of the six *E. urophylla* pollinator clones were genetically identical (*i.e.* the same clone) indicating a case of mistaken identity.

(Box 19.2 continued on next page)

Box 19.2. Applications of molecular markers in a *Eucalyptus* seed orchard. (Continued from previous page)

- 29% of the progeny were sired by pollen originating from outside the orchard; so, the 300 m buffer was not completely effective against the insect vectors from nearby plantations.
- Three of the *E. urophylla* clones sired 99% of the remaining non-contaminated progeny, and this was likely due to compatible flowering phenology with the *E. grandis* clone.
- One of the three main pollinator clones sired significantly more progeny in the selected sample than in the random control and conversely one of the three sired significantly fewer in the selected compared to non-selected samples of progeny.
- There were no selfs in the selected sample of progeny, 8.3% in the non-selected control and 80% in the sample of slow-growing runts.

5. *Applications.* The results validated the concept of the two-species seed orchard designed to produce F_1 seed of *E. urograndis* since the large majority of progeny seedlings in operational plantations were, in fact, hybrids produced by the orchard. The results also led to some management modifications to increase efficiency and realized genetic gains. First, three of the six pollinator clones that were ineffective in siring offspring were eliminated. Next, cultural practices were changed to include intensive fertilization to promote flowering and the addition of bee cages in order to reduce selfing and pollen contamination.

Finally, the authors discussed the possibility of establishing a new two-clone seed orchard composed of the *E. grandis* clone and the one *E. urophylla* clone shown to sire more offspring in the selected sample of progeny. This latter concept is particularly important because it demonstrates the potential of using molecular markers to genotype progeny growing in operational plantations, *in lieu* of progeny tests, to make retrospective (backward) selections for either roguing a seed orchard or for a breeding program.

Hybrid Breeding

There is growing interest in breeding interspecific hybrids to increase genetic gains or extend the planting zone to broader edaphoclimatic conditions (*Chapters 12* and *17*). There are at least two ways in which molecular markers can potentially be used to increase the efficiency of hybrid breeding: (1) By increasing the efficiency of backcross or other multiple-generation breeding aimed at interspecific introgression of a single trait from one species into another; and (2) By aiding mate selection when breeding for hybrid vigor.

In the first case, it is common that one of the species is well adapted and has many of the desirable attributes, while the second brings a specific set of needed characteristics (*e.g.* desirable wood quality or increased resistance to disease or cold). In this situation, the goal is to transfer favorable alleles at specific loci from the donor species into the recipient species as a form of introgression. Molecular markers associated with the specific set of desirable traits in the donor species can be monitored to enhance the transfer of alleles into the recipient species through multiple generations of breeding and testing. At the same time, molecular markers well distributed throughout the recipient species can be monitored to ensure that new selections each generation contain mostly germplasm from the desirable, recipient species except for the specific traits being introgressed. This reduces linkage drag associated with less desirable characteristics in the donor species.

The use of molecular markers in backcross breeding to facilitate introgression is probably the most prevalent application of molecular markers in crop breeding, and nearly always, the molecular markers have been used to tag a major gene for disease or insect resistance (Koebner and Summers, 2002; Barone, 2004; Francia *et al.*, 2005). The application becomes more problematic when the desirable trait(s) in the donor species are polygenic. Also, the molecular markers are almost always used in conjunction with, not instead of, classical field testing. Still, the potential clearly exists to use molecular markers in combination with classical methods to enhance hybrid breeding aimed at introgression.

The second potential use of molecular markers in hybrid breeding is to better identify mates in each of the two species to enhance genetic gains in the progeny. This has been used in animal breeding to select individuals within each of the animal breeds that are more genetically distant from each other and therefore have higher potential for hybrid vigor (Hayes and Miller, 2000; Tambasco-Talhari *et al.*, 2005). The rationale is based on hybrid vigor being caused by dominance (as opposed to overdominance) and this leading to a breed x mate interaction in which specific pairs of mates have higher hybrid specific combining ability (SCA, *Chapter 6*). Mating individuals from the two breeds that have higher pairwise genetic distances (D_m in Equation 19.7) was shown to increase heterozygosity and hybrid vigor in the progeny (Tambasco-Talhari *et al.*, 2005), and utilizing the distance information in conjunction with the within-breed general combining abilities (GCA, *Chapter 6*) maximized the total genetic merit of the progeny.

To illustrate the concept of mate selection using pairwise genetic distance in hybrid breeding, consider the example in Box 19.2. Suppose that individuals 1 and 2 are from species A, while individuals 3 and 4 are from species B. Use of individual 3 (instead of individual 4) to mate with either individual 1 or 2 results in completely heterozygous and superior progeny. Since this depends on dominance at complementary loci, the benefits will not be accumulated in a recurrent breeding program for hybrid GCA in a composite or synthetic hybrid breeding population (*Chapter 17*); however, each generation, genetic distances could be used with GCA values from tests to determine mating schemes for producing hybrid progeny for operational deployment.

Smart and Ideotype Breeding

Functional genomics (*Chapter 18*) combined with detailed studies of physiology, morphology, disease expression, wood properties and tree ontogeny have the potential to greatly increase our knowledge of: (1) Gene action (additive, dominance, epistasis and pleiotropy); (2) Gene function (the genes and underlying mechanisms controlling important phenotypic traits); and (3) Gene regulation (coordinated expression of genes through time, space, seasons and tissues of the plant). All of these advances have the potential to provide insights for the breeder to use in tree improvement programs.

Ideotype breeding was introduced (Box 17.7) as a conceptual approach to genetic improvement in which the breeder develops models or constructs that specify the desirable underlying characteristics of improved genotypes to be developed. The ultimate objective is to develop an ideotype that is well adapted to produce high yields and excellent product quality across the ranges of edaphoclimatic conditions, silvicultural treatments and product objectives. Historically, ideotype breeding has been only a conceptual model, but knowledge gained from molecular genetics and other disciplines has the potential to make it a reality.

More than ever before molecular geneticists are working with physiologists and other scientists to understand the underlying mechanisms controlling complex traits. For example, there are many component physiological, anatomical, morphological and phenological traits affecting the aggregate trait of bole volume growth per hectare (*i.e.* stand yield). Examples of component traits might be photosynthetic rate, water use efficiency, nutrient uptake efficiency, and crown architecture, and each of these component traits may, themselves, be complex polygenic traits.

The field of functional genomics (*Chapter 18*) promises to allow scientists to understand the relative importance of each of the many component traits influencing the aggregate target trait and to develop molecular marker systems for them. For example, suppose several loci are identified that affect volume growth through different underlying mechanisms, then it would be possible to focus selection on those mechanisms that produce crop ideotypes that maximize stand yield. Applications like these that combine molecular markers with knowledge of underlying physiological mechanisms are some years in the future and will require considerable research by teams of scientists working across multiple disciplines.

SUMMARY AND CONCLUSIONS

Marker-assisted selection is a form of selection applied to genetic markers that are associated with a target trait. There are many potential benefits to marker-assisted selection in forest trees but also a number of hurdles to its successful application. Benefits include: (1) Decreasing the breeding cycle time; (2) Decreasing costs; (3) Increasing selection intensity; and (4) Increasing efficiency of selection for low heritability traits. Challenges include: (1) The large task of QTL detection; (2) QTL by environment and QTL by genetic background interactions; and (3) Linkage equilibrium conditions between markers and QTLs in tree populations.

Two basic approaches to marker-assisted selection in forest trees are possible. The first is based on finding linkages between anonymous flanking markers and QTLs in segregating populations. This approach can work for marker-assisted selection within the same segregating families as the linkages were discovered, but cannot be applied to new families because of the general lack of strong linkage disequilibrium in forest tree populations.

The second approach uses association genetics to find associations between single nucleotide polymorphisms in candidate genes and the target trait. This approach can be used for both family and within-family selection because marker by trait associations are very tightly linked and will be in complete or nearly complete linkage disequilibrium in tree populations. We believe that association genetics based on single nucleotide polymorphisms may lead to applications in which marker data are combined with phenotypic measurements from progeny tests to enhance selection and breeding in tree improvement programs.

In addition to marker-assisted selection, molecular markers have many other potential uses in breeding programs, such as quality control, understanding mating systems in seed orchards, registering and protecting varieties, families and clones, and 'smart' breeding. While many of these applications are not yet operational, the future potential is great. As marker technologies evolve, costs decline and our understanding of underlying mechanisms improves, we believe that molecular markers could become integral tools in many phases of tree improvement programs.

CHAPTER 20
GENETIC ENGINEERING – *TARGET TRAITS, TRANSFORMATION AND REGENERATION*

Genetic improvement of forest trees has advanced almost entirely using selective breeding approaches (*Chapters 11-17*). In just the last few years, marker-based breeding has been developed to complement traditional methods; however, the basic principles employed are the same as in selective breeding (*Chapter 19*). Genetic engineering, however, is a fundamentally different approach than traditional or marker-based breeding and is still in its early phases of development. **Genetic engineering** (GE) is defined as the use of recombinant DNA and asexual gene transfer methods to alter the structure or expression of specific genes and traits (FAO, 2004). Other terms have been used synonymously with GE, such as genetic modification and genetic manipulation, but these terms are somewhat ambiguous with traditional breeding. Terms used to describe the product of genetic engineering are genetically modified organism (GMO), or more precisely, recombinant DNA modified organism (RDMO), but in this chapter we use GE (*i.e.* genetically engineered) tree or plant.

Genetic engineering of plants was first demonstrated in tobacco (Horsch *et al.*, 1985). Subsequently, genetic engineering of a variety of agricultural crops was accomplished (Gasser and Fraley, 1989). The first report of a GE tree was that for *Populus* (Fillatti *et al.*, 1987). This breakthrough event led to great interest in genetic engineering of forest trees and a flood of discussion papers on its potential applications followed (Sederoff and Ledig, 1985; Dunstan, 1988; Charest and Michel, 1991; Tzfira *et al.*, 1998; Ahuja, 2000; Campbell *et al.*, 2003; Tang and Newton, 2003). These papers laid out the various approaches for genetically engineering trees and the possible target traits that could be engineered. In the last 20 years significant progress has been made in the face of considerable technical and political challenges. This chapter begins with four sections that describe the technical aspects of genetic engineering: (1) *Target Traits for Genetic Engineering*; (2) *Methods for Gene Transfer*; (3) *Vector Design and Selectable Markers*; and (4) *Regeneration Methods*. We then follow with four case studies for genetic engineering in trees: (1) *Lignin Modification*; (2) *Herbicide Tolerance*; (3) *Pest and Disease Resistance*; and (4) *Flowering Control*. The final two major sections in this chapter discuss *Transgene Expression and Stability*, and *Commercialization, Regulation and Biosafety* of GE forest trees.

TARGET TRAITS FOR GENETIC ENGINEERING

A fundamental difference between GE and traditional or marker-based breeding is that foreign genes can be introduced that code for traits not naturally found in the target plant. This includes genes that could not be introduced even through artificial hybridization. A notable example in many plants, including trees, is resistance to herbicides (Fillatti *et al.*, 1987; Slater *et al.*, 2003). A gene for glyphosate resistance (aroA) was discovered and isolated from bacteria and is not found in plants (Comai *et al.*, 1985). Another common target gene for GE is the insecticidal gene *Bt* from the bacterium *Baccilus thuringiensis*

(Estruch *et al.*, 1997). Other examples of foreign genes that might be introduced are a variety of stress tolerance genes from microorganisms and animals such as freeze tolerance genes from fish.

GE can also be used to modify native genes in trees, such as those controlling traits of long standing commercial interest (stem growth, wood properties, adaptation), but the difficulties of using GE for improving such traits are great. The first challenge is that nearly all commercially important traits are polygenic and therefore quantitatively inherited. For example, stem growth and yield might be controlled by hundreds, if not thousands of genes. If all these genes act additively and each has a small effect, then GE modification of one or a few genes might have little impact on a quantitative trait. However, recent research suggests that large effects on some quantitatively inherited traits might be possible by modifying just one gene by GE (see *Lignin Modification* below).

There are a variety of ways that a native gene controlling a quantitative trait can be modified by GE. The most common approach is to alter the expression of the gene, either by down-regulating (*i.e.* reducing) or up-regulating (*i.e.* increasing) its expression. Down-regulation can be accomplished by a variety of approaches such as anti-sense RNA or RNA interference technologies (Sharp and Zamore, 2000; Baulcombe, 2002). **Anti-sense RNA** is a technique for gene silencing where a transgene produces an mRNA in the antisense direction that is complementary to an endogenous mRNA and thus prevents the mRNA from being translated and the polypeptide produced. **RNA interference** is a recently discovered phenomenon where a double-stranded RNA molecule interferes with gene expression. These approaches are fairly straight-forward because they are specifically targeted at interrupting the expression of the native gene. Up-regulation of a target gene is much more difficult to accomplish. The simplest approach is to introduce a new copy of the target gene that has been engineered to be under the control of a strong and/or inducible promoter. The expression of the native gene is not affected under this approach and therefore confounds the desired effect on the phenotype. A more effective approach uses site directed homologous recombination (Kumar and Fladung, 2001) to replace the native gene with the engineered gene, but such approaches are a long way from being developed in forest trees.

A class of target traits that might eventually be explored in forest trees is so-called domestication traits (Campbell *et al.*, 2003; Busov *et al.*, 2005). An important example in agricultural crops is that of shattering in wheat and other cereals. To ensure dispersal of seeds, panicles (seed heads) in wild ancestors of wheat shatter (release their seeds) upon maturity. It was not until varieties of non-shattering wheat were identified that large crops could be consistently harvested. This one, simply-inherited, trait led to the domestication of wheat. Likewise, rare genes having profound effects on phenotypes might be envisioned in forest trees. Utilizing such genes, trees with no stem taper or other desirable ideotypes might ultimately be designed. Tree varieties of this type will probably be easier to develop using GE than by traditional breeding methods (Campbell *et al.*, 2003).

METHODS FOR GENE TRANSFER

Methods for gene transfer can be classified into two general categories: (1) Indirect gene transfer; and (2) Direct gene transfer (Charest and Michel, 1991; Slater *et al.*, 2003). **Indirect gene transfer** requires the use of an intermediate organism such as *Agrobacterium tumefaciens* or *Agrobacterium rhizogenes* to transmit the foreign DNA into the host cell, whereas **direct gene transfer** does not involve an intermediate organism (Fig. 20.1). Indirect DNA transfer (*Agrobacterium*) has been used successfully in *Populus* species because of high transformation efficiencies (Fillatti *et al.*, 1987; Baucher *et al.*, 1996; Fladung *et al.*, 1997; Kim *et al.*, 1997; Meilan *et al.*, 2002). Initially *Agrobacterium*-mediated methods were not highly efficient in conifer species and thus direct DNA transfer methods were explored and developed. In recent years, however, efficiencies have improved with *Agrobacterium*-mediated gene transfer, which since has been widely used in conifers (Shin *et al.*, 1994; Klimaszewska *et al.*, 1997; Levée *et al.*, 1999; Wenck *et al.*, 1999). Therefore, *Agrobacterium*-mediated methods will receive more attention in this chapter, although a brief description of direct DNA transfer is also given.

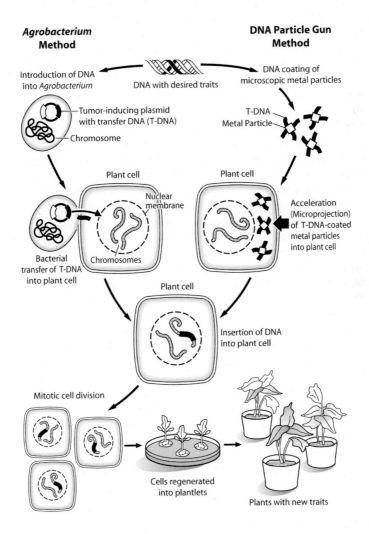

Fig. 20.1. Transformation of plant cells by *Agrobacterium*-mediated gene transfer (indirect gene transfer) shown on the left and by the biolistic (microprojection) method (direct gene transfer) shown on the right of the diagram. In both methods, the transferred foreign genes are inserted into the genome of the recipient cell. The cells carrying the foreign gene grow and differentiate into transgenic plants either by organogenesis or somatic embryogenesis. (Adapted with permission from Gasser and Fraley, 1992, p 64)

Indirect Gene Transfer

Agrobacterium tumifaciens is a soil-borne, Gram-negative bacterium that is the causal agent of crown gall disease in a number of angiosperm species. Crown gall disease is a result of the *Agrobacterium tumifaciens* having transferred a few of its genes to the host genome (horizontal gene transfer). This naturally occurring method of horizontal gene transfer was recognized as a potential approach for artificial gene transfer and hence genetic engineering.

Agrobacterium tumifaciens contains a DNA plasmid called the tumor-inducing (Ti) plasmid. A part of this plasmid, called the transfer-DNA (T-DNA) region, is the portion of *Agrobacterium tumifaciens* DNA that is actually transferred to the plant genome. The T-DNA carries a number of genes required for gall formation including genes encoding proteins in hormone biosysnthesis (auxin and cytokinin) and other metabolites such as opines and agropines. Ti plasmids have been genetically engineered to facilitate the transfer of foreign DNA without inducing gall formation in host cells (Fig. 20.2). In the section *Vector Design and Selectable Markers*, we discuss in more detail how foreign DNA is inserted into the Ti plasmid and how it is incorporated into the host plant genome.

The transfer of foreign DNA to the host plant genome is a result of several sophisticated mechanisms and is only briefly described here. The first step is to identify a suitable tissue explant for inoculation with *Agrobacterium*. In trees, a variety of explant types have been used including internodal stem or leaf segments (*Populus*) (Fillatti *et al.*, 1987; Fladung *et al.*, 1997; Han *et al.*, 1997) and somatic embryos (conifers) (Klimaszewska *et al.*, 1997; Levée *et al.*, 1997). The choice of explant type is determined primarily on the ability to regenerate a whole plant from the transformed explant.

The second step is to co-cultivate the *Agrobacterium* with the explant and allow infection to occur (Fig. 20.3). Prior to this step, however, it is often necessary to identify *Agrobacterium* strains that will infect the host plant cells (De Cleene and De Ley, 1976; Charest and Michel, 1991). Chemical inducers are often added to simulate an infection response that mimics the natural infection process. Following infection, the third step is to kill the *Agrobacterium* with an antibiotic. The infected explant is then transferred to fresh medium containing an agent to select for transformed versus untransformed cells. Kanamycin, an antibiotic, is routinely used as a selectable agent. The gene encoding resistance to

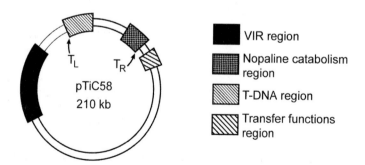

Fig. 20.2. A Ti plasmid for gene transfer in plants. The plasmid from a prescreened *Agrobacterium* strain, for example a nopaline synthesizing strain C58, can be disarmed by replacing the endogenous hormone synthesizing genes in the T-DNA region (involved in tumor formation) with recombinant genes of interest (*e.g.* pest resistance) for gene transfer. VIR is the virulence region of the Ti plasmid. (Adapted from Charest and Michel, 1991, with permission of Natural Resources Canada)

kanamycin is *nptII* and is introduced into the host along with the target genes during the transformation process. **Transformation** involves the stable introduction of foreign genes into the genome of a plant cell. Kanamycin kills the untransformed cells, leaving only transformed cells/tissue (Fig. 20.4). The final step is to regenerate the transformed explant into a whole transgenic plant following tissue culture and regeneration protocols.

Direct Gene Transfer

Agrobacterium is an elegant and sometimes highly efficient transformation system in a variety of plant species, including angiosperm tree species. One very important plant group, however, the monocots that include the cereals (corn, wheat, barley, rice, *etc.*), are not natural hosts for *Agrobacterium* and early attempts at transforming monocots with *Agrobacterium* failed. This led researchers to pursue development of alternative transformation systems motivated by the economic returns of genetically engineering these high-value crops. The conifers were also another plant group that was somewhat recalcitrant to *Agrobacterium*-mediated transformation. Thus, direct DNA gene transfer methods developed in cereals were initially tried in conifers. In the following sections, we describe a few of the more widely used direct DNA transfer methods.

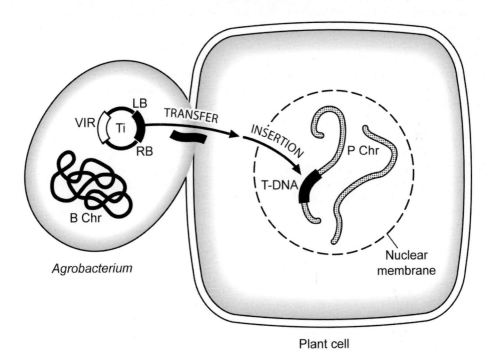

Plant cell

Fig. 20.3. Transformation of a plant cell by *Agrobacterium* carrying the Ti plasmid with the desired foreign gene. The T-DNA (between the left border (LB) and right border (RB)) in the Ti plasmid is transferred from the bacterium to the plant cell and eventually integrated into the plant's genome. The integration of the foreign gene may occur on one or several of the plant's chromosomes (PChr). The bacterial chromosome (BChr) does not play any role in the integration process. However, the virulence region (VIR) genes of the Ti plasmid are involved in the transfer of T-DNA to the plant cell. (Reproduced from Walter *et al.*, 1998a, with permission from Springer Science and Business Media)

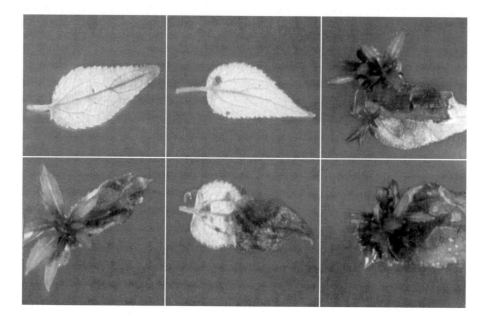

Fig. 20.4. *Agrobacterium*-mediated transformed juvenile leaves from *Populus* grown on a medium containing kanamycin. The two upper left panels show untransformed leaves: other panels show transformed leaves. Only those tissues carrying the selectable marker for kanamycin resistance gene (*NPT*) survive on this medium while untransformed tissues are killed. (Photos courtesy of R. Ahuja, Institute of Forest Genetics, Grosshansdorf, Germany (retired))

Electroporation

Electroporation has been used to deliver DNA to all the major cereals and some conifer species. Although young plant cells with thin cell walls can be used, electroporation is optimally performed using protoplasts (plant cell cultures where the cell walls have been chemically removed). In the presence of plasmids containing the foreign DNA, protoplasts are subjected to high electrical voltage, producing pores in the plasma membrane that allow the plasmid DNA to enter the cell. Subsequently, the foreign DNA integrates into the genome.

Electroporation has been effectively used in tree species to transfer and monitor transient expression of reporter genes. Electroporation has been utilized to transfer genes into protoplasts of *Populus* (Chupeau *et al.*, 1994), and protoplasts isolated from embryonic cell lines in several conifers, including *Pinus taeda* (Gupta *et al.*, 1988), *Picea glauca* (Bekkaoui *et al.*, 1988), and *Larix* x *eurolepis* (Charest *et al.*, 1991). Electroporation has been used routinely to achieve transient expression of foreign genes in conifers, but has been much less effective in the production of stable transformed plants, because of lack of efficient protoplast regeneration techniques to produce GE trees from transformed cells.

Biolistics

Biolistics, or particle bombardment, is the most widely used and efficient method for direct gene transfer in plants. Although the technique itself is rather crude (Klein *et al.*, 1987), it is capable of delivering foreign DNA directly into cells, tissues and organs. Thus, particle bombardment bypasses the limitations of *Agrobacterium* host specificity and tissue-culture related recalcitrance (Birch, 1997). To begin, tungsten or gold particles are coated with the foreign DNA to be transformed into the host plant genome (Fig. 20.1). The particles are delivered to the target plant cells at high velocity. Early biolistic methods used gun powder devices to "shoot" the particles into the target cells. A helium-driven device is now widely used. DNA on the coated particles "falls off" as the particles pass through the cell and is integrated into the host genome. Biolistics has been used for both transient (Loopstra *et al.*, 1992; Newton *et al.*, 1992; Aronen *et al.*, 1994) and stable (Ellis *et al.*, 1993; Charest *et al.*, 1996; Klimaszewska *et al.*, 1997; Walter *et al.*, 1998b) DNA transfer in conifers.

VECTOR DESIGN AND SELECTABLE MARKERS

A **recombinant gene** (also called chimeric gene) functions like a regular gene (Fig. 2.11), except in reality it consists of a patchwork of genetic sequences (*e.g.* promoter, coding sequence, terminator sequence) derived from different organisms, related or unrelated to the host. In the majority of cases, the components of recombinant genes have been obtained from viruses and bacteria, but more recently these genes have also been assembled from plants or trees for gene transfer. A recombinant gene is introduced into the host via direct or indirect gene transfer. Many silviculturally useful recombinant genes have been constructed for gene transfer in forest trees. In addition, a large number of recombinant genes from different organisms are now available for potential use.

At least three components of a recombinant gene construct are essential for the expression and detection of the transferred gene: (1) The recombinant gene of interest; (2) A promoter to regulate the expression of the recombinant gene, for example, 35S from cauliflower mosaic virus; and (3) A selectable marker gene (also called a reporter gene) to facilitate the isolation of the transferred trait gene in the transgenic plant, for example, the antibiotic resistance kanamycin gene (Fig. 20.5) (Fillatti *et al.*, 1987; Leple *et al.*, 1992; Levée *et al.*, 1999). Transformants carrying the antibiotic resistant gene survive on the kanamycin-containing medium, while non-transformants are killed. Thus, the kanamycin resistant gene acts as a selectable marker for the initial screening of transformants in gene transfer experiments. Recombinant genes with varying promoters, coding sequences, and selectable markers have been constructed and used in a large number of studies to monitor the expression of transgenes in forest trees (Charest and Michel, 1991; Fladung *et al.*, 1997; Kim *et al.*, 1997; Tang and Newton, 2003).

During infection with *Agrobacterium tumefaciens*, a piece of DNA carrying the recombinant gene construct is transferred from the bacterium to the plant cell by an unnatural form of recombination called illegitimate recombination, achieved via single strand annealing followed by ligation. The transferred segment is called transferred DNA (T-DNA), which carries the recombinant gene construct in a vector plasmid. The T-DNA is flanked by 25 base pair direct repeats, and these borders are necessary for the transfer of T-DNA from the bacterium to the plant cell (Zupan and Zambryski, 1995; Tzfira *et al.*, 2004). The integration of the T-DNA can occur in any chromosome in the genome. There may be a single insertion of T-DNA on one chromosome or several copies of the T-DNA may be inserted on the same chromosome or in several chromosomes (Fig. 20.3). Thus, it is difficult to predict the precise location(s) of the T-DNA insertion in the genome of a primary transformant. In addition, the number of copies of transferred genes may be different for each transformant, with one to several transformations occurring on one or more chromosomes. Copy number and the site of transgene insertion affect the expression of the

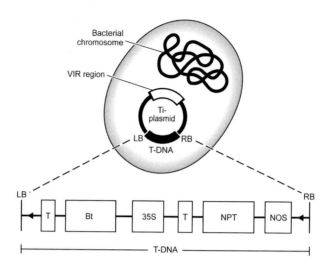

Fig. 20.5. A Ti plasmid carrying the recombinant genes in its bacterial host. The transferred DNA (T-DNA) contains a selectable marker gene for kanamycin resistance (*NPT*) and the *Bt* gene for insect resistance. The expression of the *NPT* gene is controlled by a bacterial promoter *NOS*, and the *Bt* by a 35S promoter derived from the cauliflower mosaic virus. The promoter and trait genes in recombinant gene constructs may contain DNA sequences from bacteria, viruses or plants. LB and RB are left and right borders, respectively. VIR region is the virulence region of the Ti plasmid. T is the terminal region.

transferred gene as well as homologous host genes and may be associated with various forms of transgene silencing (Hobbs *et al.*, 1993; Finnegan and McElroy, 1994; Meyer, 1995; Stam *et al.*, 1997; Ahuja, 1997; Fladung, 1999; Fagard and Vauchert, 2000; see also *Transgene Expression and Stability*). Therefore, it is necessary to optimize gene transfer methodologies that produce transgenic plants with one or relatively few copies of the transgene and that are genetically stable under greenhouse and field conditions (Ahuja, 2000).

REGENERATION METHODS

An efficient *in vitro* regeneration system is essential for GE in plants and this system needs to be both economically efficient and biologically effective for the GE plants to be deployed operationally, such as through clonal forestry (*Chapter 16*). Two approaches have been used for successful regeneration of genetically transformed cells/tissues in forest trees: organogenesis and somatic embryogenesis (Fig. 20.6). Both approaches have been employed for regeneration of transgenic plants in angiosperm trees, but in conifers, transformation and regeneration via somatic embryogenesis has been more effective since somatic tissues (buds, needles) are somewhat recalcitrant to organogenesis *in vitro*. Recent studies have shown, however, that regeneration via organogenesis from mature zygotic embryos is feasible in pines (Tang and Quyang, 1999; Tang *et al.*, 2006), but application of organogenesis in conifers is still a challenging problem.

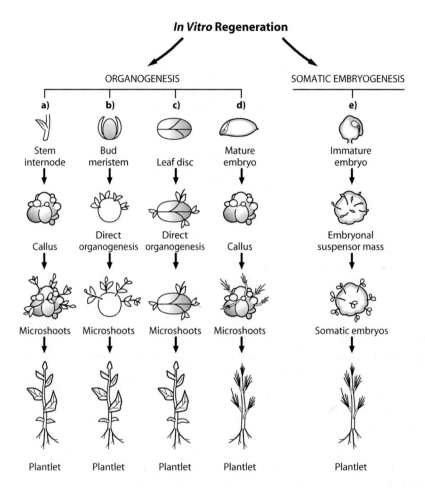

Fig. 20.6. *In vitro* regeneration of plants by organogenesis (a-d) and somatic embryogenesis (e). Organogenesis can be accomplished by culture of different types of explants, for example, internodal stem segments (a), bud meristems (b), leaf discs (c), or mature somatic embryos (d). Cultured explants may go through a callus phase (a and d), or undergo direct organogenesis. In either case, microshoots differentiate on these cultured explants on a shoot-induction medium. The microshoots are rooted in a different medium to differentiate roots for plantlet development. Organogenesis involving pathways a-c are generally used in angiosperms trees, and pathway d has been recently shown to be successful in conifers. Somatic embryogenesis (e) has been accomplished from immature zygotic embryos, mostly in conifers. The immature embryos give rise to the embryonal suspensor mass (EMS), and somatic embryos then develop in the EMS. The somatic embryos mature and finally germinate into plantlets. Organogenesis and somatic embryogenesis require different sets of culture media and growth regimes.

Organogenesis

Organogenesis (literally, the development of organs) is the production of plants through the induction of organs (shoots or roots) from bud meristems, internodal stem explants or callus cultures (Fig. 20.6). Organogenesis can occur by at least two pathways: adventitious buds/shoots on callus and axillary buds/shoots from bud meristem or internodal stem explants. The induction of shoots or roots on cultured explants depends on the appropriate culture medium used. Most frequently, variations of the ingredients in Maranshige and Skoog's (1962) medium (for example, Woody Plant Medium (Lloyd and McCown, 1981)) have been used for organogenesis in forest trees. The explants are cultured in containers (test tubes, Petri dishes, flasks or jars) under aseptic conditions to avoid microbial contamination, and maintained in artificial light (1000-3000 lux) and temperature (24-26°C) regimes for optimal growth and differentiation of organs. Growth hormones required for the differentiation of shoots include cytokinins (6,bezylaminopurine (BAP), zeatin and kinetin), while roots are generally initiated by auxins such as NAA (naphthaleneacetic acid), IAA (indol-3-acetic acid), IBA (indol-3-butyric acid) and 2,4-D (2,4-dichlorophenoxyacetic acid). Generally both cytokinins and auxins are required for the growth of tissues and the differentiation of shoots or roots is determined by the cytokinin/auxin ratio in the medium.

In order to extend tissue culture technology to large scale regeneration of transgenic plants, it is necessary to develop a micropropagation method that is relatively simple and has a high degree of reproducibility. Efficient regeneration methods via organogenesis have been developed in *Populus*, *Betula*, *Eucalyptus*, and a few other angiosperm tree species (Bonga and Durzan, 1987; Ahuja, 1993). Stock shoot cultures from bud explants or shoot tips are routinely developed and maintained for GE. Either leaf discs or internodal stem segments are co-cultivated with *Agrobacterium in vitro* for transfer of the foreign gene from the bacterium to the plant cells. Following the incubation period, tissues are grown in the medium containing a selectable marker (kanamycin or another antibiotic) for selection of the transformants that are resistant to the antibiotic. Subsequently, the tissues are grown on an antibiotic-free medium to harvest transformed shoots, which are then rooted on another medium or *ex vitro* to yield transgenic plants (Fillatti *et al.*, 1987; Fladung *et al.*, 1997; Kim *et al.*, 1997).

Somatic Embryogenesis

Somatic embryogenesis involves differentiation of embryos from somatic tissues that subsequently develop into plants. Somatic embryos derive from embryogenically competent cells in young or mature embryos (Fig. 20.6). Just like zygotic embryos, somatic embryos are bipolar having both a shoot and a root pole. In contrast to organogenesis, where shoots and roots mostly develop sequentially on different media, somatic embryogenesis is a one-step process. That is not to say that somatic embryos immediately develop from immature zygotic embryos cultured on one medium. Several steps may be required such as initiation, development of embryos, and maturation of somatic embryos each requiring different media with different concentrations of phytohormones, such as 2,4-D, BAP, kinetin, NAA and ABA (abscisic acid).

Somatic embryogenesis, however, is distinguished from organogenesis because somatic embryos develop similarly to zygotic embryos instead of sequential products of different organs. As in organogenesis, somatic embryos are regenerated in aseptic culture media under dark or light regimes. Somatic embryos have been produced in a number of hardwoods and conifer forest tree species (Becwar, 1993; Gupta and Kreitinger, 1993; Dunstan *et al.*, 1995). Stable genetic transformation has been reported in a number of conifers by either co-cultivation of embryogenic cell lines with *Agrobacterium* or by biolistic particle bombardment of embryonic tissues (Loopstra *et al.*, 1992; Charest *et al.*, 1993; Ellis *et al.*, 1993; Shin *et al.*, 1994; Klimaszewska *et al.*, 1997; Levée *et al.*, 1999; Wenck *et al.*, 1999).

APPLICATIONS OF GENETIC ENGINEERING IN FOREST TREES

The ultimate goal of genetic engineering in forest trees is to operationally deploy genetically stable transgenic trees that express or silence (or decrease the expression of) one or more target traits. Although GE research has progressed fairly well during the last two decades, there are still problems of expression of transgenes in transgenic plants. Because of long life cycles in forest trees, such concerns about transgene expression in space and time need to be fully addressed. Nevertheless, transgenic trees produced in a number of forest tree species are now undergoing field trials. Potential applications of GE in forest trees include: (1) Lignin modification; (2) Herbicide tolerance; (3) Disease and pest resistance; and (4) Flowering control. Although GE has proceeded in all four of these applications, lignin modification and flowering control are still challenging areas for domestication. GE for herbicide and disease/pest resistance has progressed to some extent in hardwoods. Transgenic trees tolerant to insects and herbicides in *Eucalyptus* (Harcourt *et al.*, 2000), herbicides in *Populus* (Meilan *et al.*, 2002) and fungal diseases in *Betula* (Pasonen *et al.*, 2004) are undergoing field evaluation for the continued, stable expression of the target traits.

Lignin Modification

Lignins are complex phenolic polymers that provide structural strength to wood and have antimicrobial properties. Lignin is cross-linked with various cell wall components and cellulose microfibrils are embedded in its dense matrix (Campbell and Sederoff, 1996). Although lignin is the most important component of wood and accounts for 25% of the global wood biomass (Leple *et al.*, 1992), it is an obstacle to efficient production of pulp and paper. The industrial removal of lignin from cellulose microfibrils is costly and residual wastes are toxic and hazardous. In order to make the pulping process less costly and more environmentally friendly, it would be desirable to reduce the quantity of lignin in individuals or change its chemical properties to make extraction easier. Of course, this must be accomplished without hurting the growth, form and stress resistance of the GE trees.

Lignin content varies among species from 15 to 36% of the dry wood mass (Higuchi, 1985). Lignin is derived from the oxidative polymerization of three different hydroxycinamyl alcohols or monolignols: p-coumaryl alcohol, coniferyl alcohol and sinapyl alcohol. These differ in their degree of methoxylation and generate the following three lignin polymers: p-hydroxyphenyl (H), guiacyl (G) and syringyl (S) units. Gymnosperm wood mainly contains G-lignin, while angiosperm wood is composed of G-S lignin (Higuchi, 1985; Campbell and Sederoff, 1996). Because of these differences in composition, lignin is relatively easier to extract from angiosperms than from gymnosperms (Campbell and Sederoff, 1996). In angiosperms there is considerable variation in the ratio of G and S units and those plants with higher S:G ratio have lignin that is more readily extractable (Chiang *et al.*, 1988; Chiang and Funaoka, 1990).

Strategies that would decrease lignin content, but still maintain the structural integrity of wood, would be highly desirable. One approach towards lowering lignin content would be to use recombinant genes that target enzyme activity in the lignin biosynthesis pathway. Lignin formation is under the control of many genes and the lignin biosynthesis pathway (Fig. 2.13) is well known (Whetten *et al.*, 1998). Recently a number of genes involved in the biosynthesis of lignin have been identified and isolated from trees, including *Populus* and *Pinus* (Allona *et al.*, 1998; Hertzberg *et al.*, 2001; Baucher *et al.*, 2003). Several genes coding for enzymes involved in the biosynthesis of lignin have been the target of lignin modification, mainly in the genus *Populus*.

An increase in the S:G ratio in wood, which leads to more readily extractable lignin, was accomplished in transgenic *Populus* plants by over-expressing the *Arabidopsis* gene encoding ferulate 5-hydroxylase under the control of a strong promoter (Franke *et al.*, 2000). The quantity of S units was effectively increased from 55 mol% in the untransformed *Populus* to 85 mol% in the transgenic *Populus*, thus effectively converting most of the G units to S units (Franke *et al.*, 2000). In addition, several other genes involved in the multi-step biosynthesis pathway leading to lignin precursors have been down regulated by antisense suppression in *Populus*.

Antisense expression of lignin biosynthesis genes, such as 4-coumarate CoA:ligase (4CL) and cinnamyl alcohol dehydrogenase (CAD), have been effective in decreasing lignin content in transgenic *Populus*. Down-regulation of the 4CL transgene not only resulted in decreased lignin content but, at the same time, increased cellulose content in transgenic *Populus tremuloides*. The transgenic *Populus* with decreased 4CL activity appeared to show better growth performance under greenhouse conditions without any fundamental change in the structural integrity of lignin (Hu *et al.*, 1999). Antisense expression of CAD resulted in modest reduction in lignin content, red coloration of wood and an increase in extractability of lignin in transgenic *Populus* (Baucher *et al.*, 1996). Field trials of transgenic *Populus* with the antisense CAD transgene also demonstrated reduced content of lignin under field conditions (Pilate *et al.*, 2002).

Lignin modification has also been investigated in *Eucalyptus*. The gene encoding cinnamate 4-hydroxylase (C4H) isolated from *Populus tremuloides* was introduced in antisense orientation into *Eucalyptus camaldulensis* via *Agrobacterium tumefaciens* (Chen *et al.*, 2001). The transfer and integration of the C4H transgene was confirmed and transgenic *Eucalyptus* is undergoing greenhouse tests for any alterations in lignin content.

Herbicide Tolerance

Herbicides are chemical compounds used to kill weeds. Tolerance to herbicides could be a useful trait in early establishment of forest tree plantations, where weeds are a major problem. One strategy is to isolate herbicide tolerance genes and genetically transfer them to trees allowing the transgenic trees to survive when adjacent weeds are sprayed with an herbicide. This has been achieved in a number of forest tree species.

A first attempt to introduce herbicide (glyphosate) tolerance in trees was carried out by Fillatti *et al.* (1987) by *Agrobacterium*-mediated transfer of the mutant *aro*A gene from a bacterium, *Salmonella typhymurium*, into *Populus* hybrid clone NC-5339 (*Populus alba* x *P. grandidentata*). In untransformed plants, glyphosate inhibits activity of 5-enolpyruvylshikimate 3-phophate (EPSP) synthase, which is a key enzyme in the biosynthesis of aromatic amino acids. The mutant gene (aroA) gene encodes an EPSP that is less sensitive to the herbicide glyphosate (Comai *et al.*, 1985). However, in the above case, the expression of the *aro*A transgene under the control of the bacterial mannopine synthetase promoter was below the expected level (Riemenschneider *et al.*, 1988). Subsequently, herbicide tolerance was improved in the same *Populus* clone by using the constitutive promoter 35S from the cauliflower mosaic virus for moderately higher expression of the herbicide tolerant *aro*A gene (Donahue *et al.*, 1994).

In addition to *aro*A, several other genes conferring herbicide tolerance have been engineered into forest tree species. These include: (1) The *bar* gene from *Streptomyces hygroscopicus* (Thompson *et al.*, 1987) that encodes phoshinothricine acetyl transferase (PAT) that inactivates the herbicide Basta in hybrid *Populus* (*Populus trichocarpa* x *P. deltoides* and *P. alba* x *P. tremula*) (DeBlock, 1990) and in *Eucalyptus camaldulensis*

(Harcourt *et al.*, 2000); and (2) A mutant *crs1-1* gene from a chlorosulfuron-herbicide-resistant line of *Arabidopsis* (Haughn *et al.*, 1988) in *Populus tremula* x *P. alba* (Brasileiro *et al.*, 1992). More recently a mutant gene CP4, which encodes an alternative form of EPSP, was isolated from *Agrobacterium* and engineered into *Populus* for improved herbicide tolerance. This transgenic *Populus* exhibited a high level of glyphosate tolerance when grown under field conditions (Meilan *et al.*, 2002).

In conifers, the *aro*A gene was introduced into *Larix decidua* and measurable tolerance to glyphosate was detected (Shin *et al.*, 1994). In addition the *bar* gene has been introduced into two species, *Pinus radiata* and *Picea abies*, and transgenic plants from these species have shown a high degree of tolerance to the herbicide Buster under greenhouse conditions (Bishop-Hurley *et al.*, 2001).

Pest and Disease Resistance

A number of insects and diseases cause considerable loss to forest productivity. Current operational practices for insect and disease control rely on classical breeding methods to develop resistant varieties or the extensive use of chemical pesticides, some of which are harmful to humans and the environment. Genetic engineering provides an alternative for controlling insects and diseases in forest trees. Two approaches have been used to engineer insect resistance in trees. One approach makes use of *Bt* genes from a bacterium, *Bacillus thuringiensis*, that encode different crystal (*cry*) toxins that are toxic to different groups of insects belonging to Lepidoptera, Diptera, and Coleoptera. Based on host-specificity and homology, these endotoxins are classified into various *cry* types, such as *cry1Aa*, *cry1Ab*, and *cry1Ac* (Estruch *et al.*, 1997). The *Bt* toxins damage the digestive mechanisms of the insects. *Bt* toxin genes have been transferred to a number of forest tree species, including *Populus* (McCown *et al.*, 1991; Wang *et al.*, 1996), *Eucalyptus* (Harcourt *et al.*, 2000), *Larix* (Shin *et al.*, 1994), *Picea* (Ellis *et al.*, 1993), *Pinus radiata* (Grace *et al.*, 2005) and *Pinus taeda* (Tang and Tian, 2003). Varying expression of *Bt* transgenes in terms of insect mortality has been observed in these forest tree species. Field trials are in progress in *Populus* and some conifers to further evaluate this form of biological control of insect damage (Wang *et al.*, 1996; Grace *et al.*, 2005).

The other approach involves transfer of a wound-inducible proteinase inhibitor II gene (*pin2*) from solanaceous plants, including potato and tomato, to forest trees. The *pin2* gene is induced in response to wounding and inhibits the protolytic activity of trypsin and chymotrypsin in the digestive system of herbivores (Klopfenstein *et al.*, 1991). Trees of *Populus* transformed with the *pin2* gene have been assayed for tolerance to willow leaf beetle, a major defoliator of *Populus* (Klopfenstein *et al.*, 1997). Larvae fed with *Populus* transgenic leaves weighed less and showed a trend towards longer development time, as compared to those fed on controls. Two other plant-derived genes, a proteinase inhibitor gene encoding oryzacystin (*ocl*) and a cystin inhibition (*cys*) gene also showed promising tolerance to beetles in *Populus* (Leple *et al.*, 1995; Delledonne *et al.*, 2001).

There have been very few studies on GE of disease resistance in forest trees. More recently, however, some progress has been made towards engineering resistance to fungal diseases in *Betula pendula*, *Populus tremula* and *P. tremula* x *P. tremuloides* hybrids. *Betula*, carrying a sugar beet chitinase IV transgene, showed increased tolerance to two fungal pathogens: leaf spot fungus (*Pyrenopeziza betulicola*) and birch rust fungus (*Melamposoridium betulinum*) in field trials of transgenic *Betula* (Pasonen *et al.*, 2004). Chitinases are produced as a response to fungal infection and are enzymes that degrade the cell walls of the invading fungus.

In another study, antifungal activity of stilbene compounds (pinosylvins and reseratol) against 16 species of pathogenic fungi was assayed in transgenic aspens carrying a pinosylvin synthase encoding gene (*STS*) from *Pinus sylvestris* (Seppänen *et al.*, 2004). Expression of the *STS* gene enhanced the resistance to the fungal pathogen, *Phellinus tremulae*, in transgenic *Populus* line H4. However, results on other transgenic lines of *Populus* indicate that the heterologous gene *STS* may be restricted in its ability to confer fungal resistance (Seppänen *et al.*, 2004) or may require a stronger promoter.

Flowering Control

Two approaches have been employed for genetic engineering of flowering in forest trees: (1) Early flowering for rapid breeding and turnover of breeding cycles; and (2) Reproductive sterility to achieve biological containment of transgenes in GE trees planted operationally. Forest trees have long generation cycles, with time to reproductive maturity extending from one year to several decades. Because of long life cycles, genetic improvement by conventional breeding strategies may be a slow and tedious process. Therefore, it would be desirable to identify genes that promote early flowering for accelerated breeding programs in forest trees. There are a large number of genes involved in the flowering process and some floral identity genes have been found (Meilan and Strauss, 1997). Most of these floral genes are conserved in higher plants with varying degree of homology. Precocious flower development in *Populus* was induced by genetic engineering of mutant floral identity genes LEAFY (*LFY*) from *Arabidopsis* (Weigel and Nilsson, 1995). Later studies with *LFY* showed that early flowering was genotype dependent and over-expression of *LFY*, and its homolog in poplar (*PTLF*), cause early flowering infrequently (Rottmann *et al.*, 2000).

Male sterility would eliminate the spread of transgenic pollen from operationally deployed GE trees. Genetic engineering strategies for reproductive sterility in forest trees include (Strauss *et al.*, 1995): (1) Ablation (prevention) of reproductive tissue; and (2) Suppression of genes necessary for fertility and flowering. These two strategies have been investigated using *Populus* as a model system because it is relatively easy to regenerate *in vitro*, can be easily transformed, and is dioecious *(i.e.* male and female flowers are borne on separate trees). A large number of putative floral identity genes, that are orthologs (Fig. 9.7) of genes in annual flowering plants, have been identified and cloned in *Populus* (Campbell *et al.*, 2003). Since *Populus* flowers under normal conditions at around 5-7 years of age, and several environmental factors, including light, day-length, and temperature, influence flowering, the results obtained in greenhouse conditions may not always hold when transgenic plants are grown under field conditions on a long-term basis.

Conifers have much longer generation cycles, with extended periods ranging from one to several decades before sexual maturity is reached. Therefore, many years of testing would be required to evaluate the fate of sterility-targeted transgenes in conifers or long-lived hardwoods. Consequently, there are many obstacles to achieving reproductive sterility in forest trees.

Ablation of reproductive tissues, using a tobacco tapetal-specific promoter TA29, which drives the expression of cytotoxin barnase from a bacterium, was effective in inducing male sterility in *Populus*. However, this and other floral prevention transgenes had a negative effect on tree growth under field conditions, but not under greenhouse conditions (Meilan *et al.*, 2001; Busov *et al.*, 2005). Antisense expression of a floral meristem identity gene to suppress the expression of genes essential for fertility has also had some success. An antisense *PTLF* (*Populus LFY*) transgene induced mutant floral phenotypes and highly reduced fertility in male *Populus* trees, but was ineffective in female trees (Rottmann *et al.*, 2000). A large repertoire (>30) of sterility transgenes, driven by endogenous and exogenous promoters, that act through different mechanisms are being targeted for reproductive sterility in *Populus* (Busov *et al.*, 2005).

TRANSGENE EXPRESSION AND STABILITY

A number of economically important traits have been engineered in several forest tree species, both hardwoods and conifers. Trees are different from annual plants in many ways; in particular, their long life cycles and the many years often required before they reach reproductive age. Therefore, stable expression of transgenes among individuals, over time in any one individual, and over generations, is a desirable goal of GE in the forest trees. Transgene stability may be affected by: (1) Gene transfer methodology; (2) Type of recombinant gene construct; and (3) Time and range of environments over which transgenics are grown. At another level, stability might be affected by the nature of recombinant genes and the promoters that drive their expression in the host tree. Finally, stability might also be affected by the long time periods that transgenes must remain active.

It appears that the *Agrobacterium*-mediated gene transfer systems produce simpler integration patterns of transgenes than direct gene transfer methods using biolistics. In transgenic *Populus*, copy number of integrated foreign genes ranged from one to several following *Agrobacterium*-mediated gene transfer (Nilsson *et al.*, 1996; Fladung *et al.*, 1997). Alternatively, biolistic transformation of conifers (*Picea glauca*, *Pinus radiata*, *Picea mariana*, and *Larix lariciana*) showed complex patterns of foreign gene integration and copy number of integrated transgenes ranged from one to more than a hundred (Ellis *et al.*, 1993; Charest *et al.*, 1996; Walter *et al.*, 1998b).

However, depending on the transformation parameters, both indirect and direct approaches of gene transfer may result in somewhat similar complex patterns of integration (Meyer, 1995; Birch, 1997). In fact, there may be considerable variation in transgene expression among different transformants of the same species using the same protocols, due to integration patterns (one or several sites) or copy number of the transgene. Multiple integrations of a transgene may cause endogenous gene inactivation and variation in gene expression. Therefore, it is important to optimize gene transfer technologies that produce preferentially simple integration patterns, preferably a single copy of the transgene.

Transgene expression may vary depending on the type of promoter used to drive its expression in a transgenic plant. For example, the most widely used promoter, 35S from cauliflower mosaic virus (CaMV), functions well in most forest tree species and provides high levels of expression of reporter genes. Recent studies show that stability of transgene expression may be increased in transgenic plants by including matrix attachment regions (MARs) to flank the recombinant gene (Han *et al.*, 1997; Allen *et al.*, 2000). MARs are AT rich regions in eukaryote genomes that bind specifically to the nuclear matrix.

The third aspect of transgene expression relates to functional utility of a transgene in time and space. For example, an herbicide tolerance transgene may remain active throughout the major part of the life of an annual crop so that it has a competitive edge over weeds when sprayed with an herbicide. On the other hand, the expression of an herbicide tolerance transgene would be required only in the first few years of tree growth when the herbicide is needed to kill competing weeds. After the initial growth period, the transgene may not be expressed in the absence of the herbicide. What might happen to the herbicide tolerant transgene during the next 10 to 50 years of the trees' life remains unknown. The transgenes for lignin modification, and for insect and disease tolerance, are required for the entire life of a tree, and their stable functionality and expression would be of paramount importance to the survival of the tree. GE for reproductive sterility would require the activation of floral identity transgenes at sexual maturity. Whether transgenes for reproductive sterility remain inactive/silent during the many years prior to sexual maturity is unclear.

COMMERCIALIZATION, REGULATION AND BIOSAFETY

Although a large number of field trials of transgenic forest tree species (both angiosperms and conifers) have been established world-wide, there are almost no operational plantations of GE forest trees. The only exception is in China, where GE poplars have been commercialized. On the other hand, annual GE crop plants have been grown commercially in many countries for nearly 10 years. Ever since the first release of a GE crop in the United States in 1994 (the "Flavrsavr" tomato that has delayed ripening characteristics), the global area of GE crops has increased more than 47-fold, from 1.7 million hectares in 1996 to 81 million hectares in 2004 (Fig. 20.7). As of 2004, the global status of approved GE food and fiber crops is (Table 20.1): (1) GE crops are planted on nearly 81 million hectares in 21 countries with more than half (47.6 million hectares) in the United States; (2) GE crops include maize, soybean, cotton, canola, rice, squash and papaya; (3) Target traits for GE crops include herbicide resistance, insect resistance and virus resistance; and (4) GE crops generate an annual global market of $4.7 billion (James, 2004).

The widespread commercialization of GE forest trees will likely not happen for some years because GE research in tree species has not progressed as far as in crop plants and there are a number of unresolved biosafety and regulatory obstacles. Commercial release of any GE tree or crop is regulated by governmental agencies for biosafety to humans and the environment. The legal authority in the United States that regulates transgenic trees and crops is the USDA Animal and Plant Health Inspection Service (APHIS) which has the mandate to protect the agricultural environment against pests and diseases (the Federal Drug Administration and the Environmental Protection agency may also be involved, depending on the trait). The regulatory criteria for evaluating a GE crop include both the process and the product. The process covers all aspects of the recombinant gene construct and how it is transferred to the host plant, and whether it is stably integrated and inherited, or affects other host genes. The product represents the phenotype of the transgenic plant, exhibiting the desired traits.

In spite of the commercial release of many GE crop plants (Table 20.1 and Fig. 20.7) and a potentially huge world market for GE forest trees, there are still obstacles to their testing and deployment, including a number of biosafety and regulatory concerns that must be addressed. These concerns are mainly based on the endogenous behavior of the transgene (its construction, stability, and interaction with other genes in the host genome in space and time) and its exogenous potential effects on forest tree plantations and forest ecosystems.

Forest trees have long life cycles, and therefore, regulatory oversight on transgenic trees should not be a simple application of the rules in place for annual crops. Some have argued, however, that the regulatory guidelines for the release of transgenic forest trees need to be assessed in view of lessons learned from crop plants (Strauss, 2003; Bradford *et al.*, 2005). The methodologies for gene transfer have improved and the trend is towards using transgenes from the same or related species (as in lignin modification and flowering control) rather than foreign genes from bacteria and viruses. If homologous genes from

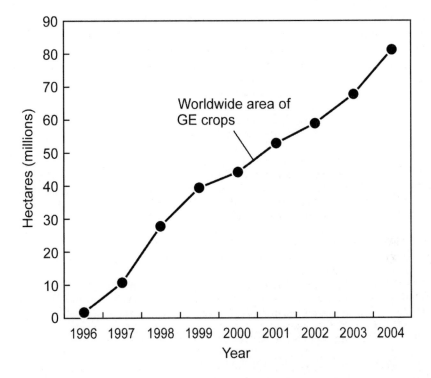

Fig. 20.7. Global area (million hectares) of genetically engineered crops, 1996 to 2004. (Reproduced from James, 2004 with permission of the International Service for the Acquisition of Agri-biotech Applications (ISAA))

Table 20.1. Global area of genetically engineered crops in 2004 (data from James, 2004)

Rank	Country	Area (millions of hectares)	GE crops
1	USA	47.6	Soybean, maize, cotton, canola, squash, papaya
2	Argentina	16.2	Soybean, maize, cotton
3	Canada	5.4	Canola, maize, soybean
4	Brazil	5.0	Soybean
5	China	3.4	Cotton
6	Paraguay	1.2	Soybean
7	India	0.5	Cotton
8	South Africa	0.5	Maize, soybean, cotton
9	Uruguay	0.3	Soybean, maize
10	Australia	0.2	Cotton
11	Romania	0.1	Soybean
12	Mexico	0.1	Cotton, soybean
13	Spain	0.1	Maize
14	Philippines	0.1	Maize
15	Colombia	<0.1	Cotton
16	Honduras	<0.1	Maize
17	Germany	<0.1	Maize
Total		81.0	

1 hectare = 2.47 acres

related species modify metabolism in the host in a similar manner to that of the donor, then potential negative impacts of gene transfer are of less concern (Strauss, 2003). Once a transgene has been found to relatively safe, the regulatory emphasis, as in conventional breeding and mutation programs, should be on the phenotype rather than the gene (Bradford *et al.*, 2005). A recent report by the US National Research Council (NRC, 2002) on biocontainment has supported the "product and not process" principle in GE crops and has further suggested that many transgenic traits will not require confinement (NRC, 2004). However, this suggestion has not been put into current governmental regulatory practice.

One major concern with commercialization of GE trees is that gene flow from transgenic plantations may have negative impacts on natural forest populations. If a transgene-carrying genome was to escape and become established in wild populations, it could conceivably displace native genotypes or lead to maladaptation. Both pollen and seed can be dispersed many kilometers from forest stands, resulting in considerable gene flow, especially among neighboring populations (*Chapter 7*; Williams, 2005). Two basic options could be used to deal with this issue: (1) The biosafety approach; and (2) The ecological approach (Williams, 2005). The biosafety approach relies on effective biosafety protocols, such as absolute sterility, that would be required before the commercial release of transgenic trees (see *Flowering Control*). Under the ecological approach, it is assumed that escape of transgenes is inevitable, so it becomes a matter of minimizing the risks; that is, by limiting the invasiveness (*i.e.* fertility and viability) of offspring from GE trees, and/or by utilizing only low-risk transgenic organisms (*e.g.* where the imported traits are functionally equivalent to those produced by conventional breeding) (DiFazio *et al.*, 2004; Williams and Davis, 2005).

In the final analysis, the benefits of GE trees must outweigh their risks if commercialization is to develop. In the current regulatory climate in much of the world, however, it is difficult to assess either the benefits or risks of employing GE in forest trees. This is because regulatory requirements and their associated costs severely limit the scale and longevity of research on transgenics (*e.g.* trees must be cut down prior to reaching reproduction age, and often harvest age) (Williams, 2005). It is clear that there are still many unknowns concerning the risks and benefits of employing GE technology in tree improvement programs, and regulatory reform may be required before these unknowns can be adequately addressed in the future

SUMMARY AND CONCLUSIONS

Genetic improvement of forest trees has largely progressed by using classical selective breeding approaches. During the past two decades, however, a new approach involving asexual transfer of genes by genetic engineering (GE) has added a new dimension for genetic modification of the tree genome. GE is fundamentally different from traditional or marker-based breeding in that foreign genes can be introduced that code for traits not normally found in the target plant.

Foreign genes can be introduced by indirect and direct approaches of gene transfer. Indirect DNA transfer requires the use of an intermediate organism such as *Agrobacterium* to transmit the foreign DNA into the host cell, whereas direct DNA transfer (for example by biolistics) does not involve an intermediate organism. Both indirect and direct approaches have been successfully used for the transfer of foreign genes into forest trees.

In order to transmit foreign genes into a tree genome, it is necessary to use recombinant DNA technology to produce a recombinant gene, a vector to transmit the recombinant gene into the host genome and an efficient *in vitro* regeneration method for the production of transgenic plants. Transgenic trees have been produced in a number of forest trees, both angiosperms and conifers.

Potential uses of GE in forest trees include: (1) Lignin modification; (2) Herbicide tolerance; (3) Disease and pest resistance; and (4) Flowering control. Some progress has been made in all four of these applications. Transgenic trees produced with desirable target traits are undergoing field trials for their future application in forestry. However, the ultimate goal of GE in forest trees is to operationally deploy stable transgenic plants that exhibit either increased or decreased expression of the target trait. Although GE has progressed fairly well during the past two decades, there are still problems of transgene expression in transgenic trees.

Because of long life cycles in forest trees, with extended vegetative phases lasting one year to several decades, concerns about transgene expression in space and time need to be fully addressed. Even in the same species using the same protocols, there may be considerable variation in the transgene expression among different transformants due to integration patterns (one or several sites) or copy number of the transgene. Multiple integration patterns may cause endogenous gene inactivation and variation in gene expression. Therefore, it is important to optimize gene transfer technologies that produce preferentially simple integration patterns in transgenic trees.

Before release of crop plants operationally, they are regulated by governmental agencies for their biosafety to humans and the environment. Forest trees would also be subject to such regulations before commercialization. However, there are number of concerns, including gene flow from transgenic trees to native forest populations, dispersal of transgenic seed, and invasiveness of transgenic trees, that must be addressed before release of transgenic trees into the market place.

REFERENCES

Aagaard, J.E., Krutovskii, K.V. and Strauss, S.H. (1998) RAPDs and allozymes exhibit similar levels of diversity and differentiation among populations and races of Douglas-fir. *Heredity* 81, 69-78.

Adams, M.D., Kelley, J.M., Gocayne, J.D., Dudnick, M., Polymeropoulod, M.H., Xiao, H., Merril, C.R., Wu, A., Olde, B.,Moreno, R., Kerlavage, A.R., McCombie, W.R. and Venter, J.C. (1991) Complementary DNA sequencing: expressed sequence tags and human genome project. *Science* 252, 1651-1656.

Adams, W.T. (1981) Population genetics and gene conservation in Pacific Northwest conifers. In: Scudder, G.G. and Reveal, J.L. (eds.) *Evolution Today, Proceedings of the 2nd International Contress of Systematic and Evolutionary Biology*. Hunt Institute for Botanical Documentation, Carnegie, Mellow University, Pittsburgh, PA, pp. 401-415.

Adams, W.T. (1983) Application of isozymes in tree breeding. In: Tanksley, S.D. and Orton, T.J. (eds.) *Isozymes in Plant Genetics and Breeding. Part A*. Elsevier Science Publishers, Amsterdam, The Netherlands, pp. 381-400.

Adams, W.T. (1992) Gene dispersal within forest tree populations. *New Forests* 6, 217-240.

Adams, W.T. and Birkes, D.S. (1989) Mating patterns in seed orchards. In: *Proceedings of the 20th Southern Forest Tree Improvement Conference*, Charleston, SC, pp. 75-86.

Adams, W.T. and Burczyk J. (2000) Magnitude and implications of gene flow in gene conservation reserves. In: Young, A., Boshier, D. and T.Boyle (eds.) *Forest Conservation Genetics: Principles and Practice*. Commonwealth Scientific and Industrial Research Organization (CSIRO) Publishing, Collingwood, Victoria, Australia, pp. 215-244.

Adams, W.T. and Campbell, R.K. (1981) Genetic adaptation and seed source specificity. In: Hobbs, S.D. and Helgerson, O.T. (eds.) *Reforestation of Skeletal Soils*. Forest Research Laboratory, Oregon State University, Corvallis, OR, pp. 78-85.

Adams, W. and Joly, R. (1980) Linkage relationships among twelve allozyme loci in loblolly pine. *Journal of Heredity* 1, 199-202.

Adams, W.T., Roberds, J.H. and Zobel B.J. (1973) Intergenotypic interactions among families of loblolly pine (*Pinus taeda* L.). *Theoretical and Applied Genetics* 43, 319-322.

Adams, W. T., Neale, D.B. and Loopstra, C.A. (1988) Verifying controlled crosses in conifer tree-improvement programs. *Silvae Genetica* 37, 147-152.

Adams, W.T., Strauss, S.H., Copes, D.L. and Griffin, A.R. (1992a) *Population Genetics of Forest Trees*. Kluwer Academic Publishers, Dordrecht, The Netherlands.

Adams, W.T., Campbell, R.K. and Kitzmiller J.H. (1992b) Genetic considerations in reforestation. In: Hobbs, S.D., Tesch, S.D., Owston, P.W., Stewart, R.E., Tappenier, J.C. and Wells, G. (eds.) *Reforestation Practices in Southwestern Oregon and Northern California*. Forest Research Laboratory, Oregon State Univiversity, Corvallis, OR, pp. 284-308.

Adams, W.T., Birkes, D.S and Erickson, V.J. (1992c) Using genetic markers to measure gene flow and pollen dispersal in forest tree seed orchards. In: Wyatt, R. (ed.) *Ecology and Evolution of Plant Reproduction*. Chapman and Hall, New York, NY, pp. 37-61.

Adams, W.T., White, T.L., Hodge, G.R. and Powell, G.L. (1994) Genetic parameter estimates for bole volume in longleaf pine: large sample estimates and influence of test characteristics. *Silvae Genetica* 43, 357-366.

Adams, W.T., Hipkins, V.D., Burczyk, J. and Randall, W.K. (1996) Pollen contamination trends in a maturing Douglas-fir seed orchard. *Canadian Journal of Forest Research* 27, 131-134.

Adams, W. T., Zuo, J., Shimizu, J.Y. and Tappeiner, J.C. (1998) Impact of alternative regeneration methods on genetic diversity in coastal Douglas-fir. *Forest Science* 44, 390-396.

Ager, A.A., Heilman, P.E. and Stettler, R.F. (1993) Genetic variation in red alder (*Alnus rubra*) in relation to native climate and geography. *Canadian Journal of Forest Research* 23, 1930-1939.

Ahuja, M.R. (ed.) (1993) *Micropropagation of Woody Plants*. Kluwer Academic Publishers, Dordrecht, The Netherlands.

Ahuja, M.R. (1997) Transgenes and genetic instability. In: Klopfenstein, N.B., Chun, W.Y.W., Kim, M.-S. and Ahuja, M.R. (eds.) *Micropropagation and Genetic Engineering and Molecular Biology of Populus*. USDA Forest Service General Technical Report RM-GTR-297, pp. 90-100.

Ahuja, M.R. (2000) Genetic engineering in forest trees: State of the art and future perspectives. In: Jain, S.M. and Minocha, S.C. (eds.) *Molecular Biology of Woody Plants, Forestry Sciences, Volume 64*. Kluwer Academic Publishers, Dordrecht, The Netherlands, pp. 31-49.

Ahuja, M.R. and Libby, W.G. (1993) *Clonal Forestry II. Conservation and Application*. Springer-Verlag, New York, NY.

Ahuja, M.R., Devey, M.E., Groover, A.T., Jermstad, K.D. and Neale, D.B. (1994) Mapped DNA probes from loblolly pine can be used for restriction fragment length polymorphism mapping in other conifers. *Theoretical and Applied Genetics* 88, 279-282.

Aitken, S.N. and Adams, W.T. (1996) Genetics of fall and winter cold hardiness of coastal Douglas-fir in Oregon. *Canadian Journal of Forest Resarch* 26, 1828-1837.

Aldrich, P. R. and Hamrick, J.L. (1998) Reproductive dominance of pasture trees in a fragmented tropical forest mosaic. *Science* 281, 103-105.

Allard, R.W. (1960) *Principles of Plant Breeding*. John Wiley & Sons, New York, NY.

Allard, R.W. and Adams, J. (1969) The role of intergenotypic interactions in plant breeding. *Proceedings of the XII International Congress of Genetics* 3, 349-370.

Allen, G.C., Spiker, S. and Thompson, W.F. (2000) Use of matrix attachment regions (MARs) to minimize transgene silencing. *Plant Molecular Biology* 43, 361-376.

Ali, I.F., Neale, D. and Marshall, K.A. (1991) Chloroplast DNA restriction fragment length polymorphism in *Sequoia sempervirens* D. Don Endl., *Pseudotsuga menziesii* (Mirb.) Franco, *Calocedrus decurrens* (Torr.), and *Pinus taeda* L. *Theoretical and Applied Genetics* 81, 83-89.

Allina, S.M., Pri-Hadash, A., Theilmann, D.A., Ellis, B.E. and Douglas, C.J. (1998) 4-coumarate: coenzyme A ligase in hybrid poplar: properties of native enzymes, cDNA cloning, and analysis of recombinant enzymes. *Plant Physiology* 116, 743-754.

Allona, I., Quinn, M., Shoop, E., Swope, K., St. Cyr, S., Carlis, J., Riedl, J., Retzel, E., Campbell, M., Sederoff, R. and Whetten, R. (1998) Analysis of xylem formation in pine by cDNA sequencing. *Proceedings of the National Academy of Science of the United States of America* 95, 9693-9698.

Almqvist, C. and Ekberg, I. (2001) Interstock and GA 4/7 effects on flowering after topgrafting in *Pinus sylvestris*. *Forest Science* 8, 279-284.

Alosi, M.C, Neale, D. and Kinlaw, C. (1990) Expression of cab genes in Douglas-fir is not strongly regulated by light. *Plant Physiology* 93, 829-832.

Alstad, D.N. (2000) *Simulations of Population Biology*. Department of Ecology, Evolution and Behavior, University of Minnesota, St. Paul, MN. (http://www.cbs.umn.edu/populus/index.html).

American Society for Testing and Materials. (2000) Standard test methods for specific gravity of wood and wood-base materials. In: *Annual Book of American Standards* Volume 4.10, ASTM International, West Conshohocken, PA, pp. 348-354.

Ampie, E. and Ravensbeck, L. (1994) Strategy of tree improvement and forest gene resources conservation in Nicaragua. In: *Forest Genetic Resources No. 22*. FAO (The Food and Agricultural Organization of the United Nations), pp. 29-32.

Anderson, E. (1949) *Introgressive hybridization*. John Wiley & Sons, Inc., New York, NY.

Anderson, R.L. and Powers. H.R., Jr. (1985) The resistance screening center - screening for disease as a service for tree improvement programs. In: Barrows-Broaddus, J. and Powers, H.R., Jr. (eds.) *Proceedings of Rusts of Hard Pines International Union of Forest Research Organizations (IUFRO), Working Party Conference*. Georgia Center for Continuing Education, University of Georgia. Athens, GA, pp. 59-63.

Anderson, V.L. and McLean R.A. (1974) *Design of Experiments: A Realistic Approach*. Marcel Dekker, Inc., New York, NY.

Andersson, E., Jansson, R. and Lindgren, D. (1974) Some Results from second generation crossings involving inbreeding in Norway spruce (*Picea abies*). *Silvae Genetica* 23, 34-43.

Araujo, J.A., Sousa, R., Lemos, L. and Borralho, N.M.G. (1996) Breeding values for growth in *Eucalyptus globulus* combining clonal and full-sib progeny information. *Silvae Genetica* 45, 223-226.

Araujo, J.A., Lemos, L., Ramos, A., Ferreira, J.G. and G. Borralho, N.M.G. (1997) The RAIZ *Euca-*

lyptus globulus breeding program: A BLUP rolling-front strategy with a mixed clonal and seedling deployment scheme. In: *Proceedings of the International Union of Forest Research Organizations (IUFRO), Conference on sobre Silvicultura e Melhoramento de Eucaliptos*. El Salvador, Brazil, pp. 371-376.

Arbuthnot, A. (2000) Clonal testing of Eucalyptus at Mondi Kraft, Richards Bay. In: *Proceedings of the International Union of Forest Research Organizations Working Party, Forest Genetics for the Next Millinnium*, Anonymous. Durban, South Africa, pp. 61-64.

Arnold, R.J. (1990) Control pollination radiata pine seed - a comparison of seedling and cutting options for large-scale deployment. *New Zealand Journal of Forestry* 35, 12-17.

Aronen, T., Haggman, H. and Hohtola, A. (1994) Transient B-glucuronidase expression in Scots pine tissues derived from mature trees. *Canadian Journal of Forest Research* 24, 2006-2011.

Askew, G.R. and Burrows, P.M. (1983) Minimum coancestry selection I. A *Pinus taeda* population and its simulation. *Silvae Genetica* 32, 125-131.

Atwood, R.A. (2000) Genetic Parameters and Gains for Growth and Wood Properties in Florida Source Loblolly Pine in the Southeastern United States. School of Forest Resources and Conservation, Institute of Food and Agricultural Sciences, University of Florida, Gainesville, FL.

Aubry, C.A., Adams, W.T and Fahey, T.D. (1998) Determination of relative economic weights for multitrait selection in coastal Douglas-fir. *Canadian Journal of Forest Research* 28, 1164-1170.

Avery, O.T., MacLeod, C.M. and McCarthy, M. (1944) Studies on the chemical nature of the substance inducing transformation of pneumococcal types. *Journal of Experimental Medicine* 98, 451-460.

Axelrod, D. (1986) Cenozoic History of Some Western American Pines. *Annals of the Missouri Botanical Garden* 73, 565-641.

Baez, M.N. and White, T.L. (1997) Breeding strategy for the first-generation of *Pinus taeda* in the northeast region of Argentina. In: *Proceedings of the 24th Southern Forest Tree Improvement Conference*. Orlando, FL, pp. 110-117.

Bahrman, N. and Damerval, C. (1989) Linkage relationships of loci controlling protein amounts in maritime pine (*Pinus pinaster* Ait.). *Heredity* 63, 267-274.

Bailey, N.T.J. (1961) *Introduction to the Mathematical Theory of Genetic Linkage*. Oxford University Press, London, UK.

Bailey, R.G. (1989) Explanatory supplement to ecoregions map of the continents. *Environmental Conservation* 16, 307-309.

Baker, L.H. and R.N. Curnow, R.N. (1969) Choice of population size and use of variation between replicate populations in plant breeding selection programs. *Crop Science* 9, 555-560.

Baker, R.J. (1986) *Selection Indices in Plant Breeding*. CRC Press, Boca Raton, FL.

Ball, S.T., Mulla, D.J. and Konzak, C.F. (1993) Spatial heterogeneity affects variety trial interpretation. *Crop Science* 33, 931-935.

Balocchi, C.E. (1990) Age trends of genetic parameters and selection efficiency for loblolly pine (*Pinus taeda* L.). Ph.D. Dissertation, North Carolina State University, Raleigh, NC.

Balocchi, C.E. (1996) Gain optimisation through vegetative multiplication of tropical and subtropical pines. In: Dieters, M.J., Matheson, A.C., Nikles, D.G., Harwood, C.E. and Walker, S.M. (eds.) *Proceedings of the QueenslandForest Research Institute-International Union of Forest Research Organizations (QFRI-IUFRO), Conference on Tree Improvement for Sustainable Tropical Forestry*. Caloundra, Queensland, Australia, pp. 304-306.

Balocchi, C.E. (1997) Radiata pine as an exotic species. In: *Proceedings of the 24th Southern Forest Tree Improvement Conference*. Orlando, FL, pp. 11-17.

Balocchi, C.E., Bridgwater, F.E., Zobel, B.J. and Jahromi, S. (1993) Age trends in genetic parameters for tree height in a nonselected population of loblolly pine. *Forest Science* 39, 231-251.

Balocchi, C.E., Bridgwater, F.E. and Bryant, R. (1994) Selection efficiency in a nonselected population of loblolly pine. *Forest Science* 40, 452-473.

Bannister, M.H. (1965) Variation in the breeding systems of *Pinus radiata*. In: Baker, H.G. and Stebbins, G.L. (eds.) *The Genetics of Colonizing Species*. Academic Press, New York, NY, pp. 353-372.

Baradat, P. (1976) Use of juvenile-mature relationships and information from relatives in combined multitrait selection. In: *Proceedings of the International Union of Forest Research Organizations (IUFRO), Joint Meeting on Advanced Generation Breeding*. Bordeaux, France, pp. 121-138.

Barber, H.N. (1965) Selection in natural populations. *Heredity* 20, 551-572.

Barbour, R.C., Potts, B.M., Vaillancourt, R.E., Tibbits, W.N. and Wiltshire, R.E. (2002) Gene flow between introduced and native *Eucalyptus* species. *New Forests* 23, 177-191.

Barnes, B.V., Bingham, R.T. and Squillace, A.E. (1962) Selective fertilization in *Pinus monticola* Dougl. II. Results of additional tests. *Silvae Genetica* 11, 103-111.

Barnes, R.D. (1986) Multiple population tree breeding in Zimbabwe. In: *Proceedings of the International Union of Forest Research Organizations (IUFRO), Conference on Breeding Theory, Progeny Testing and Seed Orchards.* Williamsburg, VA, pp. 285-297.

Barnes, R.D. 1995. The breeding seedling orchard in the multiple population breeding strategy. *Silvae Genetica* 44, 81-88.

Barnes, R.D. and Mullin, L.J. (1989) The multiple population breeding strategy in Zimbabwe: five year results. In: Gibson, G.I., Griffin, A.R. and Matheson, A.C. (eds.) *Breeding Tropical Trees: Population Structure and Genetic Improvement Strategies in Clonal and Seedling Forestry.* Oxford Forestry Institute, Oxford UK, pp. 148-158.

Barnes, R.D., Burley, J., Gibson, G.L. and Garcia de Leon, J.P. (1984) Genotype-environment interactions in tropical pines and their effects on the structure of breeding populations. *Silvae Genetica* 33, 186-198.

Barnes, R.D., Matheson, A.C., Mullin, L.J. and Birks, J. (1987) Dominance in a metric trait of *Pinus patula* Shiede and Deppe. *Forest Science* 33, 809-815.

Barnes, R.D., Mullin, L.J. and Battle, G. (1992) Genetic control of fifth year traits in *Pinus patula* Schiede and Deppe. *Silvae Genetica* 41, 242-248.

Barnes, R.D., White, T.L., Nyoka, B.I., John, S. and Pswarayi, I.Z. (1995) The composite breeding seedling orchard. In: Potts, B.M. Borralho, N.M.G. Reid, J.B., Cromer, R.N., Tibbits, W.N. and Raymond, C.A. (eds.) *Proceeding of the International Union of Forest Research Organizations (IUFRO), Conference on Eucalypt Plantations: Improving Fibre Yield and Quality.* Hobart, Australia, pp. 285-288.

Barnes, R.L. and Bengtson, G.W. (1968) Effects of fertilization, irrigation, and cover cropping on flowering and on nitrogen and soluble sugar composition of slash pine. *Forest Science* 14, 172-180.

Barone, A. (2004) Molecular marker-assisted selection for potato breeding. *American Journal of Potato Research* 81, 111-117.

Barreneche, T., Casasoli, M., Russell, K., Akkak, A., Meddour, H., Plomion, C., Villani, F. and Kremer, A. (2003) Comparative mapping between *Quercus* and *Castanea* using simple sequence repeats (SSRs). *Theoretical and Applied Genetics* 108, 558-566.

Barton, N.H. and Turelli, M. (1989) Evolutionary quantitative genetics: How little do we know? *Annual Review of Genetics* 23, 337-370.

Baucher, M., Chabbert, B., Pilate, G., Van Doorsselaere, J., Toller, M.T., Petit-Conil, M., Cornu, D., Monties, B., Van Montagu, M., Inze, D., Jouanin, L. and Boerjan, W. (1996) Red xylem and higher lignin extractability by down-regulating a cinnamyl alcohol dehydrogenase in poplar (*Populus tremula* x *P. alba*). *Plant Physiology* 112, 1479-1490.

Baucher, M., Halpin, C., Petit-Conil, M. and Boerjan, W. (2003) Lignin: genetic engineering and impact on pulping. *Critical Reviews in Biochemistry and Molecular Biology* 38, 305-350.

Baulcombe, D. (2002) RNA silencing. *Current Biology* 12, R82-R84.

Bawa, K.S. (1974) Breeding systems of tree species of a lowland tropical community. *Evolution* 28, 85-92.

Bawa, K.S. and Opler, P.A. (1975) Dioecism in tropical forest trees. *Evolution* 29, 167-179.

Bawa, K.S., Perry, D.R. and Beach, J.H. (1985) Reproductive biology of tropical lowland rain forest trees. I. Sexual systems and incompatibility mechanisms. *American Journal of Botany* 72, 331-345.

Beadle, G.W. and Tatum, E.L. (1941) Genetic control of biochemical reactions in Neurospora. *Proceedings of the National Academy of Sciences of the United States of America* 27, 499-506.

Beavis, W.D. (1995) The power and deceit of QTL experiments: lessons from comparative QTL studies. In: *Proceedings 49th Annual Corn and Sorghum Industry Research Conference.* Washington, DC.

Becker, W.A. (1975) *Manual of Quantitative Genetics.* Washington State University Press, Pullman, WA.

Becwar, M.R. (1993) Conifer somatic embryogenesis and clonal forestry. In: Ahuja, M.R. and Libby,

W.J. (eds.) *Clonal Forestry. I. Genetics and Biotechnology.* Springer Verlag, Berlin, Germany, pp. 200-223.

Bekkaoui, F. Pilon, M., Laine, E., Raju, D.S.S., Crosby, W.L. and Dunstan, D.I. (1988) Transient gene expression in electroporated *Picea glauca* protoplasts. *Plant Cell Reports.* 7, 481-484.

Bell, G.D. and Fletcher, A.M. (1978) Computer organised orchard layouts (COOL) based on the permutated neighbourhood design concept. *Silvae Genetica* 27, 223-225.

Benet, H., Guries, R., Boury, S. and Smalley, E. (1995) Identification of RAPD markers linked to a black leaf spot resistance gene in Chinese elm. *Theoretical and Applied Genetics* 90, 1068-1073.

Bennetzen, J. and Freeling, M. (1993) Grasses as a single genetic system: genome composition, collinearity and compatibility. *Trends in Genetics* 9, 259-261.

Bentzer, B.G. (1993) Strategies for clonal forestry with Norway spruce. In: Ahuja, M.R. and Libby, W.J. (eds.) *Clonal Forestry II: Conservation and Application.* Springer-Verlag. New York, NY, pp. 120-138.

Bentzer, B.G., Foster, G.S., Hellberg, A.R. and Podzorski, A.C. (1989) Trends in genetic and environmental parameters, genetic correlations, and response to indirect selection for 10-year volume in a Norway spruce clonal experiment. *Canadian Journal of Forest Research* 19, 897-903.

Berg, E.E. and Hamrick, J.L. (1995) Fine-scale genetic structure of a turkey oak forest. *Evolution* 49, 110-120.

Bergmann, F. (1978) The allelic distribution at an acid phosphatase locus in Norway spruce (*Picea abies*). *Theoretical and Applied Genetics* 52, 57-64.

Bergmann, F. and Ruetz. W.F. (1991) Isozyme genetic variation and heterozygosity in random tree samples and selected orchard clones from the same Norway spruce populations. *Forest Ecology and Management* 46, 39-47.

Bernardo, R. (2001) What if we know all the genes for a quantitative trait in hybrid crops? *Crop Science* 41, 1-4.

Bernatzky, R. and Mulcahy, D. (1992) Marker-aided selection in a backcross breeding program for resistance to chestnut blight in the American chestnut. *Canadian Journal of Forest Research* 22, 1031-1035.

Binelli, G. and Bucci, G. (1994) A genetic linkage map of *Picea abies* Karst. based on RAPD markers, as a tool in population genetics. *Theoretical and Applied Genetics* 8, 283-288.

Birch, R.G. (1997) Plant transformation: problems and strategies for practical application. *Annual Review of Plant Physiology and Molecular Biology* 48, 297-326.

Bishop-Hurley, S.L., Zubkiewicz, R.J., Grace, L.J., Gardner, R.C., Wagner, A. and Walter C. (2001) Conifer genetic engineering: transgenic *Pinus radiata* (D. Don.) and *Picea abies* (Karst.) plants are resistant to the herbicide Buster. *Plant Cell Reports* 20, 235-243.

Bjorkman, A. and Gullberg, U. (1996) Poplar breeding and selection strategies. In: Stettler, R.F., Bradshaw, H.D., Heilman, P.E. and Hinckley, T.M. (eds.) *Biology of Populus and its Implications for Management and Conservation.* NRC Research Press, National Research Council of Canada, Ottawa, Ontario, Canada, pp. 139-158.

Bjorkman, E. (1964) Breeding for resistance to disease in forest trees. *Unasylva* 18, 73-81.

Bobola, M., Smith, D. and Klein, A. (1992) Five major nuclear ribosomal repeats represent a large and variable fraction of the genomic DNA of *Picea rubens* and *P. mariana*. *Molecular Biology and Evolution* 9(1), 125-137.

Boldman, K.G., Kriese, L.A., van Vleck, L.D. and Kachman, S.D. (1993) *A Manual for Use of MTDFREML*. USDA Agricultural Research Service, Washington, D.C.

Bolsinger, C.L. and Jaramillo, A.E. (1990) *Taxus brevifolia* Nutt. In: Burns. R.M. and Honkala, B.H. (eds.) *Silvics of North America. Vol. I. Conifers.* USDA Forest Service Agricultural Handbook 654.

Bonga, J.M. (1982) Vegetative propagation in relation to juvenility, maturation and rejuvenation. In: Bongarten, B. and Durzan, D.J. (eds.) *Tissue Culture and Forestry.* Nijhoff, Boston, MA, pp. 387-412.

Bonga, J.M. and Durzan, D.J. (eds.). (1987) *Cell and Tissue Culture in Forestry, Volume 3*. Martinus Nijhoff Publishers, Dordrecht, The Netherlands.

Bongarten, B.C. and Dowd, J.F. (1987) Regression and spline methods for removing environmental variance in progeny tests. In: *Proceedings of the 19th Southern Forest Tree Improvement Con-*

ference. College Station, TX, pp. 312-319.

Bonnet-Masimbert, M. and Webber, J.E. (1995) From flower induction to seed production in forest tree seed orchards. *Tree Physiology* 15, 419-426.

Booth, T.H. (1990) Mapping regions climatically suitable for particular tree species at the global scale. *Forest Ecology and Management* 36, 47-60.

Booth, T.H. and Pryor, L.D. (1991) Climatic requirements of some commercially important eucalypt species. *Forest Ecology and Management* 43, 47-60.

Bornman, C.H. and Botha, A.M. (2000) Somatic seed: balancing expectations against achievements. In: *Proceedings of the International Union of Forest Research Organizations (IUFRO) Working Party, Forest Genetics for the Next Millinnium*. Durban, South Africa, pp. 76-79.

Borralho, N.M.G. (1992) Gains expected from clonal forestry under various selection and propagation strategies. In: *Proceedings of the Meeting on Mass Production Technology for Genetically-Improved Fast-Growing Forest Tree Species*. Bordeaux, France, pp. 327-338.

Borralho, N.M.G. (1995) The impact of Individual tree mixed models (BLUP) in tree breeding strategies. In: Potts, B.M., Borralho, N.M.G., Reid, J.B., Cromer, R.N., Tibbits, W.N. and Raymond, C.A. (eds.) *Proceedings of the International Union of Forest Research Organizations (IUFRO), Symposium on Eucalypt Plantations: Improving Fibre Yield and Quality*. Hobart, Australia, pp. 141-145.

Borralho, N.M.G. and Dutkowski, G.W. (1996) A 'rolling front' strategy for breeding trees. In: Dieters, M.J., Matheson, A.C., Nikles, D.G., Harwood, C.E. and Walker, S.M. (eds.) *Proceedings of the Queensland Forest Research Institute-International Union of Forest Research Organizations (QFRI-IUFRO), Conference on Tree Improvement for Sustainable Tropical Forestry*. Caloundra, Queensland, Australia, pp. 317-322.

Borralho, N.M.G. and Dutkowski, G.W. (1998) Comparison of rolling front and discrete generation breeding strategies for trees. *Canadian Journal of Forest Research* 28, 987-993.

Borralho, N.M.G., Almeida, I.M. and Cotterill, P.P. (1992a) Genetic control of growth of young *Eucalyptus globulus* clones in Portugal. *Silvae Genetica* 41, 100-105.

Borralho, N.M.G., Cotterill, P.P. and Kanowski, P.P. (1992b) Genetic parameters and gains expected from selection for dry weight in *Eucalytus globulus* ssp. *globulus* in Portugal. *Forest Science* 38, 80-94.

Borralho, N.M.G., Cotterill, P.P. and Kanowski, P.J. (1993) Breeding objectives for pulp production of *Eucalyptus globulus* under different industrial cost structures. *Canadian Journal of Forest Research* 23, 648-656.

Borzan, Z. and Papes, D. (1978) Karyotype analysis in *Pinus*: A contribution to the standardization of the karyotype analysis and review of some applied techniques. *Silvae Genetica* 27, 144-150.

Bos, I. and Caligari, P. (1995) *Selection Methods in Plant Breeding*. Chapman & Hall, New York, NY.

Boshier, D.H., Chase, M.R. and Bawa, K.S. (1995a) Population genetics of *Cordia alliodora* (Boraginaceae), a neotropical tree. 2. Mating system. *American Journal of Botany* 82, 476-483.

Boshier, D.H., Chase, M.R. and Bawa, K.S. (1995b) Population genetics of *Cordia alliodora* (Boraginaceae), a neotropical tree. 3. Gene flow, neighborhood, and population substructure. *American Journal of Botany* 82, 484-490.

Botstein, D., White, R.L., Skolnick, M. and Davis, R.W. (1980) Construction of a genetic linkage map in man using restriction fragment length polymorphisms. *American Journal of Human Genetics* 32, 314-331.

Bourdon, R.M. (1997) *Understanding Animal Breeding*. Prentice-Hall, Upper Saddle River, NJ.

Bowcock, A.M., Ruiz-Linares, A., Tomfohrde, J., Minch, E., Kidd, J.R and Cavalli-Sforza , L.L. (1994). High resolution of human evolutionary trees with polymorphic microsatellites. *Nature* 368, 455-457.

Bower, R.C., Ross, S.D. and Eastham, A.M. (1986) Management of a western hemlock containerized seed orchard. In: *Proceedings of the International Union of Forest Research Organizations (IUFRO), Conference on Breeding Theory, Progeny Testing and Seed Orchards*. Williamsburg, VA, pp. 604-612.

Boyer, W.D. (1958) Longleaf pine seed dispersal in south Alabama. *Journal of Forestry* 56, 265-268.

Boyle, T., Liengsiri, C. and Piewluang, C. (1990) Genetic structure of black spruce on two contrasting sites. *Heredity* 65, 393-399.

Boyle, T.J.B., Cossalter, C. and Griffin, A.R. (1997) Genetic resources for plantation forestry. In: Nambiar, E.K.S. and Brown, A.G. (eds.). *Management of Soil, Nutrients and Water in Tropical Plantation Forests*. Australian Center for International Agricultural Research, Canberra, Australia, pp. 25-63.

Bradford, K.J., Deynze, A.V., Gutterson, N., Parrott, W. and Strauss, S.H. (2005) Regulating transgenic crops sensibly: lessons from plant breeding, biotechnology and genomics. *Nature Biotechnology* 23, 439-444.

Bradshaw Jr., H. and Foster, G.S. (1992) Marker aided selection and propagation systems in trees: advantages of cloning for studying quantitative inheritance. *Canadian Journal of Forest Research* 22, 1044-1049.

Bradshaw, H.D., Jr. and Stettler, R. (1995) Molecular genetics of growth and development of *Populus*. IV. Mapping QTLs with large effects on growth, form, and phenology traits in a forest tree. *Genetics* 139, 963-973.

Bradshaw, H.D., Jr., Villar, M., Watson, B.D., Otto, K.G., Stewart, S. and Stettler, R.F. (1994) Molecular genetics of growth and development in *Populus*. III. A genetic linkage map of a hybrid poplar composed of RFLP, STS, and RAPD markers. *Theoretical and Applied Genetics* 89, 167-178.

Bramlett, D.L. (1997) Genetic gain from mass controlled pollination and topworking. *Journal of Forestry* 95, 15-19.

Bramlett, D.L. and Bridgwater, F.E. (1987) Effect of a clonal row orchard design on the seed yields of loblolly pine. In: *Proceedings of the 19th Southern Forest Tree Improvement Conference*. College Station, TX, pp. 253-260.

Bramlett, D.L. and Burris, L.C. (1995) Topworking young scions into reproductively-mature loblolly pine. In: Weir, R.J. and Hatcher, A.V. (eds.). *Proceedings of the 23rd Southern Forest Tree Improvement Conference*. North Carolina State University. Asheville, NC, pp. 234-241.

Bramlett, D.L. and Godbee, J.F., Jr. (1982) Inventory-monitoring system for southern pine seed orchards. *Georgia Forestry Research Paper 28*. Georgia Forestry Commission, Macon, GA

Brandao, L.G. (1984) The new eucalypt forest. In: *Proceedings of the Marcus Wallenberg Symposium*. Marcus Wallenberg Foundation, Falun, Sweden, pp. 3-15.

Brasileiro, A.C.M., Tourner, C., Leple, J.C., Combes, V. and Jouanin, L. (1992) Expression of the mutant *Arabidopsis thaliana* acetolacetate synthase gene confers chlorosulfuron resistance to transgenic poplar plants. *Transgenic Research* 1, 133-141.

Brewbaker, J.L. and Sun, W.G. (1996) Improvement of nitrogen-fixing trees for enhanced site quality. In: Dieters, M.J., Matheson, A.C., Nikles, D.G., Harwood, C.E. and Walker, S.M. (eds.) *Proceedings of the Queensland Forest Research Institute-International Union of Forest Research Organizations (QFRI-IUFRO), Conference on Tree Improvement for Sustainable Tropical Forestry*. Caloundra, Queensland, Australia, pp. 437-442.

Bridgwater, F.E. (1992) Mating designs. In: Fins, L., Friedman, S.T. and Brotschol, J.V. (eds.) *Handbook of Quantitative Forest Genetics*. Kluwer Academic Publishers, Boston, MA, pp. 69-95.

Bridgwater, F.E. and Bramlett, D.L. (1982) Supplemental mass pollination to increase seed yields in loblolly pine seed orchards. *Southern Journal of Applied Forestry* 6, 100-104.

Bridgwater, F.E. and Squillace, A.E. (1986) Selection indexes for forest trees. In: *Advanced Generation Breeding of Forest Trees,* Southern Cooperative Series Bulletin 309. Louisiana Agricultural Experiment Station, Baton Rouge, LA, pp. 17-20.

Bridgwater, F.E. and Stonecypher, R.W. (1979) Index selection for volume and straightness in a loblolly pine population. In: *Proceedings of the 15th Southern Forest Tree Improvement Conference*. Mississippi State, MS, pp. 132-139.

Bridgwater, F.E., Talbert, J.T. and Jahromi, S. (1983a) Index selection for increased dry weight production in a young loblolly pine population. *Silvae Genetica* 32, 157-161.

Bridgwater, F.E., Talbert, J.T. and Rockwood, D.L. (1983b) Field design for genetic tests of forest trees. In: *Progeny Testing of Forest Trees: Southern Cooperative Series Bulletin 275*. Texas A & M University, College Station, TX, pp. 28-39.

Bridgwater, F.E., Williams, C.G. and Campbell, R.G. (1985) Patterns of leader elongation in loblolly pine families. *Forest Science* 31, 933-944.

Bridgwater, F.E., Bramlett, D.L. and Matthews, F.R. (1987) Supplement mass pollination is feasible on an operational scale. In: *Proceedings of the 19ᵗʰ Southern Forest Tree Improvement Conference*. College Station, TX, pp. 216-222.

Bridgwater, F.E., Barnes, R.D. and White, T.L. (1997) Loblolly and slash pines as exotics. In: *Proceedings of the 24ᵗʰ Southern Forest Tree Improvement Conference*. Orlando, FL, pp. 18-32.

Briggs, D. and Walters, S. (1997) *Plant Variation and Evolution*. Cambridge University Press, Cambridge, UK.

Briggs, F.N. and Knowles, P. (1967) *Introduction to Plant Breeding*. Reinhold Publishing Corporation, New York, NY.

Britten, R.J. and Kohne, D.E. (1968) Repeated sequences in DNA. *Science* 161, 529-540.

Brown, A.G., Nambiar, E.K.S. and Cossalter, C. (1997) Plantations for the tropics-their role, extent and nature. In: Nambiar, E.K.S. and Brown, A.G. (eds) *Management of Soil, Nutrients and Water in Tropical Plantation Forests*. ACIAR, Canberra, Australia, pp. 1-23.

Brown, A.H.D. (1989) The case for core collections. In: Brown, A.H.D., Frankel, O.H., Marshall, D.R. and Williams, J.T. (eds.) *The Use of Plant Genetic Resources*. Cambridge University Press, UK, pp. 136-156.

Brown, A.H.D. and Allard, R.W. (1970) Estimation of the mating system in open-pollinated maize populations using isozyme polymorphisms. *Genetics* 66, 133-145.

Brown, A.H.D. and Hardner, C.M. (2000) Sampling the gene pools of forest trees for *ex situ* conservation. In: Young, A., Boshier, D. and Boyle, T. (eds) *Forest Conservation Genetics: Principles and Practice*. Commonwealth Scientific and Industrial Research Organization (CSIRO) Publishing, Collingwood, Victoria, Australia, pp. 185-196.

Brown, C.L. and Goddard, R.E. (1961) Silvical considerations in the selection of plus phenotypes. *Journal of Forestry* 59, 420-426.

Brown, G. and Carlson, J.E. (1997) Molecular cytogenetics of the genes encoding 18s-5.8s-26s rRNA and 5s rRNA in two species of spruce (*Picea*). *Theoretical and Applied Genetics* 95, 1-9.

Brown, G., Amarasinghe, V., Kiss, G. and Carlson, J. (1993) Preliminary karyotype and chromosomal localization of ribosomal DNA sites in white spruce using fluorescence in situ hybridization. *Genome* 36, 310-316.

Brown, G., Kadel III, E., Bassoni, D., Kiehne, K., Temesgen, B., van Buijtenen, J., Sewell, M., Marshall, K. and Neale, D. (2001) Anchored reference loci in Loblolly pine (*Pinus taeda* L.) for integrating pine genomics. *Genetics* 159, 799-809.

Brown, G., Bassoni, D., Gill, G., Fontana, J., Wheeler, N., Megraw, R., Davis, M., Sewell, M., Tuskan, G. and Neale, D. (2003) Identification of quantitative trait loci influencing wood property traits in loblolly pine (*Pinus taeda* L.) III. QTL verification and candidate gene mapping. *Genetics* 164, 1537-1546.

Brown, G.R., Gill, G.P., Kuntz, R.J., Langley, C.H., and Neale, D.B.. (2004) Nucleotide diversity and linkage disequilibrium in loblolly pine. *Proceedings of the National Academy of Sciences of the United States of America* 101(42), 15255-15260.

Brownie, C. and Gumpertz, M.L. (1997) Validity of spatial analyses for large field trials. *American Statistical Association and International Biometric Society Journal of Agricultural, Biological, and Environmental Statistics* 2, 1-23.

Brownie, C., Bowman, D.T. and Burton, J.W. (1993) Estimating spatial variation on analysis of data from yield trials: a comparison of methods. *Agronomy Journal* 85, 1244-1253.

Bryant, E.H. and Meffert L.M. (1996) Nonadditive genetic structuring of morphometric variation in relation to a population bottleneck. *Heredity* 77, 168-176.

Buford, M.A. and Burkhart, H.E. (1987) Genetic improvement effects on growth and yield of loblolly pine plantations. *Forest Science* 33, 707-724.

Bugos, R.C., Chiang, V.L. and Campbell, W.H. (1991) cDNA cloning, sequence analysis and seasonal expression of lignin bispecific caffeic acid/5-hydroxyferulic acid O-methyltransferase of aspen. *Plant Molecular Biology* 17, 1203-1215.

Bulmer, M.G. (1985) *The Mathematical Theory of Quantitative Genetics*. Clarendon Press, Oxford, UK.

Burczyk J., Adams, W.T. and Shimizu, J.Y. (1996) Mating patterns and pollen dispersal in a natural knobcone pine (*Pinus attenuata* Lemmon.) stand. *Heredity* 77: 251-260.

Burdon, J.J. (1987) *Diseases and Plant Population Biology*. Cambridge University Press, UK.

Burdon, R.D. (1977) Genetic correlation as a concept for studying genotype-environment interaction in forest tree breeding. *Silvae Genetica* 26, 168-175.

Burdon, R. D. (1979) Generalisation of multi-trait selection indices using information from several sites. *New Zealand Journal of Forest Science* 9, 145-152.

Burdon, R.D. (1982a) Selection indices using information from multiple sources for the single-trait case. *Silvae Genetica* 31, 81-85.

Burdon, R.D. (1982b) The roles and optimal place of vegetative propagation in tree breeding strategies. In: *Proceedings of the International Union of Forest Research Organizations (IUFRO), Meeting About Breeding Strategies Including Multiclonal Varieties.* Sensenstein, West Germany, pp. 66-88.

Burdon, R.D. (1986) Clonal forestry and breeding strategies - a perspective. In: *Proceedings of the International Union of Forest Research Organizations (IUFRO), Conference on Breeding Theory, Progeny Testing and Seed Orchards.* Williamsburg, VA, pp. 645-659.

Burdon, R.D. (1989) When is cloning on an operational scale appropriate? In: Gibson, G.I., Griffin, A.R. and Matheson, A.C. (eds.) *Proceedings of the Breeding Tropical Trees: Population Structure and Genetic Improvement Strategies in Clonal and Seedling Forestry.* Oxford Forestry Institute. Oxford, UK, pp. 9-27.

Burdon, R.D. (1990) Comment on "selection system efficiencies for computer-simulated progeny test field designs in loblolly pine" – J.A.Loo-Dinkins, C.G. Tauer and C.C. Lambeth. *Theoretical and Applied Genetics* 7981, 89-96.

Burdon, R.D. (1991) Genetic correlations between environments with genetic groups missing in some environments. *Silvae Genetica* 40, 66-67.

Burdon, R.D. (2001) Genetic diversity and disease resistance: some considerations for research, breeding, and deployment. *Canadian Journal of Forest Research* 31, 596-606.

Burdon, R.D. and Bannister, M.H. (1992) Genetic survey of *Pinus radiata.* 4: Variance structures and heritabilities in juvenile clones. *New Zealand Journal of Forest Science* 22, 187-210.

Burdon, R.D. and Namkoong, G. (1983) Multiple populations and sublines. *Silvae Genetica* 32, 221-222.

Burdon, R.D. and Shelbourne, C.J.A. (1971) Breeding populations for recurrent selection: Conflicts and possible solutions. *New Zealand Journal of Forest Science* 1, 174-193.

Burdon, R.D. and Shelbourne, C.J.A. (1974) The use of vegetative propagules for obtaining genetic information. *New Zealand Journal of Forest Science* 4, 418-425.

Burdon, R.D. and van Buijtenen, J.P. (1990) Expected efficiencies of mating designs for reselection of parents. *Canadian Journal of Forest Research* 20, 1664-1671.

Burdon, R.D., Shelbourne, C.J.A. and Wilcox, M.D. (1977) Advanced selection strategies. In: *Proceedings of the 3rd World Conference on Forest Tree Breeding.* Canberra, Australia, pp. 1133-1147.

Burley, J. (1980) Choice of species and possibility of genetic improvement for smallholder and community forests. *Commonwealth Forestry Review* 59, 311-326.

Busby, C.L. (1983) Crown-quality assessment and the relative economic importance of growth and crown characters in mature loblolly pine. In: *Proceedings of the 17th Southern Forest Tree Improvement Conference.* Athens, GA, pp. 121-130

Bush, R.M. and Smouse, P.E. (1992) Evidence for the adaptive significance of allozymes in forest trees. *New Forests* 6, 179-196.

Busov, V.B., Brunner, A.M. Meilan, R., Filichken, S., Ganio, L., Gandhi, S. and Strauss, S.H. (2005) Genetic transformation: a powerful tool for dissection of adaptive traits in trees. *New Phytologist* 167, 9-18.

Butcher, P.A., Bell, J.C. and Moran, G.F. (1992) Patterns of genetic diversity and nature of the breeding system in *Melaleuca alternifolia* (Myrtaceae). *Australian Journal of Botany* 40, 365-375.

Butcher, T.B. (1988) HAPSO development in western Australia. In: Dieters, M.J. and Nikles, D.G. (eds.) *Proceedings of the 10th Australian Forest Genetics Research Group 1.* Gympie, Queensland, Australia, pp. 145-148.

Butland, S., Chow, M. and Ellis, B. (1998) A diverse family of phenylalanine ammonia-lyase genes expressed in pine trees and cell cultures. *Plant Molecular Biology* 37, 15-24.

Butterfield, R. (1996) Early species selection for tropical reforestation: A consideration of stability.

Forest Ecology and Management 81, 161-168.

Byram, T.D. and Lowe, W.J. (1985) Longleaf pine tree improvement in the Western Gulf region. In: *Proceedings of the 18th Southern Forest Tree Improvement Conference*. Long Beach, MS, pp. 78-87.

Byram, T.D. and Lowe, W.J. (1986) General and specific combining ability estimates for growth in loblolly pine. In: *Proceedings of the International Union of Forest Research Organizations (IUFRO), Conference on Breeding Theory, Progeny Testing and Seed Orchards*. Williamsburg, VA, pp 352-360.

Byram, T.D. and Lowe, W.J. (1988) Specific gravity variation in a loblolly pine seed source study in the western gulf region. *Forest Science* 34, 798-803.

Byram, T.D., Bridgwater, F.E., Gooding, G.D. and Lowe, W.J. (1997) 45th Progress Report of the Cooperative Forest Tree Improvement Program. Texas A & M University, College Station, TX.

Byrne, M. (2000) Disease threats and the conservation genetics of forest trees. In: Young A., Boshier, D. and Boyle, T. (eds.) *Forest Conservation Genetics: Principles and Practice*. Commonwealth Scientific and Industrial Research Organization (CSIRO) Publishing, Collingwood, Victoria, Australia, pp. 159-166.

Byrne, M., Moran, G.F. and Tibbits, W.N. (1993) Restriction map and maternal inheritance of chloroplast DNA in *Eucalyptus nitens*. *The Journal of Heredity* 84(3), 218-220.

Byrne, M., Moran, G.F., Murrell, J.C. and Tibbits, W.N. (1994) Detection and inheritance of RFLPs in *Eucalyptus nitens*. *Theoretical and Applied Genetics* 89(4), 397-402.

Byrne, M., Murrell, J., Allen, B. and Moran, G. (1995) An integrated genetic linkage map for eucalyptus using RFLP, RAPD and isozyme markers. *Theoretical and Applied Genetics* 91, 869-875.

Cade, S.C. and Hsin, L.Y. (1977) Girdling: Its effect on seed-cone bud production, seed yield, and seed quality in Douglas-fir. In: Weyerhaeuser Forestry Research Technical Report 042. Western Forestry Research Center, Centralia, WA, pp. 9.

Callaham, R.Z. (1964) Provenance research: Investigation of genetic diversity associated with geography. *Unasylva* 18, 40-50.

CAMCORE. (1996) CAMCORE Annual Report. Central America & Mexico Coniferous Resources Cooperative, Department of Forestry, North Carolina State University, Raleigh, NC.

CAMCORE. (1997) CAMCORE Annual Report. Central America & Mexico Coniferous Resources Cooperative, Department of Forestry, North Carolina State University, Raleigh, NC.

Cameron, N.D. (1997) *Selection Indices and Prediction of Genetic Merit in Animal Breeding*. CAB International, Wallingford, Oxon, UK.

Cameron, J.N., Cotterill, P.P. and Whiteman, P.H. (1988) Key elements of a breeding plan for temperate eucalypts in Australia. In: Dieters, M.J. and Nikles, D.G. (eds.) *Proceedings of the 10th Australian Forest Genetics Research Working Group 1*. Gympie, Queensland,Australia, pp. 69-80.

Cameron, J.N., Cotterill, P.P. and Whiteman, P.H. (1989) Key elements of a breeding plan for temperate eucalypts in Australia. In: Gibson, G.I., Griffin, A.R. and Matheson, A.C. (eds.) *Breeding Tropical Trees: Population Structure and Genetic Improvements Strategies in Clonal and Seedling Forestry*. Oxford Forestry Institute, Oxford, UK, pp. 159-168

Campbell, M.M. and Sederoff, R.R. (1996) Variation in lignin content and composition: mechanisms of control and implications for the genetic improvement of plants. *Plant Physiology* 110:3-13.

Campbell, M.M., Brunner, A.M., Jones, H.M. and Strauss, S.H. (2003) Forestry's fertile crescent: the application of biotechnology to forest trees. *Plant Biotechnology Journal* 1, 141-154.

Campbell, R.K. (1965) Phenotypic variation and repeatability of stem sinuosity in Douglas-fir. *Northwest Science* 39, 47-59.

Campbell, R.K. (1979) Genecology of Douglas-fir in a watershed in Oregon Cascades. *Ecology* 60, 1036-1050.

Campbell, R.K. (1986) Mapped genetic variation of Douglas-fir to guide seed transfer in southwest Oregon. *Silvae Genetica* 35, 85-96.

Campbell, R.K. (1991) Soils, seed-zone maps and physiography: Guidelines for seed transfer of Douglas-fir in southwestern Oregon. *Forest Science* 37, 973-986.

Campbell, R.K. and Sorensen, F.C. (1973) Cold-acclimation in seedling Douglas-fir related to phenology and provenance. *Ecology* 54, 1148-1151.

Campbell, R.K. and Sorensen, F.C. (1978) Effect of test environment on expression of clines and on delimitation of seed zones in Douglas-fir. *Theoretical and Applied Genetics* 51, 233-246.

Campbell, R.K. and Sugano, A.I. (1979) Genecology of bud-burst phenology in Douglas-fir: Response to flushing temperature and chilling. *Botanical Gazette* 140, 223-231.

Campinhos, E. and Ikemori, Y.K. (1989) Selection and management of the basic population *E. grandis* and *E. Urophylla* established at Aracruz for the long term breeding programme. In: Gibson, G.I., Griffin, A.R. and Matheson, A.C. (eds.) *Breeding Tropical Trees: Population Structure and Genetic Improvement Strategies in Clonal and Seedling Forestry*. Oxford Forestry Institute Oxford UK, pp. 169-175.

Cannell, G.R. (1978) Improving per hectare forest productivity. In: University of Florida School of Forest Resources and Conservation (ed.) *Proceedings of the 5th North American Forest Biology Workshop*. March 13-15. Gainesville, FL, pp. 120-148.

Cannon, P.G. and Low, C.B. (1994) A computer-aided test layout for open-pollinated breeding populations of insect-pollinated trees species. *Silvae Genetica* 43, 265-267.

Cannon, P.G. and Shelbourne, C.J.A. (1991) The New Zealand eucalypt breeding programme. In: Shonau, A.P.G. (ed.) *International Union of Forest Research Organizations (IUFRO), Symposium on Intensive Forestry: The Role of Eucalypts*. Durban, South Africa, pp. 198-207.

Cannon, P.G. and Shelbourne, C.J.A. (1993) Forward selection plots in breeding programmes with insect-pollinated tree species. *New Zealand Journal of Forest Science* 23, 3-9.

Canovas, F., McLarney, B. and Silverthorne, J. (1993) Light-independent synthesis of LHC IIb polypeptides and assembly of the major pigmented complexes during the initial stages of *Pinus palustris* seedling development. *Photosynthesis Research* 38, 89-97.

Cardon, L. and Bell, J. (2001) Association study designs for complex diseases. *Nature Review Genetics* 2, 91-99.

Carlisle, A. and Teich, A.H. (1978) Analysing benefits and costs of tree-breeding programmes. *Unasylva* 30, 34-37.

Carlson, J.E., Tulsieram, L.K., Glaubitz, J.C., Luk, V.W.K., Kauffeldt, C. and Rutledge, R. (1991) Segregation of random amplified DNA markers in F1 progeny of conifers. *Theoretical and Applied Genetics* 83, 194-200.

Carney, S.E., Wolf, D.E. and Rieseberg, L.H. (2000) Hybridisation and forest conservation. In: Young A., Boshier, D. and Boyle, T. (eds.) *Forest Conservation Genetics: Principles and Practice*. Commonwealth Scientific and Industrial Research Organization (CSIRO) Publishing. Collingwood, Victoria, Australia, pp. 167-182.

Carson, M. J. (1986) Advantages of clonal forestry for Pinus radiata - real or imagined? *New Zealand Journal of Forest Science* 16, 403-415.

Carson, M. J., R. D. Burdon, S. D. Carson, A. Firth, C. J. A. Shelbourne, and T. G. Vincent. 1990. Realizing genetic gains in production forests. *In: Proceedings of the International Union of Forest Research Organizations (IUFRO) Conference on Douglas-fir, lodgepole pine, Sitka spruce and Abies spp.* Olympia, WA..

Carson, M.J., Burdon, R.D., Carson, S.D., Firth, A., Shelbourne, C.J.A. and Vincent, T.G. (1991) Realising genetic gains in production forests. *Proceedings of the 11th Regional Working Group 1 (Forest Genetics) Meeting*. Coonawarra, South Africa, pp. 170-173.

Carson, S.D. (1991) Genotype x environment interaction and optimal number of progeny test sites for improving *Pinus radiata* in New Zealand. *New Zealand Journal of Forest Science* 21, 32-49.

Carson, S.D., Garcia, O. and J.D. Hayes, J.D. (1999) Realized gain and prediction of yield with genetically improved *Pinus radiata* in New Zealand. *Forest Science* 45, 186-200.

Cato, S.A. and Richardson, T.E. (1996) Inter- and intraspecific polymorphism at chloroplast SSR loci and the inheritance of plastids in *Pinus radiata* D. Don. *Theoretical and Applied Genetics* 93, 587-592.

Cato, S.A., Gardner, R.C., Kent, J. and Richardson, T.E. (2000) A rapid PCR-based method for genetically mapping ESTs. *Theoretical and Applied Genetics* 102, 296-306.

Cervera, M.T., Gusmao, J., Steenackers, M., Peleman, J., Storme, V., Broeck, A.V., Montagu, M.V. and Boerjan, W. (1996) Identification of AFLP molecular markers for resistance against *Melampsora larici-populina* in *Populus*. *Theoretical and Applied Genetics* 93, 733-737.

Cervera, M.T., Plomion, C., Malpica, C. (2000): Molecular markers and genome mapping in woody plants. In: Jain, S.M. and Minocha, S.C. (eds.) *Molecular biology of woody plants. Forestry Sci-*

ences, Volume 64. Kluwer Academic Publishers, Dordrecht, The Netherlands, pp. 375-394.

Chagne, D., Brown, G., Lalanne, C., Madur, D., Pot, D., Neale, D. and Plomion, C. (2003) Comparative genome and QTL mapping between maritime and loblolly pines. *Molecular Breeding* 12, 185-195.

Chaisurisri, K. and El-Kassaby, Y.A. (1994) Genetic diversity in a seed production population vs natural populations of Sitka spruce. *Biodiversity and Conservation* 3, 512-523.

Chaisurisri, K., Edwards, D.G.W. and El-Kassaby, Y.A. (1992) Genetic control of seed size and germination in Sitka spruce. *Silvae Genetica* 41, 348-355.

Chaisurisri, K., Edwards, D.G.W. and El-Kassaby, Y.A. (1993) Accelerating aging in Sitka spruce seed. *Silvae Genetica* 42, 303-308.

Chaisurisri, K., Mitton, J.B. and El-Kassaby, Y.A. (1994) Variation in the mating system of Sitka spruce (*Picea sitchensis*): evidence for partial assortative mating. *American Journal of Botany* 81, 1410-1415.

Changtragoon, S. and Finkeldey, R. (1995) Patterns of genetic variation and characterization of the mating system of *Pinus merkusii* in Thailand. *Forest Genetics* 2, 87-97.

Charest, P.J. and Michel, M.F. (1991) Basics of plant genetic engineering and its potential applications to tree species. Canadian Forestry Service, Petawawa National Forestry Institute, Information Report PI-X-104, pp. 1-48.

Charest, P.J., Devantier, V., Ward, C. Jones, C., Schaffer, U. and Klimaszewska, K.K. (1991) Transient expression of foreign chimeric genes in the gymnosperm hybrid larch following electroporation. *Canadian Journal of Botany* 69, 1731-1736.

Charest, P.J., Caléro, N., Lachance, D., Datla, R.S.S., Duchêsne, L.C. and Tsang, E.W.T. (1993) Microprojectile-DNA delivery in conifer species: factors affecting assessment of transient gene expression using β-glucuronidase reporter gene. *Plant Cell Reports* 12, 189-193.

Charest, P.J., Devantier, Y. and Lachance, D. (1996) Stable genetic transformation of *Picea mariana* (black spruce) via microprojectile bombardment. *In Vitro Cellular and Developmental Biology* 32, 91-99.

Charlesworth, D. and Charlesworth, B. (1987) Inbreeding depression and its evolutionary consequences. *Annual Review of Ecology and Systematics* 18, 237-268.

Charlesworth, D., Morgan, M.T. and Charlesworth, B. (1990) Inbreeding depression, genetic load, and the evolution of outcrossing rates in a multilocus system with no linkage. *Evolution* 44, 1469-1498.

Chase, M.R., Moller, C., Kessell, P. and Bawa, K.S. (1996) Distant gene flow in tropical trees. *Nature* 383, 398-399.

Cheliak, W.M. and Pitel, J.A. (1984) *Techniques for starch gel electrophoresis of enzymes from forest tree species*. Canadian Forestry Service, Petawawa National Forestry Institute, Information Report PI-X-42.

Cheliak, W.M., Morgan, K., Strobeck, C., Yeh, F.C.H. and Dancik, B.P. (1983) Estimation of mating system parameters in plant populations using the EM algorithm. *Theoretical and Applied Genetics* 65, 157-161.

Cheliak, W.M., Dancik, B.P., Morgan, K., Yeh, F.C.H. and Strobeck, C. (1985) Temporal variation of the mating system in a natural population of jack pine. *Genetics* 109, 569-584.

Cheliak, W.M., Wang, J. and Pitel, J.A. (1988) Population structure and genic diversity in tamarack, *Larix laricina* (Du Roi) K. Koch. *Canadian Journal of Forest Research* 18, 1318-1324.

Chen, Z.-Z., Chang, S.-H., Ho, C.-K., Chen, Y.-C., Chen, Tsai, J.-B. and Chiang, V.-L. (2001) Plant production of transgenic *Eucalyptus camaldulensis* carrying the *Populus tremuloides* cinnamate 4-hydroxylase gene. *Taiwan Journal of Forest Science* 16, 249-258.

Chiang, V.L. and Funaoka, M. (1990) The difference between guaicyl and guaicyl-syringyl lignins in their response to kraft dilignification. *Holzforschung* 44, 309-313.

Chiang, V.L., Puumala, R.J., Takeuchi, H. and Eckert, R.E. (1988) Composition of softwood and hardwood kraft pulping. *TAPPI* 71, 173-176.

Christophe, C. and Birot, Y. (1983) Genetic structures and expected genetic gains from multitrait selection in wild populations of Douglas-fir and Sitka spruce. II. Practical application of index selection on several populations. *Silvae Genetica* 32, 173-181.

Chupeau, M.C., Pautot, V. and Chupeau, Y. (1994) Recovery of transgenic trees after electroporation of poplar protoplasts. *Transgenic Research* 3, 13-19.

Ciampolini, R. Moazami-Goudarzi, K.; Vaiman, D.; Dillmann, C.; Mazzanti, E.; Foulley, J.L.; Leveziel, H. and Cianci, D. (1995) Individual multilocus genotypes using microsatellite polymorphisms to permit the analysis of the genetic variability within and between Italian beef cattle breeds. *Journal of Animal Science* 75, 3259-3268

Clark, I. (1980) The semivariogram. In: *Geostatistics*. McGraw-Hill, Inc., New York, NY, pp. 17-40.

Cochran, W.G. and Cox, G.M. (1957) *Experimental Designs*. John Wiley & Sons, New York, NY. *Proceedings of the National Academy of Sciences of the United States of America* 77, 546-549.

Cockerham, C.C. (1954) An extension of the concept of partitioning hereditary variance for analysis of covariances among relatives when epistasis is present. *Genetics* 39, 859-881.

Cockerham, C.C. and Burrows, P.M. (1980) Selection limits and stratgies. *Proceedings of the National Academy of Sciences* 77(1), 546-549.

Coggeshall, M.V. and Beineke, W.F. (1986) The use of multiple breeding populations to improve northern red oak (*Quercus rubra* L.) in India. In: *Proceedings of the International Union of Forest Research Organizations (IUFRO) Conference on Breeding Theory, Progeny Testing and Seed Orchards*. Williamsburg, VA, pp. 540-546.

Coggeshall, M.V. and Beineke, W.F. (1986) The use of multiple breeding populations to improve northern red oak (*Quercus rubra* L.) in India. In: *Proceedings of the International Union of Forest Research Organizations (IUFRO) Conference on Breeding Theory, Progeny Testing and Seed Orchards*. Williamsburg, VA, pp 540-546

Colbert, S.R., Jokela, E.J. and Neary, D.G. (1990) Effects of annual fertilization and sustained weed control on dry matter partitioning, leaf area, and growth efficiency of juvenile loblolly and slash pine. *Forest Science* 36, 995-1014.

Coles, J.F. and Fowler, D.P. (1976) Inbreeding in neighboring trees in two white spruce populations. *Silvae Genetica* 25, 29-34.

Comai, L., Faciotti, D., Hiatt, W.R., Thompson, G., Ross, R. and Stalker, D. (1985) Expression in plants of a mutant *aroA* gene from *Salmonella typhymurium* confers tolerance to glyphosate. *Nature* 317, 741-744.

Comstock, R.E. (1996) *Quantitative Genetics with Special Reference to Plant and Animal Breeding*. Iowa State University Press, Ames, IA.

Conkle, M.T. (1971) Inheritance of alcohol dehydrogenase and leucine aminopeptidase isozymes in knobcone pine. *Forest Science* 17, 190-194.

Conkle, M.T. (1973) Growth data for 29 years from the California elevational transect study of ponderosa pine. *Forest Science* 19, 31-39.

Conkle, M.T. (1979) Isozyme variation and linkage in six conifer species. In: *Proceedings of the Symposium on North American Forest Trees and Forest Insects*, pp. 11-17.

Conkle, M.T. (1971) Inheritance of alcohol dehydrogenase and Leucine aminopeptidase isozymes in knobcone pine. *Forest Science* 17, 190-194.

Conkle, M.T. (Technical Coordinator) (1981a) *Proceedings of the symposium on isozymes of North American forest trees and forest insects*. USDA Forest Service General Technical Report PSW-48.

Conkle, M. (1981b) Isozyme variation and linkage in six conifer species. In: *Proceedings of the Symposium on North American Forest Trees and Forest Insects*. USDA Forest Service General Technical Report PSW-48, pp. 11-17.

Conkle, M.T. (1992) Genetic diversity - seeing the forest through the trees. *New Forests* 6, 5-22.

Conkle, M. and Critchfield, W. (1998) Genetic variation and hybridization of ponderosa pine. In: *Proceedings of Ponderosa Pine – The Species and its Management Symposium*. Washington State University, Pullman, WA, pp. 27-43.

Conkle, M.T., Hodgskiss, P.D., Nunnaly, L.B. and Hunter, S.C. (1982) *Starch gel electrophoresis of conifer seeds: A laboratory manual*. USDA Forest Service General Technical Report PSW-64.

Copes, D.L. (1974) Genetics of graft rejection in Douglas-fir. *Canadian Journal of Forest Research* 4, 186-192.

Copes, D.L. (1981) Isoenzyme uniformity in western red cedar seedlings from Oregon and Washington. *Canadian Journal of Forest Research* 11, 451-453.

Copes, D.L. (1982) Field tests of graft compatible Douglas-fir seedling rootstocks. *Silvae Genetica* 31, 183-187.

Cooper, M. and Merrill, R.E. (2000) Heterosis: its exploitation in crop breeding. In: *Hybrid Breeding and Genetics of Forest Trees. Proceedings of the Queensland Forest Research Institute/Cooperative Research Center-Sustainable Production Forestry (QFRI/CRC-SPF) Symposium*. Noosa, Queensland, Australia, pp. 316-329.

Cornelius, J. (1994) Heritabilities and additive genetic coefficients of variation in forest trees. *Canadian Journal of Forest Research* 24, 372-379.

Cotterill, P.P. (1984) A plan for breeding radiata pine. *Silvae Genetica* 33, 84-90.

Cotterill, P.P. (1986) Genetic gains expected from alternative breeding strategies including simple low cost options. *Silvae Genetica* 35, 212-223.

Cotterill, P.P. (1989) The nucleus breeding system. In: *Proceedings of the 20th Southern Forest Tree Improvement Conference*. Charleston, SC, pp. 36-42.

Cotterill, P.P. (2001) Enterprise and leadership in genetic project breeding: nucleus and cluster strategies. In: *Proceedings of the International Union of Forest Research Organizations (IUFRO), Congress on Developing the Eucalypt of the Future*. Valdivia, Chile, pp. 12-16.

Cotterill, P.P. and Cameron, J.N. (1989) *Radiata Pine Breeding Plan*, Technical Report 89/20. APM Forests Pty. Ltd., Victoria, Australia.

Cotterill, P.P. and Dean, C.A. (1988) Changes in the genetic control of growth of radiata pine to 16 years and efficiencies of early selection. *Silvae Genetica* 37, 138-146.

Cotterill, P.P. and Dean, C.A. (1990) *Successful Tree Breeding with Index Selection*. Commonwealth Science and Industrial Organization (CSIRO) Division of Forestry, Canberra, Australia.

Cotterill, P.P. and Jackson, N. (1985) On index selection. I. Method of determining economic weight. *Silvae Genetica* 34:56-63.

Cotterill, P.P. and Jackson, N. (1989) Gains expected from clonal orchards under alternative breeding strategies. *Forest Science* 35, 183-196.

Cotterill, P.P. and James, J.W. (1984) Number of offspring and plot sizes required for progeny testing. *Silvae Genetica* 33, 203-209.

Cotterill, P.P. and Zed, P.G. (1980) Estimates of genetic parameters for growth and form traits in four *Pinus radiata* D. Don progeny tests in south Australia. *Australian Forest Research* 10, 155-167.

Cotterill, P. P., Correll, R.L. and Boardman, R. (1983) Methods of estimating the average performance of families across incomplete open-pollinated progeny tests. *Silvae Genetica* 32, 28-32.

Cotterill, P.P., Dean, C., Cameron, J. and Brindbergs, M. (1989) Nucleus breeding: a new strategy for rapid improvement under clonal forestry. In: *Proceedings of the International Union of Forest Research Organizations (IUFRO), Conference on Breeding Tropical Trees: Population Structure and Genetic Improvement Strategies in Clonal and Seedling Forestry*. Pattaya, Thailand, pp. 390-451

Cown, D.J. (1978) Comparison of the pilodyn and torsiometer methods for the rapid assessment of wood density in living trees. *New Zealand Journal of Forest Science* 8, 384-391.

Cressie, N.A.C. (1993) *Statistics for Spatial Data*. John Wiley & Sons, Inc., New York, NY.

Critchfield, W.B. (1975) Interspecific Hybridization in *Pinus*: A Summary Review. *Symposium on Interspecific and Interprovenance Hybridization in Forest Trees*. In: Fowler, D.P. and C.W. Yeatman, (eds.) *14th Meeting of the Canadian Tree Improvement Association, Part 2*, Fredericton, New Brunswick, pp. 99-105.

Critchfield, W.B. (1984) Impact of the Pleistocene on the genetic structure of North American conifers. In: Lanner, R.M. (ed.) *Proceeding of the 8th North American Forest Biology Workshop*, Logan, UT, pp. 70-118.

Critchfield, W.B. (1985) The late Quaternary history of lodgepole and jack pine. *Canadian Journal of Forest Research* 15, 749-772.

Crow, J.F. and Kimura, M. (1970) *An Introduction to Population Genetics Theory*. Harper & Row, New York, NY.

Crumpacker, D.W. (1967) Genetic loads in maize (*Zea mays* L.) and other cross-fertilized plants and animals. *Evolutionary Biology* 1, 306-324.

Cubbage, F.W., Pye, J.M., Holmes, T.P. and Wagner, J.E. (2000) An economic evaluation of fusiform rust protection research. *Southern Journal of Applied Forestry* 24, 77-85.

Cullis, B.R. and Gleeson, A.C. (1991) Spatial analysis of field experiments - an extension to two dimensions. *Biometrics* 47, 1449-1460.

Cullis, C.A., Creissen, G.P., Gorman, S.W. and Teasdale, R.D. (1998a) The 25S, 18S, 5S ribosomal

RNA genes from *Pinus radiata* D. Don. Petawawa National Forestry Institute Report, Chalk River, Ontario, Canada.

Cullis, B.R., Gogel, B., Verbyla, A.P., and Thompson, R. (1998b) Spatial analysis of multi-environment early generation variety trials. *Biometrics* 54, 1-18.

Danbury, D.J. (1971) Seed production costs for radiata pine seed orchards. *Australian Forestry* 35, 143-151.

Danuscvicious, D. and Lindgren, D. (2002) Two-stage selection strategies in tree breeding considering gain, diversity, time and cost. *Forest Genetics* 9, 147-159.

Darwin, C. (1859) *The Origin of Species*. Washington Square Press, New York, NY.

David, A. and Keathley, D. (1996) Inheritance of mitochondrial DNA in interspecific crosses of *Picea glauca* and *Picea omorika*. *Canadian Journal of Forest Research* 26, 428-432.

Davis, L.S. (1967) Investments in loblolly pine clonal seed orchards: production costs and economic potential. *Journal of Forestry* 65, 882-887.

Dayanandan, S., Rajora, O.P. and Bawa, K.S. (1998) Isolation and characterization of microsatellites in trembling aspen (*Populus tremuloides*). *Theoretical and Applied Genetics* 96, 950-956.

Dean, C.A., Cotterill, P.P. and Cameron, J.N. (1983) Genetic parameters and gains expected from multiple trait selection of radiata pine in eastern Victoria. *Australian Forest Research* 13, 271-278.

Dean, C.A., Cotterill, P.P. and Eisemann, R.L. (1986) Genetic parameters and gains expected from selection in *Pinus caribaea* var. *hondurensis* in northern Queensland, Australia. *Silvae Genetica* 35, 229-236.

de Assis, T.F. (2000) Production and use of *Eucalyptus* hybrids for industrial purposes. In: *Hybrid Breeding and Genetics of Forest Trees. Proceedings of the Queensland Forest Research Institute/Cooperative Research Center-Sustainable Production Forestry (QFRI/CRC-SPF) Symposium*. Noosa, Queensland, Australia, pp. 63-74.

De Block, M. (1990) Factors influencing the tissue culture and the *Agrobacterium tumefaciens*-mediated transformation of hybrid aspen and poplar clones. *Plant Physiology* 93, 1110-1116.

De Cleene, M. and De Ley, J. (1985) The host range of crown gall. *Botanical Review* 42, 389-466.

Dekkers, J.C.M. and Hospital, F. (2002) The use of molecular genetics in the improvement of agricultural crops. *Nature Reviews Genetics* 3, 22-32.

Delledonne, M., Allegro, G., Belenghi, B. and Balestrazzi, A. (2001) Transformation of white poplar (*Populus alba* L.) with a novel *Arabidopsis thaliana* cystein proteinase inhibitor and analysis of insect pest resistance. *Molecular Breeding* 7, 35-47.

Dempster, E.R. (1955) Genetic models in relation to animal breeding. *Biometrics* 11, 535-536.

Dempster, E.R. and Lerner, I.M. (1950) Heritability of threshold characters. *Genetics* 35, 212-236.

Denison, N.P. and Kietzka, J.E. (1992) The use and importance of hybrid intensive forestry in South Africa. In: *Proceedings of the International Union of Forest Research Organizations (IUFRO), Resolving Tropical Forest Resource Concerns through Tree Improvement, Gene Conservation, and Domestication of New Species*. Cali, Colombia, pp. 348-358.

Denison, N.P. and Quaile, D.R. (1987) The applied clonal eucalypt programme in Mondi forests. *South African Forestry Journal* 142, 60-66.

deSouza, S.M., Hodge, G.R. and White, T.L. (1992) Indirect prediction of breeding values for fusiform rust resistance of slash pine parents using greenhouse tests. *Forest Science* 38, 45-60.

DeVerno, L., Charest, P. and Bonen, L. (1993) Inheritance of mitochondrial DNA in the conifer *Larix*. *Theoretical and Applied Genetics* 86, 383-388.

De-Vescovi, M.A. and Sziklai, O. (1975) Comparative karyotype analysis of Douglas-fir. *Silvae Genetica* 24, 68-73.

Devey, M.E., Jermstad, K.D., Tauer, C.G. and Neale, D.B. (1991) Inheritance of RFLP loci in a loblolly pine three-generation pedigree. *Theoretical and Applied Genetics* 83, 238-242.

Devey, M., Delfino-Mix, A., Donaldson, D., Kinloch, B. and Neale, D. (1995) Efficient mapping of a gene for resistance to white pine blister rust in sugar pine. *Proceedings of the National Academy of Sciences of the United States of America* 92, 2066-2070.

Devey, M., Bell, J., Smith, D., Neale, D. and Moran, G. (1996) A genetic linkage map for *Pinus radiata* based on RFLP, RAPD, and microsatellite markers. *Theoretical and Applied Genetics* 92, 673-679.

Devey, M., Sewell, M., Uren, T. and Neale, D. (1999) Comparative mapping in loblolly and radiata pine using RFLP and microsatellite markers. *Theoretical and Applied Genetics* 99, 656-662

Devey, M.E., Carson, S.D., Nolan, M.F., Matheson, A.C., Te Riini, C. and Hohepa, J. (2004) QTL associations for density and diameter in *Pinus radiata* and the potential for marker-aided selection. *Theoretical and Applied Genetics* 108, 516-524.

Devlin, B. and Ellstrand, N.C. (1990) The development and application of a refined method for estimating gene flow from angiosperm paternity analysis. *Evolution* 44, 248-259.

Dhillon, S. (1980) Nuclear volume, chromosome size and DNA content relationships in three species of *Pinus. Cytologia* 45, 555-560.

Dhir, N.K. and Miksche, J.P. (1974) Intraspecific variation of nuclear DNA content in *Pinus resinosa* Ait. *Canadian Journal of Genetic Cytology* 16, 77-83.

Dickerson, G.E. (1962) Implications of genetic-environmental interaction in animal breeding. *Animal Production* 4, 47-64.

Dickinson, H. and Antonovics, J. (1973) The effects of environmental heterogeneity on the genetics of finite populations. *Genetics* 73, 713-735.

Dickmann, D.I. (1985) The ideotype concept applied to forest trees. Pp. 89-101 In: Cannell, M.G.R. and Jackson, J.E. (eds.) *Attributes of Trees as Crop Plants*. Institute of Terrestrial Ecology, Huntington, England, pp. 89-101.

Dickmann, D. I. and Keathley, D.E. (1996) Linking physiology, molecular genetics, and the *Populus* ideotype. In: Bradshaw, A.D., Heilman, P.E. and Hinckley, T.M. (eds.) *Biology of Populus amd its Implications for Management and Conservation*. NRC Press, National Research Council of Canada, Ottawa, Ontario Canada, pp. 491-514.

Dieters, M.J. and Nikles, D.G. (1997) The genetic improvement of caribbean pine (*Pinus caribaea* Morelet) – building on a firm foundation. *Proceedings of the 24th Southern Forest Tree Improvement Conference*. Orlando, FL, pp. 33-52.

Dieters, M.J., White, T.L. and Hodge, G.R. (1995) Genetic parameter estimates for volume from full-sib tests of slash pine (*Pinus elliottii*). *Canadian Journal of Forest Research* 25, 1397-1408.

DiFazio, S.P., Slavov, G.T., Burczyk, J., Leonardi, S. and Strauss, S.H. (2004) Gene flow from tree plantations and implications for transgenic risk assessment. In: Walter, C. and Carson, M. (eds.) *Plantation Forest Biotechnology for the 21st Century*. Research Signpost, Travandrum, Kerala, India, pp. 405-422.

Di-Giovanni, F., Kevan, P.G. and Arnold, J. (1996) Lower planetary boundary layer profiles of atmospheric conifer pollen above a seed orchard in northern Ontario, Canada. *Forest Ecology and Management* 83, 87-97.

Dinerstein, E., Wikramanayake, E.D. and M. Forney, M. (1995) Conserving the reservoirs and remants of tropical moist forest in the Indo-Pacific region. In: Primack, R.B. and Lovejoy, T.J. (eds.) *Ecology, Conservation and Management of Southeast Asian Rainforests*. Yale University Press, London, UK, pp. 140-175.

Dobzhansky, T. (1970) *Genetics of the Evolutionary Process*. Columbia University Press, New York, NY.

Doerksen, A.H. and Ching, K.K. (1972) Karyotypes in the genus *Pseudotsuga. Forest Science* 18, 66-69.

Donahue, R.A., Davis, T.D., Michler, C.H., Riemenschneider, D.E. Carter, D.R., Marquardt, P.E., Sankhla, N., Sankhla, D., Haissig, B.E. and Isebrands, J.G. (1994) Growth, photosynthesis, and herbicide tolerance of genetically modified hybrid poplar. *Canadian Journal of Forest Research* 24, 2377-2383.

Donald, C.M. (1968) The breeding of crop ideotypes. *Euphytica* 17, 385-403.

Donald, C. M. and Hamblin, J. (1976) The biological yield and harvest index of cereals as agronomic and plant breeding criteria. *Advances in Agronomy* 28, 361-405.

Dorman, K.W. (1976) *The Genetics and Breeding of Southern Pines*. USDA Forest Service Agricultural Handbook 471.

Doudrick, R.L., Heslop-Harrison, J.S., Nelson, C.D., Schmidt, T., Nance, W.L. and Schwarzacher, T. (1995) Karyotype of slash pine (*Pinus elliotii* var. *elliotii*) using patterns of fluorescence *in situ* hybridization and fluorochrome banding. *Journal of Heredity* 86, 289-296.

Dow, B.D. and Ashley, M.V. (1996) Microsatellite analysis of seed dispersal and parentage of saplings in bur oak, *Quercus macrocarpa. Molecular Ecology* 5, 615-627.

Dow, B.D., Ashley, M.V. and Howe, H.F. (1995) Characterization of highly variable (GA/CT)$_n$ microsatellites in the bur oak, *Quercus macrocarpa. Theoretical and Applied Genetics* 91, 137-141.

Duffield, J.W. (1952) Relationships and Species Hybridization in the Genus *Pinus. Forstgenetik* 1, 93-97.

Duffield, J.W. (1990) Forest regions of North America and the world. In: Young, R.A. and Giese, R.L. (eds.) *Introduction to Forest Science*. 2nd Edition. John Wiley & Sons, New York, NY, pp. 33-61.

Dumolin, S., Demesure, B. and Petit, R.J. (1995) Inheritance of chloroplast and mitochondrial genomes in pedunculate oak investigated with an efficient PCR method. *Theoretical and Applied Genetics* 91, 1253-1256.

Dumolin-Lapegue, S., Kremer, A. and Petit, R.J. (1999) Are chloroplast and mitochondrial DNA variation species independent in oaks? *Evolution* 53, 1406-1413.

Duncan, E.A., van Deventer, F., Kietzka, J.E., Lindley, R.C. and Denison, N.P. (2000) The applied subtropical Eucalyptus clonal programme in Mondi forests, Zululand coastal region. In: *Proceedings of the International Union of Forest Research Organizations (IUFRO) Working Party, Forest Genetics for the Next Millinnium*. Durban, South Africa, pp. 95-97.

Dungey, H.S., Dieters, M.J. and Nikles, D.G. (2000) Hybrid breeding and genetics of forest trees. In: *Proceedings of the Queensland Forest Research Institute/Cooperative Research Center-Sustainable Production Forestry (QFRI/CRC-SPF) Symposium*. Department of Primary Industries, Brisbane, Australia.

Dunstan, D.I. (1988) Prospects of progress in conifer biotechnology. *Canadian Journal of Forest Research* 18, 1497-1506.

Dunstan, D.J., Tautorus, T.E. and Thorpe, T.A. (1995) Somatic embryogenesis in woody plants. In: Thorpe, T.A. (ed.) *In Vitro Embryogenesis in Plants*. Kluwer Academic Publishers, Dordrecht, The Netherlands, pp. 471-538.

Durel, C.E., Bertin, P. and Kremer, A. (1996) Relationship between inbreeding depression and inbreeding coefficient in maritime pine (*Pinus pinaster*). *Theoretical and Applied Genetics* 92, 347-356.

Dutrow, G. and Row, C. (1976) Measuring financial gains from genetically superior trees. USDA Forest Service Research Paper SO-132.

Dvorak, W.S. (1996) Integrating exploration, conservation, and utilisation: threats and remedies in the 21st century. In: Dieters, M.J., Matheson, A.C., Nikles, D.G., Harwood, C.E., and Walker, S.M. (eds.), *Proceedings of the International Union of Forest Research Organizations (IUFRO), Conference on Tree Improvement for Sustainable Tropical Forestry*, Caloundra, Australia, pp. 19-26.

Dvorak, W.S. and Donahue, J.K. (1992) CAMCORE Research Review 1980-1992. Department of Forestry, College of Forestry Research, North Carolina State University, Raleigh, NC.

Dvorak, W.S., Donahue, J.K. and Hodge, G.R. (1996) Fifteen years of *ex situ* gene conservation of Mexican and Central American forest species by the CAMCORE Cooperative. *Forest Genetics Resources, No. 24,* FAO (The Food and Agricultural Organization of the United Nations), pp. 15-21.

Dvorak, W., Jordon, A., Hodge, G. and Romero, J. (2000) Assessing evolutionary relationships of pines in the Oocarpae and Australes subsections using RAPD markers. *New Forests* 20, 163-192.

Dvornyk, V., Sirvio, A., Mikkonen, M. and Savolainen, O. (2002) Low nucleotide diversity at the pal1 locus in the widely distributed *Pinus sylvestris. Molecular Biology and Evolution* 19, 179-188.

Ebell, L.F. and McMullan, E.E. (1970) Nitrogenous substances associated with differential cone production responses of Douglas-fir to ammonium and nitrate fertilization. *Canadian Journal of Botany* 48, 2169-2177.

Echt, C.S. and May-Marquardt, P. (1997) Survey of microsatellite DNA in pine. *Genome* 40, 9-17.

Echt, C.S., May-Marquardt, P., Hseih, M. and Zahorchak, R. (1996) Characterization of microsatellite markers in eastern white pine. *Genome* 39, 1102-1108.

Eckenwalder, J. (1976) Re-evalutation of Cupresaceae and Taxodiaceae: A proposed merger. *Madrono* 23, 237-300.

Eckenwalder, J.E. (1984) Natural Intersectional hybridization between North American species of

Populus (Salicaceae) in sections Aigeiros and Tacamahaca. III. Paleobotany and evolution. *Canadian Journal of Botany* 62, 336-342.

Eckenwalder, J.E. (1996) Ch.1: Systematics and evolution of *Populus*. In: Stettler, R.F., Bradshaw, H.D., Heilman, P.E. and Hinckley, T.M. (eds.) *Biology of Populus and its Implications for Management and Conservation*. NRC Research Press, National Research Council of Canada, Ottawa, Quebec, pp.7-32.

Edwards, D.G.W. and El-Kassaby, Y.A. (1995) Douglas-fir genotypic response to seed stratification germination parameters. *Seed Science and Technology Journal* 23, 771-778.

Edwards, D.G.W. and El-Kassaby, Y.A. (1996) The biology and management of forest seeds: genetic perspectives. *Forestry Chronicle* 72, 481-484.

Edwards, M., Stuber, C. and Wendel, J. (1987) Molecular marker facilitated investigations of the quantitative trait loci in maize. I. Numbers, genomic distribution, and types of gene action. *Genetics* 116, 113-125.

Egertsdotter, U., van Zyl, L., MacKay, J., Peter, G., Whetten, R. and Sederoff, R. (2004) Gene expression profiling of wood formation: an analysis of seasonal variation. *Plant Biology* 6, 654-663.

Ehrenberg, C. (1970) Breeding for stem quality. *Unasylva* 24, 23-31.

Ehrlich, P. and Ehrlich, A. (1981) *Extinction*. Random House, New York, NY.

El Mousadik, A. and Petit, R.J. (1996) Chloroplast DNA phylogeography of the argan tree of Morocco. *Molecular Ecology* 5, 547-555.

Eldridge, K.G. (1982) Genetic improvements from a radiata pine seed orchard. *New Zealand Journal of Forest Science* 12, 404-411.

Eldridge, K., Davidson, J., Harwood, C. and van Wyk, G. (1994) *Eucalypt Domestication and Breeding*. Oxford University Press, Oxford, UK.

El-Kassaby, Y.A. (1989) Genetics of seed orchards: expectations and realities. In: *Proceedings of the 20th Southern Forest Tree Improvement Conference*. Charleston, SC, pp. 87-109.

El-Kassaby, Y.A. (1992) Domestication and genetic diversity - should we be concerned? *Forest Chronicle* 68, 687-700.

El-Kassaby, Y.A. (1999) Impacts of industrial forestry on genetic diversity of temperate forest trees. In: Matyas, C. (ed.) *Forest Genetics and Sustainability*. Kluwer Academic Publishers, Boston, MA, pp. 155-170.

El-Kassaby, Y.A. (2000) Effect of forest tree domestication on gene pools. In: Young, A., D. Boshier, D., and Boyle, T. (eds.) *Forest Conservation Genetics: Principles and Practice*. Commonwealth Scientific and Industrial Research Organization (CSIRO) Publishing, Collingwood, Victoria, Australia, pp. 197-213.

El-Kassaby, Y.A. and Jaquish, B. (1996) Population density and mating pattern in western larch. *Journal of Heredity* 87, 438-443.

El-Kassaby, Y A. and Ritland, K. (1986) The relation of outcrossing and contamination to reproductive phenology and supplemental mass pollination in a Douglas-fir seed orchard. *Silvae Genetica* 35, 240-244.

El-Kassaby, Y.A. and Ritland, K. (1996) Impact of selection and breeding on the genetic diversity in Douglas-fir. *Biodiversity and Conservation* 5, 795-813.

El-Kassaby, Y.A., Parkinson, J. and Devitt, W.J.B. (1986) The effect of crown segment on the mating system in a Douglas-fir (*Pseudotsuga menziesii* (Mirb.) Franco) seed orchard. *Silvae Genetica* 35, 149-155.

El-Kassaby, Y.A., Ritland, K., Fashler, A.M.K. and Devitt, W.J.B. (1988) The role of reproductive phenology upon the mating system of a Douglas-fir seed orchard. *Silvae Genetica* 37, 76-82.

El-Lakany, M.H. and Sziklai, O. (1971) Intraspecific variation in nuclear characteristics of Douglas-fir. *Advancing Frontiers of Plant Sciences* 28, 363-378.

Ellis, D.D., McCabe, D.E., McInnis, S., Ramachandran, R., Russel, D.R., Wallace, K.M., Martinell, B.J., Roberts, D.R., Raffa, K.F. and McCown, B.H. (1993) Stable transformation of *Picea glauca* by particle acceleration. *Bio/Technology* 11, 84-89.

Ellstrand, N.C. (1992) Gene flow among seed plant populations. *New Forests* 6, 241-256.

Endo, M. and Lambeth, C.C. (1992) Promising potential of hybrid, *Eucalyptus grandis* x *Eucalyptus urophylla*, over *Eucalyptus grandis* in Colombia. In: *Proceedings of the International Union of Forest Research Organizations (IUFRO), Resolving Tropical Forest Resource Concerns*

Through Tree Improvement, Gene Conservation, and Domestication of New Species. Cali, Colombia, pp. 366-371.

Ennos, R.A. (1994) Estimating the relative rates of pollen and seed migration among plant populations. *Heredity* 72, 250-259.

Epperson, B.K. (1992) Spatial structure of genetic variation within populations of forest trees. *New Forests* 6, 257-278.

Epperson, B.K. and Allard, R.W. (1987) Linkage disequilibrium between allozymes in natural populations of lodgepole pine. *Genetics* 115, 341-352.

Epperson, B.K. and Allard, R.W. (1989) Spatial autocorrelation analysis of the distribution of genotypes within populations of lodgepole pine. *Genetics* 121, 369-377.

Erickson, V.J. and Adams, W.T. (1989) Mating success in a coastal Douglas-fir seed orchard as affected by distance and floral phenology. *Canadian Journal of Forest Research* 19, 1248-1255.

Erickson, V.J. and Adams, W.T. (1990) Mating system variation among individual ramets in a Douglas-fir seed orchard. *Canadian Journal of Forest Research* 20, 1672-1675.

Eriksson, G. and Lundkvist, K. (1986) Adaptation and breeding of forest trees in boreal areas. In: Lindgren D. (ed.) Provenances and Forest Breeding for High Latitudes. *Proceedings of the Frans Kempe Symposium in Umeå.* Swedish University of Agricultural Sciences, Uppsala, Sweden, pp. 67-80.

Eriksson, G., Namkoong, G. and Roberds, J.H. (1993) Dynamic gene conservation for uncertain futures. *Forest Ecology and Management* 62, 15-37.

Estruch, J.J., Carrozi, N.B. Desai, N., Duck, N.B., Warren, G.W. and Koziel, M.G. (1997) Transgenic plants: an emerging approach to pest control. *Nature Biotechnology* 15, 137-141.

Evans, J. (1992a) *Plantation Forestry in the Tropics.* Clarendon Press, Oxford, UK.

Evans, J. (1992b) What to plant? In: Evans, J. (ed.) *Plantation Forestry in the Tropics.* Clarendon Press, Oxford, UK pp. 99-121.

Fagard, M. and Vauchert, H. (2000) (Tans) gene silencing in plants: how many mechanisms? *Annual Review* 51, 167-194.

Falconer, D.S. and Mackay, T.F.C. (1996) *Introduction to Quantitative Genetics.* Longman, Essex, England.

FAO (The Food and Agricultural Organization of the United Nations) (1995a) Forest Assessment 1990: Global Synthesis. Paper 124. Rome, Italy.

FAO (The Food and Agricultural Organization of the United Nations) (1995b) Forest Assessment 1990: Tropical Plantations. Paper 128. Rome, Italy.

FAO (The Food and Agricultural Organization of the United Nations) (1997) State of the World's Forests. FAO, Rome, Italy.

FAO (The Food and Agricultural Organization of the United Nations), DFSC (The Danish International Development Agency Forest Seed Center) and IPGRI (International Plant Genetic Resource Institute) (2001) Forest genetic resources conservation and management. In: *Managed Natural Forests (in situ).* International Plant Genetic Resources Institute, Rome, Italy.

FAO (The Food and Agricultural Organization of the United Nations) (2004) Preliminary review of biotechnology in forestry, including genetic modification. Forest Genetic Resources Working Paper FGR/59E. Forest Resources Development Service, Forest Resources Division. Rome, Italy.

Farjon, A. and Styles, B.T. (1997) *Pinus. Flora Neotropica Monograph 70.* New York Botanical Garden, N.Y.

Farmer, R.E., O'Reilly, G.J., and Shaotang, D. (1993) Genetic variation in juvenile growth of tamarack (*Larix laricina*) in northwestern Ontario. *Canadian Journal of Forest Research* 23, 1852-1862.

Farris, M.A. and Mitton, J.B. (1984) Population density, outcrossing rate, and heterozygote superiority in ponderosa pine. *Evolution* 38, 1151-1154.

Faulkner, R. (1975) *Seed Orchards.* Forestry Commission Bulletin No. 54. Her Majesty's Stationery Office, London, UK.

Felsenstein, J. (1989) *PHYLIP 3.2 Manual.* University of California Herbarium, Berkeley, CA.

Fernandez, G.C.J. (1991) Analysis of genotype x environment interactions by stability parameters. *HortScience* 26, 947-950.

Ferreira, M. and Santos, P.E.T. (1997) Genetic improvement of *Eucalyptus* in Brazil-brief review and perspectives. In: *Proceedings of the Conference of the International Union of Forest Re search Organizations (IUFRO), sobre Silvicultura e Melhoramento de Eucaliptos*. El Salvador, Brazil, pp. 14-33.

Fillatti, J.J., Sellmer, J., McGown, B., Haissig, B.E. and Comai, L. (1987) *Agrobacterium* mediated transformation and regeneration of *Populus*. *Molecular and General Genetics* 206, 192-199.

Finlay, K.W. and Wilkinson, G.N. (1963) The analysis of adaptation in a plant-breeding program. *Australian Journal of Agricultural Research* 14, 742-754.

Finnegan, J. and McElroy, D. (1994) Transgene inactivation: plants fight back. *Bio/Technology* 12, 883-888.

Fins, L. and Moore, J.A. (1984) Economic analysis of a tree improvement program for western larch. *Journal of Forestry* 82, 675-679.

Fins, L., Friedman, S.T., and Brotschol, J.V. (1992) *Handbook of Quantitative Forest Genetics*. Kluwer Academic Publishers, Dordrecht, The Netherlands.

Fisher, P.J., Richardson, T.E. and Gardner, R.C. (1998) Characteristics of single-and multi-copy microsatellites from *Pinus radiata*. *Theoretical and Applied Genetics* 96, 969-979.

Fisher, R.A. (1925) *Statistical Methods for Research Workers*. Oliver and Boyd, Edinburgh, UK.

Fisher, R.A. (1930) *The Genetical Theory of Natural Selection*. Clarendon Press, Oxford, UK.

Fladung, M. (1999) Gene stability in transgenic aspen (*Populus*). I. Flanking DNA sequences and T-DNA structure. *Molecular and General Genetics* 260, 1097-1103.

Fladung, M., Kumar, S. and Ahuja, M.R. (1997) Genetic transformation with different chimeric gene constructs: transformation efficiency and molecular analysis. *Transgenic Research* 6, 111-121.

Flavell, R.B., O'Dell, M., Thompson, W.F., Vincentz, M., Sardana, R. and Barker, R.F. (1986) The differential expression of ribosomal RNA genes. *Transactions of the Royal Society of London* 314, 385-397.

Florin, R. Evolution in cordaites and conifers. *Acta Horti Bergiani* 15, 285-388.

Foster, G.S. (1985) Genetic parameters for two Eastern cottonwood populations in the Lower Mississippi valley. *Proceedings of the 18th Southern Forest Tree Improvement Conference*, Long Beach, MS, pp. 258-266.

Foster, G.S. (1986) Making clonal forestry pay: breeding and selection for extreme genotypes. In: *Proceeding of the International Union of Forest Research Organizations (IUFRO), Conference on Breeding Theory, Progeny Testing and Seed Orchards*. Williamsburg,VA, pp. 582-590.

Foster, G.S. (1990) Genetic control of rooting ability of stem cuttings of lobololly pine. *Canadian Journal of Forest Research* 20, 1361-1368.

Foster, G.S. (1992) Estimating yield: beyond breeding values. In: Fins, L., Friedman, S.T. and Brotschol, J.V. (eds.) *Handbook of Quantitative Forest Genetics*. Kluwer Academic Publishers. Boston, MA, pp. 229-269.

Foster, G.S. (1993) Selection and breeding for extreme genotypes. In: Ahuja, M.R. and Libby, W.J. (eds.) *Clonal Forestry I. Genetics and Biotechnology*. Springer-Verlag, New York, NY, pp. 50-67.

Foster, G.S. and Shaw, D.V. (1988) Using clonal replicates to explore genetic variation in a perennial plant species. *Theoretical and Applied Genetics* 76, 788-794.

Foster, G.S., Campbell, R.K. and Adams, W.T. (1984) Heritability, gain, and C effects in rooting of western hemlock cuttings. *Canadian Journal of Forest Research* 14, 628-638.

Foulley, J.L. and Im, S. (1989) Probability statements about the transmitting ability of progeny-tested sires for an all-or-none trait with application to twinning in cattle. *Genetic Selection Evolution* 21, 359-376.

Foulley, J.L., Hanocq, E. and Boichard, D. (1992) A criterion for measuring the degree of connectedness in linear models of genetic evaluation. *Genetic Selection Evolution* 24, 315-330.

Fowler, D.P. (1965a) Effects of inbreeding in red pine, *Pinus resinosa* Ait., IV. Comparison with other northeastern *Pinus* species. *Silvae Genetica* 14, 76-81.

Fowler, D.P. (1965b) Natural self-fertilization in three jack pines and its implications in seed orchard management. *Forest Science* 11, 55-58.

Fowler, D.P. (1986) *Strategies for the genetic improvement of important tree species in the maritimes*. Canadian Forest Service Information Report M-X-156.

Fowler, D.P. (1987) Tree improvement strategies - flexibility required. In: Proceedings of the 21st Tree Improvement Association. Truro, Nova Scotia, pp, 85-95.

Fowler, D.P. and Lester, D.T. (1970) The genetics of red pine. USDA Forest Service Research Paper WO-8.

Fowler, D.P. and Morris, R.W. (1977) Genetic diversity in red pine: evidence for low genic heterozygosity. *Canadian Journal of Forest Research* 7, 343-347.

Frampton, L.J., Jr. and Foster, G.S. (1993) Field testing vegetative propagules. In: Ahuja, M.R. and Libby, W.J. (eds.). *Clonal Forestry I. Genetics and Biotechnology.* Springer-Verlag, New York, NY, pp. 110-134.

Francia, E., Tacconi, G., Crosatti, C., Barabaschi, D., Bulgarelli, D., Aglio, E. D. and Vale, G. (2005) Marker assisted selection in crop plants. *Plant Cell Tissue and Organ Culture* 82, 317-342.

Franke, R., McMichael, C.M., Meyer, K., Shirley, A.M., Cusumano, J.C. and Chapple, C. (2000) Modified lignin in tobacco and poplar plants over-expressing the *Arabidopsis* gene encoding ferulate 5-hydroxylase. *Plant Journal* 22, 223-234.

Frankel, O.H. (1986) Genetic resources - museum of utility. In: Williams, T.A. and Wratt. G.S. (eds). *Department of Scientific and Industrial Research, Agronomy Society of New Zealand, Plant Breeding Symposium,* Christchurch, New Zealand, pp. 3-7.

Frankel, O.H., Brown, A.H.D. and Burdon, J.J. (1995). *The Conservation of Plant Biodiversity.* Cambridge University Press, Cambridge, UK.

Frankham, R. (1995) Effective population size/adult population size ratios in wildlife: a review. *Genetics Research Cambridge* 66, 95-107.

Franklin, E.C. (1969a) Inbreeding depression in metrical traits of loblolly pine (*Pinus taeda* L.) as a result of self-pollination. *North Carolina State University School Forest Resources Technical Report 40*, 1-19.

Franklin, E.C. (1969b) Mutant forms found by self-pollination in loblolly pine. *Journal of Heredity* 60, 315-320.

Franklin, E.C. (1970) Survey of mutant forms and inbreeding depression in species of the family Pinaceae. USDA Forest Service Research Paper 61.

Franklin, E.C. (1971) Estimating frequency of natural selfing based on segregating mutant forms. *Silvae Genetica* 20, 193-195.

Franklin, E.C. (1972) Genetic load in loblolly pine. *American Naturalist* 106, 262-265.

Franklin, E.C. (1974) Pollination in slash pine: First come, first served. In: Kraus, J. (ed.) *Seed Yield from Southern Pine Seed Orchards.* Georgia Forest Research Council, Macon, GA, pp. 15-20.

Franklin, E.C. (1979) Model relating levels of genetic variance to stand development in four North American conifers. *Silvae Genetica* 28, 207-212.

Franklin, E.C. (1986) Estimation of genetic parameters through four generations of selection in *Eucalyptus grandis.* In: *Proceedings of the International Union of Forest Research Organizations (IUFRO), Conference on Breeding Theory, Progeny Testing and Seed Orchards.* Williamsburg, VA, pp. 200-209.

Franklin, E.C. (1989) Selection strategies for eucalypt tree improvement: four generations of selection in *Eucalyptus grandis* demonstrate valuable methodology. In: Gibson, G.I., Griffin, A.R. and Matheson, A.C. (eds.) *Breeding Tropical Trees: Population Structure and Genetic Improvement Strategies in Clonal and Seedling Forestry.* Oxford Forestry Institute, Oxford, UK, pp. 197-209.

Franklin, I.R. (1980) Evolutionary changes in small populations. In: Soule, M.E. and Wilcox, B.A. (eds.) *Conservation Biology: An Evolutionary-Ecological Perspective.* Sinauer Associates Inc., Publishers, Sunderland, MA, pp. 135-149.

Frewen, B., Chen, T., Howe, G., Davis, J., Rohde, A., Boerjan, W, and Bradshaw Jr., H. (2000) Quantitative trait loci and candidate gene mapping of bud set and bud flush in *Populus. Genetics* 154, 837-845.

Friedman, S.T. and Adams, W.T. (1982) Genetic efficiency in loblolly pine seed orchards. In: *Proceedings of the 16th Southern Forest Tree Improvement Conference.* Virginia Poly Tech and State University, Blacksburg, VA, pp. 213-220.

Friedman, S.T. and Neale, D.B. (1993) Biochemical and molecular markers. In: *Advances in Pollen Management.* USDA Forest Service Agricultural Handbook No. 698.

Fryer, J.H. and Ledig, F.T. (1972) Microevolution of the photosynthetic temperature optimum in relation to the elevational complex gradient. *Canadian Journal of Botany* 50, 1231-1235.

Fu, Y., Clarke, G.P.Y., Namkoong, G. and Yanchuk. A.D. (1998) Incomplete block designs for genetic testing: statistical efficiencies of estimating family means. *Canadian Journal of Forest Research* 28, 977-986.

Fu, Y., Yanchuk, A.D. and Namkoong, G. (1999a) Incomplete block designs for genetic testing: some practical considerations. *Canadian Journal of Forest Research* 29, 1871-1878.

Fu, Y., Yanchuk, A.D., G. Namkoong, G. and Clarke, G.P.Y. (1999b) Incomplete block designs for genetic testing: statistical efficiencies with missing observations. *Forest Science* 45, 374-380.

Fu, Y., Yanchuk, A.D. and Namkoong, G. (1999c) Spatial patterns of tree height variations in a series of Douglas-fir progeny trials: implications for genetic testing. *Canadian Journal of Forest Research* 29, 714-723.

Furnier, G.R. and Adams, W.T. (1986a) Geographic patterns of allozyme variation in Jeffrey pine. *American Journal of Botany* 73, 1009-1015.

Furnier, G.R. and Adams, W.T. (1986b) Mating system in natural populations of Jeffrey pine. *American Journal of Botany* 73, 1002-1008.

Furnier, G.R., Knowles, P., Clyde, M.A. and Dancik, B.P. (1987) Effects of avian seed dispersal on the genetic structure of whitebark pine populations. *Evolution* 41, 607-612.

Futuyma, D. (1998) *Evolutionary Biology*. Sinauer Associates Inc., Publishers, Sunderland, MA.

Fyfe, J.L. and Bailey, N.T. (1951) Plant breeding studies in leguminous forage crops. I. Natural cross-breeding in winter beans. *Journal of Agricultural Science* 41, 371-378.

Gabriel, W.J. (1967) Reproductive behavior in sugar maple: self-compatibility, cross-compatibility, agamospermy, and agamocarpy. *Silvae Genetica* 16, 165-168.

Gasser, C.S. and Fraley, R.T. (1989) Genetically engineered plants for crop improvement. *Science* 244, 1293-1299.

Gasser, C.S. and Fraley, R.T. (1992) Trnsgenic crops. *Scientific American* 266(6), 62-69.

Gea, L., Lindgren, D., Shelbourne, C.J.A. and Mullin, L.J. (1997) Complementing inbreeding coefficient information with status number: implications for structuring breeding populations. *New Zealand Journal of Forestry* 27, 255-271.

Geburek, T. (1997) Isoenzymes and DNA markers in gene conservation of forest trees. *Biodiversity and Conservation* 6, 1639-1654.

Geburek, T. (2000) Effects of environmental pollution on the genetics of forest trees. In: Young, A., Boshier, D. and Boyle, T. (eds.) *Forest Conservation Genetics: Principles and Practice*. Commonwealth Scientific and Industrial Research Organization (CSIRO) Publishing, Collingwood, Victoria, Australia, pp. 145-158.

Gerber, S., Rodolphe, F., Bahrman, N. and Baradat, P. (1993) Seed-protein variation in maritime pine (*Pinus pinaster* Ait.) revealed by two-dimensional electrophoresis: genetic determinism and construction of a linkage map. *Theoretical and Applied Genetics* 85, 521-528.

Gibson, G.L., Barnes, R.D., and Berrington, J. (1988) Provenance productivity of *Pinus caribaea* and its interaction with environment. *Commonwealth Forestry Review* 62, 93-106.

Giertych, M. (1965) Systematic lay-outs for seed orchards. *Silvae Genetica* 14, 91-94.

Giertych, M. (1975) Seed orchard designs. In: Faulkner, R. (ed.) *Seed Orchards, Forestry Commission Bulletin 54*. Her Majesty's Stationery Office, London, UK, pp. 25-37.

Gilmour, A.R., Thompson, R., Cullis, B.R. and Welham, S.J. (1997) *ASREML User's Manual*. Orange, Australia.

Glaubitz, J.C. and Moran, G.F. (2000) Genetic tools: The use of biochemical and molecular markers. In: Young, A., Booshier, D. and Boyle, T. (eds.) *Forest Conservation Genetics: Principles and Practice*. Commonwealth Scientific and Industrial Research Organization (CSIRO) Publishing, Collingwood, Victoria, Australia, pp.39-59.

Goddard, R.E. and Strickland, R.K. (1964) Crooked stem form in loblolly pine. *Silvae Genetica* 13, 155-157.

Godt, M.J.W., Hamrick, J.L. and Williams, J.H. (2001) Comparison of genetic diversity in white spruce (*Picea glauca*) and jack pine (*Pinus banksiana*) seed orchards with natural populations. *Canadian Journal of Forest Research* 31, 943-949.

Goff, S., Ricke, D., Lan, T., Presting, G., Wang, R., Dunn, M., Glazebrook, J., Sessions, A., Oeller, P., Varma, H., Hadley, D, Hutchison, D., Martin, C., Katagiri, F., Lange, B.M., Moughamer, T., Xia, Y., Budworth, P., Zhong, J., Miguel, T., Paszkowski, U., Zhang, S., Colbert, M., Sun W., Chen, L., Cooper, B., Park, S., Wood, T.C., Mao, L., Quail, P., Wing, R., Dean, R., Yu, Y., Zharkikh, A., Shen, R., Sahasrabudhe, S., Thomas, A., Cannings, R., Gutin A., Pruss, D., Reid,

J., Tavtigian, S., Mitchell, J., Eldredge, G., Scholl, T., Miller, R.M., Bhatnagar, S., Adey, N., Rubano, T., Tusneem, N., Robinson, R., Feldhaus, J., Macalma, T., Oliphant, A. and Briggs, S. (2002) A draft sequence of the rice genome (*Oryza sativa* L. ssp. *Japonica*). *Science* 296, 92-100.

Gonzalez-Martinez, S.C., Wheeler, N.C., Ersoz, E., Nelson, C.D. and Neale, D.B. (2006) Association Genetics in *Pinus taeda* L. I. Wood property graits. *Genetics*; published ahead of print on November 16, 2006 as doi: 10.1534/genetics.106.061127.

Goodnight, C.J. (1988) Epistasis and the effect of founder events on the additive genetic variance. *Evolution* 42, 441-454.

Goodnight, C.J. (1995) Epistasis and the increase in additive genetic variance: implications for phase 1 of Wright's shifting-balance process. *Evolution* 49, 502-511.

Gould, S.J. and Johnson, R.F. (1972) Geographic variation. *Annual Review of Ecology and Systematics* 3, 457-498.

Govindaraju, D. and Cullis, C. (1992) Ribosomal DNA variation among popualtions of a *Pinus rigida* Mill. (pitch pine) ecosystem: I. Distribution of copy numbers. *Heredity* 69, 133-140.

Govindaraju, D., Lewis, P. and Cullis, C. (1992) Phylogenetic analysis of pines using ribosomal DNA restriction fragment length polymorphisms. *Plant Systematics and Evolution* 179, 141-153.

Grace, L.J., Charity, J.A., Gresham, B., Kay, N. and Walter, C. (2005) Insect resistance transgenic *Pinus radiata*. *Plant Cell Reports* 24, 103-111.

Graham, R.T. (1990) *Pinus monticola*, Dougl. ex D. Don. In: Burns, R.M. and Honkala, B.H. (eds.). *Silvics of North America. Vol. I. Conifers.* USDA Forest Service Agricultural Handbook 654, pp. 385-394.

Grant, M.C. and Mitton, J.B. (1977) Genetic differentiation among growth forms of Engelmann spruce and subalpine fir at tree line. *Arctic and Alpine Research* 9, 259-263.

Grant, V. (1971) *Plant Speciation*. Columbia University Press, New York, NY.

Grattapaglia, D., Bertolucci, F. and Sederoff, R. (1995) Genetic mapping of QTLs controlling vegetative propagation in *Eucalyptus grandis* and *E. urophylla* using a pseudo-testcross strategy and RAPD markers. *Theoretical and Applied Genetics* 90, 933-947.

Grattapaglia, D., Ribeiro, V.J. and Rezende, G.D.S.P. (2004) Retrospective selection of elite parent trees using paternity testing with microsatellite markers: an alternative short term breeding tactic for *Eucalyptus*. *Theoretical and Applied Genetics* 109, 192-199

Gray, M.W. (1989) Origin and evolution of mitochondrial DNA. *Annual Review of Cell Biology* 5, 25-50.

Greenwood, M.S. (1977) Seed orchard fertilization: Optimizing time and rate of ammonium nitrate application for grafted loblolly pine (*Pinus taeda* L.). In: *Proceedings of the 14th Southern Forest Tree Improvement Conference*. Gainesville, FL, pp. 164-169.

Greenwood, M.S. (1983) Maximizing genetic gain in loblolly pine by application of accelerated breeding methods and new concepts in orchard design. In: *Proceedings of the 17th Southern Forest Tree Improvement Conference*. Athens, GA, pp. 290-296

Greenwood, M.S. (1987) Rejuvenation of forest trees. *Plant Growth Regulation* 6, 1-12.

Greenwood, M.S. and Hutchinson, K.W. (1993) Maturation as a developmental process. In: Ahuja, M.R. and Libby, W.J. (eds.) *Clonal Forestry I. Genetics and Biotechnology*. Springer-Verlag, New York, NY, pp. 14-33.

Greenwood, M.S., O'Gwynn, C.H. and Wallace, P.G. (1979) Management of an indoor, potted loblolly pine breeding orchard. In: *Proceedings of the 15th Southern Forest Tree Improvement Conference*. Mississippi State, MS, pp. 94-98.

Greenwood, M.S., Adams, G.W. and Gillespie, M. (1987) Shortening the breeding cycle of some northeastern conifers. In: *Proceedings of the 21st Tree Improvement Association*. Truro, Nova Scotia, pp. 43-52.

Gregorius, H.R. and Roberds, J.H. (1986) Measurement of genetical differentiation among subpopulations. *Theoretical and Applied Genetics* 71, 826-834.

Gregory, J.D., Guinness, W.M. and Davey, C.B. (1982) Fertilization and irrigation stimulate flowering and cone production in a loblolly pine seed orchard. *Southern Journal of Applied Forestry* 6, 44-48.

Griffin, A.R. (1982) Clonal variation in radiata pine seed orchards. I. Some flowering, cone and seed production traits. *Australian Journal of Forest Research* 12, 295-302.

Griffin, A.R. (1989) Sexual reproduction and tree improvement strategy - with particular reference to eucalyptus. In: Gibson, G.I., Griffin, A.R. and Matheson, A.C. (eds.) *Breeding Tropical Trees: Population Structure and Genetic Improvement Stratgies in Clonal and Seedling Forestry.* Oxford Forestry Institute, Oxford, UK, pp. 52-67.

Griffin, A.R. (1993) Potential for genetic improvement of Eucalyptus in Chile. In: Barros, S., Prado, J.A. and Alvear, C. (eds.) *Proceedings of the Los Eucaliptos en el Desarrollo Forestal de Chile.* Pucon, Chile, pp. 1-25.

Griffin, A.R. and Cotterill, P.P. (1988) Genetic variation in growth of outcrossed, selfed and open-pollinated progenies of *Eucalyptus regnans* and some implications for breeding strategy. *Silvae Genetica* 37, 124-131.

Griffin, A.R. and Lindgren, D. (1985) Effect of inbreeding on production of filled seed in *Pinus radiata* -- experimental results and a model of gene action. *Theoretical and Applied Genetics* 71, 334-343.

Griffin, A.R., Burgess, I.P. and Wolf, L. (1988) Patterns of natural and manipulated hybridisation in the genus *Eucalyptus* L'Herit – a review. *Australian Journal of Botany* 36, 41-66.

Griffin, R., Harbard, J.L., Centurion, C. and Santini, P. (2000) Breeding *Eucalyptus grandis* x *globulus* and other interspecific hybrids with high inviability - problem analysis and experience at Shell Forestry Projects in Uruguay and Chile. In: *Proceedings of the Queensland Forest Research Institute/Cooperative Research Center-Sustainable Production Forestry (QFRI/CRC-SPF) Symposium on Hybrid Breeding and Genetics of Forest Trees.* Noosa, Queensland, Australia, pp. 1-13.

Groeneveld, E., Kovac, M., and Wang, T. (1990) PEST, a general purpose BLUP package for multivariate prediction and estimation. In: *Proceedings of the 4th World Congress on Genetics Applied to Livestock*, pp. 488-491.

Grondona, M.O., Crossa, J., Fox, P.N. and Pfeiffer, W.H. (1996) Analysis of variety yield trials using two-dimensional separable ARIMA processes. *Biometrics* 52, 763-770.

Groover, A., Devey, M., Fiddler, T., Lee, J., Megraw, R., Mitchell-Olds, T., Sherman, B., Vujcic, S., Williams, C. and Neale, D. (1994) Identification of quantitative trait loci influencing wood specific gravity in loblolly pine. *Genetics* 138, 1293-1300.

Grossnickle, S.C., Cyr, D. and Polonenko, D.R. (1996) Somatic embryogenesis tissue culture for the propagation of conifer seedlings: a technology comes of age. *Tree Planters' Notes* 47, 48-57.

Gulbaba, A.G., Velioglu, E., Ozer, A.S., Dogan, B., Doerksen, A.H. and Adams, W.T. (1998) Population genetic structure of Kazdagi fir (*Abies equitrojani* Aschers. and Sint.) a narrow endemic to Turkey: Implications for *in-situ* conservation. In: Zencirci, N., Kaya, Z. Anikster, Y. and Adams, W.T. (eds.), *Proceedings of the International Symposium on in situ Conservation Plant Genetic Diversity.* Central Research Institute for Field Crops, Ankara, Turkey, pp. 271-280.

Gullberg, U. (1993) Towards making willows pilot species for coppicing production. *Forestry Chronicle* 69, 721-726.

Gupta, P.K. and Kreitinger, M. (1993) Somatic seeds in forest trees. In: Ahuja, M.R. (ed.) Micropropagation of Woody Plants. Kluwer Academic Publishers, Dordrecht, The Netherlands, pp. 107-119.

Gupta, P.K., Dandekar, A.M. and Durzan, D.J. (1988) Somatic proembryo formation and transient expression of a luciferase gene in Douglas fir and loblolly pine protoplasts. *Plant Science* 58, 85-92.

Gupta, P.K., Rustgi, S. and Kulwal, P.L. (2005) Linkage disequilibrium and association studies in higher plants: present status and future prospects. *Plant Molecular Biology* 57, 461-485.

Gurevitch, J., Scheiner, S. M. and Fox, G. A. (2002) *The Ecology of Plants.* Sinauer Associates, Inc. Publishers, Sunderland, MA.

Guries, R.P. (1984) Genetic variation and population differentiation in forest trees. In: *Proceedings of the 8th North American Forest Biology Workshop.* Logan, UT, pp. 119-131.

Guries, R.P., Friedman, S.T. and Ledig, F.T. (1978) A megagametophyte analysis of genetic linkage in pitch pine (*Pinus rigida* Mill). *Heredity* 40, 309-314.

Haapanen, M. (1996) Impact of family-by-trial interaction on utility of progeny testing methods for scots pine. *Silvae Genetica* 45, 130-135.

Hagler, R.W. (1996) The global wood fiber equation - a new world order. *Tappi Journal* 79, 51-54.

Hagman, M. (1967) Genetic mechanisms affecting inbreeding and outbreeding in forest trees: their significance for microevolution of forest tree species. In: *Proceedings of the 14ᵗʰ International Union of Forest Research Organizations (IUFRO) Congress*, Section 22, Volume III, Munchen, Germany, pp. 346-365.

Hakman, I. and von Arnold, S. (1985) Plantlet regeneration through somatic embryogenesis in *Picea abies* (Norway spruce). *Journal of Plant Physiology* 121, 149-158.

Hall, J.P. (1988) Promotion of flowering in black spruce using gibberellins. *Silvae Genetica* 37, 135-138.

Hall, P., Chase, M.R. and Bawa, K.S. (1994) Low genetic variation but high population differentiation in a common tropical forest tree species. *Conservation Biology* 8, 471-482.

Hall, P., Walker, S. and Bawa, K.S. (1996) Effect of forest fragmentation on genetic diversity and mating system in a tropical tree, *Pithecellobium elegans*. *Conservation Biology* 10, 757-768.

Hallauer, A.R. and Miranda, J.B. (1981) *Quantitative Genetics in Maize Breeding*. Iowa State University Press, Ames, IA.

Hamilton, P.C., Chandler, L.R., Brodie, A.W. and Cornelius, J.P. (1998) A financial analysis of a small scale *Gmelina arborea* Roxb. improvement program in Costa Rica. *New Forests* 16, 89-99.

Hamrick, J.L. and Godt, M.J.W. (1990) Allozyme diversity in plant species. In: Brown, A.H.D., Clegg, M.T., Kahler, A.L. and Weir, B.S. (eds.) *Plant Population Genetics, Breeding and Genetic Resources*. Sinauer Associates Inc., Publishers, Sunderland, MA, pp. 43-63.

Hamrick, J.L. and Murawski, D.A. (1990) The breeding structure of tropical tree populations. *Plant Species Biology* 5, 157-165.

Hamrick, J.L., Godt, M.J.W. and Sherman-Broyles, S.L. (1992) Factors influencing levels of genetic diversity in woody plant species. *New Forests* 6, 95-124.

Hamrick, J.L., Murawski, D.A. and Nason, J.D. (1993a) The influence of seed dispersal mechanisms on the genetic structure of tropical tree populations. *Vegetatio* 107/108, 281-297.

Hamrick, J.L., Platt, W.J. and Hessing, M. (1993b) Genetic variation in longleaf pine. In: Hermann, S.M. (ed.) *Proceedings Tall Timbers Fire Ecology Conference No. 18, The Longleaf Pine Ecosystem: Ecology, Restoration and Management*. Tallahassee, FL, pp. 193-203.

Han, K.-H., Ma, C. and Strauss, S.H. (1997) Matrix attachment regions (MARs) enhance transformation frequency and transgene expression in poplar. *Transgenic Research* 6, 415-420.

Hanocq, E., Boichard, D. and Foulley, J.L. (1996) A simulation study of the effect of connectedness on genetic trend. *Genetics Selection Evolution* 28, 67-82.

Hanover, J.W. (1966a) Inheritance of 3-carene concentration in *Pinus monticola*. *Forest Science* 12, 447-450.

Hanover, J.W. (1966b) Genetics of terpenes. I. Gene control of monoterpene levels in *Pinus monticola* Doug. *Heredity* 21, 73-84.

Hanover, J.W. (1992) Applications of terpene analysis in forest genetics. *New Forests* 6, 159-178.

Hanson,W.D. (1963). Heritability. In: Hanson,W.D. and Robinson, H.F. (eds.) *Statistical Genetics and Plant Breeding*. NAS-NRC Publication 982, Washington, D.C., pp. 125-140.

Harbard, J.L., Griffin, A.R., and Espejo, J. (1999) Mass controlled pollination of *Eucalyptus globulus* - a practical reality. *Canadian Journal of Forest Research* 29, 1457-1463

Harcourt, R.L., Kyozuka, J., Floyd, R.B., Bateman, K.S., Tanaka, H., Decroocq, V., Llewellyn, D.J., Zhu, X., Peacock, W.J. and Dennis, E.S. (2000) Insect-and herbicide-resistant transgenic eucalypts. *Molecular Breeding* 6, 307-315.

Hardner, C.M. and Potts, B.M. (1995) Inbreeding depression and changes in variation after selfing in *Eucalyptus globulus* ssp. *globulus*. *Silvae Genetica* 44, 46-54.

Hardner, C.M., Vaillancourt, R.E. and Potts, B.M. (1996) Stand density influences outcrossing rate and growth of open-pollinated families of *Eucalyptus globulus*. *Silvae Genetica* 45, 226-228.

Hare, R.C. and Switzer, G.L. (1969) Introgression with shortleaf pine may explain rust resistance in western loblolly pine. USDA Forest Service Research Note SO-88.

Harry, D.E., Temesgen, B. and Neale, D.B. (1998) Codominant PCR-based markers for *Pinus taeda* developed from mapped cDNA clones. *Theoretical and Applied Genetics* 97, 327-336.

Hart, J. (1987) A cladistic analysis of conifers: preliminary results. *Journal of the Arnold Arboretum* 68, 269-307.

Hart, J. (1988) Rust fungi and host plant coevolution: do primitive hosts harbor primitive parasites?

Cladistics 4, 339-366.

Hartl, D.L. (2000) *A Primer of Population Genetics*. Sinauer Associates Inc., Publishers, Sunderland, MA.

Hartl, D.L. and Clark, G.A. (1989) *Principles of Population Genetics*. Sinauer Associates Inc., Publishers, Sunderland, MA.

Hatcher, A.V. and Weir, R.J. (1981) Design and layout of advanced generation seed orchards. In: *Proceedings of the 16th Southern Forest Tree Improvement Conference*. Blacksburg, VA, pp. 205-212.

Hattemer, H.H. and Melchior, G.H. (1993) Genetics and its application to tropical forestry. In: Pancel, L. (ed.) *Tropical Forestry Handbook, Vol. 1*. Springler Verlag, New York, N.Y, pp. 333-380.

Haughn, G.W., Smith, J., Mazur, B. and Sommerville, C. (1988) Transformation with a mutant *Arabidopsis* acetolatesynthase gene renders tobacco resistant to sulfonylurea herbicide. *Molecular and General Genetics* 211, 266-271.

Hayes, B.J. and Miller, S. (2000) Mate selection strategies to exploit across- and within-breed dominance variation. *Journal of Animal Breeding and Genetics* 117, 347-359.

Hazel, L.N. (1943) The genetic basis for constructing selection indexes. *Genetics* 28, 476-490.

Heaman, J.C. (1986) A breeding program in coastal Douglas-fir, 1983-1985. In: *Proceedings of the 20th Meeting of the Canadian Tree Improvement Association*. Quebec City, Canada, pp. 186-188.

Heath, L., Ramakrishnan, N., Sederoff, R., Whetten, R., Chevone, B., Struble, C., Jouenne, V., Chen, D., van Zyl, L., and Grene, R. (2002) Studying the functional genomics of stress responses in loblolly pine with the Expresso microarray experiment management system. *Comparative and Functional Genomics* 3, 226-243.

Hedrick, P.A. (1985) *Genetics of Populations*. Jones & Bartlett Publishers, Boston, MA.

Hedrick, P.W., Ginevan, M.E. and Ewing, E.P. (1976) Genetic polymorphism in heterogeneous environments. *Annual Review of Ecology and Systematics* 7, 1-32.

Helms, John A. (ed.) (1998) *The Dictionary of Forestry*. Society of American Foresters, Bethesda, MD.

Henderson, C.R. (1949) Estimation of changes in herd environment. *Journal of Dairy Science* 32, 709.

Henderson, C.R. (1950) Estimation of genetic parameters. *Annals of Mathematical Statistics* 21, 309.

Henderson, C.R. (1963) Selection index and expected genetic advance. In: Hanson, W.D. and Robinson, H.F. (eds.) *Statistical Genetics and Plant Breeding. National Academy of Sciences-National Research Council (NAS-NRC) Publication No. 982*. Washington, DC, pp. 141-163.

Henderson, C.R. (1974) General flexibility of linear model techniques for sire evaluation. *Journal of Dairy Science* 57, 963-972.

Henderson, C.R. (1975) Best linear unbiased estimation and prediction under a selection model. *Biometrics* 31, 423-447.

Henderson, C.R. (1976) A simple method for computing the inverse of a numerator relationship matrix used in prediction of breeding values. *Biometrics* 32, 69-83.

Henderson, C.R. (1984) *Applications of Linear Models in Animal Breeding*. University of Guelph, Guelph, Ontario, Canada.

Hermann, R.K. and Lavender. D.P. (1968) Early growth of Douglas-fir from various altitudes and aspects in southern Oregon. *Silvae Genetica* 17, 143-151.

Hertzberg, M., Aspeborg, H., Schrader, J., Andersson, A., Erlandsson, R., Blomqvist, K., Bhalerao, R., Uhlen, M., Teeri, T., Lundeberg, J., Sundberg, B., Nilsson, P., and Sandberg, G. (2001) A transcriptional roadmap to wood formation. *Proceedings of the National Academy of Sciences of the United States of America* 98, 14732-14737.

Higuchi, T. (1985) Biosynthesis of lignin. In: Higuchi, T. (ed.) *Biosynthesis and Biodegradation of Wood Components*. Academic Press, New York, NY, pp. 141-160.

Hill, W.G. (1984) On selection among groups with heterogeneous variance. *Animal Production* 39, 473-477.

Hirayoshi, I. and Nakamura, Y. (1943) Chromosome number of *Sequoia sempervirens*. *Botanische Zoologie* 11, 73-75.

Hizume, M. and Akiyama, M. (1992) Sized variation of chromomycin A3-band in chromosomes of

Douglas fir, *Psuedotsuga menziesii. Japanese Journal of Genetics* 67, 425-435.

Hizume, M., Ishida, F. and Murata, M. (1992) Multiple locations of the rRNA genes in chromosomes of pines, *Pinus densiflora* and *P. thunbergii. Japanese Journal of Genetics* 67, 389-396.

Ho, R.H. (1988a) Gibberellin A4/7 enhances seed-cone production in field-grown Black spruce. *Canadian Journal of Forest Research* 18, 139-142.

Ho, R.H. (1988b) Promotion of cone production on White spruce grafts by Gibberellin A4/7 application. *Forest Ecology and Management* 23, 39-46.

Hobbs, S.L.A., Warkenstein, T.D. and Delong, C.M.O. (1993) Transgene copy number can be positively or negatively associated with transgene expression. *Plant Molecular Biology* 21, 17-26.

Hodge, G.R. (1985) Parent vs. offspring selection: A case study. In: *Proceedings of the 18th Southern Forest Tree Improvement Conference*. Long Beach, MS, 145-154.

Hodge, G.R. (1997) Selection procedures with overlapping generations. In: *Proceedings of the International Union of Forest Research Organizations (IUFRO), Conference on the Genetics of Radiata Pine. Forest Research Institute (FRI) Bulletin No. 203*, Rotorua, New Zealand, pp. 199-206.

Hodge, G.R. and White, T.L.. (1992a) Concepts of selection and gain prediction. In: Fins, L., Friedman, S. and Brotschol. J. (eds.) *Handbook of Quantitative Forest Genetics*. Kluwer Academic Publishers, Dordrecht, The Netherlands, pp. 140-194.

Hodge, G.R. and White, T.L. (1992b) Genetic parameter estimates for growth traits at different ages in slash pine and some implications for breeding. *Silvae Genetica* 41, 252-262.

Hodge, G.R. and White, T.L. (1993) Advanced-generation wind-pollinated seed orchard design. *New Forests* 7, 213-236.

Hodge, G.R., White, T.L., De Souza, S.M. and Powell, G.L. (1989) Predicted genetic gains from one generation of slash pine tree improvement. *Southern Journal of Applied Forestry* 13, 51-56.

Hodge, G.R., Schmidt, R.A. and White, T.L. (1990) Substantial realized gains from mass selection of fusiform rust-free trees in highly-infected stands of slash pine. *Southern Journal of Applied Forestry* 14, 143-146.

Hodge, G.R., Volker, P.W., Potts, B.M. and Owen, J.V. (1996) A comparison of genetic information from open-pollinated and control-pollinated progeny tests in two eucalypts species. *Theoretical and Applied Genetics* 92, 53-63.

Hofer, A. and Kennedy, B.W. (1993) Genetic evaluation for a quantitative trait controlled by polygenes and a major locus with genotypes not or only partly known. *Genetics Selection Evolution* 25, 537-555.

Hohenboken, W.D. (1985) Phenotypic, genetic and environmental correlations. In: Chapman, A.B. (ed.) *General and Quantitative Genetics*. Elsevier Science Publisher, NewYork, NY, pp. 121-134.

Horsch, R.E., Fry, J.R. Hoffmann, N.L., Eichholtz, D., Rogers, S.G. and Fraley, R.T. (1985) A simple and general method for transferring genes into plants. *Science* 227, 1229-1231.

Hotta, Y. and Miksche, J. (1974) Ribosomal-RNA genes in four coniferous species. *Cell Differentiation* 2, 299-305.

Howe, G., Saruul, P., Davis, J. and Chen, T. (2000) Quantitative genetics of bud phenology, frost damage, and winter survival in an F_2 family of hybrid poplars. *Theoretical and Applied Genetics* 101, 632-342.

Hu, W.-J., Kawaoka, A., Tsai, C.J., Lung, J., Osakabe, K., Ebinuma, H. and Chiang, V.L. (1998) Compartmentalized expression of two structurally and functionally distinct 4-coumarate: CoA ligase genes in aspen (*Populus tremuloides*). *Proceedings of the National Academy of Sciences of the United States of America* 95, 5407-5412.

Hu, W.-J., Harding, S.A., Lung, J., Popko, J.L., Ralph, J., Stokke, D.D., Tsai, C.-J., and Chiang, V.L. (1999) Repression of lignin biosynthesis promotes cellulose accumulation and growth in transgenic trees. *Nature Biotechnology* 17, 808-812.

Huber, D.A. (1993) *Optimal Mating Designs and Optimal Techniques for Analysis of Quantitative Traits in Forest Genetics*. Department of Forestry, University of Florida, Gainesville, Florida.

Huber, D.A., White, T.L. and Hodge, G.R. (1992) Efficiency of half-sib, half-diallel and circular mating designs in the estimation of genetic parameters in forestry: A simulation. *Forest Science* 38, 757-776.

Huber, D.A., White, T.L. and Hodge, G.R. (1994) Variance component estimation techniques compared for two mating designs with forest genetic architecture through computer simulation. *Theoretical and Applied Genetics* 88, 236-242.

Huehn, M. (1987) Clonal mixtures, juvenile-mature correlations and necessary number of clones. *Silvae Genetica* 36, 83-92.

Huehn, M. (1988) Multiclonal mixtures and number of clones. *Silvae Genetica* 37, 67-73.

Huhn, M. (1992a) Multiclonal mixtures and number of clones: II. Number of clones and yield stability (deterministic approach with competition). *Silvae Genetica* 41, 205-213.

Huhn, M. (1992b) Theoretical studies on the number of components in mixtures. III. Number of components and risk considerations. *Theoretical and Applied Genetics* 72, 211-218.

Hurme, P., Sillanpaa, E., Arjas, E., Repo, T. and Savolainen, O. (2000) Genetic basis of climatic adaptation in Scots pine by Bayesian quantitative trait analyses. *Genetics* 156, 1309-1322.

Husband, B.C. and Schemske, D.W. (1996) Evolution of the magnitude and timing of inbreeding depression in plants. *Evolution* 50, 54-70.

IHGSC (International Human Genome Sequencing Consortium) (2001) Initial sequencing and analysis of the human genome. *Nature* 409, 860-921.

Ikemori, Y. (1990) Genetic variation in characteristics of *Eucalyptus grandis* (Hill) maiden raised from micro-propagation, macro-propagation and seed. Ph.D. Dissertation, Green College, Oxford University, UK.

IPCC (Intergovernmental Panel on Climate Change) (2001) Climate change 2000: Impacts, adaptation, and vulnerability: Contribution of working group II. In: McCarthy, J.J., Canziani, O.F., Leary, N.A., Dokken, D.J. and White, K.S. (eds.) *Third Assessment Report, Intergovernmental Panel on Climate Change.* Cambridge University Press, UK.

Isaac, L.A. (1930) Seed flight in the Douglas fir region. *Journal of Forestry* 28, 492-499.

Isabel, N., Beaulieu, J. and Bousquet, J. (1995) Complete congruence between gene diversity estimates derived from genotypic data at enzyme and random amplified polymorphic DNA loci in black spruce. *Proceedings of the National Academy of Sciences of the United States of America* 99, 6369-6373.

IUCN (The World Conservation Union) (1994) *Guidelines for Protected Area Management Categories.* The World Conservation Union's Commission on National Parks and Protected Areas (**CNPPA**) with the assistance of The World Conservation Monitoring Centre (WCMC). IUCN, Gland, Switzerland and Cambridge, UK.

Jackson, N. and Turner, H.N. (1972) Optimal structure for a co-operative nucleus breeding system. In: Proceedings of the Australian Society of Animal Production 9, 55-64.

Jain, S.M. and Minocha, S.C. (2000) *Molecular Biology of Woody Plants, Forestry Sciences, Volume 64.* Kluwer Academic Publishers, Dordrecht, The Netherlands.

James, C. (2004) The global status of commercialized Biotech/GM Crops: (2004) The International Service for the Acquisition of Agri-biotech Applications (ISAAA) Brief # 32. http://www.isaaa.org

James, J.W. (1972) Optimum selection intensity in breeding programmes. *Animal Production* 14, 1-9.

James, J.W. (1977) Open nucleus breeding systems. *Animal Production* 24, 287-305.

Jansson, S. and Gustafsson, P. (1990) Type I and Type II genes for the chlorophyll a/b-binding protein in the gymnosperm *Pinus sylvestris* (Scots pine): cDNA cloning and sequence analysis. *Plant Molecular Biology* 14, 287-296.

Jansson, S. and Gustafsson, P. (1991) Evolutionary conservation of the chlorophyll a/b-binding proteins: cDNA's encoding Type I, II, and III LHC I polypeptides from the gymnosperm Scots pine. *Molecular and General Genetics* 229, 67-76.

Jansson, S. and Gustafsson, P. (1994) Characterization of a Lhcb5 cDNA from Scots Pine (*Pinus sylvestris*). *Plant Physiology* 106, 1695-1696.

Janzen, D.H. (1971) Euglossine bees as long-distance pollinators of tropical plants. *Science* 171, 203-205.

Jarvis, S.F., Borralho, N.M.G. and Potts, B.M. (1995) Implementation of a multivariate BLUP model for genetic evaluation of *Eucalyptus globulus* in Australia. In: Potts, B.M., Borralho, N.M.G., Reid, J.B., Cromer, R.N., Tibbits, W.N. and Raymond, C.A. (eds.) *Proceedings of the International Union of Forest Research Organizations (IUFRO), Symposium on Eucalypt Plantations: Improving Fibre Yield and Quality.* Hobart, Australia, pp. 212-216.

Jayawickrama, K.J. and Carson, M.J. (2000) A breeding strategy for New Zealand Radiata Pine Breeding Cooperative. *Silvae Genetica* 49, 82-90.

Jayawickrama, K.J.S., Carson, M.J., Jefferson, P.A. and Firth. A. (1997) Development of the New Zealand radiata pine breeding population. In: Burdon, R.D. and Moore, J.M. (eds.) *Proceedings of the International Union of Forest Research Organizations (IUFRO), Symposium on Genetics of Radiata Pine*. Rotorua, New Zealand, pp. 217-225.

Jeffreys, A.J., Wilson, V. and Thein, S.L. (1985a) Hypervariable 'minisatellite' regions in human DNA. *Nature* 314, 67-73.

Jeffreys, A.J., Wilson, V. and Thein, S.L. (1985b) Individual-specific 'fingerprints' of human DNA. *Nature* 316, 76-79.

Jermstad, K.D., Reem, A.M., Henifin, J.R., Wheeler, N.C. and Neale, D.B. (1994) Inheritance of restriction fragment length polymorphisms and random amplified polymorphic DNAs in coastal Douglas-fir. *Theoretical and Applied Genetics* 89, 758-766.

Jermstad, K., Bassoni, D., Wheeler, N. and Neale, D. (1998) A sex-averaged genetic linkage map in coastal Douglas-fir (*Pseudotsuga menziesii* [Mirb.] Franco var '*menziesii*') based on RFLP and RAPD markers. *Theoretical and Applied Genetics* 97, 762-770.

Jermstad, K., Bassoni, D., Jech, K., Ritchie, G., Wheeler, N. and Neale, D. (2001) Mapping of quantitative trait loci controlling adaptive traits in coastal Douglas-fir. I. Timing of vegetative bud flush. *Theoretical and Applied Genetics* 102, 1142-1151.

Jermstad, K.D., Bassoni, D.L., Jech, K.S., Ritchie, G.A., Wheeler, N.C. and Neale, D.B. (2003) Mapping of quantitative trait loci controlling adaptive traits in coastal Douglas-fir. III. QTL by environment interactions. *Genetics* 165, 1489-1506.

Jett, J.B. (1986) Reaching full production: A review of seed orchard management in the southeastern United States. In: *Proceeedings of the International Union of Forest Research Organizations (IUFRO), Conference on Breeding Theory, Progeny Testing and Seed Orchards*. Williamsburg, VA, pp. 34-58.

Jett, J.B. (1987) Seed orchard management: Something old and something new. In: *Proceedings of the 19th Southern Forest Tree Improvement Conference*. College Station, TX, pp. 160-171.

Jett, J.B. and Talbert, J.T. (1982) The place of wood specific gravity in development of advanced generation seed orchards. *Southern Journal of Applied Forestry* 6, 177-180.

John, J.A. and Williams, E.R. (1995) *Cyclic and Computer Generated Designs*. Chapman & Hall. London, UK.

Johnson, G.R. (1997) Site-to-site genetic correlations and their implications on breeding zone size and optimum number of progeny test sites for coastal Douglas-fir. *Silvae Genetica* 46, 280-285.

Johnson, G.R. and Burdon, R.D. (1990) Family-site interaction in *Pinus radiata*: Implications for progeny testing strategy and regionalized breeding in New Zealand. *Silvae Genetica* 39, 55-62.

Johnson, G.R., Sniezko, R.A. and Mandel, N.L. (1997) Age trends in Douglas-fir genetic parameters and implications for optimum selection age. *Silvae Genetica* 349-358.

Johnson, G., Wheeler, N. and Strauss, S. (2000) Financial feasibility of marker-aided selection in Douglas-fir. *Canadian Journal of Forest Research* 30, 1942-1952.

Johnson, R. (1998) *Breeding Design Considerations for Coastal Douglas-fir.* USDA Forest Service General Technical Report PNW-GTR 411.

Judd, W., Campbell, C., Kellogg, E. and Stevens, P. (1999) Plant systematics: a phylogenetic approach. Sinauer Associates, Inc., Publishers, Sunderland, MA.

Kamm, A., Doudrick, R.L., Heslop-Harrison, J.S. and Schmidt, T. (1996) The genomic and physical organization of Ty1-copia-like sequences as a component of large genomes in *Pinus elliottii* var. *elliottii* and other gymnosperms. *Proceedings of the National Academy of Sciences of the United States of America* 93, 2708-2713.

Kang, H. (1979a) Long-term tree breeding. In: *Proceedings of the 15th Southern Forest Tree Improvement Conference*. Mississippi State University, Starkeville, MS, pp. 66-72.

Kang, H. (1979b) Designing a tree breeding system. In: *Proceedings of the 17th Meeting of the Canadian Tree Improvement Association*. Gander, Newfoundland, pp. 51-66.

Kang, H. (1991) Components of juvenile-mature correlations in forest trees. *Theoretical and Applied Genetics* 81, 173-184.

Kang, H. and Namkoong, G. (1979) Limits of artificial selection under balanced mating systems. *Silvae Genetica* 28, 53-60.

Kang, II. and Namkoong, G. (1980) Limits of artificial selection under unbalanced mating systems. *Theoretical and Applied Genetics* 58, 181-191.

Kang, H. and Namkoong, G. (1988) Inbreeding effective population size under some artificial selection schemes. I. Linear distribution of breeding values. *Theoretical and Applied Genetics* 75, 333-339.

Kang, H. and Nienstaedt, H. (1987) Managing long-term tree breeding stock. *Silvae Genetica* 36, 30-39.

Kannenberg, L.W. (1983) Utilization of genetic diversity in crop breeding. In: Yeatman, C.W., Kafton, D. and Wilkes, G. (eds.) *Plant Gene Resources: A Conservation Imperative. AAAS (American Association for the Advancement of Science) Selected Symposium 87.* Westview Press, Boulder, CO, pp 93-111.

Kanowski, P.J. (1996) Sustaining tropical forestry: tree improvement in the biological and social context. In: Dieters, M.J., Matheson, A.C., D.G. Nikles, D.G., Harwood, C.E. and Walker, S.M. (eds.) *Proceedings of the Queensland Forest Research Institute-International Union of Forest Research Organizations (QFRI-IUFRO), Conference on Tree Improvement for Sustainable Tropical Forestry.* Caloundra, Queensland, Australia, pp. 295-300.

Kanowski, P. J. (2000) Politics, policies and the conservation of forest genetic diversity. In: Young, A., Boshier, D. and Boyle, T. (eds.) *Forest Conservation Genetics: Principles and Practice.* Commonwealth Scientific and Industrial Research Organization (CSIRO) Publishing, Collingwood, Victoria, Australia, pp. 275-287.

Kanowski, P.J. and Nikles, D.G. (1989) A summary of plans for continuing genetic improvement of *Pinus caribaea* var. *hondurensis* in Queensland. In: Gibson, G.I., Griffin, A.R., and Matheson, A.C. (eds.) *Breeding Tropical Trees: Population Structure and Genetic Improvement Strategies in Clonal Seedling Forestry.* Oxford Forestry Institute, Oxford, UK, pp. 236-349.

Kanowski, P.J., Savill, P.S., Adlard, P.G., Burley, J., Evans, J., Palmer, J.R. and Wood, P.J. (1992) Plantation forestry. In: Sharma, N. (ed.) *Managing the World's Forests.* Kendall/Hunt, Dubuque, IA, pp. 375-401.

Karhu, A., Hurme, P., Karjalainen, M., Karvonen, P., Karkkainen, K., Neale, D.B. and Sovalainen, O. (1996) Do molecular markers reflect patterns of differentiation in adaptive traits of conifers? *Theoretical and Applied Genetics* 93, 215-221.

Kärkkäinen, K., Koski, V. and Savolainen, O. (1996) Geographical variation in the inbreeding depression of Scots pine. *Evolution* 50, 111-119.

Karnosky, D.F. (1977) Evidence for genetic control of response to sulphur dioxide and ozone in *Populus tremuloides. Canadian Journal of Forest Research* 7, 437-440.

Kaya, Z., Ching, K.K., and Stafford, S.G. (1985) A statistical analysis of karyotypes of European black pine (*Pinus nigra* Arnold) from different sources. *Silvae Genetica* 34, 148-156.

Kaya, Z., Sewell, M. and Neale, D. (1999) Identification of quantitative trait loci influencing annual height and diameter increment growth in loblolly pine (*Pinus taeda* L.). *Theoretical and Applied Genetics* 98, 586-592.

Kearsey, M.J. and Farquar, A.G. (1998) QTL analysis in plants: Where are we now? *Heredity* 80, 137-142.

Keim, P., Paige, K.N., Whitham, T.G. and Lark, K.G. (1989) Genetic analysis of an interspecific hybrid swarm of *Populus*: occurrence of unidirectional introgression. *Genetics* 123, 557-565.

Kennedy, B.W. and Sorensen, D.A. (1988) Properties of mixed-model methods for prediction of genetic merit. In: Weir, B.S., Eisen, E.J., Goodman, M.M. and Namkoong, G. (eds.) *Proceedings of the 2nd International Conference on Quantitative Genetics.* Sinauer Associates Inc., Publishers, Sunderland, MA, pp. 91-103.

Kennedy, B.W., Quinton, M. and van Arendonk, J.A.M. (1992) Estimation of effects of single genes on quantitative traits. *Journal of Animal Science* 70, 2000-2012.

Kephart, S.R. (1990) Starch gel electrophoresis of plant isozymes: a comparative analysis of techniques. *American Journal of Botany* 77, 693-712.

Kerr, R.J. and Goddard, M.E. (1997) A comparison between the use of MAS and clonal tests in tree breeding programmes. In: Burdon, R.D. and Moore, J.M. (eds.) *Proceedings of the International Union of Forest Research Organizations (IUFRO), Conference on the Genetics of Radiata Pine.* Rotorua, New Zealand, pp. 297-303.

Kerr, R.J., Jarvis, S.F. and Goddard, M.E. (1996) The use of genetic markers in tree breeding programs. In: Dieters, M.J. Matheson, A.C. , Nickles, D.G., Hardwood, C.E. and Walker, S.M. (eds.) *Proceedings of the QueenslandForest Research Institute-International Union of Forest Research Organizations (QFRI-IUFRO), Conference on Tree Improvement for Sustainable Tropical Forestry*. Caloundra, Queensland, Australia, pp. 498-505.

Kerr, R.J., McRae, T.A., Dutkowski, G.W., Apiolaza, L.A. and Tier, B. (2001) Treeplan - a genetic evaluation system for forest tree improvement. In: *Proceedings of the International Union of Forest Research Organizations (IUFRO), Conference on Developing the Eucalypt of the Future*. Valdivia, Chile, pp. 6.

Kerr, R.J., Dieters, M.J., Tier, B. and Dungey, H.S. (2004a) Simulation of forest tree breeding strategies. *Canadian Journal of Forest Research* 34, 195-208.

Kerr, R.J., Dieters, M.J. and Tier, B. (2004b) Simulation of the comparative gains from four different hybrid tree breeding strategies. *Canadian Journal of Forest Research* 34, 209-220.

Kertadikara, A.W.S. and Prat, D. (1995) Genetic structure and mating system in teak (*Tectona grandis* L.f.) provenances. *Silvae Genetica* 44, 104-110.

Khasa, P.D., Vallee, G., Li, P., Magnussen, S., Camire, C. and Bousquet, J. (1995) Performance of five tropical tree species on four sites in Zaire. *Commonwealth Forestry Review* 74, 129-137.

Khoshoo, T.N. (1959) Polyploidy in gymnosperms. *Evolution* 13, 24-39.

Khoshoo, T.N. (1961) Chromosome numbers in gymnosperms. *Silvae Genetica* 10(1), 1-9.

Khurana, D.K. and Khosla, P.K. (1998) Hybrids in forest tree improvement. In: Mandal, A.K. and Gibson, G.L. (eds.) *Forest Genetics and Tree Breeding*. CBS Publishers and Distributors, Darya Ganj, New Delhi, India, pp. 86-102.

Kim, M.-S., Klopfenstein, N.B. and Chun, Y.W. (1997) *Agrobacterium*-mediated transformation of *Populus* species. In: Klopfenstein, N.B., Chun, Y.W., Kim, M.-S. and Ahuja, M.R. (eds.) *Micropropagation, Genetic Engineering and Molecular Biology of Populus*. USDA Forest Servic General Technical Report RM-GTR-297, pp. 51-59.

King, J.N. and Burdon, R.D. (1991) Time trends in inheritance and projected efficiencies of early selection in a large 17-year-old progeny test of *Pinus radiata*. *Canadian Journal of Forest Research* 21, 1200-1207.

King, J. and Cartwright, C. (1995) Western hemlock breeding program. In: Lavereau, J. (ed.) *Proceedings of the 25th Canadian Tree Improvement Association, Part I. CTIA/WFGA Conference on Evolution and Tree Breeding: Advances in Quantitative and Molecular Genetics for Population Improvement*. Victoria, British Columbia, Canada, pp 16-17.

King, J.N. and Johnson, G.R. (1996) Monte Carlo simulation models of breeding-population advancement. *Silvae Genetica* 42, 68-78.

King, J.N. and Wilcox, M.D. (1988) Family tests as a basis for the genetic improvement of *Eucalyptus nitens* in New Zealand. *New Zealand Journal of Forest Science* 18, 253-266.

King, J.N., Yeh, F.C., Heaman, J. Ch. and Dancik, B.P. (1988) Selection of wood density and diameter in controlled crosses of coastal Douglas-fir. *Silvae Genetica* 37, 152-157.

Kinghorn, B.P. (2000) Crossbreeding strategies to maximise economic returns. In: *Hybrid Breeding and Genetics of Forest Trees. Proceedings of the Queensland Forest Research Institute/Cooperative Research Center-Sustainable Production Forestry (QFRI/CRC-SPF) Symposium*. Noosa, Queensland, Australia, pp. 291-302.

Kinghorn, B.P., Kennedy, B.W., and Smith, C. (1993) A method for screening for genes of major effect. *Genetics* 134, 351-360.

Kinlaw, C. and Neale, D. (1997) Complex gene families in pine genomes. *Elsevier Trends Journals* 2(9), 356-359.

Kinloch, B.B., Parks, G.K. and Fowler, C.W. (1970) White pine blister rust: Simply inherited resistance in sugar pine. *Science* 167, 193-195.

Kinloch, B.B., Westfall, R.D. and Forrest, G.I. (1986) Caledonian Scots pine: origins and genetic structure. *The New Phytologist* 104, 703-729.

Kirst, M., Jonhson, A., Retzel, E., Whetten, R., Vasques-Kool, J., O'Malley, D., Baucom, C., Bonner, E., Hubbard, K. and Sederoff, R. (2003) Apparent homology of expressed genes in loblolly pine (*Pinus taeda* L.) with *Arabidopsis thaliana*. *Proceedings of the National Academy of Science of the United States of America* 100, 7383-7388.

Kirst, M., Cordeiro, C. M., Rezende, G.D.S.P. and Grattapaglia. D. (2005) Power of microsatellite markers for fingerprinting and parentage analysis in *Eucalyptus grandis* breeding populations.

Journal of Heredity 96, 1-6.

Kjaer, E.D. (1996) Estimation of effective population number in a *Picea abies* (Karst.) seed orchard based on flower assessment. *Scandinavian Journal of Forest Research* 11, 111-121.

Klein, J.I. (1987) Selection and mating strategies in second generation breeding populations of conifer tree improvement programs. *In: Proceedings of the 21st Tree Improvement Association.* Truro, Nova Scotia, pp. 170-180.

Klein, T.M., Wolf, E.D., Wu, R. and Sanford, J.C. (1987) High-velocity microprojectiles for delivering nucleic acids into living cells. *Nature* 327, 70-73.

Kleinschmit, J. (1978) Sitka spruce in Germany. *Proceedings of the International Union of Forest Research Organizations (IUFRO), Joint Meeting of Working Parties, Volume 2.* British Columbia Ministry of Forests, Information Services Branch, Victoria, BC, Canada, pp. 183-191.

Kleinschmit, J. (1979) Limitations for restriction of the genetic variation. *Silvae Genetica* 28, 61-67.

Kleinschmit, J. and Bastien, J.C. (1992) The International Union of Forest Research Organization's (IUFRO) role in Douglas-fir (*Pseudotsuga menziesii* [Mirb] Franco) tree improvement. *Silvae Genetica* 41, 161-173.

Kleinschmit, J., Khurana, D.K., Gerhold, H.D. and Libby, W.J. (1993) Past, present, and anticipated applications of clonal forestry. In: Ahuja, M.R. and Libby, W.J. (eds.) *Clonal Forestry II. Conservation and Application.* Springer-Verlag, New York, NY, pp. 9-41.

Klekowski, E.J., Jr. (1992) Mutation rates in diploid annuals - are they immutable? *International Journal of Plant Science* 153, 462-265.

Klekowski, E.J. and Godfrey, P.J. (1989) Aging and mutation in plants. *Nature* 340, 389-391.

Klekowski, E.J., Jr., Lowenfeld, R. and Hepler, P.K. (1994) Mangrove genetics. II. Outcrossing and lower spontaneous mutation rates in Puerto Rican *Rhizophora*. *International Journal of Plant Science* 155, 373-381.

Klimaszewska, K., Devantier, V., Lachance, D., Lelu, M.A. and Charest, P.J. (1997) *Larix laricina* (tamarack): somatic embryogenesis and genetic transformation. *Canadian Journal of Forest Research* 27, 538-550.

Klopfenstein, N.B., Shi, N.Q., Kernan, A., McNabb, H.S., Jr., Hall, R.B., Hart, E.R. and Thornburg, R.W. (1991) Transgenic hybrid poplar expresses a wound-inducing potato proteinase inhibitor II. – CAT gene fusion. *Canadian Journal of Forest Research* 21:1321-1328.

Klopfenstein, N.B., Allen, K.K., Avila, F.J., Heuchelin, S.A., Martinez, J., Carman, R.C., Hall, R.B., Hart, E.R. and McNabb, H.S. (1997) Proteinase inhibitor II gene in transgenic poplar: chemical and biological assays. *Biomass and Bioenergy* 12:299-311.

Knott, S., Neale, D., Sewell, M. and Haley, C. (1997) Multiple marker mapping of quantitative trait loci in an outbred pedigree of loblolly pine. *Theoretical and Applied Genetics* 94, 810-820.

Knowles, P. (1991) Spatial genetic structure within two natural stands of black spruce (*Picea mariana* (Mill.) B.S.P.). *Silvae Genetica* 40, 13-19.

Knowles, P. and Grant, M.C. (1985) Genetic variation of lodgepole pine over time and microgeographical space. *Canadian Journal of Botany* 63, 722-727.

Knowles, P., Furnier, G.R., Aleksiuk, M.A. and Perry, D.J. (1987) Significant levels of self-fertilization in natural populations of tamarack. *Canadian Journal of Botany* 65, 1087-1091.

Knowles, P., Perry, D.J. and Foster, H.A. (1992) Spatial genetic structure in two tamarack (*Larix laricina* (Du Roi) K. Koch) populations with differing establishment histories. *Evolution* 46, 572-576.

Koebner, R. and Summers, R. (2002) The impact of molecular markers on the wheat breeding paradigm. *Cellular and Molecular Biology Letters* 7, 695-702.

Kojima, K., Yamamoto, N. and Sasaki, S. (1992) Structure of the pine (*Pinus thunbergii*) chlorophyll a/b-binding protein gene expressed in the absence of light. *Plant Molecular Biology* 19, 405-410.

Komulainen, P., Brown, G.R., Mikkonen, M., Karhu, A., Garcia-Gil M.R., O'Malley, D., Lee, B., Neale, D.B. and Savolainen, O. (2003) Comparing EST-based genetic maps between *Pinus sylvestris* and *P. taeda. Theoretical and Applied Genetics* 107, 667-678.

Kondo, T., Tsumura, Y., Kawahara, T. and Okamura, M. (1998) Paternal inheritance of chloroplast and mitochondrial DNA in interspecific hybrids of *Chamaecyparis* spp. *Breeding Science* 48, 177-179.

Koski, V. (1970) A Study of Pollen Dispersal as a Mechanism of Gene Flow in Conifers, *Communi-*

cationes Instituti Forestalis Fenniae 70.4, Helsinki, Finland.

Koski, V. and Malmivaara, E. (1974) The role of self-fertilization in a marginal population of *Picea abies* and *Pinus sylvestris*. *Proceedings of the International Union of Forest Research Organizations (IUFRO), Joint Meeting of Working Parties on Population and Ecological Genetics, Breeding Theory and Progeny Testing*, 5.02.OY, Stockholm, Sweden. pp. 55-166.

Kossack, D.S. and Kinlaw, C.S. (1999) IFG, a gypsy-like retrotransposon in *Pinus* (Pinaceae), has an extensive history in pines. *Plant Molecular Biology* 39, 417-426.

Kostia, S., Varvio, S.L., Vakkari, P. and Pulkkinen, P. (1995) Microsatellite sequences in a conifer, *Pinus sylvestris*. *Genome* 38, 1244-1248.

Kozlowski, T.T. and Pallardy, S.G. (1979) *Physiology of Woody Plants*. Academic Press, London, UK.

Kremer, A., Petit, R., Zanetto, A., Fougere, V., Ducousso, A., Wagran, D. and Chauvin, C. (1991) Nuclear and organelle gene diversity in *Quercus robur* and *Q. Petraea*. In: Muller-Starck, G. and Ziehe, M. (eds.) *Genetic Variation in European Populations of Forest Trees*. Sauerlander's Verlag, Frankfurt, Germany, pp. 141-166.

Kriebel, H. (1985) DNA sequence components of the *Pinus strobus* nuclear genome. *Canadian Journal of Forest Research* 15, 1-4.

Krietman, M. (1996) The neutral theory is dead. Long live the neutral theory. *BioEssays* 18, 678-683.

Krugman, S.L., Stein, W.L. and Schmitt, D.M. (1974) Seed biology. In: Shopmeyer, C.S. (Technical Coordinator) *Seed of Woody Plants in the United States*. United States Department of Agriculture, Agricultural Handbook 450, pp. 5-40.

Krupkin, A., Liston, A. and Strauss, S. (1996) Phylogenetic analysis of the hard pines (*Pinus* subgenus *Pinus*, Pinaceae) from chloroplast DNA restriction site analysis. *American Journal of Botany* 83, 489-498.

Krutovskii, K.V. and Neale, D. B. (2003) Forest genomics and new molecular genetic approaches to measuring and conserving adaptive genetic diversity in forest trees. In: Geburek, T. and Turok, J. (eds.) *Conservation and Managementof Forest Genetic Resources in Europe*, Arbora Publishers, Zvolen.

Krutovsky K.V., Troggio, M., Brown, G.R., Jermstad, K.D., Neale, D.B. (2004) Comparative Mapping in the Pinaceae. *Genetics* 168, 447-461.

Krusche, D. and Geburek, T. (1991) Conservation of forest gene resources as related to sample size. *Forest Ecology and Management* 40, 145-150.

Krutzsch, P. (1992) The International Union of Forest Research Organization's (IUFRO) role in coniferous tree improvement: Norway spruce. *Silvae Genetica* 41, 143-150.

Kuittinen, H., Muona, O. Kärkkäinen, K. and Borzan, Z. (1991) Serbian spruce, a narrow endemic, contains much genetic variation. *Canadian Journal of Forest Research* 21, 363-367.

Kumar, L.S. (1999) DNA markers in plant improvement: An overview. *Biotechnology Advances* 17, 143-182.

Kumar, S. (2004) Effect of selfing on various economic traits in *Pinus radiata* and some implications for breeding strategy. *Forest Science* 50, 571-578.

Kumar, S. and Fladung, M. (2001) Controlling transgene integration in plants. *Trends in Plant Science* 6, 156-159.

Kumar, S. and Fladung, M. (2004) *Molecular Genetics of Breeding of Forest Trees*. Food Products Press, New York, London, Oxford, 436 pp.

Kumar, S. and Garrick, D. (2001) Genetic response to within-family selection using molecular markers in some radiata pine breeding schemes. *Canadian Journal of Forest Research* 31, 779-785.

Kummerly, W. (1973) *The Forest*. Kummerly and Frey, Geographical Publishers, Berne, Switzerland, pp. 299.

Kundo, S.K. and Tigerstedt, P.M.A. (1997) Geographical variation in seed and seedling traits of neem (*Azadirachta indica* A. Juss) among ten populations studied in growth chamber. *Silvae Genetica* 46, 129-137.

Kuser, J.E. and Ching, K.K. (1980) Provenance variation in phenology and cold hardiness of western hemlock seedlings. *Forest Science* 26, 463-470.

Kuser, J.E. and Ledig, F.T. (1987) Provenance and progeny variation in pitch pine from the Atlantic Coastal Plain. *Forest Science* 33, 558-564.

Kvarnheden, A. and Engstrom, P. (1992) Genetically stable, individual-specific differences in hypervariable DNA in Norway spruce, detected by hybridization to a phage M13 probe. *Canadian Journal of Forest Research* 22, 117-123.

Kvarnheden, A., Tandre, K. and Engstrom, P. (1995) A cdc2 homologue and closely related processed retropseudogenes from Norway spruce. *Plant Molecular Biology* 27, 391-403.

La Bastide, J.G.A. (1967) A computer program for the layouts of seed orchards. *Euphytica* 16, 321-323.

Ladrach, W.E. (1998) Provenance research: The concept, application and achievement. In: Mandal, A.K. and Gibson, G.I. (eds.) *Forest Genetics and Tree Breeding*. CBS Publishers and Distributors, New Delhi, India, pp. 16-37.

La Farge, T. (1993) Realized gains in volume, volume per acre and straightness in unrogued orchards of three southern pine species. In: *Proceedings of the 22nd Southern Forest Tree Improvement Conference*. Atlanta, GA, pp. 183-193.

La Farge, T. and Lewis, R.A. (1987) Phenotypic selection effective in a northern red oak seedling seed orchard. In: *Proceedings of the 19th Southern Forest Tree Improvement Conference*. College Station, TX, pp. 200-207.

Lagercrantz, U. and Ryman, N. (1990) Genetic structure of Norway spruce (*Picea abies*): concordance of morphological and allozymic variation. *Evolution* 44, 38-53.

Lambeth, C.C. (1980) Juvenile-mature correlations in Pinaceae and implications for early selection. *Forest Science* 26, 571-580.

Lambeth, C.C. (1983) Early testing - an overview with emphasis on loblolly pine. In: *Proceedings of the 17th Southern Forest Tree Improvement Conference*. Athens, GA, pp. 297-311.

Lambeth, C.C. and Lopez, J.L. (1994) An *E. grandis* clonal tree improvement program for Carton de Colombia. Smurfit Carton de Colombia Research Report 120, Cali, Colombia.

Lambeth, C.C., van Buijtenen, J.P., Duke, S.D. and McCullough, R.B. (1983a) Early selection is effective in 20-year-old genetic tests of loblolly pine. *Silvae Genetica* 32, 210-215.

Lambeth, C.C., Gladstone, W.T. and Stonecypher, R.W. (1983b) Statistical efficiency of row and noncontiguous family plots in genetic tests of loblolly pine. *Silvae Genetica* 32, 24-28.

Lambeth, C.C., Dougherty, P.M., Gladstone, W.T., McCullough, R.B. and Wells, O.O. (1984) Large-scale planting of North Carolina loblolly pine in Arkansas and Oklahoma: A case of gain versus risk. *Journal of Forestry* 82, 736-741.

Lambeth, C.C., Lopez, J.L. and Easley, D.F. (1989) An accelerated *Eucalyptus grandis* clonal tree improvement programme for the Andes mountains of Colombia. In: Gibson. G.I., Griffin, A.R. and Matheson, A.C. (eds.) *Breeding Tropical Trees: Population Structure and Genetic Improvement Strategies in Clonal and Seedling Forestry*. Oxford Forestry Institute, Oxford, UK, pp. 259-266.

Lambeth, C., Lee, B.-C., O'Malley, D. and Wheeler, N. (2001) Polymix breeding with parental analysis in progeny: an alternative to full-sib breeding and testing. *Theoretical and Applied Genetics* 103, 930-943.

Land, S.B., Bongarten, B.C. and Toliver, J.R. (1987) Genetic parameters and selection indices from provenance/progeny tests. In: *Statistical Considerations in Genetic Testing of Forest Trees. Southern Cooperative Series Bulletin 324*. University of Florida, Gainesville, FL, pp. 59-74.

Lande, R. (1988a) Quantitative genetics and evolutionary theory. In: Weir, B.S., Eisen, E.J., Goodman, M.M. and Namkoong, G. (eds.) *Proceedings of the 2nd International Conference on Quantitative Genetics*. Sinauer Associates Inc., Publishers, Sunderland, Massachusetts, pp. 71-84.

Lande, R. (1988b) Genetics and demography in biological conservation. *Science* 241, 1455-1460.

Lande, R. (1995) Mutation and conservation. *Conservation Biology* 9, 782-791.

Lande, R. and Thompson, R. (1990) Efficiency of marker-assisted selection in the improvement of quantitative traits. *Genetics* 124, 743-756.

Lander, E. and Botstein, D. (1989) Mapping Mendelian factors underlying quantitative traits using RFLP linkage maps. *Genetics* 121, 185-199.

Lander, E.S., Green, P., Abrahamson, J., Barlow, A., Daley, M.J., Lincoln, S.E. and Newburg, L. (1987) MAPMAKER: An interactive computer package for constructing primary genetic linkage maps of experimental and natural populations. *Genomics* 1, 174-181.

Langlet, O. (1959) A cline or not a cline? A case of Scots pine. *Silvae Genetica* 8, 13-22.

Langner, W. (1953) Eine mendelspaltung bei aurea-formen von *Picea abies* (L.) Karst als mittel zur klarung der befruchtungsverhaltnisse im Walde. *Zeitschrift für Forestgenetische Forstpflanzen-*

züchtung 2, 49-51.

Lanner, R.M. (1966) Needed: a new approach to the study of pollen dispersion. *Silvae Genetica* 15, 50-52.

Lanner, R. (1980) Avian seed dispersal as a factor in the ecology and evolution of limber and white-bark pines. In: *Proceedings of the 6th North American Forest Biology Workshop*. University of Alberta, Edmunton, Alberta, Canada, pp.15-48.

Lanner, R. (1982) Adaptions of whitebark pine for seed dispersal by Clark's nutcracker. *Canadian Journal of Forest Research* 12, 391-402.

Lanner, R. (1996) *Made for Each Other: A Symbiosis of Birds and Pines*. Oxford University Press, New York, NY.

Lantz, C.W. and Kraus, J.F. (1987) A guide to southern pine seed sources. USDA Forest Service General Technical Report SE-43, p. 34.

Latta, R.G., Linhart, Y.B., Fleck, D. and Elliot, M. (1998) Direct and indirect estimates of seed versus pollen movement within a population of ponderosa pine. *Evolution* 52, 61-67.

LeCorre, V., Dumolin-Lapegue, S. and Kremer, A. (1997) Genetic variation at allozyme and RAPD loci in sessile oak *Quercus petraea* (Matt.) Liebl.: the role of history and geography. *Molecular Ecology* 6, 519-529.

Ledig, F.T. (1974) An analysis of methods for the selection of trees from wild stands. *Forest Science* 20, 2-16.

Ledig, F.T. (1986) Heterozygosity, heterosis, and fitness in outbreeding plants. In: Soule, M.E. (ed.) *Conservation Biology*. Sinauer Associates, Sunderland, MA, pp. 77-104.

Ledig, F.T. (1988a) The conservation of diversity in forest trees: why and how should genes be conserved? *BioScience* 38, 471-479.

Ledig, F.T. (1988b) The Conservation of Genetic Diversity: The Road to La Trinidad. *Leslie L. Schaffer Lectureship in Forest Science*. University of British Columbia. Vancouver, B.C.

Ledig, F.T. (1992) Human impacts on genetic diversity in forest ecosystems. *Oikos* 63, 87-108.

Ledig, F.T. (1998) Genetic variation in *Pinus*. In: Richardson, D.M. (ed.) *Ecology and Biogeography of Pinus*. Cambridge University Press, Cambridge, UK, pp. 251-280.

Ledig, F.T. and Conkle, M.T. (1983) Gene diversity and genetic structure in a narrow endemic, Torrey pine (*Pinus torreyana parry* ex carr.). *Evolution* 37, 79-85.

Ledig, F.T. and Kitzmiller, J.H. (1992) Genetic strategies for reforestation in the face of global climate change. *Forest Ecology and Management* 50, 153-169.

Ledig, F.T. and Korbobo, D.R. (1983) Adaptation of sugar maple populations along altitudinal gradients: photosynthesis, respiration, and specific leaf weight. *American Journal of Botany* 70, 256-265.

Ledig, F.T. and Porterfield, R.L. (1982) Tree improvement in western conifers: Economic aspects. *Journal of Forestry* 80, 653-657.

Ledig, F.T. and Smith, D.M. (1981) The influence of silvicultural practices on genetic improvement: height growth and weevil resistance in eastern white pine. *Silvae Genetica* 30, 30-36.

Ledig, F.T., Jacob-Cervantes V., Hodgskiss, P.D. and Eguiluz-Piedra, T. (1997) Recent evolution and divergence among populations of a rare Mexican endemic, Chihuahua spruce, following holocene climatic warming. *Evolution* 51, 1815-1827.

Ledig, F.T., Bermejo-Velazques, B., Hodgskiss, P.D., Johnson. D.R., Flores-Lopez, C. and Jacob-Cervantes, V. (2000) The mating system and genetic diversity in Martinez spruce, an extremely rare endemic of Mexico's Sierra Madre Oriental: an example of faculative selfing and survival in interglacial refugia. *Canadian Journal of Forest Research* 30, 1156-1164.

Lehner, A., Campbell, M., Wheeler, N., Poykko, T., Glossl, J., Kreike, J. and Neale, D. (1995) Identification of a RAPD marker linked to the pendula gene in Norway spruce (*Picea abies* L. Karst. f. *pendula*). *Theoretical and Applied Genetics* 91, 1092-1094.

Leonardi, S., Raddi, S. and Borghetti, M. (1996) Spatial autocorrelation of allozyme traits in a Norway spruce (*Picea abies*) population. *Canadian Journal of Forest Research* 26, 63-71.

Lepisto, M. (1985) The inheritance of pendula spruce (*Picea abies* f. *pendula*) and utilization of the narrow-crowned type in spruce breeding. *Foundation for Forest Tree Breeding* 1, 1-6.

Leple, J.C., Brasileiro, A.C.M., Michel, M.F., Delmonte, F. and Jouanin, L. (1992) Transgenic plants: expression of chimeric genes using four different constructs. *Plant Cell Reports* 11, 137-141.

Leple, J.C., Bonade-Bottino, M., Augustin, S., Pilate, G., Le Tan, V.D., Delplanque, Λ., Cornu, D. and Jonanin, L. (1995) Toxicity to *Chrysomela tremulae* (coleopteran, chrysomelidae) of transgenic poplars expressing a cystein proteinase inhibitor. *Molecular Breeding* 1, 319-328.

Leppik, E. (1953) Some viewpoints on the phylogeny of rust fungi. I. Coniferous rusts. *Mycologia* 45, 46-74.

Leppik, E. (1967) Some viewpoints on the phylogeny of rust fungi. VI. Biogenic radiation. *Mycologia* 59, 568-579.

Lesica, P. and Allendorf, F.W. (1995) When are peripheral populations valuable for conservation? *Conservation Biology* 9, 753-760.

Levée, V., Lelu, M.A. Jouanin, L. and Pilate, G. (1997) *Agrobacterium tumefaciens*-mediated transformation of hybrid larch (*Larix kaempferi* x *L. deciduas*) and transgenic regeneration. *Plant Cell Reports* 16, 680-685.

Levée, V., Garin, K., Klimaszeweska, K. and Séguin, A. (1999) Stable genetic transformation of white pine (*Pinus strobus* L.) after cocultivation of embryogenic tissues with *Agrobacterium tumefaciens*. *Molecular Breeding* 5, 429-440.

Levin, D.A. (1984) Inbreeding and proximity-dependent crossing success in *Phlox drummondii*. *Evolution* 38, 116-127.

Levin, D.A. and Kerster, H.W. (1974) Gene flow in seed plants. In: Dobzhansky, T., Hecht, M.T. and Steere, W.C. (eds.), *Evolutionary Biology 7*. Plenum Press, New York, NY, pp. 139-220.

Lewin, B. (1997) *Genes VI*. Oxford University Press, New York, NY.

Lewontin, R.C. (1984) Detecting population differences in quantitative characters as opposed to gene frequencies. *American Naturalist* 23, 115-124.

Li, B. 1995. Aspen improvement strategies for western Canada - Alberta and Saskatchewan. *Forestry Chronicle* 71, 720-724.

Li, B. and McKeand, S.E. (1989) Stability of loblolly pine families in the southeastern U.S. *Silvae Genetica* 38, 96-101.

Li, B. and Wu, R. (2000) Quantitative genetics of heterosis in aspen hybrids. In: *Proceedings of the Queensland Forest Research Institute/Cooperative Research Center-Sustainable Production Forestry (QFRI/CRC-SPF), Symposium on Hybrid Breeding and Genetics of Forest Trees*. Noosa, Queensland, Australia. pp. 184-190.

Li, B. and Wyckoff, G.W. (1991) A breeding strategy to improve aspen hybrids for the University of Minnesota Aspen/Larch Genetics Cooperative. In: Hall. R.B., Hanover, J.W. and Nyong'o, R.N. (eds.) *Proceedings of the International Energy Agency Joint Meetings*. Grand Rapids, MN, pp. 33-41.

Li, B. and Wyckoff, G.W. (1994) Breeding strategies for *Larix decidua, L. Leptolepis* and their hybrids in the United States. *Forest Genetics* 1, 65-72.

Li, B., McKeand, S.E. and Allen, H.L. (1989) Early selection of loblolly pine families based on seedling shoot elongation characters. In: *Proceedings of the 20th Southern Forest Tree Improvement Conference*. Charleston, SC, pp. 228-234.

Li, B., Howe, G.T., and Wu, R. (1998) Developmental factors responsible for heterosis in aspen hybrids (*Populus tremuloides x P. tremula*). *Tree Physiology* 18, 29-36.

Li, B., McKeand, S.E. and Weir, R.J. (1999) Tree improvement and sustainable forestry - impact of two cycles of loblolly pine breeding in the U.S.A. *Forest Genetics* 6, 229-234.

Li, B., McKeand, S.E. and Weir, R.J. (2000) Impact of forest genetics on sustainable forestry - results from two cycles of loblolly pine breeding in the U.S. *Journal of Sustainable Forestry* 10, 79-85.

Li, L., Popko, J.L., Zhang, X.H., Osakabe, K., Tsai, C.J., Joshi, C.P. and Chiang, V.L. (1997) A novel multifunctional 0-methyltransferase implicated in a dual methylation pathway associated with lignin biosynthesis in loblolly pine. *Proceedings of the National Academy of Sciences of the United States of America* 94, 5461-5466.

Li, P. and Adams, W.T. (1989) Range-wide patterns of allozyme variation in Douglas-fir (*Pseudotsuga menziesii*). *Canadian Journal of Forest Research* 19, 149-161.

Li, W. (1997) *Molecular Evolution*. Sinauer Associates, Inc., Publishers, Sunderland, MA.

Libby, W.J. (1973) Domestication strategies for forest trees. *Canadian Journal of Forest Research* 3, 265-276.

Libby, W.J., (1982) What is a safe number of clones per plantation? In: Heybroek, H.M., Stephen,

B.R. and VonWeissenberg, K. (eds.) *Resistance to Diseases and Pests in Forest Trees. Proceedings of the 3rd International Workshop on the Genetics of Host-Parasite Interactions of Forestry*. Wageningen, The Netherlands, pp. 342-360.

Libby, W.J. (1987) Testing and deployment of genetically-engineered trees. In: Bonga, J.M. and Durzan, D.J. (eds.) *Cell and Tissue Culture in Forestry*. Nijhoff, Boston, MA, pp. 167-197.

Libby, W.J. and Ahuja, M.R. (1993) Clonal forestry. In: Ahuja, M.R. and Libby, W.J. (eds.) *Clonal Forestry II. Conservation and Application*. Springer-Verlag, New York, NY, pp. 1-8.

Libby, W.J. and Cockerham, C.C. (1980) Random non-contiguous plots in interlocking field layouts. *Silvae Genetica* 29, 183-190.

Libby, W.J. and Jund, E. (1962) Variance associated with cloning. *Heredity* 17, 533-540.

Libby, W.J. and Rauter, R.M. (1984) Advantages of clonal forestry. *Forestry Chronicle* 60, 145-149.

Lidholm, J. and Gustafsson, P. (1991) The chloroplast genome of the gymnosperm *Pinus contorta*: a physical map and a complete collection of overlapping clones. *Current Genetics* 20, 161-166.

Lincoln, S.E., Daly, M.J. and Lander, E.S. (1993) Mapping genes controlling quantitative traits using MAPMAKER/QTL Version 1.1. A tutorial and reference manual. Whitehead Institute for Biomedical Research Technical Report, Second Edition, Cambridge, MA.

Lindgren, D. (1985) Cost-efficient number of test sites for ranking entries in field trials. *Biometrics* 41, 887-893.

Lindgren, D. (1986) How should breeders respond to breeding values? In: *Proceedings of the International Union of Forest Research Organizations (IUFRO), Conference on Breeding Theory, Progeny Testing and Seed Orchards*. Williamsburg, VA, pp. 361-372.

Lindgren, D. (1993) The population biology of clonal deployment. In: Ahuja, M.R. and Libby, W.J. (eds.) *Clonal Forestry I. Genetics and Biotechnology*. Springer-Verlag, New York, NY, pp. 34-49.

Lindgren, D. and El-Kassaby, Y.A. (1989) Genetic consequences of combining selective cone harvesting and genetic thinning in clonal seed orchards. *Silvae Genetica* 38, 65-70.

Lindgren, D. and Gregorius, H.R. (1976) Inbreeding and coancestry. Pp. 49-72 In: *Proceedings of the International Union of Forest Research Organizations (IUFRO), Conference on Advanced Generation Breeding*. French Institute for Agronomy Research (INRA), Cectas, France, pp. 49-72.

Lindgren, D. and Matheson, A.C. (1986) An algorithm for increasing the genetic quality of seed from seed orchards by using the better clones in higher proportions. *Silvae Genetica* 35, 173-177.

Lindgren D. and Mullin, T.J. (1998) Relatedness and status number in seed orchard crops. *Canadian Journal of Forest Research* 28, 276-283.

Lindgren, D., Libby, W.S. and Bondesson, F.L. (1989) Deployment to plantations of numbers and proportions of clones with special emphasis on maximizing gain at a constant diversity. *Theoretical and Applied Genetics* 77, 825-831.

Lindgren, D., Gea, L. and Jefferson, P. (1996) Loss of genetic diversity monitored by status number. *Silvae Genetica* 45, 52-59.

Lindgren, D., Wei, R.P. and Lee, S.J. (1997) How to calculate optimum family number when starting a breeding program. *Forest Science* 43, 206-212.

Linhart, Y.B., Mitton, J.B., Sturgeon, K.B. and Davis, M.L. (1981) Genetic variation in space and time in a population of Ponderosa pine. *Heredity* 46, 407-426.

Lipow, S.R., St. Clair, J.B. and Johnson, G.R. (2001) *Ex situ* gene conservation for conifers in the Pacific Northwest. USDA Forest Service, General Technical Report PNW-GTR-528.

Lipow, S.R., Johnson, G.R., St. Clair, J.B and Jayawickrama, K.J. (2003) The role of tree improvement programs for *ex situ* gene conservation of coastal Douglas-fir in the Pacific Northwest. *Forest Genetics* 10, 111-120.

Lipow, S.R., Vance-Borland, K., St.Clair, J.B., Hendrickson, J.A. and McCain, C. (2004) Gap analysis of conserved genetic resources for forest trees. *Conservation Biolology* 18, 412-423.

Liston, A., Robinson, W. A., Pinero, D. and Alvarez-Buylla, E.R. (1999) Phylogenetics of *Pinus* (Pinaceae) based on nuclear ribosomal DNA internal transcribed spacer region sequences. *Molecular Phylogenetics and Evolution* 11, 95-109.

Liston, A., Gernandt, D.S., Vining, T. F., Campbell, C.S. and Piñero, D. (2003) Molecular phylogeny of Pinaceae and *Pinus*. *Proceedings of the International Conifer Conference. Acta Horticulturae*

615, 107-114.

Litt, M. and Luty, J.A. (1989) A hypervariable microsatellite revealed by *in vitro* amplification of a dinucleotide repeat within the cardiac muscle actin gene. *American Journal of Human Genetics* 44, 397-401.

Littell, R.C., Milliken, G.A., Stroup, W.W. and Wolfinger. R.D. (1996) Spatial variability. In: *SAS Institute*. Cary, NC, pp. 303-330.

Little, E.L. (1971) *Atlas of United States Trees*. USDA Forest Service.

Little, E., Jr. and Critchfield, W. (1969) *Subdivisions of the Genus Pinus (Pines)*. USDA Forest Service.

Liu, B. (1998) Stern R. (ed.) *Statistical Genomics: Linkage, Mapping, and QTL Analysis*. CRC Press LLC, Boca Raton, LA, 611 pp.

Liu, Z. and Furnier, G.R. (1993) Inheritance and linkage of allozymes and restriction fragment length polymorphisms in trembling aspen. *Journal of Heredity* 84, 419-424.

Long, A. and Langley, C. (1999) The power of association studies to detect the contribution of candidate loci to variation in complex traits. *Genome Research* 9, 720-731.

Loo, J.A., Tauer, C.G. and van Buijtenen, J.P. (1984) Juvenile-mature relationships and heritability estimates of several traits in loblolly pine (*Pinus taeda*). *Canadian Journal of Forest Research* 14, 822-825.

Loo-Dinkins, J. (1992) Field test design. In: Fins, L., Friedman, S.T. and Brotschol, J.V. (eds.) *Handbook of Quantitative Forest Genetics*. Kluwer Acadademic Publishers, Dordrecht, The Netherlands, pp. 96-139.

Loo-Dinkins, J.A. and Tauer, C.G. (1987) Statistical efficiency of six progeny test field designs on three loblolly pine (*Pinus taeda* L.) site types. *Canadian Journal of Forest Research* 17, 1066-1070.

Loo-Dinkins, J.A., Tauer, C.G. and Lambeth, C.C. (1990) Selection system efficiencies for computer simulated progeny test field designs in loblolly pine. *Theoretical and Applied Genetics* 79, 89-96.

Lopes, U.V., Huber, D.A. and White, T.L. (2000) Comparison of methods for prediction of genetic gain from mass selection on binary threshold traits. *Silvae Genetica* 49, 50-56.

Lott, L.H., Lott, L.M., Stine, M., Kubisiak, T.L. and Nelson, C.D. (2003) Top grafting longleaf x slash pine F1 hybrids on mature longleaf and slash pine interstocks. In: *Proceedings of the 27th Southern Forest Tree Improvement Conference*. Stillwater, OK, pp. 24-27.

Loveless, M.D. (1992) Isozyme variation in tropical trees: patterns of genetic organization. *New Forests* 6, 67-94.

Lowe, W.J. and van Buijtenen, J.P. (1981) Tree improvement philosophy and strategy for the Western Gulf Forest Tree Improvement Program. *Proceedings of the 15th North American Quantitative Forest Genetics Group Workshop*. Coeur d'Alene, ID, pp. 43-50.

Lowe, W.J. and van Buijtenen, J.P. (1986) The development of a subline system in an operational tree improvement program. In: *Proceedings of the International Union of Forest Research Organizations (IUFRO), Conference on Breeding Theory, Progeny Testing and Seed Orchards*. Williamsburg, VA, pp. 96-106.

Lowe, W.J. and van Buijtenen, J.P. (1989) The incorporation of early testing procedures into an operational tree improvement program. *Silvae Genetica* 38, 243-250.

Lowenfeld, R. and Klekowski, J.E. (1992) Mangrove genetics. I. Mating system and mutation rates of *Rhizophora mangel* in Florida and San Salvador Island, Bahamas. *International Journal of Plant Science* 153, 394-399.

Lowerts, G.A. (1986) Realized genetic gain from loblolly and slash pine first generation seed orchards. In: *Proceedings of the International Union of Forest Research Organizations (IUFRO), Conference on Breeding Theory, Progeny Testing and Seed Orchards*. Williamsburg, VA, pp. 142-149.

Lowerts, G.A. (1987) Tests of realized genetic gain from a coastal Virginia loblolly pine first generation seed orchard. In: *Proceedings of the 19th Southern Forest Tree Improvement Conference*. College Station, TX, pp 423-431.

Lloyd, G. and McCown, B. (1981) Commercially feasible micropropagation of mountain laurel (*Kalmia latiflora*) by use of shoot tip culture. *Proceedings of the International Plant Propagation Society* 30, 421-427.

Loopstra, C.A., Weissinger, A.K. and Sederoff, R.R. (1992) Transient gene expression in differentiating pine wood using microprojectile bombardment. *Canadian Journal of Forest Research* 22, 993-996.

Lstiburek, M., Mullin, T.J., Lindgren, D. and Rosvall, O. (2004a) Open-nucleus breeding strategies compared with population-wide positive assortative mating. I. Equal distribution of testing effort. *Theoretical and Applied Genetics* 109, 1196-1203.

Lstiburek, M., Mullin, T.J., Lindgren, D. and Rosvall, O. (2004b) Open-nucleus breeding strategies compared with population-wide positive assortative mating. II. Unequal distribution of testing effort. *Theoretical and Applied Genetics* 109, 1169-1177.

Lubaretz, O., Fuchs, J., Ahne, R., Meister, A. and Schubert, I. (1996) Karyotyping of three Pinaceae species via fluorescent *in situ* hybridization and computer-aided chromosome analysis. *Theoretical and Applied Genetics* 92, 411-416.

Lush, J.L. (1935) Progeny test and individual performance as indicators of an animal's breeding value. *Journal of Dairy Science* 18, 1-19.

Lynch, M. (1996) A quantitative-genetic perspective on conservation issues. In: Avise, J.C. and Hamrick, J.L. (eds.) *Conservation Genetics: Case Histories from Nature*. Chapman and Hall, New York, NY, pp 471-501.

Lynch, M. and Walsh, B. (1998) *Genetics and Analysis of Quantitative Traits*. Sinauer Associates Inc., Publishers, Sunderland, MA.

Mackay, J.J., O'Malley, D.M., Presnell, T., Booker, F.L., Campbell, M.M., Whetten, R.W. and Sederoff, R.R. (1997) Inheritance, gene expression, and lignin characterization in a mutant pine deficient in cinnamyl alcohol dehydrogenase. *Proceedings of the National Academy of Sciences of the United States of America* 94, 8255-8260.

MacPherson, P. and Filion, G.W. (1981) Karyotype analysis and the distribution of constitutive heterochromatin in five species of *Pinus*. *The Journal of Heredity* 72, 193-198.

Madalena, F.E. and Hill, W.G. (1972) Population structures in artificial selection programmes: simulation studies. *Genetic Research* 20, 75-99.

Madalena, F.E. and Robertson, A. (1975) Population structure in artificial selection: studies with *Drosophila melongaster*. *Genetical Research* 24, 113-126.

Magnussen, S. (1989) Determination of optimum selection ages: A simulation approach. In: *Proceedings of the 20th Southern Forest Tree Improvement Conference*. Charleston, SC, pp. 269-285.

Magnussen, S. (1990) Application and comparison of spatial models in analyzing tree-genetics field trials. *Canadian Journal of Forest Research* 20, 536-546.

Magnussen, S. (1993) Design and analysis of tree genetic trials. *Canadian Journal of Forest Research* 23, 1144-1149.

Magnussen, S. and Yanchuk, A.D. (1994) Time trends of predicted breeding values in selected crosses of coastal Douglas-fir in British Columbia: A methodological study. *Forest Science* 40, 663-685.

Mahalovich, M.F. and Bridgwater, F.E. (1989) Modeling elite populations and positive assortative mating in recurrent selection programs for general combining ability. In: *Proceedings of the 20th Southern Forest Tree Improvement Conference*. Charleston, SC, pp 43-49.

Mandal, A.K. and Gibson, G.L. (1998) *Forest Genetics and Tree Breeding*. CBS Publishers & Distributors, New Delhi, India.

Mandel, J. (1971) A new analysis of variance model for non-additive data. *Technometrics* 13, 1-18.

Mangold, R.D. and Libby, W.J. (1978) A model for reforestation with optimal and suboptimal tree populations. *Silvae Genetica* 27, 66-68.

Mantysaari, E.A., Quaas, R.L. and Grohn, Y.T. (1991) Simulation study on covariance component estimation for two binary traits in an underlying continuous scale. *Journal of Dairy Science* 74, 580-591.

Marco, M.A. and White, T.L. (2002) Genetic parameter estimates and genetic gains for *Eucalyptus grandis* and *E. dunnii* in Argentina. *Forest Genetics* 9, 211-220.

Marques, C.M., Araujo, J.A., Ferreira, J.G., Whetten, R., O'Malley, D.M., Liu, B.-H. and Sederoff, R. (1998) AFLP genetic maps of *Eucalyptus globulus* and *E. tereticornis*. *Theoretical and Applied Genetics* 96, 727-737.

Marques, C., Brondani, R., Grattapaglia, D. and Sederoff, R. (2002) Conservation and synteny of

SSR loci and QTLs for vegetative propagation in four *Eucalyptus* species. *Theoretical and Applied Genetics* 103, 474-478.

Marshall, D.R. (1990) Crop genetic resources: current and emerging issues. In: Brown, A.H.D., Clegg, M.T., Kahler, A.L. and Weir, B.S. (eds.). *Plant Population Genetics, Breeding and Genetic Resources*. Sinauer Associates Inc., Publishers, Sunderland, MA, pp. 367-388.

Marshall, D.R. and Brown, A.H.D. (1975) Optimum sampling strategies in genetic conservation. In: Frankel, O.H. and Hawkes, J.G. (eds.) *Crop Genetic Resources for Today and Tomorrow*. Cambridge University Press, Cambridge, UK, pp. 21-40.

Martin, T.A., Johnsen, K.H. and White, T.L. (2001) Ideotype development in southern pines: rationale and strategies for overcoming scale-related obstacles. *Forest Science* 47, 21-28.

Matheson, A.C. (1989) Statistical methods and problems in testing large numbers of genotypes across sites. In: G. G.I., A. R. Griffin, and A. C. Matheson, A.C. (eds.). *Breeding Tropical Trees: Population Structure and Genetic Improvement Strategies in Clonal and Seedling Forestry*. Oxford Forestry Institute, Oxford, UK, 93-105.

Matheson, A.C. and Harwood, C.E. (1997) Breeding tropical Australian acacias. In: *Proceedings of the 24th Southern Forest Tree Improvement Conference*. Orlando, FL, pp. 69-80.

Matheson, A.C. and Lindgren, D. (1985) Gains from the clonal and the clonal seed-orchard options compared for tree breeding programs. *Theoretical and Applied Genetics* 71, 242-249.

Matheson, A.C. and Raymond, C.A. (1984a) Provenance x environment interaction: Its detection, practical importance and use with particular reference to tropical forestry. In: Barnes, R.D. and Gibson, G.L. (eds.) *Provenance and Genetic Improvement Strategies in Tropical Forest Trees*. Commonwealth Forestry Institute, Oxford, UK, pp. 81-117.

Matheson, A.C. and Raymond, C.A. (1984b) The impact of genotype x environment interactions on Australian *Pinus radiata* breeding programs. *Australian Forest Research,* 14, 11-25.

Matheson, A.C. and Raymond, C.A. (1986) A review of provenance x environment interaction: its practical importance and use with particular reference to the tropics. *Commonwealth Forestry Review* 65, 283-302.

Matheson, A.C., Spencer, D.J., and Magnussen, D. (1994) Optimum age for selection in *Pinus radiata* using basal area under bark for age:age correlations. *Silvae Genetica* 43, 352-357.

Matheson, A.C., White, T.L. and Powell, G.L. (1995) Effects of inbreeding on growth, stem form and rust resistance in *Pinus elliottii*. *Silvae Genetica* 44, 37-45.

Matos, J.A. and Schaal, B.A. (2000) Chloroplast evolution in the *Pinus montezumae* complex: A coalescent approach to hybridization. *Evolution* 54, 1218-1233.

Matyas, C. (1999) *Forest Genetics and Sustainability*. Kluwer Academic Publishers, Boston, MA.

Matze, A.J.M. and Chilton, M.D. (1981) Site-specific insertion of genes into T-DNA of the *Agrobacterium* tumor-inducing plasmid. An approach to genetic engineering of higher plant cells. *Journal of Molecular and Applied Genetics* 1, 39-49.

McClenaghan, Jr. L.R. and Beauchamp, A.C. (1986) Low genetic differentiation among isolated populations of the California fan palm (*Washingtonia filifera*). *Evolution* 40, 315-322.

McClure, M.S., Salom, S.M. and Shields, K.S. (2001) *Hemlock Woolly Adelgid. Forest Health Technology Enterprise Team*. USDA Forest Service.

McCown, B.H., McCabe, D.E., Russell, D.R., Robison, D.J., Barton, K.A. and Raffa, K.F. (1991) Stable transformation of *Populus* and incorporation of pest resistance by electric discharge particle acceleration. *Plant Cell Reports* 9, 590-594.

McKeand, S.E. (1988) Optimum age for family selection for growth in genetic tests of loblolly pine. *Forest Science* 34, 400-411.

McKeand, S.E. and Beineke, F. (1980) Sublining for half-sib breeding populations of forest trees. *Silvae Genetica* 29, 14-17.

McKeand, S.E. and Bridgwater, F.E. (1986) When to establish advanced generation seed orchards. *Silvae Genetica* 35, 245-247.

McKeand, S.E. and Bridgwater, F.E. (1992) Third-generation breeding strategy for the North Carolina State University-Industry Cooperative Tree Improvement Program. In: *Proceedings of the International Union of Forest Research Organizations (IUFRO), Symposium on Resolving Tropical Forestry Resource Concerns Through Tree Improvement, Gene Conservation, and Domestication of New Species*. Cali, Colombia, pp. 234-240.

McKeand, S.E. and Bridgwater, F. (1998) A strategy for the third breeding cycle of loblolly pine in the Southeastern USA. *Silvae Genetica* 47, 223-234.

McKeand, S.E. and Raley, F. (2000) Interstock effect on strobilus initiation in topgrafted loblolly pine. *Forest Genetics* 7, 179-182.

McKeand, S.E. and Svensson, J. (1997) Sustainable management of genetic resources. *Journal of Forestry* 95, 4-9.

McKeand, S.E., Mullin, T.J., Byram, T.D. and White, T.L. (2003) Deployment of genetically improved loblolly and slash pines in the South. *Journal of Forestry* 101, 32-37.

McKenney, D.W., van Vuuren, W. and Fox, G.C. (1989) An economic comparison of alternative tree improvement strategies: A simulation approach. *Canadian Journal of Agricultural Economics* 37, 211-232.

McKinley, C.R. (1983) Objectives of progeny tests. In: *Progeny Testing of Forest Trees,* Southern Cooperative Series Bulletin 275. Texas A & M University, College Station, TX, pp. 2-13.

McKinnon, G.E., Vaillancourt, R.E., Jackson, H.D. and Potts, B.M. (2001) Chloroplast sharing in the Tasmanian eucalypts. *Evolution* 55, 703-711.

Megraw, R.A. (1985) *Wood Quality Factors in Loblolly Pine: The Influence of Tree Age, Position in Tree, and Cultural Practice on Wood Specific Gravity, Fiber Length and Fibril Angle.* TAPPI Press, Atlanta, GA.

Meilan, R. and Strauss, S.H. (1997) Poplar genetically engineered for reproductive sterility and accelerated flowering. In: Klopfenstein, N.B., Chun, W.Y.W., Kim, M.-S. and Ahuja, M.R. (eds.) Micropropagation, *Genetic Engineering and Molecular Biology of Populus.* USDA Forest Service Technical Report RM-GTR-297, pp. 212-219.

Meilan, R., Brunner, A.M., Skinner, J.S. and Strauss, S.H. (2001) Modification of flowering in transgenic trees. In: Morohoshi, N. and Komamine, A. (eds.) *Molecular Breeding of Woody Plants.* Elsevier Science Publishers, New York, NY, pp. 247-256.

Meilan, R., Han, K.-H., Ma, C., DiFazio, S.P., Eaton, J.A., Hoien, E.A., Stanton, B.J., Crockett, R.P., Taylor, M.L., James, R.R., Skinner, J.S., Jouanin, L., Pilate, G. and Strauss, S.H. (2002) The *CP4* transgene provides high levels of tolerance to Roundup herbicide in field-grown hybrid poplars. *Canadian Journal of Forest Research* 32, 967-976.

Mejnartowicz, M. (1991) Inheritance of chloroplast DNA in *Populus. Theoretical and Applied Genetics* 82, 477-480.

Mendel, G. (1866) Versuche über pflanzen hybriden Verh. naturforsch. Verein Brünn 4, 3-47. Sinnott, E.W., Dunn, L.C. and Dobzhansky, T. (1958) *Principles of Genetics* (Abstract). McGraw-Hill Book Company, Inc., New York, NY.

Menzies, M.I. and Aimers-Halliday. J. (1997) Propagation options for clonal forestry with *Pinus radiata.* In: Burdon, R.D. and Moore, J.M. (eds.) In: *Proceedings of the International Union of Forest Research Organizations (IUFRO), Conference on the Genetics of Radiata Pine.* Rotorua, New Zealand, pp. 256-263.

Mergen, F. and Thielges, B. (1967) Intraspecific variation in nuclear volume in four conifers. *Evolution* 21, 720-724.

Merkle, S.A. and Adams, W.T. (1988) Multivariate analysis of allozyme variation patterns in coastal Douglas-fir from southwest Oregon. *Canadian Journal of Forest Research* 18, 181-187.

Merkle, S.A., Feret, P.P., Bramlett, D.L. and Queijo, D.L. (1982) A computer program package for use with the southern pine seed orchard inventory-monitoring system. *School of Forestry and Wildlife Publication No. FWS-2-82.* Virginia Polytechnic Institute, Blacksburg, VA.

Meskimen, G. (1983) Realized gain from breeding *Eucalyptus grandis* in Florida. *United States Department of Agriculture, Pacific Southwest Experiment Station, General Technical Report RSW-69.* Berkeley, CA, pp. 121-128

Meskimen, G.F., Rockwood, D.L. and Reddy, K.V. (1987) Development of *Eucalyptus* clones for a summer rainfall environment with periodic severe frosts. *New Forests* 3, 197-205.

Meyer, K. (1985) Maximum likelihood estimation of variance components for a multivariate mixed model with equal design matrices. *Biometrics* 41, 153-165.

Meyer, K. (1989) Restricted maximum likelihood to estimate variance components for animal models with several random effects using a derivative-free algorithm. *Genetics Selection Evolution* 21, 317-340.

Meyer, P. (1995) Variation in transgene expression in plants. *Euphytica* 85, 359-366.

Michelmore, R., Paran, I. and Kesseli, R. (1991) Identification of markers linked to disease-resistance genes by bulked segregant analysis: a rapid method to detect markers in specific genomic regions by using segregating populations. *Proceedings of the National Academy of Sciences of the United States of Americ.* 88, 9828-9832.

Miglani, G.S. (1998) *Dictionary of Plant Genetics and Molecular Biology.* Food Products Press, Binghamton, NY.

Miksche, J. (1967) Variation in DNA content of several gymnosperms. *Canadian Journal of Genetics and Cytology* 9, 717-222.

Miksche, J. (1968) Quantitative study of intraspecific variation of DNA per cell in *Picea glauca* and *Pinus banksiana. Canadian Journal of Genetics and Cytology* 10, 590-600.

Miksche, J. (1971) Intraspecific variation of DNA per cell between *Picea sitchensis* (Bong.) Carr. provenances. *Chromosoma* 32, 343-352.

Miksche, J.P. and Hotta, J. (1973) DNA base composition and repetitious DNA in several conifers. *Chromosoma* 41, 29-36.

Millar, C.I. (1983) A steep cline in *Pinus muricata. Evolution* 37, 311-319.

Millar, C.I. (1989) Allozyme variation of bishop pine associated with pygmy-forest soils in northern California. *Canadian Journal of Forest Research* 19, 870-879.

Millar, C., (1993) Impact of the Eocene on the evolution of *Pinus* L. (1993) *Annals of the Missouri Botanical Garden* 80, 471-498.

Millar, C. (1998) Early evolution of pines. In: Richardson, D.M. (ed.) *Ecology and Biogeography of Pinus.* Cambridge University Press, Cambridge, UK, pp. 69-91.

Millar, C. and Kinloch, B. (1991) Taxonomy, phylogeny, and coevolution of pines and their stem rusts. In: Hiratsuka, Y., Samoil, J., Blenis, P., Crane, P. and Laishley, B. (eds.) *Rusts of Pine. Proceedings of the 3rd International Union of Forest Research Organizations (IUFRO), Rusts of Pine Working Party Conference.* Forestry Canada Northwest Region, Northern Centre, Edmonton, Alberta, Information Report NOR-X-317, pp. 1-38.

Millar, C.I. and Libby, W.J. (1991) Strategies for conserving clinal, ecotypic, and disjunct population diversity in widespread species. In: Falk, D.A. and Holsinger, K.E. (eds.) *Genetics and Conservation of Rare Plants.* Oxford University Press, New York, NY, pp. 149-170.

Millar, C.I. and Marshall, K.A. (1991) Allozyme variation of Port Orford-cedar (*Chamaecyparis lawsoniana*): Implications for genetic conservation. *Forest Science* 27, 1060-1077.

Millar, C.I. and Westfall, R. (1992) Allozyme markers in forest genetic conservation. *New Forests* 6, 347-371.

Miller, C., Jr. (1976) Early evolution in the Pinaceae. *Review of Palaeobotany and Palynology* 21, 101-117.

Miller, C., Jr. (1977) Mesozoic conifers. *The Botanical Review* 43, 217-280.

Miller, C., Jr. (1982) Current status of Paleozoic and Mesozoic conifers. *Review of Palaeobotany and Palynology* 37, 99-114.

Mirov, N. (1956) Composition of turpentine of lodgepole x jack pine hybrids. *Canadian Journal of Botany* 34, 443-457.

Mirov, N. (1967) *The genus Pinus.* The Ronald Press Company, New York, NY.

Mitton, J.B. (1998a) *Apparent overdominance in natural plant populations. Concepts and Breeding of Heterosis in Crop Plants.* CSSA Special Publication No. 25, Crop Science Society of America, Madison, WI, pp. 57-69.

Mitton, J.B. (1998b) Allozymes in tree breeding research. In: Mandal, A.K. and Gibson, G.L. (eds.) *Forest Genetics and Tree Breeding.* CBS Publishers and Distributors, Daryaganj, New Delhi, India, pp. 239-251.

Mitton, J.B., Sturgeon, K.B. and Davis, M.L. (1980) Genetic differentiation in ponderosa pine along a steep elevational transect. *Silvae Genetica* 29, 100-103.

Mitton, J.B., Latta, R.G. and Rehfeldt, G.E. (1997) The pattern of inbreeding in washoe pine and survival of inbred progeny under optimal environmental conditions. *Silvae Genetica* 46, 215-219.

Mode, C.J. and Robinson, H.F. (1959) Pleiotropism and the genetic variance and covariance. *Biometrics* 15, 518-537.

Mogensen, H.L. (1975) Ovule abortion in *Quercus* (Fagaceae). *American Journal of Botany* 62, 160-165.

Moran, G.F. (1992) Patterns of genetic diversity in Australian tree species. *New Forests* 6, 49-66.

Moran, G.F. and Adams, W.T. (1989) Microgeographical patterns of allozyme differentiation in Douglas-fir from southwest Oregon. *Forest Science* 35, 3-15.

Moran, G.F. and Hopper, S.D. (1983) Genetic diversity and the insular population structure of the rare granite rock species *Eucalyptus caesia* Benth. *Australian Journal of Botany* 31, 161-172.

Moran, G.F., Bell, J.C. and Eldridge, K.G. (1988) The genetic structure and the conservation of the five natural populations of *Pinus radiata*. *Canadian Journal of Forest Research* 18, 506-514.

Moran, G.F., Muona, O. and Bell, J.C. (1989) *Acacia mangium*: a tropical forest tree of the coastal lowlands with low genetic diversity. *Evolution* 43, 231-235.

Morgante, M. and Salamini, F. (2003) From plant genomics to breeding practice. *Current Opinion in Biotechnology* 14, 214-219.

Morgante, M., Vendramin, G.G. and Rossi, P. (1991) Effects of stand density on outcrossing rate in two Norway spruce (*Picea abies*) populations. *Canadian Journal of Botany* 69, 2704-2708.

Morgante, M., Vendramin, G.G., Rossi, P. and A.M. Olivieri, A.M. (1993) Selection against inbreds in early life-cycle phases *in Pinus leucodermis*. *Heredity* 70, 622-627.

Morgenstern, E.K. (1962) Note on chromosome morphology in *Picea rubens* Sarg. and *Picea mariana* (Mill.) B.S.P. *Genetica* 11, 163-164.

Morgenstern, E.K. (1978) Range-wide genetic variation of black spruce. *Canadian Journal of Forest Research* 8, 463-473.

Morgenstern, E.K. (1996) *Geographic Variation in Forest Trees*. UBC Press, Vancouver, BC.

Morgenstern, E.K. and Teich, A.H. (1969) Phenotypic stability of height growth of jack pine provenances. *Canadian Journal of Genetic Cytology* 11, 110-117.

Morgenstern, E.K., Corriveau, A.G., and Fowler, D.P. (1981) A provenance test of red pine in nine environments in eastern Canada. *Canadian Journal of Forest Research* 11, 124-131.

Morton, N.E., Crow, J.F. and Muller, H.J. (1956) An estimate of the mutational damage in man from data on consanguineous marriages. *Proceedings of the National Academy of Sciences of the United States of America* 42, 855-863.

Moss, E. (1949) Natural pine hybrids in Alberta. *Canadian Journal of Forest Research Section C* 27, 218-229.

Mosseller, A., Innes, D.J. and Roberts, B.R. (1991) Lack of allozymic variation in disjunct Newfoundland populations of red pine (*Pinus resinosa*). *Canadian Journal of Forest Research* 21, 525-528.

Mosseller, A., Egger, K.N. and Hughes, G.A. (1992) Low levels of genetic diversity in red pine confirmed by random amplified polymorphic DNA markers. *Canadian Journal of Forest Research* 22, 1332-1337.

Mosteller, F. and Tukey, J.W. (1977) *Data Analysis and Multiple Regression*. Addison-Wesley Publishers, London, UK.

Mousseau, T.A. and Roff, D.A. (1987) Natural selection and the heritability of fitness components. *Heredity* 59, 181-197.

Mrode, R.A. (1996) *Linear Models for the Prediction of Animal Breeding Values*. CAB International, Wallingford, Oxon, UK.

Mukai, Y., Tazaki, K., Fujii, T. and Yamamoto, N. (1992) Light-independent expression of three photosynthetic genes, cab, rbcS and rbcL, in coniferous plants. *Plant and Cell Physiology* 33(7), 859-866.

Müller, G. (1976) A simple method of estimating rates of self-fertilization by analyzing isozymes in tree seeds. *Silvae Genetica* 25, 15-17.

Müller, G. (1977) Short note: cross-fertilization in a conifer stand inferred from enzyme gene-markers in seeds. *Silvae Genetica* 26, 223-226.

Müller-Starck, G., Baradat, P. and Bergmann, F. (1992) Genetic variation within European tree species. *New Forests* 6, 23-47.

Mullin, T.J. and Park, Y.S. (1992) Estimating genetic gains from alternative breeding strategies for clonal forestry. *Canadian Journal of Forest Research* 22, 14-23.

Mullin, T.J. and Park, Y.S. (1994) Genetic parameters and age-age correlations in a clonally replicated test of black spruce after 10 years. *Canadian Journal of Forest Research* 24, 2330-2341.

Mullis, K.B. (1990) The unusual origin of the polymerase chain reaction. *Scientific American* 262(4), 56-65.

Muona, O. (1990) Population genetics in forest tree improvement. In: Brown, H.D., Clegg, M.T.,

Kahler, A.L. and Weir, B.S. (eds.) *Plant Population Genetics, Breeding and Genetic Resources.* Sinauer Associates Inc., Publishers, Sunderland, MA, pp. 282-298.

Muona, O. and Harju, A. (1989) Effective population sizes, genetic variability, and mating system in natural stands and seed orchards of *Pinus sylvestris. Silvae Genetica* 38, 221-228.

Muona, O. and Schmidt, A.E. (1985) A multilocus study of natural populations of *Pinus sylvestris.* In: Gregorius, H.R. (ed.) *Population Genetics in Forestry.* Lecture Notes in Biomathematics, Springer Verlag, New York, NY, pp. 226-240.

Murashige, T. and Skoog, F. (1962) A revised medium for rapid growth and bioassays with tobacco cultures. *Physiologia Plantarum* 15, 473-497.

Murawski, D.A. and Hamrick, J.L. (1991) The effect of the density of flowering individuals on the mating systems of nine tropical tree species. *Heredity* 67, 167-174.

Murawski, D.A. and Hamrick, J.L. (1992a) The mating system of *Cavanillesia platanifolia* under extremes of flowering-tree density: A test of predictions. *Biotropica* 24, 99-101.

Murawski, D.A. and Hamrick, J.L. (1992b) Mating system and phenology of *Ceiba pentandra* (Bombacaceae) in Central Panama. *Journal of Heredity* 83, 401-404.

Murawski, D.A., Hamrick, J.L., Hubbell, S.P. and Foster, R.B. (1990) Mating systems of two Bombacaceous trees of a neotropical moist forest. *Oecologia* 82, 501-506.

Murawski, D.A., Nimal Gunatilleke, I.A.U. and Bawa, K.S. (1994) The effects of selective logging on inbreeding in *Shorea megistophylla* (Dipterocarpaceae) from Sri Lanka. *Conservation Biology* 8, 997-1002.

Nakamura, R.R. and Wheeler, N.C. (1992) Pollen competition and parental success in Douglas-fir. *Evolution* 46, 846-851.

Nambiar, E.K.S. (1996) Sustaining productivity of forests as a continuing challenge to soil science. *Soil Science Society of America Journal* 60, 1629-1642.

Namkoong, G. (1966a) Statistical analysis of introgression. *Biometrics* 22, 488-502.

Namkoong, G. (1966b) Inbreeding effects on estimation of genetic additive variance. *Forest Science* 12, 8-13.

Namkoong, G. (1969) Nonoptimality of local races. *Proceedings of the 10th Southern Forest Tree Improvement Conference.* Houston,TX, pp. 149-153.

Namkoong, G. (1970) Optimum allocation of selection intensity in two stages of truncation selection. *Biometrics* 26, 465-476.

Namkoong, G. (1976) A multiple-index selection strategy. *Silvae Genetica* 25, 199-201.

Namkoong, G. (1979) *Introduction to Quantitative Genetics in Forestry.* United States Department of Agriculture Technical Bulletin No.1588.

Namkoong, G. (1984) A control concept of gene conservation. *Silvae Genetica* 33, 160-163.

Namkoong, G. (1997) A gene conservation plan for loblolly pine. *Canadian Journal of Forest Research* 27, 433-437.

Namkoong, G. and Conkle, M.T. (1976) Time trends in genetic control of height growth in ponderosa pine. *Forest Science* 22, 2-12.

Namkoong, G. and Kang, H. (1990) Quantitative genetics of forest trees. *Plant Breeding Reviews* 8:139-188.

Namkoong, G. and Roberds, J.H. (1982) Short-term loss of neutral alleles in small population breeding. *Silvae Genetica* 31, 1-6.

Namkoong, G., Snyder, E.B. and Stonecypher, R.W. (1966) Heritability and gain concepts for evaluating breeding systems such as seedling orchards. *Silvae Genetica* 15, 76-84.

Namkoong, G., Usanis, R.A. and Silen, R.R. (1972) Age-related variation in genetic control of height growth in Douglas-fir. *Theoretical and Applied Genetics* 42, 151-159.

Namkoong, G., Kang, H.C. and Brouard, J.S. (1988) *Tree Breeding: Principles and Strategies.* Springer-Verlag, New York, NY.

Nance, W.L., Tuskan, G.A., Nelson, C.D. and Doudrick, R.L. (1992) Potential applications of molecular markers for genetic analysis of host-pathogen systems in forest trees. *Canadian Journal of Forest Research* 22, 1036-1043.

Nason, J.D. and Hamrick, J.L. (1997) Reproductive and genetic consequences of forest fragmentation: two case studies of neotropical canopy trees. *Journal of Heredity* 88, 264-276.

Nason, J.D., Herre, E.A. and Hamrick, J.L. (1996) Paternity analysis of the breeding structure of strangler fig populations: evidence for substantial long-distance wasp dispersal. *Journal of Bio-*

geography 23, 501-512.

Nason, J.D., Aldrich, P.R. and Hamrick, J.L. (1997) Dispersal and the dynamics of genetic structure in fragmented tropical tree populations. In: Laurance, W.F. and Bierregaard, R.O. (eds.) *Tropical Forest Remnants: Ecology, Management and Conservation of Fragmented Communities.* University Chicago Press, Chicago, IL, pp. 304-320.

Nason, J.D., Herre, E.A. and Hamrick, J.L. (1998) The breeding structure of a tropical keystone plant resource. *Nature* 391, 685-687

Natarajan, A.T., Ohba, K. and Simak, M. (1961) Karyotype Analysis of *Pinus silvestris. Heredity* 47, 379-382.

National Research Council (1991) *Managing Global Genetic Resources: Forest Trees.* National Academy Press, Washington, DC.

National Research Council (2002) *Environmental Effects of Transgenic Plants. The Scope and Adequacy of Regulation.* National Academy Press, Washington, D.C..

National Research Council (2004) *Biological Confinement of Genetically engineered Organisms.* National Academy Press, Washington, D.C.

Neale, D.B. (1985) Genetic Implications of shelterwood regeneration of Douglas-fir in southwest Oregon. *Forest Science* 31, 995-1005.

Neale, D.B. and Adams, W.T. (1985a) The mating system in natural and shelterwood stands of Douglas-fir. *Theoretical and Applied Genetics* 71, 201-207.

Neale, D.B. and Adams, W.T. (1985b) Allozyme and mating-system variation in balsam fir (*Abies balsamea*) across a continuous elevational transect. *Canadian Journal of Botany* 63, 2448-2453.

Neale, D.B. and Harry, D.E. (1994) Genetic mapping in forest trees: RFLPs, RAPDs and beyond. *AgBiotech News and Information* 6, 107N-114N.

Neale, D.B. and Savolainen, O. (2004) Association genetics of complex traits in conifers. *Trends in Plant Science* 9, 325-330.

Neale, D.B. and Sederoff, R.R. (1989) Paternal inheritance of chloroplast DNA and maternal inheritance of mitochondrial DNA in loblolly pine. *Theoretical and Applied Genetics* 77, 212-216.

Neale, D.B. and Williams, C.G. (1991) Restriction fragment length polymorphism mapping on conifers and applications to forest tree genetics and tree improvement. *Canadian Journal of Forest Research* 21, 545-554.

Neale, D.B., Wheeler, N. and Allard, R.W. (1986) Paternal inheritance of chloroplast DNA in Douglas-fir. *Canadian Journal of Forest Research* 16, 1152-1154.

Neale, D., Marshall, K. and Sederoff, R. (1989) Chloroplast and mitochondrial DNA are paternally inherited in *Sequoia sempervirens* D. Don Endl. *Proceedings of the National Academy of Sciences of the United States of America* 86, 9347-9349.

Neale, D., Marshall, K. and Harry, D. (1991) Inheritance of chloroplast and mitochondrial DNA in incense-cedar (*Calocedrus decurrens*). *Canadian Journal of Forest Research* 21, 717-720.

Nei, M. (1987) *Molecular Evolutionary Genetics.* Columbia University Press, New York, NY.

Nei, M. and Kumar, S. (2000) *Molecular Evolution and Phylogenetics.* Oxford University Press, New York, NY.

Nei, M., Maruyama, T. and Chakraborty, R. (1975) The bottleneck effect and genetic variability in populations. *Evolution* 29, 1-10.

Nelson, C., Nance, W. and Doudrick, R. (1993) A partial genetic linkage map of slash pine (*Pinus elliottii Englem. var. elliottii*) based on random amplified polymorphic DNAs. *Theoretical and Applied Genetics* 87, 145-151.

Neter, J. and Wasserman, W. (1974) *Applied Linear Statistical Models.* Richard D.Irwin and Company, Homewood, IL.

Newton, R.J., Vibrah, H.S., Dong, N., Clapman, D.H., and Von Anold, S. (1992) Expression of an abscisic acid responsive promoter in *Picea abies* (L.) Karst. following bombardment from an electric discharge particle accelerator. *Plant Cell Reports* 11, 188-191.

Newton, R., Wakamiya, I. and Price, H.J. (1993) *Handbook of Plant and Crop Stress.* Marcel Dekker, Inc., New York, NY.

Nicholas, F.W. (1980) Size of population required for artificial selection. *Genetical Research* 35, 85-105.

Niklas, K. (1997) *The Evolutionary Biology of Plants.* The University of Chicago Press, Chicago, IL.

Nikles, D.G. (1986) Strategy and rationale of the breeding programmes with *Pinus caribaea* and its hybrids in Australia. In: *Proceedings of the International Union of Forest Research Organizations (IUFRO), Conference on Breeding Theory, Progeny Testing and Seed Orchards.* Williamsburg, VA, pp. 298.

Nikles, D.G. (1992) Hybrids of forest trees: the bases of hybrid superiority and a discussion of breeding methods. *Proceedings of the International Union of Forest Research Organizations (IUFRO), Conference Resolving Tropical Forest Research Concerns Through Tree Improvement, Gene Conservation, and Domestication of New Species.* Cali, Colombia., pp. 333-347.

Nikles, D.G. (2000) Experience with some *Pinus* hybrids in Queensland, Australia. In: *Hybrid Breeding and Genetics of Forest Trees. Proceedings of the Queensland Forest Research Institute/Cooperative Research Center-Sustainable Production Forestry (QFRI/CRC-SPF). Symposium.* Noosa, Queensland, Australia, pp. 27-43.

Nikles, D.G. and Robinson, M.J. (1989) The development of *Pinus* hybrids for operational use in Queensland. In: Gibson, G.I., Griffin, A.R. and Matheson, A.C. (eds.) *Breeding Tropical Trees: Population Structure and Genetic Improvement Strategies in Clonal and Seedling Forestry.* Oxford Forestry Institute, Oxford, UK, pp. 272-282

Nilsson, O., Little, C.H.A., Sandberg, G. and Olsson. O. (1996) Expression of two heterologous promoters *Agrobacterium rhizogenes rolC* and cauliflower mosaic virus 35S, in the stem of transgenic hybrid aspen plants during the annual cycle of growth and dormancy. *Plant Molecular Biology* 31, 887-895.

Nirenberg, M.W. and Matthei J.H. (1961) The dependence of cell-free protein synthesis in *E. coli* upon naturally occurring or synthetic polyribonucleotides. *Proceedings of the National Academy of Sciences of the United States of America* 47, 1588-1602.

Oak, S.W., Blakeslee G.M., and Rockwood, D.L. (1987) Pitch canker resistant slash pine identified by greenhouse screening. In: *Proceedings of the 19th Southern Forest Tree Improvement Conference.* College Station, TX, pp. 132-139.

Ohba, K. (1993) Clonal forestry with sugi (*Cryptomeria japonica*). In: Ahuja, M.R. and Libby, W.J. (eds.) *Clonal Forestry II: Conservation and Application.* Springer-Verlag, New York, NY, pp. 66-90.

Ohba, K., Iwakawa, M., Okada, Y. and Murai, M. (1971) Paternal transmission of a plastid anomaly in some reciprocal crosses of Sugi, *Cryptomeria japonica* D.Don. *Silvae Genetica* 20(4), 101-107.

Ohta, T. (1996) The current significance and standing of neutral and nearly neutral theories. *BioEssays* 18, 673-677.

Ohta, T. and Cockerham, C.C. (1974) Detrimental genes with partial selfing and effects on a neutral locus. *Genetical Research, Cambridge* 23, 191-200.

Oldfield, S., Lusty, C., and MacKinven, A. (1998) *The World List of Threatened Trees.* World Conservation Press, Cambridge, UK.

O'Malley, D. and McKeand, S. (1994) Marker assisted selection for breeding value in forest trees. *Forest Genetics* 1, 207-218.

O'Malley, D.M. and Whetten, R. (1997) Molecular markers and forest trees. In: Gaetano-Anolles, G. and Gresshoff, P.M. (eds.) *DNA markers: Protocols, Applications and Overviews.* John Wiley & Sons, New York, NY.

O'Malley, D.M., Porter, S. and Sederoff, R.R. (1992) Purification, characterization, and cloning of cinnamyl alcohol dehydrogenase in loblolly pine (*Pinus taeda* L.). *Plant Physiology* 98, 1364-1371.

Orr-Ewing, A.L. (1957) A cytological study of the effects of self-pollination on *Pseudotsuga menziesii* (Mirb.) Franco. *Silvae Genetica* 6, 179-185.

Osorio, L.F. (1999) Estimation of genetic parameters, optimal test designs and prediction of the genetic merit of clonal and seedling material of *Eucalyptus grandis*. School of Forest Resources and Conservation, University of Florida, Gainesville, FL.

Osorio, L.F., Wright, J.A. and White, T.L. (1995) Breeding strategy for *Eucalyptus grandis* at Smurfit Carton de Colombia. In: Potts, B.M., Borralho, N.M.G., Reid, J.B., Cromer, R.N., Tibbits, W.N. and Raymond, C.A. (eds.) *Proceedings of the International Union of Forest Research Organizations (IUFRO), Conference on Eucalypt Plantations: Improving Fibre Yield and Quality.* Hobart, Australia, pp. 264-266.

Ostrander, E.A., Jong, P.M, Rine, J. and Duyk, G. (1992) Construction of small insert genomic libraries highly enriched for microsatellite repeat sequences. *Proceedings of the National Academy of Sciences of the United States of America* 89, 3419-3423.

Owens, J.N. and Blake, M.D. (1985) Forest Tree Seed Production. Information Report PI-X-53, Petawawa National Forestry Institute, Canadian Forestry Service, Agriculture Canada. Victoria, BC, Canada.

Owens, J.N., Colangeli, A.M., Morris, S.J. (1991) Factors affecting seed set in Douglas-fir (*Psuedotsuga menziesii*). *Canadian Journal of Botany* 69, 229-238.

Palmer, E.J. (1948) Hybrid oaks of North America. J*ournal of Arnold Arbor* 29, 1-48.

Pancel, L. (1993) Species selection. In: Pancel, L. (ed.) *Tropical Forestry Handbook, Vol. 1.* Springler Verlag, New York., N.Y, pp. 569-643.

Papadakis, J.S. (1937) Methode statistique pour des experiences sur champ. *Bulletin de l'Institut de l'Amelioration des Plantes Thessalonique, 23.*

Park, Y.S. and Fowler, D.P. (1981) Provenance tests of red pine in the maritimes. Canadian Forest Service Information Report M-X-131.

Park, Y.S. and Fowler, D.P. (1982) Effects of inbreeding and genetic variances in a natural population of tamarack [*Larix laricina* (Du Roi) K. Koch] in eastern Canada. *Silvae Genetica* 31, 21-26.

Parker, S.R., White, T.L., Hodge, G.R. and Powell, G.L. (1998) The effects of scion maturation on growth and reproduction of grafted slash pine. *New Forests* 15, 243-259.

Pasonen, H.-L., Seppänen, S.-K., Degefu, Y., Rytkönen, A., Von Weissenberg, K., and Pappinen, A. (2004) Field performance of chitinase transgenic silver birches (*Betula pendula*): resistance to fungal disease. *Theoretical and Applied Genetics* 109, 562-570.

Patterson, H.D. and Thompson, R. (1971) Recovery of interblock information when block sizes are unequal. *Biometrika* 58, 545-554.

Patterson, H.D. and Williams, E.R. (1976) A new class of resolvable incomplete block designs. *Biometrika* 63, 83-92.

Patterson, H.D., Williams, E.R. and Hunter, E.A. (1978) Block designs for variety trials. *Journal of Agricultural Science* 90, 395-400.

Paul, A.D., Foster, G.S., Caldwell, T., and McRae, J. (1997) Trends in genetic and environmental parameters for height, diameter, and volume in a multilocation clonal study with loblolly pine. *Forest Science* 43, 87-98.

Pederick, L.A. (1967) The structure and identification of the chromosomes of *Pinus radiata* D. Don. *Silvae Genetica* 16, 69-77.

Pederick, L.A. (1968) Chromosome inversions in *Pinus radiata*. *Silvae Genetica* 17, 22-26.

Pederick, L.A. (1970) Chromosome relationships between *Pinus* species. *Silvae Genetica* 19, 171-180.

Pederick, L.A. and Griffin, A.R. (1977) The genetic improvement of radiata pine in Australasia, In: *Proceedings of the 3rd World Consultation on Forest Tree Breeding*. Canberra, Australia, 561-572.

Perry, D. and Bousquet, J. (1998) Sequence-tagged-site (STS) markers of arbitrary genes: development, characterization and analysis of linkage in black spruce. *Genetics* 149, 1089-1098.

Perry, D.J. and Dancik, B.P. (1986) Mating system dynamics of lodgepole pine in Alberta, Canada. *Silvae Genetica* 35, 190-195.

Perry, D.J. and Furnier, G.R. (1996) *Pinus banksiana* has at least seven expressed alcohol dehydrogenase genes in two linked groups. *Proceedings of the National Academy of Sciences of the United States of America* 93, 13020-13023.

Perry, D.J. and Knowles, P. (1990) Evidence of high self-fertilization in natural populations of eastern white cedar (*Thuja occidentalis*). *Canadian Journal of Botany* 68, 663-668.

Peters, G.B., Lonie, J.S. and Moran, G.F. (1990) The breeding system, genetic diversity and pollen sterility in *Eucalyptus pulverulenta*, a rare species with small disjunct populations. *Australian Journal of Botany* 38, 559-570.

Pfeiffer, A., Olivieri, A. and Morgante, M. (1997) Identification and characterization of microsatellites in Norway spruce (*Picea abies* K.). *Genome* 40, 411-419.

Pigliucci, M., Benedettelli, S. and Villani, F. (1990) Spatial patterns of genetic variability in Italian chestnut (*Castanea sativa*). *Canadian Journal of Botany* 68, 1962-1967.

Pilate, G., Guiney, E., Holt, K., Petit-Conil, Michel, Lapierre, C., Leplé, J.-C., Pollet, B., Mila, I.,

Webster, E.A., Marstorp, H.G., Hopkins, D.W., Jouanin, L., Boerjan, W., Schuch, W., Cornu, D. and Halpin, C. (2002) Field and pulping performances of transgenic trees with altered lignification. *Nature Biotechnology* 20, 607-612.

Pilger, R. (1926) *Pinus. Die Naturlichen Pflanzenfamilien* 13, 331-342.

Pimm, S.L., Russell, G.J., Gittleman, J.L. and Brooks, T.M. (1995) The future of biodiversity. *Science* 269, 347-350.

Plessas, M.E. and Strauss, S.H. (1986) Allozyme differentiation among populations, stands, and cohorts in Monterey pine. *Canadian Journal of Forest Research* 16, 1155-1164.

Plomion, C. and Durel, C. (1985) Estimation of the average effects of specific alleles detected by the pseudo-testcross QTL mapping strategy. *Genetics of Self Evolution* 28, 223-235.

Plomion, C., Costa, P., Bahrman, N. and Frigerio, J.M. (1997) Genetic analysis of needle proteins in maritime pine. *Silvae Genetica* 46, 2-3.

Pollack, J. and Dancik, B. (1985) Monoterpene and morphological variation and hybridization of *Pinus contorta* and *P. banksiana* in Alberta. *Canadian Journal of Botany* 63, 201-210.

Ponoy, B., Hong, Y.-P., Woods, J., Jaquish, B. and Carlson, J. (1994) Chloroplast DNA diversity of Douglas-fir in British Columbia. *Canadian Journal of Forest Research* 24, 1824-1834.

Porterfield, R.L. (1975) Economic aspects of tree improvement programs. In: Thielges, B.A. (ed.), *Forest Tree Improvement-The Third Decade. Proceedings of the 24th Annual Louisiana State University Symposium.* Louisiana State University, Baton Rouge, LA, pp. 99-117.

Porterfield, R.L. and Ledig, F.T. (1977) The economics of tree improvement programs in the northeast, pp. 35-47. In: *Proceedings of the 25th Northeast Forest Tree Improvement Conference.* Orono, ME, pp. 35-47.

Potts, B.M. and Dungey, H.S. (2001) Hybridisation of *Eucalyptus*: key issues for breeders and geneticists. In: *Proceedings of the International Union of Forest Research Organizations (IUFRO), Conference on Developing the Eucalypt of the Future.* Valdivia, Chile, pp. 34.

Powell, G.L. and White, T.L. (1994) Cone and seed yields from slash pine seed orchards. *Southern Journal of Applied Forestry* 18, 122-127.

Powell, J.R. and Taylor, C.E. (1979) Genetic variation in ecologically diverse environments. *American Scientist* 67, 590-596.

Powell, M.B. and Nikles, D.G. (1996) Performance of *Pinus elliottii* var. *elliottii* and *P. Caribaea* var. *hondurensis*, and their F_1, F_2 and backcross hybrids across a range of sites in Queensland. In: Dieters, M.J., Matheson. A.C., Nikles, D.G., Harwood, C.E. and Walker, S.M. (eds.) *Proceedings of the Queensland Forest Research Institute-International Union of Forest Research Organizations (QFRI-IUFRO), Conference on Tree Improvement for Sustainable Tropical Forestry.* Caloundra, Queensland, Australia, pp. 382-383.

Powell, M.B., Borralho, N.M.G., Wormald, N. and Chow, E. (1997) What X-A program to optimise selection and mate allocation in tree breeding. In: *Proceedings of the International Union of Forest Research Organizations (IUFRO), Conference on sobre Silvicultura e Melhoramento de Eucaliptos.* El Salvador, Brazil, pp 427-432.

Powell, W., Morgante, M., Andre, C., McNicol, J.W., Machray, G.C., Doyle, J.J., Tingey, S.V. and Rafalski, J.A. (1995) Hypervariable microsatellites provide a general source of polymorphic DNA markers for the chloroplast genome. *Current Biology* 5, 1023-1029.

Pravdin, L.F., Abaturova, G.A. and Shershukova, O.P. (1976) Karyological analysis of European and Siberian spruce and their hybrids in the USSR. *Genetica* 25, 89-94.

Price, H.J., Sparrow, A.H. and Nauman, A.F. (1973) Evolutionary and developmental considerations of the variability of nuclear parameters in higher plants. I. Genome volume, interphase chromosome volume and estimated DNA content of 236 gymnosperms. *Brockhaven Symposia in Biology* 25, 390-421.

Price, R., Olsen-Stojkovich, J. and Lowenstein, J. (1987) Relationships among the genera of Pinaceae: An immunological comparison. *Systematic Botany* 12, 91-97.

Price, R., Liston, A. and Strauss, S. (1998) Phylogeny and systematics of *Pinus*. In: Richardson, D.M. (ed.) E*cology and Biogeography of Pinus.* Cambridge University Press, Cambridge, UK, pp.49-68.

Pryor, L.D. and Johnson, L.A.S. (1981) Eucalyptus, the universal Australian. In: Keast, A. (ed.) *Ecological Biogeography of Australia.* Junk, The Hague, The Netherlands, pp. 499-536.

Purnell, R.C. and Kellison, R.C. (1983) A tree improvement program for southern hardwoods. In:

Proceedings of the 17th Southern Forest Tree Improvement Conference. Athens, GA, pp. 90-98.

Qiu, Y-L. and Parks, C.R. (1994) Disparity of allozyme variation levels in three Magnolia (Magnoliaceae) species from the Southeastern United States. *American Journal of Botany* 81, 1300-1308.

Radetzky, R. (1990) Analysis of mitochondrial DNA and its inheritance in *Populus*. *Current Genetics* 18, 429-434.

Radhamani, A., Nicodemus, A., Nagarajan, B. and Mandal, A.K. (1998) Reproductive biology of tropical tree species. In: Mandal, A.K. and Gibson, G.L. (eds.) *Forest Genetics and Tree Breeding.* CBS Publishers and Distributors, Darya Ganj, New Delhi, India, pp. 194-204.

Rafalski, A. (2002) Applications of single nucleotide polymorphisms in crop genetics. *Current Opinion in Plant Biology* 5, 94-100.

Rajora, O.P. and Dancik, B.P. (1992) Chloroplast DNA inheritance in *Populus*. *Theoretical and Applied Genetics* 84, 280-285.

Rajora, O.P., DeVerno, L.L. Mosseller, A. and Innes, D.J. (1998) Genetic diversity and population structure of disjunct Newfoundland and central Ontario populations of eastern white pine (*Pinus strobus*). *Canadian Journal of Botany* 76, 500-508.

Randall, W.K. (1996) *Forest Tree Seed Zones for Western Oregon.* Oregon Department of Forestry, Salem, OR.

Rapley, L.P., Allen, G.R., and Botts, B.M. (2004a) Genetic variation in *Eucalyptus globulus* in relation to susceptibility from attack by the southern eucalypt leaf beetle, *Chrysophtharta Agricola*. *Australian Journal of Botany 52, 747-756.*

Rapley, L.P., Allen, G.R., and Botts, B.M. (2004b) Genetic variation of *Eucalyptus globulus* in relation to autumn gum moth *Mnesampela* private *Lepidoptera*: Geometridae) oviposition preference. *Forest Ecology and Management* 194, 169-175.

Ratnam, W., Lee, C.T., Muhammad, N. and Boyle, T.J.B. (1999) Impact of logging on genetic diversity in humid tropical forests. In: Matyas, C. (ed.) *Forest Genetics and Sustainability.* Kluwer Academic Press, Boston, MA, pp. 171-182.

Raymond, C.A., Owen, J.V., Eldridge, K.G. and Harwood, C.E. (1992) Screening eucalypts for frost tolerance in breeding programs. *Canadian Journal of Forest Research* 22, 1271-1277.

Redmond, C.H. and Anderson, R.L. (1986) Economic benefits of using the Resistance Screening Center to assess relative resistance to fusiform rust. *Southern Journal of Applied Forestry* 10, 34-37.

Reddy, K.V. and Rockwood, D.L. (1989) Breeding strategies for coppice production in a *Eucalyptus grandis* base population with four generations of selection. *Silvae Genetica* 38, 148-151.

Reed, D.H. and Frankham, R. (2001) How closely correlated are molecular and quantitative measures of genetic variation? A meta-analysis. *Evolution* 55, 1095-1103.

Rehfeldt, G.E. (1983a) Genetic variability within Douglas-fir populations: Implications for tree improvement. *Silvae Genetica* 32, 9-14.

Rehfeldt, G.E. (1983b) Seed Transfer Guidelines for Douglas-fir in Central Idaho. USDA Forest Service Research Note INT-337.

Rehfeldt, G.E. (1983c) Adaptation of *Pinus contorta* populations to heterogeneous environments in northern Idaho. *Canadian Journal of Forest Research* 13, 405-411.

Rehfeldt, G.E. (1985) Genetic variances and covariances in *Pinus contorta*: estimates of genetic gains from index selection. *Silvae Genetica* 34, 26-33.

Rehfeldt, G.E. (1986) Adaptive variation in *Pinus ponderosa* from intermountain regions. I. Snake and Salmon River basins. *Forest Science* 32, 79-92.

Rehfeldt, G.E. (1988) Ecological genetics of *Pinus contorta* from the Rocky Mountains (USA): a synthesis. *Silvae Genetica* 37, 131-135.

Rehfeldt, G.E. (1989) Ecological adaptations in Douglas-fir (*Pseudotsuga menziesii* var. *glauca*): a synthesis. *Forest Ecology and Management* 28, 203-215.

Rehfeldt, G.E. (2000) Genes, Climate and Wood. *The Leslie L. Schaffer Lectureship in Forest Science.* University of British Columbia, Vancouver, BC.

Rehfeldt, G.E., Hoff, R.J. and Steinhoff, R.J. (1984) Geographic patterns of genetic variation in *Pinus monticola*. *Botanical Gazette* 145, 229-239.

Rehfeldt, G.E., Wykoff, W.R., Hoff, R.J. and Steinhoff, R.J. (1991) Genetic gains in growth and simulated yield of *Pinus monticola*. *Forest Science* 37, 326-342.

Reilly, J.J. and Nikles, D.G. (1977) Analysing benefits and costs of tree improvement: *Pinus cari-*

baea. In: *Proceedings of the 3rd World Consultation on Forest Tree Breeding.* Canberra, Australia.

Riemenschneider, D.E., Haissig, B.E., Selmer, J. and Fillatti, J.J. (1988) Expression of a herbicide tolerance gene in young plants of transgenic hybrid poplar clone. In: Ahuja, M.R. (ed.) *Somatic Cell Genetics of Woody Plants.* Kluwer Academic Publishers, Dordrecht, The Netherlands, pp. 73-80.

Remington, D.L., Wu, R.L., MacKay, J.J., McKeand, S.E. and O'Malley, D.M. (1998) Average effect of a mutation in lignin biosynthesis in loblolly pine. *Theoretical and Applied Genetics* 99, 705-710.

Remington, D., Whetten, R., Liu, B. and O' Malley, D. (1999) Construction of an AFLP genetic map with nearly complete genome coverage in *Pinus taeda. Theoretical and Applied Genetics* 98, 1279-1292.

Remington, D., Thornsberry, J., Matsuoka, Y., Wilson, L., Whitt, S., Doebley, J., Kresovich, S., Goodman, M. and Buckler IV, E. (2001) Structure of linkage disequilibrium and phenotypic associations in the maize genome. *Proceedings of the National Academy of Sciences of the United States of America* 98, 11479-11484.

Retief, E.C.L. and Clarke, C.R.E. (2000) The effect of site potential on eucalypt clonal performance in coastal Zululand, South Africa. In: *Proceedings of the International Union of Forest Research Organizations (IUFRO), Working Party, Forest Genetics for the Next Millennium.* Durban, South Africa, pp. 192-196.

Ridley, M. (1993) *Evolution.* Blackwell Scientific, Boston, MA.

Riemenschneider, D.E. (1988) Heritability, age-age correlations and inferences regarding juvenile selection in jack pine. *Forest Science* 34, 1076-1082.

Risch, N. (2000) Searching for genetic determinants in the new millennium. *Nature* 405, 847-856.

Ritland, K. and El-Kassaby, Y.A. (1985) The nature of inbreeding in a seed orchard of Douglas fir as shown by an efficient multilocus model. *Theoretical and Applied Genetics* 71, 375-384.

Ritland, K. and Jain, S. (1981) A model for the estimation of outcrossing rate and gene frequencies using *n* independent loci. *Heredity* 47, 35-52.

Roberds, J.H. and Bishir, J.W. (1997) Risk analyses in clonal forestry. *Canadian Journal of Forest Research* 27, 425-432.

Roberds, J. and Conkle, M.T. (1984) Genetic structure in loblolly pine stands: allozyme variation in parents and progeny. *Forest Science* 30, 319-329.

Roberds, J.H., Friedman, S.T. and El-Kassaby, Y.A. (1991) Effective number of pollen parents in clonal seed orchards. *Theoretical and Applied Genetics* 82, 313-320.

Robertson, A. (1960) A theory of limits in artificial selection. *Proceedings of the Royal Society Series B-Biological Sciences* 153, 234-249.

Robertson, A. (1961) Inbreeding in artificial programmes. *Genetic Research* 2, 189-194.

Rockwood, D.L. and Kok, H.R. (1977) Development and potential of a longleaf pine seedling seed orchard. In: *Proceedings of the 14th Southern Forest Tree Improvement Conference.* Gainesville, FL, pp. 78-86.

Rockwood, D.L., Warrag, E.E., Javenshir, K. and Kratz, K. (1989) Genetic improvement for *Eucalyptus grandis* for Southern Florida. In: *Proceedings of the 20th Southern Forest Tree Improvement Conference.* Charleston, SC, pp. 403-410.

Rockwood, D. L., Dinus, R.J., Kramer, J.M., McDonough, T.J., Raymond, C.A., Owen, J.V. and DeValerio, J.T. (1993) Genetic variation for rooting, growth, frost hardiness and wood, fiber, and pulping properties in Florida-grown *Eucalyptus amplifolia.* In: *Proceedings of the 22nd Southern Forest Tree Improvement Conference.* Atlanta, GA, pp. 81-88.

Roff, D.A. (1997) *Evolutionary Quantitative Genetics.* Chapman & Hall, New York.

Roff, D.A. and Mousseau, T.A. (1987) Quantitative genetics and fitness: Lessons from *Drosophila. Heredity* 58, 103-118.

Rogers, D.L., Stettler, R.F. and Heilmann, P.E. (1989) Genetic variation and productivity of *Populus trichocarpa* and its hybrids. III. Structure and pattern of variation in a 3-year field test. *Canadian Journal of Forest Research* 19, 372-377.

Rogers, S. and Bendich, A. (1987) Ribosomal RNA in plants: variability in copy number and in the intergenic spacer. *Plant Molecular Biology* 9, 509-520.

Rogstad, S., Patton II, J. and Schaal, B. (1988) M13 repeat probe detects DNA minisatellite-like sequences in gymnosperms and angiosperms. *Proceedings of the National Academy of Sciences of the United States of America* 85, 9176-9178.

Rogstad, S., Nybom, H. and Schaal, B. (1991) The tetrapod "DNA fingerprinting" M13 repeat probe reveals genetic diversity and clonal growth in quaking aspen (*Populus tremuloides*, Salicaceae). *Plant Systematics and Evolution* 175, 115-123.

Rosvall, O., Lindgren, D. and Mullin, T.J. (1998) Sustainability robustness and efficiency of a multi-generation breeding strategy based on within-family clonal selection. *Silvae Genetica* 47, 307-321.

Rottmann, W.H., Meilan, R. Sheppard, L.A., Brunner, A.M., Skinner, J.S., Ma, C., Cheng, S., Jouanin, L., Pilate, G. and Strauss, S.H. (2000) Diverse effects of overexpression of *LEAFY* and *PTLF*, a poplar (*Populus*) homolog of *LEAFY/FLORICAULA*, in transgenic poplar and *Arabidopsis*. *Plant Journal* 22, 235-246.

Rudin, D. and Ekberg, I. (1978) Linkage studies in *Pinus sylvestris L.*: using macro gametophyte allozymes. *Silvae Genetica* 27, 1-12.

Ruotsalainen, S. and Lindgren, D. (1997) Predicting genetic gain of backward and forward selection in forest tree breeding. *Silvae Genetica* 47, 42-50.

Ruotsalainen, S. and Lindgren, D. (2001) Number of founders for a breeding population using variable parental contribution. *Forest Genetics* 8, 57-67.

Rushton, B.S. (1993) Natural hybridization within the genus *Quercus* L. *Annales des Sciences Forestieres* 50, 73s-90s.

Russell, J.H. and Libby, W.J. (1986) Clonal testing efficiency: The trade-offs between clones tested and ramets per clone. *Canadian Journal of Forest Research* 16, 925-930.

Russell, J.H. and Loo-Dinkins, J.A. (1993) Distribution of testing effort in cloned genetic tests. *Silvae Genetica* 42, 98-104.

Saintagne, C., Bodenes, C., Barreneche, T., Pot, D., Plomion, C. and Kremer, A. (2004) Distribution of genomic regions differentiating oak species assessed by QTL detection. *Heredity* 92, 20-30.

Sakai, A., Scarf, S., Faloona, F., Mullis, K.B., Horn, G.T., Erlich, H.A. and Arnhiem, N. (1985) Enzymatic amplification of beta-globin genomic sequences and restriction site analysis for diagnosis of sickle-cell anemia. *Science* 230, 1350-1354.

Sanger, F., Nicklen, S. and Coulson, A.R. (1977) DNA sequencing with chain terminating inhibitors. *Proceedings of the National Academy of Sciences of the United States of America* 74, 5463-5467.

Santamour, F.S. (1960) New chromosome counts in *Pinus* and *Picea*. *Silvae Genetica* 9, 87-88.

SAS Institute Inc. (1988) *SAS/STAT User's Guide, Release 6.03*. SAS Institute Inc., Cary, NC.

SAS Institute Inc. (1996) *SAS/STAT Software: Changes and Enhancements*. SAS Institute Inc., Cary, NC.

Savill, P.S. and Evans, J. (1986) *Plantation Silviculture in Temperate Regions*. Oxford University Press, Oxford, UK.

Savolainen, O. (1994) Genetic variation and fitness: conservation lessons from pines. In: Loeschcke, V., Tomiuk, J. and Jain, S.-K. (eds.) *Conservation Genetics*. Birkhauser Verlag, Basel, Switzerland, pp. 27-36.

Savolainen, O. and Hedrick, P. (1995) Heterozygosity and fitness: no association in Scots pine. *Genetics* 140, 755-766.

Savolainen, O. and Kärkkäinen, K. (1992) Effects of forest management on gene pools. *New Forests* 6, 329-345.

Savolainen, O., Karkkainen, K. and Kuittinen, H. (1992) Estimating numbers of embryonic lethals in conifers. *Heredity* 69, 308-314.

Sax, K. (1923) The association of size differences with seedcoat pattern and pigmentation in *Phaseolus vulgaris*. *Genetics* 8, 552.

Sax, K. and Sax, H.J. (1933) Chromosome number and morphology in the conifers. *Journal of the Arnold Arboretum* 14, 356-375.

Saylor, L.C. (1961) A karyotypic analysis of selected species of *Pinus*. *Silvae Genetica* 10, 77-84.

Saylor, L.C. (1964) Karyotype analysis of *Pinus*-group *Lariciones*. *Silvae Genetica* 13, 165-170.

Saylor, L.C. (1972) Karyotype analysis of the genus *Pinus* - subgenus *Pinus*. *Silvae Genetica* 21, 155-163.

Scalfi, M., Troggio, M., Piovani, P., Leonardi, S., Magnaschi, G., Vendramin, G.G. and Menozi, P.

(2004) A RAPD, AFLP and SRR linkage map, and QTL analysis in European beech (*Fagus sylvatica* L.). *Theoretical and Applied Genetics* 108, 433-441.

Schemske, D.W. and Lande, R.L. (1985) The evolution of self-fertilization and inbreeding depression in plants. II. Empirical observations. *Evolution* 39, 41-52.

Schena, M., Shalon, D., Davis, R. and Brown, P. (1995) Quantitative monitoring of gene expression patterns with a complementary DNA microarray. *Science* 270, 467-470.

Schenk, P., Kazan, K., Wilson, I., Anderson, J., Richmond, T., Somerville, S. and Manners, J. (2000) Coordinated plant defense responses in *Arabidopsis* revealed by microarray analysis. *Proceedings from the National Academy of Sciences of the United States of America* 97, 11655-11660.

Schlarbaum, S.E. and Tzuchiya, T. (1975a) The chromosome study of giant sequoia, *Sequoiadendron giganteum. The Journal of Heredity* 66, 41-42.

Schlarbaum, S.E. and Tsuchiya, T. (1975b) Chromosomes of incense cedar. *Silvae Genetica* 33, 56-62.

Schmidtling, R.C. (1999) Revising the seed zones for southern pines. *Proceedings of the 25th Southern Forest Tree Improvement Conference.* New Orleans, LA, pp. 152-154.

Schmidtling, R.C. and Hipkins, V. (1998) Genetic diversity in longleaf pine (*Pinus palustris* Mill.): Influence of historical and prehistorical events. *Canadian Journal of Forest Research* 28, 1135-1145.

Schmidtling, R.C., Carroll, E. and LaFarge, T. (1999) Allozyme diversity of selected and natural loblolly pine populations. *Silvae Genetica* 48, 35-45.

Schnabel, A. and Hamrick, J.L. (1990) Organization of genetic diversity within and among populations of *Gleditsia triacanthos* (Leguminosae). *American Journal of Botany* 77, 1060-1069.

Schnabel, A. and Hamrick, J.L. (1995) Understanding the population genetic structure of *Gleditsia triacanthos* L: the scale and pattern of pollen gene flow. *Evolution* 49, 921-931.

Schnabel, A., Laushman, R.H. and Hamrick, J.L. (1991) Comparative genetic structure of two co-occurring tree species, *Maclura pomifera* (Moraceae) and *Gleditsia triacanthos* (Leguminosae). *Heredity* 67, 357-364.

Schoen, D.J. and Stewart, S.C. (1986) Variation in male reproductive investment and male reproductive success in white spruce. *Evolution* 40, 1109-1120.

Schofield, E.K. (1989) Effects of introduced plants and animals on island vegetation: examples from the Galapagos Archipelago. *Conservation Biology* 3, 227-238.

Scholz, F., Gregorius, H.R. and Rudin, D. (1989) *Genetic Aspects of Air Pollutants in Forest Tree Populations.* Springer-Verlag, New York, NY.

Scotti-Saintagne, C., Bodenes, C., Barreneche, T., Bertocchi, E., Plomion, C. and Kremer, A. (2004) Detection of quantitative trait loci controlling bud burst and height growth in *Quercus robur* L. *Theoretical and Applied Genetics* 109, 1648-1659.

Schuster, W.S.F. and Mitton, J.B. (1991) Relatedness within clusters of a bird-dispersed pine and the potential for kin interactions. *Heredity* 67, 41-48.

Searle, S.R. (1974) Prediction, mixed models, and variance components. In: *Reliability and Biometry.* SIAM Publishing, Philadelphia, PA, pp. 229-266.

Searle, S.R. (1987) *Linear Models for Unbalanced Data.* John Wiley & Sons, New York, NY.

Searle, S.R., Casella, G. and McCulloch, C.E. (1992) *Variance Components.* John Wiley & Sons, Inc., New York, New York.

Seavey, S.R. and Bawa, K.S. (1986) Late-acting self-incompatibility in angiosperms. *The Botanical Review* 52, 195-219.

Sedjo, R.A. and Botkin, D.B. (1997) Using forest plantations to spare natural forests. *Environment* 39, 14-20.

Sederoff, R.R. (1999) Tree genomes: what will we understand about them by the year 2020? In: Matyas, C. (ed.) *Forest Genetics and Sustainablity.* Kluwer Academic Publishers, Dordrecht, The Netherlands.

Sederoff, R.R. and Ledig, F.T. (1985) Increasing forest productivity and value through biotechnology. In: *Forest Potential, Productivity and Value. Weyerhaueser Science Symposium, Volume 4.* Weyerhaueser Company, Tacoma, WA, pp. 253-276

Seppänen, S.-K., Syrjälä, L., von Weissenberg, K., Teeri, T.H., Paajanen, L. and Pappinen, A. (2004) Antifungal activity of stilbenes in vitro bioassays and transgenic *Populus* expressing a gene encoding pinosylvin synthase. *Plant Cell Reports* 22, 84-593.

Sewell, M., and Neale, D. (2000) Mapping quantitative traits in forest trees. In: Jain, S., and Minocha, S. (eds.) *Molecular Biology of Woody Plants, Forestry Sciences, Volume 64*. Kluwer Academic Publishers, Dordrecht, The Netherlands, pp. 407-423.

Sewell, M., Que, Y-L, Parks, C. and Chase, M. (1993) Genetic evidence for trace paternal transmission of plastids in *Liriodendron* and *Magnolia* (Magnoliaceae). *American Journal of Botany* 80(7), 854-858.

Sewell, M., Sherman, B. and Neale, D. (1999) A consensus map for loblolly pine (*Pinus taeda* L.). I. Construction and integration of individual linkage maps from two outbred three-generation pedigrees. *Genetics* 151, 321-330.

Sewell, M., Bassoni, D., Megraw, R., and Wheeler, N.C. and Neale, D.B. (2000) Identification of QTLs influencing wood property traits in loblolly pine (*Pinus taeda* L.). I. Physical wood properties. *Theoretical and Applied Genetics* 101, 1273-1281.

Sewell, M., Davis, M.F., Tuskan, G.A., Wheeler, N. C., Elam, C.C., Bassoni, D.L. and Neale, D.B. (2002) Identification of QTLs influencing wood property traits in loblolly pine (*Pinus taeda* L.). II. Chemical wood properties. *Theoretical and Applied Genetics* 104, 214-222.

Shapcott, A. (1995) The spatial genetic structure in natural populations of the Australian temperate rainforest tree *Atherosperma moschatum* (Labill.) (Monimiaceae). *Heredity* 74, 28-38.

Sharma, N.P. (1992) *Managing the World's Forests: Looking for Balance Between Conservation and Development*. Kendall/Hunt Publishing Company, Dubuque, IA.

Sharp, P.A. and Zamore, P.D. (2000) RNA interference. *Science* 287, 2431-2433.

Shaw, D.V. and Allard, R.W. (1982) Estimation of outcrossing rates in Douglas-fir using isoenzyme markers. *Theoretical and Applied Genetics* 62, 113-120.

Shaw, D.V., Kahler, A.L. and Allard, R.W. (1981) A multilocus estimator of mating system parameters in plant populations. *Proceedings of the National Academy of Sciences of the United States of America* 78, 1298-1302.

Shaw, G. (1914) *The Genus Pinus*. Arnold Arboretum Publications, The Murray Printing Company, Forage Village, MA.

Shea, K.L. (1987) Effects of population structure and cone production on outcrossing rates in Engelmann spruce and subalpine fir. *Evolution* 4, 124-136.

Shearer, B.L. and Dillon, M. (1995) Susceptibility of plant species in *Eucalyptus marginata* forest to infection by *Phytophthora cinnamomi*. *Australian Journal of Botany* 43, 113-134.

Shelbourne, C.J.A. (1969) Tree breeding methods. In: *Forest Research Institute Technical Paper 55*. New Zealand Forest Service.

Shelbourne, C.J.A. (1991) Genetic gains from different kinds of breeding population and seed or plant production population. In: *Proceedings of the International Union of Forest Research Organizations (IUFRO), Symposium on Intensive Forestry: The Role of Eucalypts*. Durban, South Africa, pp. 300-317.

Shelbourne, C.J.A. (1992) Genetic gains from different kinds of breeding population and seed or plant production population. *South African Forestry Journal* 160, 49-65.

Shelbourne, C.J.A. and Low, C.B. (1980) Multi-trait index selection and associated genetic gains of *Pinus radiata* progenies at five sites. *New Zealand Journal of Forest Science* 10, 307-324.

Shelbourne, C.J.A. and Thulin, I.J. (1974) Early results from a clonal selection and testing programme with radiata pine. *New Zealand Journal of Forest Science* 4, 387-398.

Shelbourne, C.J.A., Burdon, R.D., Carson, S.D., Firth, A. and Vincent, T.G. (1986) Development plan for radiata pine breeding. New Zealand Forest Service.

Shelbourne, C.J.A., Carson, M.J. and Wilcox, M.D. (1989) New techniques in the genetic improvement of radiata pine. *Commonwealth Forestry Review* 68, 191-201.

Shen, H.H., Rudin, D. and Lindgren, D. (1981) Study of the pollination pattern in a Scots pine seed orchard by means of isozyme analysis. *Silvae Genetica* 30, 7-14.

Shin, D.-I., Podila, G.K., Huang, Y. and Karnosky, D.F. (1994) Transgenic larch expressing genes for herbicide and insect resistance. *Canadian Journal of Forest Research* 24, 2059-2067.

Shortt, R.L., Hawkin, B.J. and Woods, J.H. (1996) Inbreeding effects on the spring frost hardiness of coastal Douglas-fir. *Canadian Journal of Forest Research* 26, 1049-1054.

Shukla, G.K. (1972) Some statistical aspects of partitioning genotype-environmental components of varibility. *Heredity* 29, 237-245.

Sierra-Lucero, V., McKeand, S.E., Huber, D., Rockwood, D.L. and White, T.L. (2002) Performance differences and genetic parameters for four coastal provenances of loblolly pine in the south-eastern United States. *Forest Science* 48(4), 732-742.

Silen, R.R. (1966) A simple, progressive, tree improvement program for Douglas-fir. USDA Forest Service Research Note PNW-45.

Silen, R.R. (1978) Genetics of Douglas-fir. USDA Forest Service Research Paper, WO-35.

Silen, R. and Osterhaus, C. (1979) Reduction of genetic base by sizing of bulked Douglas-fir seed lots. *Tree Planters' Notes* 30, 24-30.

Silen, R.R. and Wheat, J.G. (1979) Progressive tree improvement program in coastal Douglas-fir. *Journal of Forestry* 77, 78-83.

Silva, J.C., Dutkowski, G.W. and Gilmour, A.R. (2001) Analysis of early tree height in forest genetic trials is enhanced by including a spatially correlated residual. *Canadian Journal of Forest Research* 31, 1887-1893.

Simon, J.P., Bergeron, Y. and Gagnon, D. (1986) Isozyme uniformity in populations of red pine (*Pinus resinosa*) in the Abitibi Region, Quebec. *Canadian Journal of Forest Research* 6, 1133-1135.

Simmons, A.J. (1996) Delivery of improvement for agroforestry trees. In: Dieters, M.J., Matheson, A.C., Nikles, D.G., Harwood, C.E. and Walker, S.M. (eds.) *Proceedings of the Queensland Forest Research Institute-International Union of Forest Research Organizations (QFRI-IUFRO), Conference on Tree Improvement for Sustainable Tropical Forestry*. Caloundra, Queensland, Australia, pp. 391-400.

Skrøppa, T. and Tho, T. (1990) Diallel crosses in *Picea abies*. I. Variation in seed yield and seed weight. *Scandinavian Journal of Forest Research* 5, 355-367.

Slater, A., Scott, N.W. and Fowler, M.R. (2003) *Plant Biotechnology: The Genetic Manipulation of Plants*. Oxford University Press, London, UK, 346p.

Slatkin, M. (1987) Gene flow and the geographic structure of natural populations. *Science* 236, 787-792.

Slatkin, M. and Barton, N.H. (1989) A comparison of three indirect methods for estimating average levels of gene flow. *Evolution* 43, 1349-1368.

Slavov, G.T., Howe, G.T., Yakoulev, I., Edwards, K.J., Krutouskii, K.V., Tuskan, G.A., Carlson, J.E., Strauss, S.H. and Adams, W.T. (2004) Highly variable SSR markers in Douglas-fir: Mendelian inheritance and map locations. *Theoretical and Applied Genetics* 108, 373-390.

Slavov, G.T., Howe, G.T. and Adams, W.T. (2005) Pollen contamination and mating patterns in a Douglas-fir seed orchard as measured by single sequence repeat markers. *Canadian Journal of Forest Research* 35, 1492-1603.

Sluder, E.A. (1980) A study of geographic variation in loblolly pine in Georgia: 20th-year results. USDA Forest Service Research Paper SE-213.

Smalley, E.B. and Guries, R.P. (1993) Breeding elms for resistance to Dutch elm disease. *Annual Review of Phytopathology* 31, 325-352.

Smith, C. (1969) Optimum selection procedures in animal breeding. *Animal Production* 11, 433-442.

Smith, C.C., Hamrick, J.L. and Kramer, C.L. (1988) The effects of stand density on frequency of filled seeds and fecundity in lodgepole pine (*Pinus contorta* Dougl.). *Canadian Journal of Forest Research* 18, 453-460.

Smith, D.B. and Adams, W.T. (1983) Measuring pollen contamination in clonal seed orchards with the aid of genetic markers. In: *Proceedings of the 17th Southern Forest Tree Improvement Conference*. Athens, GA, pp. 69-77.

Smith, D. and Devey, M. (1994) Occurrence and inheritance of microsatellites in *Pinus radiata*. *Genome* 37, 977-983.

Sneath, P.A. and Sokal, R.R. (1973) *Numerical Taxonomy*. W.H. Freeman and Company, San Francisco, CA.

Snedecor, G.W. and Cochran, W.G. (1967) *Statistical Methods*. Iowa State University Press, Ames, IA.

Sniezko, R.A. (1996) Developing resistance to white pine blister rust in sugar pine in Oregon. In: Kinloch, B.B., Marosy, M. and Huddleston, M.E. (eds.) *Symposium Proceedings of the California Sugar Pine Management Committee*. Publication No. 3362, University of California, Division of Agriculture and Natural Resources. Davis, CA, pp. 171-178.

Sniezko, R.A. and Zobel, B.J. (1988) Seedling height and diameter variation of various degrees of inbred and outcross progenies of loblolly pine. *Silvae Genetica* 37, 50-60.

Snyder, E.B. (1972) *Glossary for Forest Tree Improvement Workers*. USDA Forest Service.

Sohn, S. and Goddard, R.E. (1979) Influence of infection percent on improvement of fusiform rust resistance in slash pine. *Silvae Genetica* 28, 173-180.

Soltis, D.E. and Soltis, P.S. (1989) *Isozymes in Plant Biology*. Dioscorides Press, Portland, OR.

Sorensen, F. (1969) Embryonic genetic load in coastal Douglas-fir, *Pseudotsuga menziesii* var. *menziesii. American Naturalist* 103, 389-398.

Sorensen, F. (1971) Estimate of self-fertility in coastal Douglas-fir from inbreeding studies. *Silvae Genetica* 20, 115-120.

Sorensen, F.C. (1973) Frequency of seedlings from natural self-fertilization in coastal Douglas-fir. *Silvae Genetica* 22, 20-24.

Sorensen, F.C. (1982) The roles of polyembryony and embryo viability in the genetic system of conifers. *Evolution* 36, 725-733.

Sorensen, F.C. (1983) Geographic variation in seedling Douglas-fir (*Pseudotsuga menziesii*) from the western Siskiyou mountains of Oregon. *Ecology* 64, 696-702.

Sorensen, F.C. (1987) Estimated frequency of natural selfing in Lodgepole pine (*Pinus contorta* var. *murrayana*) from Central Oregon. *Silvae Genetica* 36, 215-221.

Sorensen, F.C. (1994) Frequency of seedlings from natural self-fertilization in Pacific Northwest ponderosa pine (*Pinus ponderosa* Dougl. Ex Laws.). *Silvae Genetica* 43, 100-108

Sorensen, F.C. (1997) Effects of sib mating and wind pollination on nursery seedling size, growth components, and phenology of Douglas-fir seed-orchard progenies. *Canadian Journal of Forest Research* 27, 557-566.

Sorensen, F.C. (1999) Relationship between self-fertility, allocation of growth, and inbreeding depression in three coniferous species. *Evolution* 53, 417-425.

Sorensen, F.C. and Adams, W.T. (1993) Self fertility and natural selfing in three Oregon Cascade populations of lodgepole pine. In: Lindgren, D. (ed.) *Proceedings* of the *Pinus contorta - from Untamed Forest to Domesticated Crop. Meeting of the International Union of Forest Research Organizations (IUFRO) WP2.02.06 and Frans Kempe Symposium.* Department of Genetics and Plant Physiology, Swedish University of Agricultural Sciences, Umea Report 11, 358-374.

Sorensen, F.C. and Campbell, R.K. (1997) Near neighbor pollination and plant vigor in coastal Douglas-fir. *Forest Genetics* 4, 149-157.

Sorensen, F.C. and Miles, R.S. (1982) Inbreeding depression in height, height growth, and survival of Douglas-fir, ponderosa pine, and noble fir to 10 years of age. *Forest Science* 28, 283-292.

Sorensen, F.C. and White, T.L. (1988) Effect of natural inbreeding on variance structure in tests of wind-pollination Douglas-fir progenies. *Forest Science* 34, 102-118.

Sorensen, F.C., Franklin, J.F. and Woollard, R. (1976) Self-pollination effects on seed and seedling traits in noble fir. *Forest Science* 22, 155-159.

Soule, M E. (1980) Thresholds for survival: maintaining fitness and evolutionary potential. In: Soule, M.E. and Wilcox, B.A. (eds.) *Conservation Biology: An Evolutionary-Ecological Perspective.* Sinauer Associates Inc., Publishers, Sunderland, MA, pp. 119-133.

Soule, M.E. (1986) *Conservation Biology: The Science of Scarcity and Diversity.* Sinauer Associates Inc., Publishers, Sunderland, MA.

Southern, E.M. (1975) Detection of specific sequences among DNA fragments separated by gel electrophoresis. *Journal of Molecular Biology* 98, 503-517.

Sprague, J.R., Talbert, J.T., Jett, J.B. and Bryant, R.L. (1983) Utility of the pilodyn in selection for mature wood specific gravity in loblolly pine. *Forest Science* 29, 696-701.

Squillace, A.E. (1966) Geographic variation in slash pine. *Forest Science Monograph 10*.

Squillace, A.E. (1971) Inheritance of monoterpene composition in cortical oleoresin of slash pine. *Forest Science* 17, 381-387.

Squillace, A.E. (1973) Comparison of some alternative second-generation breeding plans for slash pine. In: *Southern Forest Tree Improvement Conference.* Baton Rouge, LA, pp. 2-13.

Squillace, A.E. (1974) Average genetic correlations among offspring from open-pollinated forest trees. *Silvae Genetica* 23, 149-156.

Squillace, A.E. and Silen, R.R. (1962) Racial variation on ponderosa pine. *Forest Science Monograph 2*.

St. Clair, J.B. (1994) Genetic variation in tree structure and its relation to size in Douglas-fir. I. Biomass partitioning, foliage efficiency, stem form, and wood density. *Canadian Journal of Forest Research* 24, 1226-1235.

Stacy, E.A., Hamrick, J.L., Nason, J.D. Hubbell, S.P., Foster, R.B. and Condit, R. (1996) Pollen dispersal in low-density populations of three neotropical tree species. *American Naturalist* 148, 275-298.

Stam, M., Mol, J.N.M. and Kooter, J.M. (1997) The silence of genes in transgenic plants. *Annals of Botany* 79, 3-12.

Stam, P. (1993) Construction of integrated genetic linkage maps by means of a new computer package: JOINMAP. *Plant Journal* 3, 739-744.

Stam, P. and van Ooijen, J.W. (1995) JOINMAP™ version 2.0: software for the calculation of genetic linkage maps. CPRO-DLO, Wageningen, The Netherlands.

Stasolla, C., van Zyl, L., Egertsdotter, U., Craig, D., Liu, W. and Sederoff, R. (2003) The effects of the polyethylene glycol on gene expression of developing white spruce somatic embryos. *Plant Physiology* 131, 49-60.

Stebbins, G.L. (1948) The chromosomes and relationships of *Metasequoia* and *Sequoia*. *Science* 108, 95-98.

Stebbins, G.L. (1950) *Variation and Evolution in Plants*. Columbia University Press, New York, NY.

Stebbins, G. (1959) The role of hybridization in evolution. *Proceedings of the American Philosophical Society* 103, 231-251.

Stebbins, G.L. (1971) *Chromosomal Evolution in Higher Plants*. Edward Arnold, London, UK.

Steinhoff, R.J. (1974) Inheritance of cone color in *Pinus monticola*. *The Journal of Heredity* 65, 60-61.

Steinhoff, R.J., Joyce, D.G. and Fins, L. (1983) Isozyme variation in *Pinus monticola*. *Canadian Journal of Forest Research* 13, 1122-1132.

Sterky, F., Regan, S., Karlsson, H., Hertzberg, M., Rohde, A., Holmberg, A., Amini, B., Bhalerao, R., Larsson, M., Villarroel, R., van Mantagu, M., Sandberg, G., Olsson, O., Teeri, T., Boerjan, W., Gustafsson, P., Uhlen, M., Sundberg, B. and Lundeberg, J. (1998) Gene discovery in the wood-forming tissues of poplar: analysis of 5,692 expressed sequence tags. *Proceedings of the National Academy of Sciences of the United States of America* 95, 13330-13335.

Stettler, R.F., Fenn, R.C., Heilman, P.E. and Stanton B.J. 1988. *Populus trichocarpa* x *Populus deltoides* hybrids for short rotation culture: variation patterns and 4-year field performance. *Canadian Journal of Forest Research* 18, 745-753.

Stettler, R.F., Zsuffa, L. and Wu, R. (1996) The role of hybridization in the genetic manipulation of *Populus*. In: Stettler. R.F., Jr., Bradshaw, H.D., Heilman, P.E. and Hinckley, T.M. (eds.) *Biology of Populus and its Implications for Management and Conservation*. NRC Research Press, National Research Council of Canada. Ottawa, Ontario, pp. 87-112.

Stewart, S.C. (1994) Simultaneous estimation of pollen contamination and pollen fertilities of individual trees. *Theoretical and Applied Genetics* 88, 593-596.

Stirling, B., Newcombe, G., Vrebalov, J., Bosdet, I. and Bradshaw, Jr., H. (2001) Suppressed recombination around the *MXC3* locus, a major gene for resistance to poplar leaf rust. *Theoretical and Applied Genetics* 103, 1129-1137.

Stoehr, M.U. and El-Kassaby, Y.A. (1997) Levels of genetic diversity at different stages of the domestication cycle of interior spruce in British Columbia. *Theoretical and Applied Genetics* 94, 83-90.

Stoehr, M.U., Orvar, B.L., Vo, T.M., Gawley, J.R., Webber, J.E. and Newton, C.H. (1998) Application of a chloroplast DNA marker in seed orchard management evaluations of Douglas-fir. *Canadian Journal of Forest Research* 28, 187-195.

Stonecypher, R.W. and McCullough, R.B. (1986) Estimates of additive and non-additive genetic variances from a clonal diallel of Douglas-fir *Pseudotsuga mensiesii* (Mirb.) Franco. *Proceedings of the International Union of Forest Research Organizations (IUFRO), Conference on Breeding Theory, Progeny Testing and Seed Orchards*. Williamsburg, Virginia, pp. 211-227.

Stonecypher, R.W., Piesch, R.F., Heilman, G.G., Chapman, J.G. and Reno, H.J. (1996) Results from genetic tests of selected parents of Douglas-fir (*Pseudotsuga menziesii* [Mirb.] Franco) in an applied tree improvement program. *Forest Science Monographs* 32.

Strauss, S.H. (2003) Genomics, genetic engineering, and domestication of crops. *Science* 300, 61-62.

Strauss, S.H. and Critchfield, W.B. (1982) Inheritance of β-pinene in xylem oleoresin of knobcone X Monterey pine hybrids. *Forest Science* 28, 687-696.

Strauss, S. and Doerksen, A. (1990) Restriction fragment analysis of pine phylogeny. *Evolution* 44, 1081-1096.

Strauss, S.H. and Howe, G. (1990) An investigation of somatic variability for ribosmal RNA gene number in old-growth Sitka spruce. *Canadian Journal of Forest Research* 20, 853-856.

Strauss, S.H. and Tsai, C.-H. (1988) Ribosomal gene number variability in Douglas-fir. *Journal of Heredity* 79, 453-458.

Strauss, S.H., Palmer, J.D., Howe, G.T. and Doerksen, A.H. (1988) Chloroplast genomes of two conifers lack a large inverted repeat and are extensively rearranged. *Proceedings of the National Academy of Sciences of the United States of America* 85, 3898-3902.

Strauss, S.H., Bousquet, J., Hipkins, V.D. and Hong, Y.-P. (1992a) Biochemical and molecular genetic markers in biosystematic studies of forest trees. *New Forests* 6, 125-158.

Strauss, S., Lande, R. and Namkoong, G. (1992b) Limitations of molecular-marker-aided selection in forest tree breeding. *Canadian Journal of Forest Research* 22, 1050-1061.

Strauss, S.H., Hong, Y.-P. and Hipkins, V.D. (1993) High levels of population differentiation for mitochondrial DNA haplotypes in *Pinus radiata, muricata*, and *attenuata. Theoretical and Applied Genetics* 86, 605-611.

Strauss, S.H., Rottmann, W.H., Brunner, A.M. and Sheppard, L.A. (1995) Genetic engineering of reproductive sterility in forest trees. *Molecular Breeding* 1, 5-26.

Streiff, R., Labbe, T., Bacilieri, R., Steinkellner, H., Gloessl, J. and Kremer, A. (1998) Within population genetic structure in *Quercus robur* L. and *Quercus petraea* (Matt.) Liebl. assessed with isozymes and microsatellites. *Molecular Ecology* 7, 317-328.

Streiff, R., Ducousso, A., Lexer, C., Steinkellner, H., Gloessl, J. and Kremer, A. (1999) Pollen dispersal inferred from paternity analysis in a mixed oak stand of *Quercus robur* L. and *Q. petraea* (Matt.) Liebl. *Molecular Ecology* 8, 831-841.

Stroup, W.W., Baezinger, P.S. and Mulitze D.K. (1994) Removing spatial variation from wheat yield trials: a comparison of methods. *Crop Science* 86, 62-66.

Stukely, M.J.C. and Crane, C.E. (1994) Genetically based resistance of *Eucalyptus marginata* to *Phytophthora cinnamomi. Phytopathology* 84, 650-656.

Subramaniam, R., Reinold, S., Molitor, E.K. and Douglas, C.J. (1993) Structure, inheritance, and expression of hybrid poplar (*Populus trichocarpa* x *Populus deltoides*) phenylalanine ammonia-lyase genes. *Plant Physiology* 102, 71-83.

Surles, S.E., Hamrick, J.L. and Bongarten, B.C. (1989) Allozyme variation in black locust (*Robinia pseudoacacia*). *Canadian Journal of Forest Research* 19, 471-479.

Sutton, B.C.S., Flanagan, D.J., Gawley, J.R., Newton, C.H., Lester, D.T. and El-Kassaby, Y.A. (1991) Inheritance of chloroplast and mitochondrial DNA in *Picea* and composition of hybrids from introgression zones. *Theoretical and Applied Genetics* 82, 242-248.

Sweet, G.B., Dickson, R.L., Donaldson, B.D. and H. Litchwark. (1992) Controlled pollination without isolation - A new approach to the management of radiata pine seed orchards. *Silvae Genetica* 41, 95-99.

Swofford, D. (1993) PAUP: Phylogenetic analysis using parsimony. Version 3.1.1. Distributed by the Illinois Natural History Survey, Champaign, IL.

Syvanen, A. (2001) Accessing genetic variation: genotyping single nucleotide polymorphisms. *Nature Review Genetics* 2, 930-942.

Szmidt, A., Alden, T. and Hallgren, J.-E. (1987) Paternal inheritance of chloroplast DNA in *Larix. Plant Molecular Biology* 9, 59-64.

Szmidt, A.E., Wang, X.R. and Lu, M.Z. (1996) Empirical assessment of allozyme and RAPD variation in *Pinus sylvestris* (L.) using haploid tissue analysis. *Heredity* 76, 412-420.

Talbert, C.B. (1985) Two-stage early selection: A method for prioritization and weighting of traits. In: *Proceedings of the 18th Southern Forest Tree Improvement Conference*. Long Beach, MS, pp. 107-116.

Talbert, C.B. (1986) Multi-criterion index selection as a tool for operational tree improvement. In: *Proceedings of the International Union of Forest Research Organizations (IUFRO), Conference on Breeding Theory, Progeny Testing and Seed Orchards*. Williamsburg, VA, pp. 228-238.

Talbert, C.B. and Lambeth, C.C. (1986) Early testing and multi-stage selection. In: *Advanced Gen-*

eration Breeding of Forest Trees, Southern Cooperative Series Bulletin No. 309. Louisiana Agricultural Experiment Station, Baton Rouge, LA, pp. 43-52.

Talbert, C.B., Ritchie, G.A. and Gupta, P. (1993) Conifer vegetative propagation: an overview from a commercialization perspective. In: Ahuja, M.R. and Libby, W.J. (eds.) *Clonal Forestry I. Genetics and Biotechnology*. Springer-Verlag., New York, NY, pp. 145-181.

Talbert, J.T. (1979) An advanced-generation breeding plan for the North Carolina State University-Industry pine tree improvement cooperative. *Silvae Genetica* 28, 72-75.

Talbert, J.T., Weir, R.J. and Arnold, R.D. (1985) Costs and benefits of a mature first generation loblolly pine tree improvement program. *Journal of Forestry* 83, 162-165.

Tambasco-Talhari, D., Mello de Alencar, M., Paro de Paz, C.C., da Cruz, G.M., de Andrade Rodrigues, A., Packer, I.U., Coutinho, L.L. and Correia de Almeida Regitano, L. (2005) Molecular marker heterozygosities and genetic distances as correlates of production traits in F_1 bovine crosses. *Genetics and Molecular Biology* 28, 218-224.

Tang, W. and Newton, R.J. 2003. Genetic transformation of conifers and its application in forest biotechnology. Plant Cell Rep. 22:1-15.

Tang, W. and Quyang, F. (1999) Plant regeneration via organogenesis from six families of loblolly pine. *Plant, Cell, Tissue and Organ Culture* 58, 223-226.

Tang, W. and Tian, Y. (2003) Transgenic loblolly pine (*Pinus taeda* L.) plants expressing a modified δ-endotoxin gene from *Bacillus thuringiensis* with enhanced resistance to *Dendrolimus punctatus* Walker and *Crypyothelea formosicola* Staud. *Journal of Experimental Botany* 54, 835-844.

Tang, W., Newton, R.J. and Charles, T.M. (2006) Plant regeneration through adventitious shoot differentiation from callus cultures of slash pine (*Pinus elliottii*). *Journal of Plant Physiology* 163, 98-101.

Tanksley, S., Ganal, M. and Martin, G. (1995) Chromosome landing: a paradigm for map-based gene cloning in plants with large genomes. *Trends in Genetics* 11, 63-68.

Tauer, C.G. (1975) Competition between selected black cottonwood genotypes. *Silvae Genetica* 24(2/3), 44-49.

Tauer, C., Hallgren, S. and Martin, B. (1992) Using marker-aided selection to improve tree growth response to abiotic stress. *Canadian Journal of Forest Research* 22, 1018-1030.

Taylor, F.W. (1981) Rapid determination of southern pine specific gravity with a pilodyn tester. *Forest Science* 27, 59-61.

Teich, A.H. and Holst, M.J. (1974) White spruce limestone ecotypes. *Forestry Chronicle* 50, pp. 110-111.

Temesgen, B., Neale, D.B. and Harry, D.E. (2000) Use of haploid mixtures and heteroduplex analysis enhance polymorphisms revealed by denaturing gradient gel electrophoresis. *BioTechniques* 28, 114-122.

Temesgen, B., Brown, G.R., Harry, D.E., Kinlaw, C.S., Sewell, M.M. and Neale, D.B. (2001) Genetic mapping of expressed sequence tag polymorphism (ESTP) markers in loblolly pine (*Pinus taeda* L.). *Theoretical and Applied Genetics* 102, 664-675.

Teoh, S.B. and Rees, H. (1976) Nuclear DNA amounts in populations of *Picea* and *Pinus* species. *Heredity* 36, 123-137.

Terborgh, J. (1986) Keystone plant resources in the tropical forest. In: Soule, M.E. (ed.) *Conservation Biology: The Science of Scarcity and Diversity*. Sinauer Associates Inc., Publishers, Sunderland, MA, pp. 330-344.

Thamarus, K., Groom, K., Bradley, A., Raymond, C. A., Schimleck, L. R., Williams, E. R. and Moran, G. F. (2004) Identification of quantitative trait loci for wood and fibre properties in two full-sib families of *Eucalyptus globulus*. *Theoretical and Applied Genetics* 109, 856-864.

The *Arabidopsis* Genome Initiative (TAGI). (2000) Analysis of the genome sequence of the flowering plant *Arabidopsis thaliana*. *Nature* 408, 796-815.

Thoday, J. (1961) Location of polygenes. *Nature* 191, 368-378.

Thomas, G. and Ching, K.K. (1968) A comparative karyotype analysis of *Pseudotsuga menziesii* (Mirb.) Franco, and *Pseudotsuga wilsoniana* (Hayata). *Silvae Genetica* 17, 138-143.

Thompson, C.J., Movva, N.R., Tizard, R., Crameri, R., Davies, J.E., Lauwereys, M. and Botterman, J. (1998) Characterization of the herbicide-resistance gene *bar* from *Streptomyces hygroscopicus*. *EMBO Journal* 6, 2519-2523.

Thompson, J. (1994) *The Coevolutionary Process*. University of Chicago Press, Chicago, IL.

Thompson, R.S., Anderson, K. and Bartlein, P. (1999) Atlas of relations between climatic parameters and distributions of important trees and shrubs in North America - introduction and conifers. *United States Geological Survey Professional Paper 1650-A*.

Thomson, T.A., Lester, D.T. and Martin, J.A. (1987) Marginal analysis and cost effectiveness in seed orchard management. *Canadian Journal of Forest Research* 17, 510-515.

Thomson, T.A., Lester, D.T., Martin, J.A. and Foster, G.S. (1989) Using economic and decision making concepts to evaluate and design a corporate tree improvement program. *Silvae Genetica* 38, 21-28.

Tibbits, W.N., Hodge, G.R. and White, T.L. (1991) Predicting breeding values for freezing resistance in *Eucalyptus globulus*. In: *International Union of Forest Research Organizations (IUFRO), Symposium on Intensive Forestry: The Role of Eucalypts*. Durban, South Africa, pp. 330-333.

Tibbits, W.N., Boomsma, D.B. and Jarvis, S. (1997) Distribution, biology, genetics, and improvement programs for *Eucalyptus globulus* and *E. nitens* around the world. *Proceedings of the 24th Southern Forest Tree Improvement Conference*. Orlando, FL, pp. 81-95.

Toda, R. (1964) A brief review and conclusions of the discussion on seed orchards. *Silvae Genetica* 13, 1-4.

Tomback, D. and Linhart, Y. (1990) The evolution of bird-dispersed pines. *Evolutionary Ecology* 4, 185-219.

Tsai, L.M. and Yuan, C.T. (1995) A practical approach to conservation of genetic diversity in Malaysia: Genetic resource areas. In: Bayleard, T.J.B and Boontawee, B. (eds.) *Measuring and Monitoring Biodiversity in Tropical and Temperate Forests*. Center for International Forestry Research (CIFOR). Bogor, Indonesia, pp. 207-217.

Tsumura, Y., Yoshimura, K., Tomaru, N. and Ohba, K. (1995) Molecular phylogeny of conifers using RFLP analysis of PCR-amplified specific chloroplast genes. *Theoretical and Applied Genetics* 91, 1222-1236.

Tsumura, Y., Suyama, Y., Yoshimura, K., Shirato, N. and Mukai, Y. (1997) Sequence-tagged-sites (STSs) of cDNA clones in *Cryptomeria japonica* and their evaluation as molecular markers in conifers. *Theoretical and Applied Genetics* 94, 764-772.

Tulsieram, L., Glaubitz, J, Kiss, G and Carlson, J. (1992) Single tree genetic linkage mapping in conifers using haploid DNA from megagametophytes. *BioTechnology* 10, 686-690.

Turnbull, K.J. and Griffin, A.R. (1986) The concept of provenance and its relationship to infraspecific classification in forest trees. In: Styles, B.T. (ed.) *Infraspecific Classification of Wild and Cultivated Plants*. Clarendon Press, Oxford, UK, pp. 157-189.

Tzfira, T., Zuker, A. and Altman, A. (1998) Forest tree biotechnology: genetic transformation and its application to future forests. *Trends in Biotechnology* 16, 439-446.

Tzfira, T., Li, J., Lacroix, B. and Citovsky, V. (2004) *Agrobacterium* T-DNA integration: molecules and models. *Trends in Genetics* 20, 375-383.

van Buijtenen, J.P. (1971) Seed orchard design, theory and practice. In: *Proceedings of the 11th Southern Forest Tree Improvement Conference*. Atlanta, GA, pp. 197-206.

van Buijtenen, J.P. (1976) Mating designs. In: *Proceedings of the International Union of Forest Research Organizations (IUFRO), Joint Meeting on Advanced Generation Breeding*. Bordeaux, France, pp. 11-27.

van Buijtenen, J.P. (1978) Response of "lost pines" seed sources to site quality. *Proceedings of the 5th North American Forest Biology Workshop*. University of Florida, Gainesville, FL.

van Buijtenen, J.P. (1992) Fundamental genetic principles. In: Fins, L., Friedman, S.T. and Brotschol, J.V. (eds.) *Handbook of Quantitative Forest Genetics*. Kluwer Academic Publishers, Boston, MA, pp. 29-68.

van Buijtenen, J.P. and Bridgwater, F. (1986) Mating and genetic test designs. In: *Advanced Generation Breeding of Forest Trees*, Southern Cooperative Series Bulletin 309. Louisiana Agricultural Experiment Station, Baton Rouge, LA, pp. 5-10.

van Buijtenen, J.P. and Burdon, R.D. (1990). Expected efficiencies of mating designs for advanced generation selection. *Canadian Journal of Forest Research* 20, 1648-1663.

van Buijtenen, J. P. and Lowe, W.J. (1979) The use of breeding groups in advanced generation breeding. In: *Proceedings of the 15th Southern Forest Tree Improvement Conference*. Mississippi State University, Mississippi State, MS, pp. 59-65.

van de Ven, W.T.G. and McNicol, R.J. (1996) Microsatellites as DNA markers in Sitka spruce. *Theoretical and Applied Genetics* 93, 613-617.

Van Doorsselaere, J., Baucher, M., Feuillet, C., Boudet, A.M., Van Montagu, M. and Inze, D. (1995) Isolation of cinnamyl alcohol dehydrogenase cDNAs from two important economic species: alfalfa and poplar. Demonstration of a high homology of the gene within angiosperms. *Plant Physiology and Biochemistry* 33, 105-109.

Van Ooijen, J.W. MapQTL, Software for the mapping of quantitative trait loci in experimental populations. Kyazma B.V., Wageningen, The Netherlands.

van Vleck, L.D. (1993) *Selection Index and Introduction to Mixed Model Methods*. CRC Press, Boca Raton, FL.

van Vleck, L.D., Pollak, E.J. and Oltenacu, E.A. (1987). *Genetics for the Animal Sciences*. W.H. Freeman and Company, New York, NY.

Van Wyk, G. (1981) Pollen management for eucalypts. In: *Pollen Management Handbook. United States Department of Agriculture Handbook No. 587*, Washington DC, pp. 84-88.

Van Wyk, G. (1985a) Genetic variation in wood preservation of fast grown *Eucalyptus grandis*. *South Africa Forestry Journal* 135, 33-39.

Van Wyk, G. (1985b) Tree breeding in support of vegetative propagation of *Eucalyptus grandis* (Hill) Maiden. *South African Forestry Journal* 132, 33-39.

Vander Wall, S.B. (1992) The role of animals in dispersing a "wind-dispersed" pine. *Ecology* 73, 614-621.

Vander Wall, S.B. (1994) Removal of wind-dispersed pine seeds by ground-foraging vertebrates. *Oikos* 69, 125-132.

Vander Wall, S.B. and Balda, R.P. (1977) Coadaptations of the Clark's Nutcracker and the pinon pine for efficient seed harvest and dispersal. *Ecological Monographs* 47, 89-111.

Van Zyl, L., von Arnold, S., Bozhkov, P., Chen, Y., Egertsdotter, U., MacKay, J., Sederoff, R., Shen, J., Zelena, L. and Clapham, D. (2002) Heterologous array analysis in *Pinaceae*: hybridization of *Pinus* taeda cDNA arrays with cDNA from needles an embryogenic cultures of *P. taeda, P. sylvestris, or Picea abies. Comparative and Functional Genomics* 3, 306-318.

Vargas-Hernandez, J. and Adams, W.T. (1992) Age-age correlations and early selection for wood density in young coastal Douglas-fir. *Forest Science* 38, 467-478.

Vendramin, G.G., Lelli, L., Rossi, P. and Morgante, M. (1996) A set of primers for the amplification of 20 chloroplast microsatellites in Pinaceae. *Molecular Ecology* 5, 595-598.

Venter, C., Adams, M., Myers, E., Li, P., Mural, R., Sutton, G., Smith, H., Yandell, M., Evans, C., Holt, R., Gocayne, J., Amanatides, P., Ballew, R., Huson, D., Wortman, J., Zhang, Q., Kodira, C., Zheng, X., Chen, L., Skupski, M., Subramanian, G., Thomas, P., Zhang, J., Gabor, G., Miklos, Nelson, C., Broder, S., Clark, A., Nadeau, J., McKusick, V., Norton Zinder, Levine, A., Roberts, R., Simon. M., Slayman, C., Hunkapiller, M., Bolanos, R., Delcher, A., Dew, I., Fasulo, D., Flanigan, M., Florea, L., Halpern, A., Hannenhalli, S., Kravitz, S., Levy, S., Mobarry, C., Reinert, K., Remington, K., Abu-Threideh, J., Beasley, E., Biddick, K., Bonazzi, V., Brandon, R., Cargill, M., Chandramouliswaran, I., Charlab, R., Chaturvedi, K., Deng, Z., Di Francesco, V., Dunn, P., Eilbeck, K., Evangelista, C., Gabrielian, A., Gan, W., Ge, W., Gong, F., Gu, Z., Guan, P., Heiman, T., Higgins, M., Ji, R., Ke, Z., Ketchum, K., Lai, Z., Lei, Y., Li, Z., Li, J., Liang, Y., Lin, X., Lu, F., Merkulov, G., Milshina, N., Moore, H., Naik, A., Narayan, V., Neelam, B., Nusskern, D., Rusch, D., Salzberg, S., Shao, W., Shue, B., Sun, J., Wang, Y., Wang, A., Wang, X., Wang, J., Wei, M., Wides, R., Xiao, C., Yan, C., Yao, A., Ye, J., Zhan, M., Zhang, W., Zhang, H., Zhao, Q., Zheng, L., Zhong, F., Zhong, W., Zhu, S., Zhao, S., Gilbert, D., Baumhueter, S., Spier, G., Carter, C., Cravchik, A., Woodage, T., Ali, F., An, H., Awe, A., Baldwin, D., Baden, H., Barnstead, M., Barrow, I., Beeson, K., Busam, D., Carver, A., Center, A., Cheng, M., Curry, L., Danaher, S., Davenport, L., Desilets, R., Dietz, S., Dodson, K., Doup, L., Ferriera, S., Garg, N., Glueksmann, A., Hart, B., Haynes, J., Haynes, C., Heiner, C., Hladun, S., Hostin, D., Houck, J., Howland, T., Ibegwam, C., Johnson, J., Kalush, F., Kline, L., Koduru, S., Love, A., Mann, F., May, D., McCawley, S., McIntosh, T., McMullen, I., Moy, M., Moy, L., Murphy, B., Nelson, K., Pfannkoch, C., Pratts, E., Puri, V., Qureshi, H., Reardon, M., Rodriguez, R., Rogers, Y., Romblad, D., Ruhfel, B., Scott, R., Sitter, C., Smallwood, M., Stewart, E., Strong, R., Suh, E., Thomas, R., Tint, N., Tse, S., Vech, C., Wang, G., Wetter, J., Williams, S., Williams, M., Windsor, S., Winn-Deen, E., Wolfe, K., Zaveri, J., Zaveri, K., Abril, J.,

Guigó, R., Campbell, M., Sjolander, K., Karlak, B., Kejariwal, A., Mi, H., Lazareva, B., Hatton, T., Narechania, A., Diemer, K., Muruganujan, A., Guo, N., Sato, S., Bafna, V., Istrail, S., Lippert, R., Schwartz, R., Walenz, B., Yooseph, S., Allen, D., Basu, A., Baxendale, J., Blick, L., Caminha, M., Carnes-Stine, J., Caulk, P., Chiang, Y., Coyne, M., Dahlke, C., Mays, A., Dombroski, M., Donnelly, M., Ely, D., Esparham, S., Fosler, C., Gire, H., Glanowski, S., Glasser, K., Glodek, A., Gorokhov, M., Graham, K., Gropman, B., Harris, M., Heil, J., Henderson, S., Hoover, J., Jennings, D., Jordan, C., Jordan, J., Kasha, L., Kagan, L., Kraft, C., Levitsky, A., Lewis, M., Liu, X., Lopez, J., Ma, D., Majoros, W., McDaniel, J., Murphy, S., Newman, M., Nguyen, T., Nguyen, N., Nodell, M., Pan, S., Peck, J., Rowe, W., Sanders, R., Scott, J., Simpson, M., Smith, T., Sprague, A., Stockwell, T., Turner, R., Venter, E., Wang, M., Wen, M., Wu, D., Wu, M., Xia, A., Zandieh, A., Zhu , X. (2001) The sequence of the human genome. *Science* 291, 1304-1351.

Vergara, P.R. and Griffin, A.R. (1997) Fibre yield improvement program (FYIP) of *Eucalyptus globulus* Labill. in Santa Fe group, Chile. In: *Proceedings of the Conference of the International Union of Forest Research Organizations (IUFRO), sobre Silvicultura e Melhoramento de Eucaliptos.* El Salvador, Brazil, pp. 206-212.

Verhaegen, D., Plomion, C., Gion, J., Poitel, M., Costa, P. and Kremer, A. (1997) Quantitative trait dissection analysis in Eucalyptus using RAPD markers: 1. Detection of QTL in interspecific hybrid progeny, stability of QTL expression across different ages. *Theoretical and Applied Genetics* 95, 597-608.

Ver Hoef, J.M. and Cressie, N. (1993) Spatial statistics: analysis of field experiments. In: Scheiner, S.M. and Gurevitch, J. (eds.) *Design and Analysis of Ecological Experiments.* Chapman and Hall, Inc., New York, NY, pp. 319-341.

Verryn, S.D. (2000) *Eucalyptus* hybrid breeding in South Africa. In: *Hybrid Breeding and Genetics of Forest Trees. Proceedings of the Queensland Forest Research Institute/Cooperative Research Center-Sustainable Production Forestry (QFRI/CRC-SPF) Symposium.* Noosa, Queensland, Australia, pp. 191-199.

Vicario, F., Vendramin, G.G., Rossi, P., Lio, P. and Giannini, R. (1995) Allozyme, chloroplast DNA and RAPD markers for determining genetic relationships between *Abies alba* and the relic population of *Abies nebrodensis. Theoretical and Applied Genetics* 90, 1012-1018.

Villar, M., Lefevre, F., Bradshaw, J. and Teissier du Cros, E. (1996) Molecular genetics of rust resistance in poplars (*Melampsora larici-populina Kleb/Populus sp.*) by bulked segregant analysis in a 2 x 2 factorial mating design. *Genetics* 143, 531-536.

Vincent, T.G. (1997) Application of flowering and seed production research results to radiata pine seed production in New Zealand. In: Burdon, R.D. and Moore, J.M. (eds.) *Proceedings of the International Union of Forest Research Organizations (IUFRO), Conference on Genetics of Radiata Pine.* Rotorua, New Zealand, pp. 97-103.

Visscher, P.M., Thompson, R. and Hill, W.G. (1991) Estmation of genetic and environmental variances for fat yield in individual herds and an investagation into heterogeneity of variance between herds. *Livestock Production Science* 28, 273-290.

Volker, P.W., Dean, C.A., Tibbits, W.N. and Ravenwood, I.C. (1990) Genetic parameters and gains expected from selection in *Eucalyptus globulus* in Tasmania. *Silvae Genetica* 39, 18-21.

Vollmann, J., Buerstmayr, H. and Ruckenbauer, P. (1996) Efficient control of spatial variation in yield trials using neighbour plot residuals. *Experimental Agriculture* 32, 185-197.

Voo, K.S., Whetten, R.W., O'Malley, D.M. and Sederoff, R.R. (1995) 4-Coumarate: coenzyme A ligase from loblolly pine xylem: Isolation, characterization, and complementary DNA cloning. *Plant Physiology* 108, 85-97.

Vos, P., Hogers, R., Bleeker, M., Reijans, M., van de Lee, T., Hornes, M., Frijters, A., Pot, J., Peleman, J., Kuiper, M. and Zabeau, M. (1995) AFLP: a new technique for DNA fingerprinting. *Nucleic Acids Research* 23, 4407-4414.

Wackernagel, H. and Schmitt, M. (2001) Statistical interpolation models. In: von Storch, H. and Floser, G. (eds.) *Models in Environmental Research.* Springer-Verlag, New York, NY, pp. 185-201.

Wagner, D.B., Furnier, G.R., Saghai-Maroof, M.A., Williams, S.M., Dancik, B.P. and Allard, R.W. (1987) Chloroplast DNA polymorphisms in lodgepole and jack pines and their hybrids. *Proceedings of the National Academy of Sciences of the United States of America* 84, 2097-2100.

Wagner, D.B., Dong, J., Carlson, M.R. and Yanchuk, A.D. (1991a) Paternal leakage of mitochondrial DNA in *Pinus*. *Theoretical and Applied Genetics* 82, 510-514.

Wagner, D., Sun, Z-X, Govindaraju, D.R. and Dancik, B. (1991b) Spatial patterns of chloroplast DNA and cone morphology variation within populations of a *Pinus banksian-Pinus contorta* sympatric region. *The American Naturalist* 138, 156-170.

Wagner, D.B., Nance, W.L., Nelson, C.D., Li, T., Patel, R.N. and Govindaraju, D.R. (1992) Taxonomic patterns and inheritance of chloroplast DNA variation in a survey of *Pinus echinata, Pinus elliottii, Pinus palustris,* and *Pinus taeda. Canadian Journal of Forest Research* 22(5), 683-689.

Wakamiya, I., Newton, R., Johnston, J.S. and Price, H.J. (1993) Genome size and environmental factors in the genus *Pinus. American Journal of Botany* 80(11), 1235-1241.

Wakamiya, I., Price, H.J, Messina, M. and Newton, R. (1996) Pine genome size diversity and water relations. *Physiologia Plantarum* 96, 13-20.

Wakasugi, T., Tsudzuki, J., Ito, S., Nakashima, K., Tsudzuki, T. and Sugiura, M. (1994a) Loss of all *ndh* genes as determined by sequencing the entire chloroplast genome of the black pine *Pinus thunbergii. Proceedings of the National Academy of Science of the United States of America* 91, 9794-9798.

Wakasugi, T, Tsudzuki, J, Ito, S., Shibata, M. and Sugiura, M. (1994b) A physical map and clone bank of the black pine (*Pinus thunbergii*) chloroplast genome. *Plant Molecular Biology Reporter* 12(3), 227-241.

Walter, C., Carson, S.D., Richardson, M.T. and Carson, M. (1998a) Review: Application of biotechnology to forestry – molecular biology of conifers. *World Journal of Microbiology and Biotechnology* 14, 321-330.

Walter, C., Grace, L.J., Wagner, A., White, R., Walden, A.R., Donaldson, S.S., Hinton, H., Gardner, D. and Smith R. (1998b) Stable transformation and regeneration of transgenic plants of *Pinus radiata* D. *Plant Cell Reports* 17, 460-469.

Wang, C., Perry, T.O. and Johnson, A.G. (1960) Pollen dispersion of slash pine (*Pinus elliottii* Engelm.) with special reference to seed orchard management. *Silvae Genetica* 9, 78-86.

Wang, G., Castiglione, S., Chen, Y., Li, L., Han, Y., Tian, Y., Gabriel, D.W., Han, Y., Mang, K. and Sala, F. (1996) Poplar (*Populus nigra* L) plants transformed with a Bacillus thuringiensis toxin gene: insecticidal activity and genomic analysis. *Transgenic Research* 5:280-301.

Wang, X. R. and Szmidt, A. E. (1994) Hybridization and chloroplast DNA variation in a *Pinus* species complex from Asia. *Evolution* 48, 1020-1031.

Wang, X.R., Szmidt, A.E., Lewandowski, A. and Wang, Z.R. (1990) Evolutionary analysis of *Pinus densata* (Masters), a putative tertiary hybrid. *Theoretical and Applied Genetics* 80, 635-647.

Watson, J.D. and Crick, F.H.C. (1953) Molecular structure of nucleic acids: A structure for Deoxyribase Nucleic Acid. *Nature* 171, 737-738.

Watt, M.S., Garnett, B.T. and Walker, J.C.F. (1996) The use of the pilodyn for assessing outerwood density in New Zealand radiata pine. *Forest Products Journal.* 46, 101-106.

Webb, D.B., Wood, P.J., Smith, J.P., and Henman, G.S. (1984) *A Guide to Species Selection for Tropical and Sub-Tropical Plantations. Tropical Forestry Papers No. 15.* Commonwealth Forestry Institute, University of Oxford, UK.

Weber, J.C. and Stettler, R.F. (1981) Isoenzyme variation among ten populations of *Populus trichocarpa* Torr. et Gray in the Pacific Northwest. *Silvae Genetica* 30, 82-87.

Weber, J.L. and May, P. (1989) Abundant class of human DNA polymorphisms which can be typed using the polymerase chain reaction. *American Journal of Human Genetics* 44, 388-396.

Wei, R.P. and Lindgren, D. (1995) Optimal family contributions and a linear approximation. *Theoretical Population Biology* 48, 318-332.

Weigel, D. and Nilsson, O. (1995) A developmental switch sufficient for flower initiation in diverse plants. *Nature* 377, 495-500.

Weir, R.J. and Zobel, B.J. (1975) Managing genetic resources for the future: a plan for the North Carolina State Industry Cooperative Tree Improvement Program. In: *Proceedings of the 13th Southern Forest Tree Improvement Conference.* Raleigh, NC, pp. 73-82.

Weiss, K. and Clark, A. (2002) Linkage disequilibrium and the mapping of complex human traits. *Trends in Genetics* 18, 19-24.

Wells, O.O. (1969) Results of the southwide pine seed source study through 1968-69. *Proceedings of* *the 10th Southern Forest Tree Improvement Conference.* pp. 117-129.

Wells, O.O. (1983) Southwide pine seed source study-Loblolly pine at 25 years. *Southern Journal of Applied Forestry* 7, 63-71.

Wells, O.O. (1985) Use of Livingston Parish, Louisiana loblolly pine by forest products industries in the Southeast. *Southern Journal of Applied Forestry* 9, 180-185.

Wells, O.O. and Lambeth, C.C. (1983) Loblolly pine provenance test in southern Arkansas: 25th year results. *Southern Journal of Applied Forestry* 7, 71-75.

Wells, O.O. and Snyder, E.B. (1976) Longleaf pine half-sib progeny test. *Forest Science* 22, 404-406.

Wells, O.O. and Wakeley, P.C. (1966) Geographic variation in survival, growth, and fusiform rust infection of planted loblolly pine. *Forest Science Monograph 11.*

Wells, O.O. and Wakeley, P.C. (1970) Variation in shortleaf pine from several geographic sources. *Forest Science* 11, 415-423.

Wendel, J. (2000) Genome evolution in polyploids. *Plant Molecular Biology* 42, 225-249.

Wenck, A.R., Quinn, M., Whetten, R.W. Pullman, G. and Sederoff, R. (1999) High-efficiency *Agrobacterium*-mediated transformation of Norway spruce (*Picea abies*) and loblolly pine (*Pinus taeda*). *Plant Molecular Biology* 39, 407-416.

Westfall, R.D. (1992) Developing seed transfer zones. In: Fins, L., Friedman, S.T. and Brotschol, J.V. (eds.) *Handbook of Quantitative Forest Genetics.* Kluwer Academic Publishers, Boston, MA., pp. 313-398.

Westfall, R.D. and Conkle, M.T. (1992) Allozyme markers in breeding zone designation. *New Forests* 6, 279-309.

Welsh, J. and McClellend, M. (1990) Fingerprinting genomes using PCR with arbitrary primers. *Nucleic Acids Research* 18, 7213-7218.

Wheeler, N.C. and Bramlett, D.L. (1991) Flower stimulation treatments in a loblolly pine seed orchard. *Southern Journal of Applied Forestry* 15, 44-50.

Wheeler, N.C. and Guries, R.P. (1982) Population structure, genic diversity, and morphological variation in *Pinus contorta* Dougl. *Canadian Journal of Forest Research* 12, 595-606.

Wheeler, N.C. and Guries, R.P. (1987) A quantitative measure of introgression between lodgepole and jack pines. *Canadian Journal of Botany* 65, 1876-1885.

Wheeler, N. and Jech, K. (1986) Pollen contamination in a mature Douglas-fir seed orchard. In: *Proceedings of the International Union of Forest Research Organizations (IUFRO), Conference on Breeding Theory, Progeny Testing and Seed Orchards.* Williamsburg, VA, pp. 160-171.

Wheeler, N.C. and Jech, K.S. (1992) The use of electrophoretic markers in seed orchard research. *New Forests* 6, 311-328.

Wheeler, N.C., Jech, K.S., Masters, S.A., O'Brien, C.J., Timmons, D.W., Stonecypher, R. and Lupkes, A. (1995) Genetic variation and parameter estimates in *Taxus brevifolia* (Pacific yew). *Canadian Journal of Forest Research* 25, 1913-1927.

Whetten, R.W. and Sederoff, R.R. (1992) Phenylalanine ammonia-lyase from loblolly pine: purification of the enzyme and isolation of complementary DNA clones. *Plant Physiology* 98, 380-386.

Whetten, R.W., MacKay, J.J. and Sederoff, R.R. (1998) Recent advances in understanding lignin biosynthesis. *Annual Review of Plant Physiology and Plant Molecular Biology* 49, 587-609.

Whetten, R., Sun, Y. and Sederoff, R. (2001) Functional genomics and cell wall biosynthesis in loblolly pine. *Plant Molecular Biology* 47, 275-291.

Whitaker, D., Williams, E.R. and John, J.A. (2002) *CycDesigN: A Package for the Computer Generation of Experimental Designs.* Commonwealth Scientific and Industrial Research Organization (CSIRO) Forestry and Forest Products, Canberra, Australia.

White, T.L. (1987a) Drought tolerance of southwestern Oregon Douglas-fir. *Forest Science* 33, 283-293.

White, T.L. (1987b) A conceptual framework for tree improvement programs. *New Forests* 4, 325-342.

White, T.L. (1992) Advanced-generation breeding populations: size and structure. In: *Proceedings of the International Union of Forest Research Organizations (IUFRO), Resolving Tropical Forest Resource Concerns Through Tree Improvement, Gene Conservation, and Domestication of New Species..* Cali, Colombia, pp. 208-222.

White, T.L. (1996) Genetic parameter estimates and breeding value predictions: Issues and implications in tree improvement programs. In: *Proceedings of the International Union of Forest Research Organizations (IUFRO), Symposium on Tree Improvement for Sustainable Tropical Forestry*. Queensland, Australia, pp. 110-117.

White, T.L. (2001) Breeding strategies for forest trees: Concepts and challenges. *South African Forestry Journal* 190, 31-42.

White, T.L. (2004) Breeding theory and genetic testing. Pp. 1551-1561 In: Burley, J., Evans, J. and Youngquist, J.A. (eds.) *Encyclopedia of Forest Sciences*. Elsevier Academic Press, New York, NY, pp. 1551-1561.

White, T.L. and Carson, M.J. (2004) Breeding programs of conifers. In: Walter, C. and Carson, M.J. (eds.) *Plantation Forest Biotechnology for the 21st Century, 2004*. Research Signpost. Kerala, India, pp. 61-85.

White, T.L. and Ching, K.K. (1985) Provenance study of Douglas-fir in the Pacific Northwest region. IV. Field performance at age 25 years. *Silvae Genetica* 34, 84-90.

White, T.L. and Hodge, G.R. (1989) *Predicting Breeding Values with Applications in Forest Tree Improvement*. Kluwer Academic Publishers, Dordrecht, The Netherlands.

White, T.L. and Hodge, G.R. (1992) Test designs and optimal age for parental selection in advanced-generation progeny tests of slash pine. *Silvae Genetica* 41, 293-302.

White, T.L. and Rockwood. D.L. (1993) A breeding strategy for minor species of Eucalyptus. In: Barros, S., Prado, J.A. and Alvear, C. (eds.) *Proceedings of Los Eucaliptos en el Desarrollo Forestal de Chile*. Pucon, Chile, pp. 27-41.

White, T.L., Lavender, D.P., Ching, K.K. and Hinz, P. (1981) First-year height growth of southwestern Oregon Douglas-fir in three test environments. *Silvae Genetica* 30, 173-178.

White, T.L., Hodge, G.R., Powell, G.L., Kok, H.R., De Souza, S.M., Blakeslee, G. and Rockwood, D. (1988) 13th Progress Report, Cooperative Forest Genetics Research Program. Department of Forestry, University of Florida, Gainesville, FL.

White, T.L., Hodge, G.R. and Powell, G.L. (1993) Advanced-generation breeding strategy for slash pine in the southeastern United States. *Silvae Genetica* 42, 359-371.

White, T.L., Matheson, A.C., Cotterill, P., Johnson, R.G., Rout, A.F. and Boomsma, D.B. (1999) A nucleus breeding plan for radiata pine in Australia. *Silvae Genetica* 48, 122-133.

White, T.L., Huber, D.A. and Powell, G.L. (2003) Third-cycle breeding strategy for slash pine by the Cooperative Forest Genetics Research Program. In: *Proceedings of the 27th Southern Forest Tree Improvement Conference*. Stillwater, OK, pp. 17-29.

Whittemore, A. and Schaal, B. (1991) Interspecific gene flow in sympatric oaks. *Proceedings of the National Academy of Sciences of the United States of America* 88, 2540-2544.

Wilcox, P., Amerson, H., Kuhlman, E., Liu, B., O'Malley, D. and Sederoff, R. (1996) Detection of a major gene for resistance to fusiform rust disease in loblolly pine by genomic mapping. *Proceedings of the National Academy of Sciences of the United States of America* 93, 3859-3864.

Wilcox, P., Carson, S., Richardson, T., Ball, R., Horgan, G. and Carter, P. (2001) Benefit-cost analysis of DNA marker-based selection in progenies of *Pinus radiata* seed orchard parents. *Canadian Journal of Forest Research* 31, 2213-2224.

Williams, C.G. (1988) Accelerated short-term genetic testing for loblolly pine families. *Canadian Journal of Forest Research* 18, 1085-1089.

Williams, C. (2005) Framing the issues on transgenic trees. *Nature Biotechnology 23, 530-532.*

Williams, C.G. and de Steiguer, J.E. (1990) Value of production orchards based on two cycles of breeding and testing. *Forest Science* 36, 156-168.

Williams, C.G. and Hamrick, J.L. (1996) Genetic diversity levels in an advanced-generation *Pinus taeda* L. program measured using molecular markers. *Forest Genetics Research* 23, 45-50.

Williams, C. and Neale, D. (1992) Conifer wood quality and marker-aided selection: a case study. *Canadian Journal of Forest Research* 22, 1009-1017.

Williams, C.G. and Savolainen, O. (1996) Inbreeding depression in conifers: implications for breeding strategy. *Forest Science* 42, 102-117.

Williams, C., Hamrick, J.L. and Lewis, P.O. (1995) Multiple-population versus hierachical conifer breeding programs: a comparison of genetic diversity levels. *Theoretical and Applied Genetics* 90, 584-594.

Williams, E.R. and Matheson, A.C. (1994) *Experimental Design and Analysis for Use in Tree Improvement*. Commonwealth Scientific and Industrial Research Organization (CSIRO) Cataloguing-in-Publication Entry. East Melbourne, Victoria.

Williams, E.R. and Talbot, L.M. (1993) *Alpha+: Experimental Designs for Variety Trials, Version 1.0. Design User Manual*. Commonwealth Scientific and Industrial Research Organization (CSIRO) Publishing, Canberra, Australia.

Williams, E.R., Matheson, A.C., and Harwood, C.E. (2002) *Experimental Design and Analysis for Tree Improvement*. Commonwealth Scientific and Industrial Research Organization (CSIRO) Publishing, Collingwood, Australia, pp. 214.

Williams, J.G.K., Kubelik, A.R., Livak, J., Rafalski, J.A. and Tingey, S.V. (1990) DNA polymorphisms amplified by arbitrary primers are useful as genetic markers. *Nucleic Acids Research* 18, 6531-6535.

Willis, J.H. (1992) Genetic analysis of inbreeding depression caused by chlorophyll-deficient lethals in *Mimulus guttatus*. *Heredity* 69, 562-572.

Wilson, B.C. (1990) Gene-pool reserves of Douglas-fir. *Forest Ecology and Management* 35, 121-130.

Wilson, E.O. (1992) *The Diversity of Life*. Harvard University Press, Cambridge, MA.

Wilusz, W. and Giertych, M. (1974) Effects of classical silviculture on the genetic quality of the progeny. *Silvae Genetica* 23, 127-130.

Woods, J.H. and Heaman, J.C. (1989) Effects of different inbreeding levels on filled seed production in Douglas-fir. *Canadian Journal of Forest Research* 19, 54-59.

Woolaston, R.R. and Jarvis, S.F. (1995) The importance of breeding objectives in forest tree improvement. In: Potts, B.M., Borralho, N.M.G., Reid, J.B., Cromer, R.N., Tibbits, W.N. and Raymond, C.A. (eds.) *Proceedings of the International Union of Forest Research Organizations (IUFRO), Symposium on Eucalypt Plantations: Improving Fibre Yield and Quality*. Hobart, Australia, pp. 184-188.

Woolaston, R.R., Kanowski, P.J. and Nikles, D.G. (1990) Genetic parameter estimates for *Pinus caribaea hondurensis* in coastal Queensland, Australia. *Silvae Genetica* 39, 21-28.

Woolaston, R.R., Kanowski, P.J. and Nikles, D.G. (1991) Genotype-environment interactions in *Pinus caribaea* var. *hondurensis* in Queensland, Australia. II. Family x site interactions. *Silvae Genetica* 40, 228-232.

Wright, J.A. (1997) A review of the worldwide activities in tree improvement for *Eucalyptus grandis*, *Eucalyptus urophylla* and the hybrid *urograndis*. In: *Proceedings of the 24th Southern Forest Tree Improvement Conference*. Orlando, FL, pp. 96-102.

Wright, J.W. (1952) Pollen dispersion of some forest trees. USDA Forest Service Station Paper 46.

Wright, J.W. (1961) Progeny tests or seed orchards? *Recent Advances in Botany* 2, 1681-1687.

Wright, J.W. (1976) *Introduction to Forest Genetics*. Academic Press, Inc., New York, NY, 463 pp.

Wright, S. (1931) Evolution in Mendelian populations. *Genetics* 16, 97-159.

Wright, S. (1969) *Evolution and the Genetics of Populations. Vol. 2, The Theory of Gene Frequencies*. University of Chicago Press, Chicago, IL.

Wright, S. (1978) *Evolution and the Genetics of Populations. Volume 4. Variability Within and Among Populations*. University of Chicago Press, Chicago, IL.

World Resources Institute (1994) *World Resources 1994-95: A Report by the World Resources Institute in Collaboration with the United Nations Environment Programme and the United Nations Development Programme*. Oxford University Press, New York, NY.

Wu, H.X. (2002) Study of early selection in tree breeding 4. Efficiency of marker-aided early selection (MAES). *Silvae Genetica* 51, 261-269.

Wu, H.X. and Ying, C.C. (1997) Genetic parameters and selection efficiencies in resistance to western gall rust, stalactiform blister rust, needle cast, and sequoia pitch moth in lodgepole pine. *Forest Science* 43, 571-581.

Wu, J., Krutovskii, K. and Strauss, S. (1999) Nuclear DNA diversity, population differentiation, and phylogenetic relationships in the California closed-clone pines based on RAPD and allozyme markers. *Genome* 24, 893-908.

Xie, C.Y., Yeh, F.C., Dancik, B.P. and Strobeck, C. (1991) Joint estimation of immigration and mating system parameters in gymnosperms using the EM algorithm. *Theoretical and Applied Genetics* 83, 137-140.

Yamada, Y. (1962) Genotype by environment interaction and genetic correlation of the same trait under different environments. *Japanese Journal of Genetics* 37, 498-509.

Yamamoto, N., Mukai, Y., Matsuoka, M., Kano-Murakami, Y., Tanaka, Y., Ohashi, Y., Ozeki, Y. and Odani, K. (1991) Light-independent expression of cab and rbcS genes in dark-grown pine seedlings. *Plant Physiology* 95, 379-383.

Yamamoto, N., Tada, Y. and Fujimara, T. (1994) The promoter of a pine photosynthetic gene allows expression of a beta-glucuronidase reporter gene in transgenic rice plants in a light-independent but tissue-specific manner. *Plant and Cell Physiology* 35(5), 773-778.

Yanchuk, A.D. (2001) A quantitative framework for breeding and conservation of forest tree genetic resources in British Columbia. *Canadian Journal of Forest Research* 31, 566-576.

Yanchuk, A.D. and Lester, D.T. (1996) Setting priorities for conservation of the conifer genetic resources of British Columbia. *Forestry Chronicle* 72, 406-415.

Yang, R. C., Yeh, F.C. and Yanchuk, A.D. (1996) A comparison of isozyme and quantitative genetic variation in *Pinus contorta* ssp. *latifolia* by FST. *Forest Genetics* 142, 1045-1052.

Yazdani, R. and Lindgren, D. (1991) The impact of self-pollination on production of sound selfed seeds. In: Fineshi, S., Malvolti, M.E., Cannata, F. and Hattemer, H.H. (eds.), *Biochemical Markers in the Population Genetics of Forest Trees.* SPB Academic Publishing, The Hague, The Netherlands, pp. 143-147.

Yazdani, R., Muona, O., Rudin. D. and Szmidt, A.E. (1985) Genetic structure of a *Pinus sylvestris* L. seed-tree stand and naturally regenerated understory. *Forest Science* 31, 430-436.

Yeh, F.C. (1988) Isozyme variation of *Thuja plicata (*Cupressaceae*)* in British Columbia. *Biochemical Systematics and Ecology* 16, 373-377.

Yeh, F.C., Cheliak, W.M., Dancik, B.P., Illingworth, K., Trust, D.C. and Pyrhitka, B.A. 1985. Population differentiation in lodgepole pine, P*inus contorta* ssp. *latifolia*: a discriminant analysis of allozyme variation. *Canadian Journal of Genetic Cytology* 27, 210-218.

Yim, K.B. (1963) Karyotype analysis of *Pinus rigida. Heredity* 47, 274-276.

Young, A.G. and Boyle, T.J. (2000) Forest fragmentation. In: Young, A, Boshier, D. and Boyle, T. (eds.) *Forest Conservation Genetics: Principles and Practice.* Commonwealth Scientific and Industrial Research Organization (CSIRO) Publishing, Victoria, Australia, pp. 123-134.

Young, A., Boshier, D. and Boyle, T. (2000) *Forest Conservation Genetics: Principles and Practice.* Commonwealth Scientific and Industrial Research Organization (CSIRO) Publishing, Collingwood, Victoria, Australia.

Young, N. (1999) A cautiously optimistic vision for marker-assisted breeding. *Molecular Breeding* 5, 505-510.

Yu, J., Hu, S., Wang, J., Wong, G., Li, S., Liu, B., Deng, Y., Dai, L., Zhou, Y., Zhang, X., Cao, M., Liu, J., Sun, J. Tang, J., Chen, Y., Huang, X., Lin, W., Ye, C., Tong, W., Cong, L., Geng, J., Han, Y., Li, L., Li, W., Hu, G., Huang, X., Li, W., Li, J., Liu, Z., Li, L., Liu, J., Qi, Q., Liu, J., Li, L., Li, T., Wang, X., Lu, H., Wu, T., Zhu, M., Ni, P., Han, J., Dong, W., Ren, X., Feng, X., Cui, P., Li, X., Wang, H., Xu, X., Zhai, W., Xu, Z., Zhang, J., He, S., Zhang, J., Xu, J., Zhang, K., Aheng, X., Dong, J., Zeng, W., Tao, L., Ye, J., Tan, J., Ren, X., Chen, X., He, J., Liu, D., Tian, W., Tian, C., Xia, H., Bao, Q., Li, G., Gao, H., Cao, T., Wang, J., Zhao, W., Li,, P., Chen, W., Wang, X., Zhang, Y., Hu, J., Wang, J., Liu, S., Yang, J., Zhang, G., Xiong, Y., Li, Z., Mao, L., Zhou, C., Zhu, Z., Chen, R., Hao, B., Zheng, W., Chen, S., Guo, W., Li, G., Liu, S., Tao, M., Wang, J., Zhu, L., Yuan, L. and Yang, H.. (2002) A draft sequence of the rice genome (*Oryza sativa* L. ssp. *indica*). *Science* 296, 79-92.

Zavarin, E., Critchfield, W. and Snajberk, K. (1969) Turpentine composition of *Pinus contorta* x *Pinus banksiana* hybrids and hybrid derivatives. *Canadian Journal of Botany* 47, 1443-1453.

Zhang, J., Steenackers, M., Storme, V., Neyrinck, S., Van Montagu, M., Gerats, T. and Boerjan, W. (2001) Fine mapping and identification of nucleotide binding site/leucine-rich repeat sequences at the *MER* locus in *Populus deltoides* "S9-2". *The American Phytopathology Society* 91, 1069-1073.

Zhang, X.H. and Chiang, V.L. (1997) Molecular cloning of 4-coumarate: coenzyme A ligase in loblolly pine and the roles of this enzyme in the biosynthesis of lignin in compression wood. *Plant Physiology* 113, 65-74.

Zheng, Y. and Ennos, R.A. (1997) Changes in the mating systems of populations of *Pinus caribaea* Morelet var. c*aribaea* under domestication. *Forest Genetics* 4, 209-215.

Zheng, Y.-Q., Andersson, E.W. and Lindgren, D. (1998) A model for infusion of unrelated material into a breeding population. *Silvae Genetica* 47, 94-101.

Zimmerman, D.L. and Harville, D.A. (1991) A random approach to the analysis of field-plot experiments and other spatial experiments. *Biometrics* 47:223-239.

Zimmerman, R. H., Hackett, W.P. and Pharis, R.P. (1985) Hormonal aspects of phase change and precocious flowering. In: Pharis, R.P. and D. M. Reid, D.M. (eds.) *Encyclopedia of Plant Physiology*. Vol. II. Springer-Verlag. New York, NY, pp. 79-115.

Zobel, B. (1953) Are there natural loblolly-shortleaf pine hybrids? *Journal of Forestry* 51, 494-495.

Zobel, B.J. (1993) Clonal forestry in the eucalypts. In: Ahuja, M.R. and Libby, W.J. (eds.) *Clonal Forestry II: Conservation and Application*. Springer-Verlag. New York, NY, pp. 139-148.

Zobel, B.J. and Jett, J.B. (1995) *Genetics of Wood Production*. Springer-Verlag, New York, NY.

Zobel, B.J. and Talbert, B.J. (1984) *Applied Forest Tree Improvement*. John Wiley & Sons, New York, NY, pp. 448.

Zobel, B.J. and van Buijtenen, J.P. (1989) *Wood Variation: Its Causes and Control*. Springer-Verlag, New York, NY.

Zobel, B.J., Van Wyk, G. and Stahl, P. (1987) *Growing Exotic Forests*. John Wiley & Sons. New York, NY.

Zsuffa, L., Sennerby-Forsse, L., Weisgerber, H. and Hall, R.B. (1993) Strategies for clonal forestry with poplars, aspens, and willows. In: Ahuja, M.R. and Libby, W.J. (eds.) *Clonal Forestry II: Conservation and Application*. Springer-Verlag, New York, NY, pp. 91-119.

Zupan, J.R. and Zambryski, P. (1995). Transfer of T-DNA from *Agrobacterium* to the plant cell. *Plant Physiology* 107, 1041-1047.

INDEX

Note: page numbers in **bold** refer to major topics emboldened in text. Page numbers in *italics* refer to figures, tables, and boxes.